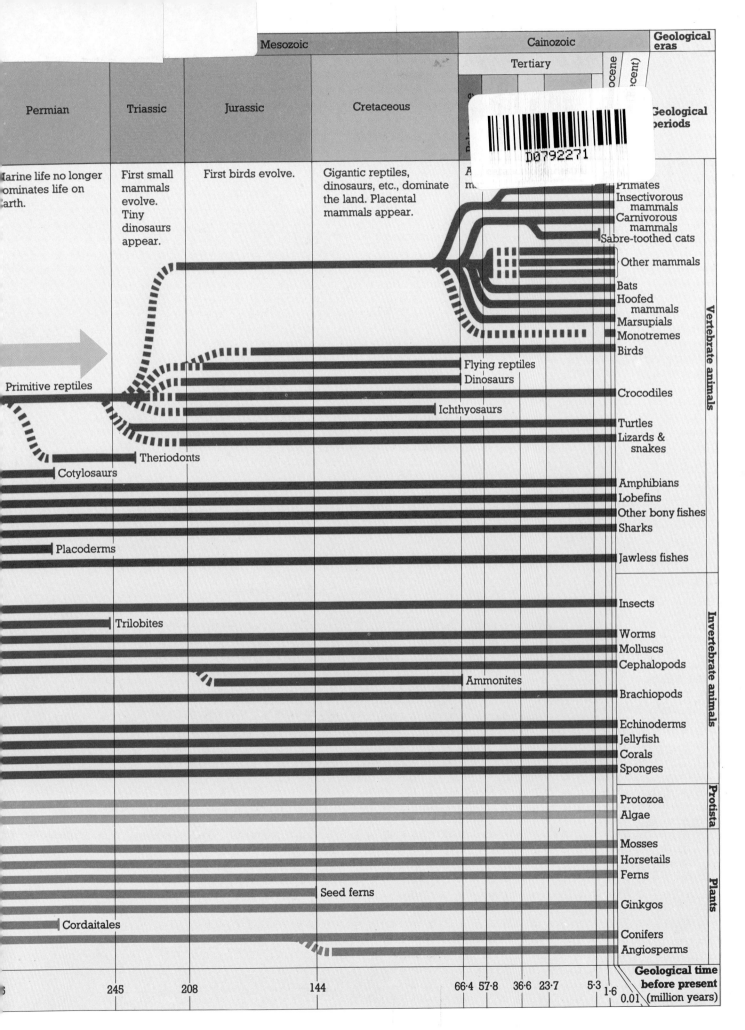

OXFORD
ILLUSTRATED ENCYCLOPEDIA

Volume 2
THE NATURAL WORLD

OXFORD
ILLUSTRATED ENCYCLOPEDIA

General Editor Harry Judge

Executive Editor Anthony Toyne

Volume 2
THE NATURAL WORLD

Volume Editor Malcolm Coe

Consultant Editor Sir Richard Doll

OXFORD
OXFORD UNIVERSITY PRESS
NEW YORK   MELBOURNE
1985

Oxford University Press, Walton Street, Oxford, OX2 6DP

Oxford New York Toronto
Delhi Bombay Calcutta Madras Karachi
Kuala Lumpur Singapore Hong Kong Tokyo
Nairobi Dar es Salaam Cape Town
Melbourne Auckland

and associated companies in
Beirut Berlin Ibadan Mexico City Nicosia

Oxford is a trademark of Oxford University Press

Text © Oxford University Press 1985
Captions, design, and illustrations © The Rainbird Publishing Group Limited 1985

British Library Cataloguing in Publication Data

Oxford illustrated encyclopedia.
Vol. 2: The natural world.
1. Encyclopedias and dictionaries
I. Judge, Harry   II. Toyne, Anthony
III. Coe, Malcolm
032   AE5

ISBN 0-19-869134-3

Library of Congress Cataloging in Publication Data

Oxford illustrated encyclopedia.

Contents: v. 1. The physical world/volume editor,
Sir Vivian Fuchs—v. 2. The natural world/volume
editor, Malcolm Coe.
1. Encyclopedias and dictionaries.   I. Judge,
Harry George.   II. Toyne, Anthony.
AE5.O94   1985   032   85-4876
ISBN 0-19-869134-3 (v. 2)

Designed and produced by The Rainbird Publishing Group Limited, London,
on behalf of Oxford University Press

Database prepared by Alphabyte Limited, Cheltenham, Gloucestershire, England
Illustrations originated by Gilchrist Brothers Limited, Leeds, England
Printed and bound by Royal Smeets Offset BV, Weert, The Netherlands

General Preface

As the worlds that we inhabit become more complicated, so encyclopedias become more and not less important. Fortunately, as that need increases so also does human skill and ingenuity in producing major works of reference. The ancient Greeks would have been fascinated both by the growth of the circle of the arts and sciences and by the new methods of exploring and describing all that lies within the circle.

The *Oxford Illustrated Encyclopedia* stands now at the end of a noble tradition and, like its predecessors, has the mission of offering to all readers the very best account of modern knowledge. It does so in a particular way.

Each of the volumes is arranged alphabetically around a certain theme. This means that each of these volumes can be enjoyed with profit on its own and studied either intensively or at leisure. One of the great delights of encyclopedias is that they serve equally those who wish to dip into a particular topic (or settle a doubt or argument), and those who want to acquire a rounded understanding of a major field of human knowledge.

More important still, the volumes have been carefully arranged in such a way that, taken together, they provide a clear and authoritative map of contemporary knowledge — spanning the sciences and the arts, modern society and its long history, the life of the imagination and the shape of politics, the achievement of men and women, and the expanding world of space.

An arrangement by volumes devoted to broad themes has other advantages. It means that responsibility for each volume can be assumed by a scholar of acknowledged international reputation in that field. The objectives have been twofold: first, to provide a scholarly basis for a general work (for an excyclopedia which offers half-truths and unfinished stories can be worse than useless) and, second, to ensure broad coverage, fairness, and objectivity for an international audience. Thus the volumes have been written by teams of expert contributors who have made every effort to ensure not only that the information is accurate and up to date but also that their articles can be readily understood by those who wish to learn, at whatever age, and by those who want to help them, whether as parents or as teachers. Thus also they have taken a worldwide view, rather than bearing a particular country or continent in mind, so that their contributions are international in the fullest sense. Preparation of the text has been backed by the full resources of the Oxford University Press.

One decision had to be taken in the early days of planning these volumes. Should we aim to produce a relatively small number of longer articles, or a much richer collection of shorter entries? We decided that most of those consulting the Encyclopedia would welcome immediate access to the information for which they were seeking and have therefore decided upon short entries that are concise and to the point, many of them containing cross-references for the user who wishes to explore more widely.

To be able to see and understand is, for many topics within the circle of knowledge, at least as important as to be able to read and understand. For the presentation of items which can best be appreciated visually, much research by the staff of the Rainbird Publishing Group has gone into the selection of illustration, and the use of colour is judicious. The intention has been not so much to decorate, although the pictures are handsome, as to supplement and expand the information given in the text.

I am grateful alike to the editors, consultants, contributors, and illustrators whose common aim has been to provide, within the space available, the best and most reliable representation that can be achieved in explaining and enriching the world.

HARRY JUDGE

Foreword

From the time when our first human ancestors evolved three or more million years ago, they must have been aware of their surroundings and classified the various natural phenomena that impinged upon their lives. Their observations would have encompassed the distinction between animals and plants, and between organisms that were edible or poisonous, harmless or predatory. Even so, it was not until the time of the first great modern taxonomist, Carl Linnaeus, that we received the first classification of living things that could be used everywhere, irrespective of language or culture. This advance came with the publication of Linnaeus's *Species Plantarum* in 1753, which provided us with a descriptive tool based upon the sexual features of plants. Then, in the nineteenth century, Charles Darwin set classification in an evolutionary perspective: on 1 July 1858, he and Alfred Russel Wallace made the first public statement of their theory of evolution by natural selection before the Linnean Society of London.

Today we have advanced far beyond these early ideas, for not only do we recognize evolution as the fundamental biological process, but also we now have a much clearer idea of the mechanisms involved. The underlying process forms the subject-matter of the science of genetics, which helps explain how variation in species leads to the origin of new forms, whether this be by gradual process of selection, by rather sudden jumps, or by a combination of both. The fundamental unit of these changes is the gene, and it was the discovery of the helical structure of DNA (deoxyribonucleic acid) that made it possible to explain one of the most difficult aspects of the way in which the genetic mechanism functions. With this knowledge, we are now able to view living organisms not just in a descriptive mode but in a functional way as well.

To put together a volume describing the natural world has not been easy, for living organisms alone comprise over five million known species (of which over 90 per cent are animals), with possibly twice as many as this number still undescribed. In choosing representatives to cover this vast array of species, we have tried to include all the major animal and plant groups from as wide a geographical range as possible. Thus, while readers in one part of the world will not be able to locate every common bird that occurs in their region, they should be able to find a representative of the group to which it belongs. Here the cross-reference system has been particularly important, for the scientific name may be more widely known than a more restricted local name. In cases where there are no generally accepted common names, we have used the scientific name only. Wherever possible the contributors have included references to the family or other major group to which the organism belongs as a guide to the reader who wishes to follow the topic further. In drawing up our plant entries, we have tried to strike a balance between describing the commoner species or groups encountered in the countryside, including those that are rare or have unusual habits, and those that have in many cases been used by man for thousands of years as food or for medicinal purposes.

In addition to living organisms, we have included entries on a number of the more important fossil and extinct forms of animals and plants; these forms are central to our understanding of the history and origins of the organisms that survive today. Moreover, those who find the classification of living organisms something of a mystery will find a number of entries that explain the manner in which biologists group together the animal and plant species they study.

Natural science today does not consist of just a list of plant and animal names, for there is now a wide variety of new disciplines that study the ways in which living organisms work. These new disciplines include the science of ecology, which looks at the interrelationships between living organisms and their non-living environment, and the sciences of biochemistry and physiology, which tell us how the chemical processes such as respiration, nutrition, and reproduction of animals and plants work. Most of the major subdivisions of natural science that are included in this volume are interrelated at a classificatory or functional level; nowhere is this more obvious than in the field of medical science—for the study of human anatomy has close affinities with, and origins in, the study of the gross structure of vertebrates, and in particular of mammals, while the origins of disease may commonly be connected with the animals or plants that cause it, whether these be viruses, bacteria, roundworms, or fungi. Thus our medical headings have been chosen to represent the commoner medical conditions and the basic anatomical and morphological features of the human species.

The study of the natural world, like all other aspects of human knowledge, is related ultimately to the people who have pursued it. In this context we have included biographical details of the major contributors to the advancement of the various fields of natural science, from the Greek philosophers to the natural scientists of the twentieth century.

The extent to which we have achieved in this volume a balanced treatment of this very large field has been to a large degree due to the care with which the thirty-four specialist contributors have prepared their entries, and to the efforts taken to find illustrations that will provide the reader with as wide a visual representation as is possible in the confines of a volume of 384 pages. It is hoped that those using *The Natural World*, whether they be amateur naturalists, students, or members of the informed public, will derive as much pleasure from it as the editors have had in preparing it.

MALCOLM COE

ASSISTANT VOLUME EDITORS

Sandra Raphael   Keith Porter

CONTRIBUTORS

Ailsa Allaby

Rachel Banister

R. C. Beatty

J. K. Burras

B. T. Clark

Dr Malcolm Coe

Peter Coxhead

Jane Ellen Dockray

Martin Dockray

Professor A. C. Dornhorst

Dr S. K. Eltringham

Dr R. B. Freeman

Linda Gamlin

Dr G. M. Goodwin

Dr Roger Hall

Dr T. S. Kemp

Dr C. J. McCarthy

Dr D. J. Mabberley

R. A. S. Melluish

Dr Marion Nixon

Dr Jennifer Owen

Dr C. M. Perrins

Dr W. D. Phillips

Dr Keith Porter

Sandra Raphael

Sylvia Remington

Dr Mark Ridley

Dr B. Stonehouse

Dr Christopher Ward

Dr Sarah Watkinson

Professor A. G. M. Weddell

A. C. Wheeler

Ailsa White

Dr Pat Willmer

A User's Guide

This book is designed for easy use, but the following notes may be helpful to the reader.

ALPHABETICAL ARRANGEMENT   The entries are arranged in A–Z order of their headwords up to the first comma (thus **Logan, James** comes before **Loganberries**). Names beginning with 'St.' are placed as if spelt 'Saint' (**St. John's Wort** comes before **Saithe**).

HEADWORDS   Entries are usually placed under the keyword in the title (the surname, for instance, in a group of names); if there is no obvious keyword, an entry appears under the word that most users are likely to look up first. Where an animal or plant is known by both a common name and a popular derivative of its scientific name, the entry is usually placed under the common name with a single-line cross-reference from the derived name—for example there is a cross-reference from **Nematodes** to **Roundworms**. In the case of large families of plants, many individual species are mentioned under the name of the most obvious member of that family (thus members of the primrose family such as cowslip and polyanthus are mentioned under **Primrose**). If the animal or plant you are interested in does not have an individual entry you may find something about it under the common family name or the name of a closely related species. Headwords generally take the singular form of a name if a single species or subject is meant, the plural form if more than one; but professional usage sometimes omits the final 's' from a plural form. A combination of these factors occasionally results in an entry's appearing in a different alphabetical position from the one that might be expected—for example the entry for bats is given under **Bats** not **Bat** and therefore follows **Bateleurs** and **Bateson**.

NAMES   The common names of plants and animals used throughout this volume are supplemented by scientific names if these are necessary to distinguish between unrelated organisms bearing the same common name or between related species known by variations of a particular common name. Scientific names are also introduced in order to identify organisms that are known by different popular names in different countries. The common names adopted reflect popular usage in the English-speaking world and some allowance is made for organisms with two popular names in common use (thus the small bird **Chickadee** is included with a cross-reference to **Tits**). Separate entries on **Phylum**, **Orders**, **Families**, **Genera**, and **Species** explain the group names used in biological classification.

CROSS-REFERENCES   An asterisk (*) in front of a word denotes a cross-reference and indicates the headword of the other entry to which attention is being drawn. Cross-references in the text appear only in places where reference is likely to amplify or increase understanding of the entry you are reading. They are not given automatically in all cases where a separate entry can be found, so if you come across a name or a term about which you would like to know more it is worth looking for an entry in its alphabetical place even if no cross-reference is marked.

WEIGHTS AND MEASURES   Both metric measures and their American equivalents are given throughout, and in the scientific entries SI units are used. Large measures (such as the weights of the biggest animals or the heights of trees) are generally rounded off, partly for the sake of simplicity and occasionally to reflect the difficulty of precise measurement.

TERMINOLOGY   Wherever possible, technical terms are explained in their context. Those not explained can be found in any good standard dictionary. The terms 'tropical', 'subtropical', and 'temperate' are used to refer to the geographical divisions of the world along lines of latitude, rather than to precise climatic conditions: tropical regions lie between the Tropic of Cancer (23° 27′ N.) and the Tropic of Capricorn (23° 27′ S.); subtropical regions border on the tropics and extend as far as 30° N. and S.; temperate regions extend from the subtropics to the Arctic and Antarctic circles (66° 33′ N. and S.). The term 'Old World' refers to Europe, Africa, and Asia, and 'New World' to the American continents.

ILLUSTRATIONS   Pictures and diagrams almost invariably occur on the same page as the entries to which they relate or on a facing page, and where this is not the case the position of an illustration is indicated in the text. Picture captions supplement the information given in the text and indicate in bold type the title of the relevant entry. The endpaper diagrams illustrating the evolution and classification of life on earth act as a ready guide to the major groupings of living organisms covered in this volume.

RELATIONSHIP TO OTHER VOLUMES   This survey of the natural world is part of a series, the *Oxford Illustrated Encyclopedia*, to be published over the period 1985–8 and comprising eight thematic volumes:

1 *The Physical World*
2 *The Natural World*
3 *The History of Mankind: Evolution to Revolution*
4 *The History of Mankind: The Old World and the New*
5 *The Arts*
6 *The World of Technology*
7 *Human Society*
8 *The Universe*

The volume is self-contained, having no cross-references to any of its companions, and is therefore usable on its own. Some aspects of certain topics are, however, the subjects of other volumes. Many chemical compounds referred to in this volume will be found as entries in Volume 1. Specialist applications of biological concepts, such as biotechnology and genetic engineering, will be found in Volume 6. The history and socio-economic aspects of agriculture will be found in volumes 3, 4, and 7 (which last will also include entries on living persons). Such items appearing in more than one volume will be linked by a general index in Volume 8.

A

Aardvark, or Cape ant-eater: an African mammal quite unrelated to the South American ant-eaters. It is found in all habitats except dense forest, from Ethiopia to the Cape of Good Hope. Growing up to 1·4 m. (4 ft. 6 in.) in length, it has a round body, tapering tail, and short, thick legs with powerful claws. The long, narrow head has large, donkey-like ears, and an elongated snout. The aardvark digs extensive burrows, large enough for a small man to enter. Feeding at night, it quickly demolishes termite nests up to 1·8 m. (6 ft.) in height, and consumes the termites with its sticky tongue, which is 30 cm. (12 in.) long. A single young is born and is cared for by the mother alone.

Aardwolf: a solitary, nocturnal, African mammal, related to the hyenas. Unlike other members of the hyena family, it feeds on termites. It lives on sandy plains or in bush country. Up to 50 cm. (20 in.) high at the shoulder, but spindly and slender, it weighs up to 27 kg. (60 lb.). The yellow-grey coat has black stripes and the lower parts of the limbs are black; the bushy tail has a black tip and the muzzle is black and hairless. Termites are swept up by

Aardvark means 'earth pig' in Afrikaans and this frontal shot of the animal shows its adaptations for locating and digging out insect prey. When foraging, the aardvark keeps its long snout close to the ground and its tubular ears pointed forwards. Once a termites' or an ants' nest is located, it uses its long, spoon-shaped claws to dig down and break open the nest.

its long, sticky tongue. A litter of two to four young, blind at birth, is produced each year in November or December, after a gestation period of 90–110 days.

Abalones are marine *molluscs related to limpets and snails, with flat, ear-shaped shells that have a series of holes to one side, providing the exit for the water currents that allow them to breathe. They live in shallow coastal waters and their shells are designed for minimum resistance to water movements. Abalones produce particularly good mother-of-pearl on the inner shell, hence their frequent use as ornamental ash-trays. They are also used as food in some areas, especially on the western coast of North America and on southern Pacific islands.

Abdomen is a term used to describe that part of the body, or trunk, of an animal which is distinct from the head or thorax. In mammals the abdomen contains the organs of digestion, excretion, and reproduction. It is separated from the thorax, containing the heart and lungs, by a muscular sheet called the diaphragm. The muscular walls of the mammalian abdomen are flexible, allowing for expansion of the intestines during feeding and respiratory movements of the diaphragm. The membrane lining the abdomen is called the peritoneum. In most other vertebrates there is no clear subdivision between thorax and abdomen, although fishes, reptiles, and amphibians do have a division between the heart and the other organs.

In arthropods, such as insects, crabs, or spiders, the abdomen is one of three major subdivisions of the body, forming the posterior, and often largest, part of the animal. In common with most invertebrates, the arthropod abdomen is divided into segments as well as containing the organs of digestion, excretion, and reproduction.

Abortion is the loss of a foetus before it is able to survive outside the mother. Spontaneous abortion, or miscarriage, is common early in pregnancy. Many such aborted foetuses have serious malformations or chromosomal abnormalities. Repeated spontaneous abortion may be due to a lethal genetic combination or a defect in the mother's reproductive system. Deliberate abortion is easily and safely carried out early in the pregnancy, but with increasing difficulty and risk later than the thirteenth week.

Abortion in animals, such as infective abortion in cows, can be an important cause of loss to farmers. Some animals, such as rabbits, resorb non-viable foetuses rather than abort them. They use this natural method of birth control when food or some other resource is scarce.

Abscesses are localized collections of pus in the body tissues. The usual cause is the presence of harmful bacteria, but a sterile abscess may follow the injection of something causing local tissue damage. Pus consists of white blood cells, whose main task is to destroy bacteria, but which release enzymes that can cause further tissue damage. An abscess can cause illness until it works its way to a surface and discharges, or is surgically drained. Healing usually follows adequate drainage. The term empyema is used when pus collects in a preformed body cavity, such as the gall bladder. An abscess within the skin is called a boil.

Acacias are tropical and subtropical trees (wattles), bushes, and climbers of the pea family. They are particularly common in the drier regions of Africa, Australia, and South America with about 400 species. Many of them are armed with hooks or large spines, and in some the thorns have an inflated base inhabited by ants, which deter other insects and also climbing plants. Wind blowing over the ants' entrance holes on certain African species gives rise to an eerie whistling noise.

Acacias may be the principal trees of some vegetation zones such as the dry plains of India. One African species, *Acacia albida*, produces leaves when all other trees are leafless, making it a valuable fodder tree. Others yield timber, gum arabic, tanning materials, the dye for the original khaki cloth, and flowers used in perfumery. The florists' mimosa, or silver wattle, *A. dealbata*, has yellow heads of tiny flowers typical of the genus. The true *Mimosa* is a separate genus which includes the *sensitive plant. The false acacia is a species of *Robinia*.

Acanthocephala ('spiny-headed worms') is a phylum of some 500 species of worm-like animals which are endo-parasites, usually 1–2 cm. ($\frac{1}{4}$–$\frac{3}{4}$ in.) long. The juvenile acanthocephalans are parasitic initially within insects, and are passed on to their primary vertebrate host when their insect host is eaten by the vertebrate. Inside the vertebrate host the larvae hatch out and attach themselves to the intestine wall using a strongly spined proboscis. These

worms have no gut and feed by direct absorption of the host's fluids. Females may produce millions of eggs, which pass out of the vertebrate host in faeces.

Acanthus is the generic name from the Greek *acantha*, thorn, in allusion to the spiny leaves and bracts of a group of fifty or so species of perennial herbs, chiefly native to the Mediterranean region. They belong to the family Acanthaceae, which also includes black-eyed Susan, *Thunbergia alata*. The shapely leaves were the inspiration for certain ornaments in classical architecture, including the capital of the Corinthian column. Some species, notably *Acanthus spinosus* and *A. mollis*, are grown as garden plants for their spikes of white, pink, or purple flowers, as well as for their decorative leaves.

Accentors make up the family Prunellidae, which includes thirteen species of sparrow-sized birds which are typically brownish, streaked with black, although some species have touches of pink, red, or yellow. They are confined to the Old World, North Africa, and the Middle East. Almost all species live in rocky or mountainous areas. The dunnock or hedge sparrow, *Prunella modularis*, is atypical in being widespread at low altitudes and is a common garden bird in Europe. This species has a complex family life, with territories being occupied by a single pair, two males and a female, two females and a male, or two pairs. All species build concealed nests of twigs, fibres, and hairs, and lay two to five light blue eggs. They feed on insects, fruits, and small seeds.

Acne is a skin disorder caused by excessive secretion and blocking of the sebaceous glands. These glands are associated with hair follicles and normally serve to keep the skin supple. They are stimulated by sex hormones and tend to react excessively at puberty. The condition usually settles down in early adult life, but by that time may have caused disfiguring scars as well as much distress. Modern treatment can lessen the effects of the condition.

Aconites are a group of widely distributed plants of the genus *Aconitum*, part of the buttercup family. All contain the very poisonous alkaloid aconitine, once used to poison wolves, a quality from which the common name of wolfsbane was derived. The name monkshood for *A. napellus* indicates the shape of the attractive blue flower. As a decorative herbaceous plant, *A. napellus* has long been grown in European gardens. The yellow-flowered winter aconite, *Eranthis hyemalis*, is also a member of the buttercup family.

Acorn worms, named for their acorn-like proboscis, are soft-bodied, worm-like animals which are closely related to vertebrates. They form a class within the phylum hemichordata and are usually 9–45 cm. (4–18 in.) long, although a Brazilian species, *Balanoglossus gigas*, can reach 1·5 m. (5 ft.) in length. They live in sandy sea-beds, with only the front of the body emerging to feed; and for this they use gill-slits, acting as filters. These structures are also found in sea-squirts, lancelets, and fishes, and explain why hemichordates were once considered as a subphylum of the Chordata (which includes all vertebrates).

Acupuncture is a medical treatment, long practised in China, in which sharp needles are inserted and rotated at specific points on the body for the relief of pain. It is not

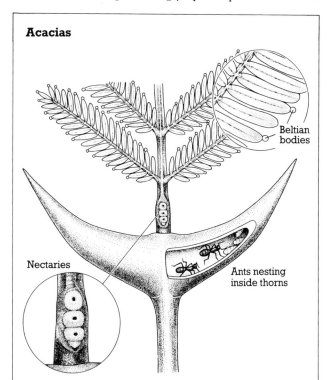

Acacias

Beltian bodies

Nectaries

Ants nesting inside thorns

The bull's horn acacias of the Central American forests have evolved an intricate relationship, symbiosis, with ants of the genus *Pseudomyrmex*. The ants nest in holes gnawed in the horn-shaped thorns and feed from nectaries at the base of the leaves; modified leaf tips, Beltian bodies, also provide protein and fat-rich food for their larvae. New leaves appear all year round ensuring a reliable food supply. The ants in turn remove fungi from leaves, drive away other insects, and destroy any plant that grows too close to their tree, thus protecting it from defoliation, shading, and competition from other plants. Without the ants the bull's horn acacia rarely survives to produce seed.

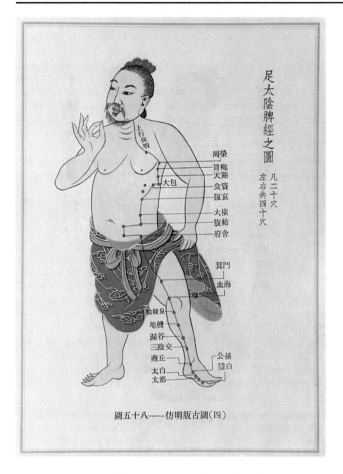

足太陰脾經之圖

凡二十穴
左右共四十穴

上行俠咽

周榮
胃鄉
天谿
食竇
腹哀

大包

大橫
腹結
府舍

箕門

血海

陰陵泉
地機
漏谷
三陰交
商丘
太白
太都

公孫
隱白

圖五十八——仿明版古圖（四）

Acupuncture needles are inserted at points along imaginary lines called meridians, as shown in this medieval Chinese chart. The meridians are believed to represent energy pathways in the body. The needles interrupt this flow to affect sensations such as pain.

certain how the method works but it is widely used in the Far East as an alternative to anaesthetics during major surgery. Western doctors accept that it may have a place in treating chronic headache or backache, in part because psychologically it seems to be of non-specific benefit to patients, perhaps by a mechanism akin to hypnotism.

Adanson, Michel (1727–1806), French botanist, developed the system of plant classification devised by Linnaeus, placing plants in natural orders on the basis of their overall affinity. His *Familles naturelles des Plantes* (1763) was notable for its predictive value. The *baobab genus *Adansonia* was named after him.

Adaptation in nature is the process by which plants and animals have evolved to fit and operate efficiently in a particular ecological niche. Successful adaptation ensures that an organism will be able to feed, grow, reproduce, and tolerate climatic and other fluctuations within the environment. To this end, adaptation may involve modification of structure, function, and, in animals, behaviour. Adaptation arises through *natural selection, which acts upon inheritable variation (arising from *mutation) within a species to remove less effective individuals. Thus, the species as a whole becomes increasingly specialized to fit its particular role or niche. For example, plants in dry habitats have extensive roots and mech-

anisms to reduce water loss and to store water, or they grow and flower rapidly only when rain falls.

In a similar way adaptive 'radiation' can occur, whereby new species evolve from one common ancestor to fit a range of microhabitats, or niches. On the Galapagos Islands Darwin's finches, or *Galapagos finches, probably arose from a single mainland species which first colonized these volcanic islands.

Adder: a species of snake belonging to the *viper family, with a wide distribution in Europe. It is found in a wide variety of habitats, including marshes, heaths, moors, hedgerows, and open woodland. It is Britain's only poisonous snake and, although venomous, its bite is rarely fatal to man. Adults grow up to 65 cm. (2 ft. 2 in.) in length. Males tend to be greyish in colour, females brown or reddish; both sexes generally have a dark zigzag stripe down the back, but it is not uncommon for entirely black individuals to occur.

Addison, Thomas (1793–1860), English physician, was the first to recognize Addison's Disease, a syndrome due to defective functioning of the *adrenal glands and often characterized by a bronzy pigmentation of the skin. He was also the first to describe appendicitis adequately and to explain the features of pernicious *anaemia.

A female **adder** and young. The gestation period normally lasts several weeks, the young being born during late August and early September. During cold summers, and in the northern part of the adder's range, birth can be delayed until the spring of the following year.

Adenoids are part of a ring of lymphoid tissue (*lymph glands) surrounding the openings of the mouth and nose into the gullet. They are found in all mammals and protect against the entrance of infection. In man, they are larger in children than in adults, and may either grow to restrict the passage of air through the nose, or block the tubes taking air to the middle ear, causing deafness. In such cases they have to be removed surgically.

Adenosine triphosphate (ATP) is found in all living organisms and is made up of a natural, organic nitrogenous compound called adenine, the five-carbon sugar, ribose, and a chain of three phosphate (PO_4 groups. The second and third phosphates are attached by high-energy bonds. When these bonds are broken to give adenosine diphosphate (ADP) or adenosine monophosphate (AMP), this energy is released. ATP acts as a *coenzyme and its breakdown is coupled to other reactions so that the energy released is used to perform work, for example in protein synthesis, ion transport, and muscle contraction. ATP is used in light production, as in fireflies, and this property was once used as an assay of the amount of ATP present in a solution. An enzyme called luciferinase was obtained from fireflies and when mixed with ATP produced a certain quantity of light in proportion to the amount of ATP. The cell's supplies of ATP are continually renewed by *respiration, and stored as a source of energy.

Adler, Alfred (1870-1937), Viennese psychologist, proposed that human striving is a need for power and is a reaction against inferiority. This theory led to the term 'inferiority complex'.

Adrenal glands in mammals consist of a pair of small *endocrine glands found near the kidneys. Each gland has two parts, the inner medulla and an outer cortex. Its outer part releases aldosterone, a steroid *hormone which regulates an animal's fluid and salt content, and glucocorticoids, which assist in co-ordinating the daily patterns of sugar and fat usage. The inner part of the gland is formed of tissue specialized to produce and store large amounts of noradrenaline and adrenaline, hormones that increase the breakdown of body fat and increase the blood sugar level. Their actions raise the blood pressure by increasing the power and rate of the heartbeat and regulating the diameter of blood vessels. In other vertebrate groups, the adrenal gland is a diffuse organ located among the other body contents.

African lily *Lily.

Agamids are a *lizard family comprising about 300 species. Some forms resemble lizards of the iguana family in external appearance and habits, but the two families occur in different parts of the world. Agamids occur in warmer parts of the Old World, including Australia. They are typical lizards, having well-developed limbs, and most species have a fairly long tail. Their life-styles are diverse; terrestrial, rock-dwelling, and tree-dwelling forms occur, including flying lizards and frilled lizards within the last category. They are usually easy to find in areas where they exist, due to their being active in the daytime and the dramatic display behaviour of some species. They include species such as the bearded lizard, *Amphibolorus barbatus*, the frilled lizard, *Chlamydosaurus kingii*, and the glass snake, *Ophisaurus apodus*.

Agarics, or gill fungi, are *fungi belonging to the order Agaricales, which produce *mushrooms or toadstools as their spore-bearing structures; examples are blewits, death caps, fly agarics, and the cultivated mushroom. The spores are produced in millions from gills or pores underneath the cap and are carried away in air currents. The cap of an agaric, left overnight on clean paper, will leave a 'spore print'. Most agarics are *saprophytes and play an important part in the breakdown of dead wood and leaves.

Agaves, or century plants, are all species native to Central and South America. They form a family of some 700 species of monocotyledons. Their common name alludes to the long period before the plant is mature enough to flower (in reality this is between seven and forty years, depending on the species). The fleshy, sword-like leaves form large rosettes—in some species up to 3 m. (10 ft.) across—which are expended in the production of a flowering stem often several metres in height. *A. sisalana* produces the fibre sisal, while the Mexican liquors pulque and tequila are distilled from the sap obtained by cutting away the developing flower stem of *A. americana*.

Ageing is the gradual but inevitable process that leads to the deterioration of the cells and tissue components of an organism. Its biochemical basis is still poorly understood. Ageing tissues are more susceptible to diseases than those

To ensure pollination, **agaves** often remain in flower for long periods of up to five months. These specimens are *Agave scabra*, a Mexican species, in bloom in the Chihuahuan Desert in the north of the country.

of other age classes. Characteristic features of 'normal' ageing include impaired muscular strength and activity; dryness and thinning of the skin, causing wrinkles; loss of hair and of hair colour; thinning of bones and limitation of joint movements; and deficiencies in vision and hearing. Ageing cells are unable to respond to hormonal signals, with a resulting loss of fertility and ability to withstand stress.

Aggression among animals is attacking, offensive behaviour, which is common in animals for the defence of self and territory, and the establishment of social hierarchies. Nearly all vertebrates and a number of invertebrates have a territory for all or part of their lives, and intruders may be met by a threat display. The bared teeth, snarling, and raised hackles of the domestic dog are typical. If this fails to deter the intruder, a fight is likely, sometimes, as in gannet colonies, to the death. An intruder of the same species may avoid a fight by giving out appeasement signals and adopting submissive postures.

In many instances the dominance of animals within their social group is determined by aggressive display or by fighting. In a flock of chickens, fights establish a 'pecking' order, for food and space, that will subsequently be maintained by display alone. This status quo persists until new individuals enter or some annual event, such as rutting in deer, occurs and then the dominance structure has to be redefined.

Aggression, particularly in vertebrates where a male may mate with several females, often ensures that the fittest, strongest males of the species father most offspring. This confers consequent benefit to subsequent generations in that the *genes which enabled their parents to compete for food and to reproduce successfully are passed on.

Agoutis are a genus (*Dasyprocta*) of eleven species of South and Central American *rodents that look like large, rather long-legged guinea-pigs with hoof-like claws and a short or rudimentary tail. They are characterized by long,

coarse hair on the rump, which is often conspicuously coloured. The hairs are erected when the animal is alarmed. Agoutis live on the floor of tropical evergreen forests or in savannah. They feed on fruit, roots, seeds, and succulent plants. They make up the family Dasyproctidae together with two species of acouchis (*Myoprocta* species) and are distinct from the family Agoutidae, which includes the two species of *paca.

AIDS, an acronym for Acquired Immune Deficiency Syndrome, appears generally as a new disease, though it is probably endemic in central Africa. It results in the progressive failure of cellular *immunity, leaving the sufferer susceptible to lethal infections. Many victims also have Kaposi's sarcoma, an otherwise rare vascular tumour. Transmission of this disease requires direct contact between body fluid and infected blood or serum. The disease is most common among young homosexuals but also occurs in patients receiving blood donated by carriers of the disease.

Albacore *Tuna.

Albatrosses are long-winged oceanic birds with a wingspan of 2–3.5 m. (6–12 ft.), and with a distinctive gliding and soaring flight. The albatross family contains nine Southern Hemisphere species inhabiting temperate and Antarctic waters, and three tropical species of the central Pacific. They all feed on surface-caught fish and squid. They breed on windswept islands, laying single eggs like their closest relatives the petrels.

Alcoholism is addiction to alcohol. Physical dependency is shown by withdrawal symptoms (shaking, sweating,

A wandering **albatross**, *Diomedea exulans*, and its single chick on South Georgia. Albatrosses spend most of their life at sea, coming to land only to breed, in the case of this species, only every two years.

and nausea) on stopping drinking for even a short time. Associated with these symptoms is severe anxiety and a craving for alcohol. Dependence is common in cases of a daily consumption of more than 100 g. (3·5 oz.), which is equivalent to 1 litre (2 US pints) of wine, half a bottle of spirits, or 4 litres (8 US pints) of beer. The health consequences are physical damage to the stomach, liver, and nervous system.

Alder flies are medium to large insects with two pairs of smoky wings, each with a net-like arrangement of veins. They are unrelated to true flies and are part of the order Megaloptera, which, with the *snake flies, includes some 200 species worldwide. The adult alder flies are found near still or slow-moving water. The larvae are predatory and aquatic, with strong, biting jaws and seven pairs of feathery gills fringing the tapering abdomen. A North American species of alder fly, *Corydalis* species, is known as the Dobson fly. The adult can have a wing-span of up to 16 cm. (6 in.) and has larvae up to 7·5 cm. (3 in.) long.

Alders comprise about thirty-five species of deciduous trees and shrubs native to north temperate zones and the Andes. They form part of the *birch family, along with hazels and hornbeams. They are readily distinguished from the allied birches, *Betula* species, as they have small, woody cones which persist after the seeds have been dispersed. The common alder, *Alnus glutinosa*, of Eurasia and North Africa, is a plant of wet places in woods and stream-sides. The wood of the North American red alder, *A. rubra*, has been used for furniture. The common alder has a soft, light timber used in the production of charcoal and plywood.

Alfalfa is one of the most important forage crops in the world, cultivated widely in temperate and warm temperate areas, principally in the USA. Known as 'lucerne' in Europe, South Africa, and Australia, this protein-rich *legume, originating in the Middle East, was the principal fodder of Roman cavalry- and chariot-horses. It is a perennial, deep-rooting, bushy plant with purple, sometimes yellow, flowers; the young leaves are used as a vegetable in China and elsewhere.

Algae are one of the major groups of living organisms and are often referred to as simple plants because they make their own food by *photosynthesis. The 13,000 or so species classed as algae include all the marine *seaweeds and vast numbers of microscopic, single-celled species which are major components of *plankton. Algae vary in size from single-celled forms such as Euglena, through ribbon-like, or filamentous species such as Spirogyra, to the huge kelps which may reach 60 m. (197 ft.) in length. Algae are cryptogams; they reproduce by means of spores. The spores move with the use of tiny hairs, called flagellae, that work like paddles.

As planktonic organisms in both marine and freshwater habitats, algae often undergo spectacular outbursts of growth. These phases, known as algal blooms, occur in response to an increase in nutrient levels in their lake, river, or sea. The 'red tides' of some parts of the world are caused by blooms of algae with red photosynthetic pigments. The green water so characteristic of river or lake *eutrophication is due to tremendous numbers of planktonic algae. The foul smell associated with algal blooms is caused by algal cells using up all the oxygen in the water and releasing gases such as hydrogen sulphide.

The classification of algae has always presented a problem to biologists as many 'algae' have features in common with both animals and plants. Many species lack a cellulose cell wall, considered as basic to all plant cells, while others possess the ability to photosynthesize, yet also ingest food particles. Thus algae have been grouped along with plants by some, or as a separate group by others. They are now regarded as part of the kingdom Protista, which also includes the bacteria, fungi, and Protozoa. The organisms known as *blue-green algae are more closely related to bacteria than to algae.

Alimentary canal: a muscular tube, starting at the mouth and ending at the anus. It is concerned with the passage, digestion, and absorption of nutrients. It varies in complexity depending upon the animal and its food. In an earthworm it is a straight tube. In birds it contains specialized parts, such as the crop and gizzard, in which food is broken up. In mammals the tract differs in herbivores, omnivores, and carnivores.

Alkaloids are organic nitrogen-containing compounds, which are found in the stems, roots, bark, or leaves of plants. Many alkaloids, such as morphine, cocaine, quinine, and caffeine, are valuable drugs with useful pharmacological properties; others, including curare, strychnine, and colchicine are poisons.

Allergies are excessive reactions of the body to some substance. Examples are *asthma, provoked by the inhalation of pollen, and *urticaria, provoked by certain foods. The usual mechanism is that the offending substance combines with a special type of *antibody, which is attached to certain cells. These then release histamine and other compounds to produce the reaction. Allergy is an important cause of adverse reactions to drugs. Members of some families have a propensity to produce larger than normal amounts of the antibody concerned, becoming allergic to many substances. Although allergic reactions are unpleasant and sometimes dangerous, they are thought to be part of the body's natural defence, particularly against parasitic worms.

Alligator bugs *Lantern flies.

Alligators are the only members of the crocodilian order of *reptiles confined to temperate zones. There are two species of alligators: the American alligator, *Alligator mississipiensis*, inhabits wetlands in the southeastern USA, whereas the Chinese alligator, *A. sinensis*, is restricted to the lower Yangtze valley. They form the family Alligatoridae, which also includes the *caymans.

The American alligator has been known, exceptionally, to attain a length of 5·8 m. (19 ft. 2 in.). The female deposits her eggs in a nest constructed of mud and rotting vegetation and guards them until they hatch. Although hide-hunting and habitat loss have reduced the range of the American alligator, protective legislation has contributed to a recent recovery in numbers. The endangered Chinese alligator is smaller, secretive in its habits, and spends much of its time in burrows close to water. Alligators are distinguished from crocodiles by having the upper teeth lying outside the lower teeth when the mouth is closed. Their head is broader and shorter than that of the crocodile.

An American **alligator** lies up in its water-hole in the Everglades swamp, Florida. Its log-like appearance provides a useful disguise when it is hunting. Alligators are inactive for much of the time, remaining immobile in order to ambush passing prey, to digest their food, or to raise their body temperature by basking in the sun.

Allopathy is a term used, principally by homoeopathists, to denote the orthodox use of drugs, as opposed to *homoeopathy.

Almonds are deciduous, ornamental fruit trees belonging to the *cherry subfamily (Prunoideae), and represented by a single species. Almond trees grow up to 8 m. (27 ft.) tall, producing pink blossom in spring. Originating in the Near East, they are naturalized in southern Europe and introduced and cultivated in many other parts of the world, including California, South Australia, and South Africa. The bitter almond, a subspecies of the common almond, contains prussic acid and is the chief source of almond oil. The common, or sweet, almond is the type grown for its oval, edible nuts.

Aloes are a large group of African succulents with fleshy leaves and, in some species, trunk-like stems up to 7 m. (23 ft.) in height. They belong to the lily family, along with tulips, onions, and others. Bitter aloes is a pharmaceutical product of the plant sap. The flowers of aloes are bell-like, or tubular, and produced on long stalks.

Alopecia means baldness. The common type affecting young men is a condition which may be inherited through either parent, but is only expressed under the influence of

the male sex hormone. Other types of premature hair loss are comparatively rare, and often temporary. Most men and many women tend to go bald in old age.

Alpaca: a domesticated South American *ruminant, related to the llama. It has been known since 200 BC but nothing precise is recorded about its domestication. It is bred as a beast of burden as well as for its wool and meat. Its long fleece, often reaching to the ground, is of uniform black, reddish-brown, or white, or is sometimes piebald. It is sheared, every two years, and may provide up to 5 kg. (11 lb.) of wool each time. The alpaca thrives best at altitudes of 4,000–4,800 m. (13,000–16,000 ft.) in the high Andes and grows to about 90 cm. (3 ft.) tall at the shoulders. The female has a gestation period of eleven months and foals in the rainy season. The centre for breeding has, since time immemorial, been on the high plateau around Lake Titicaca.

Alpine plants, or arctic-alpines, are adapted to habitats with low temperatures, seasonal snow cover, and strong winds. They are found at various altitudes from sea-level in Arctic and Antarctic zones, to several thousand metres in the tropics and subtropics. Usually dwarf and compact in habit, they may be shrubs, bulbs, or mat-forming species. All are adapted to a short growing season, either by flowering after the snow melts, or by reproducing by vegetative means. Most are waxy or hairy so that water loss is minimized.

Alternation of generations   *Life cycles.

Alternative medicine, or fringe medicine, is a term used to describe the many unorthodox systems such as *homoeopathy, osteopathy, and *acupuncture. Each system proposes its own theory of disease and its treatment. Some are based partly on ideas already accepted in orthodox medicine, others on ideas totally estranged from accepted theories. Some unorthodox treatment may be effective.

Amaryllis, or belladonna lily: a plant genus with a single species. Related to the daffodil, the true amaryllis, *Amaryllis belladonna*, is native to South Africa. It has a bulb and a cycle of growth and rest. The large, sweetly scented, rose-pink flowers are produced at the end of the dry season and they are followed by strap-like leaves which grow during the rainy period. The name is wrongly applied to the South American bulbous plant *Hippeastrum*, which is often grown as an indoor plant.

Ambrosia beetles, or pin-hole borers, are small, dark, cylindrical *bark beetles, The young adults carry fruiting bodies of certain fungi into their egg tunnels, in which the developing hyphae form food for the beetles' larvae. There are two families of beetles which have species known as ambrosia beetles, the Scolytidae and the large tropical family, the Platypodidae.

American chameleons   *Anoles.

Amino acids are natural organic compounds which are the subunits of *proteins. There are twenty different amino acids found in living organisms, although some are rarely found. Each amino acid molecule consists of a central carbon atom with four bonds attaching it to hydrogen

(—H), amino (—NH₂), and carboxyl (—COOH) groups, and also to a variable '—R' group known as the '—R' group. The —R group is made up of carbon, hydrogen, oxygen and, in some amino acids, sulphur or nitrogen. It has a different structure in each of the twenty amino acids and determines their individual chemical properties. Amino acids are joined together like the coaches of a train by chemical bonds (peptide bonds) formed between the carboxyl group of one, and the amino group of the next. Short chains of two to twenty amino acids are called peptides; longer chains are called polypeptides. In proteins, the chains are twisted and folded to give a three-dimensional shape, which is determined by the order of the individual amino acids.

In the human diet, ten of the twenty different types, called 'essential' amino acids, must be present; the remaining 'non-essential' types can be manufactured by the liver. Amino acids cannot be stored in the body for more than four hours, so all of the essential ones must be eaten together and in the correct proportions. Excess amino acids in the body, obtained from the digestion of proteins, are 'deaminated': the amino group is removed and converted into *urea (in man and most mammals) for excretion. In addition to building proteins, amino acids are used for making hormones, certain vitamins, coenzymes, antibodies, and transmitters of impulses between nerve cells.

Ammonites are the fossil remains of a group of *cephalopod molluscs, now extinct but which flourished in the seas during the Mesozoic Era (245–66.4 million years ago). They consist of flat, tightly coiled shells which both protected the animal and probably contained air to help it to float. In some respects, these animals were probably like present-day species of *nautilus*.

Amnesia means loss of memory. The commonest cause is damage to the brain, following a blow to the head. Memory of events immediately before the injury, as well as those afterwards, may be lost, and the victim may be unconscious for a time. Memory usually returns after a period ranging from minutes to months, depending on the severity of the damage.

An **ammonite**, *Psiloceras planorbis*, from Watchet in Somerset, England. The colour and shining quality of the fossil are due to the pearly shell having been partially preserved. The fossil dates from the Jurassic Period, 180 million years ago.

Amoebae are classic examples of unicellular *protozoa. They are usually highly asymmetric creatures which move by means of flowing pseudopodia (foot-like protrusions of the flexible cell wall). Within the one-celled body a large nucleus and many microscopic organelles occur, yet there is usually little detail to be seen even in giant amoebae, which are up to 0.5 cm. (⅕ in.) long. Amoebae live in seas, fresh water, and soils, or occasionally as parasites (causing amoebic dysentery). Besides the 'naked' forms there are amoebae with shells, which live in small ponds and among mosses. These shells are made of silica or chitin, often elaborately sculpted, with one opening where the pseudopodia can be protruded to catch food. Amoebae are closely related to *foraminifera and *radiolaria.

Amphibians are a class of vertebrate animals which can live on land and in water and have eggs which are externally fertilized. Their class includes frogs, toads, newts, salamanders, sirens, and caecilians. They are modern descendants of fish-like animals which made the transition from water on to land, an event which probably happened in the Devonian period, some 350 million years ago. The change from fish-like ancestor to present-day amphibian involved the development of limbs, lungs, and a way of changing from an aquatic larva to an adult amphibian. This latter change is called metamorphosis.

The skin is particularly important to an amphibian, since it acts as a respiratory (air-breathing) surface, although it needs to be kept moist to function in this way. The lungs, present in most though not all amphibians, are simple sac-like structures which connect via a tube to the oral cavity. An efficient excretory system removes the waste products of metabolism, while controlling salt and water losses from the body. Frogs and toads have a true *ear and a more complex heart than their fish ancestors. These features, which first appeared among vertebrates in the amphibians, together with a variety of further structural and behavioural adaptations, have enabled the amphibians to colonize a wide variety of habitats in most continents of the world, except Antarctica.

Modern amphibians (approximately 4,000 species) belong to three distinct major orders: the Caudata or tailed amphibians (newts, salamanders, and sirens); the Anura or tail-less amphibians (frogs and toads); and the Gymnophiona or Apoda (the limbless, worm-like caecilians).

Amphioxus *Lancelets.

Amphipods are an order of *crustaceans with over 4,600 species, including the very common sandhoppers, *Orchestia*, and the freshwater 'shrimps', such as *Gammarus pulex*, as well as many marine shrimp-like animals and the elongate 'skeleton shrimps', *Caprella*. Most amphipods are specialized to live at the bottom of ponds and streams or along sea-shores; they have different sets of legs for swimming, walking, and jumping. Amphipods are generally flattened laterally and look hunch-backed as they scavenge on mud or sand, sometimes even swimming on their sides. Some species are parasitic and have evolved bizarre body shapes. Any stream or pond should yield examples of the commonest freshwater species, *G. pulex*, often found in mating pairs. The male may carry the female beneath him for days before fertilizing the eggs in her brood-pouch.

Amphistaenians, or Amphisbaerids, are *reptiles that resemble earthworms, being exceedingly specialized for a

Freshwater **amphipods**, *Gammarus pulex*, mating. A female *Gammarus* will produce four to six broods, each of between seven and forty young, during her life.

burrowing life-style. They are limbless, with the exception of three Mexican species which have fore-limbs only. Their blunt heads and strong skulls are well adapted for ramming through soil. They comprise about 130 species and, in spite of sometimes being referred to as worm lizards, are quite distinct from both lizards and snakes. In the Americas they range from Florida to Patagonia; in the Old World they occur in Africa, western Asia, Spain, and Portugal.

Anaconda: a species of snake, belonging to the boa family, that occurs in northern South America. It is the largest living snake; there are credible records of specimens up to 11·4 m. (37 ft. 5 in.) long. Found usually in pools and streams, or sometimes basking in trees or bushes on an adjacent bank, it feeds mainly on mammals and birds that come to the water to drink, though aquatic animals such as fishes, turtles, and small caymans are also taken. A smaller, closely related, species, the southern anaconda, *Eunectes notaeus*, occurs in Paraguay. Anacondas kill by constriction, effectively suffocating the prey animal.

Anaemia is a decrease in the concentration of *haemoglobin in the blood, usually in association with a reduced number of red blood cells. Among the most common causes are deficiency of iron, an essential constituent of haemoglobin, and hereditary diseases such as sickle-cell anaemia and *thalassaemia, both of which are associated with abnormal haemoglobin. People with sickle-cell anaemia have red blood cells that collapse into a sickle shape from lack of oxygen. This condition offers some defence against attacks by malarial parasites.

Anaesthesia is the state induced by a variety of drugs, known as anaesthetics. It may be local, as in dentistry, when a drug is injected close to the nerves in the mouth to stop them transmitting pain signals; or it may be gen-

eral, as in major operations, when patients are made unconscious by injecting a drug, which acts on the brain, dulling all sensation of pain. Anaesthetists administer the drugs and maintain unconsciousness in the patient by using anaesthetic gases; they also monitor and support the vital functions of the body during long operations and so play an important role in modern surgery.

Anatomy as a science is the study of the shape, form, consistency, and organization of plants and animals. Comparative anatomy deals with differences in these aspects between humans and other animals and is important in the study of evolution.

Study of the external features of living organisms is called morphology and covers all external detail down to that visible only under an electron microscope. Anatomy embraces morphology, and applies also to the internal features of organs, skeletons, and even cellular detail. Dissection is its most powerful tool and its historical basis, while modern non-destructive means of studying internal structures include body-scanners and X-ray. The study of tissues is known as histology, and that of cells as cytology.

Anchovies are several species of slender fishes, related to the herring, with a clear green back, silvery sides, and a very long upper jaw. The European anchovy, *Engraulis encrasicolus*, forms large schools near the surface of the sea round European coasts. Other species of the anchovy family (Engraulidae) live in warm and temperate waters around the world; many are important commercial fishes.

Anemones, of which there are twenty or so wild species, mostly native to the Northern Hemisphere, include woodland and sun-loving kinds. They belong to the buttercup family and have either a tuberous corm or a fibrous root. Several have been developed as garden plants of great merit, including *Anemone coronaria*, the cut-flower of the florist. The common name of wind-flower is used for the small, white-flowered *A. sylvestris*, native to Europe.

Anemones, sea *Sea anemones.

Aneurysms are localized bulgings of an artery following a weakening in its wall. They are inherently unstable, and

Students attend an **anatomy** lecture at the Barber-Surgeons' Hall, London in 1581. At this time, the major application of anatomy was in the field of amputations.

tend to enlarge until they rupture. Many aneurysms affect the aorta, the main artery of the body, of elderly people, and are a consequence of *arteriosclerosis. Aortic aneurysms due to syphilis used to be common but are now rare. Small aneurysms, due to weakness in blood vessels supplying the brain, can cause one type of *stroke. Such weaknesses, if they occur, are commonly present at birth.

Angel-fish is a name used for two distinct and unrelated groups of fishes. The freshwater angel-fishes, *Pterophyllum* species, are South American *cichlids with very deep bodies and long rays in the dorsal, anal, and pelvic fins. Several species are recognized, all small, and all barred with dark vertical stripes. They are popular tropical aquarium fishes. Marine angel-fishes, *Pomacanthus* species, live mainly on coral reefs in all tropical seas. They are brightly coloured and deep-bodied and all have a sharp, backwardly pointing spine on the lower edge of the gill cover. They belong to the same family (Chaetontidae) as the marine butterfly fishes.

Angelica, although occurring naturally in northern Europe, Greenland, and Iceland, is cultivated and naturalized in many parts of Europe. A large biennial herb, it is a member of the carrot family, Umbelliferae. Pieces of the young stem and leaf-stalk are crystallized in sugar and used in confectionery as flavouring and decoration.

Angina means 'strangling', and is still sometimes used to denote certain types of sore throat, but usually angina pectoris is implied—a painful constricting sensation in the chest. The cause is typically a narrowing of one of the arteries supplying the heart. The symptom is provoked when the heart is required to increase its work and is unable to increase its blood supply correspondingly. The sensation thus appears at a certain level of exertion, and subsides quickly on its discontinuance.

Angiosperms are the dominant group of land plants in the world today and include all flowering plants. Thus most trees, all grasses, herbaceous plants, herbs, shrubs, and some aquatic plants are termed angiosperms. They exhibit a diversity of form and ecology greater than has any other group of plants at any point in earth's history. They probably arose in the Cretaceous Period from a group of seed ferns, and they differ from other seed-bearing plants in usually having closed carpels, which protect the seeds from desiccation and attacks from animals. They also have a unique form of fertilization whereby, in addition to the egg, other accessory cells are fertilized to give the endosperm, which later nourishes the developing plant embryo. The alternative group of plants to angiosperms is the group *gymnosperms.

Anglerfishes form an order of bony fishes in which the first ray of the dorsal fin is modified to form a flexible 'fishing-rod' with a fleshy bait, which is used to lure fishes close to the mouth of the angler. They all have large jaws with very sharp and often long teeth, and feed almost entirely on fishes. They are all marine, and include the shallow-water angler, *Lophius piscatorius*, of the eastern Atlantic, and the North American goosefish, *L. americanus*. Many of the deep-sea anglers have luminous lures, and bizarre body shapes. Females are always larger than males and in at least four families the males are parasitic on the adult females.

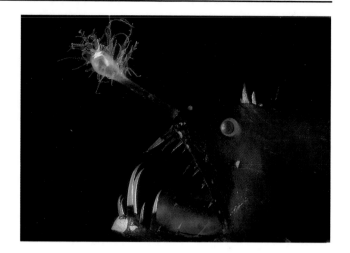

A deep-sea **anglerfish**, *Linophryne polypagon*, shows its huge gape and multi-branched luminous lure, which attracts prey. Adult males of this species are unknown but are believed to be parasitic on the female's body.

Animal behaviour is the way an animal responds to its environment and to its own and other species. It may ensure mating and continuation of the species, or survival of individuals through *adaptation to the environment. Behaviour may be inherited or learned, and both types play a part in survival. Even simple organisms show complex inherited behaviour patterns, often referred to as instinctive behaviour. These may facilitate feeding, as in the web-building of many spiders, or ensure successful reproduction, as in courtship behaviour. Inherited behaviour may provide communication between social animals, such as the waggle dances of hive *bees which tell others of the whereabouts of food. Inherited behaviour patterns tend to be inflexible and usually consist of a series of actions, each of which requires a particular signal or reaction from another individual. More sophisticated organisms, such as mammals, also show inherited behaviour, for example when human infants cling instinctively to their mothers. In addition, such organisms are able to pass on learned behaviour from one generation to the next. For example, lion cubs learn to hunt by watching and copying their parents.

The well-developed brain of humans, and other *primates, allows them to learn a wide range of behaviour patterns. These include speech, social interaction, and manipulative and mental skills. As with any other feature of an animal, behaviour patterns are subject to *natural selection, and many species rapidly adapt to new situations.

Animals are organisms which rely upon preformed food, as distinct from *autotrophic organisms. One of the main distinguishing features of many species of animals is their ability to move freely through air or water. This mobility, which extends from the smallest flagellate protozoa to the largest mammals, is effected by muscles. At a cellular level, animal cells are bounded by flexible *membrane systems, unlike the rigid cellulose walls of plant cells.

With the exception of *coelenterates, such as jellyfish, and *echinoderms, most animals are bilaterally symmetrical. This means that their body can be divided into similar halves along a line drawn from head to tail. Associated with this body-plan is the development of a head (cephalization). This is the first part of a mobile animal to reach new stimuli in its environment, and hence contains a vast array of sensory organs. The high activity levels of most animals, and their large multicellular bodies, necessitated the evolution of efficient circulatory systems to move oxygen and food to tissues, and waste products away. Larger bodies also need good excretory and nervous systems.

The link between animals and plants is revealed in the basic chemical reactions which sustain life. Processes such as *respiration, and the enzymes used in many similar biochemical pathways, are common to both kingdoms. Modern biology does not classify organisms as either animals or plants, but recognizes that there are 'grey' areas in between the two. Viruses, fungi, bacteria, and blue-green algae often defy categorization. Some major groupings such as protozoa are now classified according to their autotrophic or heterotrophic tendencies.

Anise, or aniseed, a member of the hemlock family, is an annual herb, introduced long ago from China to Europe, Asia, and North America. The small, greyish-brown, aromatic fruits are used for flavouring, giving the characteristic flavour to the liqueur 'anisette' as well as to various beverages, cakes, and sweets. Oil of anise, distilled from the fruits, is used in cough medicines and lozenges.

Annelids are the advanced and diverse phylum of worms, over 8,700 species of which show segmentation, being built up from many similar ring-shaped body-sections. The annelids are divided into three main classes according to their pattern of body hairs, or appendages. The polychaetes form the bulk of the phylum with over 5,300 species of marine worms, such as bristle worms and lugworms. The oligochaetes, which lack the segmental appendages of the last class, include the common earthworm (which has stiff bristles on each segment) and live on land or in fresh water. Among the 3,100 species of oligochaete are the giant Australian earthworms, which reach 3 m. (10 ft.) in length. The final class is the hirudinea, which contains over 500 species of smooth, segmented worms called leeches.

The annelid body is generally soft and cylindrical, with an outer wall and an inner gut, separated by the *coelom, which acts as the hydrostatic (fluid-filled) skeleton. Small bristles or longer appendages anchor the animal into sand or mud; in free-swimming predatory species these appendages are larger, supporting the lateral paddles that assist locomotion. Annelids also have a proper blood-system with a long, tubular, dorsal heart, unlike most other 'worms'; the blood vessels may be seen through the transparent skin, with the gills or even the whole body appearing reddish. They also have a simple brain, and can show learned behaviour.

Annuals (plants) germinate from seeds, develop, flower, and fruit within a single year. The field poppy is an example. Species which produce several short-lived generations in a year, such as groundsel, are called ephemerals.

Anoa   *Buffaloes.

Anoles belong to the iguana family of lizards, though they are sometimes, misleadingly, called American chameleons. There are 165 species in the genus *Anolis*. They range from the southeastern USA to Brazil, being par-

ticularly diverse in the West Indies. They live mainly in trees and bushes; their fingers and toes are adapted for climbing, bearing pads, rather like those of some geckos. More than one species often occurs in the same place, but complex territorial behaviour generally ensures that competition is minimized. Males display by head-bobbing and by extending a throat flap which is often coloured bright red or yellow. They are small lizards ranging from 13 cm. (5 in.) to 46 cm. (18 in.) in length.

Anorexia is loss of appetite, and is a symptom of many illnesses. Little is known about the control of appetite, and attempts to induce anorexia by drugs in the treatment of obesity have been largely unsuccessful. Anorexia nervosa is a psychological illness, occurring usually in adolescent girls. It often starts with a desire to lose weight which becomes an obsession. There is considerable weight loss, and treatment is directed towards restoring a healthy physique.

Ant birds are a large family of about 230 species. They are largely confined to South America, although about thirty-five occur in Central America. Ranging in size from that of a sparrow to that of a large thrush, they are mostly coloured black, brown, and white, though a few are strikingly patterned and some have bright blue patches of skin around the eyes. They live on or near the ground in scrub or tropical rain forest. They build a simple cup-shaped nest and lay two pale, speckled eggs. They are largely insectivorous, and some species follow columns of army ants and catch the insects which these flush out—hence the name of the family.

Ant-eaters are relatives of the sloths, armadillos, and pangolins. The four species of ant-eater are native to South and Central America. The giant ant-eater, *Myrmecophaga tridactyla*, is 2·4 m. (8 ft.) long with an enormously elongated face terminating in an extremely small mouth. It has a long hairy coat and a long bushy tail and lives in forests and savannah. The front legs have powerful claws, the middle one being much enlarged and sabre-like, used in tearing open ant nests or occasionally termite nests. As the insects swarm out they are collected by the long, smooth, whip-like tongue, which often extends 22 cm. (9 in.) out of its mouth and is moistened with a sticky saliva. In captivity the giant ant-eater has lived for up to twenty-six years. A single offspring is born each year, the female carrying it on her back; the youngster is not fully independent for almost two years. The giant ant-eater, now reduced to a fraction of its former range, is an endangered species, but some populations live in conservation areas.

The remaining three species of ant-eater: the northern tamandua, *Tamandua mexicana*, southern tamandua, *T. teradactyla*, and the silky ant-eater, *Cyclopes didactylus*, live in trees, where they climb and cling by their hind-legs, so leaving the fore-legs free to dig for ants. The snout of the silky ant-eater is much shorter than that of the others.

Antelopes do not form a single zoological grouping: the term is vague and is used for several diverse lines of the family Bovidae, which also includes sheep, cattle, and goats. The 'typical' antelopes are mostly tall, slender, graceful, swift-moving animals with smooth hair and upward-pointing hollow horns. Examples of antelopes include gazelles, gnus, hartebeest, klipspringers, kudus, nilgai, oryxes, and springboks.

Anthrax is an acute infection of sheep and cattle, transmissible to man through infected animal hair, hides, and excrement, but it is now very rare. It is due to a specific type of bacteria, which may be present in the gut of sheep. The organism has the capacity to form spores, which may lie dormant in soil, but remain potentially infectious, for many years. The disease attacks the lungs or skin in man and is fatal if untreated.

Antibiotics are substances extracted from micro-organisms such as fungi, and used to destroy or inhibit the growth of others, such as infective bacteria and fungi. Only those with few side-effects can be used in medicine. Inappropriate or inadequate use often leads to the development of resistant strains of the damaging micro-organism. Many infective bacteria are now resistant to early forms of the first antibiotic, penicillin, a drug to which some people are allergic.

The most commonly used antibiotics include cephalosporins, streptomycin, and tetracycline, as well as the penicillins. These natural extracts are often chemically modified to increase their stability and specificity. The exploitation of antibiotics, along with synthetic drugs, has revolutionized medicine in developed countries, robbing acute infections of their terrors, and reducing tuberculosis from a scourge to a nuisance.

Antibodies are an important group of proteins, called immunoglobulins, which defend vertebrate animals against infection. They inactivate viruses and bacterial toxins and recruit various migratory body-cells to kill and remove foreign material. Antibodies, attached to the surface of white blood cells called lymphocytes, bind themselves to substances foreign to the body. These substances are called antigens, and may be particles such as pollen, or chemicals. The highly selective reaction of an antigen with its antibody activates the parent lymphocyte to multiply into special cells able to produce and release into the blood very large quantities of soluble antibody protein. The reaction forms an important component of *immunity to specific diseases.

Antirrhinums are commonly known as snapdragons, in allusion to the shape of the flower. There are about forty perennial, herbaceous species native to the Northern Hemisphere. They belong to the foxglove family and are related to figwort and speedwell. Many garden varieties in a large range of colours and sizes have been developed from *Antirrhinum majus*.

Antlers are the horns of deer. They are usually found only in the male, but are present in both sexes of the reindeer. They are bony outgrowths and are shed each year. After the old antlers are cast, new ones begin to grow almost immediately and, to satisfy the craving for minerals, especially calcium phosphate, the animal will gnaw the shed antlers. In successive years, there is an increase in size and complexity of the antlers, with progressively more branches, or tines, as the animal grows older. Full development is reached at about six years of age, after which there may be some decline in the form of the antlers. A stag of the red deer whose antlers have twelve tines is known as royal.

Ant lions are a family of some 1,200 species of dragonfly-like insects with wing-spans of up to 12·5 cm. (5 in.). They are mainly tropical insects, and part of the same insect order as the *lacewings. Their larvae excavate conical pits in sand and lie at the bottom. When another insect, often an ant, passes over the top they flick sand at it to make it fall in. They have strong, sickle-shaped jaws with which they hold their prey, while sucking out its juices.

Ants are insects which belong to the same order as bees, wasps, and sawflies. They are found in almost all terrestrial habitats. There are about 14,000 species, most occurring in the tropics. Ants, whose sizes range from 1 mm. to 4 cm. ($\frac{1}{25}$ to $1\frac{1}{2}$ in.), live in colonies made up of different funtional types, or castes. These include workers, soldiers, and reproductive castes. At certain seasons, winged males and females are produced synchronously by adjacent colonies of a species. These leave their nests in swarms and mate, after which the males die. The fertilized females shed their wings and usually start a new colony. Queens of some species join existing colonies, a few infiltrate colonies of different species and kill the queen, and others raise workers which raid nests of different species for 'slaves'. Colony cohesion and organization are maintained by the queen's *pheromones and by exchange of regurgitated food and saliva.

Most ants excavate subterranean nests, but some build nests in the branches of trees, or excavate tree-trunks. Tailor or weaver ants, such as *Decophylla smaragdina*, use their silk-producing larvae to sew leaves into a chamber. The majority are voracious predators, although some species also avidly eat sugary fluids, such as nectar and honeydew. A few store and eat seeds, and leaf-cutter ants of tropical America (*Atta* species) cut small sections off leaves and use these as a basis for fungal 'gardens', which are harvested for food.

Ape is a term applied in popular usage to any tail-less monkey but which should be reserved for *primates of the family Pongidae. The living apes include chimpanzees and gorillas from Africa, and gibbons, siamangs, and orang-

A leaf-cutter **ant**, *Atta* species, from Trinidad, one of the 200 species that inhabit the New World tropics, where they are serious economic pests. They cut leaves for use as compost, on which to grow the fungi that constitute their food. This habit is shared with certain species of termites and wood-boring beetles.

utans from Asia. Instead of walking along the branches of trees as monkeys do, apes swing by their powerful, muscular arms, which are longer than their legs. However, gorillas rarely swing by their arms owing to their large size. In all apes, both hands and feet are efficient grasping organs. Most of the apes are vegetarians but chimpanzees occasionally eat meat.

The apes have highly developed social organization and systems of communication, except the orang-utan, which is solitary. The chimpanzee is capable of producing some thirty distinct sounds and its facial musculature allows a wide range of expressions, such as anger, pleasure, laughter, and surprise. The gestation period of apes is up to nine months in some species, and their life-span can be up to forty years in the cases of the chimpanzee, gorilla, and orang-utan.

Aphasia, or dysphasia, is impairment of the use of language, as distinct from dysarthria, which is impairment of pronunciation. Aphasia way be limited to difficulty in finding appropriate words, but more commonly there is also difficulty in comprehending spoken and written speech. It is usually due to damage to a special area of the cortex of one hemisphere of the brain, usually the left.

Aphids (singular: aphis) are small insects which belong to the homopteran *bug family Aphididae with over 3,000 species worldwide. They all have sucking mouthparts and feed on the sap of plants. Their life cycles are often extremely complex, including parthenogenetic and sexual generations, winged or wingless adults, and alternation of food plants. In the course of feeding, they excrete excess sugars as honeydew; this is exploited by ants. Many species, commonly known as greenfly, blackfly, or plant lice, are important pests of crops.

Appendicitis is inflammation of the appendix, and is a common condition in childhood and early adult life. The risk of the dangerous development of *peritonitis is largely avoided by early surgical removal of the appendix while the inflammation is still localized.

Appendix (anatomy): the name given to a blind-ended tube, up to 10 cm. (4 in.) long in man, which protrudes from the inner side of the caecum, or the first part of the large *intestine. The appendix is often considered vestigial in man, but it is also present in many other mammals. In herbivorous mammals it is a well-developed elongated organ which assists in the digestion of cellulose.

Apples are small trees, growing up to 6 m. (20 ft.) tall, represented by some twenty-five species in north temperate regions. The wild or crab apple, *Malus sylvestris*, has small sour fruits; breeding and selection from it has yielded thousands of varieties of cider, cooking, and dessert apples. Other species of *Malus* are grown as ornamental trees, especially for their colourful spring blossom, as are many of their relatives in the rose family. Some cultivated apples are self-fertile but most require pollination by another compatible variety for good fruit production.

'Appleseed, Johnny' (1774–1847), was an American orchardist, whose real name was John Chapman. From his Pennsylvania nursery he distributed apple saplings and seeds to settlers moving into the Ohio valley. Later he himself followed, planting seeds and tending older trees.

Apraxia is a disorder characterized by the inability of an otherwise normal person to carry out correctly a voluntary movement like lighting a match. It may be due to failure to activate, within brain centres, the necessary sequence of muscular movements.

Apricots are cultivated mainly in California, China, Japan, and northern Africa, as they are susceptible to frosts in cooler climates. The medium-sized trees, up to 10 m. (33 ft.) tall, produce white, or occasionally pink, blossom. As a member of the cherry subfamily, apricots produce edible fruit similar to that of the peach, both species originating from China. The wild apricot is the parent of all cultivated varieties.

Aquilegias are perennial herbaceous plants from the mountains and temperate zones of the Northern Hemisphere. They are relatives of the delphinium, clematis, and others of the buttercup family. The common name columbine indicates the dove- or bird-like appearance of the flowers, which have long, nectar-bearing spurs attractive to bumble-bees.

Arachnids are one of the most numerous classes within the *arthropods, being second in number only to the insects. There are at least 60,000 species, including scorpions, spiders, mites, and ticks. They range in size from a fraction of a centimetre in some mites, to 18 cm. (7 in.) in the African scorpion, *Pandirus*. Arachnids succeed in all habitats from tundra to desert, mostly as carnivores, pursuing or trapping other small animals. Their body plan consists typically of two main parts: the front part, or 'head', bears eyes, jaws, *pedipalps, and eight walking legs, while the abdomen houses most of the organs.

Arachnid jaws cannot bite strongly, so poisons are used to subdue prey, or silk to enwrap it, before it is chewed. Being so aggressive, many spiders and scorpions have to be very careful when mating; males are usually smaller, and may be eaten by their lovers! The males use ritual

White-flowered **aquilegias**, *Aquilegia fragrans*, grow in typical scattered clumps beside an alpine lake in Kashmir at 3,650 m. (12,000 ft.).

dancing during courtship, bringing gifts to the female or wooing her with instinctive caresses. They deposit packages of sperm, generally within her body, and then retreat hastily. The females produce eggs which may be held inside the body until hatching, laid singly, or laid collectively in batches. Development and growth necessitate a series of skin changes, or moults.

Arapaima: a tropical South American fish and the largest freshwater fish known, growing up to a length of 4 m. (13 ft. 3 in.). It is long and slender, with a flattened, scaleless head and low dorsal and anal fins. It breeds in April to May in clear areas of the Amazon river system, in which a nest about 15 cm. (6 in.) deep and 50 cm. (20 in.) wide is hollowed out for eggs. The nest is guarded by the male. It is an important food-fish for the aboriginal inhabitants of the Amazon, but has become scarce in many areas. In Brazil, it is called pirarucu.

Archaeopteryx: the oldest known fossil bird, found in parts of Germany in rocks about 150 million years old. Five fossil specimens exist. The fossils have bird-like features, such as the imprints of feathers and wings, but also many reptilian characteristics, notably teeth and a long tail formed of vertebrae. Archaeopteryx is considered to be a link between living birds and their reptilian ancestor, probably a small dinosaur.

A beautifully preserved fossil of **archaeopteryx** from the limestone quarries of Solenhofen, Bavaria. Though bird-like, it differed from modern birds in several respects. For example, the bones of the 'fingers' were complete, not reduced in number, and the metacarpal bones in the hands and metatarsal bones in the feet were separate and not fused. *Archaeopteryx* was as large as a medium-sized crow and had a wing-span of 58 cm.(23 in.).

Archer fishes are a family of four species which is widespread in the lowland rivers and estuaries of Southeast Asia and northern Australia. The common archer fish, *Toxotes jaculatrix*, has a relatively deep body, especially towards the tail, and a pointed snout. Inside the mouth, the palate has a lengthwise groove through which water can be ejected by sudden compression of the tongue and gill covers. By this means it can spit drops of water between 1 and 3 m. (3–10 ft.) away, accurately hitting insects which fall into the water and are then eaten.

Argali: the largest of the wild sheep of Eurasia, found on the central Asian plateau. It weighs around 158 kg. (350 lb.) and has long horns with a 1·5 m. (5 ft.) curve. Herds live in the valleys in winter and graze up to 5,500 m. (18,000 ft.) or above in summer. It will strike the ground sharply if alarmed to warn other members of the herd. Individuals may live for up to six years.

Argonauts (invertebrates)   *Paper nautiluses.

Armadillos are twenty species of armoured mammals found in South, Central, and southern parts of North America. The largest, the giant armadillo, *Priodontes maximus*, reaches 1·5 m. (5 ft.) in length, while the lesser fairy armadillo, *Chlamyphorus truncatus*, is only 12 cm. (5 in.) long. The armour takes the form of horny bands and plates, which are modifications of the skin, connected by flexible tissue. The narrow flexible bands on the back serve to break the rigidity, and the number of these bands is given in the common names of the species such as the Brazilian three-banded armadillo, *Tolypeutes tricinctus*, and the nine-banded, or common long-nosed armadillo, *Dasypus novemcinctus*. Many can draw their legs and feet beneath the shell and a few can roll into a ball. They dig burrows with their powerful claws, which are also used when foraging for insects, arachnids, eggs, worms, and small reptiles, or leaves and shoots. Armadillos exhibit polyembryony, which is the production of two or more genetically identical offspring (*clones) from a single fertilized egg. Most species have four young, but up to twelve may be produced. Armadillos belong to the same order as ant-eaters, sloths, and pangolins.

Arms (anatomy), peculiar to man and primates, are held by muscles to the shoulder-bones (scapulae), which in turn are held by muscles to the sides of the chest wall. The collar-bones (clavicles) hold the joints of the shoulder-bone and the arm from the side of the body, forming shoulders, and allow the arms to be far more mobile than the fore-limbs of all other mammals. They can be positioned forwards, backwards, above the head, and across the body in front and behind.

Strictly speaking, arms consist of the arm-bones (humerus) stretching from shoulder to elbow, where they join the forearms. The forearms stretch from the elbows to the wrists. Each forearm contains two bones: one (the ulna) which is hinged to the elbow, and the other (the radius) which rotates round it so that the wrist can be moved.

Army worms are the caterpillars of several species of moths of the family Noctuidae (*owlet moths), which move *en masse* in search of food-plants when their previous supply is exhausted. Species such as the African army worm, *Spodoptera exempta*, are migratory as adult moths and can occur in most tropical parts of the world. Army

Armadillo

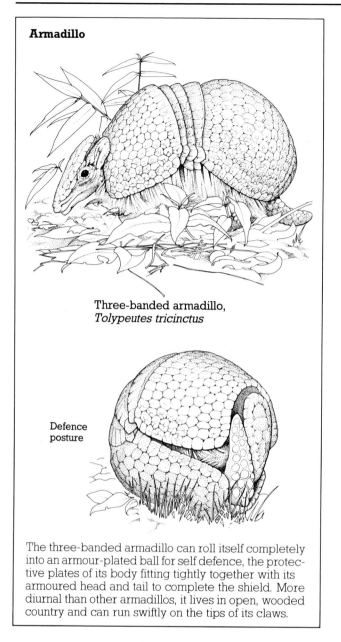

Three-banded armadillo,
Tolypeutes tricinctus

Defence posture

The three-banded armadillo can roll itself completely into an armour-plated ball for self defence, the protective plates of its body fitting tightly together with its armoured head and tail to complete the shield. More diurnal than other armadillos, it lives in open, wooded country and can run swiftly on the tips of its claws.

worms are usually pests of crops such as cereals, tobacco, maize, cotton, and potatoes.

Arrau turtle *Greaved turtle.

Arrow-poison frogs are used by the Choco Indians of Colombia to poison their blow-gun darts. They obtain the poison by impaling the frog on a stick, sometimes using the heat of a camp-fire to extract the poison. One species, *Phyllobates terribilis*, described in 1978, is so toxic that the Indians need only to rub the dart tip along the back of the living frog to obtain the poison. Approximately seventy-six species of arrow-poison frogs are known, most of them highly coloured and restricted to the tropical regions of South and Central America. They are an interesting family of frogs as the male carries the fertilized eggs, and eventually the tadpoles, on his back until they are ready to be released in a pool and lead independent lives. The smallest known species of frog, *Sminthillus limbatus*, measuring a mere 1 cm. ($\frac{1}{2}$ in.) in length, occurs in this family.

The bright colours of this male **arrow-poison frog**, *Dendrobates granuliferus*, warn predators of its poisonous flesh. Note the tadpole carried on its back.

Arrowroot is a tropical, herbaceous perennial whose swollen rhizomes yield a highly digestible fine-grained starch, particularly useful in infant and invalid diets. The commercial cultivation of the arrowroot, *Maranta arundinacea*, is essentially limited to the West Indian island of St. Vincent, although the plant is native to northern South America. It is also a source of the coating of carbonless paper, used for computer printouts.

Arrow worms are a phylum of some sixty-five species of marine *planktonic animals, with dart-like bodies, usually only 1–5 cm. ($\frac{1}{2}$–2 in.) long. The adults are swift, predatory creatures, their heads bearing eyes, tiny teeth, and spines. Stabilized by their lateral fins, they swim by rapid muscle contractions in pursuit of other plankton, and they change depth daily or seasonally as the planktonic masses migrate up and down. The developmental stages of their embryo show them to be part of the same major group of animals as the vertebrates, despite the somewhat worm-like appearance of the adults.

Artemisias are aromatic shrubs and herbs of the sunflower family. There are many species in the genus *Artemisia*, mainly from the Northern Hemisphere, and most possess hairy, silvery leaves, rendering the plants very tolerant of arid conditions. The flowers are small, yellow, and button-shaped. Some species have medicinal properties and yield stimulants and worming compounds (hence the common name wormwood for *A. absinthium*). The shrub southernwood is *A. abrotanum*.

Arteries are the blood vessels of vertebrates which carry oxygen-rich blood from the heart to the body tissues. Transfer of oxygen and other nutrients is by diffusion across thin-walled capillaries. Arteries spread through all tissues and pass eventually into *veins via tubules called capillaries.

Arteriosclerosis is a medical term describing the hardening of the arteries, which occurs with increasing age and which is not in itself a serious matter. The term is also used with reference to the narrowing or blocking of arteries by patchy deposition of fats in their walls (atheroma); this can lead to *thrombosis. The effects of this process on the heart and brain are the major cause of death in developed countries. Among the risk factors implicated are cigarette smoking, high blood pressure, too much dietary fat, and lack of exercise.

Arthritis is inflammation of the joints. It may be due to local infection, but more commonly it is a response to a degeneration of the *cartilage (osteoarthritis), or an aberration of *immunity (rheumatoid arthritis). The former is predominantly a disorder of later life, whereas the latter may affect younger people as well. Both conditions are common causes of disability, and neither is well understood. Rheumatoid arthritis is characterized by the production within the joints of an *antibody active against a constituent of other antibodies, the ensuing interaction being responsible for the inflammation. Many diseases can cause arthritis; a disease of the metabolism of uric acid leads to the form known as gout.

Arthropods are the largest phylum of animals and comprise some 750,000 species. Although enormously diverse in form and habits, they are quite easy to recognize. They share two critical features: a tough segmented covering or *exoskeleton (forming a waterproof cuticle in insects and spiders, or a thick shell in crabs), and a number of jointed legs (usually one pair per segment) used for swimming, walking, or food-handling.

Arthropods can be found in every conceivable habitat. *Crustaceans live mainly in the seas and fresh water. On land, *arachnids (such as spiders and mites), *myriapods (such as centipedes), and *insects abound. Winged insects also thrive in aerial habitats. Arthropod success in all these ecological niches is due to several features: the protective exoskeleton and its associated sense organs; speedy and nimble locomotion, legs being mechanically superior to the hydrostatic system of many other invertebrates; diversity of cuticular mouthparts, which can exploit almost any food; and (on land especially) elaborate courtship and mating strategies and the production of resistant eggs. The small size of athropods (necessitated by the cuticle) is also useful, as it allows them to inhabit small spaces among vegetation, within crevices, or in soils, which are unavailable to larger creatures.

Artichokes are members of the sunflower family, or Compositae, and include the globe artichoke, *Cynara scolymus*, and Jerusalem artichoke, *Helianthus tuberosus*. The fleshy flower-heads of the globe artichoke, a plant of Mediterranean origin, have been used as a vegetable since ancient Greek and Roman times. The green to purple flower-heads make the plant a handsome addition to gardens in temperate regions. The Jerusalem artichoke of North America is a tall plant up to 2 m. (6 ft. 6 in.) in height and is a close relative of the sunflower; it is grown for its sweet fleshed under-ground tubers.

Arum lilies are members of the Araceae, a family with a distinctive flower structure consisting of a spathe, or petal-like sheath, enclosing a spadix, or club-like column, with male and female flowers attached to the base. In the florists' varieties, the spathe may be white, yellow, or

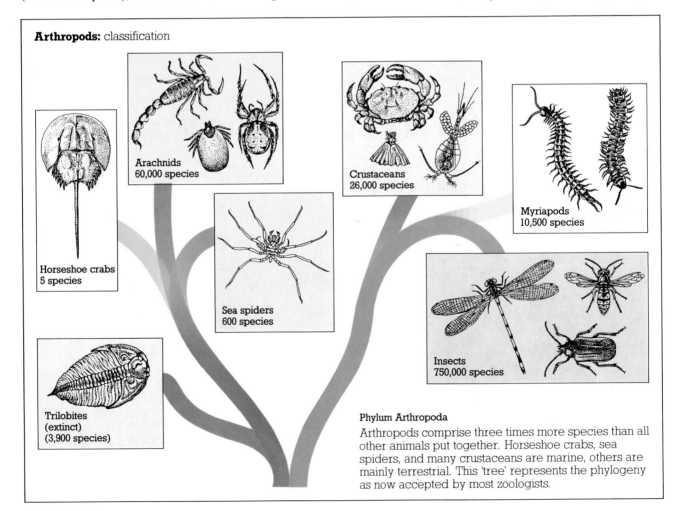

Arthropods: classification

Arachnids
60,000 species

Crustaceans
26,000 species

Myriapods
10,500 species

Horseshoe crabs
5 species

Sea spiders
600 species

Insects
750,000 species

Trilobites
(extinct)
(3,900 species)

Phylum Arthropoda

Arthropods comprise three times more species than all other animals put together. Horseshoe crabs, sea spiders, and many crustaceans are marine, others are mainly terrestrial. This 'tree' represents the phylogeny as now accepted by most zoologists.

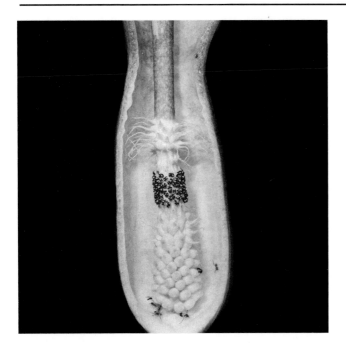

The downward-pointing hairs in the spadix of the **arum lily**, *A. maculatum*, (seen here in section), trap small flies carrying pollen from other plants.

rose-coloured. The flowers of several species smell of carrion and are attractive to and pollinated by flies.

Ascaris: a genus of *roundworms which are intestinal parasites of man, dogs, cats, pigs, and many other vertebrates. The human ascarid, *Ascaris lumbricoides*, is similar in appearance to a smooth, pale earthworm and feeds as an adult on the host's intestinal contents. The adult female can reach 35 cm. (14 in.) in length, and her eggs pass out in faeces to be transmitted to new hosts in contaminated food or drink. Infestation is very common in tropical and subtropical regions, and damage is caused by the migration of *Ascaris* larvae into other organs of the host body.

Ash (tree) is the name for a few of the seventy species of *Fraxinus* native to the Northern Hemisphere. They belong to the same family as olive and lilac and have pinnate or, more rarely, simple leaves and winged fruits. Some are small shrubs, but most are trees. The common ash of Europe, *F. excelsior*, has very strong, pale wood, long used for tool handles, bows, and hockey sticks. Its flowers generally have no petals, though those of the manna ash, *F. ornus*, of southern Europe have slender whitish petals.

A few unrelated trees, like the rowan, are sometimes called ash.

Asparagus is a member of the lily family which has been cultivated in Europe since the time of the ancient Greeks. The cylindrical young green or purple-tipped shoots or 'spears', with scale-like leaves clustered near the tip, develop in late spring and summer and are harvested when 20–30 cm. (8–12 in.) high.

Asphodels are members of the lily family, mainly from the Mediterranean region. They are perennial herbaceous plants with fleshy roots and racemes of white or yellow starry flowers. Many species of the genera *Tofieldia* and *Narthecium* grow in wet meadows or bogs. The related

Asphodeline was the asphodel of ancient Greece, the flower of the dead.

Aspidistras are evergreen stemless plants of the lily family, and include eight species of the genus *Aspidistra* which are native to eastern Asia. The large, decorative, green leaves have made them popular as indoor plants. The flowers, which are produced at ground level, are pollinated by small flies.

Assassin bugs, or cone-nosed bugs, are medium to large predatory *bugs 1–4 cm. (½–1½ in.) long, belonging to the family Reduviidae. They are a diverse group with over 3,000 species, each roughly oval in shape with a black or brownish flattened body, an elongate, narrow head and an abdomen which is widest at the middle. The piercing, sucking mouthparts form a pointed 'snout' of three segments that fits into a groove on the underside of the thorax when not in use. This is used to pierce and suck out the juices of prey, using a pair of sliding needles, or stylets. A few species are bloodsuckers that inflict painful bites on man, and several American species of *Triatoma* are vectors of the trypanosome (parasite) that causes Chaga's disease. One species, *Rhodnius prolixus*, has been widely used in the laboratory investigation of the hormonal control of growth, moulting, and metamorphosis in insects.

Asses belong to the same genus as the horse. Both species of wild asses—the African, *Equus africanus*, and Asiatic, *E. hemionus*—live in steppes, semi-deserts, and even in the deserts of Africa and Asia. The largest is the kiang, a subspecies of the Asiatic ass, of Tibet and the Himalayas. The Asiatic asses live in herds of up to fifteen in number, led by a stallion, with mares and their young. Foals are born during the summer months. The African ass is an endangered species found in the northeast corner of the continent; the *donkey is the domesticated form of it.

Asthma is the difficulty experienced in breathing due to excessive contraction of the involuntary muscle in the walls of bronchial tubes leading into the lungs, with consequent narrowing of the tubes. The muscle reacts excessively to a wide range of stimuli such as infections, exertion, and most importantly allergens, the substances which cause *allergic reactions. If the attack is prolonged it is complicated by plugging of the small airways by abnormal secretion, and it is this that can make asthma a threat to life. Asthma commonly starts in childhood, and

A large **assassin bug** sucks the body fluids from a Peruvian rain-forest grasshopper.

about half those affected improve or recover around puberty. The condition can be alleviated by drug treatment.

Athlete's foot *Ringworm.

Atlas moths are a genus of *silk moths found only in East and Southeast Asia. The atlas moth, *Attacus atlas*, and Edwards' atlas moth, *A. edwardsi*, are among the largest moths in the world, with wing-spans up to 30 cm. (12 in.). They have hooked wing-tips and bold brown and cream markings, and in some countries are protected by law. Their caterpillars, which feed on the leaves of a wide range of trees and which may reach 10 cm. (4 in.) in length, are green and brown with large tubercules on each segment.

Atwater, Wilbur Olin (1844–1907), American chemistry professor, demonstrated the validity of the law of conservation of energy for humans, and drew up a table of the calorific value of various foods. Moreover, he discovered that free atmospheric nitrogen is assimilated by plants of the pea family.

Aubergine or egg plant: a relative of the tomato and potato, belonging to the genus *Solanum*, and originating in tropical Asia. It is a herbaceous, branched perennial cultivated widely as an annual and is a particular favourite in India (where it is known as the brinjal), the West Indies, the USA, and Mediterranean countries. A somewhat spiny plant with big leaves, it bears egg- or sausage-shaped fruits which are large—up to 15 cm. (6 in.) long—and smooth, varying in colour from the more common deep purple to green, red, or white.

Aubrieta is a genus of trailing evergreen plants named in honour of Claude Aubriet, a French flower-painter. About fifteen species are known; they are mostly alpine and are distributed from Italy to Asia Minor. They are close relatives of the wallflower and other species of the mustard family. *Aubrieta deltoidea* is the parent of many garden varieties, mostly with purple, lilac, or pink flowers.

Audubon, John James (1785–1851), was a Haiti-born American naturalist and painter whose magnificent work *The Birds of America* (1827–38) attempted to depict every known American bird in its habitat. The accompanying text, *Ornithological Biography* (5 vols., 1831–9), includes first-hand descriptions of frontier adventures during his years of travel. Further source material is in his *Journal* (1929) and *Letters* (1930), both published after his death.

Auenbrugger, Leopold (1722–1809), Austrian physician, introduced percussion of the chest as a diagnostic procedure. He found that if the chest was tapped firmly with the fingers, then sounds of varying resonance were heard which related to the location and state of health of the underlying organs.

Auks make up a family of some twenty-two species of mainly black and white Northern Hemisphere seabirds with distinctive whirring flight. The family includes razorbills, little auks, the now extinct great auks, guillemots, auklets, murres, and puffins. Auks nest in colonies, mainly on high cliffs and steep-sided islands. Some make burrows. They dive and swim powerfully under water using their webbed feet and short wings, and feed on fish, molluscs, and crustaceans. Most species lay one or two eggs. In some

J. J. **Audubon** in an engraving based on a contemporary painting by F. Cruikshank. Audubon was one of the first and most influential American wildlife conservationists.

species the young birds leave the nests before they can fly, and swim to their feeding grounds.

Auriculas *Primrose.

Auroch is the name for a now extinct type of wild ox which was the ancestor of domestic cattle. It first appeared about 230,000 years ago and became extinct in the Bronze Age, 3,600–1,000 years ago. It occurred across much of Europe and western Asia. Its decline was probably due to persecution by man. A number of Stone Age cave paintings depict this animal, showing its importance to early man.

Australian copperhead *Copperhead.

Autism is a rare and severe mental illness of childhood. Failure to develop normal interpersonal relationships or normal speech patterns or play can occur. Repetitive solo play with objects is often preferred to playing with others. Mental handicap is sometimes present.

Autonomic nervous system: the part of the nervous system that controls the vegetative functions of the body such as the circulation of the blood, intestinal activity and secretion, and the production of chemical 'messengers' *hormones that circulate in the blood. The system is subdivided into the sympathetic and parasympathetic nervous systems. Each is controlled by areas within the central nervous system, which lie in the brain and/or spinal cord, and are linked to organs under involuntary control by *nerves. The outflow from the autonomic centres is not under direct conscious control, although it can be influenced by deliberate relaxation or meditation. The separate systems tend to have opposite effects: the parasympathetic vagus nerve acts to slow heart-rate, while the sympathetic nerves speed it up. Sympathetic nervous activity increases in response to fear, producing increases in heart-rate and blood pressure, diversion of blood flow to the muscles in readiness for action, and increasing energy production. Anxiety and *neurosis are often associated with enhanced autonomic nervous activity.

Autotrophic organisms are those which manufacture complex organic molecules, such as carbohydrates and proteins, from simple inorganic substances, including water, carbon dioxide, and mineral salts. The most common and important autotrophs are the green *plants which use the sun's energy to fuel the manufacturing process, *photosynthesis. Other autotrophs, mostly bacteria, make use of energy released by various chemical reactions to convert inorganic molecules into food. This process is known as chemosynthesis. Non-autotrophic organisms are termed *heterotrophic.

Autumn crocus is the common name for *Colchicum*, a genus of plants which are unrelated to the true crocus. These plants are part of the lily family, and as such are related to tulips, onions, and leeks. The bulb-like corms normally produce their flowers in autumn before the leaves appear. They may be identified as distinct from the crocus by the presence of six stamens instead of three. The poisonous chemical colchicine is extracted from the corms and seeds, especially from *C. autumnale*. An alternative name is meadow saffron.

Avicenna, or Ali ibn-Sina (980–1037), Persian physician and philosopher, travelled widely, and wrote about a hundred books on philosophy, poetry, mathematics, and medicine, his most famous being the *Canon of Medicine*. Comprising five books, it included detailed descriptions of many diseases, and of the composition and preparation of medicines. It was regarded as an authoritative medical textbook not only in the Arab world, but also in the West, following its translation into Latin.

Avocado pears are small to medium-sized tropical evergreen trees belonging to the family Lauraceae which also includes bay laurel and cinnamon. The green, pear-shaped fruit is a rich source of energy in tropical diets, as it contains protein- and fat-rich, creamy flesh, and is rich in vitamins A, B, and E. Cultivated for at least 9,000 years in its Central American homeland, it has only relatively recently been grown widely in tropical and subtropical areas; it is now an important cash crop in Florida, California, Hawaii, Israel, and tropical Australia and South America. The old name of alligator pear refers to the thick, rough skin.

Avocets are *wading birds in the family Recurvirostridae, with slender, upcurved beaks which they sweep from side to side in the shallow water when feeding. The pied Palearctic avocet, *Recurvirostra avosetta*, now breeds in Britain after a century's absence. Three other species inhabit North America, the Andes, and Australia. *Stilts are members of the same family.

Axolotl: a well-known species of *mole salamander, *Ambystoma mexicanum*, from Mexico, which remains a feathery-gilled larva all of its life in the natural state. The axolotl is capable of reproducing, as it becomes sexually mature in the larva state, a condition known as neotenous. Two types are known, a black form and a white form with

An **aye-aye** foraging on the forest floor. Its large eyes seek out the burrows of insect larvae, and fruit, while its prominent ears probably help it to detect sounds made by larvae hidden under the bark of dead trees.

pink gills. They resemble newts and feed upon aquatic invertebrates such as insect larvae, worms, and crustaceans, but they may also attack small, wounded fishes.

Aye-aye: a species of small *primate related to *lemurs, but forming its own unique family, the Daubentoniidae. In common with lemurs, the aye-aye is found only in Madagascar. It has a body length of 40 cm. (16 in.) with an additional 55 cm. (22 in.) of bushy tail. It has large, forward-facing eyes, large bat-like ears, quite a small snout, and a thick, dark brown coat. A nocturnal animal, it inhabits coastal rain forests, where it lives high in the larger branches of the trees. Unlike lemurs, the aye-aye has specialized as an insect feeder; the large upper and lower incisor teeth grow continuously throughout its life, like those of rodents, and its third finger is exceptionally thin and elongated, enabling it to dig out insect larvae from trees; it also occasionally eats fruit. The other digits also have claws which help it to cling to tree trunks. It builds nests in the forks of trees and produces one offspring at each birth. The young aye-aye is carried on the mother's back for some time and shares her nest. Thought to be extinct, it has recently been rediscovered and rescued by the establishment of a reserve. A similar, but slightly larger, species of aye-aye existed on Madagascar some 2,000–3,000 years ago, but this is now extinct.

Ayurvedic medicine is an ancient system of medicine based on the use of natural plant materials. It is practised widely in India but less so in Western countries, where it is regarded as a form of *alternative medicine.

Azaleas   *Rhododendrons.

B

Babblers are a large family of about 280 species of birds which occur throughout the warmer areas of Asia, some of the species extending to Africa and Australia. Ranging from sparrow-sized to jay-sized, they are a very diverse family. Most species are brown, but some have brightly coloured patches. The Pekin robin, *Leiothrix lutea*, is greyish above and red and yellow in the underparts and on the wings. The white-crested laughing thrush, *Garrulax leucolophus*, one of about forty-five laughing thrushes, is mainly brown but with a striking white head and chest, a black eye-stripe, and bright red eyes. Other groups within the family include parrotbills and the African rockfowl.

Most are insectivorous, catching their food with a long, thin beak. Some of the tree-babblers have strongly curved beaks which they use for probing into soft wood and leaves. Many species live in groups of six to twelve and defend a joint territory.

Babirusas   *Pigs.

Baboons are five species of Old World monkeys that have evolved from tree-dwelling ancestors to become terrestrial, walking on all four limbs. Typical open-country monkeys, they are found all over the savannah, semi-desert, and lightly forested regions of Africa south of the Sahara Desert. (The species of baboons known as the *mandrill and the drill, however, live in more forested habitats.) The face is elongated and rather dog-like, and the jaw carries a long row of grinding molar teeth. Baboons feed on the ground, eating seeds, tubers, grass, and insects, and this makes them vulnerable to predators. Troops of baboons will often associate with a herd of ungulates (hoofed mammals) such as impala, which are alert and give warning of approaching predators. The association is of mutual benefit, as baboons are powerful animals and give protection to the impala from smaller predators. The hamadryas baboon, *Papio hamadryas*, is 76 cm. (30 in.) tall with a tail 61 cm. (24 in.) long; the females have brown hair and the males have grey hair with a long mane. They live in highly organized societies of twenty-five to thirty animals, and occasionally up to 200. The society is usually hierarchical, and the males defend females with young. A single offspring is born and it is carried by its mother for several months. Other species are the common, or savannah, baboon, *P. cynocephalus*, and the gelada baboon, *Theropithecus gelada*.

Bacilli are *bacteria having the shape of short rods.

Bacteria are a large group of mainly unicellular microorganisms, which along with *blue-green algae form one of the three major groupings of living organisms, the *Protista. They are characterized by a simple *nucleus without a bounding membrane, a single *chromosome formed into a ring, and the lack of *mitochondria. Their cell wall, formed largely of a polysaccharide sugar, may be spherical, rod-like, or spiral, or individual cells may collect as filamentous colonies. Each bacterium is microscopic, between 0·0005 and 0·005 mm. (up to 0·0002 in.) in size.

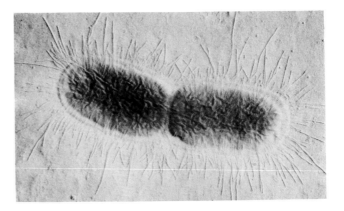

Multiplication of **bacteria** by simple division of the parent cell is shown in this *Salmonella* species, which causes a digestive upset in man. The visible whip-like appendages, the flagella, are used for movement.

They reproduce by simple division of cells or by sexual *reproduction.

Many bacteria are involved in the breakdown of organic matter, while others can survive on inorganic materials such as carbon dioxide. A minority of species are parasitic, or infective agents, of animals or plants. They cause problems when they produce *toxins which cause illness. In man, many diseases including cholera, typhoid, and diphtheria are caused by bacterial infections.

Bactrian camel   *Camels.

Badgers are large, stocky mammals of the weasel family, and are related to weasels, skunks, otters, and the ratel. The eight species of badger are found throughout Eurasia and North America. These are: the Eurasian badger, *Meles meles*; the hog badger, *Arctonyx collaris*, from Southeast Asia; the Malayan stink badger, or teledu, *Mydaus javanensis*; the Palawan stink badger, *Suillotaxus marchei*; the American badger, *Taxidea taxus*; and three species of Asian ferret badgers, *Melogale* species. The stink badgers and the ferret badgers produce a pungent secretion from the anal glands, which they can discharge after the manner of a skunk. All badgers burrow and have similar omnivorous habits, with the American badger being more carnivorous than other species.

Baer, Karl Ernst von (1792–1876), Estonian biologist, discovered the mammalian ova; this was the chief of his many contributions to embryology. He formulated a principle that, in the developing embryo, general characters appear before special ones. Baer's studies were used by Darwin in his theory of evolution.

Bagworm moths are some 800 species of *moths in the family Psychidae. Their name comes from their caterpillars, which enclose themselves in a silken bag, often incorporating fragments of leaves and twigs. Adult male moths have dull unmarked wings, up to 25 mm. (1 in.) across, and are swift-flying; females are wingless, never leaving the bag, and in many species lack legs, antennae, eyes, and mouthparts.

Bain, Alexander (1818–1903), Scottish philosopher, was an exponent of a system of psychology which traces psychological phenomena to the nerves and brain.

Baird, Spencer Fullerton (1823–87), American author of a *History of North American Birds* (5 vols., 1875–84), began the influential 'Baird school' of accurate ornithological description. He helped also to compile the *Catalogue of North American Mammals* and made contributions to the study of fishes and plants.

Bakewell, Robert (1725–95), English pioneer in livestock breeding and husbandry, produced pedigree herds of sheep and cattle from his Leicestershire farm. He showed that, with proper irrigation, grassland could produce four crops of grass a year, and that improved feeding contributed greatly to the quality of carcasses.

Bamboos are a widely distributed group of evergreen woody-stemmed plants which are members of the *grass family. The 200 or so species are divided into several genera and they may vary in height from a few centimetres (an inch) to 30 m. (100 ft.) or more. The hard-skinned stems are very durable and have many uses, including building, furniture making, basketry, and as garden canes. The flowering of bamboos is very unusual. It usually occurs at indefinite and lengthy intervals of up to 120 years, and often results in the death of the plant.

Bananas originated in Southeast Asia from two wild species, *Musa acuminata* and *M. balbisiana*. These large-leaved, perennial, herbaceous plants, reaching up to 10 m. (33 ft.) in height, are now grown throughout the tropics, predominantly in the humid lowland areas, and produce the most widely consumed tropical fruit, with an estimated world yield of 35,000,000 tonnes (38,580,000 US tons) per annum. Central America and the West Indies export large quantities of the sweet, dessert types to North America and Europe. A cooking type known as plantain, *M. paradisiaca*, with a higher starch content and flesh too hard and indigestible to be eaten raw, is used for local consumption. Other varieties are also known as 'plantains', in particular in India and East Africa. Cultivated banana plants are sterile, ensuring seedless fruits, and are propagated vegetatively from suckers. They are relatively long-lived—fifty years or more.

Bandicoots are Australasian *marsupials about the size of a rabbit, which they resemble in their reproductive rate (the highest of any marsupial) and elongated hind legs. With their long, flexible, rodent-like muzzles, they root in the soil for their insect food. The pouch opens towards the tail. Once widespread and numerous, some of the nineteen species are now very rare. The rabbit-eared bandicoots comprise two species often known as bilbies.

Banks, Sir Joseph (1743–1820), English naturalist and explorer, brought back many plants and insects from his first expedition to Labrador and Newfoundland. He accompanied Captain James Cook in the voyage of the *Endeavour*, sent by the Royal Society to observe the transit of Venus in 1769, which set a pattern for geographical and scientific exploration of the Pacific. Banks collected plants and seeds wherever they went and found so many in a part of New South Wales that Cook named the region Botany Bay.

He helped to establish the Botanic Gardens at Kew, near London, not only as a repository of thousands of living specimens from all over the world, but as a centre for the introduction of plants to new regions, including

When the **banana** plant is mature, a bud emerges from the leaf crown and produces many flower clusters. Only the upper clusters bear fruit. Each cluster is known as a hand, each fruit as a finger.

An engraving of Sir Joseph **Banks**, from a painting by J. Phillips in the Royal Society. Banks became President of the Royal Society in 1778 and greatly improved its status. It was he, also, who was responsible for the purchase of Linnaeus's collection and the subsequent founding of the Linnean Society.

*breadfruit and tea. He also imported merino sheep from Spain and sent them on to Australia. His herbarium and library in London became a centre of taxonomic research, freely available to scholars from all over the world, and after his death it became part of the British Museum.

Banksias *Proteas.

Bank swallows *Sand martins.

Bank voles are a species of *vole which occurs over much of Europe and western Asia in most vegetation types, provided there is sufficient cover. The tail is long for a vole, being about half the length of the 9-cm. (4-in.) long body, so that this vole resembles a mouse. It is mainly nocturnal, breeding throughout the summer and, if food is abundant, sometimes throughout much of the year. Predators of the bank vole include hawks, owls, weasels, and stoats. The bank vole belongs to the genus *Clethrionomys*, which includes a further seven species commonly called red-backed voles, some native to North America.

Banting, Sir Frederick Grant (1891–1941), Canadian physiologist, with C. H. Best discovered the hormone insulin and used it to treat *diabetes in 1922. An extract of part of the pancreas was purified and used successfully to treat a fourteen-year-old diabetic patient.

Banyan: a species of *fig, *Ficus benghalensis*, which is a sacred tree in India. It is notable for its long aerial roots which develop on the trunk and branches and then take root in the ground, forming pillars which make an extensive thicket—one tree is said to have sheltered the army of Alexander the Great. Banyan is so called because Europeans noted that certain traders (banians) always sat beneath such a tree.

A **baobab** tree, *Adansonia digitata*, in southern Africa during its dormant season. Its soft trunk is occasionally excavated by the Africans and used as a house. The bark yields a fibre for rope and cloth.

Baobab: a species of deciduous tree, *Adansonia digitata*, which grows in arid environments in tropical and southern Africa. A characteristic feature is its massive, barrel-shaped trunk, growing up to 30 m. (100 ft.) in circumference and reaching a height of 17 m. (56 ft.). This huge trunk is adapted for water storage and gives the impression of being too large in proportion to the crown. The baobab is part of the family Bombacaceae, which also includes the balsa, silk cotton or kapok, and durian trees in tropical regions.

Barbary ape: a species of *Macaque monkey found in northwest Africa, and in a small colony on the Rock of Gibraltar. It is the only monkey that is native to the Mediterranean lands and is more closely related to the monkeys of India and Southeast Asia than to African monkeys living south of the Sahara Desert. It is sociable and lives in small groups which travel through rocky woodland searching for their food of fruit, leaves, roots, and insects.

Barbel: a fish, *Barbus barbus*, of the *carp family, with a wide distribution in Europe. It prefers deep water with moderate currents, and lives close to the river-bed. Related species in the same genus, also often called barbels, live in Europe, Asia, and Africa; in the last two continents there are many small, brightly coloured species known as barbs, which are often kept as aquarium fishes.

Barberries are shrubs in the family Berberidaceae, and are distributed in temperate regions of the Northern

Hemisphere and of South America. There are around 450 species both evergreen and deciduous, the leaves of the latter turning to bright colours in autumn. All species bear small yellow or orange flowers, which are followed by red, yellow, blue, or black fleshy berries, according to species. The common barberry, *Berberis vulgaris*, found in Europe, and northern Asia, and introduced to North America, is the host at a stage in the life cycle of wheat black-rust disease.

Barbets are a family of seventy-eight species of birds allied to the toucans and woodpeckers. They vary in length from about that of a sparrow to that of a largish thrush, and are stocky, large-headed, stout-beaked birds. Most are brightly coloured with patches of red, blue, and yellow on a basic plumage of bright green, brown, or black. They live in wooded areas of the tropics feeding on fruits and insects, and are non-migratory. They nest in holes, usually in trees, but sometimes in the ground, and lay two to four white eggs.

Bark is the outer covering of the stems of woody, perennial land plants and is derived from the dead cells of *phloem. The bark protects the phloem and other living parts of the tree, and thus it often contains preservatives, some of medicinal and tanning importance. The bark of certain trees is harvested for cork (oaks) and a kind of cloth from the lacebark tree, *Lagetta linteraria*.

Bark beetles are small beetles of the subfamilies Scolytinae and Platypodinae whose larvae feed on the wood of living trees or shrubs. They are included within the *weevil family. The young adults bore through the bark and make egg tunnels, usually between the bark and the wood, but sometimes into the wood. The males usually make the tunnels and the females lay eggs in them. The larvae make tunnels at right angles to the egg tunnel; they pupate at the ends of their tunnels, and the newly emerged adults bore their way out. The larvae of *ambrosia beetles feed on species of fungi brought in by their parents.

The oriental green **barbet**, *Megalaima zeylanica*, is an Asian species. Its streaked head and breast feathers distinguish it from the other green-backed barbets. This pose is characteristic, as barbets spend much time sitting motionless in the uppermost branches of trees.

Barley was first cultivated in western Asia at least 9,000 years ago. This annual grass of temperate regions provides the most important grain (malting barley) used in brewing beer; grain not used for this purpose is fed to livestock. The two commonly cultivated species are the six-row barley, *Hordeum vulgare*, and the two-row type, *H. distichon*. Both species, although typically associated with light soils, are grown in a wide range of areas and about 10,000,000 hectares (38,000 square miles) per annum are cultivated worldwide. Europe is the largest producer, and Britain achieves the highest yields. Like wheat, barley can be sown in temperate countries in the autumn or spring. Pearl barley is produced by grinding away the outer husk of the grains.

Barnacle geese are easily distinguished from other *geese by the white face contrasting with the black neck and breast. They breed in colonies on Arctic cliffs and rocks, and winter in northwestern Europe. In medieval times people believed that they were hatched from goose-barnacles (molluscs). They are similar in many respects to Canada geese.

Barnacles are a familiar sight on rocky sea-shores worldwide. Despite their calcified mollusc-like shells, they are actually *crustaceans, though very unusual ones. They have stiff, jointed legs like their relatives, but are sessile, sitting upside-down within the shell and kicking food particles into their mouths. This sweeping leg-action can easily be watched in rock-pools. Most of the 900 or so species of barnacles live cemented on to rocks or other animals' shells, but some have stalks and a few are parasitic. Since they are sessile, mating presents problems, so each barnacle (for they are *hermaphrodites) has a long penis which reaches into a neighbour's shell. The resulting larvae are an important component of *plankton.

Barn owls belong to a family containing ten species of closely related *owls with distinctive pale, heart-shaped faces and slender legs. All, including the widely distributed common barn owl, *Tyto alba*, are nocturnal predators of small mammals, and have acute hearing. They often nest in barns and other farm buildings.

Barracouta *Snoek.

Barracudas (*see over*) are marine fishes which live in tropical and warm temperate seas, including the southern Mediterranean. They make up a family of some twenty species. Long and slender-bodied, with a pointed head, the lower jaw being the longer, they have large, fang-like teeth. They live in shallow coastal waters and in lower estuaries. They are voracious predators, eating all kinds of fishes which they capture with a high-speed charge. The largest species, the great barracuda, *Sphyraena barracuda*, which is found in the western Atlantic, grows to 1·8 m. (5 ft. 9 in.). It occasionally attacks bathers, especially in the West Indies.

Barramundis, of which there are two species, are Asiatic and Australian relatives of the *arapaima. The spotted barramundi, *Scleropages formosus*, is found in Southeast Asia and the Fitzroy river system of Australia, and the northern barramundi, *S. leichardti*, in northern Australia and New Guinea. They live in slow-flowing rivers and creeks, feeding on fishes, frogs, crustaceans, and insects, and grow to

A school of **barracudas** in the Red Sea. The formation of schools provides them with an efficient means of exploiting limited food resources. Wide scanning of the environment by a large group of fishes is more likely to reveal a source of food than is the more restricted hunting of a solitary individual.

about 1 m. (3 ft. 3 in.) in length. The females brood the eggs and hatch the young in a pouch within the throat. The name barramundi is also used to describe the giant perch, *Lates calcifer*, of the Indo-Pacific region. This is regarded as a good food fish in Australia.

Barrel-fish: a species of the family Centrolophidae, known by this name because the young fishes swim in company with floating wood at the surface of the sea. They also accompany jellyfishes and salps. The barrel-fish, *Hyperoglyphe perciforma*, found throughout the North Atlantic, grows to 90 cm. (3 ft.) long, and lives in deep water when adult.

Bartram, John and William, father and son, were both pioneering American botanists. John (1699-1777) gathered specimens for his Philadelphia gardens during his travels from the Catskills to Florida and from Pennsylvania to Ontario. By exchanging seeds and bulbs with European botanists, and by cross-fertilizing, he developed many new *hybrids.

His son William (1739-1823) accompanied him and described his father's botanical journeys vividly in *Travels* (1791), a book which attracted European attention. William was also an ornithologist and produced a list of 215 native birds.

Basil is a member of the mint family, Labiatae. It is a slightly hairy, aromatic annual, 40 cm. (1 ft. 4 in.) tall, and is grown as a culinary herb in many areas of the world, including tropical Asia, Africa, and some Pacific islands. The production of basil oil is centred around the Mediterranean. A close relative is the sacred bush of the Hindus, grown in front of their houses and temples.

Basilisks are lizards of the *iguana family found in forests of Central America and northern South America. They live in trees and bushes overhanging rivers. When disturbed they plunge into the water and may rear up on to their hind legs and run away across the water's surface. They achieve this remarkable feat with the aid of flaps on

their toes which open as the foot strikes the water; young, lighter individuals can water-walk with greater ease than larger lizards can.

Basket stars *Brittle stars.

Basking shark: the second-largest living species of fish (the largest is the whale shark), which can be 11 m. (36 ft.) in length. It lives in cool, temperate seas, and in summer swims near the surface, feeding on plankton which it sieves from the water with its gill rakers. Its teeth are numerous but minute. In winter it migrates to warmer areas or may simply live inactive close to the sea-bed of the cool, temperate sea.

Bass is a name given to two major groups of fishes. The sea basses, a family of some 400 species, are typified by the torpedo-shaped fish, *Dicentrarchus labrax*, which has strong first dorsal fin spines, a green-grey back, and brilliant silvery sides. It lives in the coastal waters of Europe, often in turbulent water near reefs, and in the breakers, but it also ascends estuaries. It eats crustaceans and a wide range of schooling fishes. It is slow-growing and some large specimens may be at least twenty years old. In North America a similar species, the striped bass, *Morone saxatilis*, is native to the Atlantic coast and has been successfully introduced to the Pacific coast.

The second group of bass comprises six species of the genus *Micropterus*, which are part of the largely North American sunfish family.

Bateleurs are long-winged African harrier-*eagles, distinguished by their short tail and scavenging habit. Mostly carrion feeders, they soar on extended wings over the plains in search of dead mammals, and descend either to feed directly or to harry other birds into disgorging food.

Bates, Henry Walter (1825-92), English entomologist and naturalist, explored the upper Amazon region of South America (1848-59), collecting and describing over 8,000 new animal species. His travels are recounted in his classic book, *Naturalist on the Amazon* (1863). He first described what is now known as Batesian *mimicry.

Bateson, William (1861-1926), English geneticist, continuing the work of Mendel studied discontinuity in variation by the experimental breeding of animals and plants. He ascribed these deviations from the laws of heredity to the interaction of genes, and was the first to use the term genetics. He published a classical book on genetics, *Materials for the Study of Variation* (1894).

Bats are the only mammals capable of true flight as opposed to gliding. The wings are formed from skin stretched between the fingers and the flight surface is often extended between the hind legs. Bats hang by their feet while at rest, with the wings wrapped around the body. Most of the 951 species are nocturnal and spend the day roosting in large numbers in trees, caves, or old buildings.

Among mammals, bats are second only to rodents in the number of species. The order Chiroptera, to which bats belong, is divided into two unequal groups, the large *flying foxes with around 173 species, and the smaller but more numerous insectivorous bats. The latter use ultrasound to locate their insect prey in flight, rather like human radar. Flying foxes, or fruit bats—which may have a wing-span of 2 m. (6 ft. 6 in.)—do not need to echo-locate, and lack this facility, except for one species which roosts in caves. The folds of skin around the snout and the huge ears, which give such a bizarre appearance to the insectivorous bats, serve to receive the 'radar' signals.

Bats have a curious physiology in that they do not maintain a constant body temperature but allow it to drop at night, and in winter, when most bats hibernate. Mating occurs in the autumn but the sperm is stored within the female's body and fertilization does not take place until the following spring.

Bay duck is the name used in North America for diving ducks such as the redhead, canvasback, scaup, and goldeneyes. Although bay ducks frequent freshwater habitats in summer, they commonly winter in estuaries and sheltered coastal bays. In Europe some of these species are called *pochards.

Bay laurel, known also as sweet bay, is a white-flowered evergreen tree indigenous to the Mediterranean and grow-ing to 20 m. (65 ft.) in height. Its aromatic leaves are used in cooking for flavouring. The leaves of bay laurel were woven into wreaths and used to crown victors in ancient Mediterranean civilizations.

Bay trees, from which oil of bay is distilled, belong to the family Myrtaceae which includes myrtles, eucalypts, and cloves. The principal species from which bay oil is extracted, *Pimenta racemosa*, is a tropical tree indigenous to the West Indies.

Beach fleas *Sandhoppers.

Beaded lizard *Gila monster.

Beaked whales belong to a family of eighteen species of *toothed whales and have a snout elongated into a beak. There are usually only one or two pairs of functional teeth at the tip of the lower jaw; in some species there are a number of small teeth which never break through the gums. Among the best-known members of the family are

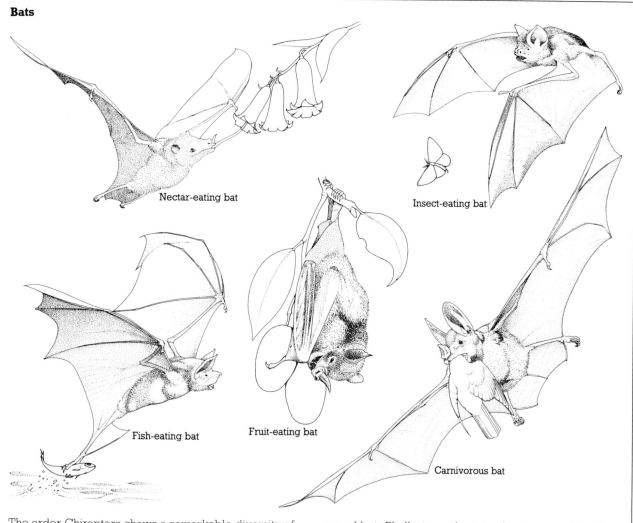

Bats

Nectar-eating bat

Insect-eating bat

Fish-eating bat

Fruit-eating bat

Carnivorous bat

The order Chiroptera shows a remarkable diversity of feeding habits. Some pollen- and nectar-eating species, such as the spear-nosed long-tongued bat, *Glossophaga soricina*, are important agents of flower pollination, whilst many fruit-eaters, for example *Artibeus* species, aid seed dispersal. Carnivorous bats including the greater spear-nosed bat, *Phyllostomus hastatus*, feed on a variety of birds, smaller bats, small rodents, and insects. Other meat-eaters are more specialized: the fishing bulldog bat, *Noctilio leporinus*, is entirely a fish-eater, whilst the horseshoe bats, *Rhinolophus* species, are insect-eaters.

Cuvier's beaked, *Ziphius cavirostris*, northern bottle-nosed, *Hyperoodon planifrons*, and Sowerby's, *Mesoplodon biden*. They are between 4 and 12 m. (13 and 40 ft.) long and feed on squid and fishes.

Beans are an extremely important group of *legumes, grown principally as annual plants in the warm temperate and subtropical areas of the world for their large, protein-rich seeds and/or immature seed pods. The genus *Phaseolus*, particularly *P. vulgaris*, is native to South and Central America and is known to have been, together with maize, the staple diet of the indigenous Indians from at least 3000 BC. The protein of these beans complements that of the maize to form a full and balanced human diet.

The haricot, snap, string, green, and French or kidney beans are all varieties of *P. vulgaris*. The seeds, or 'beans', of this species differ markedly in colour according to the variety. The white haricot bean is particularly well known as the basis of the commercial 'baked beans'. The green immature pods of the French or kidney bean are a popular vegetable in Europe, as are those of the vigorous, climbing, scarlet runner bean, *P. coccineus*. A close relative, *P. lunatus*, the butter or Lima bean, has the largest seeds of the group. The coloured types of bean can contain poisonous glycosides that liberate prussic acid when chewed. Indeed, the wild varieties of all of this group contain these substances, but the process of selective breeding has largely eliminated them from cultivated varieties. Other cultivated beans include broad beans and soya beans.

Beard-fishes are a family of deep-water fishes found in the Atlantic, North Pacific, and parts of the Indian oceans. The Atlantic species, *Polymixia nobilis*, has been found at Madeira, and from Newfoundland to the Caribbean. It grows to 25 cm. (10 in.) in length and lives at depths of about 500 m. (1,625 ft.). Its long chin-barbels suggest that it feeds on the ocean floor. This family is thought to contain six or so species, but these may all be subspecies of a single true species.

Bears are large terrestrial mammals all with a similar and familiar appearance. Characterized by their size and heavy build, they have thick limbs, a diminutive tail, small ears and eyes, a large black nose, and a somewhat dog-like head. The bears and dogs arose from the same ancestors in the Miocene period of geological time. The thick, coarse fur is dark in colour in all except the polar bear. They have large feet, the entire soles of which rest on the ground with each step they walk, giving a slow, ponderous gait. The seven species of bears are widely distributed in the Northern Hemisphere but only three species extend into the Southern Hemisphere. One of these, the spectacled bear, *Tremarctos ornatus*, is found beyond the Andes as far south as Bolivia. Most of the bears reach a length of 1·77 m. (5 ft. 10 in.), while the grizzly of Alaska, Canada, and western North America is the largest, reaching 2·87 m. (9 ft. 5 in.) and weighing up to 770 kg. (1,700 lb.).

Bears have the reputation of being particularly ferocious but most species are usually timid and peaceful, becoming formidable only when wounded or disturbed suddenly. They are omnivores, eating both animals and vegetables with the obvious exception of the *polar bear. Some will kill large terrestrial mammals, others will kill seals or scoop fish from streams; fruit and grass are also eaten. The spectacled bear feeds only on grass, fruit, and roots. The sun bear, *Helarctos malayanus*, occurs in forests of Southeast

Asia, where it eats fruit, bees, and termites, as well as rodents, birds, and eggs. The gestation period lasts six to nine months in most bear species, and the cubs when born are very small, perhaps between 0·45 and 1·8 kg. (1 and 4 lb.). An age of forty-seven years has been reached in captivity by a grizzly bear.

Bear's breeches   *Acanthus.

Beaumont, William (1785–1853), American army surgeon, pioneered gastric physiology. In 1822 at a frontier post he tended and studied a young soldier with a gastric fistula, a gunshot wound in the stomach which refused to close. His exhaustive report of his findings, *Experiments and Observations on the Gastric Juice and the Physiology of Digestion* (1833) remains a classic in its field.

Beavers are two species of aquatic *rodents, one, *Castor canadensis*, found in North America and the other, *C. fiber*, in much lower numbers, in Europe and Asia. Most American, but few Eurasian, beavers construct nests or lodges made of sticks, often plastered with mud, which freezes in the winter into a hard roof, strong enough to deter predators. The entrance to the lodge is under water and so remains open if the pond, which forms upstream from the lodge, freezes. The pond is usually formed or enlarged by damming the exit streams with stones or branches taken from trees felled by the beaver with its strong front teeth. For most of the year beavers feed on plants or tree herbage, but in winter they rely upon branches stored underwater during the summer. A family of beavers, consisting of parents and several generations of young, shares the lodge and shows a social life unique among rodents. Beavers can radically alter their habitat but their activities are generally beneficial in improving the diversity of the vegetation.

Bedbugs are oval, flattened insects of the *bug order which hide by day in beds and in crevices in furnishings and walls, emerging at night to suck the blood of vertebrates, including man. A common species throughout Europe and North America is *Cimex lectularius*; its close relative *C. rotundatus* is found in southern Asia and Africa. They produce a characteristic smell and their bites are irritating, but they transmit no disease. Related species are parasites of birds and bats.

A **beaver** adds a stick to a dam across a stream, using its strong incisors. It builds dams to raise the water-level of a pool, before constructing its living quarters or lodge. Beavers can carry sticks underwater without drowning because the back of their tongue can block off their throat to prevent water from going in.

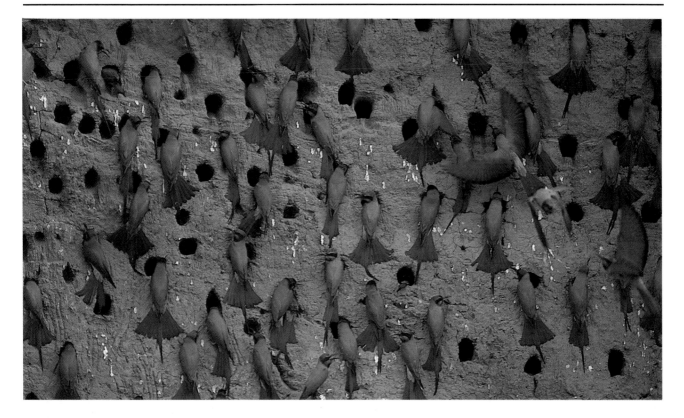

Beebe, Charles William (1877–1962), American naturalist, explorer, and author, described his expeditions to many parts of the world in books such as *Jungle Days* (1925). In *Half Mile Down* (1934), he told the story of his undersea investigation of ocean fauna in a bathysphere.

Beech is the common name used for two distinct genera: the ten species of deciduous *Fagus* of the north temperate zone and Mexico, and the thirty-six species, both evergreen and deciduous, of *Nothofagus* (known as southern beeches) of the Southern Hemisphere. They all belong to the family Fagaceae along with oaks and sweet chestnuts. Beeches cast a dense shade, permitting little undergrowth to flourish. They produce good timber, that of the European common beech being used for furniture. The nuts, or mast, are edible and yield an oil. The copper or purple beeches are variants with dark red leaves and are now planted as ornamental trees.

Bee-eaters are brilliantly coloured, small birds. Many, but not all, of the twenty-five species are bright green in colour with patches of red, blue, or yellow. All have longish, slightly curved beaks. They form a distinct family and are largely confined to the warmer areas of the Old World. The European bee-eater, *Merops apiaster*, is strongly migratory, spreading in summer from the tropics into Europe and temperate Asia. All species feed on large insects such as dragonflies and bees, often swooping down on them from perches. They nest in holes in banks or in the ground, laying two to four white eggs. *M. apiaster* is sometimes a nuisance to bee-keepers, though the seriousness of its depredations is often exaggerated.

Bee-flies are true *flies with over 2,000 species in the family Bombyliidae. Most are tropical species; they all have a long proboscis, and are often stout and hairy, with a superficial resemblance to small bumble-bees. Their

A colony of carmine **bee-eaters**, *Merops nubicus*, a gregarious species from the arid bush and savannah country of central and southern Africa. This species commonly follows grass fires to feed on escaping insects, and will also perch on other animals while waiting for the insects that they disturb.

larvae parasitize and ultimately kill developmental stages of mining bees, parasitic wasps and flies, caterpillars, or grasshopper eggs, depending upon species.

Bees are insects which, together with ants, wasps, and sawflies, make up the order Hymenoptera. The bees, which comprise some 2,000 species in the superfamily Apoidea, have a body covered with feathery hairs, two pairs of wings, the hind pair of which are smaller and linked to the front pair with a row of hooks, and mouthparts adapted for collecting pollen and sucking nectar. Most bees are solitary; only a few, the *honey-bees and *bumble-bees and their relatives, are social, with a worker caste collecting food for their colonies. The solitary bees include species which construct nests in hollow stems, or dig their own nest burrows in wood or soil. Only females build, provision, and occasionally tend their nests; males, or drones, live only to mate. The cells of a nest can be formed from plant material, mud, or wax secreted by glands on the underside of the body. Each cell contains one or more eggs and, before being sealed, is provisioned with pollen and honey, which is nectar matured in the mother's honey stomach. The young of some species are fed as they grow, while others are parasitic on other kinds of bees.

Beetles are *insects of the order Coleoptera. The front pair of their wings is thickened and hardened, forming protective covers (elytra) for the delicate, folded hind pair, and for much of the abdomen. The elytra are extended in flight but give little more than lift, the hind wings giving most lift and thrust. All beetles go through a complete *metamorphosis between egg and adult.

There are more described species of beetle than of any other order of insects, with some 300,000 species known. The smallest is smaller than the largest single-celled animal at 0·25 mm. (0·01 in.), and the largest, such as the *Hercules beetle, is larger than the smallest mammal. The great number of species is due to the variety of life-styles followed, especially by the larvae. Beetles occur worldwide except in oceans and near the poles. Most species live on land, although there are many in fresh water and a few on the sea-shore. All have biting mouthparts. Some, such as the ground beetles, are carnivorous both as adults and as larvae. The majority of species feed on plants, including fungi, with many having special adaptations for feeding on leaves, fruit, seeds, and living or dead wood. Consequently many are pests of agriculture and forestry. These include weevils, leaf beetles, and longhorns.

Beetroot is a variety of *Beta vulgaris* which also includes sugar-beet, mangel-wurzels, spinach beet, and seakale beet. The characteristic swollen, red root of beetroot is grown as an annual for a vegetable or as a perennial for fodder.

Begonias are plants with a wide distribution in sub-tropical and tropical areas of the world. The 750 or so species belong to the genus *Begonia* and most have rather succulent, jointed stems; many are tuberous rooted or rhizomatous, a few are climbers with aerial roots, and others have woody stems. All species have leaves that are asymmetrical in shape, and male and female flowers (both on the same plant) which differ in appearance. Several species have been hybridized and selectively bred for either showy red, yellow, pink, or white flowers, or ornamental foliage in reds, greens, or silver. Some species contain medically valuable drugs.

Behaviourism is an approach to psychology that concentrates on observable behaviour rather than on how the subject may feel or think. It is founded on the scientific study of animal behaviour and the ways in which this can be modified by reward or punishment. Behavioural treatments are of value in a number of mental illnesses such as phobias.

Belemnites are the fossilized shells of a group of extinct *cephalopod molluscs of the Mesozoic Era (245–66·4 million years ago). They are shaped like bullets and are actually reduced shells that lay entirely within the body of the animal and are equivalent to the cuttle-bone of the present-day cuttlefish. They may have been air-filled to assist buoyancy.

Bell, Sir Charles (1774–1842), Scottish anatomist, worked on the nervous system and discovered that there are two types of *nerve filaments, sensory and motor, along which nerve impulses can be transmitted in only one direction.

Belladonna lily   *Amaryllis.

Bell-bird is the common name for four species belonging to the *cotinga family, and which are confined to South and Central America. About the size of a largish thrush,

the males of two species are almost completely white, and in both the others have large amounts of white in their plumage. The females of all four species are yellowish-green. The birds live in rain forest and are famous for the powerful, bell-like calls given by the males. The single egg is laid in a very flimsy nest built on a tiny branch, presumably to make it difficult for predators to reach it. The female looks after the egg and young by herself.

The Australian birds, the crested bell-bird and the bell-minder, are not related.

Bellflowers   *Campanulas.

Bell magpie is an alternative name for three species of Australian magpies, now usually referred to as currawongs. They are crow-sized and shiny black with patches of white in the tail, at the base of the tail, or in the wings; all have striking, yellow eyes. Their powerful beaks are used in taking small animals (including baby birds), fruits, and seeds. In winter many move out of forests and into parks and gardens in their search for food.

Beluga   *Sturgeon.

Bernard, Claude (1813–78), French physiologist, is regarded as the founder of experimental medicine, a field in which physical and chemical methods are used to examine the functions of body tissues. He studied the formation of glycogen in the liver, the chemical action of pancreatic juice in digestion, and the possibility that the nervous system may act to control blood flow.

Berries are fleshy *fruits containing seeds which are not encased in a tough layer, or endocarp. The fruits of tomatoes, papayas, bananas, and oranges are berries. In common parlance, however, 'berry' is used for any small juicy fruit irrespective of its precise structure. The strawberry is not a berry, but a 'false' fruit, and the mulberry is not strictly a berry, but a coalescence of the stalks of many flowers and their fruits. The blackberry, raspberry, and related species have fruits which are clusters of *drupes. Berries are often brightly coloured and sweetly flavoured, making them attractive to birds and mammals, which disperse the seeds, usually in their droppings.

Bewick's swan is a European species resembling the whooper swan, but is smaller and shorter in the neck, and less vocal. It breeds in Siberia, wintering in central Asia and Europe, and takes its name from Thomas Bewick (1753–1828), an English artist who made wood-engravings of birds and animals. Their New World cousins are the *whistling swans.

Biceps are muscles having two distinct points of origin, and are present in the arms and thighs of most vertebrates. In man those that lie in the front of the arm are attached at one end to the shoulder-bone, and at their lower end to the top of the radius bone of the forearm. The muscles both flex the forearms and twist them to allow the hands to face forwards, movements which are used particularly by, for example, wrestlers, carpenters, and violinists.

Bichat, Marie François Xavier (1771–1802), French anatomist and physiologist, was one of the pioneers of the study of cells and tissues, or histology. Without using a microscope, he examined body tissues, dividing them into twenty-one different types. He proposed that tissues rather than organs were the elementary biological units for study.

Bichirs, or reedfish, of which there are twelve or so species, are African freshwater fishes. They are considered to be among the most primitive of the *bony fishes, with elongate, almost snake-like bodies covered with hard, shiny, rectangular, ganoid *scales. They live in the heavily weeded margins of rivers and lakes, and in swamps, and breathe air by means of a lung-like, double-chambered swimbladder. The Nile bichir, *Polypterus bichir*, grows to a length of 70 cm. (28 in.).

Biennials (plants), of which a typical example is the foxglove, take two growing seasons to complete their life-cycle, from seed germination and growth in the first year to flowering and fruiting in the second.

Bighorn sheep are a species of wild *sheep found in the mountains of western North America. Their horns often form more than one full turn, the record length being 1·25 m. (4 ft. 2 in.); the ewes have smaller horns, 37 cm. (15 in.) long. The species, named *Ovis canadensis*, is swift and agile, and there are few cliffs that it will not surmount, often at a gallop; its feet have pads to absorb the shock of its bouncing gait. The sexes live in separate herds except during the mating season. Another species, the snow sheep, *O. nivicola*, is sometimes known as the Siberian big-horn sheep.

Bilberry: along with cranberries and blueberries, forms part of a genus within the heather family. This low shrub grows on the heaths and moors of Europe and northern Asia and is known for its round, juicy bluish-black fruits. Also known as whortleberries, the raw fruit is too acid to be palatable.

Bile is an alkaline fluid containing yellow pigments derived from the breakdown of *haemoglobin. It also contains salts, acids, and *cholesterol, and is produced continuously by liver cells in most vertebrates. During digestion of food, the bile flows directly into the small intestine; at other times it accumulates in the gall bladder. The stored bile is expelled from the gall bladder by a hormone released when food enters the intestine from the stomach. The bile salts aid the digestion and absorption of fat.

Bilharzia, or schistosomiasis, is a tropical disease caused by *blood flukes of the genus *Schistosoma* which live in veins of the intestine. Their eggs escape via the bladder or rectum, causing bleeding and inflammation. It is very widespread in Africa and Asia. The larvae develop in freshwater snails into free-swimming forms, called cercaria, which can penetrate intact animal skin. Because of the participation of snails, intensive irrigation favours spread of the parasite. This disease is seriously debilitating, often lethal, and is one of the three greatest scourges of mankind, next to malaria and hookworm.

Bindweed is the common name for the widely distributed, scrambling, climbing, and non-climbing species of the plant genera *Convolvulus* and *Calystegia*. They belong to the same family as morning glory and sweet potato. The common bindweed, *Convolvulus arvensis* and related species, are persistent and troublesome weeds of

The common **bindweed** occurs in fields and by roadsides. It twines round the stems of other plants—always in an anti-clockwise direction—and kills them by blocking out the light.

gardens, spreading by means of whitish rhizomes, of which the smallest portions will produce new plants. Some *Convolvulus* species are used as garden plants for their large trumpet-shaped flowers. The name bindweed is also used for a member of the rhubarb family: black bindweed, *Bilderdykia convolvulus*.

Binet, Alfred (1857–1911), French psychologist, experimented on the measurement of *intelligence. The Binet–Simon tests placed a child on a scale according to his 'intelligence quotient' or 'IQ'.

Binturong *Civets.

Biochemistry is the study of the chemical processes taking place in living organisms. Individual cells contain thousands of different chemicals, and sophisticated techniques are needed for their identification and study. Such techniques show that the cells of all organisms contain four groups of very large molecules, or macromolecules. These are the two nucleic acids, *DNA and *RNA, *proteins, *carbohydrates, and *lipids. Biochemistry has also shown that all organisms share the same basic molecules of life.

Biogeochemical cycle: a term used to describe the series of events in which a chemical element circulates between living organisms and the non-living environment. The most important examples are the *oxygen cycle, the *carbon cycle, and the *nitrogen cycle.

Biogeography is the study of the geographical distribution of living things, and includes phytogeography (plants) and zoogeography (animals). Its first object was to collect information about plant and animal distribution, and to identify distinct patterns. Phytogeographers and zoogeographers have each divided the terrestrial world into major regions, generally continents or groups of continents considered to possess a characteristic flora or fauna. These two sets of regions do not have precisely the same boundaries.

The faunal system based on the divisions proposed by A. R. *Wallace and P. L. Sclater is probably the best known. In this system six faunal realms are recognized: Nearctic (most of North America); Neotropical (South and Central America); Palearctic (Europe and Asia north of the tropics, and parts of northern Africa); Ethiopian (Africa, except its northern fringe); Oriental (tropical Asia and its associated islands); and Australasian (Australasia, New Guinea, Oceania, and parts of Southeast Asia). The Nearctic and Palearctic regions are sometimes combined to form the Holarctic. The divisions used by phyto-geographers are more concerned with continental regions, although using the major division of the world into the New World (the Americas) and the Old World (mainly Europe, Asia, and Africa).

Geographical isolation can influence evolution. The theory of island biogeography is an important concept in *ecology, and postulates that the number of species in any isolated community is determined by the balance between the rate of arrival of new species, and the rate of extinction of existing ones.

Biological control of pests: a technique first adopted in the control of insect pests, using a suitable predatory insect or parasitoid (*parasite). Parasitic flies and wasps have proved useful in this respect, and one such wasp, *Encarsia formosa*, has been introduced to control the glass-house whitefly, which became a pest in British green-houses. Other successes include the introduction of the small Australian ladybird, *Rhodalia cardinalis*, to control the cottony cushion *scale-insect, which was accidentally carried from Australia to California around 1870 and caused severe damage to citrus orchards before the in-troduction of this ladybird. Such methods do not always work: over 200 pest species have been tackled in this way but half of the attempts have failed completely. However, in cases where biological control methods have proved successful these have been a very cheap and non-toxic solution to the problem.

Modern methods of biological control include spraying with fatal doses of *hormones, and releasing large num-bers of sterile males to reduce the breeding rate of females. Environmental manipulation, such as destroying habitats where pests such as mosquitoes breed, reduces the pest impact.

Biology is the science of living things. It is divided into specialist fields according to the kinds of organism studied: botany for plants, entomology for insects, and so on. It is further split into levels of study, ranging from molecular biology, the subject matter of *biochemistry, to whole communities of plants and animals, which form the subject-matter of *ecology. Biology also includes the his-tory and evolution of life on earth.

Biophysics is a general term of recent origin for the application of the theories and techniques of physics to biology. It can mean either of two things. Firstly, it can refer to the study of natural phenomena, such as the elec-trical conduction of nerve impulses, which are related to topics studied in physics itself. Secondly, it can refer to the investigation of any aspect of biology using complicated physical techniques. In either case it is necessary to under-stand something of modern physics in order to perform the research.

Biophysics, which aims to be an exact, quantitative subject, has made major contributions to many areas of biology. These include theories of nerve and muscle action,

the physical properties of cell membranes, and the struc-tures of *DNA and *protein molecules (worked out using X-ray techniques). Many of these discoveries are of great medical value.

Birch is the name for the forty species of *Betula*, a genus of trees and shrubs of the north temperate zone and Arctic region. Together with *alders (*Alnus*), *hazels (*Corylus*), and hornbeams (*Carpinus*), they make up the birch family, Betulaceae. They reach the polar limits of flowering plants, and several species are important constituents of *taiga and *tundra. In lower latitudes, birches are colonist trees, invading herbaceous or shrubby vegetation and growing rapidly. They are short-lived and after death are replaced by other species. The pale wood is used mainly for charcoal and plywood. An oil extracted from the trees is used in tanning Russian leather. The bark often peels off in long, pale strips, and that of the paper birch, *B. papyrifera*, was used in the construction of canoes, tents, and huts by North American Indians.

Bird of Paradise flowers are plants related to the banana, and native to South Africa. They bear flowers with spikes of yellow and purple, shaped like a bird's head. These plants, in the genus *Strelitzia*, are pollinated by the feet of nectar-seeking birds.

Birds are vertebrates of the class Aves. They are thought to be warm-blooded descendants of small *dinosaurs. About 9,000 living species are known. The earliest known bird, the archaeopteryx, had the most striking charac-teristic of birds: feathers and flight. The power of flight has enabled birds to reach most parts of the world and to migrate to and from breeding-grounds as the seasons change.

Birds range in weight from about 2 g. ($\frac{1}{14}$ oz.) in the case of the vervain hummingbird, *Mellisuga minima*, to 90 kg. (200 lb.) in the case of the ostrich. The largest birds are flightless, the upper limit to the weight of flying birds being about 15–20 kg. (35–45 lb.). Many of the charac-teristics of birds reflect the great need to minimize weight in order to make flight as energetically economic as poss-ible. These adaptations include loss of teeth, hollowed bones, and the fusing of many parts of the skeleton. The large breast muscles provide the source of the power needed to flap the wings. The main senses used by birds are vision and hearing, in contrast to many mammals, for whom vision may be much less important and sense of smell much more important.

Birds have been used widely by man, especially for the production of eggs and meat. They have also been popular as cage-birds on account of their beautiful songs; for sport, whether as prey or hunter (as in falconry); and, in the case of pigeons, for carrying messages.

Bird's nest fern is a species of *fern native to the Old World tropics; it grows as an epiphyte with a rosette of fronds, each up to 1·2 m. (4 ft.) in length. The plant forms a nest-like structure which traps detritus, from which its roots obtain nourishment.

Birds' nests are structures, built by almost all birds, in which they lay their eggs and raise their young. Nests vary from simple scrapes in the ground made by wading birds, through simple cups of pebbles or vegetation made by gulls and ducks, to enormous nests of twigs and branches which may last for many seasons, such as are made by birds of prey. They may be built in holes excavated in the ground or in tree trunks, on the ground or cliff ledges, in forks of trees, or suspended from the tips of branches. Some birds build floating nests of vegetation; and others nests of mud which hardens. Nests may serve to keep the eggs together during incubation, to help insulate them against adverse weather, and to conceal them from predators. In many species the young leave the nest on hatching; in others the nest provides protection for them until they are ready to fly. Some young birds continue to roost in the nest after this stage, and in some species the parents may roost in the nest throughout the year.

Birds of Paradise belong to the family Paradisaeidae; this includes about forty species, most of which are found in New Guinea, although a few occur in eastern Australia and other nearby areas. Noted for their brilliant colours and ornamentation, the males display communally; the females come to the display grounds for mating, and look after the eggs and young by themselves. The feathers were once much valued for ladies' hats and are still used for local tribal head-dresses. The earliest skins sent to Europe were poorly prepared and without their legs, giving rise to the fanciful idea that the birds never came down to earth—hence birds of paradise.

Bird-song is composed of the sounds produced through a complex structure in the *trachea, called the syrinx. Song is mainly used by male birds and is thought to serve two main functions: to defend the territory against rival males by announcing the owner's presence; and to attract potential mates. Songs of each species tend to be charac-

A beautiful male **birdwing butterfly**, *Ornithoptera chimaera*, on a hibiscus flower. This species is native to the montane forest of Papua New Guinea, from 1,000 to 2,500 m. (3,280–8, 200 ft.). After fertilization, the male plugs the female's genital tract with a frothy secretion to prevent other males from copulating with her.

A **bird's nest fern,** *Asplenium nidus*, growing on the bark of a tree in a tropical rain forest in Java. The true species has smooth fronds; this variety, with crinkly fronds, is commonly grown as a house plant.

teristic, enabling the experienced ornithologist to identify a species without seeing it.

The songs vary from quite simple vocalizations to very complex songs. The latter tend to be given by the so-called perching birds of the order Passeriformes, particularly the subgroup Oscines, the songbirds. Some species mimic the calls of other species in their song, and in a few species the male and female may produce different parts of the same song (antiphonal singing, or duetting), their split-second synchronization making it almost impossible for a listener to realize that more than one bird is involved.

Birdwing butterflies, found in Southeast Asia and Australasia, are among the largest and most beautiful butterflies; some are protected by law. The thirty-one species in the genera *Triodes*, *Trogonoptera*, and *Ornithoptera* are known as birdwings on account of their size and brilliant colours. Those species such as the Rajah Brooke's birdwing, *Trogonoptera brookiana*, are strikingly patterned green and black. This probably advertises the toxicity of the adult, as its caterpillars feed on the poisonous Aristolochiaceae plants, like other members of the swallowtail family. The largest known butterfly is Queen Alexandra's birdwing, *Ornithoptera alexandrae*, of New Guinea, the female of which has a wing-span of up to 28 cm. (11 in.).

Bison are two species of large *cattle. The American bison, *Bison bison*, once roamed the plains of North America; a larger subspecies, the wood bison, still does in a remote region of Canada. The European bison, *B. bonasus*, once lived in the forests of Europe and Asia but is now restricted to captivity; wild populations became extinct in 1919. The European bison is 1·8 m. (6 ft.) in height at the shoulder, and the American bison is rather smaller at 1·5 m. (5 ft.). Both species have large heads with short, stout horns curving outwards and upwards. Behind the head are the huge, humped shoulders and the surprisingly narrow hindquarters. The coat is a rich, dark brown, becoming almost black on the head and shoulders. Bulls fight in the rutting season for supremacy of the herd, and battles may last two days in succession. One calf is usual but two are occasionally born. After the birth, the young is ready to stand as soon as the mother has completed cleaning it from head to toe.

Biting housefly *Stable fly.

Biting lice, or bird lice, are small, wingless, flattened insects which comprise the order Mallophaga with over 2,800 species. The majority are external parasites of birds, although some species occur on mammals. They feed on dead cells and, in heavy infestations, may damage feathers or hair. Some are active runners; others move slowly. As some species are confined to certain parts of the body, a given host may carry several species at any one time. The eggs are attached to the feathers or hairs and the young resemble the adults.

Bitterns are marsh birds belonging to the same family as herons. There are twelve species found throughout the world, and all of them frequent reed-beds. When disturbed, they will thrust their head and neck upwards, facing the danger, so that their elongated shape and streaky plumage merge with the reeds. The males' booming calls in the breeding season are audible up to 5 km. (3 miles) away. These medium-sized birds construct a nest of vegetation in reed-beds and incubate four to six eggs.

Bivalves are a class of *molluscs with a two-piece hinged shell, and include many familiar sea-shore animals. Clams, scallops, cockles, mussels, and oysters all belong to this group. Most of the 20,000 or so species are specialized as sedentary, often burrowing, filter-feeders. Within the valves (the two halves of the shell) are large flattened gills covered with tiny, beating cilia; these create water currents supplying oxygen, and carrying countless *plankton, which are filtered out and eaten. Burrowing bivalves have tubular siphons to carry water in and out. They leave tell-tale depressions on sandy beaches.

Bivalves are among the commonest molluscs, often forming enormous beds in estuaries and shallow seas. They are renowned as edible-shellfish, for pearl manufacture, and sometimes as destructive burrowers into rock (piddocks) or man's marine artefacts of wood (shipworms).

Black beetles *Cockroaches.

Blackberries are members of the rose family. They fall within the general species name of *Rubus fruticosus* but are

The Eurasian **bittern**, *Botaurus stellaris*, takes up its defence posture, making it appear almost indistinguishable from its reedy surroundings.

a highly variable and complex species-group, or aggregate. The varieties grown commercially, or in gardens, are derived from plants indigenous to Britain and other temperate countries. The fruit is really a cluster of small, black berries and is sweetly flavoured. Most individual kinds are prickly, climbing or sprawling plants that produce long, arching canes, capable of rooting and producing new plants where their tips reach the soil. Thornless plants have been developed in recent years, including the variety Oregon.

Blackberries can be exceedingly vigorous in growth and are weeds in many temperate regions of the world, including those to which they have been introduced, such as South America and Australia.

Blackbird is a name used primarily for some thrushes of the genus *Turdus* and for the New World orioles. In the Old World the term primarily relates to *Turdus merula*, a common species which occurs throughout most of Europe and eastern Asia and is also found on high ground in northern India and Southeast Asia. About 25 cm. (10 in.) long, the male is completely black with a bright yellow eye-ring and yellow beak, while the female and the young are brown. They build cup-shaped nests in bushes and lay three to five eggs, which are blue with brownish speckles. The parents may raise three broods in a season. Blackbirds feed primarily on earthworms and other invertebrates, and many of those breeding in the northern parts of the Northern Hemisphere migrate to milder areas for winter.

Blackcap *Warblers.

Blackcock *Grouse.

Black currant: a common garden and commercially-grown, bushy shrub belonging to the saxifrage family, along with red currants and gooseberries. The purplish-black fruit is used in jam and tarts. It is a European native, growing wild over a wide area, although many improved varieties have been produced, differing in fruit size, flavour, and season of ripening. Established plants can become infected by a virus spread by mites, causing big-bud disease, and it is important that virus-free stock is planted.

Black-eyed Susan *Acanthus.

Blackflies, or buffalo gnats, are small, dark, thickset *flies of the family Simulidae, occurring in all parts of the world, especially the north temperate and sub-Arctic regions. Females have a vicious bite and feed on the blood of vertebrates; they are often so abundant as to be serious pests, even causing deaths of livestock and occasionally of man. In Central America and Africa they are vectors for a nematode worm that causes onchocerciasis or *river blindness. The larvae live in running water, attached to stones or to other animals by a sucker, filtering their food out of the water. Control measures include the addition of specific insecticides to the head-waters of streams.

Blackthorn, or sloe, is a deciduous, spiny shrub of hedge-rows and forest edges in Europe. It is a member of the cherry genus, *Prunus*, and has small, white flowers and blue-black fruits which have a sour, astringent taste. In Britain, these are used in preserves and in flavouring a kind of gin, and the wood is used for walking-sticks. The fruits are eaten by birds, badgers, and foxes, which disperse the seeds in their droppings.

Black widow spiders are found in many parts of the world, and are one of the few genera of *spiders with a venom dangerous to man. This venom can affect the nerves to produce muscular spasms and respiratory difficulty. The several species of these black, shiny spiders, often with a red hour-glass mark on the back of the nearly spherical abdomen, build rather untidy webs and hang beneath them, wrapping their catches up in silk trusses. Females carry large egg-cases at certain times of the year.

Bladder: the muscular bag into which urine passes from the kidneys via tubes called ureters. This organ is found in the majority of vertebrates with the exception of birds. It swells up as it fills with urine, and urination occurs when the ring of muscle, or sphincter, guarding its outlet relaxes. This is under voluntary control in adult mammals, but in the young voiding occurs automatically, being dependent on the amount of urine to be secreted and the speed at which the bladder fills. The duct through which urine passes from the bladder is called the urethra. In the male, the sphincter is surrounded by the prostate gland, which tends to enlarge in old age and interfere with its action. Hence, old male animals and men tend to become incontinent.

A female **black widow spider**, *Latrodectus mactans*, with egg sac. This North American species is commonly found in houses and outbuildings and is responsible for many bites. Female spiders are about 12–16 mm. ($\frac{1}{2}$–$\frac{3}{5}$ in.) long. Males are much smaller and flatter, with long legs. Other venomous 'widows', not necessarily black in colour, occur in different parts of the world including Australia and the Mediterranean region.

Bladderwort is the common name for some 120 species of *Utricularia*, a genus of *carnivorous plants which trap, and use as a food source, small aquatic animals such as *Daphnia*. They may be rootless aquatics or marginal bog-plants found throughout most of the world. The traps are small bladders produced on the leaves and runners, which are triggered by contact, sucking in their prey with a rush of water. They give their name to the bladderwort family, Lentibulariaceae, which also includes *butterworts.

Lesser **bladderwort**, *Utricularia minor*, a submerged aquatic, showing the bladders with trigger hairs on both young and mature shoots together with the branched, antler-like leaves.

Blennies are mostly marine fishes living in tropical and subtropical seas. A few live in fresh water. Fishes of several families are called blennies, including the comb-tooth blennies of the family Blennidae, the clinids (Clinidae), pricklebacks (Stichaediae), and gunnels (Pholididae). Their bodies are scaleless, the head is usually blunt, and the jaws have numerous, closely packed, small teeth. They have two long, slender rays in each pelvic fin, and often have flaps of skin above the eyes or elsewhere on the head. They usually live in shallow water, even in shore pools. There are over 100 species within the family groups known as blennies.

Blewits *Agarics.

Blind fishes occur in a variety of families but are mainly catfishes, cyprinids (carp family), or cave fishes (a North American family of five species). They are found over a wide geographical range, and are mostly associated with caves and underground systems, although there are a few marine species. In all of them the eyes are much reduced in size, being covered with skin but rarely totally absent. Blind fishes also lack all pigment in the skin and are pink in colour. These apparently fundamental features are a result of living permanently in underground, dark situations. Several deep-sea fishes, which live in the dark, have greatly reduced eyes and some are virtually blind. Most blind fishes belong to groups in which there are abundant alternative sense organs. The catfishes, for example, all have well-developed barbels, and the cave fishes have highly elaborate sense organs on the head and body. The blind characin, *Anoptichthys jordani*, from Mexico, has a normally coloured and sighted relative, *A. mexicanus*, in the surrounding rivers with which it will interbreed.

Blindness is the inability to see. It may be due to damage to that part of the brain involved with vision, or to the optic nerve, but much more commonly it is due to the eyes themselves. It has been estimated that there are about ten million people in the world without effective vision, and that for most of them the loss of sight could have been prevented. Specific causes of blindness are trachoma, a form of chronic *conjunctivitis, leprosy, and onchoceriasis (*river blindness). In developed countries the causes are more commonly *glaucoma, diabetes, and hereditary and congenital conditions. Restoration of sight is possible only when blindness is due to clouding of the translucent lens or cornea. In that case, removal of the lens or corneal grafting is effective.

Blind snakes include about 200 burrowing species in the family Typhlopidae, and they are distributed world-wide in tropical and subtropical regions. They are completely harmless to man and rather small, although a few species are known to exceed 61 cm. (2 ft.) in length. Their eyes are tiny and probably capable only of distinguishing between light and dark. Their bodies are covered with smooth, shiny scales which have the property of resisting soil adhesion. Their diet consists mainly of ants and other small soil creatures and they have a markedly reduced dentition; the upper jaw bears teeth, whereas the lower jaw is toothless. Snakes of the unrelated Leptotyphlopidae family are also known as blind snakes.

Blister beetles are mostly blackish and soft-bodied *beetles up to 2 cm. (¾ in.) long. If they are crushed and rubbed on human skin, a chemical in their blood, cantharidin, causes blisters, a characteristic which has led to their use in medicine as counter-irritants. Their larvae are parasitic, mostly on solitary bees. Along with the *oil beetles, they form the family Meloidae.

Blood is a fluid containing cells which circulates in the body of all animals with the exception of a few phyla of lower invertebrates. It supplies tissues with oxygen and foodstuffs and transports waste products, which are removed from the body by the excretory system. The blood of many invertebrates may be simply a solution of nutrients which bathe the tissues, for they do not need to have an active blood flow, their needs being met by diffusion of wastes out of, and nutrients into, cells. Larger invertebrates, such as crabs or insects, have a tubular heart which pumps blood from one end of the body to the other. This creates an open flow system. Other functions of invertebrate blood include the transmission of force, as in the locomotion of earthworms, and the transfer of heat from one part of the body to another.

With increases in body size, diffusion is no longer efficient, and an active flow system is needed. The vertebrate blood system, with veins, arteries, and the powerful muscular pump, the heart, supplies this need. Waste products, nutrients, and chemical messengers called hor-

Red **blood** cells seen under the scanning electron microscope. These cells are among the smallest in the human body, with a diameter of only 7 micrometres (0·007 mm.). There are about five million red cells in every cubic millimetre of blood.

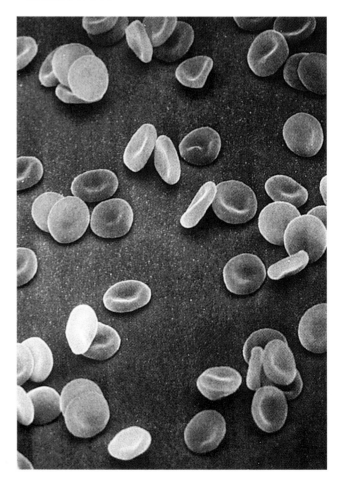

mones, are transported in the blood. Mammalian blood consists mainly of red cells, white cells, and platelets suspended in *plasma. Some of the white blood cells combat infections directly; other white cells manufacture *antibodies and strengthen the body's immunity towards unwanted foreign proteins. Red blood cells convey oxygen to tissues and remove carbon dioxide to the lungs. The plasma conveys foodstuffs to tissues and also takes away waste products. Arteries carry oxygen-rich blood away from the heart, while veins return blood rich in carbon dioxide to the heart and lungs. Capillaries are the smallest blood vessels and form a branching network connecting arterioles (the fine end-branches of arteries) and venules.

In adult vertebrates, blood is manufactured in the bone marrow.

Blood flukes are *flatworms belonging to the genus *Schistosoma* and are parasites of man throughout Africa and Asia. There are several species, each capable of causing the disease *bilharzia. They are unusual among flatworms in having separate sexes; the more slender but longer female, up to 1 cm. ($\frac{1}{2}$ in.) long, lives within a groove on the underside of the male. Eggs released into the host's bloodstream penetrate the wall of the large intestine or bladder and leave the host's body. A tiny larva, called a miracidium, hatches and bores into one of several species of freshwater snail. Here it develops into a free-swimming form of larva called a cercaria, this boring into the skin of humans as they bathe in streams or pools infested with snails. Once inside the new host's body the larva enters the bloodstream and eventually reaches the intestinal veins, via the lungs and liver.

Blood groups are systems defined for the typing of human blood for transfusion. If two bloods are incompatible, the transfused cells clump together and are then destroyed. This is caused by a reaction of *antibodies in the recipient's *plasma against antigen on the surface of red blood cells. The two most important of the thirty or so systems are the ABO and the rhesus factor (Rh). The ABO system occurs because two different antigens, A_1 and B_1, may be present on the surface of red blood cells. Individuals of group AB possess both antigens, those in blood group A or B possess a single type, and people with group O blood have neither. Group O persons have in their blood plasma antibodies to both A and B antigens so they can safely receive transfusions only from members of their own group. Persons who have any particular antigen on their red blood cells necessarily lack antibodies to it; otherwise their blood would clot all the time. Thus, people with blood group AB do not have antibodies to these antigens in their blood and can receive blood from any other group.

In the Rh system, 85 per cent of people possess the Rh antigen; the remainder, known as Rh-negative, do not. Rh-negative individuals receiving a transfusion of Rh-positive blood, or Rh-negative women pregnant with an Rh-positive child, slowly develop Rh antibodies. The antibodies may lead to severe reactions to transfusions with Rh-positive blood and may handicap the development of Rh-positive children in subsequent pregnancies. Both Rh and ABO blood groups are inherited in a straightforward genetic manner.

Blood pressure results from the force driving the blood round the closed system of blood vessels supplying an animal's body. At each beat of the heart, blood is pumped into the arteries, creating a brief peak of high pressure. The pressure declines between beats to about two-thirds of this high peak of pressure. The difference between these values is a useful indicator of the condition of the heart and arteries. A raised blood pressure (hypertension) may lead to heart disease. Blood pressure can be estimated in the limb arteries of man and other animals by compression, using an inflatable cuff and a column of mercury whose height records the pressure in the cuff.

Bloodworms *Midges.

Blowflies are true *flies in the genus *Calliphora*, and this name is often used for the 'bluebottles' such as *C. vomitoria* and similar species. The term blowfly refers to the females' habit of laying their eggs in carrion or excrement which the larvae consume. In doing so, the larvae often liquefy the carrion which is then known as 'blown'. Related species include the greenbottles of the genus *Lucilla*, and species such as the screw-worm fly, *Cochliomyia homnivorax*, of North America, which lays eggs in wounds or orifices on livestock; other species parasitize nestling birds and toads. Blowfly maggots have been used medically to clean wounds. Adult blowflies feed indiscriminately on excrement or sweet substances which they liquefy with saliva and regurgitated food. As a consequence they may contaminate food with pathogens causing dysentery, typhoid, or cholera.

Bluebells *Scilla.

Blueberries are the fruits of several species of the genus *Vaccinium*, shrubs native to North America, where they are grown commercially. They are related to the European bilberry, but bear larger, sweeter fruit. The bluish-black berries are about 1 cm. ($\frac{1}{2}$ in.) across. The high-bush blueberries may grow as tall as 4·5 m. (15 ft.); the hardier low-bush species are only about 30 cm. (12 in.) tall.

Bluebirds are three species of the thrush family and belong to the genus *Sialia*. All occur in North America and Mexico. They are rather larger in size than a sparrow, and the males of all species are bright blue on top. The mountain bluebird, *S. currucoides*, is entirely blue, while the other species are rusty red underneath. The females of all species are dull grey in colour. All inhabit open woodlands or orchards and nest in holes in trees, laying four to six blue eggs. They eat a wide range of insects and fruits.

Bluebottles *Blowflies.

Blue-green algae were probably among the first living organisms to evolve on Earth, their fossilized remains having been discovered in Pre-cambrian rocks 1,500 million years old. They are unicellular organisms once considered to be a sort of simple algae but have more features in common with bacteria. They are found in aquatic habitats and are often the cause of blue-green coloured 'algal blooms'. They lack a well-defined nucleus and propagate by budding or splitting into two halves. Other names for the group include Cyanophyceae and Myxophyceae.

Blues (butterflies) are a family of some 6,000 species of butterfly. This family, the Lycaenidae, also includes the *hairstreaks and *coppers. They are a predominantly

tropical family with a wide variety of life-styles. The upperside of the male is blue, sometimes brilliantly so, that of the female more often being brown. Several species have caterpillars that feed on the grubs of ants, while some, including the Australian mistletoe butterflies, *Orygis* species, are protected by ants. The Chinese blue butterfly, *Gerydus chinensis*, has a caterpillar that eats aphids. Many Lycaenids feed on the leaves of tropical trees and shrubs. Their larvae are typically woodlouse- or slug-shaped, unlike the normal shape of a caterpillar.

Blue whale: a species of *whalebone whale, and the largest animal that has ever lived. It can grow to a length of 33 m. (109 ft.) and weigh some 120 tonnes (132 US tons). It is distributed worldwide and migrates between high and low latitudes. It can swim at speeds of up to 36 km./hour (22 m.p.h.). In contrast with its size, its food consists of krill. This is collected on the fibrous fringe of the baleen plates which hang in its mouth; the efficiency of this system is indicated by the finding of 2 tons of krill in the stomach of a single whale.

Pairing is thought to occur in June and July and the gestation period lasts ten or eleven months. The young are born in the comparatively warm waters of lower latitudes and may be as long as 7·5 m. (24 ft. 6 in.) and weigh up to 2,790 kg. (6,150 lb.) at birth. The young grow rapidly on the rich milk, increasing from 7 to 15 m. (23 to 50 ft.) in length in seven months. Females conceive every two or three years and pregnant females of thirty years of age have been found. Blue whales may live for as long as sixty-five years. Populations of the blue whale have been severely reduced by overfishing and it is now an endangered species. Two subspecies are recognized: the blue whale, *Balaenoptera musculus musculus*, and the pygmy blue whale, *B. m. brevicauda*.

Boars are members of the *pig family. The word boar has two meanings: it is used of all uncastrated male pigs,

A pair of large **blues**, *Maculinea arion*, on wild thyme, the larval food-plant. After hatching, the caterpillar feeds on the flowers until after the third moult when it leaves the plant. It is then picked up by the ant, *Myrmica sabuleti*, which strokes the gland on the caterpillar's back and drinks the sweet droplet that exudes from the gland. The ant then carries the caterpillar to its nest where the caterpillar feeds on ant larvae. It pupates in the late spring and three weeks later the adult blue butterfly finally emerges.

and also refers universally to the five species of the genus *Sus*, including the European wild boar, *S. scrofa*, which still ranges over a wide area of the forested regions of Europe, and east into Asia. Other members of the genus are found throughout Asia and south into New Guinea. The European wild boar may stand 90 cm. (3 ft.) at the shoulder, and is almost 150 cm. (5 ft.) long. Dusky or greyish-brown to black, its coat is of long coarse hair with scattered bristles. It has large, sharp, strong tusks that may be 30 cm. (1 ft.) long. The sow often has two litters per year, each of four to six young.

Boas, like some other primitive *snakes, retain vestiges of a pelvic girdle; their hind-limbs are retained in the form of claw-like cloacal spurs. They resemble pythons in many respects but occur in some areas, such as the Americas and Madagascar, where pythons are absent. Some small species are found in the Old World (sand boas), and dwarf boas occur in New Guinea and some South Pacific Islands. All boas kill, suffocating their prey by constriction.

They exist in a variety of habitats; there are arboreal forms (tree boas), burrowing forms (such as rosy boa, rubber boa, sand boa), and partly aquatic forms (anaconda). The boa constrictor, the rainbow boas, and the ground boas (of the West Indies) are mainly surface-dwellers. There are records of the American boa constrictor reaching 5·6 m. (18 ft. 6 in.) in length.

One of the best-known species of **boa** is the boa constrictor, *Boa constrictor*, of tropical America. It is associated with humid forests but also lives in most other habitats including arid scrub.

Boat-billed herons, or boatbills, are a species of bird found in Central and South America. They are similar to night herons, but have a beak which is flat and broad and which is used to catch and eat reptiles and small mammals. This species is largely nocturnal and solitary in habit.

Boat flies   *Water boatmen.

Bobolink: a sparrow-sized bird of the New World *oriole family. The male in breeding plumage is black with a white rump, white flashes on the wing, and a creamy-buff nape; outside the breeding season he is sparrowy plumaged, as is the female at all times. Bobolinks breed in the northern half of North America, in open grassy areas, and feed on insects. They spend the northern winter in central South America after migrating across the Gulf of Mexico, a distance of 1,900–2,400 km. (1,200–1,500 miles).

Body fluids of terrestrial and aquatic animals consist mostly of water, which acts as a buffer against temperature changes and as a solvent for nutrients and waste products.

About one-third of the total body water in man is outside cells; it includes the fluid that bathes cell surfaces, the *lymph, and the blood *plasma. The remaining two-thirds make up the bulk of cellular mass. The two fluids, inside and outside cells, are in osmotic (*osmosis) balance, and are not excessively acidic or alkaline. They possess a high electrical conductivity due to dissolved mineral salts. Fluids outside cells contain more sodium and chloride and much less potassium than does that within cells.

Boerhaave, Hermann (1668–1738), Dutch physician, made Leiden the centre of European medical teaching in his time. He insisted on the need for bedside instruction and for direct post-mortem observations. He wrote several comprehensive medical textbooks which were widely accepted and translated into several languages. A disciple of Hippocrates, he taught that the aim of medicine was to cure the patient, using all available knowledge.

Boils are small *abscesses around hair follicles which are due to infection with the bacterium *Staphylococcus aureus*. This organism often persists inside the nostrils for a considerable time without causing local trouble but providing a source of infection. For this reason boils may occur in succession until the source is treated.

Bones form the vertebrate skeleton, giving it structural strength, while acting as attachment points for muscles, storage places for mineral salts, and sites for the pro-

This cross-section of human **bone** tissue from the tibia has been stained and magnified (× 27). This is compact bone. The concentric-circle structures each consist of bone cells around a canal which contains blood and lymph vessels.

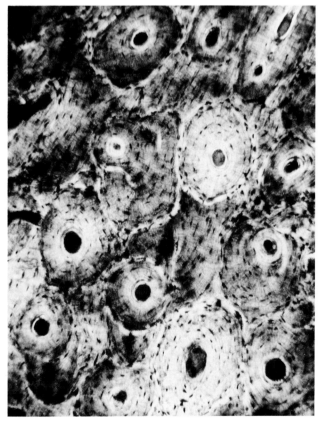

duction of blood cells. They are made of a relatively soft, rubbery substance called collagen, reinforced by calcium carbonate and calcium phosphates. The hard outside shell is called compact bone, while the interior is formed of a network of bony plates, called spongy bone. Cavities help to lighten bones and provide a site for bone marrow. In adult mammals, bones are 80 per cent calcium, which is secreted by osteocytes (bone cells) in such a way that the final shape and internal structure of each bone is best able to resist the stresses and strains it encounters. Bones are living structures and, as they grow, they are constantly being changed in shape and form. This occurs by the interaction of osteocytes, which build them up, with osteoclasts, which eat them away. The marrow of some large bones, such as the femur in adult humans, produces the red blood cells. The non-functional marrow of other bones consists of fat.

Bonito is a widely-used name for small members of the *tuna or mackerel family. In the eastern Atlantic the name is used for both the skipjack tuna, *Katsuwonus pelamis*, and Atlantic bonito, *Sarda sarda*. The Atlantic bonito ranges across the Atlantic from Scandinavia to Nova Scotia and South Africa to Brazil, and is an open-sea predator which swims in small schools, feeding on fishes and squids near the surface. In appearance it is relatively slender, with a pointed head, and a steel-blue back with numerous oblique, black lines across it. It grows to 90 cm. (3 ft.) in length.

Bonnet, Charles (1720–93), Swiss naturalist, made many contributions to entomology and botany, despite being virtually blind and deaf from an early age. He was one of the first to study *photosynthesis; and he discovered parthenogenesis (reproduction without fertilization), which is common among insects and plants.

Bony fishes are the largest class of fishes and represent some 20,000 species of marine and freshwater fishes. They all have a skeleton based upon *bones, as opposed to the boneless skeleton of the other two classes, the *jawless fish and the *cartilaginous fishes. The bony fishes are further subdivided into four main taxonomic groups: the lobefins, represented by the *coelacanth; the *lungfishes; the *bichirs; and the largest group by far, the rayfinned fishes, which include the vast majority of bony fish species.

Boobies are large seabirds with sharp, conical bills, which plunge into the sea to catch squid and small fish at the ocean surface. They are tropical equivalents of the *gannet and belong to the genus *Sula*, which contains six species. Three species are wide-ranging, two (the white booby, *S. dactylatra*, and the red-footed booby, *S. sula*) occurring in all tropical oceans. The remaining three species have restricted ranges over parts of South America.

Book lice are minute, wingless, pale yellow insects, around 1 mm. ($\frac{1}{25}$ in.) in length; they are found in house dust and often in old books. They comprise over 2,000 species in the order Psocoptera, and feed by scraping the surface of their foodstuff, causing much damage if they are numerous. They occur naturally on tree trunks and fungi, under bark, and in other places where fragments of vegetable matter collect. Some of the males make tapping noises by banging their heads; this sound has been confused with that of the death watch beetle. Some species

have winged females or, more rarely, males which can swarm under certain conditions.

Bootlace worms *Ribbon-worms.

Borage *Forget-me-nots.

Boston fern is the most familiar of several subspecies of the sword fern, *Nephrolepis exaltata*, a native of tropical regions. The name is derived from the fact that it was first discovered in a shipment of ferns sent to Boston. The Boston fern, *N. e. bostoniensis*, and other members of its genus are unusual among ferns in being able to reproduce vegetatively by means of runners.

Botanic(al) gardens were first planted in Italy—at Pisa in 1543 and Padua in 1545. They began as centres of teaching at a time when plants were the main source of medicinal drugs. The first English botanical garden was founded in Oxford in 1621, the first North American one, the Elgin Botanic Garden in New York, in 1801. The purpose of these gardens did not restrict their contents, for any new plant was considered worthy of study in case it turned out to be useful. A flood of plants, new to European gardens, came from sea voyages of exploration from the late seventeenth century onwards. By the mid-nineteenth century the Royal Botanic Gardens at Kew had become the main channel for the introduction of new plants to English gardens. Kew also became an important centre for the transfer of food and other economic plants from their native lands to similar habitats elsewhere. There are now hundreds of similar gardens, from the small ones to the giants like Kew and the Missouri Botanical Garden. Their curators still sponsor plant-hunting expeditions to relatively unexplored regions.

Although the study and cultivation of plants, native or exotic, remain the central purpose of these gardens, they have become places of entertainment for those interested in plants.

Botany is the study of plants; its origins go back to around 300 BC, when Theophrastus wrote about the form and functions of plants. This early interest in plants was largely a result of their use in medicine. Plant anatomy developed in the seventeenth century, aided by the invention of the microscope, whereas physiology, the study of processes essential to life, began a century later. Nowadays botany covers all aspects of the biology of plants.

Although plants and animals are fundamentally similar in terms of their biochemistry and cell biology, botany reflects many differences in emphasis from zoology. For example, botanists often find it more difficult than zoologists to explain the reasons for structural variations, such as the different shapes of leaves. Moreover, plants are particularly sensitive to subtle variations in climate and soil, and their *ecology cannot be understood without very careful measurements of these factors. Modern botany has contributed much to the development of agriculture and horticulture.

Bot flies, or warble flies, are a family of large *flies, often bee-like, which as larvae parasitize hoofed mammals. The fat, spiny larvae develop in the gut of horses, in the throat and windpipe of deer, and in the nasal cavities and sinuses of sheep. The eggs or tiny larvae are deposited around the mouth or nostrils, or attached to body hairs. The larvae then enter their host's body through the lips or nostrils, or by burrowing through the skin. The species of the genus *Hypoderma* are known as warble flies; their larvae form sores, or 'warbles', on the backs of cattle. Larval migration and feeding inside the host cause loss of condition by the livestock. The adult flies, often called gadflies or breeze flies, do not feed, but their presence distresses the animals.

Bottlebrushes are evergreen small trees or shrubs of the genus *Callistemon*, which is native to Australia and New Caledonia. They belong to the family Myrtaceae, which also includes myrtles, *Eucalyptus*, and cloves. All twelve or so species of bottlebrushes have narrow, leathery leaves and colourful red or yellow flowers in dense, cylindrical clusters round the stem. The stamens protrude well beyond the petals and give the flowers the appearance of the spiky brushes once used to clean bottles—hence their common name.

Botulism is a rare and dangerous type of food poisoning. It results from a toxin produced by the bacterium *Clostridium botulinum*. Resistant spores produced by this micro-organism are not killed by ordinary cooking, and can develop in home-bottled vegetables, which are responsible

The scientific name for **bottlebrushes**, *Callistemon*, derives from the Greek *kalos* (beautiful) and *stemon* (stamen). This species, *C. rigidus*, is found in New South Wales. It can grow to 5 m. (16 ft.) in height.

A satin **bower-bird**, *Ptilonorhynchus violaceus*, adds to his display. Most of the objects collected are blue, a colour similar to that of rival males. When displaying, the male picks up one of these objects and shakes it vigorously in front of the watching female. This display, together with the vocal mimicry of other birds, may continue over a period of several months until the female eventually becomes receptive.

for most of the outbreaks. The organism does not infect the body, but produces an extremely potent toxin which can kill by causing paralysis that affects the heart and lungs.

Bougainvillea is a genus of eighteen species of climbing shrubs, or *lianas, native to South America. They have been hybridized and are widely grown as garden ornamentals in the tropics and warm temperate zones. Each of the 'flowers' consists of a group of three inconspicuous blossoms surrounded by three large, showy bracts, which are white, bluish, lilac, or orange in colour. Bougainvillea form part of the chiefly tropical family Nyctaginaceae.

Bowel *Intestine.

Bower-birds form a family of eighteen species occurring in New Guinea and Australia; they range in size from that of a thrush to that of a small crow. Although a few are brightly coloured, many are brownish, with colourful, erectile crown or neck feathers. They have a stout beak for feeding on fruits, berries, and large insects.

They are particularly famous for their display grounds or bowers, which are built by the males. These are intricately woven walls of twigs, decorated with flowers, berries and leaves. Some of the gardener bower-birds, *Amblyornis* species, may concentrate their decorations around the base of a small sapling. The females come to the display grounds for mating and subsequently care for the eggs and young by themselves. Several of the New Guinea species are rare, though one, the yellow-fronted gardener bower-bird, *A. flavifrons*, previously known only from three of four skins collected in the nineteenth century, has recently been rediscovered and exists in some hundreds.

Bowfin: a species of freshwater fish of eastern North America, which lives in sluggish or still waters. It can breathe air, using its swimbladder as a lung. It breeds in spring, when the male makes a saucer-shaped hollow in the lake or river-bed in which the eggs are laid. The bowfin grows to 90 cm. (3 ft.) in length.

Boxes are evergreen shrubs or small trees of the genus *Buxus*, native to northern temperate regions and Central America. There are about thirty species, three of which are found in Europe. The common box, *B. sempervirens*, is a slow-growing, bushy, evergreen tree of western Europe. It is most widely grown as a hedging plant and as an edging plant for flower beds. The finely grained, hard, yellow wood is valued for inlay work, rules, and fine carving.

Box-fishes   *Trunk-fishes.

Boylston, Zabdiel (1679–1766), Boston physician, introduced the practice of inoculation into America against bitter opposition during the smallpox epidemic of 1721. A violent mob threatened his life; but only 6 of his 245 inoculated patients died, and in the next epidemic (1729) other doctors adopted the practice.

Brachiopoda   *Lamp shells.

Bracken is a very widespread species of *fern, belonging to the genus, *Pteridium*, which spreads by means of underground stems called rhizomes. These enable it to invade existing plant communities and spread over large areas, particularly on acid soils. Its fronds grow up to 2 m. (6 ft.) high, and it is a particular nuisance on hill farms, where it replaces grass and is poisonous to livestock. If eaten by cattle or horses, it causes bone-marrow damage with associated internal bleeding and high fever.

Bracket fungi are the spore-bearing structures of many *saprophytic and parasitic fungi, which appear on trunks and stumps of infected trees or timber. The hemispherical brackets are supplied with food by microscopic fungal threads (hyphae) growing in the wood. They belong to the fungal family Polyporaceae, often called polypores, and occur throughout much of the northern temperate zone. Familiar species in both Europe and North America include the birch bracket, *Piptoporus betulinus*, dryad's saddle, *Polyporus squamosus*, and beefsteak fungus, *Fistulina hepatica*.

Brain: the part of the central *nervous system that is within the skull in man and found in virtually all animals. Brain tissue is composed of *nerve cells and their supporting cells, the glia. Simple animals such as threadworms have brains with as few as a hundred nerve cells; the human brain contains more than ten billion. Within the brain each nerve cell is connected to many others, up to sixty thousand in the case of humans, and has links to and from the rest of the nervous system.

The human brain has a very complex structure and can be divided into three main parts: the hindbrain, the midbrain, and the forebrain. The hindbrain includes the

The **bracket fungus** *Polyporus squamosus* commonly infects elm trees, producing a white, stringy rot in the wood. Its large fruiting bodies, the brackets (seen here), can grow as large as 1 m. (3 ft.) across.

cerebellum which, in man and other vertebrates, is responsible for the unconscious muscular control of balance. The midbrain is part of the brainstem, an area through which nerves enter and leave the brain; it connects the brain to the top of the spinal cord. The forebrain, or cerebrum, is the largest portion of the human brain and consists of a left and a right hemisphere. Each hemisphere includes 'grey matter', where the nerve cells are situated, and 'white matter', where the nerve fibres lie. The outer cortex of grey matter has a crucial role in higher intellectual functions, sensory perception, and voluntary movement. The motor and sensory cortex of one hemisphere is connected with the opposite side of the body. The dominant hemisphere, usually the left, controls most aspects of language.

Brassicas are members of the *mustard family. The genus *Brassica* includes a number of familiar and valuable vegetables, such as cabbage, turnip, beetroot, and swede. Other species, such as rape, are grown for their oil-rich seeds; still others, such as mustard, are used as condiments.

Brassicas are native to Europe and the Mediterranean region, and the large number of local races within each species makes their classification very difficult. They tend to be either annual or biennial and to have a strong taproot. Some biennial species, like the swede and turnip, develop a swollen root in their first year and are food for

Brains

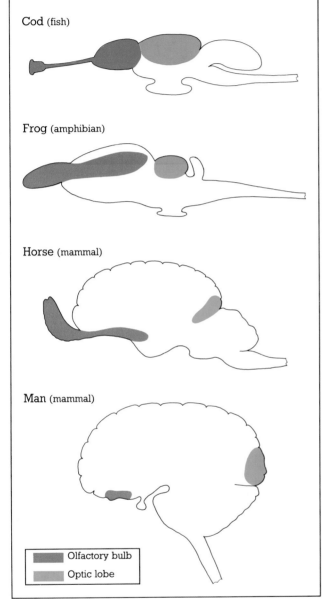

Diagrammatic sections through the brain of a typical fish, amphibian, and two mammals, showing the relative site and size of the olfactory (smell) and optic (sight) senses – relative size correlates with acuteness of, and reliance on, that sense.

Cod (fish)

Frog (amphibian)

Horse (mammal)

Man (mammal)

■ Olfactory bulb
■ Optic lobe

humans and cattle. Kohlrabi develops a swollen stem base, which is also used as a vegetable. Many wild brassicas taste bitter but centuries of selection have led to greatly reduced levels of the bitter compound.

Brazil nuts come from a 40-m. (130-ft.) tree, *Bertholletia excelsa*, a native of South American rain forests. This tree is a member of a family with some 450 species spread throughout the tropics. It produces large, round, woody fruits, up to 15 cm. (6 in.) in diameter, inside which are up to twenty-four Brazil 'nuts'. These are really *seeds, rich in oil, and not true *nuts in the botanical sense.

Breadfruits, or jackfruits, are tropical trees of the genus *Artocarpus* which are native to parts of Southeast Asia.

They belong to the family Moraceae which also includes figs, mulberries, hops, and hemp (cannabis). The species to which the name breadfruit is most commonly given is *A. altilis* which, although native to eastern Malaysia, is cultivated throughout Southeast Asia, the Pacific islands, and parts of South America. Its oval green fruits, each around 20 cm. (8 in.) in diameter, consist of 30–40 per cent carbohydrate; the bread-like flesh of this fruit forms the staple diet in some areas. Other bread-fruits include the giant jackfruit, *A. scortechinii*.

Bream is a species of fish in the *carp family, widely distributed in Europe. It is a deep-bodied fish with an exceptionally long anal fin, and thrives in slow-flowing deep rivers, canals, and lowland lakes. It feeds, often in schools, in a head-down posture. Its prey are mainly bottom-living invertebrates, which burrow into the mud, from which the bream's protrusible mouthparts are well equipped to extract them.

Breasts   *Mammary glands.

Brent geese   *Geese.

Bright, Richard (1789–1858), English physician, based his clinical work on post-mortem studies. In 1827 he was the first to recognize what is now called Bright's disease, or nephritis: an inflammation of the kidney due to a variety of causes. Bright described tuberculosis of the larynx, and did important work on abdominal tumours.

Brill is a species of *flat-fish with a thickset, large head and both eyes on the left side. Native to the seas around Europe, it is most common on sandy bottoms, from the shoreline down to 70 m. (230 ft.). It feeds on fishes such as sand eels and whiting, and less often on crustaceans.

Brimstone butterflies, found mainly in Europe and temperate Asia, are six species of bright yellow butterfly belonging to a part of the *white butterfly family. Their colour may be the source of the name 'butterfly'.

Bristle flies, or tachinid flies, are stout, true *flies, the size of houseflies or larger. Although at first glance they resemble houseflies, unlike them they have many strong, black bristles, particularly on the abdomen. The cylindrical larvae develop as internal parasites of arthropods, usually other insects, and particularly of the larvae of moths, butterflies, sawflies, and beetles. A single host may have one or several maggots within it, and is ultimately killed by these parasites. The adults most commonly feed from flowers. The 1,500 or so species make up the family Tachinidae.

Bristle-tails   *Silverfish.

Bristle worm is a name sometimes given to the class of *annelid worms called polychaetes. In narrower usage, bristle worms are species of the subclass Errantia, which includes free-swimming predatory worms and a few parasitic species. They are generally found on sandy sea-shores and species such as the ragworms (*Nereis*) are used as fishing bait. Most are flat, bristly, and usually between 5 and 10 cm. (2 and 4 in.) long. The predatory species have many lateral paddles to help them swim and have eyes and jaws as an aid to catching small invertebrate prey.

A brightly marked **bristle worm**, or fireworm, *Hermodice carunculata*, feeding on sea slug eggs, on Lanzarote, Canary Islands. The white bristles along the sides of the worm contain toxins and can be ejected as a defence.

Many show bright coloration, especially greens and reds, and some are iridescent. When sexually mature, many bristle worms change, the rear segments growing large nuptial paddles, allowing them to swim strongly to seek mates.

Brittle stars, a subclass of star-shaped *echinoderms, are closely related to the familiar *starfish, but have distinctive longer, mobile, and brittle arms, radiating from the disc-like body. They move by flexing the arms and are aided by rather sticky, stalked suckers called tube-feet along the arms which grip the sea-bed. Some brittle stars have respiratory pouches at the base of each arm, which are also used to brood the young. They are the most

A **brittle star** on the Great Barrier Reef, Australia. The numerous spines are pushed against the substrate to give leverage and assist locomotion.

numerous echinoderms, with some 2,000 species occurring worldwide. A group of species known as basket-stars have arms that branch to form a circular mass of tendrils.

Broad beans have been cultivated in the Old World, particularly the Mediterranean region, since ancient times. They belong to the genus *Vicia*, unlike other beans, and are far hardier. Small-seeded varieties, commonly called horse beans and tick beans, are used to feed livestock.

Broadbills are a family of some fourteen species of squat, brightly coloured birds, the Eurylaimidae, which occur in Africa and Southeast Asia. Mainly bright green and sparrow- to thrush-sized, they live in forests and open woodland. Their beaks are broad, blunt, and partially covered in feathers, and used for eating a wide range of fruits, insects, and other small animals. They build hanging nests of leaves and grass, often suspended over water.

Broccoli   *Cabbages.

Bromeliads are monocotyledonous plants of the pineapple family (Bromeliaceae), which are distributed from southern North America to central South America. Some species are small, terrestrial shrubs with functional root systems; species of the Andean *Puya*, for example, may reach 3 m. (10 ft.) in height. Most bromeliads, however, are short-stemmed *epiphytes with stiff, usually spiny leaves with coloured bases. The leaves form a rosette which creates a funnel or tank in which water accumulates. The adventitious roots are used mainly for attaching the plant to its support, rather than for nutrient gathering. The water held within the rosette of the plant, often as much as 5 litres (over 1 US gall.), can be the home of other plants, such as bladderworts, and many types of small animal.

The flowers are typically adapted for insect or bird pollination and are carried in the axils of coloured bracts, making species such as *Aechmea*, *Billbergia*, and *Bromelia* attractive pot plants. An unusual species of bromeliad is Spanish moss, *Tillandsia usneoides*, which is native to subtropical America, where its long, grey, moss-like shoots festoon trees.

Bronchitis is inflammation of the bronchial tubes of the *lungs. Acute bronchitis is usually due to a viral or bacterial infection. Chronic bronchitis is a misnomer. It is not primarily an inflammatory condition, but the persistent overproduction of mucus as a result of prolonged irritation, particularly by cigarette smoke. It causes a persistent cough, and often increasing shortness of breath, due to an associated narrowing of the tubes that lead into the lungs.

Brontosaurus, or thunder-lizard, was a member of the reptile-hipped *dinosaurs. It reached enormous size, about 20 m. (22 yards) in length and 30 tonnes (33 US tons) in weight, and had a very long neck and four massive legs. The fossilized remains of this dinosaur have been found in rocks 150 million years old in America. It had tiny teeth, suggesting that it was a *herbivore which may have lived in swamps. Despite its bulk, brontosaurus had the smallest brain, relative to its body size, of any living or extinct vertebrate.

Broom is the name given to a number of shrubby, yellow-flowered members of the *pea family with reduced leaves. Common broom, *Cytisus scoparius*, is native to western and central Europe but widely cultivated. It and related species are short-lived perennials bearing yellow, white, or purple flowers, and poisonous to livestock.

Broomrapes, or *Orobanche*, form a widely distributed genus of about ninety species of annual and perennial parasitic plants, without leaves or chlorophyll. They have normal flowers and produce light, wind-blown seeds. The seedlings attach themselves to the root-systems of suitable host plants by means of suckers and draw sustenance from them. They may have a wide host range or be specific to one particular host.

Brown, Robert (1773–1858), Scottish botanist and protégé of Sir Joseph *Banks, was appointed by Banks in 1801 as naturalist to the Flinders expedition to Australasia. He returned in 1805 with nearly 4,000 species of dried plants, many of them new to science, and devoted the rest of his long life to botanical studies.

Browns (butterflies), or satyrs, belong to the subfamily Satyrinae and include some 1,100 species found worldwide in many types of habitat. The upperside of most species is some shade of brown, often with a few white-centred eye-spots; the underside often has prominent eye-spots, especially near the wing margins. A few species, such as the marbled white, *Melanargia galathea*, are very differently coloured. Flight in some species is feeble, and tropical species are more active towards dusk or fly in the shade of forests. Their eye-spots act as deflection marks to protect their vulnerable bodies from attacks by birds. A few tropical species have wet- and dry-season forms or *mimic unpalatable butterflies. The caterpillars feed on grasses and related plants. The browns are often given

Bromeliads, such as this *Vriesia* species in south Mexico, tend to grow near the top of host trees, where their brightly coloured bracts and flower spikes are very conspicuous within the forest canopy.

family status, but modern classification places them as part of the *brush-footed butterflies.

Bruising is the result of the leaking of blood from small vessels in and under the skin. It normally occurs after minor injury or unusual pressure. Some people, particularly the elderly, bruise more easily than others. Only when bruises are spontaneous or unusually large should an underlying blood disorder be suspected.

Brush-footed butterflies, or nymphalids, include over 5,600 species of *butterfly in the family Nymphalidae. The name is applied to all species in which the front pair of legs is useless for walking, being short, held close to the body, and clothed in long brush-like hairs. This family contains some of the most colourful butterflies from several subfamilies, including Heliconiinae, Acraeinae, Charaxinae, as well as the *Morphos and *browns. They range in size from 2·5 cm. (1 in.) in wing-span, in the Central American *Dynamine theseus*, to over 20 cm. (8 in.) in *Morpho hecuba*. Among the more familiar species are the tortoiseshell butterflies and the Camberwell beauty or mourning cloak, *Nymphalis antiopa*, both of Eurasia and North America. The large tortoiseshell, *Nymphalis polychloros*, occurs throughout Eurasia, while Milbert's tortoiseshell, *N. milberti*, is found in North America.

Brussels sprouts are a variety of cabbage and are a valuable garden and commercial vegetable in temperate regions. They arose as a mutant strain of *Brassica olearacea* in Belgium and are characterized by the large number of swollen buds, resembling miniature cabbages, that develop, tightly packed, on the tall, main stem.

Bryony can be one of two distinct and unrelated plants. The white bryony, *Bryonia cretica*, and other *Bryonia* species are annual and perennial herbs climbing by means of tendrils and bearing male and female flowers on separate plants. They belong to the *pumpkin family. Black bryony, *Tamus communis*, is a twining climber native to Europe and the Mediterranean region. The sexes are separate, and the female plant bears red berries, which are very poisonous. It belongs to the *yam family.

Bryophytes are a subdivision of *cryptogam plants and include mosses and liverworts. They have no true roots, stems, or leaves, possessing instead structures which resemble those parts of other plants. They usually have an alternation of generation (see *life cycle) with a sexual (gametophyte) plant and a non-sexual (sporophyte) type.

Bryozoans, or moss animals, are a phylum of very common creatures (about 4,000 marine and freshwater species), so tiny that they are rarely known to the layman. Each animal, living in its own box-like skeletal casing, measures less than 1 mm. ($\frac{1}{25}$ in.), but being colonial they form encrusting masses on rocks, pilings, and ship-hulls. Many form flat, mossy sheets, often covering seaweeds. Others form branching colonies like tiny trees or fans, and their papery remains washed up on beaches are regularly dismissed as 'dead seaweed'. When these colonies are living, countless crowns of ciliated tentacles emerge, as each animal feeds on passing plankton.

Buckthorns are deciduous and evergreen shrubs belonging to the family Rhamnaceae, and are native to north temperate regions. They include species of the genus *Rhamnus*, such as the common or purging buckthorn, *R. cathartica*, of Eurasia, and *R. purshiana* of North America, which is the source of the purgative cascara sagrada. A close relative, once considered part of the genus *Rhamnus*, is the alder buckthorn, *Frangula alnus*.

In addition to the buckthorns, the family includes the jujube or Chinese date, *Zizyphus jujuba*, the Californian lilacs, *Ceanothus* species, and the spiny *Paliurus spina-christi*, which is supposed to have been used to make Christ's crown of thorns.

The sea buckthorns of the genus *Hippophae* are unrelated, and belong to the oleaster family.

Buckwheat is not a cereal but a member of the same family as rhubarb and sorrel. The common buckwheat, *Fagopyrum esculentum*, is an erect, fast-growing annual from central Asia. Cultivated in southern China for over 1,000 years, it is now also grown to a limited extent in cool, temperate regions, especially in the USSR, and occasionally in parts of the tropics. The greyish-brown, three-cornered, angular seeds resemble minute beech-nuts. Despite being 40 per cent husk by weight, the seeds yield a flour that can be used to make bread, pancakes, or a kind of porridge. Useful also as fodder for stock, the crop's main advantage is its ability to grow well in soils of low fertility, as it is particularly efficient at extracting phosphorus from the soil.

Budd, William (1811–80), English physician, as a young man nearly died of typhoid fever; the study of its origin and transmission was his chief life-work. It led him to examine contaminated water and sewer air, and to investigate a cholera epidemic in Bristol, where he lived. He was also interested in the contagious diseases of animals, such as rinderpest in cattle. The preventive measures he proposed were often resisted at the time but became the basis of his fame.

Buddleias are deciduous or evergreen small trees and shrubs of the genus *Buddleia*, forming part of the family Loganiaceae along with strychnine, *Strychnos* species. Buddleias are native to South America, South Africa, and parts of Asia. Most species have stems that are angular in

A herd of African **buffaloes** dries in the sun after wallowing in a water-hole in Amboseli National Park, Kenya.

cross-section, and have leaves with downy surfaces. The butterfly bush, *B. davidii*, from central and western China, has spikes of fragrant, lilac flowers, which are much visited by butterflies. This popular garden shrub has become naturalized in many parts of the world.

Budgerigars are possibly the world's best-known parrots. They are now bred in a wide range of colours, including blue, yellow, white, grey, and green. The wild budgerigar, *Melopsittacus undulatus*, is green underneath and has a mottled grey-green back and a yellow face. It is about 20 cm. (8 in.) in length, including the long tail. The wild birds are restricted to the drier areas of Australia, being absent from the wetter coastal districts, and sometimes occur in flocks of many thousands. They are nomadic, their migrations prompted in part by the effect of rain upon the supply of seeds. They nest in hollow trees, laying four to six white eggs.

Buffaloes are members of the *cattle family. The name is used for two distinctive species: the water buffalo of Asia, *Bubalus arnee*, and the African buffalo, *Synceros caffer*. Related species include the anoas, *B. depressicornis* and *B. quarlesi*, and tamarau, *B. mindorensis*, of the islands of Sulawesi and the Philippines respectively. Buffaloes have a rounded forehead, horns which drop and sweep widely outwards before curving upwards, and large, horizontally-held ears. Short hair leaves the skin almost naked. Their coat is black, sometimes a little brownish. All of them wallow in mud; this protects them against ticks and other insects.

The African buffalo is found all over the African continent; the largest subspecies is the Cape buffalo, which is some 2·75 m. (9 ft.) long. It is the carrier of the blood parasite that causes sleeping sickness in man, and which is transmitted by tsetse flies. The water buffalo is a native of India and Sri Lanka. Although a dangerous beast, it has been successfully domesticated and is used as a draught animal. It provides rich milk and butter and the dark meat is eaten; the skin makes a very strong leather. The name buffalo is also used when referring to the North American and European *bison.

Buffalo gnats *Blackflies.

Buffalo weavers are birds belonging to the same family as sparrows and other weavers. There are two species in the subfamily to which buffalo weavers belong: the white-billed buffalo weaver, *Bubalornis albirostris*, which is black, and the white-headed buffalo weaver, *Dinemellia dinemelli*, which is predominantly white, with a brown back and red under and above the tail. Both occur in tropical Africa and are the size of large sparrows. They are particularly noted for their communal nests; in *D. dinemelli* the individual nests are slightly separate from one another, but in *B. albirostris* they are in contact with each other and there is a common roof to all the nests in the colony.

Buffon, Georges Louis Leclerc, Comte de (1707–88), French naturalist, compiled a 44-volume *Histoire naturelle* (1749–1804), which was completed by his contemporaries after his death. This work, and especially its volumes on the animal kingdom, had great influence on the study of natural history.

Bugle (botany): a perennial creeping plant with short spikes of blue or purple flowers. It belongs to the genus *Ajuga*, which is part of the mint family. There are several species, spread over northern temperate areas. The kinds with variegated leaves are grown in gardens as ground-cover plants.

Bugs are popularly any sort of insect, but correctly they are the 68,000 or so species in the order Hemiptera, characterized by modification of the mouthparts into a piercing and sucking beak, which is usually held horizontally beneath the body when not in use. Typically, they have two pairs of wings, but these are absent in bed-bugs, many aphids, and other species. *Metamorphosis is incomplete, the young, or nymphs, being miniatures of the adults, though without wings. Although some are predatory animals, the majority feed on plants, and these include many agricultural pests, which not only inflict direct damage but also transmit viruses.

The order is divided into two suborders. In the Heteroptera, the tips of the forewings are flexible, the bases leathery. When at rest, the forewings lie flat over the folded membranous hindwings. The beak, or rostrum, rises from the front of the head. Some members of this group, such as shieldbugs, feed on plants; others, such as assassin bugs, are predatory; and many, such as water boatmen, are aquatic. The second suborder is the Homoptera, which includes cicadas and scale-insects; they are all plant-feeders. The forewings are uniform in texture, though they may be stiff, and at rest they are held roof-like over the body. The beak is underneath the head.

Bulbs are underground storage organs of plants; they are composed of fleshy leaves, usually acting to carry the plant through times of climatic stress, such as winter in the case of onions and daffodils, or drought in some desert plants. Many provide food for animals including man, especially onions in the Old World and quamash (*Camassia* species) in America. The term bulb is commonly, but botanically incorrectly, extended to other swollen underground organs, such as corms and rhizomes.

Bulbuls belong to a large family of 120 species of sparrow- to thrush-sized birds. They are mostly brown, grey, or olive-green, though some have brightly coloured patches of yellow or red under the tail. They are found in Africa and throughout most of the warmer areas of Asia,

occurring primarily in wooded country, usually in small flocks. They eat mainly fruits, but also take insects.

Bulimia is a term applied to the episodes of gross over-eating that sometimes punctuate the course of *anorexia nervosa. Injury to the part of the central nervous system which controls eating, or appetite, may also be the cause.

Bulldog bats are also known as fish-eating bats, for they are among the very few bats that catch fish. They catch them with their hind feet, which they drag through the water, presumably after the prey has been detected by echo-location. Their name comes from their jowl-like upper lip. There are two species in the genus *Noctilio*, both found around fresh water in Central and South America.

Bullfinches are six species of the *finch family belonging to the genus *Pyrrhula*. All occur in the Old World. They are sparrow-sized birds, 15 cm. (6 in.) long. The best-known species is the common bullfinch, *Pyrrhula pyrrhula*, which occurs over a wide area of Europe and Asia. The male is grey above, with a white rump, and bright red below; the female is olive-green, also with a white rump. They live in open wooded country, feeding on seeds. In spring, when these are scarce, the birds take flower buds and can be a considerable pest in orchards.

Bullfrog is the name applied to any moderate to large frog that has a loud or noisy call. It is specifically applied to the American bullfrog, *Rana catesbeiana*, which is native to the United States, although introduced elsewhere. It has an unmistakable 'jug o' rum' call. The African bull-frog, *Pyxicephalus adspersus*, grows up to 20 cm. (8 in.) long, and is capable of eating fully-grown mice.

Bullheads, or sculpins, are marine and freshwater fishes mainly confined to the Northern Hemisphere. They are most abundant in the North Pacific and in fresh water in North America, but also occur in Europe and northern Asia. The name bullhead is used in Europe to cover the family Cottidae with 130 species, but in North America this family is known as the sculpins. Usually not longer than 30 cm. (12 in.), they are squat, heavy-headed fishes with numerous spines on the head, and scaleless bodies, although the skin has prickles in it.

The freshwater bullhead, or Miller's thumb, *Cottus gobio*, is common in Britain and northern Europe. It lives mainly in rivers with a moderate current and stony bed, usually hiding beneath stones or among dense plant cover. Active mostly at night, it feeds on crustaceans and bottom-living insect larvae. Its broad, flattened head is said to resemble the shape of a miller's thumb developed from constantly assessing the fineness of ground meal. The name bullhead is used in North America to describe a group of small catfish of the genus *Ictalurus*.

Bumble-bees, or humble-bees, are large social bees of the Northern Hemisphere and mountains of tropical regions. They are often colourful in blacks, browns, whites, and yellows. Nests are built underground, or, in the case of carder bees, on the surface and are lined with grass or moss. Bumble-bees form highly organized colonies, with non-reproductive females acting as 'workers' to provision and maintain the nest. In autumn some grubs develop as males, or drones, and females; these mate and the potential queen then hibernates over winter. Bumble-bees are unusual in that they can raise their body temperature by 'shivering'; they function best at around 30 °C (86 °F).

Bunions are swellings of the joint at the base of the big toe. They are formed when the big toes are pushed laterally from their proper position. Prolonged shoe pressure causes this painful swelling, by developing a bursa, a small fluid-filled sac overlying the joint. Avoidance of high-heeled, narrow-toed shoes is extremely important. Surgery is necessary for severe symptoms.

Buntings belong to the family Emberizidae, which also includes the American sparrows and finches and are sparrow-sized birds. In most species the females are dull brown, though often with white edges to their tails. The males in breeding plumage may be strikingly coloured, with black and white patterned heads and yellow underparts. The male snow bunting, *Plectrophenax nivalis*, is largely white with black wing-tips. Most of the forty-seven species occur in the Old World and only about six in

A pair of narrow-bordered, five-spot **burnet moths**, *Zygaena lonicerae*, mating near the female's cocoon. The males of this species are efficient in tracking down the females, often only minutes after they have emerged from the cocoon.

North America. They live in a wide variety of habitats, from open country such as grassland and reed-beds, to open forest. Many species migrate long distances. They have smallish, stout beaks and eat seeds, although they may catch and eat insects in the summer.

Burbot: the only fish of the cod family to live in fresh water. It is distributed across the cold and temperate land masses of the Northern Hemisphere, from the Netherlands to the North American Great Lakes, but is extinct in England. It lives in lowland rivers and lakes, and feeds, at dawn and dusk, mainly on other bottom-living fishes, insect larvae, and crustaceans.

Burnet moths are brightly coloured day-flying moths which contain chemicals similar to cyanide and are distasteful to predators. They comprise the subfamily Zygaeninae with some 100 species in north temperate regions. Most are patterned blue and red, but the foresters, such as *Adscita* species, are brilliant metallic green. These bright colours are equalled by the yellow and black of their caterpillars (which are also poisonous).

Burns are tissue injuries from heat, electricity, or chemicals. They may be partial or involve the full depth of skin, depending on whether the germinal layer is destroyed. A severe burn may be less painful than a superficial one because sensitive nerve endings have been destroyed. Severity depends on the percentage of body surface area burned, 50 per cent resulting in a 50 per cent chance of death. Even a 10–15 per cent burn causes massive fluid loss and collapse or shock, requiring immediate transfusion and specialized hospital care to prevent infection and promote healing with minimal scarring and skin contracture. Deep burns may also require skin grafts.

Bursae are sac-like cavities of fibrous tissue which contain a thick lubricating fluid (synovial fluid). They are found in parts of the bodies of vertebrates, including man, where friction between skin, muscle, or ligaments, and bone occurs, and they help to reduce it. Typically they are found around joints, or where tendons pass over bones of the legs or arms. Infection can lead to inflammation of the bursa, causing complaints such as housemaid's knee and tennis elbow. Such inflammation is called bursitis.

Burying beetles, or sexton beetles, are medium-sized *carrion beetles, often with red and black wing-covers. They are attracted by smell to the corpses of small birds and mammals, and a male and female will dig the earth from under the corpse so that the female may lay her eggs on it. Some species show parental care by guarding the eggs, and the female may even feed the young larvae until they are old enough to fend for themselves.

Bush babies, or galagoes, are six species of primate belonging to the loris family, which includes the pottos. Bush babies live in the forests of Africa south of the Sahara. The smallest, the dwarf bush baby, *Galago demidouii*, is mouse-sized, and the largest, the thick-tailed bush baby, *G. crassicaudatus*, is the size of a domestic cat. They are extremely attractive animals, with thick woolly fur and a bushy tail. The head is round, and the very large eyes, which face forwards, are a rich, translucent brown. Their long, slender fingers and toes have pads that act like suction cups, allowing the animal to run up a tree without

using its claws. Bush babies are agile and can leap some 6 m. (20 ft.) from tree to tree. On the ground, they leap like kangaroos. They are nocturnal and emerge at sunset to rush around in quest of food, uttering a haunting cry at intervals; their food is insects, fruits, seeds, and birds' eggs.

Bush cricket   *Longhorn grasshopper.

Bush dog: a species of dog found in the tropical regions of South and Central America, where it lives in holes on river banks. Its shape resembles that of a badger more than a dog, with a long body and short legs. The body, legs, and short tail are dark brown in colour, while the large head and shoulders and small ears are yellowish, white, or buff. It will feed on any animal that it can catch and kill, sometimes hunting in packs of up to ten.

Bush pig, or red river hog: a species of pig found south of the Sahara Desert. It has a subspecies on the island of Madagascar, where it is the only even-toed, hoofed mammal native to that island. An inhabitant of dense bush, reed-beds, and heavily forested regions, the bush pig is seldom seen. Though a wary animal, it is fierce and tough, and does great damage to crops. It stands about 60 cm. (2 ft.) high at the shoulder and weighs some 90 kg. (200 lb.). It has a coat of reddish-brown hair, with white face patches. It has small eyes, pointed ears ending with tufts of hair, and a pair of warts in front of the eyes and another, smaller pair behind.

Bustards are a family of twenty-four species of chicken- to turkey-sized birds, the males often much larger than the females. Most are light brown and grey on top, and may have pale-coloured or black undersides. Some have black and white patterns on the head, and a few are crested. Bustards have longish necks and long, powerful legs; being ground-dwelling birds, they tend to run rather

A lesser **bush baby**, *Galago senegalensis*, emerges from its daytime hiding place. This is the most widely distributed species, found in forests and savannah throughout much of Africa. Gum from leaves and bark provides the basic food in dry periods.

than fly when disturbed. All occur in open country in Africa, southern Europe, southern Asia, and Australia. Many species have been severely reduced in number by man, and several are endangered. The nest is a small depression in the ground, and most species lay just two eggs. These hatch into well-feathered young, which leave the nest within a few hours.

Butcher birds are six species of the family Cracticidae from New Guinea and Australia. In the same family as the bell magpies, butcher birds are crow-sized with black, black and white, or black-grey and white markings. They have stout, powerful beaks with a hooked tip, and feed mainly upon small animals and other birds, although they also eat fruit. The name 'butcher bird' is also given to some *shrikes which, like the true butcher birds, impale their prey on thorns for later use.

Butcher's broom is a small shrub, although it belongs to the *lily family. It is native to Europe and the Mediterranean region, and forms thickets of dark green stems up to 75 cm. (30 in.) in height. It is botanically interesting because of its leaf-like spiny branches which bear small green flowers, usually of one sex, on the upper surfaces (*see over*).

Buttercups are a group of plants belonging to the genus *Ranunculus*. About 200 species are known; they are distributed throughout the temperate and alpine zones of the Northern Hemisphere, including North America. Buttercups are chiefly plants of arable lands, often preferring

A close-up of **butcher's broom** shows the flattened leaf-like branches ending in a spine, and the tiny, star-like flowers, only 3 mm. ($\frac{1}{10}$ in.) across, which appear in the spring and are pollinated by insects.

damp places. The flowers are mostly yellow but may be white or red. The common names of several species, such as mountain, bulbous, creeping, and meadow buttercup, indicate the plant's habitat or other characteristics. The buttercup family, Ranunculaceae, also includes clematis, delphiniums, and many other ornamental garden species. The whole family contains some 1,800 species, concentrated mainly in the Northern Hemisphere.

Butterflies are usually brightly coloured *insects which fly by day and rest with the wings held together. With the moths, they comprise the insect order Lepidoptera, and can be distinguished from moths by their antennae, which typically end in a knob. Like moths, they feed by sucking nectar and other liquids through a long proboscis, which is coiled when not in use. They are distributed worldwide, but the majority of the 20,000 species occur in the tropics. Among the more familiar families are the *blues, *browns, *brush-footed butterflies, *swallowtails, and *whites. The caterpillars of a few, such as the citrus swallowtail, *Papilio demodocus*, of Africa and the ubiquitous cabbage white, are regarded as pests. Butterflies range in size from the tiny blues which are only 1·5 cm. ($\frac{1}{2}$ in.) in wing-span, to the birdwing butterflies of New Guinea, which have a wing-span of 28 cm. (11 in.).

The two pairs of wings have a powdery covering of scales, which may contain pigments, or be so structured microscopically that they either reflect light of a certain wavelength or break white light into its constituent colours by interference. The conspicuous compound eyes distinguish colours, and the striking wing patterns are used in sexual recognition and often as protection. Many species that are toxic or unpalatable are vividly coloured, while other species use camouflage or some form of eye-spot or *flash coloration.

Butterfly fishes are a family of marine fishes found mainly in the tropical Indo-Pacific, but also in the Caribbean. Their bright coloration merges with the background colour, and many of the 150 known species have an eyespot near the tail, which confuses predators. Their deep, disc-shaped bodies enable them to slip between the crevices of the coral, and their long snouts and minute teeth are designed for feeding among the coral. The family also includes species known as marine angel-fishes.

Butterfly flower   *Schizanthus.

Butternut is a term used for two distinct trees and their seeds: the white walnut, *Juglans cinerea*, of eastern North America, and the Brazilian *Caryocar nuciferum* and allied species. The first resembles the common walnut, the second has a *drupe with two to four hard-shelled seeds within. Both trees are a source of oils. The timber of *Caryocar nuciferum* is valued in ship-building.

Butterworts are small stemless, often rootless, plants of boggy places. They are *carnivorous plants and have rosettes of sticky-surfaced leaves to which insects are attracted, before being trapped and utilized as a food source. About forty species are known, chiefly from the Northern Hemisphere but also extending along the Andes in the Southern Hemisphere. They are close relatives of the *bladderworts.

Button quails form a family of fourteen species which occur in Africa, Southeast Asia, and Australia. They resemble tiny partridges, and are brown, grey, and black in colour and well camouflaged. They lay two to four eggs in a grassy nest built on the ground, and the young leave the nest soon after hatching.

Buzzard is the European name for large *hawks of the genus *Buteo*. In the USA the same name is applied to certain vultures and falcons. The genus *Buteo* is widely distributed throughout Eurasia and the New World and includes twenty-four species. Buzzards are large-winged, slow-flying diurnal predators; although capable of soaring, they usually hunt from spotting-posts on rocks or trees, dropping swiftly on to their prey. They often occur on marshy or open ground, but may also hunt along the forest edge or in the canopy of a rain forest.

C

Cabbages, along with Brussels sprouts, broccoli, calabrese, cauliflower, kale, and kohlrabi, are derived from a single species, the sea cabbage, *_Brassica oleracea_, that grows wild on cliffs of northwest Europe and the Mediterranean region. This wild cabbage, known to have been grown by the Saxons and Celts in northern Europe, has few leaves and a woody stem, and is very different from the modern vegetables. Cabbages have been selected for their greatly enlarged, single, terminal bud, with its large number of tightly packed leaves which form the familiar cabbage 'head'. A range of varieties exists, each variety being distinguished by its shape (round-headed or conical), colour (red, white, or green), and season of use (spring, summer, or winter).

Cabbage whites are two species of *white butterflies with black wingtips. The most cosmopolitan species is the small white, _Pieris rapae_, which although native to Eurasia has been introduced into North America and Australia. As caterpillars they all feed on cruciferous plants, particularly brassicas, so they are regarded as pests. Caterpillars of the large whites, _P. brassicae_, are black and yellow, and gregarious; those of small whites are green, and solitary.

Cacao, or cocoa, is a small, spreading, evergreen tree up to 8 m. (26 ft.) tall, from the South and Central American rain forests. One of thirty species in its genus, the cacao, _Theobroma cacae_, is now cultivated principally in Ghana, Nigeria, and the Ivory Coast. The large, oval pods, which grow up to 30 cm. (12 in.) long, are borne on the trunk and contain twenty to sixty seeds embedded in a pink pulp. These are scooped out, fermented naturally, and then dried in the sun to produce brown cocoa beans. Fifty to sixty per cent of the cocoa bean consists of a pale yellow fat, cocoa butter, which is used as a base for chocolate manufacture, being mixed with other roasted and ground cocoa beans, sugar, and flavourings. The residue, after extraction of the butter, is used to make cocoa powder.

Cacomistle *Ring-tailed cat.

Cactus is an ancient Greek name for a prickly plant which Linnaeus later used for a family of plants, most of them spiny and some bearing short-lived flowers. Except for the African genus _Opuntia_ *prickly pear, cacti are indigenous to the hot, dry regions of North and South America, but some have been introduced to other places, where they flourish as weeds. They vary in shape and size, lacking leaves and *photosynthesizing in green stems and branches. The reduction of leaves is an adaptation for conserving water, which plants normally lose through transpiration via leaves. Many cacti have large root systems, allowing them to collect water from deep underground. Some species produce edible fruit, notably the prickly pear, while others are grown as house plants.

Caddis flies are not true flies, but comprise some 5,000 species of insects with hairy-winged adults and aquatic larvae. They comprise the order Trichoptera which has

A **caddis fly** larva, _Limnephilus flavicornis_, with a covering of empty pea mussel shells. Other species of _Limnephilus_ make cases, usually of a cylindrical shape, out of plant stems, stones, or sand grains. Most caddis fly larvae feed on detritus and algae.

a worldwide distribution. Caddis flies have a complete *metamorphosis, the eggs normally being laid in water and surrounded by a jelly-like mass. The larva breathes by gills and is generally vegetarian, though some are carnivorous. Many species build a protective case with bits of twig, shell, stone, and weed, which they bind together with silk. The pupa is unusual among insects in having movable jaws and legs. When the time comes for it to emerge, it bites its way out of the larval case and crawls or swims upwards to the surface of the water.

The adults have hairy wings, dark to pale in colour, and occasionally patterned, and are usually moth-like. Most species cannot eat, as they have no jaws, but some can sip fluids such as nectar. A bright light from a window will readily attract them.

Caecilians are probably the least known, most poorly understood family of *amphibians. They are limbless and look more like worms than amphibians, having a series of ring-like folds in the skin. They possess rudimentary eyes, sensory tentacles beneath the eyes, and a shark-like mouth with jaws and teeth. Some species have minute scales embedded in the skin. Approximately 170 species are known; they inhabit the wet tropical forest areas of South America, Africa, and Asia. Most of them burrow in soft earth, though one South American group is apparently largely aquatic. Some lay eggs; others give birth to live young. They are the only family in the order Apoda.

Caird, Sir James (1816–92), Scottish agriculturalist, travelled widely and carried out investigations into all aspects of farming. He published _English Agriculture in 1850–51_, _Prairie Farming in America_ (1859), and _Our Daily Food_ (1868), among numerous other writings.

Calabrese *Cabbages.

Calamites, or giant horsetails, are fossil plants of the genus *Calamites*, which were important tree-like plants in the swamps of the Carboniferous Period, 300 million years ago. They grew in dense stands from spreading, underground stems and the many-branched trunks could reach a height of 30 m. (98 ft.) with a diameter of 40 cm. (1 ft. 6 in.). Like their smaller living relatives the *horsetails, *Equisetum*, they bore branches and leaves in whorls around a jointed stem. The most common fossil is the grooved internal cast of the stem, but coaly compressions of leaves are sometimes found in the shales above coal seams.

Calceolaria is the plant name derived from the Latin *calceolus*, meaning 'slipper', and refers to the shape of the lower petal of the flower. About 200 species are known in the genus *Calceolaria*, which belongs to the foxglove family. They are chiefly New World plants from Central and South America and New Zealand. Several species are popular garden and glasshouse plants.

Californian poppy   *Eschscholtzia.

Calorie: a measure of energy based upon the definition that one calorie is the quantity of energy needed to raise the temperature of 1 gram of water by 1 °C. The kilocalorie (kcal) is equal to 1,000 calories, and is sometimes referred to as the large calorie when indicating the value of foods. This is interchangeable with the unit of mechanical work, the kilojoule (1 kcal = 4·2 kJ). Man's energy consumption is deduced from his heat output or his oxygen consumption. On an average diet of protein, carbohydrate, and fat, each litre of oxygen used releases 4·8 kcal. The daily expenditure of energy for an adult European is about 2,600 kcal, and even at rest he requires up to 2,000 kcal per day.

Camberwell Beauty   *Brush-footed butterflies.

Camellias are evergreen shrubs native to the Far East and widely cultivated since ancient times by the Chinese, and latterly by Europeans, for their showy spring flowers. There are about eighty species in the genus *Camellia*, which belongs to a family of trees and shrubs confined to the tropics and subtropics. *C. sinensis* is the plant from which tea is derived.

Camels are *ruminants and are among the largest of the even-toed ungulates (hoofed mammals). They form a family of six species which includes the two species of 'camel' as well as the llama, alpaca, guanaco, and vicuña. The two species commonly called camels, the dromedary, *Camelus dromedarius*, and the Bactrian camel, *C. bactrianus*, can be as tall as 2·3 m. (7 ft. 6 in.) at the shoulder. The long head is set on an arched neck. The eyes have long lashes to keep sand out, the ears are small, and the nostrils are set high above the fleshy lips. The legs are long and slender and the feet have two toes united by a web of skin, allowing the toes to spread sideways thus enabling the animal to walk on soft sand.

The two-humped, or Bactrian, camel of Asia survives in the wild in the Gobi Desert and has been domesticated since the third or fourth century BC. The one-humped, or Arabian, camel of North Africa and the Near East has been domesticated since 4000 BC and is not known in the wild today. Where the Arabian and the Bactrian camels live side by side, as in the southern USSR and Syria, they frequently cross-breed. The offspring are known as tulus.

A herd of dromedary **camels** browses desert vegetation in Arabia. Camels eat thorns and saltbush that other mammals avoid. Their thick fur insulates them against extreme heat and provides warmth during cold nights.

Camels can survive for six to eight days in the desert and can travel for 1,000 km. (620 miles) without drinking. When water does become available they can drink up to 90 litres (23 US gallons) to rehydrate their tissues. The large hump of fat provides an energy store when food is in short supply. The thick, dense fur acts as a substantial barrier to heat. By allowing their body temperature to rise during the day, camels reduce the need to sweat, thus preventing water loss. Another aid to water conservation is the concentrated nature of their urine.

Camomiles   *Chamomiles.

Camouflage (in nature) is the way in which an animal resembles another or blends into its background. Some have developed *mimicry patterns, others *protective coloration, both of which help in defence or attack. Protective camouflage is generally effective only so long as the creature remains still. Colour can be used to break up outlines, to harmonize with the background, or to provide countershading. The stripes of the tiger help it to blend into the shadows and sunspots of dense jungle, while the pure white of the polar bear matches that of its surroundings. Many fishes have a dark upperside and a paler underside to counteract the effect of light from above. Some creatures, such as the chameleon, have developed a mechanism which changes their colour to match their background. Insects use shape to a remarkable degree in concealing themselves from predators: leaf insects look like real leaves and stick insects are perfect copies of real twigs. Other animals have strange outgrowths from their bodies, such as the sea-dragon, a relative of the sea-horse, which has a body bearing membraneous streamers to conceal it perfectly in seaweed.

Campanulas are plants often called bell-flowers in allusion to the bell-like shape of the flower. At least 250 species are known in the genus *Campanula*, mostly from the

Mediterranean region, Europe, Asia, and the Caucasus. They give their name to the family Campanulaceae, which contains over 600 species, including other types of bell-flower, such as the pendulous bell-flowers *Symphyandra* species, giant bell-flowers, *Ostrowskia* species, and the salad vegetable rampion, *C. rapunculus*. Campanulas range from tiny alpine plants to herbaceous species over 2 m. (6 ft. 6 in.) in height. The attractive flowers, which are chiefly blue or white, make the frost-hardy species valuable as garden plants. *C. rotundifolia*, which is found throughout Europe, is commonly known as harebell; but in Scotland this species is often called 'bluebell'.

Campions belong to several species in the plant genus *Silene*, of which the white campion, *S. alba*, and the red campion, *S. dioica*, are the best known in Europe. Both are biennial or short-lived perennial herbs belonging to the pink family. The former is found from Europe to western Asia as a wild plant of grassy verges, banks, and hedgerows (it is introduced in North America). The usual habitat of the red campion is woodlands and clearings. Both species bear male and female flowers on different plants and, where they grow together, hybrids occur between them with intermediate characteristics. Garden plants in the genus *Lychnis* are sometimes called campions.

Canada geese are a North American species which was introduced to Europe 300 years ago. Typical large 'black'

The alpine **campanula**, *C. cochleariifolia*, or fairy's thimble, grows among rocks and on screes. Its cushion-like form provides protection from wind and desiccation. Its flower stems grow to 20 cm. (8 in.) in height.

Canada geese are gregarious outside the breeding season; this winter party is on a lake in the English Midlands. They are the largest geese found in Europe, growing up to 1 m. (40 in.) in length.

*geese, they breed regularly in parts of Europe as well as North America. The white cheek, whitish-brown (not black) breast, and brownish-grey body distinguish them from the smaller barnacle geese. There are twelve sub-species of Canada geese.

Canary is part of the name given to ten species of the genus *Serinus* of the finch family. The common canary, *S. canaria*, was first brought to Europe from the Canary Islands (its native country) in the fifteenth century. This small bird, which in nature is greenish above and yellow below, is the ancestor of all of the domestic strains of this very common cage-bird.

Most bred canaries are basically yellow. The roller varieties are particularly valued for their beautiful song, while others are bred more for their build and shape. The range of forms has been increased by cross-breeding canary stock with other finch species, the progeny being known as mules; some of the canaries with red or orange in their plumage probably have the hooded siskin, *Carduelis magellanica*, in their ancestry.

Canary creeper *Nasturtium.

Cancer is not a disease but a group of conditions differing greatly in their clinical features and response to treatment. The common factors are the uncontrolled proliferation of some type of cell, and the infiltration thereby of normal tissues. The cause is an alteration, probably involving two distinct steps, in the *DNA of the cell. This may be brought about by certain chemicals, by radiation such as X-rays, or by virusues. The latter act by inserting a foreign piece of DNA into the *chromosomes of a host cell. Virtually all alterations to DNA cause malfunction or death of individual cells. Only when an alteration leads to uncontrolled cell division will a cancerous tumour be formed.

The incidence of different types of cancer varies greatly in different countries, but is largely independent of races. The environment in which we live is thought to be the main factor in the differing incidence of cancer. Cigarette smoking accounts for nearly all lung cancers, and a sexually transmitted virus is almost certainly implicated in one type of cancer of the uterus. Exposure to sunlight is the principal cause of skin cancer. Forms of radiation (such as X-rays and cosmic rays) can lead to cancers in any part of the body, but are responsible for only a small proportion of all cancers.

The treatment of cancer is often difficult because its spread to distant parts of the body may render surgical removal impossible. Drugs and X-rays are then used to try to destroy the abnormal cells without causing too much damage to normal tissues. Progress in treatment is likely to come from the development of more selective drugs, which avoid damage to the body's defences.

Candidiasis   *Thrush (disease).

Candytuft is an annual plant, *Iberis amara*, up to 30 cm. (12 in.) high, with heads of small white flowers, and is native to western Europe. It belongs to the mustard family and is related to plants such as the cabbage, wallflower, and charlock. The genus *Iberis* contains several species, some of them perennial, which are native to Eurasia. The decorative, coloured varieties bred from *I. amara* are grown in gardens around the world.

Cane rats are two species of African *rodent related to porcupines and guinea-pigs; they resemble a rat because of their long, scaly tail. Both species are strictly vegetarian, as the alternative name of grass-cutter indicates. They are widely used as food, particularly in West Africa, and experiments have shown that they can easily be domesticated. The larger of the two species, *Thryonomys swinderianus*, can reach a body length of 60 cm. (2 ft.) and weigh 9 kg. (20 lb.).

Cannabis is both the tall annual plant cultivated for its stem fibre, *hemp, and for its oil-rich seeds, and the narcotic, otherwise known as marijuana or hashish, that is derived from the leaves, stems, and particularly flowers of this species. Deliberate cultivation for this purpose is prohibited in most countries. Certain varieties grown in India yield the powerful narcotic, ganja, from the dried female flowers, and this is used medicinally.

The seed may contain up to 22 per cent protein and 32 per cent oil, the latter being a substitute for linseed oil in paints. The seeds are a useful livestock feed, particularly for poultry.

Cantaloupes   *Melons.

Cape hunting dog   *Hunting dog.

Capercaillie is the name applied to two species of the grouse family, but particularly to the Eurasian species, *Tetrao urogallus*, a bird found primarily in the conifer belt across Europe and west and central Asia. The turkey-sized male is dusky grey with a glossy blue-green sheen on the neck and breast, and has a shaggy 'beard' of loose feathers. The female is much smaller, and is speckled brown with pale underparts. They are vegetarians, feeding on a wide range of leaves, pine needles, seeds, and berries. Because of their large size and tasty flesh they are hunted by man over large parts of their range. They were exterminated in Scotland and Ireland in the late 1700s, but a massive programme of reintroduction with Swedish stock in the mid-1800s led to their re-establishment.

Capers are plants of the genus *Capparis* and members of the Capparaceae, a family which includes over 700 species

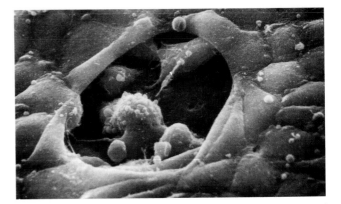

Cancer on the move: the large, granular melanoma (skin cancer) cell is escaping through a gap in a wall of normal cells to set up a cancerous growth or metastasis in another part of the body (× 1,000).

of herbaceous plants, trees, shrubs, and some lianas, distributed throughout the tropics and subtropics. The common caper, *C. spinosa*, whose pickled flower buds are used as a condiment, is native to the Mediterranean regions. It is a small, spiny shrub with conspicuous white or lilac flowers, each with a large number of long stamens.

Capsicums are a genus of plants with some fifty species native to tropical America and the West Indies. Although they are commonly referred to as peppers (for example, red, paprika, chilli, green, cayenne, and *sweet peppers), the true *pepper is a completely different plant, native to India. Capsicums are members of the potato family and are related to the aubergine, the tomato, and tobacco. They were first cultivated in the Americas in pre-Columbian times.

Capsids   *Mirid bugs.

Capuchin monkeys are perhaps the best known of the New World *monkeys, with the four species being found in Central and South America. Common and conspicuous, they live in fairly large troops of up to thirty individuals, feeding mostly on fruit. They are medium-sized: 38–53 cm. (15–21 in.) long, with a long prehensile tail covered in hair; both this and the body hair is short and brown. The name of these monkeys comes from the 'cowl' of thick hair on the crown of the head. The most distinctive species is the tufted, or brown, capuchin, *Cebus apella*, which has tufts of hair on the forehead. Other capuchins include the white-faced or -throated, *C. capucinus*, the white-fronted, *C. albifrons*, and the weeper capuchin, *C. nigrivittatus*.

Capybaras, of which there is one, or possibly two, species, are the largest *rodents in the world, weighing up to 45 kg. (99 lb.) and measuring over 1 m. (3 ft. 3 in.) in length. The capybara, *Hydrochoerus hydrochaeris*, has several subspecies throughout South America in marshy areas, one of which, the so-called Panama capybara, may be a true species or simply a subspecies of *H. hydrochaeris*. Capybaras are semi-aquatic and leave the water only to graze and are related to porcupines and guinea-pigs.

Caracaras are nine species of *hawk belonging to the falcon family. All are confined to South and Central

America except one, the common caracara, *Polyborus plancus*, which extends into North America. They are medium-sized birds of prey, 60 cm. (24 in.) long, and are brown or black with white markings. Several species have stout, powerful beaks which are used in their role of scavengers to tear carcasses. Non-scavenging species hunt their prey by stalking it on the ground. Two, the yellow-throated caracara, *Daptrius ater*, and the red-throated caracara, *D. americanus*, take quite small prey, such as frogs and wasp larvae, and also eat fruits.

Carbohydrates are one of the major classes of natural organic compounds with the general formula $C_x(H_2O)_y$. They include *sugars, *starch, and *cellulose. The simplest carbohydrates are sugars such as ribose and *glucose. These are called monosaccharides because their molecular structure is based upon a single ring of five (ribose) or six (glucose) carbon atoms. Monosaccharides are the basic units from which all other carbohydrates are built. When two of these simple sugars are bonded together, they form compounds called disaccharides; these include sucrose (common table sugar) and maltose (malt sugar). Compounds known as polysaccharides are formed by joining together ten or more monosaccharides (often hundreds). These are very important biological compounds and include storage materials, such as starch and *glycogen, and structural substances such as cellulose.

All carbohydrates are formed from simple sugars (especially glucose), which are produced by plants during *photosynthesis. Once produced, the simple sugars are converted into disaccharides and move from the leaves (*translocation) to be used elsewhere or stored. Carbohydrates are an essential component of the diet of animals and may be used as a source of energy (*respiration) or modified by being combined with fats or proteins. These more complex compounds include glycoproteins (sugar + protein), which form some enzymes and the rigid cell walls of bacteria, and glycolipids (sugar + fat), which include the insulating material (myelin) of nerves.

Carbon cycle is the name given to one of the most important *biogeochemical cycles, in which the element carbon circulates between living organisms and the non-living environment. Carbon dioxide forms 0·03 per cent of the atmosphere and is used by plants for *photosynthesis,

during which they 'lock' carbon into *carbohydrates. These compounds can be stored by the plant or used in respiration, releasing some of the carbon back into the atmosphere as carbon dioxide. The remaining carbon is trapped in the plants' tissue until it either dies or is eaten. Animals use carbon compounds obtained from plants both for respiration and for building up their own tissues until they too·die or are eaten.

The carbon compounds in dead plants and animals are broken down by decomposers such as bacteria or fungi. These organisms degrade the complex carbon compounds into simple ones, which they then use in respiration, ultimately releasing carbon dioxide or methane (CH_4). Under certain chemical or physical conditions, some of the organic material escapes the action of decomposers and forms *fossil fuels, such as coal or oil. The carbon they contain is returned to the atmosphere either by natural erosion of the deposits, or by man burning them. The carbon contained in organisms whose bodies have been incorporated in sedimentary rock, such as limestone, is released by erosion or volcanoes.

Carbuncle: an infection of skin and subcutaneous tissue with the bacterium which causes *boils. It differs from a boil in the extent of tissue destruction, which may involve an area of several square inches.

Cardamom is a tropical, perennial shrub native to Sri Lanka and India. It is a member of the ginger family, and has aromatic seeds that are highly flavoured and have been prized for chewing in those countries for centuries. It is best known as a spice, being a constituent of curry powder or paste.

Cardinal beetles are a family (Pyrochroidae) of some 100 species of medium to large *beetles, occurring mainly in northern temperate regions. They have long antennae and are often bright red—hence their name. The larvae of all species live under bark; the adults of some species visit flowers.

Cardinal fishes are a family of fishes living in shallow tropical and subtropical seas. About 170 species are known, most of them growing only up to 10 cm. (4 in.) in length. Many are uniformly reddish; others have length-

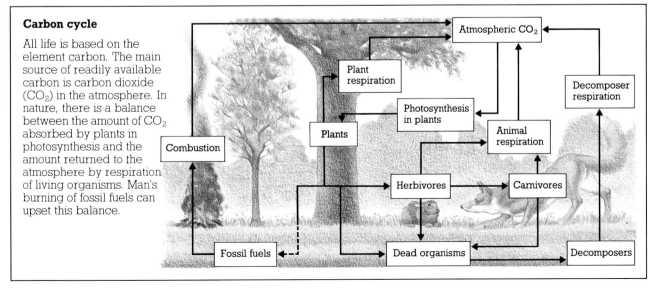

Carbon cycle

All life is based on the element carbon. The main source of readily available carbon is carbon dioxide (CO_2) in the atmosphere. In nature, there is a balance between the amount of CO_2 absorbed by plants in photosynthesis and the amount returned to the atmosphere by respiration of living organisms. Man's burning of fossil fuels can upset this balance.

Atmospheric CO_2

Plant respiration

Photosynthesis in plants

Decomposer respiration

Combustion

Plants

Animal respiration

Herbivores

Carnivores

Fossil fuels

Dead organisms

Decomposers

A pair of golden **cardinal fishes**, *Apogon diversa*, on the Great Barrier Reef, Australia. Cardinal fishes are especially numerous in the Indo-Pacific.

wise black lines along the body. They are most numerous on coral reefs, where many are nocturnal, spending daylight hours in crevices in the coral. The males of many species incubate the eggs in their mouth.

Cardinals (birds) are members of the genera *Paroaria*, *Cardinalis*, and *Pyrrhuloxia*, all of which belong to the bunting family. The best known is the red or common cardinal, *Cardinalis cardinalis*, of North and Central America. This species is about 20 cm. (8 in.) long; the male is a striking red with a red crest and a black mark around the beak, and the female is olive-green with a full red tail and crest. It breeds in thick cover, and builds an untidy cup-shaped nest of grasses and thin twigs, in which are laid three to four pale eggs speckled with dark brown and grey.

Caribou   *Reindeer.

Carnations   *Pinks.

Carnivores are, broadly speaking, meat-eating animals, but the term is used more specifically to describe an order of mammals. Although principally predators, some also eat plant material; only one member of the order, the giant panda, is wholly vegetarian. There are seven distinct families, of which four are dog-like and three cat-like. They are Ursidae (bears), Canidae (dogs, foxes), Procyonidae (racoons), Mustelidae (stoats, badgers, otters), Viverridae (mongooses, genets), Hyaenidae (hyenas), and Felidae (cats). The seals, sea-lions, and walruses used to be included among the carnivores, but now they are placed in

a separate order. Distinctive features of carnivores include the pointed canine teeth and the scissor-like cheek teeth. The claws may be blunt, as in bears and dogs, or very sharp, as in cats. All carnivores are covered with fur, which preserves their body heat.

Carnivores tend only to take prey that are about their own size, unless several animals band together in packs so that they can overpower much larger prey. The large cats usually kill their prey by suffocation; the smaller carnivores tend to have a killing bite; but many, such as wolves and hyenas, kill by tearing their prey to pieces. Many of the small carnivores are also insectivorous or eat

Carnivores

Cheetah's skull

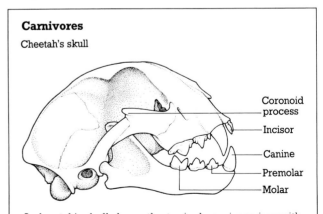

A cheetah's skull shows the typical carnivore jaw with a large bony structure, coronoid process, for the attachment of the powerful temporal muscles. Movement of the jaw is in one plane and has none of the rotary movement of other mammals. The incisor teeth are well developed for piercing, the canines for tearing, and the molars for cutting.

fruit. Even the grizzly bear may subsist for much of the year on a purely vegetarian diet of fruit and grass. All carnivores scavenge and some, such as the hyenas, are specialized scavengers.

Carnivorous plants are those which depend upon animal tissue for their sustenance. They include some 450 species spread over several plant families and genera. Their distribution is worldwide and their habitats range from acid bog lands to semi-deserts. Carnivorous plants attract and ensnare a range of insects and other invertebrates by means of traps. These may be of the pitfall (or passive) kind, or active and able to move as an aid to the capture of prey. The prey items provide the plant with nutrients, such as nitrogen, which are deficient in their habitats. The most active trappers include the *Venus's fly-trap and the *sundews, while the passive kinds are typified by the *pitcher-plant of North and South America. Among non-flowering plants, certain species of fungi catch and digest eelworms by means of a noose of fungal filaments (hyphae).

Carobs are the seedpods of *Ceratonia siliqua*, a tree legume native to the Mediterranean region. They are rich in sugars and gums, and are sometimes used as a substitute for chocolate. Carobs were probably 'the husks that the swine did eat' and the 'locusts' eaten by John the Baptist, mentioned in the Bible. The seeds may be the source of the jewellers' measure, the carat. A second species of *Ceratonia* has recently been discovered in Arabia.

Carp are fishes which give their name to the large family of freshwater fish, the Cyprinidae, which includes some 1,500 species. They all lack teeth; two pairs of barbels protruding from their lips help them to find food. Carp are omniverous, eating mainly plants and also insect larvae and snails. The European carp, *Cyprinus carpio*, is native to the Danube and the rivers of the Black Sea basin. It has been introduced to many parts of the world and is used as a food-fish in Europe. It is now also kept as an angling and as an ornamental fish. Carp prefer to live in deep, slow-flowing rivers and in lakes, especially well-vegetated ones.

Carpenter bees are large, blue-black, solitary *bees of the family Xylocopidae. They resemble dark bumble bees and have a loud, deep buzz. The queens dig long tunnels into living timber with their jaws, and build their cells of leaves, laying a single egg in each cell. Carpenter bees are mostly tropical, although a few species do occur in Europe.

Carpenter worms   *Goat moths.

Carpet moths belong to the moth family Geometridae, whose caterpillars are sometimes called *loopers. They belong to a number of genera, and are distributed worldwide. Most have mottled, marbled, or intricately barred fore-wings, and rest by day with their wings pressed flat against tree-trunks or walls.

The name is often associated with the genus *Xanthorhoe*, which includes the garden carpet moth, *X. fluctuata*, common throughout Eurasia and North Africa. Its fore-wings are white, dark at the base, and with variable dark markings in the centre. It lands head uppermost on a fence or tree-trunk, then turns head-down before settling, using its camouflage pattern to blend into its surroundings. Several

A male **carpenter bee**, *Xylocopa tabaniformis*, from northern California. The bee is using its strong mouth-parts to chew through the base of a *Diplacus* flower to steal the nectar and avoid fertilizing the flower.

similar species are common throughout the world.

The caterpillars of carpet moths eat a wide variety of plants, but confusingly the name is occasionally also used for one of the unrelated species of clothes moths.

Carpet sharks are members of a family of *cartilaginous fishes, containing about twenty-five species of small sharks, rarely longer than 4 m. (13 ft.). Most of them live in the tropical Indo-Pacific, but one, the nurse shark, *Gingly mostoma cirratum*, occurs in the tropical Atlantic. Carpet sharks are thickset and heavy-bodied with rather broad fins, and they have a pair of large barbels in front of the mouth. They are often boldly patterned with dark markings. All are bottom-living and feed on invertebrates; they are harmless to man.

Carrion beetles feed, as larvae or adults, on dead animal material. They may belong to one of several different groups, but the name is used specifically for the 200 or so species of the beetle family Silphidae, which are most common in northern temperate regions. These include roving carrion beetles, such as *Silpha*, and *burying beetles, such as *Necrophorus*.

Carrion crow: a species of the crow family which has two main subspecies. The common carrion crow is black all over, with a glossy blue sheen; the so-called hooded crow has the back and underparts grey rather than black. The common form occurs in Europe, while the hooded form extends over much of Asia. The birds are about 50 cm. (20 in.) long. They build bulky nests of twigs in the tops of trees or on rock ledges and eat a wide range of food, varying from carrion, to other birds and mammals, and to fruits and seeds.

Carrots give their name to the carrot family, the Umbelliferae, most members of which are characterized by large, disc-shaped heads of tiny white, pink, or yellow flowers. This family includes vegetables and herbs such as parsnips, angelica, fennel, celery, and hemlock. The cultivated carrot is derived from the European wild carrot,

Daucus carota, which produces a white or purplish, rather woody tap-root and flowers in its second year. The edible roots of cultivated carrots may be cylindrical or conical, and vary in length. They contain the orange pigment carotene which is a rich source of vitamin A.

Carrying capacity of a *habitat: the maximum number of individuals, of a particular species, that it is able to support. This is an important concept in *ecology and is determined by a number of factors, including the availability of food, space, and light, and the degree of competition, disease, predation, and accumulation of wastes. Such factors inhibit the population of a species from increasing beyond a certain point within the habitat, and when it reaches that point it stabilizes, fluctuating in numbers within narrow limits. Any increase in the limit may then depend upon whether the animal or plant can reduce its body size. Competition for food and space in an over-crowded habitat frequently results in smaller individuals which produce fewer offspring.

Cartilage is a structural material found in animals and exists in three forms: articular, elastic, and fibrous. Articular cartilage is the gristle at the ends of bones, forming movable joints. It has a translucent, smooth surface which, lubricated by synovial fluid, allows the ends of the bones to glide smoothly over one another. Elastic cartilage forms much of the skeleton of the larynx, epiglottis, and the external ear. It contains numerous elastic fibres and is a springy, rubber-like material. Fibrous cartilage forms discs between the spinal vertebrae, and between other bones that move only slightly over one another.

Cartilaginous fishes are one of the three main classes of fishes and include all those species which lack true bone in their skeleton. Instead, they have a skeleton made of cartilage, hardened by the deposition of calcium but still partially flexible. Their skin is covered in tooth-like (placoid) *scales, and they lack the swim bladder of the *bony fishes. The cartilaginous fishes are subdivided into three orders: the sharks; the skates, rays, and allied families; and the ratfishes, or chimaeras. The sharks, skates, and rays are sometimes classed together as the elasmobranchs.

Carver, George Washington (1864–1943), American agricultural chemist, was born into slavery. At the age of thirty-two he became director of agricultural research at the Tuskegee Institute for Negroes in Alabama and thenceforward worked to diversify farming in the American South, developing many products from soya beans, sweet potatoes, and peanuts, including peanut butter.

Case-bearer moths are tiny moths of the family Coleophoridae, whose caterpillars enclose themselves in characteristically shaped cases about 6 mm. (¼ in.) long. The 400 or so species are commonest in the Northern Hemisphere. *Coleophora* species have long, narrow, fringed wings. Young larvae are leaf-miners; older ones feed on leaves or seeds from within their attached cases, which are made of hollowed-out seeds or of silk, incorporating fragments of leaves and other plant debris. One common species, introduced to the USA from Europe, eats clover and is a minor pest.

Case-bearer moths are not unique in making cases: caterpillars of other families, including a clothes moth and bagworms, also do so.

Cashew nuts are produced by a small, evergreen tree, *Anacardium accidentale*, which is related to the mango and native to Central and South America. It is now widely cultivated throughout the world, often growing on soils too poor to support other crops. The 'nuts', which are really *seeds, are unusual in that they are not formed within the fruit, which is called the cashew apple, but are attached to it at one end and hang beneath it. They contain up to 50 per cent oil and 20 per cent protein.

Cassava is a woody shrub belonging to the *spurge family, and growing to 3 m. (9 ft. 9 in.) in height. It originated in South and Central America, but in the last few hundred years it has been introduced throughout the tropics, where it is particularly valuable, in drier areas with poor or exhausted soils.

The large, starchy, underground tubers of cassava contain a certain amount of prussic acid, a toxic chemical. The bitter types, which are the most nutritious, are cut into pieces, boiled, and squeezed to expel the poisonous sap. Types without prussic acid can be eaten without such treatment. The bitterness of all parts of the plant makes it virtually immune to locust attack and even to the ravages of baboons, factors often of vital importance in a crop of famine areas. Apart from its use as a food crop in the tropics, cassava starch enters world trade as tapioca, a form that can be stored and exported after heat treatment.

Cassowaries are large, heavily built, flightless birds which occur in New Guinea and Australia. Large individuals stand 1·5 m. (5 ft.) high. They are mainly covered with rather hair-like greyish-black plumage. The neck and head, which are featherless, are often coloured bright blue or red. They live mainly in forest and can run swiftly through the undergrowth. The head has a heavy shield, or casque, which may protect it while the bird is running. The female lays three to eight eggs in a scrape on the ground and leaves the male to incubate them. They eat mainly fruits and berries.

Castor-bean ticks *Sheep-ticks.

Casuarinas are a genus of about forty-five species of very unusual trees confined to Southeast Asia and to Australia, where they are known as she-oaks. Their slender green branchlets have tiny scale leaves, thereby resembling *horsetails. They are adapted to survive in very dry, hot conditions, and the shoots have been used as emergency fodder for animals. One species is common on sea-shores. The wood of many species is very hard.

Catalpas, or Indian bean-trees, are eleven species of deciduous tree legumes native to eastern Asia, North America, and the Caribbean. They all have very large leaves and long bean-like pods, some up to 45 cm. (18 in.) long, containing winged seeds. Most widely planted as an ornamental is *Catalpa bignonioides* of the southeastern United States: its timber is valued as well as its showy, scented white flowers.

Cataracts of the eye are restrictions to vision brought about by loss of translucency of the lens. There are many causes but the most common is an age-related change in one of the special lens proteins. Clear sight may be restored by removal of the lens. A strong spectacle lens, or plastic contact lens, is then required to achieve focus.

An adult double-wattled **cassowary**, *Casuarius casuarius*. Adults are extremely similar in appearance; males are largely responsible for rearing the chicks.

Catarrh is a mild inflammation of an internal surface of the body associated with increased mucus secretion. The term is usually applied to the nose or air passages. It may be due to a mild infection or to an allergy.

Catbird is a name used for members of three separate families of birds. The first is in respect of two species of mocking bird: the catbird, *Dumetella carolinensis*, and the black catbird, *Melanoptila glabrirostris*, both of which occur in Central and North America. The second family is that of the babblers, with the Abyssinian catbird, *Parophasma galinieri*, found in high mountains in Ethiopia. The third family is that of the bower-birds from Australia and New Guinea, which has three species of *Ailuroedus* called catbirds, and one species of *Scenopoectes*.

Caterpillars are the *larvae of butterflies, moths, and sawflies. In shape they resemble segmented worms but possess several pairs of legs, and strong jaws. They are adapted to a life of eating, and store food in preparation for the development of the adult insect. Some moths do not feed as adults and rely entirely upon the foodstores laid down as fat by the caterpillar.

The caterpillar's body is composed of ring-shaped segments, the first three of which carry pairs of clawed legs for grasping leaves. The remaining segments carry five pairs of fleshy prolegs in moths and butterflies, and between six and eight in sawflies. These prolegs, which often have finely hooked 'soles', are used to grip the stems of plants, and are also used in walking. Breathing openings, called spiracles, occur on each segment and connect with tubes which branch throughout the body. The head has

simple eyes (a single pair in sawfly larvae, but several in moths and butterflies), and tiny tubes called spinnerets which extrude silken strands. The most important structures on the head are sideways-working jaws which connect to a straight digestive tube or intestine.

The caterpillar's skin is covered with hairs, sometimes long and sometimes so short that the body looks bare. Moulting occurs at regular intervals, usually four to six times in the life of the caterpillar, allowing room for growth. Most caterpillars move slowly, and rely on *protective coloration for defence. Some have unpleasant scent organs or spray poison; others carry large horns or spines, while yet others live in a protective casing.

Catfishes are a group of fishes which typically have several whisker-like barbels around the mouth. Some 2,000 species are known; most of them are freshwater fishes but about fifty, in two families, are sea fishes. They are widely distributed in all tropical and temperate continents. Most of them have scaleless skins but many South American species have a covering of hard, large scales. Catfishes range in size from the South American candiru, *Vandellia cirrhosa*, only 2·5 cm. (1 in.) long, to the European wels, *Silurus glanis*, which is 3 m. (10 ft.) long.

Catkins are spikes of stalkless, simple flowers, separated by bracts, and typically the inflorescences of trees such as birches, willows, and poplars. Most catkins have no scent and are wind-pollinated, but willow catkins are insect-pollinated and produce nectar.

Cats are *carnivores belonging to the family Felidae, which includes pumas, ocelots, leopards, lions, tigers, and jaguars. The cats are the most specialized of the carnivores, and are well adapted to a hunting life. The ears, eyes whiskers, and nose are well developed as organs of sense. The teeth are of the specialized carnivore pattern; even

some of the rear molars (the carnassials) have developed into cutting blades.

In order to pursue prey, cats make use of scent, sight, and even such obscure clues as footmarks. They are masters in the art of leaping; from a running, walking, standing, or sitting position they can catapult into the air to hit their prey with stunning impact. They land with jaws wide open, teeth bared, and claws extended ready to sink into the throat and flesh of their prey. Cats have the sharpest claws of all mammals, claws that can be withdrawn into a sheath by all species except the cheetah. In this way, the claws are protected and the cat can move silently on its pads. Some of the patterns of behaviour associated with hunting may be observed in the play of kittens.

Little is known of the origin of the domestic cat but there are a number of similarities with the European *wild cat. Many breeds of the domestic cat have been developed with long or short coats.

Cattell, James McKeen (1860-1944), American professor of psychology, developed 'mental tests' concerned with sensory discrimination and co-ordination of movement.

Cattle are one of the most economically important members of the animal kingdom. Twelve species of cattle belong to the subfamily Bovinae, and include bison, buffalo, and yak, as well as all the domesticated breeds. This subfamily includes the genus *Bos*, which comprises five species of so-called 'true' cattle, such as the yak, gaur, and the cow of domestication. Both sexes of all cattle have horns which are never shed. They are *ruminant mammals, adapted to feed on grasses by having the teeth of the upper jaw fused into a pad. They lack incisor or canine teeth, and those of the lower jaw work with a grinding motion against the pad of the upper jaw. All cattle have cloven hoofs; that is, they are even-toed.

Domesticated cattle, *Bos primigenius*, are derived from the now extinct *auroch, and have been part of agricultural life for thousands of years. Some breeds have become draught animals; others provide man with milk, cream, and cheese, as well as meat, hide, horn, and bone. Females mature at eighteen months to three years and are referred to as heifers until the first calf is born, after which they are called cows. A cow will continue to breed for more than ten years; she will normally produce milk only when her calf is small, but lactation can be prolonged by showing her calf to her and by the act of milking. Many cows can produce over 9,000 litres (2,000 US gallons) per year. Much food is needed to maintain this output: a cow will eat about 70 kg. (150 lb.) of grass in a day, eight hours being spent eating and the remaining sixteen hours resting and chewing the cud.

Cauliflower: a relative of the *cabbages. It is grown for its succulent, immature flower-head, which is a large, roundish mass of white or creamy-white flower-buds.

Cave bears lived in Europe and Asia during the Pleistocene Ice Age (270,000-20,000 years ago). They were much larger than any living brown bear, and had long front feet and a domed forehead. Huge quantities of

A **cat** falls on its feet by reflex action. Signals from the eyes and ears cause neck muscles to rotate the head to a position with which the rest of the body rapidly aligns.

their bones have been found in caves, where they seem to have lived in family groups. Remains of their teeth suggest they were vegetarian. They became extinct during the last period of glaciation in Europe.

Cave fishes   *Blind fishes.

Cave lions are extinct members of the cat family and lived in Europe during the Middle and Upper Pleistocene Epoch (370,000–10,000 years ago). They were one-third larger than any living lion. Cave paintings indicate that they had a shaggy coat but no mane or tuft at the end of the tail. They died out at the end of the last Ice Age.

Cavies are some fourteen species of small South American *rodents of which the domestic *guinea-pig is the most familiar example. All are very similar in appearance and habits, although they live in a variety of habitats, including open grassland, forest edges, and rocky ground. Strictly vegetarian, they are mainly nocturnal and spend the day in burrows. Other species include the mara, or Patagonian hare, *Dolichotis patagonum*; the rock cavy, *Keradon rupestris*; and the desert cavies, *Microcavia*.

Caymans are tropical relatives of *alligators; their five species are found from Mexico southward through Central and South America. They have bony plates in their belly skins, which in some forms, especially smooth-fronted caymans, *Paleosuchus* species, are so well developed that their hides are of little use to the leather industry. The black cayman, *Melanosuchus niger*, and the broad-nosed cayman, *Caiman crocodilus*, are commercially valuable because of their more pliable skins, and they have been subject to overexploitation.

Cedars are four species of evergreen conifers of the genus *Cedrus*, extending from the Mediterranean region to the Himalayas. However, the name has long been used for trees with wood of similar qualities, as in Spanish cedar, *Cedrela odorata*, which is used for cigar boxes. The true cedars have tufts of rather small needles, and cones consisting of papery scales which carry the seeds. The Cedars of Lebanon, native in the eastern Mediterranean region, have been reduced to a few pathetic stands, but the tree is widely planted elsewhere. Cedars are members of the pine family, and closely related to larches.

Celery is a member of the *carrot family. Wild celery, *Apium graveolens*, is locally distributed in Europe, western Asia, and Africa. Cultivated celery is a relatively new crop, selected for its edible, swollen, leaf bases. The traditional commercial types are grown in rich soils, in trenches, as the stems need to be blanched by covering them with earth. A vegetable also derived from the wild celery is celeriac, of which the swollen root-like base of the stem is the part eaten.

Cells (biology) were named by Robert Hooke (1635–1703), who used the term to describe the compartments he observed in cork sections, using one of the earliest microscopes. The cell is now recognized as the basic unit of structure and function in all living things. Today cytologists (cell biologists) define cells as units of biological activity, each surrounded by a selectively permeable membrane and capable of reproducing itself independently.

Cells are usually microscopic, though some, such as the ova of vertebrates, are considerably larger. Their shapes include cubes, spheres, discs, spindles, rods, and numerous other variations. Some cells are amoeboid: that is, they have the ability to change their shape. Although differing in magnitude and shape, all cells have a similar internal organization, consisting of *cytoplasm and various structures known as organelles. In all cells except those of bacteria and blue-green algae, the major organelle is the *nucleus, which controls all activities within the cell. Other vital organelles include *mitochondria and ribosomes (where proteins are synthesized). The primitive cells of bacteria and blue-green algae are called prokaryotic cells. The cells of all other organisms are called eukaryotic cells. The differences between cells arise because they have become adapted to perform specific functions. Photosynthetic cells, for example, contain *chlorophyll, which is usually contained within organelles called chloroplasts, whereas spermatozoa are specialized cells possessing flagella for propulsion.

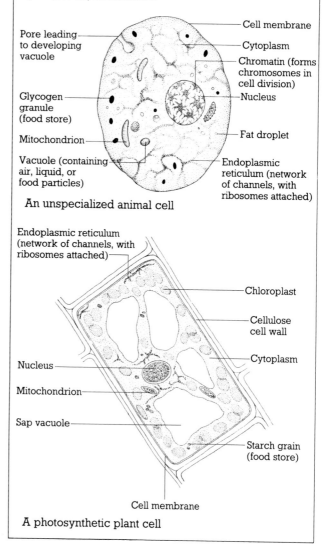

Cells

Semi-diagrammatic representation of a typical animal and plant cell revealing their incredible complexity, here magnified many thousand times — a human tissue cell may be just 0.0025 mm. (1/10,000 in.) in diameter.

Cell membrane
Cytoplasm
Chromatin (forms chromosomes in cell division)
Nucleus
Fat droplet
Endoplasmic reticulum (network of channels, with ribosomes attached)
Pore leading to developing vacuole
Glycogen granule (food store)
Mitochondrion
Vacuole (containing air, liquid, or food particles)

An unspecialized animal cell

Endoplasmic reticulum (network of channels, with ribosomes attached)
Chloroplast
Cellulose cell wall
Cytoplasm
Nucleus
Mitochondrion
Sap vacuole
Starch grain (food store)
Cell membrane

A photosynthetic plant cell

Cellulose is the structural material of plant cell walls. Its molecules, which comprise very long chains of glucose molecules (polysaccharide), are insoluble in water, unlike other carbohydrates. Typically, the cellulose chains lie parallel, forming tough microfibrils which may be embedded in lignin for greater strength, as in the wood of trees. The digestion of cellulose by all multicellular herbivores, from termites to ruminants, depends on the presence of certain types of bacteria or protozoa in the digestive system. These are the only organisms able to produce the enzymes necessary to digest cellulose.

Centaury is a name applied in Europe to several representatives of the gentian family, including *Centaurium erythraea*, a small annual plant with pink or red flowers, which is often found at the edges of roads and on dryish grassy slopes. In America the name is applied to the related plant, the rose pink, *Sabatia angularis*.

Centipedes are generally seen only when rocks or soils are disturbed. Then they scuttle to the nearest shelter, for they need cool, moist habitats, and are normally active only at night. They are *myriapods, belonging to the class Chilopoda with some 3,000 species. They have a single pair of jointed legs, of the typical *arthropod type, on each segment; this distinguishes them from millipedes, which have two pairs per segment. They can also be distinguished from millipedes by the stout poison claws beneath the head, which are used for nocturnal predation. Usually centipedes are cryptically coloured, but some tropical forms reaching 26 cm. (10 in.) long display brilliant coloration. Sperm is transferred, after courtship, in silken packages; and the females of many species brood over their eggs.

Century plants *Agaves.

Cephalopods are probably the most peculiar and the most advanced *invertebrates. They are a class of *molluscs, yet they have many unique features. Among the living 650 or so species are the familiar squid, octopus, cuttlefish, and nautilus. They move by jet propulsion, squirting water from directed muscular jets. They have large, focusing eyes, complex brains, and a remarkable learning capacity. They use smoke-screens of ink and extraordinary skin-colour changes to confuse predators, and they are active carnivores. The typical cumbersome molluscan shell is usually reduced or *vestigial.

Cephalopod means 'head–foot', as the foot of other molluscs has evolved in cephalopods into a ring of tentacles around the mouth, bearing suckers to capture prey. Although the cephalopods are successful as intelligent marine carnivores, they have never adapted to other ways of life, and the modern species are only a fraction of the 7,500 fossil species which existed before the evolution of competing fish.

Cereals belong to the very large family of *grasses, *Graminae*. Relatively few species have ever been cultivated as cereal crops, but these provide the staple diet of the vast majority of the world's population and have done so for the last 9,000 years. They are all annual grasses that flower and seed in their first year, and have been selected for tough stems and seed heads that do not shed the grain, once ripe. Some, such as *wheat, have been selected to give naked grain, free of chaff. Familiar cultivated cereals include maize, oats, rye, rice, sorghum, wheat, and millet.

The 'seeds' or grains of cereals are really a specialized type of fruit. They are easy to sow, easy to harvest, and easy to store, three vital characteristics. They contain about 70 per cent by weight of carbohydrate and 9–14 per cent protein. Regarded principally as a source of carbohydrate, they also provide about 70 per cent of the world's protein supply by virtue of the quantities eaten. Bread can be made only from those cereals containing large amounts of the protein gluten, and bread wheat is the outstanding example of their type.

Cerebral palsy is an effect of brain damage during or before birth. The most common effect is spastic paralysis, involving poor muscular co-ordination, and stiffness, due to sustained involuntary contraction of the muscles. Choreoathetosis is another disability caused by cerebral palsy; here involuntary writhing movements are prominent. Other effects include defective senses and some impairment of intelligence. Assessment of the child's intellectual ability is of great importance, but is often rendered difficult by his difficulty in expressing himself.

Chafers are *beetles of the family Scarabaeidae, especially those of the subfamilies Cetoniinae (2,600 species) and Melolonthinae (9,000 species). The first group includes large beetles with metallic-coloured elytra (wing-cases) and characteristic fan-shaped antennae. The second group includes the cockchafer, *Melolontha melolontha*, whose subterranean larvae can do great damage to roots of plants. The majority of species are found in the tropics and are related to scarab beetles.

Chaffinch: two species of the genus *Fringilla* from the finch family. Both are sparrow-sized and have conspicuous white wing-bars. The blue, or Canary Island, chaffinch, *F. teydea*, is confined to the Canary Islands; the male is a blue-grey colour, while the female is a duller, more greenish colour. The other species, the common chaffinch, *F. coelebs*, is widespread in Europe and western Asia. In some areas the females may migrate further than the males and, as a result, flocks of predominantly male or female birds may be found in some wintering areas.

Chaga's disease *Trypanosomiasis.

Chalcids are small to minute *wasps with reduced veins on the wings, and are usually black or metallic in colour. The larvae of most species are parasitic on the larvae of other insects, eventually killing the host. Other species are found in galls (abnormal growths of tissue on plants), being parasites on the larvae which have caused the gall to form, or even causing galls themselves. The chalcids form one of the largest superfamilies of Hymenoptera (bees, wasps, and ants) and comprise an estimated 25,000 species! They include the smallest insects, fairy-flies (*Mymaridae*), which have a body length of 0·2 mm. ($\frac{1}{127}$ in.).

Chameleons are an Old World family of lizards with about eighty species, the majority of which occur in Africa and Madagascar. They are adapted to life in trees and bushes, having fingers and toes that grasp. Chameleons have a prehensile tail, able to wrap around a twig and stabilize the arboreal forms. They capture their food, such as insects, by shooting out their long tongue, which may be extended as far as the length of their body. The prey is

A Namib **chameleon**, *Chamaeleo namaquensis*, shows an interesting use of colour. By turning black it absorbs heat to warm up quickly in the early morning sun.

carried back into the mouth on the tongue-tip to be crushed in the chameleon's jaws. Their famous ability to change colour is affected by light, temperature, and stress.

Chamois: a goat-like mountain-dweller belonging to the same family as the wild goat and sheep. It is found in the Alps, the Apennines, and east through the Carpathians into Asia Minor, living high above the tree-line. An elusive, swift-footed animal, it is recognized by its round, short horns (in both sexes) which rise perpendicularly from its head to turn backwards and downwards at the tip to form a hook. It is a slender animal, up to 81 cm. (32 in.) tall at the shoulders, with a soft, tawny-coloured coat in summer, changing to blackish-brown in winter. It has cup-shaped depressions in its feet that enable it to rest on ledges no larger than a man's hand.

Chamomiles, or camomiles, are strongly scented herbs of the genus *Anthemis*, which belong to the sunflower family. They occur throughout Europe and southwest Asia. The daisy-like flowers of the prostrate species, such as *A. nobilis*, are cropped for their medicinal properties as a tonic and to reduce fevers. These species, particularly the non-flowering varieties, are also used as a lawn-grass substitute on dry soils.

Chanterelles are *fungi belonging to the family Cantharellaceae, and are found in woodland. Most of them are brown or yellow, with funnel-shaped fruit bodies; the spores are borne on the underside of the funnel, which is folded into ridges. The common chanterelle, *Cantharellus cibarius*, is regarded as a delicacy, but the very similar-looking but unrelated Jack o'lantern fungus, *Clitocybe olearia*, is poisonous.

Chaparral is a type of vegetation, consisting of low trees and shrubs with tough, leathery leaves. It is found growing in southern California, where short, cool, wet winters and long, hot summers prevail. The term is sometimes extended to vegetation of similar aspect in the similar conditions of the Mediterranean (*maquis*), the Cape of South Africa, and parts of southern Australia.

Characins are freshwater fishes of the large family Characidae, with over 500 species, which occur in both warm-temperate and tropical America and in tropical Africa. They include the predatory piranhas, *Serrasalmus* species, of South America and the tigerfish, *Hydrocynus goliath*, of Africa, as well as the aquarists' brightly coloured *tetras. Characins are usually slender, with fully scaled bodies, a rayed dorsal fin, and an adipose dorsal fin. Their teeth are usually highly developed.

Charr is a species of fish of the salmon family, distributed in cool, fresh waters of the Northern Hemisphere, and in Arctic seas. In the north it is migratory, feeding in the sea on crustaceans and fishes and returning to rivers to spawn. Elsewhere it is found in mountain lakes in which it has been isolated since the Ice Ages. In lakes it is very variable in size, coloration, and body form.

Chat is a name used for birds of several different families, but mainly for Australian chats, which are four species of small birds of the genus *Ephthianura* that live in very dry areas in Australia, and for a number of groups of species in the thrush family. Most of the latter are small, brightly coloured species which live in open country, but some are secretive and live among rocks or scrub. Four members of the family of New World *warblers are also called chats, the best known being the yellow-breasted chat, *Icteria virens*, which is a summer visitor to North America. It is a heavily built bird for a warbler, brown above, bright yellow below, and with a rather large bill.

Chatterer is a name now usually applied to the scaly chatterer, *Turdoides aylmeri*, a member of the *babbler family; it is a greyish-brown bird with paler edgings to its feathers. This species occurs in East Africa. Chatterer has also in the past been applied to other members of this family, and to *cotingas, *waxwings, and other species.

Cheese-mites, which are related to other *mite species, such as grain mites, that attack man's stored products, are especially fond of foods with high fat and protein content, so that cheese is inevitably at risk. In some European countries these tiny mites are deliberately introduced into ripening cheeses in order to impart a characteristic fragrance and appearance.

Cheese-skipper: a species of *fly, *Piophila casei*, whose larvae live in maturing and stored cheese. If disturbed, the larva takes its 'tail' in its mouth, tenses its body, and then suddenly releases the tail, resulting in a skipping movement. It belongs to a small family of true flies which normally breed in carrion.

Cheetah: a species of big *cat with long, slender legs; it is the fastest land animal on earth, able to reach a speed of 72 km./hour (45 m.p.h.) in two seconds from a standing

A mother **cheetah** and cubs at Masai Mara National Park, Kenya. The young separate from their mother when they are thirteen to twenty months old, though siblings remain together for several months longer.

start. The top speed of this cat is 112 km./hour (70 m.p.h.); this pace, however, can only be maintained over some 450 m. (1,500 ft.) or so. It has for long been associated with man and trained to hunt like a dog. Unlike many cats, it hunts during daylight, using sight rather than smell. In Asia it was once found from the Caspian Sea to Sumatra but it is becoming a rarity; in Africa it is disappearing with the encroachment of man.

Chemofossils are organic compounds found in ancient sedimentary rock. They were originally formed by biological activity and have survived virtually unchanged, perhaps billions of years after the living organism that made them perished. Because they indicate the presence of life they are also known as 'biological markers', and their presence in Pre-Cambrian rocks has suggested that life existed more than 3,000 million years ago.

Cherries are small deciduous trees, typical of the subfamily Prunoideae, and part of the much larger rose family. Their nearest relatives include almonds, apricots, damsons, peaches, and plums. The cultivated cherry has originated from the wild cherry or gean, *Prunus avium*, found in woodlands in Europe and Asia. The fruits are borne in clusters on relatively long flower stalks and vary in colour from pale yellow to dark red. They are very susceptible to bird damage, and commercial orchards are concentrated in certain areas, such as Kent in England, to minimize losses. The genus *Prunus*, to which the cultivated cherry belongs, contains some 200 species growing in tem-

perate regions worldwide; not all of them have edible fruits. Flowering cherry trees, first developed in Japan, include varieties with more clustered, white to pink single or double flowers.

Chestnut trees are of two distinct types. The first, typified by the sweet or Spanish chestnut, *Castanea sativa*, contains ten to twelve species native to southern Europe, Asia, and North America. The second type includes the horse chestnuts, or buckeyes (*Aesculus* species), of southeast Europe, North America, and northeast Asia. They include twenty-five species which are quite unrelated to the sweet chestnuts. The sweet chestnut has simple, toothed leaves, bears edible seeds, and is related to beech. It is widely planted for its excellent timber, and coppices well, producing wood for stakes and fencing. The common horse chestnut, *A. hippocastanum*, is related to litchi and is native to the Balkans. It has been much planted and its inedible seeds are the conkers of children's games. Other species and hybrids of *Aesculus* are cultivated for their flowers.

Chevrotains, or 'mouse deer', are neither mouse nor deer but are even-toed hoofed mammals, placed in a separate family, Tragulidae. There are four species of these small animals; one species approaches a hare in size. The males have tusk-like upper canine teeth that project below the mouth. Most common in low country in rain forests and jungles, they may also live up in mountains. The Indian chevrotain or spotted mouse deer, *Tragulus meminna*, is the smallest species, and lives in Sri Lanka and India. There are two species of slightly larger Malayan chevrotain: the lesser mouse deer, *T. javanicus*, and larger mouse deer, *T. napu*. Both live in tropical rain forest. The African water chevrotain, *Hyemoschus aquaticus*, is somewhat larger and takes to the water at the least provocation. It occurs over West and Central Africa.

Chickadee   *Tits.

Chicken: a wide variety of domesticated forms of the red jungle-fowl, *Gallus domesticus*, which is a member of the pheasant family and, in the wild state, occurs in Southeast Asia. This species was probably first domesticated in India, probably as early as 3000 BC, and was gradually transported to other parts of the world. There is good evidence that it was in the Middle East by at least 700 BC. It was probably taken to the New World via East Asia and Peru or Ecuador, before Europeans reached North America. Initially it seems to have been of value as much as a sacrificial animal as it was for food.

The modern poultry industry developed on a large scale in the late nineteeth century and led to a great proliferation of breeds of different sizes and colours. Many were developed for their egg-laying potential or for meat, but some, such as the silkies and other bantams, were bred largely for decorative purposes. In eastern Asia, especially Japan, a number of other decorative breeds were produced, including the Phoenix and Yokohama strains, which have central tail feathers 6 m. (20 ft.) or more long. These are the longest feathers known.

Chicken pox is a highly infectious virus infection, producing a characteristic rash of small blisters. Most people are infected in childhood and suffer only a trivial illness, after which they are immune. Infection in adults can be more severe. The *Herpes zoster* virus, involved in chicken

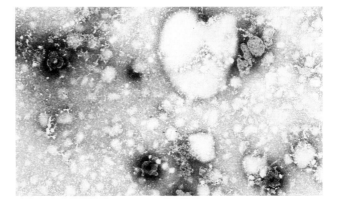

The virus that causes **chicken pox** (large irregular object, *top right*) measures less than a thousandth of a millimetre across. It spreads from person to person by airborne droplets or by direct contact with the patient.

pox, has the capacity of surviving in a dormant form in the body for many years, and becoming active again to produce an attack of *shingles.

Chick-peas are leafy, much-branched annual *legumes. Of western Asian origin, they are widely grown in northern India, throughout the Middle East, and in parts of southern Europe. In India they are known as gram, and are an important food crop, as are all lentils. Their short swollen pods contain one or two whitish-brown, beaked seeds between 5 and 7 mm. ($\frac{1}{8}$ and $\frac{1}{4}$ in.) in diameter.

Chicory is a widespread native plant of Europe and Asia. A relative of the daisy, it is an easily-grown salad vegetable which is eaten as 'greens' or forced and blanched. Certain varieties are grown for their large roots, which when dried, ground, and roasted may be blended with coffee.

Chiffchaff   *Warblers.

Chilblains are patches of swelling, redness, pain, and irritation on fingers or toes due to inflammation following prolonged exposure to cold. There is much individual variation in susceptibility, which is not well understood.

Chilli: a bushy perennial *capsicum with much smaller fruits than those of its close relative, the sweet pepper. The fruits can be as short as 1 cm. ($\frac{1}{2}$ in.). There is a range of forms, with the red fruits differing in their pungency. All taste considerably hotter than the other capsicums, and some can cause irritation to the human skin. Chillies are a constituent of curry powder and tabasco sauce. Cayenne pepper is made from the powdered, dried fruits of some species.

Chimaeras, or ratfish, are members of the family Chimaeridae, part of the class of *cartilaginous fishes. The family is widely distributed in the cooler regions of the North Atlantic, Indian, and Pacific oceans. They are allied to the sharks by virtue of their cartilaginous skeletons, but they have a single opening on each side for the gills, covered by a hard flap, similar to the condition in *bony fishes. Many species live in deep oceans.

Chimpanzee: the smaller of the two Old World *apes living in Africa, the gorilla being the larger. It is found

Chimpanzees are one of the few animals, apart from humans, to use tools. This young chimpanzee uses a stick to enlarge the hole made by tree-dwelling ants, which are then removed on the stick and licked off.

from lowland forest up to altitudes of 3,300 m. (11,000 ft.), where the gorilla is absent. Chimpanzees have a human-like face with long features and a prominent jaw. The ears are large and often pale. The rather sparse body hair is black, and the skin changes from flesh colour in juveniles to bronze and then black in adults. The forelimbs are longer than the hind-limbs, and the chimpanzee walks quadrupedally. Its time is spent equally between walking on the ground and climbing trees. Males are somewhat larger than females and can weigh up to 50 kg. (110 lb.).

Chimpanzees are mainly fruit-eaters, but they also eat leaves, nuts, bark, ants, termites, and even meat. They are very sociable animals, with large and complex societies of up to eighty individuals. They communicate by a wide variety of sounds and gestures. The home range may occupy an area of 20 sq. km. (8 sq. miles) in rich forest land, but will be greater in more frugal places. The groups divide into smaller bands such as nursery, all-male, and mixed adults. A prolific food source is advertised by hooting, screaming, and drumming, to draw other bands to the site. Each night the animals build fresh sleeping nests high in the trees. A single young is born after a gestation period of 225 days and the infant at first is carried on its mother's belly. Mother and infant stay together for up to a year after the birth of a second offspring.

China-mark moths are small, delicately marked, brown and white moths, which are said to resemble porcelain. They belong to two genera in the family Pyralidae. Their caterpillars feed on submerged or floating aquatic plants; some have gills on the sides, others trap a layer of air in body hairs. They construct webs or cases of silk, usually incorporating leaf fragments. The genus *Nymphula* contains several species of china-mark moths throughout the Northern Hemisphere. The genus *Paraponyx* has several species found worldwide.

Chinchilla rats are neither chinchillas nor rats but two species of cavy-like *rodent from the Andes. They are rat-sized, and have soft, dense fur resembling that of chin-

chillas, whereas their large rounded ears, pointed snouts, and long tails give them the appearance of rats. They are colonial and live in underground burrows. Their diet is entirely herbivorous.

Chinchillas are two species of South American *rodents characterized by long hind legs, a bushy tail, and soft, luxuriant fur, which is highly valued in the fur trade. The chinchillas, *Chinchilla brevicaudata* and *C. lanigera*, inhabit the Andes of western South America, where they may now be almost extinct. They are still common in commercial fur farms. The family Chinchillidae also includes the *viscachas. All members of this family are vegetarians, eating grasses and low-growing plants.

Chipmunks are *ground squirrels, that occur in North America and northern Asia. They usually have alternate light and dark stripes running down the body. There are twenty-four species in the genus *Tamias*, found in a variety of habitats from open grassland to mountain forests. They are noted for their habit of hoarding food, chiefly nuts and seeds, which they eat during the winter at intervals when they awake from hibernation. As they do not always find each store, this habit helps the germination of trees by the scattering of the seeds.

Chives are closely related to the onion, garlic, and leek, and belong to the lily family. They are widespread across the Northern Hemisphere and also occur in Asia Minor and the Himalayas. They produce dense tufts of bright green, tubular leaves, up to 25 cm. (10 in.) long, which are used in salads, and bear attractive round purple or pink flower-heads.

Chlorophyll is any one of several closely related green pigments that are used by most plants in *photosynthesis. Chlorophyll molecules have a flat, lollipop shape and each contains a magnesium atom fixed at its centre. During photosynthesis, the structure absorbs energy from red and blue light, but it reflects green, giving the chlorophyll its colour. The molecules, 'excited' by the absorbed light, transfer electrons to energy-carrier substances in the cell, thus providing energy for making carbohydrates.

There are four main types of chlorophyll, differing from each other in the number of carbon atoms attached to a small part of the molecule. Chlorophyll-a and -b are found in all green plants; chlorophyll-c is found in brown algae and diatoms; chlorophyll-d occurs in red algae. In organisms coloured red, brown or purple, the green colour of chlorophyll is modified by other light-trapping pigments which can be yellow (carotenoids), or blue or red (phycobilins).

Cholera is an acute infection of the small intestine with a specific bacterium, *Vibrio cholerae*. The organism adheres to the gut lining, multiplying but not invading the tissues. The bacterium produces a toxin which causes a great loss of fluid from the sufferer's body. The loss of fluid may be so great as to kill the patient in a few hours. The infection is contracted from food or, more commonly, water which is contaminated with the excrement of infected people. Fluid replacement is successful in curing the ailment, and drug treatment may hasten the recovery.

Cholesterol, a neutral fat of animal origin, is a *steroid found in blood and body tissues. An essential component of cell *membranes, it modifies the physical properties of the water-repellent layers. It also acts as the raw material on which steroid hormones, bile salts, and acids are based. Much of the body's cholesterol is made in the liver, and some can be absorbed from diets rich in animal fat. People with high levels of cholesterol in their blood are thought by some experts to be prone to circulatory diseases, such as *arteriosclerosis.

Choughs are two species of the crow family. Both are about 38 cm. (15 in.) long and have glossy blue-black plumage and red legs. The red-beaked, or common, chough, *Pyrrhocorax pyrrhocorax*, has a long, sharply curved red beak, whereas the alpine chough, *P. graculus*, has a shorter, straighter, and stouter, yellow beak. Both species occur in Europe and southern parts of west and central Asia, the red-billed chough living primarily at low altitudes, the alpine chough at high altitudes. They feed on small insects and nest in cliff crevices.

Chromosomes are the structures in living organisms that carry the hereditary molecules (*DNA). Bacteria do not possess chromosomes; their DNA is carried in a simpler form in the cytoplasm. In the cells of all higher organisms they are visible under a microscope, and are found in the cell nucleus. Chromosomes consist of a mixture of *nucleic acids and *proteins. Some of the proteins support the DNA molecule, while others act as *enzymes, to catalyse its various functions.

Most species have more than one chromosome. In man there are twenty-three different chromosomes, and each kind is present in two copies in every cell except gametes, making forty-six in all. The condition of having two copies of each kind is called diploidy and is found in most kinds of animal. One pair of chromosomes, called the sex chromosomes, determine the sex of the organism.

Chrysalis: the *pupae of some butterflies, the term being also used occasionally for the pupae of moths and other insects. In a true chrysalis, the wings and legs are closely stuck to the body, and there may be a silken girdle round the middle that attaches it to a stick or plant, and sometimes there are patches of gold and silver colour. The chrysalids of some species can wriggle the abdomen section, but the rest is stationary. Chrysalids are not contained in a cocoon.

Chromosomes within the actively dividing root cells of an onion plant, at various stages of the cell division process, photographed here under the light microscope (× 600).

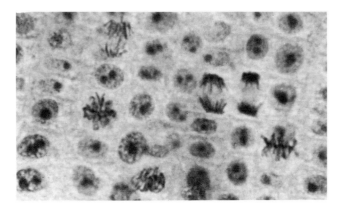

Chrysanthemums are annual or perennial plants, chiefly of herbaceous habit but occasionally shrub-like. About 100 species are known within the genus *Chrysanthemum*, and include plants such as feverfew, pyrethrum, and tansy. Their distribution includes Europe, Asia, America, and Africa. Several species have been developed as garden flowers, and the exhibition or florists' varieties are the result of centuries of selective breeding from early Chinese and Japanese kinds; their colours and shapes are now of infinite variety, ranging from the large mop-headed kinds to the tiny pompom types. Their flower shape reveals them as members of the sunflower family and relatives of the daisy.

Chub, or chubb, are fishes of the *carp family, typified by the European chub, *Leuciscus cephalus*, which grows up to 61 cm. (22 in.) long. They are broad across the head and back, and have a strongly concave edge to the anal fin. Most abundant in rivers but also living in lakes, they are schooling fishes when young, becoming solitary with age. There are some seventeen species of North American chub belonging to the genus *Hybopsis*.

Cicadas are large insects of the true *bug family, found mainly in tropical regions. The adults feed on the sap of trees and shrubs, but the larvae live underground on roots, and may live for many years before completing their life

A **cicada**, *Fidicina* species, dries off and hardens its exoskeleton after emerging from its larval skin in a Central American rain forest.

cycle. The females are silent, but the males produce loud, high-pitched, persistent songs from organs on the underside of the body. Each buzzing song is characteristic of the species and is usually made by day.

Cichlids are a very large family of freshwater fishes which live mostly in Africa and in South and Central America. A few species occur naturally as far north as Texas, and in parts of Asia, but many are found in the African Great Lakes. In general they are perch-like in body form, with spiny dorsal and anal fins, and the head and body are fully scaled. They have a single nostril each side. There are about 700 species, occupying a wide range of micro-habitats; some are even adapted to life in bodies of water that regularly dry out.

Cinchonas are a genus of some forty species of evergreen trees of South America, belonging to the gardenia family and related to coffee. They are the source of natural quinine, formerly much used in the treatment of malaria and in flavouring tonic water and other drinks. The active principle is concentrated in the bark. Plantations were established in India and Java, but nowadays the natural substance is of less importance due to the adoption of synthetic antimalarial drugs.

Cinnamon is a spice obtained from the dried bark of a species of evergreen trees belonging to the laurel family, *Cinnamomum zeylanicum*. It is native to Sri Lanka, and its exploitation was the incentive for the successive Por-

On plantations the **cinnamon** tree is pruned so that it grows only as a bush. This specimen in the wild displays the characteristic young red leaves.

tuguese, Dutch, and British colonizations of that country. Cinnamon oil, used in medicine, is extracted from the bark residue, while an oil distilled from the green leaves is used as a substitute for clove oil. Much of the cinnamon consumed in the USA today comes from a Chinese and Indonesian species with a stronger flavour. Camphor is obtained, in a similar manner to cinnamon oil, from *Cinnamomum camphora*, a closely related species.

Cirrhosis is a condition resulting from extensive damage to the cells of the *liver. This leads to changes in the structure and composition of the liver and impairment of function in this vital organ. Blood flow through the liver is restricted by the changes and can lead to increased blood pressure in abdominal veins, leading to internal haemorrhage. Causes include viral infections such as *hepatitis, an excess of iron or copper in the liver, and, most commonly in developed countries, alcoholism. Damage caused is irreversible and can be halted but not cured.

Citronellas are species of tropical grass, *Cymbopogon* species, which are extensively grown in Sri Lanka, Burma, and Java for distillation of the scented oil, much used in soap manufacture. They are not to be confused with the South American genus *Citronella*, which are small trees or bushes belonging to the family Icacinaceae.

Citrus fruits, strictly speaking, come from plants of the genus *Citrus*, which give their name to the citrus family, the Rutaceae. Also in this family are similar fruits such as kumquats (*Fortunella*) and the inedible trifoliate orange *Poncirus trifoliata*. All, except the trifoliate orange, are evergreen trees with simple leaves, and usually with spines at the leaf axils of the younger shoots. The majority of citrus fruits are of Asian origin, although the grapefruit originated in Barbados. Citrus fruits were cultivated in China at least 5,000 years ago.

The fruits are a special type of berry, with juice-filled segments surrounded by a white, spongy tissue, and a tougher, outer peel. All are typically orange- or lemon-shaped but vary in size from the massive *shaddock, up to 30 cm. (12 in.) in diameter, to the lime, which is 4-6 cm. ($1\frac{1}{2}$-$3\frac{1}{2}$ in.) in diameter. The sweetness of the juice, which is rich in vitamin C, varies according to its proportion of sugar to acid. Sweet orange has about 12 per cent sugar and 1 per cent citric acid, whereas lemons and limes contain 4 per cent and 6 per cent acid respectively.

Classification of the various species and cultivars is difficult, particularly as hybrids are common. Artificial hybrids include the tangor (mandarin × sweet orange), tangelo (mandarin × grapefruit), lemonime (lemon × lime), and limequat (lime × kumquat) among others. Citrus fruits are usually propagated by a form of grafting known as budding. A bud of the chosen cultivar is grafted on the rootstock of the sour orange (for heavy soils) or rough lemon (for light soils).

Civets are carnivores of the mongoose family, found throughout Africa and tropical Asia. There are nineteen species including the *palm civets and 'true' civets. All are medium-sized animals, rather bigger than a large cat, with elongated bodies and snouts, long bushy tails, and short legs. They prey upon rodents, birds, and insects, and they also take fruit. They have well-developed anal scent glands whose secretions, although evil-smelling, can be refined for use in the perfume industry. The true civets are part of the same subfamily as genets and linsangs. The best-known and largest species of true civet is the African civet, *Civetticus civetta*, which occurs in a wide range of habitats, from tropical rain forest to arid grassland. The binturong, *Arctictis binturong*, of Southeast Asia is an unusual civet in that it is one of the few mammals in the Old World to have a prehensile tail.

Clams, sometimes used as a term for all *bivalves, more specifically refers to the burrowing hard-shell clams (or quahogs), such as *Mercenaria mercenaria*, and soft-shell clams (or gapers), such as *Mya arenaria*, both types common beneath sandy and muddy beaches. Both groups include large, rather ovoid bivalves, but the soft-shell clam is endowed with long, stout siphons and can burrow down to 50 cm. (20 in.). Other molluscs commonly called clams include the Venus clams (*Clausinella*), razor clams (*Ensis*), and the giant clams (*Tridacna*) of the Indo-Pacific, which are over 1 m. (3 ft.) across and weigh 1,100 kg. (1·2 US tons). All claims feed exclusively on tiny plankton.

Clam worms   *Bristle worms.

Classes, in the classification of animals and plants, rank below *phyla and include orders of animals or plants. If the number of orders in a class is large, it may be divided into smaller units called suborders. Members of each class show characteristics indicating common evolutionary descent, but the limits which define the class are often the subject of disagreement among biologists.

Clavicles, or collar-bones, are jointed at their mid-point to the upper part of the breast bone and laterally to the shoulder bones just above the shoulder joint. They keep the shoulder joints away from the chest wall and are longer and larger in man than in any other mammal. The arrangement in man permits a greater degree of movement of the arm at the shoulder joint than in other animals.

Claws are formed on the tips of the digits of vertebrates. They are keratinized epidermal structures of skin and the simplest form may partly embrace the digit. The claw is V-shaped in section and pointed at its tip, and may be used for holding prey, grooming, and clinging to the ground or a tree. Those of the cat can be retracted into a sheath, allowing it to approach its prey silently. Mammals may have nails or hoofs in place of claws. The nail, essentially the same type of structure as a claw, but broadened and flattened, is found in primates. Hoofs are characteristically a development of ungulates, mammals that walk on the tips of their digits. These include horses, cattle, sheep, goats, and antelopes. The terminal bone, or phalanx, is broadened and the claw becomes modified to surround it, while a pad forms below. All these keratinized structures grow continually outward from a cell layer beneath or at its base. Keratin, which is also the structural protein of hair, is a typical fibrous protein.

Clearwing moths are a large family of small day-flying moths, which is distributed worldwide. They have extensive transparent patches on the wings, the scales from these areas having been shed on the first flight. Many species resemble wasps, bees, or ichneumon flies in shape and coloration, and are avoided by predators because of this resemblance. The maggot-like caterpillars of clear-

The **clearwing moth**, *Sesia apiformis*, resting on a tree trunk looks remarkably like a hornet. The adults are most often seen in this position in late spring and early summer. The caterpillars burrow in aspen and poplar wood for three years before pupating.

wings feed by burrowing in tree trunks, stems, or roots. The currant clearwing, *Synanthedon salmachus*, can be a pest of currant bushes in Europe and North America.

Cleft palate is a birth defect caused by the failure of the tissues forming the roof of the mouth to join along the midpoint. It occurs in about one in a thousand births and can be associated with a cleft or hare lip. Undiagnosed, the baby has difficulty feeding and later with speech development. Treatment consists of plastic surgery at about one year of age, special dental care, and speech therapy.

Clegs   *Horseflies.

Clematis is a genus of some 250 species of plants, most of them climbers, with opposite, usually compound leaves. They are members of the buttercup family. The climbing species attach themselves by bending their leaf-stalks round the support. The flowers are conspicuous because of their coloured sepals, there being no petals, while the fluffy fruits are single-seeded with a long fluffy 'parachute', which promotes wind-dispersal. Many species are cultivated, the most commonly seen being the large-flowered hybrids known collectively as *C.* × *jackmannii*.

Clementine is a *citrus fruit thought to be a hybrid between the mandarin and the sweet orange. It is intermediate between its parents in size, colour, and the looseness of peel. The most important areas for commercial production are in North Africa.

Click beetles are a large family of some 7,000 species of *beetle found worldwide. They include *fireflies such as *Pyrophorus noctiluca* and many species common in grassland. Most species are elongate and oval in shape. When they fall on their backs they can right themselves by a special

Climbing plants

Various adaptations developed for climbing through, and clinging to, other plants.

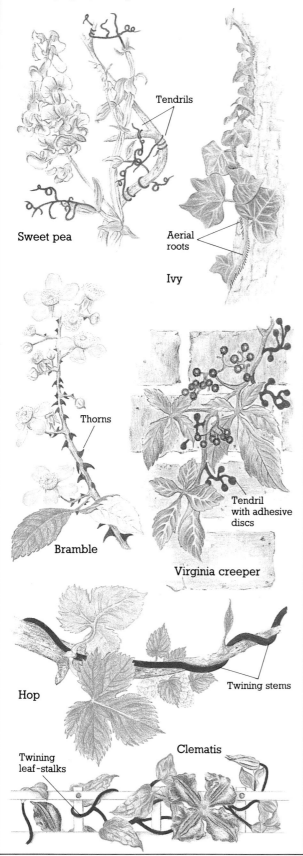

Sweet pea

Tendrils

Aerial roots

Ivy

Thorns

Bramble

Tendril with adhesive discs

Virginia creeper

Hop

Twining stems

Twining leaf-stalks

Clematis

mechanism on the thorax, which throws them into the air, with an audible click. The larvae of some species are known as wireworms and these do tremendous damage to plant roots in some parts of the world. The wireworm larvae may take up to five years to develop fully.

Climbing ferns belong to the genus *Lygodium*, which includes about twenty-five species, mostly native to the tropics and subtropics. One species, *L. palmatum*, is native to the USA from Florida to Massachusetts. These ferns coil round other vegetation, using their leaf stalks to climb, and can reach 4·5 m. (15 ft.).

Climbing perch: a species of Asiatic freshwater fish which ranges from southern China to Sri Lanka. It lives in large rivers, canals, and ponds, often where oxygen levels are low, and it breathes air by means of respiratory organs in a cavity above each set of gills. It can travel overland from one patch of water to another, pulling itself slowly along by means of its pelvic fin spines and spiky gill covers, but its climbing ability has been much exaggerated.

Climbing plants are those which attach themselves to supports (often other plants) in order to grow towards the light to display their leaves, flowers, and fruits. The adaptations for climbing are numerous and include aerial roots, hooked prickles, twining leaf-stalks, branchlets modified as tendrils (often with the addition of adhesive discs), and leaf-tip tendrils. Moreover, climbers include simple scrambling plants, woody twining kinds, and the *lianas. Those with twining stems regularly spiral in one direction, either to the right or left, and this is constant for each species: for example, convolvulus turns to the left, whereas the common hop spirals to the right.

Clingfishes are members of the family Gobiesocidae, which contains some 100 species. They are mostly small, scaleless fishes with a maximum length of 30 cm. (12 in.), and with the pelvic fins adapted into a powerful sucking disc, by means of which they cling to rocks. They are mostly marine fishes of temperate and tropical shallow waters, but a few live in fresh water in Central America. The shore clingfish, *Lepadogaster lepadogaster*, is common on rocky shores in western Europe.

Clitoris is the name given to the female mammalian equivalent of the male penis, but it is not transversed by the urethra. It has a body ending in a conical head called a glans, all formed of erectile tissue. It is situated just below the pubis and above the opening of the urethra. Like the nipple, the clitoris is supplied with numerous nerves and is very sensitive to stimulation.

Clones are organisms or cells that have been produced from one parent by asexual *reproduction. The individual organisms are precise copies of the parent and, most importantly, are genetically identical to it. Clones are found naturally among single-celled organisms, a few invertebrates, such as corals, and some asexually reproducing plants. They are also produced artificially, especially from useful cells, such as bacteria, possessing particular chemical abilities. The techniques of genetic engineering, where selected pieces of *DNA are inserted into the chromosome of bacteria, can give these micro-organisms the ability to produce valuable drugs or hormones. Artificial cloning has enabled the commercial production of substances such as the human growth hormone somatostatin, and human insulin. Clones of farm animals have also been produced experimentally.

Clothes moth is a name used for mostly small, dark coloured moths, whose caterpillars eat furs, woollen goods, and other textiles. In the wild, they are scavengers on carcasses and in birds' nests and mammal burrows, but some have spread worldwide in association with man. They get insufficient nutrients from clean textiles, and prefer raw or soiled wool. Some species have become less abundant with the introduction of man-made fibres; others eat these too, and chew through polythene. The common clothes moth, *Tineola bisselliella*, originated in warm countries but spread elsewhere with the efficient heating of houses. These moths are among the 2,400 or so species in the family Tineidae.

Clouded leopard: a species of big *cats, some 90 cm. (3 ft.) in length. They are found in Southeast Asia in the densest forests. The coat is of soft, thick fur and is decorated with spots and stripes. The canine teeth are relatively longer than those of other cats. These powerful animals are nocturnal hunters and can kill deer, though they usually prey upon smaller mammals.

Clovers are herbaceous annual or perennial *legumes. Their generic name, *Trifolium*, refers to the appearance of their leaves, each consisting of three leaflets. They are a vital component of pastures in temperate regions of the world; the creeping, white clover, *T. repens*, is particularly valuable for sheep grazing, for example in New Zealand, while the taller, red clover, *T. pratense*, is useful in hay meadows. Their vital role in agriculture is in maintaining soil fertility by converting atmospheric nitrogen into a form suitable for plant growth. An annual species of Mediterranean areas, burrowing clover, *T. subterraneum*, is interesting in that, after flowering and setting seed, its flower stalks grow downwards to bury its own seed; it has been introduced successfully to other parts of the world with similar climates, such as south Australia, where it is now the most important pasture legume.

Cloves come from *Syzygium aromaticum*, a tree native to the Moluccas, and, together with nutmeg, were responsible for their being named the Spice Islands. The two spices were so prized that in the seventeenth century the Dutch deliberately destroyed trees in all but three of the islands to maximize the price obtained in Europe. The cloves are the unopened flower-buds, picked and dried in the sun. They were used in China as long ago as 300 BC and were ordered to be used as breath-sweeteners by courtiers addressing the Emperor. Along with eucalyptus and allspice, they belong to the myrtle family.

Club foot is a defective development of the foot, noticeable at birth and able to be corrected if treatment is given in infancy. If untreated, normal growth is impossible and the adult foot appears stunted and lumpy.

Clubmosses are *cryptogam plants that bear spores in spore cases which are usually aggregated into club-like cones. They are closely related to the ferns, and the 1,250 or so species are placed in the class Lycopisda; they include the genera *Lycopodium* and *Selaginella*, which are usually

The beautifully patterned coat of the **clouded leopard**, made up of pale 'sunspots' and dark blotches, breaks up the animal's body outline as it rests during the day amongst forest foliage. Note the large paws, which are used effectively to swat and capture smaller prey either in the trees or on the ground.

small plants with tufts of branching stems clothed with very small leaves. Clubmosses have roots and woody tissue, and can be considered as perennial plants. Like horsetails, they once included giant tree-like forms which were numerous in the Carboniferous Period, 300 million years ago, and contributed towards coal formation. The fossil clubmoss, *Lepidodendron*, appears to have been about 30 m. (100 ft.) high. Many present-day species are found among the ground flora of tropical rain forest, and some grow as *epiphytes.

Cluster-flies are true *flies which resemble houseflies and belong to the same family as *blowflies. The adults of *Pollenia rudis* hibernate as adults and are sometimes found clustering in crevices or in the corners of attics. Their eggs are laid in soil, and the larvae are parasitic in earthworms, eventually killing their host.

Alpine **clubmoss**, *Diaphiastrum alpinum*, a European species found on heaths and moors. The long green cones are the spore cases, typical of this plant group.

Coalfish is a species of fish known as the saithe pollack in North America, and is a North Atlantic member of the cod family. Its back and upper sides are very dark green in colour. It is a schooling fish that lives from the surface waters down to 200 m. (650 ft.) and feeds on small crustaceans and fishes. It is an important commercial fish.

Coatis are four species of slim, long-tailed South American carnivores related to raccoons. Their most striking feature is the long, flexible snout with which they root for insects in the ground or in dead logs. They are omnivorous feeders, foraging by day in packs of up to a dozen animals. Males are more solitary but join the pack when a female is on heat. The ring-tailed coati, *Nasua nasua*, grows up to 65 cm. (25 in.) and, as its name implies, has an equal length banded tail.

Cobras are venomous *snakes, well known for their tendency, when alarmed, to rear up and spread a hood, which is formed in the neck region by loose skin. True cobras of the genus *Naja* occur in Africa, the Middle East, and Asia. The venom fangs of cobras are carried permanently erect within the mouth, in contrast to the foldable fangs of vipers. The discharge orifices of the fangs are modified in some species, such as the ringhole or spitting cobra, *Naja nigricollis*, to enable them to spray venom into the eyes of an aggressor. This causes great pain and even temporary blindness if it reaches the eyes. The king cobra, or hamadryad, *N. hannah*, is famous for preying on other, non-venomous snakes. The Indian cobra, *N. naja*, is one of the world's most dangerous snakes, killing a great many people every year.

Coccyx: the four fused vertebral bones lowermost on the spine of man and other tail-less mammals. It is the bony part of an abbreviated tail which is not visible in man, and does not contain any part of the spinal cord. In old age it often fuses with the sacrum, the large triangular bone at the base of the spine. It can be felt just above the cleft of the buttocks and can be bent backwards and forwards.

Cochlea: an organ of the inner *ear, consisting of a fluid-filled, spiral-shaped tube, inside which runs a membrane sensitive to sound. It is concerned with the breakdown of sound into its component frequencies before sending the information to the brain along the cochlear nerve. Sound waves pass along the cochlea and different frequencies cause vibrations to be set up in different parts of the cochlear membrane. High frequencies are detected in the region nearest the middle ear, low frequencies are detected at the tip of the cochlea. This organ is best developed in mammals and is present in a non-spiral form in birds.

Cockatoos belong to the parrot family, and occur in Australia, New Guinea, and some adjacent islands. They are about 40–65 cm. (16–26 in.) in length. All have powerful beaks; that of the palm cockatoo, *Probosciger aterrimus*, is exceptionally powerful and is used for opening large nuts. In plumage the eighteen species of cockatoo are basically either black, like the red-tailed cockatoo, *Calyptorhyncus magnificus*, or white, like the sulphur-crested cockatoo, *Cacatua galerita*, though a few are pink, like the galah, *Eolophus roseicapillus* which is grey above with deep pink underneath and a pale pink crown; all have erectile crests. Cockatoos are noisy birds and live in groups ranging from a few to many birds. They feed primarily on seeds, and the galah and the little corella, *Cacatua sanguinea*, sometimes do considerable damage to grain and rice crops. They nest in large holes in trees, laying two to three round white eggs. In captivity many of them live a very long time, fifty to sixty years having been recorded.

Cockchafers *Chafers.

Cockles are *bivalve molluscs that live just below the surface of sandy beaches in vast numbers. The edible cockle, *Cardium edule*, occurs between tide-marks, filtering fine food particles from the sea through short siphons that barely protrude above the sand. Other species, such as the prickly cockle, *C. echinata*, live further down the beach, permanently covered by water. All cockles have rather plump shells, usually heavily ribbed, perhaps to improve anchorage in the sand.

Cock-of-the-rock: two species of birds from the cotinga family, belonging to the genus *Rupicola*. They are the Guianan cock-of-the-rock, *R. rupicola*, and the Andean cock-of-the-rock, *R. peruviana*. They live in tropical forests of northern South America. The males are a brilliant orange with black wings and tail; the orange feathers on the crown are stiffened to form a permanent crest. The females are olive-brown. The males display in communal groups, or leks, each bird having a separate court (an area of ground cleared of leaves). The females attend the leks for mating and then leave and look after the eggs and young by themselves.

A male Andean **cock-of-the-rock** shows his brilliant plumage and crest. When displaying, the males show little animation, performing curiously static dances, holding their position for several minutes at a time.

The **coconut** palm, *Cocos nucifera*, may produce 200 nuts in one year. Once fallen, coconuts can survive in sea water and float from island to island, but in the Pacific region man has been an agent in their spread.

Cockroaches are usually large, brown or black insects which feed on decaying vegetable material. Together with *praying mantids, they make up the order Dictyoptera, which includes some 4,000 species of cockroaches. The adults have a flattened shape and the front wings (if present) are thickened and slippery to the touch. Cockroaches produce their eggs in a tough case called an ootheca, which the female usually carries around with her until the nymphs are ready to hatch. The young resemble the adults. Cockroaches are mainly tropical, but there are a few small species in temperate countries. The best known are those which are common household pests, especially the black beetle or oriental cockroach, *Blatta orientalis*, and the German and American cockroaches, *Blattella germanica* and *Periplaneta americana*. In the tropics, a few species are colourful and live on flowers.

Cockscomb (plant)   *Love-lies-bleeding.

Cocoa   *Cacao.

Coconut: a feathery-leaved, tropical palm, up to 24 m. (80 ft.) in height. Its nut consists of a single hard-shelled seed, surrounded by a thick, fibrous, outer husk. The white-fleshed kernel has a central cavity containing coconut milk, which is a refreshing drink. Ripe nuts are harvested by climbing the tree (or by training monkeys to do so) or by allowing them to fall naturally.

The tree grows well in coastal areas of the tropics and subtropics, with a steady supply of nuts being produced throughout the year if the temperature and rainfall are fairly evenly distributed. It is reasonably tolerant of salt and the fruits are able to float and survive in seawater. As they are carried in ocean currents, the species has been dispersed throughout the coastal tropics. The fluid contained in the seed allows it to germinate successfully, even in dry sand.

The husk of the coconut (coir) is used to make mats, floor coverings, upholstery filling, and ropes, and the shells are useful as fuel. The dried, white flesh of the kernel, is known as *copra.

Coconut-crabs, or robber crabs, reveal their relationship to hermit crabs in their asymmetrical abdomen, but, unlike most hermit crabs, the adults leave the sea to live on land. They use their stout, spiky claws to climb palm trees, though they have to descend backwards. They eat most types of vegetation, but are especially keen on fallen coconuts. Like all land *crabs, they nevertheless rely on the sea for spawning, and drink copiously to replace water lost by evaporation.

Cod: a fish which lives in the cool waters of the North Atlantic at depths ranging from near the shoreline to 600 m. (1,950 ft.). It is a schooling fish that prefers to swim at depths of 30–80 m. (98–260 ft.) above the sea-bed, although it feeds on the bottom. The individuals keep in contact with each other by means of sounds which are produced and amplified by the swim bladder. The cod feeds on a large variety of smaller fishes and crustaceans, and also eats molluscs and brittle stars. An immensely valuable food-fish, it is caught mainly by trawling, but many are taken on lines. The cod family contains fishes such as haddock, *Melanogrammus aeglefinus*, burbot, *Lota lota*, and true cod of the genus *Gadus*.

Codlin moth: a species of small moth found worldwide. It feeds as a caterpillar by burrowing through fruits such as apples and peaches. The caterpillars are notorious pests of fruit. Related species include the pea moth, *Cydia nigricana*, the oriental fruit moth, *C. molesta*, and the jumping bean moth, *C. saltitans*. All are part of the family Tortricidae.

Coelacanth: a species of fish that is the sole living representative of the crossopterygian fishes which were abundant in the Devonian and Cretaceous periods, 400–90 million years ago. This group of fishes is well known as fossils, and was thought to be extinct until a live specimen was caught in December 1938 off East London, South Africa. Since then many specimens have been caught around the Comoro Archipelago, northwest of Madagascar. It grows to 1·9 m. (6 ft.) in length, is dark blue in colour, and lives at depths of 150–400 m. (500–1,300 ft.) on near-vertical, underwater rock cliffs. The Coelacanth gives birth to well-developed young in litters of up to nineteen, and feeds on fishes.

Coelacanth

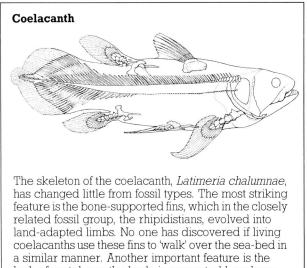

The skeleton of the coelacanth, *Latimeria chalumnae*, has changed little from fossil types. The most striking feature is the bone-supported fins, which in the closely related fossil group, the rhipidistians, evolved into land-adapted limbs. No one has discovered if living coelacanths use these fins to 'walk' over the sea-bed in a similar manner. Another important feature is the lack of vertebrae; the body is supported by a large, stiff rod of cartilage, called the notochord.

Coelom: the fluid-filled body-cavity in animals that separates the skin from the gut, allowing organs to develop independently; it also provides a circulating fluid. Animals which lack a coelom are small and usually simple, while coelomate animals (such as annelids, arthropods, echinoderms, and chordates) are larger and more specialized. Those lacking stiff skeletons use their coelom for hydrostatic support, especially when their body is segmented, as in annelid worms.

Coenzymes are relatively small, non-protein organic molecules that are essential for the correct functioning of some *enzymes. The coenzyme fits into the active site of the enzyme along with the substance upon which the enzyme acts. The coenzyme is chemically changed after use and is converted back to its original form by further enzyme-controlled reactions. Among the most important examples are *adenosine triphosphate (ATP) and coenzyme-A, which plays an important role in the *Krebs' cycle.

Coffee was first brewed from the roasted and ground seeds of the small evergreen tree *Coffea arabica* by the Arabs of Yemen. A native of Ethiopia, this species grows up to 5 m. (16 ft.). It now provides most of the world's coffee, the bulk of it being grown in South and Central America, particularly Brazil, although it is also an important crop in parts of East Africa. The white, scented flowers produce red berries containing two grey-green seeds. These are separated from the pulp, washed repeatedly, fermented for one or two days, and sun-dried for a week before the outer skin or parchment is removed to leave the well-known coffee bean. These processes require great skill and care to protect the quality of the bean, the full flavour and aroma being further enhanced by roasting. Along with bedstraws and gardenias the genus *Coffea* belongs to the Rubiaceae family, one of the largest flowering plant families with over 7,000 species.

Most of the coffee produced in Africa is from *C. canephora*, native to the rain forests of central Africa. This 'robusta' coffee is cheaper to produce and is being increasingly used as a blend with 'arabica'. Robusta is also used extensively in the manufacture of instant coffee. A

hardy, native species of Liberia, *C. liberica* produces an inferior coffee of bitter flavour that is the least important of the cultivated species.

Coke, Thomas William (1752-1842), English landowner, was the foremost exponent of British agricultural improvement in the eighteenth century. He inherited Holkham Hall with its large estate in Norfolk and here began scientific farming, transforming land which had 'but one blade of grass, and two rabbits fighting for it' into rich pasture. He also experimented in stockbreeding. At the age of eighty-five he was created Earl of Leicester.

Colds *Common colds.

Colitis is inflammation of the *colon resulting from infection by micro-organisms, some of which cause *dysentery. Symptoms include diarrhoea, often with the passage of blood and mucus, and pain in the lower abdomen. Similar symptoms, acute or chronic, are produced by ulcerative colitis, a disease of obscure origin which also affects the rectum. Forms of colitis include amoebic colitis, caused by the amoeba *Entamoeba histolytica*, and the more common infective colitis, which is caused by bacteria.

Colon (anatomy): the general name for the main part of the large *intestine in mammals. In man the small intestine, which leads from the stomach, enters it just above the appendix, in the right lower part of the abdomen. The colon passes up towards the liver, crosses towards the spleen, and then drops down into the rectum. It is concerned chiefly with the absorption of water in food, following digestion. In herbivores, bacteria break down cellulose in the colon and the resulting sugars are absorbed.

Colorado beetle: a species of leaf beetle, up to 1 cm. ($\frac{3}{8}$ in.) in length, which is oval in shape and has striped yellow and black wing-covers. It is a serious pest of potato crops across North America and Europe. The beetle probably fed originally on wild species of potato in the Andes, but spread to the Rockies and, in the twentieth century, invaded the potato-growing countries of the world. It is regularly accidentally introduced into Britain, and a great effort is made to prevent it from becoming established.

Colour blindness, in which certain colours cannot be distinguished, is most commonly due to an inherited condition. About 8 per cent of European males and about 0·5 per cent of females may be affected. Total colour blindness is very rare. Cones (colour-sensitive receptors) containing single visual pigments selective for red, green, and blue light are present in the normal human eye. Disturbances of colour vision will occur if the amount of pigment per cone is reduced or if one or more of the three cone systems are absent. Problems in distinguishing reds and greens are the most common.

Colubrid snakes form a family that includes about two-thirds of living *snake species. They form the main proportion of the snake species in all major regions of the world except Australia, where front-fanged (proteroglyphous) snakes predominate.

Although most colubrids are non-venomous, many possess enlarged (often grooved) teeth on the rear of the upper jaw to enable toxic secretions to be introduced into the body of a prey or an enemy. These forms are generally

Larvae and adult **Colorado beetles**, feeding on potato leaves, show boldly marked, warning coloration. Although notorious as the scourge of potato crops, this insect also feeds on other members of the potato family, including tobacco and deadly nightshade.

A Malayan flying lemur or **colugo** and its young, peering out from beneath the gliding membrane or patagium, which also serves as a soft, warm pouch. The young, like those of marsupials, are born in a poorly developed state and are carried by the mother until weaned.

not dangerous, but the African boomslang, *Dispholidus typus*, and the African twig snake, *Thelotornis kirtlandi*, are exceptions, and human deaths have resulted from their bites.

The family includes sand snakes, *Psammophis* species, whip snakes, *Coluber* species, tree snakes, *Chrysopelea* species, rat snakes, *Elaphe* species, and king snakes, *Lampropeltis* species.

Colugos, or flying lemurs, are insectivorous mammals. In spite of their common name, the two species, the Malayan colugo, *Cynocephalus variegatus*, and Philippine colugo, *C. volans*, are not lemurs and cannot fly. They can, however, glide by means of a 'wing' or patagium. This is a membrane between the fore- and hind-limbs, and including the toes and tail, that allows it to move from tree to tree, often gliding for some 70 m. (230 ft.). The Malayan colugo lives in dense tropical forests and is found not only in Malaya but also in southern China, Indonesia, and neighbouring islands. A nocturnal animal, it sleeps by day, hanging by its feet from a branch. It wakes at dusk to forage for food. A single young is born and clings to its mother's breast or belly. The Philippine colugo, as its name suggests, is restricted to the forests of the Philippines.

Columbine　*Aquilegia.

Coma is an imprecise medical term for any condition in which consciousness is lost and from which patients cannot be fully aroused. The common causes are head injury, drug overdose, diabetes, strokes, and infections. Many patients in coma recover more or less completely under intensive medical care, but if the brain is extensively damaged the coma becomes permanent and the patient may be described as 'brain-dead'.

Comb jellies　*Sea gooseberries.

Comfrey is the common name for several species of perennial herbaceous plant of the genus *Symphytum*, which is part of the forget-me-not family, the Boraginaceae. The common comfrey, *S. officinale*, has a fleshy, branching root-system and large, hairy leaves. The flowering stem, 30–120 cm. (12–48 in.) high, bears drooping clusters of creamy-yellow or purplish blooms. It is native to temperate Asia and Europe. The leaves and roots have long been considered to have medicinal properties.

Commensalism means 'eating at the same table' and is used in biology to refer to an association between organisms of different species in which one (the commensal) gains from the relationship, and the other (the host) neither benefits nor loses. A well-known example involves the hermit crab, whose adopted shell frequently has, attached to it, colonies of *hydra-like hydroids. The hydroid gains food particles and a firm yet mobile anchorage from the crab. The crab receives nothing from the association, as far as we know. An association in which both partners benefit is called *symbiosis.

Common colds are infections by one of a large number of related *viruses that cause inflammation of the mucous membranes of nasal passages and throat. The resulting illness is mild, but bacterial infection of nasal sinuses or bronchi may follow. Colds are transmitted through viral particles in the air after sneezing or coughing by the sufferer. There is at present no effective treatment of the viral infection, but bacterial complications respond to *antibiotics.

Community (ecology): a natural assemblage of plants and animals, living within a particular area or *habitat. Although a community may comprise organisms in several habitats, the term usually refers to a specific habitat such as a woodland or grassland site. The organisms are usually interdependent, either directly, through *food chains and webs, or indirectly, as one species modifies the environment to the advantage or detriment of others. In most communities, relatively few species are present in large numbers; rather, many species occur in small numbers. The community may be named after the predominant species, as in an oakwood community or a reed-bed community, and it may be either stable, persisting over many years, or changing, through *succession.

Complexes (psychoanalysis) are ideas or beliefs which are often repressed and largely subconscious, yet influence our behaviour. They may arise either unconsciously, like the Oedipus complex from conflicts in childhood, or consciously, like the inferiority complex.

Compost is a natural product of the *decomposition of dead plant material. The most familiar compost occurring naturally is that present in mature, deciduous woodland as leaf-mould on the forest floor. Compost is used by gardeners to provide a supply of *humus and nutrients for the garden and can be produced in quantity from vegetable peelings, grass cuttings, and dead leaves. The garden compost heap also requires oxygen, and the correct decomposing micro-organisms. The process of making garden compost mimics the natural breakdown of all dead plant material, and the activity of the decomposers produces an appreciable quantity of heat.

Concussion is temporary impairment or loss of consciousness following a blow on the head. Recovery is usually complete within a few minutes, or at the most a few hours. The mechanism of concussion is not fully understood, but it is suspected that sudden acceleration of the head can affect the brainstem, where consciousness is controlled. Although a full recovery is normally made, concussion can produce permanent brain damage.

Condors are two species of American *vultures. Andean condors, *Vultur gryphus*, with wings spanning almost 3 m. (10 ft.), occur mainly high up in the western mountains of South America, generally in the tropics, but are also widely distributed along the deserts and wooded cordilleras further south. Californian condors, *Gymnogyps californianus*, which are slightly smaller, are found only in inland California, where their tiny population is protected.

Coneflowers are the thirty or so species of the plant genus *Rudbeckia*, which is native to North America. They are perennial, herbaceous members of the sunflower family, with showy flowers consisting of a ring of yellow ray-petals, surrounding a vertical cone of brown-to-purple tube-shaped florets.

Conger eels are marine fishes which grow to a considerable size, up to 2·75 m. (9 ft.) in length. The European conger eel, *Conger conger*, is abundant on underwater wrecks and reefs, but also lives in deep shore-pools and in crevices in harbour walls. It feeds mainly on bottom-living crustaceans, octopuses, and fishes. A similar species, *C. oceanus*, lives on the Atlantic coast of North America, and about twenty other species occur in the Atlantic and Pacific oceans. Conger eels spawn in deep water, and their young, called leptocephalus larvae, are transparent and flattened, and drift towards the shore, where they change into tiny eels.

Congo eels are amphibians which make up the smallest of the *salamander families, comprising only three, mainly aquatic, species, restricted to the southeastern USA. They are eel-like, although they have tiny, useless fore- and

The Andean **condor** is the heaviest flying bird, weighing up to 12 kg. (26 lb. 8 oz.). It ranges from western Venezuela to as far south as Tierra del Fuego, and feeds mainly on carrion, detected by sight from great heights.

hind-limbs. Their larvae have external gills and meta-morphose into adults that retain a pair of gill slits. Up to 200 eggs are laid by a single female, in a shallow muddy depression, and are guarded by her.

Conies *Hyraxes, *Pikas.

Conifers are trees which comprise the majority of living *gymnosperms and are characterized by their small pollen-bearing male cones and larger seed-bearing female cones. Most are evergreen, the needle-like leaves living up to ten years, though a few species, such as the larch, are deciduous. Conifers are the dominant trees over large areas of the cool temperate zones, but a few species, such as the pines *Araucaria* and *Agathis*, are native to the tropics. The tallest of any living species of plant, conifers are the most widely used of plantation trees, being of enormous importance in the timber and paper industries.

Conjunctivitis is inflammation of the membrane cover-ing the white of the eye and lining the inside of the lids. It may be caused by infection, or by exposure to irritant chemicals or ultraviolet light. It is not usually a serious condition, but trachoma, a specific chronic form, may damage the cornea and cause blindness.

Consciousness is the totality of a person's waking ex-perience, thought, and feeling. It is regarded as the de-fining characteristic of the human mind and thus the highest functional property of the human mind. It is lost in sleep, anaesthesia, coma, and death.

Clouding or alteration of consciousness is produced by some brain diseases such as epilepsy, by drug intoxications, and by hypnosis. Patients are then seen to be not fully accessible to either the outside world or their own usual internal reactions.

Conservation of nature is the management of natural resources to avoid destruction of species and habitats. It involves maintaining and protecting habitats, controlling the harvesting of natural populations, and seeking to re-duce pollution and other man-made threats to other or-ganisms. Conservation is important because man himself forms part of the living world, and damage to the stability and resilience of *ecosystems will ultimately damage man's well-being.

Clearance of forests to create agricultural land has dam-aged many habitats and, in South America particularly, it has promoted erosion and loss of condition of the soil. Moreover, the haphazard destruction of the flora and fauna of tropical rain forest has led to the extinction of many organisms potentially useful to man as medicines or as food, or for crop breeding. Some governments provide protected areas, such as wildlife parks and nature reserves, and impose bans on hunting endangered species. Many regulate their fishing industries, but several major fisheries have collapsed in recent years through overfishing. The reintroduction of captive-bred species may be used to re-build a wild population, as in the case of the *Hawaiian goose. Zoos can be important: the Arabian oryx is now thought to survive only in captivity.

Constipation is a less frequent movement of the bowels than is normal. Only if it is of recent onset is it of medical significance, and it may then be a symptom of intestinal disease. The usual treatment is adjustment of the diet to include more fibre or, less desirably, the administration of laxatives.

Contagion *Infectious diseases.

Contraception is the use of physical or chemical means to prevent unwanted pregnancy. Many parents wish to plan the timing and number of children in a family, and most countries now aim to keep population growth in proportion to available resources. Various methods of con-traception are effective and readily available.

Barrier methods work by preventing sperm reaching the cervix or neck of the *uterus. The sheath or condom, worn by the man, is the most popular method of con-traception; the equivalent method in women is the dia-phragm or cap. Oral contraceptives containing steroid hormones are very effective in preventing pregnancy. The main type of pill contains a combination of oestrogen and progestogen, which prevents the normal release of the egg. The 'mini-pill' contains only progestogen and alters the receptivity of the cervical mucus and uterine lining. Intra-uterine devices are made of plastic, often with metal, and are inserted into the uterus. They probably prevent the fertilized egg from attaching itself to the wall of the uterus, and are not quite as effective as the combined pill. The rhythm method is successful for those couples who can identify the fertile period around ovulation, and during it abstain from intercourse. Permanent contraception can be provided by sterilization.

Convolvulus *Bindweed.

Coolabah *Eucalyptus.

Cooper, Sir Astley Paston (1768–1841), English anat-omist, became the most famous surgeon of his day. He was an indefatigable dissector, using the resurrectionists for the supply of exhumed bodies, and insisted on post-mortem examinations. One of his most celebrated oper-ations was tying the aorta for *aneurysm.

Coots are a widespread group of *rails with lobed toes. The Old World and Australian common coot, *Fulica atra*, is a stocky black bird, larger than a moorhen, and with a white bill and forehead. Flocks of coots are often seen in winter with ducks, swimming and diving on open water. There are nine species of coot within the genus *Fulica*; with the exception of *F. atra*, most are South American.

Copperhead: a stout-bodied venomous snake of the *pit-viper group that occurs in the eastern and southern USA. Its common name derives from its head coloration which is often coppery-red; the colour of the body is pink-ish or coppery-orange with darker cross-bands. Maximum body length is about 1·4 m. (4 ft. 6 in.). Its bite is painful but rarely fatal to man. The unrelated Australian cop-perhead of southeastern Australia, Tasmania, and some Bass Strait islands, is also fairly stout-bodied; but the color-ation ranges from reddish-brown to greyish or black and the maximum length is 1·7 m. (5 ft. 7 in.). It is often found in marshy areas, particularly in thick grass and piles of stones. It is dangerously venomous but human casualties are uncommon.

Coppers (butterflies) typically have metallic, coppery-coloured wings, and together with *blues and *hairstreaks

A purple-shot **copper** butterfly, *Heodes alciphron*, basking on bramble buds. This attractive species ranges from western Europe to Iran. Adults are found in open countryside with flowery banks and meadows. The caterpillars feed on docks, *Rumex* species.

belong to the family Lycaenidae. Most of the 500 or so species of coppers are north temperate in distribution. Often only the male is truly copper-coloured, and it may have a purple, metallic sheen in addition. Like other Lycaenid butterflies, coppers have a quick, darting flight and are regular visitors to flowers.

Copra is produced from the white, oil-rich (60–65 per cent) kernel of the coconut palm. Cup copra is produced when freshly harvested nuts are split open and either sun- or kiln-dried. Coconut oil, used in the manufacture of soaps, cooking fats, and margarine is extracted from this copra. The major producers are the Philippines and Indonesia; lesser amounts come from India, Sri Lanka, and Mexico. Ball copra is produced when unhusked nuts are stored in the shade for up to twelve months.

Coral fossils belong to three main marine groups. The rugose corals (so called because of their wrinkled look) were the first to appear, during the Cambrian Period (570–505 million years ago). They lived either singly or in colonies and the fossilized skeletons of the solitary ones often have a distinctive horn shape. The tabulate corals arose in the Ordovician Period, 500 million years ago. These were colonial species, forming loosely branching or compact bun-shaped masses. Both corals contributed to Palaeozoic reef limestones and both became extinct at the end of the Palaeozoic Era, 245 million years ago. They were followed by the scleractinian, or stony, corals which still survive today and are major limestone-reef builders.

Corals are tiny animals which belong to the same class (Anthozoa) as *sea anemones and *sea fans, all part of the phylum Coelentrata. Every individual secretes its own supporting skeleton, and in many species these join together as a co-ordinated structure. Tiny vulnerable animals thereby become massive colonies. There are two main groups of living corals: the stony corals, which may be solitary or colonial, and the octocorallian corals, which are always colonial.

Most stony corals have an external calcium-carbonate skeleton, deposited by photosynthesizing microorganisms. This dependence restricts them to depths to which light can penetrate. These corals form most of the large tropical reefs, their surfaces pitted with small cup-shaped cavities where the separate *polyps live and feed, sharing nutrients through lateral cellular connections. The colony produces new polyps by budding, the specific pattern of budding determining whether the coral becomes branched (as in staghorn coral) or a massive and furrowed dome (as in brain coral). A few stony corals are solitary, especially in cold or deep waters. Octocorallian corals have a skeleton laid down within their own bodies, so the surface is living tissue; they include the soft corals and gorgonians. Such types always have eight feathery tentacles per polyp.

Corals feed on small fish and invertebrate plankton, using stinging-cells on outstretched tentacles, often with mucous nets, to filter finer particles. As well as growing by budding, they also reproduce sexually, producing tiny planktonic larvae, which disperse and colonize new areas.

Coral snakes are brightly coloured snakes that may belong to one of several families in Asia, Australia, southern Africa, and the Americas. They are poisonous, and have *cobra-style fangs. Their striking colours (often combinations of red, yellow, and black) apparently serve as a warning to would-be predators. True coral snakes hunt at night and are inactive during the daytime; they normally eat lizards and other snakes. In some areas, harmless snakes appear to gain protection by mimicking the patterns of the venomous species. An example of this is the milk snake, or false coral snake, *Lampropeltis doliata*, which is really a species of king snake, native to southeast North America.

Cork (botanical) is a waterproof layer produced on the outer surface of young shoots of trees before the formation of bark. The cork of commerce, however, is derived from the bark of the cork oak, *Quercus suber*, of southern Europe, notably Portugal, and North Africa.

Cormorants are slender-billed coastal and riverine birds, usually dark brown or black. The thirty-two species of the family occur worldwide from the tropics to both polar regions. They are foot-swimmers and strong divers, with webbed feet and stout legs, and often hunt for fish, shrimps, and other aquatic food in small or large groups. Before diving they wet their plumage thoroughly, and after hunting they fly to a rock or branch to dry out, extending their wings in a characteristic pose. They breed colonially in untidy twig, seaweed, and guano nests, laying two to four eggs.

The green cormorant, *Phalacrocorax aristotelis*, is also known as the shag.

Corms are underground storage organs which carry plants through adverse growing conditions, such as winter or drought. They are derived from flattened *stems, bearing lateral buds, or small corms, and scale-like leaves. They occur in plants such as *Gladiolus* and *Crocus* and often last just one year. The flowering stem is produced from an apical bud with the exhaustion of the old corm which, in turn, is replaced by one or more of the lateral ones. There is no sharp distinction to be drawn between corms and *rhizomes or stem *tubers.

Corn is a term used, as in peppercorn, to describe a grain or seed of a plant, particularly cereals. Alternatively, the growing cereal plant or crop can be referred to as 'corn'. In Britain, for example, all cereal crops are referred to loosely as 'corn'. In the USA it is a term used specifically to describe maize and hence is also used for its products such as corn syrup, corn whiskey, and cornflakes.

Corn buntings *Buntings.

Corncrake *Rails.

Cornflowers are plants of the genus *Centaurea*, which also includes other members of the sunflower family, such as knapweeds and star thistles. The wild cornflower, *C. cyanus*, which is native to Europe, has deep blue flowers and is the parent of the many coloured garden varieties.

A **coral snake**, *Micrurus nigrocinctus*, from the rain forest of Costa Rica, showing its bright warning coloration. Here, as an additional defence, the snake shows tail-wriggling behaviour to direct attention away from the vulnerable head.

Corns are painful, pea-sized lumps in the skin of the feet. They grow over toe joints and between toes, where friction causes thickening of the horny layer of the skin into an inverted pyramid that presses down. Treatment consists of softening agents, or attention by a chiropodist.

Corn salad, or lamb's lettuce, is an annual plant belonging to the valerian family, and with leaves like a forget-me-not. It is native to Europe, and is found in arable fields, on hedgebanks, and on roadsides. It is cultivated for its leaves, which are used in salads.

Coronary thrombosis is the blocking by a blood clot of an artery supplying the heart. The affected artery is nearly always narrowed at the site of the blockage by fat deposits. The usual outcome is a heart attack, resulting in the destruction of a part of the heart muscle. This may cause death or serious disability, or may, if small, heal without leaving the patient any the worse.

The African white-necked **cormorant** is a race of the common European species, *Phalacrocorax carbo*. This individual from Lake Nakuru, Kenya, is shown in typical wing-drying pose.

Cotingas are birds of the family Cotingidae, comprising about eighty species which are mostly native to rain forest in South and Central America. The group includes the bell-bird, cock-of-the-rock, and umbrella birds. Although one or two species are only sparrow-sized, most are thrush-to crow-sized. The males of some species are dull brown or green, but most are strikingly coloured—red, purple, and blue being common colours; they often have bare patches of skin on the face. In many species the males are polygamous, displaying at leks and taking no part in raising the young. The main diet of cotingas is fruit. Most build very flimsy inconspicuous nests and lay only one or two eggs. Some species are known only in a few local areas, and are threatened by forest clearance.

Cotoneasters are deciduous and evergreen shrubs and small trees, and include a few climbers and scramblers. They are members of the rose family, like thorns and rowans, and form a genus with some fifty or so species. They are native to mountainous areas of the Old World temperate zone. Many are cultivated as ground-cover and wall plants, notably *Cotoneaster horizontalis* of Japan. Their seeds are dispersed by birds, attracted by the red or black colour of the fruits, which they eat.

Cotton is a plant of the genus *Gossypium*, a relation of the mallow and hollyhock. This rather complex genus of annual and perennial plants is grown throughout the world in subtropical regions for the valuable fibre it produces. The United States is the world's leading producer, followed by the USSR, China, and India.

Cotton plants on a plantation in Madagascar show it to be a woody shrub which reaches an average height of 2 m. (6 ft. 6 in.). Grown as an annual, it needs warm weather for germination, ample sunshine for growth and fruiting, and dry weather at harvest time.

The fruits are capsules known as bolls that split open when ripe, to reveal a mass of white fibres. This is largely 'lint', and consists of cellulose hairs, up to 50 mm. (2 in.) long, growing from the seed-coat. As the hairs dry, they twist, enabling a fine, strong thread to be spun. Usually the bolls are picked by hand and the lint is torn or pulled from the seed-coat by machines known as gins; mechanical harvesters are common only in the USA. A fuzz of hairs, too short to be spun, is also extracted and is used for cotton wool. A cooking or salad oil is extracted from the seed, and the protein-rich seedcake is a valuable livestock food and can even be used as a nitrogenous fertilizer.

Cotton spinners *Sea cucumbers.

Cotton stainers *Red bugs.

Cotyledons are the first leaves to appear as a plant seed germinates. Dicotyledons are plants that have a pair of leaves wrapped round the embryo within the seed, mono-cotyledons have one.

Coucals are birds of the genus *Centropus*, which belongs to the cuckoo family. They are found in parts of Africa, India, Southeast Asia, some tropical islands, and Australia. Most of the nine species are brownish or black and are long-tailed and short-winged; they are weak fliers. Unlike many other members of the family, coucals do not parasitize other birds, but incubate their eggs and raise their own young. They build untidy, domed nests with a side entrance, usually fairly low down in thick vegetation. They eat insects and other small animals.

Couch grass is a particularly troublesome weed of cultivated ground. It is also known by several other names, such as scutch or twitch-grass, all applying to the species *Agropyron repens*. It spreads rapidly by means of underground stems or rhizomes. Cultivation techniques, particularly with mechanical equipment, tend to divide and propagate the plant. As a wild grass species, it is widely distributed throughout Europe, northern Africa, Siberia, and North America. Other members of the genus *Agropyron* have couch as part of their common name.

Coué, Emile (1857–1926), French psychologist, advocated the power of auto-suggestion, especially as a cure for disease. He was famous for his theme, 'Day by day and in every way I am getting better and better.'

Cougar *Puma.

Courgettes *Marrows.

Coursers are birds closely related to the sandpipers and plovers. They occur in the Old World, where they live in open, dryish country or deserts. The best-known of the seven species is the cream-coloured courser, *Cursorius cursor*, of semi-desert in North Africa and the Middle East. This species is about 20 cm. (8 in.) long, and pale sandy coloured, with a greyish crown and striking black and white stripes behind the eye. Coursers feed mainly on insects and lay their eggs (usually two) in a scrape on bare ground.

Courtship is the behaviour which, in many types of animal, helps bring together opposite sexes of the same

Courtship

Courtship sequence of the European smooth newt, *Triturus vulgaris*

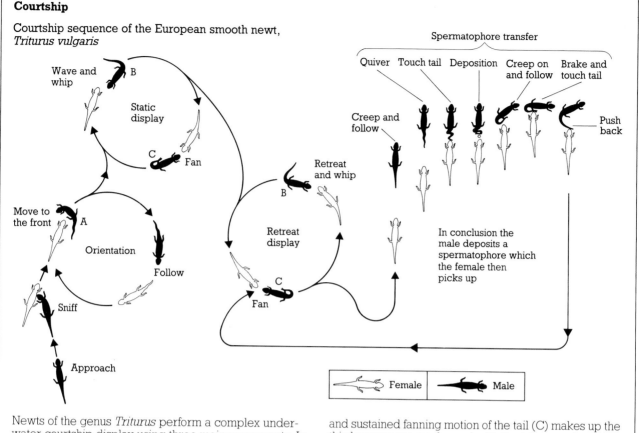

Newts of the genus *Triturus* perform a complex underwater courtship display using three main movements. In the first movement the male gently waves his tail and aligns himself to show off his full breeding coloration (A). This is followed by a whipping movement produced by the tail being lashed against his body (B), sending a powerful current of water towards the female. A delicate and sustained fanning motion of the tail (C) makes up the third movement, performed with the tail curved against the flank and vibrating the tip. This sends a stream of water containing a chemical secretion from the male's body towards the female. The female must respond to all stages of the display in order for courtship to be successful.

species at the right time and place, and in a suitable position for successful mating. Sound, scent, touch, and *display may all be used in courtship, and the pattern and combination of signals is unique to each species. This usually prevents individuals of different species from trying to mate.

Many animals spend most of their lives avoiding or driving off others of the same species. Courtship, much of which is controlled by instinct, helps overcome these solitary and aggressive tendencies, so that mating can take place without danger to the participants. Some aspects of courtship help the individual to select a good mate. The male kingfisher presents a fish to the female during courtship, thus demonstrating that he would be capable of helping to feed the young. Gifts may sometimes have another purpose. In some species of spider the male presents a fly to the female to keep her busy during copulation so that he is not himself eaten.

Cowbirds are seven species of birds in the family Icteridae, which also includes the New World orioles or blackbirds. They vary in size from that of a thrush to that of a small crow. Virtually all come from subtropical and tropical America. Some of the species are parasitic, laying their eggs in the nests of other birds. The brown-headed cowbird, *Molothrus ater*, parasitizes a wide range of small birds in North America, while the bay-winged cowbird,

Molothrus badius, steals nests built by other species but raises its young itself.

Cowries are beautifully shaped and often brightly coloured gastropod *molluscs. The regular outer shell conceals an inner shell which is spiralled just like that of related snails. The opening to the shell is toothed, and

A tiger **cowrie**, *Cypraea tigris*, browses on seaweed in seas around Mauritius. The mantle, a two-lobed muscular flap, partially covers the shell. When fully extended, the two lobes meet midway to form a fleshy covering.

from this the soft body is extended in a fold over the top of the shell, covering most of the shell in the living animal. Cowries feed upon colonial marine invertebrates such as bryozoans and sea squirts. Large tropical species may be spectacularly marked with spots and banding, and some are common enough for the shells to be used as currency in certain island communities.

Cows   *Cattle.

Cowslip   *Primrose.

Coyote: a species of wild dog confined to North America. It is very similar to a wolf in appearance but considerably smaller, being about 60 cm. (24 in.) at the shoulder. Its distribution range overlaps that of wolves, but it can be distinguished from a wolf by the downward carriage of the tail when running, by the differing skull dimensions, and from the fact that the coyote does not form packs. It feeds on small mammals, birds, frogs, and fish as well as carrion, but its favourite prey is rabbits and hares. Although not a serious pest, it is known as a sheep-killer in certain regions. Alterations of the habitat and the near extermination of wolves have led to a considerable extension of the coyote's range. In the eastern USA, it has hybridized with domestic dogs.

Coypu: a South American species of large *rodent. Coypus look like huge rats but are related to guinea-pigs. They are aquatic but feed on waterside vegetation. Their soft, dense underfur, known as nutria, is of commercial value, and they were once farmed in Britain, America, and elsewhere. Many escapes occurred and they are now widely established as a feral species. They have become a pest, not only because they eat root crops but also because they cause flooding by burrowing into the banks of streams and drainage ditches.

Crab apples   *Apples.

Crabs, probably the best-known *crustaceans, are found throughout the shallow seas and tidal zone. They share with other large crustaceans, or decapods, such as prawns, crayfish, and lobsters, the possession of ten main walking-legs, the first pair of which are enlarged as pincers. Three front pairs of limbs are modified as mouthparts, giving a complex array of feeding structures. Crabs can be recognized by the large carapace (shell), which is expanded laterally to cover the respiratory gills; the remaining abdominal segments are tucked underneath this, giving the crab its squat, oval shape. Most are also characterized by sideways locomotion, some ghost crabs achieving 6·5 km./hour (4 m.p.h.) over sand.

Crabs range from a few millimetres (tenths of an inch) long in species such as pea crabs, *Dissodactylus*, which live inside bivalves' shells, to the dimensions of spider crabs, *Macrocheira kaempferi*, the largest of all *arthropods, which has a span across the legs of 4 m. (13 ft.). Crabs live in oceans, coastal waters, and rivers, and a few have been successful on land. Most are bottom-dwelling, though a few can swim powerfully. Many have curious and fascinating habits: the decorator crabs, such as *Oregonia gracilis*, deliberately plant other organisms on their own shells; the *hermit crabs inhabit other animals' cast-off shells; some land crabs climb trees, and the 'cleaner-crabs' offer a clean-up service to fish. All crabs grow by casting,

or shedding, their old *exoskeleton to reveal a larger but temporarily soft new 'skin' beneath.

Cranberries are the fruits of evergreen species of the genus *Vaccinium*, part of the heather family. They are native to North America, Europe, and northern Asia, and are closely related to bilberries and blueberries. The low, creeping shrubs bear small, bright red berries, about 1 cm. ($\frac{1}{2}$ in.) across, which are so acid that they need to be cooked before being eaten.

Crane flies, or daddy-long-legs, are slender-bodied true *flies with very long legs and an elongate body. Most have simple thread-like antennae and a conspicuous pair of halteres (drumstick-like balancing organs which replace the hind-wings of true flies). The elongate, cylindrical larvae, some of which are aquatic or found in damp wood, also include destructive leather-jackets, which eat the roots and lower stems of a variety of plants, especially grasses. The adults feed on nectar, if at all, and despite the sting-like appearance of the female's ovipositor, are harmless.

Cranes are ground-living birds, looking rather like herons but related to rails and bustards. Cranes make up a family of some fourteen species with representatives in most parts of the world; they include demoiselle and Stanley cranes. The largest are 1·5 m. (5 ft.) tall, and they all have long necks and legs but short beaks. They are mainly grey or white, with elongated wing-feathers trailing over their 'tails', and are well known for their resonant calls; the call of the common crane of Europe, *Grus grus*, is a trumpeting 'grooh'. They perform ceremonial dances, and possibly pair for life. Many species migrate in large flocks, flying in lines or wedge formations.

Cranesbills are species of the genus *Geranium*, a large group of perennial or occasionally annual plants, native to the temperate areas of the world. The leaves of most species are deeply lobed or dissected, and the flowers, which may be white, blue, pink, or red, are often large and attractive. When mature, the sections of the fruit split explosively, expelling the carpels and seeds. They are part of the family Geraniaceae, which has some 750 species distributed throughout temperate and subtropical regions.

Crawfish, also known as spiny lobsters or rock lobsters, are deep-water marine *crustaceans, rarely found inshore. They lack lobster-like pincers, and defend themselves instead with a very heavy and spiny carapace and by flailing sharp-edged antennae. Crawfish are caught for food in some countries.

Crayfish live in fresh, flowing water, especially streams in chalk and limestone areas. Related to, and closely resembling, the *lobsters, they have similar habits. They are mostly nocturnal, feeding on snails, tadpoles, and insect larvae, and generally hide beneath stones during the day. The female, which may live for twenty years, carries the eggs until they hatch in spring, and the youngsters cling on to her for some time while undergoing early moults. Most crayfish stay rather small, but some Australian species can reach 50 cm. (20 in.) long.

Creeper is a name used for several unrelated groups of arboreal birds. The New Zealand creeper, *Finschia no-*

vaeseelandiae, a member of the family of Australian *warblers, is a small brown bird 15 cm. (6 in.) long, which lives in the forests of the South Island, New Zealand. The Philippine family of creepers contains two species in the genus *Rhaldornis*, while the Hawaiian creeper, *Loxops maculata*, is a member of the family of Hawaiian honeycreepers. Other birds with 'creeper' as part of their name include the *treecreepers and *woodcreepers.

Cress is a salad vegetable eaten in its seedling stage. If allowed to grow it develops into a short, strong-smelling annual with tiny white flowers. Cress, *Lepidium sativum*, originated in Asia and belongs to the *mustard family.

Crested swifts are birds allied to the swifts, and include three species occurring in Southeast Asia and some Pacific Islands. They are long-winged, short-tailed, and greyish to bronzy green, and they catch insects in flight. They build extremely small nests of bits of bark or moss, held together with saliva, and usually positioned on the side of small branches of trees. The female lays a single white egg. The nest is too frail to support her weight, so she perches on the branch while incubating.

Crèvecoeur, J. Hector St. John de (1735–1813), French-born immigrant to America, settled in New York State and in *Letters from an American Farmer* (1732) vividly portrayed life on the soil. He introduced the system of

A Nile **crocodile** hatches after a three-month incubation period. Before hatching the young emit high-pitched cries which stimulate the mother to dig open the nest and expose the eggs. She then carries the young to the water and protects them for up to two months. Despite the high degree of parental care, a large proportion of young crocodiles are taken by predators such as monitor lizards, marabou storks, goliath herons, and African crows, before they are even a year old.

clover crops and several new crop plants, including the leguminous *alfalfa, or lucerne, for fodder.

Crickets are black or brown, flattened insects belonging to the same order (Orthoptera) as grasshoppers. There are large ovipositors in the females and sound-producing organs on the front wings of the males. The house cricket, *Acheta domesticus*, which is pale brown, often lives with man in buildings, and sings mostly at night. Field crickets, such as the European *Gryllus campestris*, are black and live in burrows, the males singing by day at the mouths of the burrows. The related mole crickets, of the family Gryllotalpidae, are almost entirely subterranean and cannot jump. There are some 2,300 species of true cricket throughout the world, and some fifty species of mole cricket.

Crinoids *Sea-lilies.

Crocodiles are members of the archosaurian group of *reptiles, and are apparently more closely related to birds and extinct groups such as dinosaurs than they are to other reptiles. They have a number of adaptations for an aquatic life-style: their nostrils, for instance, are situated on top of the snout and can be closed by muscular action, and there are internal flaps in the throat which can be closed to prevent the lungs from flooding, when the mouth is open under water.

There are three main categories of living crocodilians: true crocodiles; *alligators (together with caymans); and the *gharial. True crocodiles comprise some thirteen species found in the tropical Americas, Africa, Asia, the East Indies, and Australasia. The largest living crocodilians belong to this group; they are the estuarine crocodiles, *Crocodylus porosus*, ranging from Asia to Australasia, of which there are reports of specimens up to 8·1 m. (30 ft.) long. At the other end of the scale is the Congo dwarf crocodile, *Osteolaemus* species, from West Africa which

reaches a maximum length of only 1·14 m. (3 ft. 9 in.). Some crocodiles exhibit parental care: the female Nile crocodile, *C. niloticus*, once its eggs have incubated, digs open its nest and carries the hatchlings to the water.

Crocodiles are carnivorous reptiles and attack and kill any animal smaller than themselves. Members of some of the larger species have been known to develop a taste for human flesh.

Crocuses are small, monocotyledonous plants with underground *corms and adventitious roots. There are about eighty species, native to central and southern Europe, North Africa, and western Asia. They belong to the iris family and often grow at high altitudes. Crocuses have marked cycles of growth and rest, corresponding to wet and dry seasons. The attractive flowers open only in sunlight, or on bright days. The many colourful garden varieties are derived from any one of several species of *Crocus*. From the stigmas of *C. sativus* the yellow culinary dye saffron is produced.

Crossbills are four species of bird of the genus *Loxia*, which in turn is part of the finch family. They are native to the coniferous forests of the Northern Hemisphere. They are the size of large sparrows; the males are red, the females olive-green. Their most distinctive feature is that the two halves of the bills, or beaks, are elongated and twisted, crossing over each other at the tips, hence their name. This shape is adapted for levering up and twisting over the scales of conifer cones in order to extract the seeds, which form their staple diet.

Cross-fertilization occurs when *gametes from two different organisms fuse (*fertilization). Most animals and some plants are dioecious, that is they have separate sexes. They can reproduce only by means of cross-fertilization, thus ensuring that *genes from different organisms mix, and creating genetic variation (or individuality). Monoecious organisms, which include the majority of plants, have both male and female organs on the same individual and usually make use of physiological or anatomical mechanisms to increase the chances of cross-fertilization.

Crown of thorns starfish are *echinoderms with particularly long spines, which give them their name. They can occur in very large numbers, up to fifteen adults per square metre (seventeen adults per square yard) in some parts of the Red Sea and off the Australian coast, where they seriously damage the reefs by eating *corals. These numbers result from population explosions that seem to have a natural rhythm, and occur every seventy years or so. Under natural conditions they may be self-regulating before irrevocable destruction results; though man's modification of the environment may affect this balance.

Crows are birds of the crow family, Corvidae, a large family of about 103 species with a worldwide distribution; it includes many familiar birds, such as jays, rooks, jackdaws, magpies, ravens, choughs, and nutcrackers. They are medium to large birds, 30–40 cm. (12–16 in.) long; a few, such as the magpies, are much larger than this because they have very long tails. The Chinese red-billed blue magpie, *Urocissa erythrorhyncha*, is 70 cm. (28 in.) long, of which 40 cm. (16 in.) is tail.

More narrowly, the word crow refers to the thirty-nine species of the genus *Corvus*. Most of these are black,

with powerful black beaks, which are used for scavenging, catching small animals, and opening seeds. The best known species include the carrion crow, *C. corone*, the Indian house crow, *C. splendens*, and the North American crow, *C. brachyrhynchos*.

Crucian carp: a robust, rather deep-bodied fish, olive-green in colour, which, unlike the common carp, has no barbels around its mouth. Widely distributed in European lakes and the backwaters of slow-flowing rivers, it lives in heavily overgrown waters, can survive well where there is little oxygen, and can tolerate extreme cold. It feeds on plants and insect larvae, and spawns in May or June, the golden eggs adhering to plants.

Crustaceans are one of the three main classes of *arthropods (the others being insects and arachnids), and are particularly numerous in the sea. There are about 26,000 known species, which are distinguished from other arthropods by having calcified external skeletons and two pairs of antennae (insects have one pair; arachnids none).

Crustaceans have a three-part body: head, thorax, and abdomen, each made up of segments which bear a pair of limbs. The limbs have two parts, the upper part usually forming a gill. In primitive forms the limbs are all similar, and the lower part is used for swimming and simultaneous filter-feeding. The brine shrimps, fairy shrimps, and *Daphnia* (collectively called branchiopods) are of this type. Most of the familiar crustaceans, such as prawns, true shrimps, crabs, lobsters, and the like, are more advanced, and their legs are greatly modified. Some remain oar-like for swimming; others are specialized for crawling, grasping, jumping, sperm transfer, brooding eggs, or capturing and handling large food items.

Crustaceans are as ubiquitous in the sea as insects are on land, and some planktonic copepods, of which there are 7,500 species, may well be the single most abundant life-form, especially species such as *Calanus finmarchius*. Some crustaceans also live in estuaries and fresh water (where *amphipods and *Daphnia* are often dominant), and a few forms are amphibious (shore crabs) or fully terrestrial (coconut-crabs and woodlice).

Cryptic coloration *Camouflage in nature.

Cryptogams are plants and fungi that disperse themselves by microscopic *spores rather than seeds. They include fungi, algae, hornworts, liverworts, mosses, clubmosses, horsetails, and ferns. The term implies 'hidden sexuality' and was given to these plants before their sexual life cycles had been discovered; it is not much used by modern botanists. All cryptogamic plants have an alternation of generations. This means that the spore germinates to produce a plant (called the gametophyte) that bears the female and male structures which produce egg cells and swimming sperms respectively. After fertilization the egg cell in the female sexual structure (archegonium) develops into a new, different type of plant, called the sporophyte, which will bear the spores. In some groups of plants the gametophyte is the main stage. For example, it forms the green, leafy plant of mosses; the sporophyte is the spore-containing capsule and its stalk. But in ferns the gametophyte is small (called the prothallus) and the main plant is the sporophyte.

The fossil record shows that seed-bearing plants, or *phanerogams, evolved from cryptogams. Larger spores, destined to produce female gametophytes, were retained on the parent sporophyte, and smaller spores, containing the very early stages of the male gametophyte, came to be dispersed in the form of pollen grains. The ovary of a seed plant contains minute female gametophytes and its pollen minute male ones, the main plant body being a highly evolved sporophyte. Sporophytes are usually diploid (twice the basic number of chromosomes) and gametophytes haploid (having half the diploid number).

Cuchia: a species of freshwater, eel-like fish, widely distributed in southern and eastern India and Burma. It lives in swamps, ditches, and backwaters of rivers, and although these are often oxygen-deficient, it survives by breathing air into lung-like sacs in the roof of the gill chamber. Locally, it is an important food-fish which grows to 70 cm. (28 in.) in length. Despite its body shape, it is not an eel and belongs to a distinct order, often called swamp eels.

Cuckoo bees are small, yellow and black bees, which belong to the same superfamily as the true bees. They have no pollen baskets and some species resemble wasps in appearance. The females lay their eggs in the cells of other bees, where their larvae feed on the pollen stored for the host's larvae.

Cuckoos are widespread in both Old and New Worlds. Varying from sparrow-sized to crow-sized, they are mostly coloured grey, brown, and black, though some of the smaller ones are bronze-green. Although many of them build an ordinary cup-nest and raise their young themselves, some lay their eggs in the nests of other birds and allow the host species to raise the young cuckoos. Most famous of these is the European cuckoo, *Cuculus canorus*, whose females lay eggs of various colours. Those of each female closely resemble those of the particular species which they parasitize. The young cuckoo hatches before the eggs of the host and ejects them from the nest, so leaving itself as the only mouth to feed. Other groups in the family include anis, roadrunners, and coucals. The cuckoo family encompasses some 130 species.

Cuckoo-shrikes are birds which are neither shrikes nor cuckoos, and occur throughout the warmer areas of Africa, India, Southeast Asia, Australia, and many Pacific islands. They vary in size from that of a sparrow to that of a large thrush. Cuckoo-shrikes and trillers, which belong to the same family of seventy-two species, tend to be brownish, grey, or black, often with pale or white underparts. Minivets, also part of the same family, are usually more brightly coloured, with glossy blue-black upperparts and bright red or yellow underparts. The cuckoo-shrike feeds on insects and small fruits and builds a very small cup nest in twigs at the tip of a branch.

Cuckoo-spit   *Frog-hoppers.

Cuckoo wasps, or ruby tails, are *wasps of the family Chrysididae. They have a body of metallic red, blue, or green colours and a conspicuous ovipositor. The females lay their eggs in the cells of solitary wasps or bees, where the young feed on the larvae of the hosts and sometimes on food stored for them. They are a large family including over 1,000 species in the genus *Chrysis*. This includes the widespread European species, *C. ignita*.

A European **cuckoo** sits begging for food from its foster-parent, a reed warbler. Despite the young cuckoo's large size and the lack of competition from siblings, its bright red gape acts as a further stimulus for the host birds to gather more and more food; it is fed for seven to eight weeks before becoming fledged.

Cucujo: a species of *click beetle of the tropical New World, which is the brightest, most luminous animal known. The adults produce light from spots on the thorax as well as from the tip of the abdomen. The eggs and the larvae are also luminous.

Cucumbers belong to the pumpkin family and are most closely related to melons. Thought to be of Asian origin, they have been cultivated for centuries throughout Europe, being known to the ancient Greeks and Romans. The fruit of the plant is edible, usually eaten raw. A relative, the gherkin, *Cucumis anguria*, native to tropical and subtropical America, produces smaller, prickly fruits. Both

A portrait of Nicholas **Culpeper**, whose herbal treatments were popular amongst the poor of his time but were never taken up by the medical profession. He initiated the idea that distinctive marks on plants indicated their medicinal properties.

are trailing or climbing plants with large, angular, toothed leaves and separate male and female flowers.

Culpeper, Nicholas (1616–54), English physician and herbalist, wrote many books on medicine and astrology, including an English translation of the College of Physicians' Latin pharmacopoeia. He is chiefly remembered for his *English Physician* (1652), which was later revised and enlarged and became known as 'Culpeper's Herbal'.

Curassows are birds which form the family Cracidae, some of which are called guans. Related to other game-birds, curassows are large, 50–90 cm. (20–36 in.) long, somewhat turkey-like, birds. They are glossy black or brown, often with white areas on the body. Primarily South American, some occur in Central America, and one, the chachalaca, *Ortalis vetula*, is found in the southern United States. They build flimsy nests of twigs on the branches of trees and lay two or three white eggs. In appearance they resemble the *archaeopterix.

Curlews are large, streaky brown, buff, and white *wading birds with long, downcurved beaks. The eight species of their genus (*Numeniys*) include the whimbrel, *N. phaeopus*, and the Eskimo curlew, *N. borealis*, but this last species may be extinct. The curlew, *N. arquata*, has a loud, ringing call 'cour-lee' and a remarkable bubbling song.

Currants are hardy, deciduous shrubs belonging to the saxifrage family, especially *black currants and *red-currants. The term currant is also applied to the small, dried, black fruit of a dessert grape grown in Greece.

Currawong   *Bell magpies.

Cuscuses   *Phalangers.

Cushion star   *Starfish.

Custard apple   *Soursop.

Cuticle is a term for the outer layer of some animals and plants which acts to prevent water loss. In a plant it consists of a waxy layer which covers the stems and leaves. The cuticle of arthropods, especially insects, is a layer of waterproof material secreted by the underlying epidermal cells. In insects, it consists of multiple layers of a *carbohydrate (polysaccharide), structural material called chitin. This is set in a matrix of softer protein, rather like fibre glass and resin. A new layer is added daily to the insect cuticle and the whole is shed when the animal moults. The cuticle of other arthropods may be of a collagen-type material.

Cuttlefish are *cephalopods, with an internal shell (the cuttlebone) which provides buoyancy, lateral fins for gentle swimming, and jet propulsion for escaping. Their skin undergoes remarkably rapid colour and pattern changes, especially when they are mating. Eight arms and two longer tentacles are used for feeding, mainly on crustaceans. These prey can be hunted out by the cuttlefish squirting water at the sand which conceals them. Male cuttlefish (like other male cephalopods) have one arm modified as a sperm-transferring organ when mature.

Cutworms are species of *owlet moths of the family Noctuidae. They have destructive caterpillars, which feed by night on roots and shoots, and by day hide in soil or beneath stones. Many species, such as the dark sword-grass, or black cutworm, *Agrotis ipsilon*, which feed on crops and grasses, are serious pests.

Cuvier, G. L. C. F. Dagobert, Baron, (1769–1832), French naturalist, founded the science of palaeontology with a study of fossil elephants. Guided by the principle of correlation of parts (that organs are not separate entities, but related to their functions), he believed that findings of fossil feathers showed that there were very ancient animals which could fly. Pioneering also in the comparative anatomy of living animals, he developed an impressive system of zoological classification.

Cyanophyceae   *Blue-green algae.

Cycads make up a group of some 100 species of *gymnosperm. They look quite unlike conifers and resemble palms in their stout-stemmed habit and large, fern-like leaves. Present-day cycads are classified into three families, largely according to the vein pattern of the leaves: one includes about twenty species of *Cycas* from Madagascar and tropical Asia; the second is made up of a single species of *Stangeria* native to southern Africa; and the third family comprises some eighty species in eight genera native to tropical and warm, temperate zones. These are relics of a

A **cuttlefish** (*above*), *Sepia* species, 'glides' over a reef in the Philippines by gentle undulations of its lateral fins. The siphon, used for jet propulsion, is below the head. Cuttlefish hunt at night in shallow water and bury themselves in the sand during the day.

Eastern Cape **cycads** (*below*), *Encephalartos altensteinii*, one of some thirty species from South and tropical Africa. The remains of leaf bases cause the knobbled appearance of the trunk. Cycads are slow growing; some specimens are thought to be over 1,000 years old.

formerly much more widespread and numerous group, which reached its zenith 100–150 million years ago. The pith of some *Cycas* species is made into sago in Asia.

Cyclamens are plant species belonging to the primrose family. The sixteen known species within the genus *Cyclamen* are Mediterranean in origin and all bear attractive pinkish or white flowers with upturned petals. *C. persicum* is the species from which the large-flowered varieties grown as pot plants are derived. In Italy and southern France, the tuberous rootstocks or corms are eaten by pigs, hence the common name sowbread.

Cylinder snakes, or pipe snakes, are harmless, rather primitive, burrowing snakes which occur in tropical areas of Asia (nine species) and northern South America (one species). The New World form, *Anilius*, has very reduced eyes and is a *coral snake mimic, with a pattern of red and black bands. The Malaysian pipe snake has a bright red underside to its tail, which it displays when threatened.

Cypresses are evergreen conifers of the tree genera *Cupressus* and *Chamaecyparis*, though trees in other genera may be called cypresses too. For example, the tree known as swamp cypress (*Taxodium*) is a relative of the *redwoods. In *Cupressus*, found in temperate and subtropical regions of the world, the leaves are reduced to tiny scales pressed close to the shoots. The Monterey cypress, *Cupressus macrocarpa*, is a familiar example since it is widely planted for timber and shade. *Chamaecyparis*, found only

in North America and Japan, has larger leaves, and this genus includes Lawson's cypress, *C. lawsoniana*, which produces a useful timber. The Leyland cypress is a fast-growing hybrid between these two genera, and is widely used for hedging.

Cystic fibrosis is one of the commonest hereditary diseases. It alters the secreted products of the pancreas, sweat glands, and glands of the intestine. It can seriously affect the bronchial tubes of the lungs. The salt content of sweat is increased by this disorder, a symptom which aids diagnosis. Sufferers are vulnerable to chronic chest infections and may be severely disabled as a result.

The *gene controlling the expression of cystic fibrosis is of a type known as recessive. Many people may 'carry' it without knowing, and pass it on to their children. The disorder is expressed only when both parents are carriers, even then only in an average of one birth in four.

Cystitis is inflammation of the urinary *bladder, usually due to infection by the bacterium *Escherichia coli*. It creates an urge to pass urine more frequently than normal, usually with a burning sensation when doing so. Many anti-bacterial drugs pass through our body and are concentrated in the urine, which makes elimination of the infection easy.

Cysts are small sacs or closed cavities filled with a fluid. They may be caused by a blockage preventing secretion from a gland, or form part of a tumour. Parasitic cysts are part of the life cycle of certain worms.

Cytoplasm is the content of a cell, which surrounds the nucleus. It consists of a water-based liquid with dissolved salts, proteins, and other nutrients. Also included are larger particles and cell components, or organelles. These include *mitochondria, Golgi body, and other structures essential to the functioning of the cell.

Dab: a species of small *flat-fish, extremely common in shallow coastal waters around Europe at depths of 2–40 m. (6–130 ft.). It prefers to live on sandy bottoms, but is also found on mud, and in estuaries. Its upperside is warm brown in colour, and its scales have rough edges.

Dabchick is another name for the little grebe, *Tachybaptus ruficollis*, and a few other small *grebes. Little grebes are the commonest European species, distinguished by having a thicker neck than other species, coloured brown and chestnut, without the crest or ear-tufts that many grebes have in summer. They nest on ponds, lakes, and rivers.

Dace is a small, silvery, freshwater fish of wide distribution in Britain and Eurasia. Related to the *chub, it is more slender and has a concave anal fin. It lives in rivers with a moderately fast current, feeding on insects at the surface and in the water. It spawns in early spring on shallow, gravelly bottoms. The name dace is also used for several North American freshwater fishes of the genus *Lepidomedea*.

Daddy-long-legs   *Crane flies, *Harvestmen.

Daffodils are varieties of *Narcissus*, a genus of bulbous plants whose the petals are fused or united into a large tube or trumpet. They include the wild *N. pseudonarcissus* of the poet Wordsworth. Selective breeding, particularly since the late nineteenth century, has resulted in the introduction of many fine varieties with a range of colour and shape of flower. They belong to the same family as snowdrop and amaryllis. This contains some 1,100 species native to warm temperate, or subtropical regions.

Dahlias are tuberous-rooted perennials, chiefly of herbaceous habit. About twelve species are known, all of Mexican origin, and include the tree-like *Dahlia excelsa*, which will grow to a height of 6–7 m. (19–23 ft.). The development of the many decorative garden varieties began with the introduction to Europe of the first plants from Mexico in 1789. Today they are among the most popular of garden plants because of the variety of plant size and the colour range of their flowers. They belong to the sunflower family, along with plants such as daisies, chrysanthemums, and many other familiar species.

Daisies, of which there are a great many kinds, are sometimes identified by prefixes such as dog, ox-eye, Michaelmas, and Swan River. All are members of the *sunflower family, the Compositae, which includes many familiar garden plants such as asters, calendulas, pyrethrums, and ragworts. *Bellis perennis* is the daisy which occurs as a lawn weed in Europe. Its American counterpart is *B. integrifolia*.

Damselflies are predatory insects with two pairs of long, net-veined wings. Along with *dragon-flies they form the order Odonata, which includes some 5,000 species. Unlike

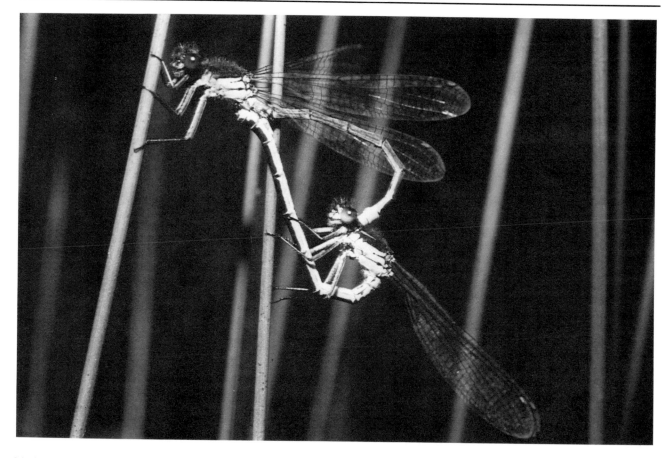

Mating red-eyed **damselflies**, *Erythromma najas*, a Eurasian species with a wing-span of 4·5 cm. (1½ in.). They are found on still water and frequently come to rest on water-lily leaves.

dragon-flies, they are slender, their flight is weak and fluttery, and the eyes are widely separated. They are always found near water, often perched on plants with the wings vertically together. They insert their eggs in the stems of waterplants and the aquatic larvae have three paddle-like external gills at the tip of their abdomen.

Damsons are small deciduous trees, somewhat spiny, closely related to the cherries. They are sometimes classified as a subspecies of the European plum, sometimes as a separate species. Native to western Asia, they bear oval, blue-black fruit about 2 cm. (1 in.) long.

Dandelions are yellow-flowered perennial plants, which belong to the genus *Taraxacum*, with fleshy thong-like roots and a white latex sap. The common name is said to be derived from the French, *dent de lion*, lion's tooth, in reference to the deeply toothed leaves. The species, though native to Europe and Asia, is now widely spread throughout the temperate regions, partly as a result of its use as a salad plant and because of its wind-borne seed distribution. Dandelions belong to a complex group of species with many intermediate varieties and hybrids. The genus *Taraxacum* is part of the sunflower family.

Daphne is a genus of some seventy species of Old World temperate or subtropical shrubs and small trees, many of them evergreen. They are part of the family Thymelaeaceae, which includes over 500 species, mainly native to Africa. Some species of *Daphne* are grown in gardens for their sweetly scented flowers, notably *D. mezereum* of Europe. The bark of *D. bholua* and other species from the Himalayas is used locally to make paper.

Daphnia, or water fleas, are small, freshwater *crustaceans found worldwide, which form the suborder Cladocera. They are enclosed in a transparent carapace, with a beaked head, long antennae, prominent eyes, and a terminal spine. The antennae beat to propel the animal jerkily upwards and act as parachutes during descent. The legs filter-feed within the carapace. *Daphnia* brood their eggs, producing only females by parthenogenesis in summer, then in autumn producing male and female offspring which mate to produce eggs which will survive the winter.

Darters are a family of four species of water-birds, found in Africa, India, Australia, and the Americas. They resemble cormorants and are alternatively called anhingas, water turkeys, or snakebirds. They all belong to the genus *Anhinga*. The movement of the narrow head and thin kinked neck, which are often all that is exposed above water, is strongly suggestive of a snake.

Darters (fish) are members of the *perch family and comprise some 120 species of North American freshwater fishes. They are mostly small, up to 8 cm. (3 in.) in length, and rather slender, with two dorsal fins, the first of which is spiny. They have adapted to a wide range of habitats. Males become brightly coloured in the breeding season.

Darwin, Charles Robert (1809–82), English scientist, revolutionized biological thinking at the end of the nineteenth century. At the age of twenty-three he joined the

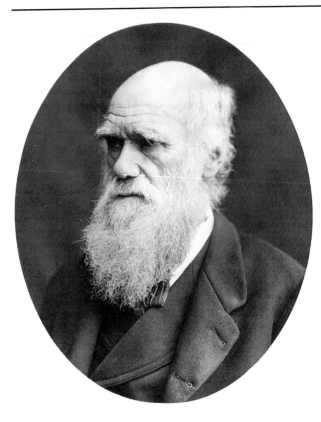

A portrait of Charles **Darwin** taken in about 1877. Apart from his famous work on evolution, he wrote numerous treatises on his travels and on the plant and animal species he encountered, including an account of orchids and their pollination by insects.

naval survey ship *Beagle*, as geologist and naturalist, for a five-year voyage round the world. In South America he studied the plants, animals, and geology of Patagonia before travelling up the west coast to make a detailed study of the geology of the Andes, and propounding a theory of their origin. His observations on coral islands led to his theory of reef formation. A mass of data and specimens of plants, animals, and fossils collected from islands such as the Galapagos strongly influenced his future thinking.

Darwin continually compared living animals with fossil finds, and also compared the differences in species which occurred from place to place. The ultimate result of his experiences, and of his practical knowledge of variation and breeding gained from his own garden plants and animals, was his now famous book, *The Origin of Species by Means of Natural Selection, or the Preservation of Favoured Races in the Struggle for Life* (1859), usually referred to simply as *Origin of Species*. It set out to show that organisms tend to produce offspring varying slightly from their parents, that the process of *natural selection tends to favour the survival of those best adapted to their environment. The book caused great and lasting controversy, which was particularly heated on the question of human evolution. His other classic works include *The Formation of Vegetable Mould through the action of Worms, with observations on their habits* (1881) and a monograph on barnacles.

Date palms are tall trees, growing up to 30 m. (98 ft.) with a high crown of large leaves typical of many members of the palm family. The yellowish to reddish-brown fruits of the date palm are either harvested unripe (soft dates) or allowed to dry (dried dates). Date palms flourish in the hot, dry regions of Arabia, the Middle East, and North Africa. A vital food plant in these regions, the fruits contain up to 70 per cent sugar. All parts of the tree are useful: the leaves and stems are used in house construction, leaf fibres are used for mats, baskets, saddles, and ropes, and date seeds are fed to livestock.

Datura   *Thorn apple.

Deafness, the loss of hearing in the *ear, has many causes. Loss of response to high tones is an almost inevitable accompaniment of ageing, and often becomes severe enough to interfere with the understanding of speech. Unless the loss is great, most deaf people can be helped by hearing-aids. Abnormalities of the inner ear are the usual cause of congenital deafness. Middle-ear deafness as a result of infection was previously much more common than it is today.

Death is the cessation of function of the major organs of the body required to sustain life. Failure of the heart, lungs, liver, kidneys, or brain can lead to death of the whole body unless compensatory medical treatment is possible.

The permanent loss of function of the brainstem, which controls respiration and other vital reflexes, is called brain death. This is the modern accepted point at which death can be diagnosed, even if the rest of the body still functions.

Death cap is one of many species of deadly poisonous *agaric fungi belonging to the family Amanitaceae. The death cap, *Amanita phalloides*, has a fruit body with a greenish-white cap, white gills and stalk, a white ring round the stalk, and a membraneous cup-like volva at the base. It is found in woods and pasture land. Symptoms of poisoning appear only some hours after it has been eaten, and it is usually fatal. Other deadly poisonous species include the destroying angel, *A. virosa*, the browning amanita, *A. brunnescens*, of North America, and fly agaric, *A. muscaria*.

Death's-head hawk moths   *Hawk moths.

Death watch beetles are beetles whose larvae bore into mature timber, potentially causing immense damage. The adult males strike their heads against the wood as a mating call, making sounds which can be heard on still nights. *Book lice can make similar, but fainter, noises. Death watch beetles belong to the family Anobiidae, which also includes the common woodworm, *Anobium punctatum*. Others of the 1,100 species in this family attack stored foods, tobacco, and even drugs.

Deciduous forests are comprised of trees which, under normal conditions of growth, regularly shed all their leaves. They are generally found where there is an annual climatic change such as a cold winter or prolonged dry period. At these times, the majority of tree species in deciduous forests will be leafless. They occupy large areas of the tropics, subtropics, and temperate zones, the most familiar being the north temperate forests with their spectacular leaf colours as they synchronously shed them in the autumn (fall). In the tropics they are known as monsoon forests or evergreen tropical rain forests depending upon the local climate. Deciduous trees are not always broad-leaved, especially in the tropics, where many species of broad-leaved tree are evergreen, while some *conifers are deciduous.

Decomposition (biological) ensures that nutrients are continuously recycled in any *ecosystem. It is the breakdown of dead plants and animals by organisms such as bacteria and fungi. The process of decomposition is helped by many invertebrates such as earthworms, snails, and beetles which break down the dead organic material into a form suitable for micro-organisms to act upon. Decomposition frees nutrients, which are then available to other organisms, particularly *autotrophs, and produces *humus. Compost heaps, biogas units, and sewage works all harness these decomposer organisms under artificial conditions.

Deep-sea fishes include a wide range of taxonomic groups. They can be divided into mid-water (bathypelagic) and bottom-living forms (benthic). The latter range down to the very deepest of ocean trenches. For example, a deep-sea cod of the family Eretmophoridae was caught at 8,000 m. (26,500 ft.) in the Puerto Rico Trench.

Fishes bearing light-organs, such as *lantern fishes, are common in the twilight zone of the deep sea, down to about 1,000 m. (3,250 ft.). Many species have large eyes to make the most of this living light. Species living below this level often have small eyes which in some cases are covered with skin. Most deep-sea fishes are black or red in colour. The deep sea is relatively poor in sources of food and consequently many of its inhabitants have huge mouths, as in gulper eels, and massive teeth, as in black

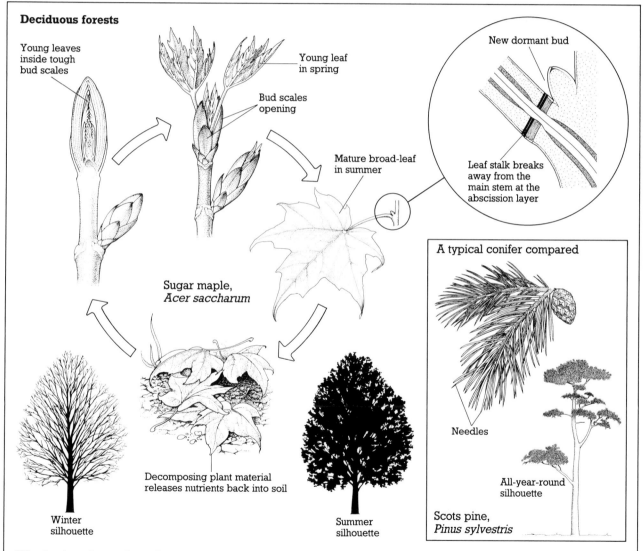

Deciduous forests

Young leaves inside tough bud scales

Young leaf in spring

Bud scales opening

Mature broad-leaf in summer

New dormant bud

Leaf stalk breaks away from the main stem at the abscission layer

Sugar maple, *Acer saccharum*

A typical conifer compared

Needles

All-year-round silhouette

Scots pine, *Pinus sylvestris*

Winter silhouette

Decomposing plant material releases nutrients back into soil

Summer silhouette

Woody plants have adopted several strategies to adapt to climates that are drier than the swamps from which they evolved. Deciduous trees solve the problem by producing hormones that stop the supply of water and nutrients to the leaves as the dry season (winter) approaches. This is called abscission. Deprived of water, the leaves die, change colour, and fall. Temperate regions do not have a pronounced dry season but water in the soil may freeze in winter and be unavailable to plants. Most broad-leaved trees native to temperate regions are deciduous but a few species such as holly, *Ilex* species, retain their leaves and have a waxy coating to minimize water loss. In contrast, conifers have needle-like leaves which reduce water loss. To compensate for their small size, conifer leaves have more highly concentrated chlorophyll than the leaves of broad-leaved trees and are therefore usually a darker green in colour. Most conifers retain their leaves through the winter.

dragon fish. In addition, most deep-sea fish have highly extensible stomachs. The benthic forms find most of their food buried in the bottom mud and they tend to have downward-facing mouths, long snouts, and elaborate sensory organs in the *lateral line of the head. They also frequently have long tails and anal fins and swim by undulating the body, allowing a head-down feeding posture. Some deep-sea angler-fishes, in addition to having luminous fishing lures, large mouths, and huge teeth, have a unique means of ensuring breeding success in that the males are parasites on the much larger females.

Deer are some thirty-six species of large, even-toed, hoofed, land mammals. They make up the family Cervidae and have been common since the Pliocene as browsing animals of the forest. They are spread throughout the Northern Hemisphere over Asia, Europe, and North America; in the Southern Hemisphere they are almost entirely limited to South America.

The weight of the body is carried by the third and fourth toes, protected by hoofs, while on either side of the feet are the small second and fifth toes, the dew claws, which do not touch the ground. Deer are characterized by the presence of *antlers or tusk-like canine teeth in the males. The muzzle is naked and the ears large. Below the eyes are scent glands, which in the male are odoriferous. The tail, usually short, is held erect at moments of alarm. Deer are *ruminants, browsers or grazers, with cud-chewing habits. They come out at dawn and dusk to feed on leaves, grass, berries, shoots, ferns, root crops, and even bark, before settling in a safe, quiet place to chew the cud. All deer are swift and have highly developed senses and social organization, which helps them to avoid danger. They live in herds based on the supremacy of the leading male, whose position is maintained or overthrown by fights, seldom fierce, with male rivals.

Deer-flies *Horse flies.

Deer mice *Woodmice.

Deficiency disease is any disorder due to lack of an essential dietary factor, such as a *vitamin, protein, or a *trace element. It is most likely to occur when the amount of food is adequate, but limited in variety. For example, peoples dependent upon maize as a staple food suffer from pellagra, a disease caused by a shortage of the B vitamin called nicotinamide, which is lacking in maize. Similarly, a lack of thiamin (vitamin B_1) causes beriberi amongst people dependent upon polished rice. Such diseases are easily cured by redressing the deficiency in the diet.

Delphiniums are annual, biennial, and perennial plants, which belong to the same family as the buttercup. About 200 species are known and they are chiefly from the temperate zones of the Northern Hemisphere, including Europe and North America. Blue is the predominant colour but there are also red, pink, and white species. *Delphinium hybridum* and *D. grandiflorum* with tall flower spikes are popular garden plants. The larkspur, *Consolida ambigua*, is an annual species with many coloured varieties, and is closely related to the delphiniums.

Dementia is the global reduction in mental functions, such as the loss of memory and reasoning. It is not an integral aspect of ageing but always results from a specific pathological process, most commonly Alzheimer's disease. Sufferers lose track of recent events and may forget where they are, as well as the date and time of day, while memory of the distant past may remain good. Behaviour may become impulsive, with other changes in mood such as irritability and lack of personal care.

Demoiselle crane: a species of bird, *Anthropoides virgo*, which is one of the smallest members of the *crane family. It has a wide breeding range from southeastern Europe to central Asia, wintering in northeastern Africa, India, and China. The head and neck are black with large white ear-tufts while the rest of the body is ashy grey.

Denaturation is the change that occurs in proteins when they are exposed to temperature and acidity conditions outside their normal range or to some other chemicals. These changes affect the three-dimensional shape of the protein molecule which is so important to its normal function. In the case of enzyme denaturation, the precise shape of the active site is altered so that the substrate molecules no longer fit and the catalytic action of the enzyme is destroyed. Denaturation is often a reversible process, as many proteins can re-fold into their working shape if their chemical environment is correct. However, in the most familiar form of denatured protein, heated egg white, where the albumen protein chains become entangled, the change is not reversible.

Deodar *Cedars.

Depression is a mental state of sadness, hopelessness, and despair. It may become so severe that sufferers stop eating and sleeping normally, lose interest and pleasure in everything, and may make determined efforts to commit suicide. After treatment with appropriate drugs, about 80 per cent of such cases recover.

Dermatitis is inflammation of the skin. There are many types, with causes varying from direct skin contact with an irritant substance to being part of an allergic reaction. The irritant may be one to which all skins will react or may be unique to that person. Some allergic types of dermatitis are identical to *eczema.

Desert animals live in regions where rainfall is less than 25 cm. (10 in.) a year. The dryness is frequently accentuated by high temperatures during the day. Without exception, desert animals show adaptation to the arid conditions. Most important is their impermeable covering which reduces desiccation. Insects have a waxy cuticle; reptiles, thick horny scales; and mammals, cornified skin impregnated with oily sebum produced from the sweat glands. Water loss is further reduced in many desert mammals, such as the kangaroo rats, by excretion of highly concentrated urine. The structure of the kidneys of desert vertebrates is adapted for efficient water reabsorption. Reptiles, birds, and insects excrete crystalline uric acid, losing little water. Kangaroo rats, like many small desert animals, hide during the day in their burrows, where it is cooler and the humid atmosphere reduces respiratory water loss. Camels are adapted to allow their body temperature to rise during the day and lose heat at night, thus avoiding sweating. They metabolize fat (stored in their hump) for energy, which releases water and their tissues can tolerate dehydration and rapid rehydration.

Desert birds have to shade their eggs from the excesses of heat. Lizards control their temperatures by adjusting their orientation to the sun, often climbing off the ground to escape the intense heating effect at soil level. Aestivation, or dormancy, during adverse conditions is also common among animals in seasonally arid deserts.

Desert dormouse *Dormice.

Desert plants follow one of two different strategies for survival: some tolerate desert conditions all year round, and others avoid extreme desert conditions by germinating, growing, flowering, and dying during rainy spells. Plants in the second group are similar in their structure to 'normal' plants, differing only in the speed with which they grow, flower, and set seed. Those in the former group, which includes *succulents and *cacti, have a wide pith and cortex layer which acts as an efficient insulator against heat. These plants are thus protected from the full temperature extremes of the desert.

Devil-fish *Manta ray.

Devil's apple *Mallow.

Dhole is a species of red hunting dog found in the forests of Southeast Asia. It hunts in packs, at first by scent and then by sight. After bringing its prey, often a deer, to bay, the pack encircles it. The lead dog waits for an unguarded

Ephemeral **desert plants**, such as these poppies, can lie dormant for years as seeds. A sudden downpour, such as happened here in the Kalahari Desert, can stimulate growth within the space of a few days.

moment to leap at the quarry's throat, whereupon the whole pack helps to kill it. The dhole resembles a wolf and has ten distinct subspecies, including the east and west Asian dholes.

Diabetes (mellitus) is a disorder in which insulin, an internal secretion of the *pancreas, is deficient in amount or activity. This causes the level of a sugar called glucose in the blood to be higher than normal. The defect may be partial, in which case modified diet may be adequate treatment, or severe, and require injections of insulin. The severe type can cause sufficient metabolic upset to induce coma, and even threaten life. Unless carefully treated, both types can, after many years, cause serious damage to eyes or kidneys.

Diana monkeys are a species of Old World monkey found in the Ivory Coast region of Africa. It was so named by the famous naturalist Linnaeus, because of its attractiveness. They have a black head, legs, and tail, a grey body, red-brown rump, white neck and breast, and a white band on the forehead. Rarely at rest, they move gracefully through the trees, using their long legs and slender body to best advantage. They move generally in troops of up to twelve members and hunt during the day for fruit, birds' eggs, and insects.

Dianthus *Pinks.

Diaphragm *Lungs, *Abdomen.

Diarrhoea is the frequent movement of the bowels, usually with the passage of soft or liquid faeces. Ingestion of many types of micro-organism can cause acute diarrhoea,

which is a symptom of intestinal inflammation. Prolonged periods of diarrhoea will lead to dehydration and loss of salts and nutrients.

Dicotyledons comprise the larger of the two subclasses of *angiosperms and include the great majority of all trees, shrubs, and plants. They are far more numerous than the other subclass, the *monocotyledons, with around five times as many families. The dicotyledons have two cotyledons, or seed-leaves, and a network of branched veins on the leaves. Their floral parts (petals, sepals, etc.) occur in fives or multiples thereof. There are individual exceptions to these general characteristics. In temperate regions most food crops, other than cereals, are dicotyledons.

Diet (food) is the habitual food consumption of an individual. An animal's diet should contain the several varieties of chemical compounds (nutrients) essential for existence. Special diets are designed to improve physical fitness for sport, or to overcome medical problems such as obesity, diabetes, or genetic defects. In a balanced diet, the fuel *calorie value should match an individual's energy requirements. Other essential items in diets include proteins, fats, carbohydrates, water, vitamins, salts, and *trace elements. These must be supplied in the correct relative proportions, or various nutritional disorders may develop.

Severe lack of protein causes impaired physical and mental performance and susceptibility to infection. Some intake of fat is necessary to provide essential fatty acids and fat-soluble vitamins, but not more than 30 per cent of calories should be met from fats. Foods of plant origin supply water-soluble vitamins and trace elements. Carbohydrates, consumed in large amounts, provide calories and adequate dietary fibre.

Digestion is the process of breaking down food into a form capable of being absorbed into the body of an animal. It usually involves two processes, a mechanical one, and a chemical one. The mechanical part includes tearing and chewing by teeth, crushing and grinding by the mouth, and churning by the *stomach. Chemical aspects of digestion often begin in the mouth when *enzymes in the saliva begin to break down the food into simple molecules of sugars, fats, or proteins.

The digestion of food by chemical means may begin outside the body in animals such as spiders. In vertebrates the major part of chemical digestion begins in the *alimentary canal. Humans and carnivores rely largely on a supply of special juices from the stomach (gastric juice), pancreas (pancreatic juice), and liver (bile). In *ruminants, with additional stomachs, bacterial and protozoan fermentation of vegetation occurs as well. As the food particles are propelled from mouth to stomach, and then, in solution, pass to the elongated, coiled tube of the intestines, they are exposed to a succession of different chemical treatments under the control of hormones and the nervous system.

In many vertebrates, when food is seen, smelt, and tasted, reflexes ensure that a plentiful flow of strongly acid juice is secreted by the stomach to sterilize the food as it arrives and to provide enzymes (pepsins) which begin the splitting of proteins. Acid solutions from the stomach are forced into the duodenum at intervals and mix there with alkaline pancreatic juice and bile. Pancreatic enzymes require a neutral medium to split starches, protein de-

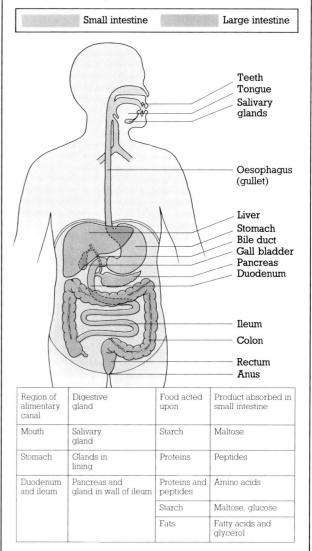

Digestion

Diagram to show the human alimentary canal and a chart summarizing the chemical breakdown of food by the digestive enzymes.

| Small intestine | Large intestine |

Teeth
Tongue
Salivary glands

Oesophagus (gullet)

Liver
Stomach
Bile duct
Gall bladder
Pancreas
Duodenum

Ileum
Colon
Rectum
Anus

Region of alimentary canal	Digestive gland	Food acted upon	Product absorbed in small intestine
Mouth	Salivary gland	Starch	Maltose
Stomach	Glands in lining	Proteins	Peptides
Duodenum and ileum	Pancreas and gland in wall of ileum	Proteins and peptides	Amino acids
		Starch	Maltose, glucose
		Fats	Fatty acids and glycerol

rivatives, and fats into simpler compounds. These are then absorbed into the blood stream via the intestinal walls.

Digitalis *Foxgloves.

Dikkops *Stone-curlews.

Dill is a hardy, aromatic, annual herb growing up to 70 cm. (2 ft. 4 in.). It is a member of the same family as carrot, carroway, and fennel and is a native of southern Europe and western Asia. The fruits are used whole or ground. The young feathery leaves, though far less aromatic, are also used as a flavouring.

Dingo: a species of wild dog that is the only wild carnivore of Australia, and is not found elsewhere in the world. It is not known how or when the dingo arrived in Australia, a land otherwise occupied by marsupials. Its fossil remains are known from the Pleistocene Era 40,000 years ago, at which time it may have associated with man. It is found in the open forests and on plains, where it hunts singly or in a pack. The dingo preys upon kangaroos, sheep, and other small animals. It is the size of a large dog with soft fur, usually rusty red or fawn in colour, the toes and tip of the tail being white. Four to eight pups are born in an underground burrow or hollow tree. It does not bark but utters a series of yelps and a plaintive howl.

Dinosaurs are extinct reptiles which dominated the Earth during the Mesozoic Era, 245–66 million years ago. There are two separate groups of dinosaurs, the reptile-hipped dinosaurs, such as *brontosaurus and *tyrannosaurus, and the bird-hipped dinosaurs like *iguanodon and *triceratops. These divisions represent skeletal differences, which in turn reflect their origin and evolution. Many dinosaurs reached enormous size although others were quite small. They were very diverse in habits as can be deduced from their dentition, with both herbivores and carnivores. They are believed to have included species which walked on two legs and those which used four. This is once again based upon studies of their fossilized skeletons, and by drawing comparisons with the skeletons of present-day animals. All the dinosaurs disappeared abruptly about 65 million years ago. The reason for this mass extinction is still unknown but is usually attributed to a cooling of the world's climate. Their nearest living descendants are the birds.

Dioscorides, Pedanius (*c.* 30–80), Cilician botanist and physician in the Roman army, compiled one of the first pharmaceutical texts, prescribing some 500 herbs for use in medicine.

Diphtheria is a highly contagious disease caused by the bacterium *Corynebacterium diphtheriae*. It usually affects the throat, and infection is acquired by inhaling droplets derived from another sufferer or a healthy carrier. The organism produces a toxin which can kill by damaging the heart. Its action on nerves causes paralysis. Inoculation with modified toxin gives excellent protection, and the condition is now rare in Western countries.

Diplodocus is a now extinct species of the reptile-hipped *dinosaurs, which inhabited what is now North America in the late Jurassic Period, some 150 million years ago. It was quadrupedal and vegetarian in habit, with a long neck and tail which made it the longest of the dinosaurs at up to 25 m. (80 ft.), and with a weight of about 24 tonnes (26·5 US tons). The fossilized remains of *Diplodocus* are known only from Wyoming and Colorado in the USA.

Dippers are a family of birds which resemble giant wrens. They are habitually aquatic and frequent fast-flowing streams. They plunge or wade in, swim both on and under the water, and even walk on the bottom hunting for food. Like wrens, they build domed nests. There are five species of dipper, two native to Eurasia, two to South America, and one to North and Central America.

Discs (anatomy), or intervertebral discs, are flat plates of a tough, fibrous cartilage which connect the *vertebrae in the backbone of vertebrates. The centre of each disc is softer than the rest due to the presence of fluid.

Discs act as fluid-containing cushions which dampen vibrations from the legs or skull that might otherwise pass along the backbone. Because man stands upright, fluid

from the disc centres is gradually squeezed into the bloodstream in daytime and re-forms at night. Young people, with healthy, fluid-filled elastic discs, are measurably taller in the morning. With increasing age the amount of fluid decreases as the centres of the discs contract and harden, and old people decrease in stature. Also, the holes through which spinal nerves pass tend to get smaller, nipping the nerves and causing pain and stiffness.

If the spine is suddenly deformed in a fall or accident, the outer edge of the disc may rupture with the result that fluid on one side forms a blister. This causes either lumbago or sciatica depending on the particular spinal nerve nipped. A so-called 'slipped disc' is in fact a condition in which the disc has burst and fragments press on ligaments and nerves.

Disc-winged bats belong to the family Thyropteridae and are so called because they have a suction disc on each

A **dingo** suckles her three-day-old pups in a hollow tree used as a den. Dingoes readily interbreed with feral dogs, so pure-bred examples are rare. They are regarded as serious pests in many parts of Australia because of their attacks on sheep and poultry.

wrist and ankle. These allow the bat to cling to smooth surfaces. The two species occur in Central and northern South America. The Madagascar disc-winged, or sucker-footed, bat, *Myzopoda aurita*, belongs to a separate family, not closely related to the true disc-winged bats.

Disease is any impairment of the normal state of health of a plant or animal. There are two main groups of diseases, pathogenic and non-pathogenic.

Pathogenic diseases are those caused by invading organisms (pathogens) which may be regarded as *pests or *parasites. These may or may not be host-specific—for instance all mammals are susceptible to the rabies virus. Certain bacteria cause diseases such as tuberculosis and cholera in man and galls and cankers in plants. Fungal diseases of plants can have serious consequences: potato blight can destroy a potato crop in four weeks. Diseases caused by animal pathogens include *malaria, *bilharzia, and *elephantiasis.

Non-pathogenic diseases are common and include *deficiency diseases caused by poor nutrition; radiation and chemical poisoning; and genetic diseases such as *haemophilia and sickle-cell *anaemia. Man is also susceptible to mental illness such as *depression. Some diseases like cancer, which has various causes, are difficult to classify. Allergies may have both genetic and environmental components.

The courtship **display** of the peacock, *Pavo cristatus*, focuses on the train, made up of iridescent tail coverts with purplish and black-centred, coppery 'eyes'. This stunning visual impact is heightened by the strutting movements and the rustling noise of the train.

Display is the presentation of colour, pattern, and movements as part of the behaviour which identifies a species and, often, the sex of an individual. The display may be permanent, seasonal, or intermittent. It may serve to initiate courtship or maintain *pair-bonding. Many birds such as the lyrebird have colourful displays; others, such as the blackbird, rely on a sequence of postures and calls. In apes, monkeys, man, and many other vertebrates, display establishes and maintains a social hierarchy and will often minimize potentially damaging aggression. Threat displays occur not only within species but also between species. Competitors or predators are deterred by *warning coloration and *flash coloration, which accompany aggressive or defensive displays.

Divers, called loons in North America, are grebe-like water birds of the family Gaviidae, with sharp bills, long bodies, and webbed feet. Though clumsy on land, they fly strongly and are expert at swimming and diving. When alarmed they often swim with only their head and neck above the water, like submarines. Their eerie calls resemble a child's shriek of pain. They breed in the northern temperate regions on islands and the shores of lakes, and winter off temperate sea-coasts. There are five species, of which the great northern diver, *Gavia immer*, and white-billed diver, *G. adamsii*, are the largest. Both are black and white with boldly chequered upper parts.

Diverticulitis is inflammation of the diverticula, or pouches, of the *colon. It is thought to be associated with a low fibre intake and occurs commonly among people used to a Western diet. The inflammation may be sudden and resemble an attack of acute appendicitis, from which

it is distinguished by being on the left-hand side of the abdomen, and by pain that is usually less severe and more prolonged.

Diving ducks   *Bay ducks.

DNA (deoxyribonucleic acid) is a *nucleic acid and the hereditary material of living organisms; it occurs in every cell and determines the characteristics of the organism by controlling the manufacture of *proteins. In higher plants and animals, the DNA is contained within the nucleus of the cell. DNA molecules are composed of four *nucleotides: each contains a sugar and a phosphate group and one of the nitrogenous bases adenine, guanine, cyto-sine, and thymine. The nucleotides join in any order to make very long strands. Molecules of DNA comprise two such strands, which run parallel to give a ladder-like structure. This is then twisted into a right-handed spiral shape called a double helix. The two strands are held together by chemical bonds between the two strands at points where the nitrogenous bases occur. These bonds can form only between adenine and thymine, or between cytosine and guanine. The DNA molecule can copy itself, or replicate, by unzipping, and using each strand as a template for the formation of a new strand. Replication precedes cell division so that identical copies of the DNA, which lie in *chromosomes, are passed to each of the daughter cells.

DNA

The double helix structure of DNA was proposed by James Watson and Francis Crick in 1953. The main strands of the double helix consist of alternating sugar and phosphate groups, and to each sugar group is bonded a nitrogenous base – adenine (A), cytosine (C), guanine (G), or thymine (T). The strands are held together by weak chemical bonds, called hydrogen bonds, that form between the bases (shown by dotted lines in the diagram below). These weak bonds make it possible for the two strands of the DNA molecule to separate in order to replicate themselves.

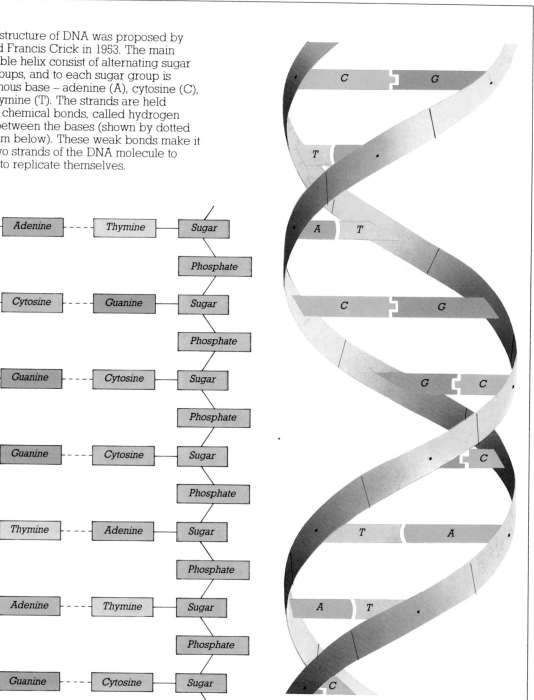

The sequence of nitrogenous bases on one of the strands acts as a code for making proteins. The code is copied by a molecule of *RNA and then transferred from the nucleus to the cytoplasm of the cell, where the assembly of protein takes place. This is often referred to as the *genetic code.

Dobsonflies *Alder flies.

Docks (plants) are perennial plants of the genus *Rumex*, which includes 150 species in north temperate regions. All species are wind-pollinated. They are related to buckwheats and rhubarb, and sorrel. They are troublesome to farmers and gardeners because of the prodigious quantities of seed they produce and their deep taproots. The curled dock, *R. crispus*, is a particularly serious weed on newly turned soils.

Dodders are twining plants of the widespread genus *Cuscuta*. Lacking leaves and chlorophyll, they are parasitic on other plants and possess suckers (haustaria) which enter the tissues of the host plant, drawing sustenance from them. They belong to the Convolvulaceae family and are related to morning glory and sweet potato. Individual species may grow on a range of hosts within a family or they may be restricted to a single host. Some are troublesome to crops such as clover and flax.

Dodo: an extinct species of large, swan-sized bird (related to the pigeons), which once lived on the island of Mau-

An engraving of a **dodo** by J. Pass, dated 1801, depicts it as a dumpy bird about the size of a turkey. Paintings of live birds show that the overall colour was brown with yellow on the wings and tail.

ritius. It had lost the power of flight for its wings were very reduced, and it was therefore easy prey to sailors landing on the island for provisions. By the middle of the seventeenth century it had consequently been rendered extinct. A number of skeletons still exist in museums, and the fragmentary one in the University Museum, Oxford, inspired Lewis Carroll to use the dodo as a character in *Alice in Wonderland*.

Dogfish are small *sharks belonging either to the family Scyliorhinidae or to the family Squalidae. The former are known in Europe as dogfish, but in North America as catsharks. Scyliorhinid sharks are rarely more than 1·5 m. (5 ft.) long; they are bottom-living, usually solitary, heavily patterned in colour, and lay eggs. Squalid sharks, which are known as spur-dogs in Britain and dogfish sharks in North America, are active mid-water sharks, which feed on fishes and squids. They are mostly grey coloured, with a long spine in front of each dorsal fin, and no anal fin. They give birth to live young.

Dog-rose *Rose.

Dogs form a family of thirty-five species within the great order of *carnivores. They appeared very early in the evolution of carnivores and those living today have changed relatively little from their fossil ancestors of 50 million years ago. The dog family, which includes wolves, foxes, dingo, and dhole, is found throughout the world. All dogs have elongated jaws with forty-two to forty-four teeth; the canines are especially long and dagger-like to hold onto the prey; the carnassial teeth are for cutting and shearing flesh and sinews; and in some species the molars are used for grinding bones. Their claws are not retractile. The female dogs, or bitches, give birth to between two and sixteen pups which, blind at birth, open their eyes at ten days of age. Growth is rapid and they are weaned at six weeks.

Wild species of dog include the bush dog and the Cape hunting dog. Wild dogs can swim, and a few can climb trees, although they are adapted to life on the ground. The muzzle is narrow, the ears large and erect, the limbs slender, and the tail long and bushy. The sense of smell is exceptionally well developed and this, with their good eyesight and keen hearing, allows them to follow a trail through a forest on a dark night. Domestic dogs now include many breeds, which may be classified into groups according to their employment: hounds for hunting, gun dogs, spaniels, non-sporting dogs, sheepdogs, watchdogs, draught animals, and pet dogs.

Dog's-tooth violets are small, spring-flowering, monocotyledonous plants with colourful flowers (the recurved petals are said to resemble dog's teeth). They belong to the lily family, and are totally unrelated to the true violets (*Viola* species). About twenty species are known, all from northern temperate zones. They are popular garden plants.

Dogwoods, or cornels, are deciduous shrubs or trees native to the United States, Asia, and Europe. They belong to a family of some 100 species, forty of which belong to the genus *Cornus*. The common European dogwood, *C. sanguinea*, has greenish white flowers and dark purple berries. Its branches are dark red, and provide attractive winter colour in gardens. The wood of species such as the

Pacific white-sided **dolphins**, *Lagenorhynchus obliquidens*, seen off southern California. This species is noted for its graceful surface displays and its large schools which may include other dolphin species.

Cornelian cherry, *Cornus mas*, and the flowering dogwood, *C. florida*, is sometimes used to make skewers (once known as 'dogs') or shuttles for the textile industry.

Dolphins belong to a family of *toothed whales and are typically beaked. The common dolphin, *Delphinus delphis*, is one of thirty-two species, and is an inhabitant of the warmer parts of the Atlantic and Mediterranean. Dolphins are schooling animals with a highly developed social organization; schools may include up to 200 individuals, and many will leap some 3 m. (9 ft. 9 in.), almost synchronously, clear from the water. Some species can maintain a speed of 36 km./hour (23 m.p.h.) for several hours. Mating probably occurs in spring and summer, the gestation period lasting eleven months; the single young is 75–90 cm. (29–36 in.) long at birth. An adult common dolphin may be as long as 2·4 m. (8 ft.). A vocal mammal, it produces a series of clicks that are audible to man.

Other species include the widespread bottle-nosed dolphin, *Tursiops truncatus*, and the killer whale, *Orcinus orca*. The last species hunts in packs, or pods, of twenty or more individuals and will encircle or trap shoals of fishes.

Donkey: a relative of the horse, derived from the wild *ass. Its name is thought to originate from the dun colour of its coat. It was probably domesticated during the New Stone Age some 12,000 years ago; the Egyptians used it extensively, as far back as 3000 BC, in their monument-building. From Egypt the donkey spread through Asia Minor, Turkey, and the Balkans, before reaching the rest of Europe in the ninth century; it was brought to the USA in 1868.

The donkey is a placid animal and makes a better pack and draught animal than the horse, to which it is also superior when carrying loads and negotiating narrow mountain paths. It has a life expectancy of up to fourteen years. A single foal is usually produced after a gestation period of nine months.

Dor beetles are large, heavily built beetles, with noisy flight. They comprise about 300 species in the family Geotrupidae. The males of some species have horns on their heads and many species are metallic blue, purple or green in colour. Their larvae are sedentary and feed on carrion, or dung. Dung-feeding is the most common life-style for dor beetles.

Dormice belong to the *rodent families Gliridae and Seleviniidae; the first contains ten species found throughout Europe, Asia, and Africa, and the second has a single species, the desert dormouse. More like voles than mice in appearance, dormice have long, bushy tails and soft fur. Their common name derives from the deep hibernation undertaken in the northern parts of their range. The common dormouse of Britain, more properly known as the hazel dormouse, *Muscardinus avellanarius*, extends as far east as the USSR and Asia Minor. As the name suggests, this dormouse is especially associated with hazel bushes. The edible dormouse, *Glis glis*, was fattened for food by the Romans in *gliraria*, hence the family name. Most species eat a variety of plant products or small animals such as insects and earthworms.

Douglas firs are five species of *conifer belonging to the pine family. Two occur in North America, and three in the Far East. Their nearest relatives are the hemlocks and they are not true firs. A Douglas fir, *Pseudotsuga menziesii*, felled in British Columbia in 1895 was the tallest living tree known, at 133 m. (440 ft.).

Douglass, William (1691–1752), Scottish-born physician, emigrated to America and was then the only doctor in Boston with a medical degree. He is credited with the first adequate clinical description of diphtheria (although he later confused it with scarlet fever in an epidemic of that disease).

Douroucouli *Night monkey.

Dover sole *Soles.

Doves are birds belonging to any of several groups of the pigeon family. The term 'dove' does not describe any particular taxonomic unit, but most of the species tend to be small ones. The Barbary dove, *Streptopelia risoria*, is kept

extensively for its pleasant song; its origin is obscure, but it is thought to be a long-domesticated form of the African collared dove, *S. roseogrisea*. The white dove of peace is a domesticated form of the rock pigeon, *Columba livia*.

Down's syndrome is a congenital condition which occurs once in every 600 births, usually when the mother is aged over thirty-eight. Affected children have a facial appearance similar in the shape of the eyes and nose to those of the Mongolian race. The syndrome is caused by a *chromosome abnormality characterized by the presence of three particular chromosomes instead of two. The condition can be detected before birth but cannot be corrected at any stage. Mental handicap is usually present to a variable degree, and a range of physical illnesses can shorten life expectancy.

Dragonets are spiny fishes of the family Callionymidae. The most common European dragonet, *Callionymus lyra*, grows to 30 cm. (12 in.) in length. This species is a bottom-living fish which prefers muddy or sandy bottoms at depths of 20–100 m. (65–325 ft.). The males are brilliantly coloured by comparison with the females and there is an elaborate display of colours during breeding in early spring. Other kinds of dragonet are abundant in the Indo-Pacific region, and also occur on the American coast.

Dragonfishes are deep-sea fishes with large mouths and long teeth—a combination of which earns them their name. Two families are recognized: scaly dragonfishes, the Stomiatidae family, and scaleless black dragonfishes, the Melanostomiatidae family. Both families have species with

An edible **dormouse** suckles her young. Both this species and the garden dormouse, *Eliomys quercinus*, produce one litter per year; other species may have up to three. Female dormice have from two to nine young, though four is the average litter size in all species.

long bodies, luminous organs on the underside, and a barbel on the chin. They are found in deep-water oceans throughout the world.

Dragon-flies are powerful aerial hunters of other insects, able to hover and fly backwards, and like *damselflies have legs placed so as to form a 'basket' beneath the mouth for catching prey. They hunt by sight; the enormous compound eyes cover most of the head and can detect movement in a full circle around the insect. Many are brilliantly coloured, and they rest, unlike damselflies, with the wings outspread. They are most often found near still or slow-moving water, although adults may range over several kilometres.

The aquatic, carnivorous nymphs, unlike those of damselflies, have gills in the rectum and pump water in and out of the anus, sometimes using this for 'jet propulsion'. As in damselflies, the mouth of the nymph with its massive jaws is hidden by the expanded, hinged lower lip, or mask, which ends in a pair of hooks and shoots out to grasp its prey.

Dragon-flies and damselflies have an unusual approach to reproduction. Preliminary to mating, the male grasps the female's neck with claspers at the end of his abdomen, and pairs are often seen flying in tandem. She curves her abdomen forward to receive sperm stored in pockets at the front of his abdomen. Eggs are then dropped into water in flight, or inserted into stems of water plants; in damselflies the male may support the female while she oviposits.

Dreams are the visions and sensations that occur during sleep. They are most common and vivid during phases of light sleep, when the eyes and body may move jerkily. During dreaming, heartbeat and breathing rate are elevated and the brain shows increased activity. In *psychoanalysis the content of dreams is examined to try to shed light on the patient's innermost worries, but this does not prove that dreams always have special meaning. They may be best understood as a by-product of the way the brain sorts and records the impressions and events of our waking hours. Many sleeping animals show eye movements and brain activity patterns similar to those in humans. This suggests that animals dream too.

Drill (mammal) *Mandrill.

Drills (invertebrates) are *molluscs, related to whelks; they use a special secretion from their foot, together with rasping teeth, to drill holes through oyster shells and so eat the soft inner tissues. The oyster drill, *Urosalpinx cinerea*, is a common pest in America, and was unintentionally introduced to parts of Europe where it now severely affects oyster fisheries.

Drinker moths *Eggar moths.

Dromedary *Camels.

Drongos are a family of twenty or so bird species which occur in Africa, India, Southeast Asia, many of the Pacific islands, and Australia. Predominantly black, they are mostly thrush-sized, though one, the racket-tailed drongo, *Dicrurus paradiseus*, has two greatly elongated tail feathers, making it some 63 cm. (25 in.) in overall length. They live chiefly in wooded country and feed mainly on insects.

Drugs include any substance used to modify the bodily function of a living organism but which is neither a normal constituent of the body nor food. Vitamins and hormones are not normally classed as drugs because they are essential for health occur naturally in our bodies or food. A few drugs are given in large doses of several grams and act in a simple manner. For example, alkalis such as magnesium carbonate are used to counteract gastric acidity. The great majority, however, are active in doses of milligrams or less. Such drugs react with specific regions, called receptors, on the surface of cells, causing a chain-reaction within.

Effective drug treatment is a recent phenomenon: the list of drugs available in the 1930s contained very few that were really useful. The more active the drug, the easier it is to produce unwanted side-effects. Occasionally, patients genetically less able to dispose of a drug may react excessively when it is first used; others may develop allergies to drugs which were initially well tolerated. Addictive drugs are those upon which the body gradually comes to depend physically or psychologically; removal of these drugs results in withdrawal symptoms. Physical dependency occurs as a result of the biochemical effects of the drug. Psychological dependency, although often causing severe social problems, is not accompanied by biochemical changes.

Drupes are *fruits with a single seed in a hard case (endocarp) usually enveloped in a juicy edible layer (mesocarp) and coated with a brightly coloured skin, as in the plum (*Prunus domestica*). However, the mesocarp may be fibrous, as in the coconut (*Cocos nucifera*), where it is the coir, or coconut matting, of commerce, or it may be woody, as in the walnut (*Juglans* species). Aggregates of drupes include such 'berries' as blackberries, raspberries, and loganberries.

Dry rot fungus causes dry rot of timber in buildings. It is able to spread from infected timber into mortar, and

Fruiting bodies of **dry rot fungus** emerge from woodwork and produce millions of spores. The main body of the fungus spreads by a branching network, or mycelium, which can create its own humid conditions by transporting water to new sites of infection.

therefore across walls. Unlike wet rot (a form of chemical decomposition of wet wood), it can attack relatively dry wood, provided that some moisture is present in the wood. It can be prevented by keeping structural timber dry and well ventilated.

Dublin Bay prawns, also called Norway lobsters, and scampi, are related to true *lobsters but have slender bodies up to 20 cm. (8 in.) long, and are coloured pink or red rather than the blue-grey of a real lobster. They live along sandy coastlines, usually in deeper water, staying clear of rock-pools. They are caught for food in Europe.

Ducks are closely related to swans and geese, but they are smaller, with shorter necks, and beat their wings faster in flight. They are much more aquatic than geese and the sexes differ in appearance. Except when moulting (in eclipse), the drakes (males) are generally more brightly coloured than the females. There are over 110 species, distributed worldwide. Some, like the sea-ducks, sawbills (mergansers), and pochards, dive for their food. Others, like mallard, teal, and widgeon, are surface-feeders that dabble or up-end. Ducks are adapted to an aquatic way of life in having webbed feet and a water-repellent plumage that gives them added buoyancy.

Duckweeds are some fifteen species of small floating or submerged perennial plants in the genus *Lemna*, which forms part of the worldwide family Lemnaceae. Most species, such as the common duckweed, *L. minor*, occur in all parts of the world with the exception of the polar regions and parts of the tropics. They often form a green carpet, covering the surface of still or slow-moving water. The closely related genus of *Wolffia* includes the smallest known species of flowering plant. All duckweeds have tiny male and female flowers, which are separate but usually on the same plant.

Dugong, or sea cow, is an entirely aquatic species of mammal, living in warm coastal waters in shallow bays and estuaries from the Red Sea throughout the Indian and Pacific oceans. It is highly adapted for life in water, the streamlined body having a layer of thick blubber. The fore-limbs are large and the digits joined to form a paddle. There are no hind-limbs and the terminal tail, like that of the whale, is horizontal. The dugong can remain submerged for up to ten minutes at a time, its nostrils closed by valves, while feeding on green seaweed and other vegetation. The dugong has poor sight but good hearing. A single pup is born in the water, and the mother carries it in her flippers.

Duikers are small antelopes found in forest and bush country of Africa south of the Sahara Desert. The seventeen species have short legs and stand about 60 cm. (2 ft.) at the shoulder. In most species, both sexes have horns; these are short and spiked with a tuft of hair between them. There are several species, including the duikerbok, or common duiker, *Sylvicapra grimmia*, in which only the male has horns. The zebra duiker, *Cephalophus zebra*, has a striking coat of orange-red with tiger-like black stripes on the back. Females give birth to one or two young, generally in May.

Dung beetles are beetles whose larvae feed on dung, especially that of herbivorous animals. They include some

The **dye plant** Himalayan indigo, *Indigofera heterantha*, a deciduous shrub, is one of several species used as a source of indigo, though today the dye is manufactured on a large scale from chemicals.

3,200 species of scarab and beetles of the subfamily Aphodiinae, in addition to *dor beetles. In many cases they are large and black, the males having horns. The adults dig chambers beneath the dung, in which they lay their eggs, or simply place eggs on the dung. Some species show maternal care for the young, and all serve a useful function in ecosystems by recycling dung.

Dunlins are small wading birds, resembling sandpipers, chestnut and black in summer but grey and white in winter. They are often to be seen on mudflats with other waders, feeding in large flocks. In flight, compact parties twist and wheel in the air with an amazing unity of purpose. Several subspecies of dunlin occur throughout northern temperate zones.

Dunnock   *Hedge-sparrow.

Duodenal ulcers   *Ulcers.

Durian: the fruit of a large tree that flourishes in the Asian rain forests and is cultivated in Malaysia, Thailand, and Indonesia. It belongs to the same family as the *baobab. The large yellowish fruits up to 25 cm. (10 in.) in diameter are coarsely spiny with a creamy inner pulp. The smell of the durian has been described as resembling bad drains with a touch of garlic, but some people find the sweet-fleshed fruit delicious. The locals build shelters in the trees in order to collect the ripe fruits the moment they drop. This is partly because the flesh deteriorates rapidly but also because the fruit is peculiarly attractive to wild animals such as monkeys and elephants.

Durmast oak   *Oaks.

Dutchman's pipe: a species of woody, deciduous climbing plant, native to the eastern USA. It belongs to the family Aristolochiaceae and, along with some 500 other species, forms the genus *Aristolochia*. Although most species are climbers, some plants in this genus are herbaceous. Relatives of the Dutchman's pipe occur throughout temperate and tropical Eurasia and America. The Dutchman's pipe itself, *A. macrophylla*, has heart-shaped leaves, and grows to height of 9 m. (30 ft.). Its yellowish-green, tubular, inflated flowers are bent into a pipe shape (hence the common name), and they are fly-pollinated.

Dye plants have been used since prehistoric times for colouring skin, and later for colouring fabrics and clothing. Until the nineteenth century, when it became possible chemically to synthesize various colouring compounds such as anilines, all dyeing used natural products obtained from plants. These natural dyes were extracted from species such as *woad (blue), wild mignonette (yellow), Egyptian privet (*henna), and *Indigofera tinctoria* (indigo).

Annatto is a red dyestuff obtained from the seeds of the Central American shrub *Bixa orellana*. It is used today in lipsticks and for colouring foods but was once used by Caribbean Indians to colour their bodies, and the resulting effect probably led to the description 'Red Indians'.

Dysentery is an infection of the intestine, caused by the bacterium *Shigella* or the amoeba *Entamoeba histolytica*. These micro-organisms lead to inflammation, ulceration, and often bleeding of the gut. This disease leads to a condition called *colitis. The parasitic amoeba which causes dysentery is spread by contamination of food or drinking water infected by sewage. The bacterial type of dysentery can also be spread by these means as well as by person-to-person contact. Severe attacks can lead to serious dehydration and loss of blood from the gut. Amoebic dysentery is restricted to the tropics and sub-tropics.

Dyslexia is a disorder which makes learning to read and to spell a difficulty. It occurs in normal children who have received instruction in reading and writing and who have not been neglected socially or deprived in other ways. It is characterized by slowness in handling and recognizing sequences of letters. Its cause is not understood but the condition can be treated with some success.

Dysphasia   *Aphasia.

Dyspraxia   *Apraxia.

Eagle owls are large, horned *owls of the genus *Bubo* with eleven species. They are native to parts of Europe, Asia, and Africa. The European eagle owl, *B. bubo*, is typical with its dark brown barred plumage, feathered legs and feet, and golden-yellow eyes. They are easily recognized on account of their eartufts, 4–5 cm. (about 2 in.) long, and their deep hooting calls. Females weigh up to one and a half times as much as males. They frequent mountains and open woodland, hunting birds, small mammals, and occasionally fish, reptiles, and frogs. They lay two to four eggs in nests on the ground.

Eagles are some fifty-six species of large *hawks of worldwide distribution. Most have long, pointed wings, spanning up to 1·5 m. (5 ft.), which are perfectly adapted to flapping, soaring, and gliding flight. The legs are strong, the feet heavily taloned and in some species feathered. The beak is strongly curved, and used for tearing prey. Eagles are generally powerful predators, catching their prey on the ground or snatching it from treetops; a few are carrion-feeders. Live prey includes reptiles, birds, and small mammals. Their plumage is usually brown or black and combined with white.

Eagles generally nest high in trees or on cliff faces, sites which give good vantage-points with updraughts to assist take-off. Their nests are untidy heaps of sticks accumulated from year to year and most species lay one or

The thick wing-coverts of this **eagle owl**, *Bubo bubo*, muffle any flight sound; this helps in surprise attacks on prey. It usually hunts low over the ground with a wavering flight, and will take prey as large as roe deer fawns, as well as hawks and other owls.

two eggs. Parents hunt over extensive territories, returning to the nest with prey before tearing it up for the chicks. They are diurnal hunters with remarkable eyesight, and can air-lift prey of considerable size. In pastoral areas they occasionally attack lambs and kids, but attacks on babies are legendary though not improbable.

Ears are sense organs which interpret vibrations in water and sound waves in air. In vertebrates the ear is also the origin of balance. Many invertebrates have sense organs which can be called ears; those of crickets and grasshoppers are good examples. The vertebrate ear is a complex structure consisting of three fluid-filled, bony tubes (semicircular canals), and a variety of chambers or additional tubes. Ears always occur in pairs, and usually have some connection with the outside world, via a flexible membrane, called a typanum, which covers a cavity.

In fishes, the ears are most important as organs of position or motion. The movement of small stones, or otoliths, within hollow chambers, enable fishes to 'feel' acceleration and judge their position relative to gravity. The ears of amphibians and reptiles are a stage more complex than those of fishes, and some species have an opening from the inner ear to the throat. The ears of birds are among the most sensitive of all, especially to different frequencies of sound.

The mammalian ear, which includes that of humans, has three main parts, the outer, middle, and inner ears. The outer ear consists of a fleshy flap, which concentrates directional sounds, and a tube lined with hairs and wax glands, leading to the eardrum (tympanum). The middle ear is an air-filled cavity, connected to the throat, which is crossed by a series of three bones, known commonly as the anvil, hammer, and stirrup bones. They transmit vibrations from the eardrum to a similar membrane closing the fluid-filled tubes of the inner ear. Nerves from the inner ear transmit the information on frequency and volume of the sound to the brain. Some vertebrates, such as man, are receptive to a wide range of frequencies, 50–20,000 Hz in the case of man.

Fruiting bodies of the **earthstar** fungus, *Astraeus hygrometicus*, a European fungus of sandy, woodland soil. The spore-producing body is the brown, oval structure, 1–5 cm. ($\frac{1}{2}$–2 in.) across.

Earthstars, like puffballs to which they are related, are woodland fungi whose spores are contained in a spherical paper-like fruit-body. The group is represented by about thirty European species. In the young stages the fruit-body, or inner peridium, is surrounded by a fleshy outer peridium and resembles a tulip bulb. At maturity the outer peridium splits into several parts that bend outwards in a star-like formation to reveal the fruit-body. The apex of this has a small pore through which the spores are puffed out of the fruit-body when it is fully mature.

Earthworms are oligochaete *annelids, with the familiar segmented body and tiny bristles to give anchorage. The thickened saddle towards the featureless front end is

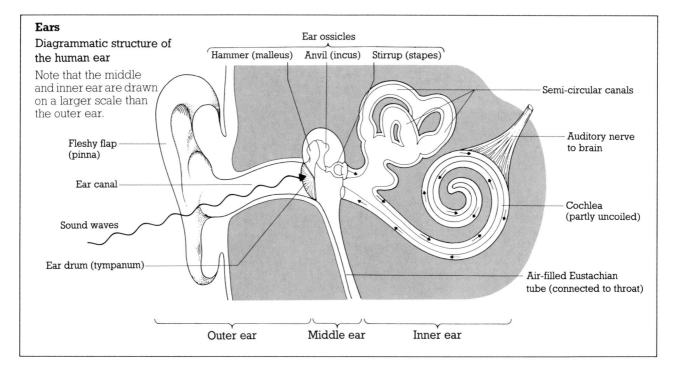

Ears

Diagrammatic structure of the human ear

Note that the middle and inner ear are drawn on a larger scale than the outer ear.

Ear ossicles
Hammer (malleus)   Anvil (incus)   Stirrup (stapes)

Fleshy flap (pinna)

Ear canal

Sound waves

Ear drum (tympanum)

Semi-circular canals

Auditory nerve to brain

Cochlea (partly uncoiled)

Air-filled Eustachian tube (connected to throat)

Outer ear   Middle ear   Inner ear

characteristic; it lies near the reproduction organs, and secretes the small cocoons into which earthworm eggs are laid for protection.

Earthworms are *hermaphrodites and mate by exchanging sperm. This usually occurs on damp nights, when two worms lie side by side outside their burrows. At other times they are rarely seen, spending their lives below ground, tunnelling and pulling down decaying vegetation, thus beneficially aerating and mixing the soils. They move through the soil by pressing tiny bristles into the walls of their tunnels and setting up waves of muscle contraction along their body. In earthworms, as in all annelids, the muscles act antagonistically against the hydrostatic pressure within their *coelom.

Earwigs are elongate, brownish insects which have at the tip of the abdomen a pair of forcep-shaped appendages, curved in the males and straight in the females. The front pair of wings is small and thickened, and when at rest the hind pair is folded like a fan beneath them. Earwigs eat both plant and animal material and can be destructive to garden flowers. The females lay their eggs in a batch, often in winter, in soil or under stones, and tend them, constantly 'licking' them to remove destructive fungi, until they hatch as miniature adults. Earwigs make up the order Dermaptera with some 1,200 species found throughout the world.

Echidnas, or spiny ant-eaters, are monotremes, belonging to the group of primitive egg-laying mammals from Australia and New Guinea. There are two species, both with a covering of spines which usually project through the fur. When disturbed, the echidna rolls up into a ball like a hedgehog. They are mainly nocturnal and spend the day in crevices or rotten logs. They are powerful diggers, whether tunnelling down into ant nests in search of these insects, or making sleeping burrows. They can run, swim, and climb trees.

Echinoderms, though familiar as sea-urchins and starfish on sea-shores, are a very peculiar and enigmatic phylum of animals. The 5,300 or so species have many unique features which set them apart from other invertebrate groups, such as their five-rayed symmetry; their internal skeletal plates, often bearing spines; their tiny, hydraulically operated tube-feet, used for feeding and locomotion; and their internal circulatory systems, quite unlike other more conventional blood-systems. In some respects they resemble chordates (a major division of the animal kingdom which includes vertebrates), and they may have arisen from the same ancestors as them. But echinoderms are in many respects primitive, and though successful in the sea have never colonized other habitats.

Echinoderms, which include the strange *sea cucumbers, are unique among animal groups in having no parasitic members. Some of the classes within this phylum were once numerous in the seas of earlier geological eras. Particularly plentiful fossil echinoderms included *sea-lilies and sea-urchins.

Ecology is the study of the relationship between living organisms and their environment, the word being derived from the Greek *oikos*, meaning house or home. There are two main approaches: autecology is the study of the ecology of a single species, and synecology is the study of the ecology of complete *ecosystems. Ecologists consider

population dynamics, interaction between species (such as competition, predation, and feeding), the environment (including the availability of nutrients), and energy flow through the ecosystem. Patterns of distribution and *succession and the impact of man are also examined. Ecological study leads to an understanding of the working of natural systems, and it enables *conservation of nature to be practised. The basic units of ecological study are the *species and the space, or niche, which they occupy in an ecosystem.

Ecosystems are the basic functional units of *ecology, since each consists of a biological *community and its

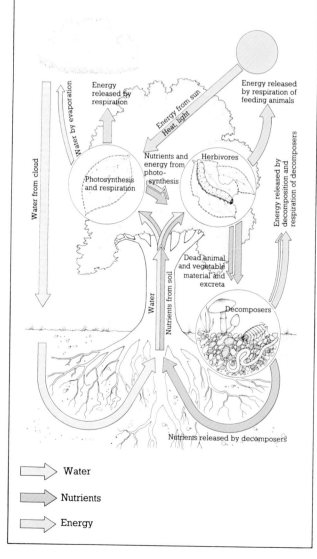

Ecosystem: energy and nutrient flow

A single ecosystem can consist of as little as a tree and its immediate surroundings. A tree uses the sun's energy together with minerals from the soil to manufacture its own tissues by photosynthesis. Energy is released from the tree during respiration. Herbivorous animals such as caterpillars convert the tree's tissues into their own, at the same time giving off energy by respiration. Decomposing soil fungi and bacteria use plant debris for their own life processes, at the same time releasing minerals which the tree requires for growth.

non-living environment. An ecosystem can be as small as a garden pond, or as large as the planet Earth, but the term usually describes an area of discrete habitat such as a rain forest or an oak wood. The essential point is that an ecosystem is a self-contained unit in all respects except for the input of energy. All nutrients such as nitrogen, carbon, water, and oxygen are recycled. All ecosystems are directly, or ultimately, dependent upon a continuous input of energy from the sun.

The radiant energy in sunlight is trapped by *autotrophic organisms, typically green plants, and converted into chemical energy which is stored in complex organic molecules. These are used as food by organisms higher in the *food chain and are eventually lost to decomposers. At each stage along the food chain a proportion of the original energy trapped from sunlight is converted into heat and lost to space. In terms of geological time, ecosystems are constantly exchanging materials, but in the commonly used sense of the term they are stable features.

Sometimes there is an import or export of energy (organic material), as in river ecosystems. The border between two ecosystems, such as the intertidal sea-shore, is called an ecotone.

Eczema is an allergic inflammation which causes a skin rash. Often the provoking substance is not identified, but it may be something ingested, such as cow's milk. When specific external contact is responsible, the condition is called dermatitis. Eczema often occurs in patients with other allergic disorders such as asthma or hay fever.

Edelweiss: a silvery-grey *alpine plant, *Leontopodium alpinum*, belonging to the sunflower family, and native to the mountains of Europe. Great symbolic value is attached to the plant by mountain guides in the European Alps, who traditionally wear a flowering sprig of the plant in their hats.

Eel-grasses are submerged aquatic plants of the genus *Zostera*, with long, dark green, ribbon-like leaves. Male and female flowers may be borne on separate, or the same plants. The eel-grasses form a family of eighteen species, all of which are found in salt or brackish water in coastal areas of temperate seas. Their pollen has the same specific gravity as salt water and is distributed by water currents.

The name is sometimes applied to submerged freshwater aquatic plants of the genus *Vallisneria*.

Eels are all long, slender fishes primarily adapted for living in crevices or burrows, although some live in the deep sea and are free-swimming. Over 600 different species of eels are known in nineteen families. Most are marine fishes; the freshwater eels all belong to the family Anguillidae, but even they return to the sea to spawn. Because of their burrowing habit their fins are poorly developed, although all have a low dorsal and anal fin which is united around the tail. Eels have no pelvic fins, and the moray eel, in addition, has no pectoral fins. The gill opening is also greatly reduced to a slit, sometimes placed low down on the throat. All eels are predatory, but their diet varies widely from group to group. Moray eels are fish-eaters; most have long, pointed fangs, although some, which eat crustaceans, have blunt teeth. Conger eels have densely packed, triangular teeth, suitable for both soft- and hard-skinned prey, while snipe-eels have minute teeth and feed on planktonic crustaceans.

The European freshwater eel, *Anguilla anguilla*, and its North American counterpart, *A. rostrata*, grow to about 1 m. (3 ft. 3 in.) in length. Both breed in the general area of the Sargasso Sea, and their leptocephalus larvae (often called glass-fish) migrate across the Atlantic. Although the American eels reach their coastline after one year, it takes three years for the European eels to reach fresh water. They travel far upstream in rivers as *elvers and can move into lakes (even travelling overland at night). All species of freshwater eels feed on insect larvae, crustaceans, and fishes, and scavenge on the river-bed.

Eelworms are familiar to farmers and gardeners as the tiny but incredibly numerous white, worm-like soil pests which eat plant roots, especially potato tubers and bulbs. These featureless *roundworms live free in the soil, but tunnel into any roots they encounter, eating as they go, and often causing cysts or galls to form on their food plants.

Egg is a name used for two similar structures: for a female germ cell or *ovum, or for a complex structure which protects and nourishes the true ovum. The eggs of birds and reptiles are examples of what are popularly known as 'eggs', having a tough protective shell. These are adapted to life on land, as the true germ cell is provided with a food store, in the form of yolk; a protective cushion and water, in the albumen; and a bacterial barrier in the membranes lying under the shell. Each egg is a self-contained life-support unit for the development of a young bird or reptile. The external horny (reptiles) or brittle (birds) shell serves to give support to the contents some degree of mechanical protection.

Eggar moths are species of large, furry-bodied moths, which are part of the family Lasiocampidae. The 1,000 or so species are mainly tropical, and have caterpillars which are clothed in hairs which can be irritant. Pupation occurs within tough, usually egg-shaped cocoons. Examples include the drinker moth, *Philudoria potatoria*, which is common and widespread over Eurasia, and the oak eggar, *Lasiocampa quercus*, which occurs in Europe and North Africa. The lackey and lappet moths also belong to this family.

Ego *Id.

Egrets comprise thirteen species of birds belonging to the genus *Egretta* and *Bubulcus* and related to the heron. Many have all-white plumage, like the little egret, *E. garzetta*. The ubiquitous and familiar cattle egret, *B. ibis*, habitually follows grazing animals to feed on the insects they disturb. Egrets are found throughout the world and are of similar size to herons and bitterns.

Ehrlich, Paul (1854–1915), German-born physician, was the founder of chemotherapy (the prevention or treatment of disease by the use of chemicals). He developed methods of staining microscopic specimens, helped to standardize diphtheria antitoxin, and proposed a chemical theory to explain immunity.

Eider is the common name for four species of mainly black and white sea ducks, which thrive off the cold northern coasts of America, Europe, and Asia. The common eider, *Somateria mollissima*, is a hefty-looking bird, 60 cm.

A plumed **egret**, *Egretta intermedia*, at Bharatapur, northwest India, shows the spear-like beak and long, powerful neck, adaptations for fishing. Both sexes will develop graceful plumes during the breeding season.

(2 ft.) long, with an odd triangular profile to its beak and head. Breeding in colonies, it is well known for the softness of the down which, like other ducks, the female plucks from her breast to line the nest. In Iceland the birds are farmed for this down and its collection does them no harm.

Elands are two species of antelopes with straight, spirally twisted horns. The giant eland, *Taurotragus derbianus*, is the largest of the antelopes, being almost 1·8 m. (6 ft.) tall at the shoulder; its horns may be up to 99 cm. (39 in.) long, though those of the female are more slender. It lives in the open country and forests of East Africa and once ranged as far south as the Cape. The common eland, *T. oryx*, once inhabited much of southern, central, and East Africa, but is now restricted to reserves.

Elders are deciduous shrubs or trees from temperate regions of the Old and New Worlds. Members of the same family as honeysuckle and snowberry, all elders belong to the genus *Sambucus*, with about twenty species. Most have large heads of white flowers followed by blue-black, red, or in some species white or yellow fruits much liked by birds. The flowers and fruits of the common elder, *S. nigra*, are used in wine-making. Other species, especially the American elder, *S. canadensis*, are widely grown for their spectacular flowers.

The plant called *ground elder is an unrelated species of the carrot family.

Elecampane is a handsome, sunflower-like plant, *Inula helenium*, with solitary, bright yellow flowers. Though native to Europe and northern Asia, it is now widely distributed in the eastern USA, where it has escaped from gardens. It belongs to the sunflower family and, like its relative the dahlia, has a starch-like chemical called inulin in its roots. This was formerly used as a medicinal plant for coughs and other lung infections. The root may be candied as confectionery.

Electric fishes are found in both the sea and fresh water and include several hundred different species of fishes from several families. Most produce only a weak electrical current but there are several kinds which produce one powerful enough to cause a severe shock to man or large animals. In the sea, the only example of the latter are the twenty or so species of electric rays or torpedoes, but all skates and rays can generate weak currents. Torpedoes, especially *Torpedo nobiliana*, use their powers to stun the fishes on which they prey, but the skates probably employ theirs for species recognition. Another group of marine fishes using electricity possibly to detect oncoming prey are the stargazers, a family of bony fish with some twenty-four species.

Some of the most powerful electric fishes are found in fresh water, and include the tropical South American electric eel, *Electrophorus electricus*, which can produce currents of 1 amp at 500 volts, and the tropical African catfish, *Malapterurus electricus*. Relatives of the electric eel, the gymnotids of South America and the African mormyrid fishes, such as the elephant-snout fish, produce weak currents. Both live in heavily sedimented rivers and swamps and use their electric powers to locate one another, avoid predators, or in navigation to avoid obstacles.

Electricity (in nature) is a form of energy used by all organisms with a *nervous system in that they produce minute electric currents by which signals are sent along nerves. Few species produce electric currents for any other purpose, except for several species of *electric fish which use them to navigate or to stun prey.

Electric-light bugs   *Giant water-bugs.

Electron transport chain, or hydrogen transport, is one of three stages in the biochemical pathway of *respiration. In this stage, electrons or hydrogen atoms are transferred by coenzymes from the other two stages, namely *glycolysis and *Krebs' cycle, and are used to produce molecules of *adenosine triphosphate (ATP). The electron transport chain consists of a series of carrier molecules, including cytochrome proteins, fixed in the membranes of *mitochondria. Within the carrier system, hydrogen atoms are split into their component protons and electrons. The protons are released to the surrounding medium, but the electrons are passed from carrier to carrier along the chain. This transfer is accompanied by the release of energy which is used for making ATP. At the end of the chain, the enzyme cytochrome oxidase transfers electrons and protons to oxygen, forming water. This final step is the only reaction in the whole of respiration which uses oxygen. If oxygen is absent the hydrogen transport pathway and Krebs' cycle are both blocked, which means that only glycolysis can continue, resulting in anaerobic respiration, or *fermentation.

Elephant grass is a name in general use for several tall-growing African grass species and for the bulrush, *Typha elephantum*. One species is a constituent of the floating islands of sudd on the Nile. It is burned to produce an impure salt of carbonate of soda for culinary use.

Elephantiasis is gross swelling of the legs and other parts of the body, with thickening of the skin due to blockage of the vessels of the *lymph system by parasitic *roundworms of the genus *Filaria*. The adult worms live in the lymphatic channels of the abdomen, and their tiny larvae circulate in the blood of their verbebrate host (including man) at night, when they may be taken up by mosquitoes. After further development inside the mosquito, they are transferred to other hosts. The most common type, caused by the Asian *Wuchereria bancrofti*, affects the legs, breasts, or scrotum.

Elephants are the largest and most powerful of the land mammals, once found in tropical jungles and grassy plains over much of Africa and Asia. The African or bush elephant is the larger of the two species living today; the smaller subspecies of the African elephant, the forest el-

Protective herding behaviour shown by **elephants** in the Amboseli National Park, Kenya. Females live together, led by a mature individual. Note the cattle egrets, feeding on insects flushed out by the elephants.

ephant, also lives in Africa. The Asiatic or Indian elephant is smaller than the bush elephant and is classified in four subspecies: the Indian, Ceylon, Sumatran, and Malaysian elephants. The trunk, characteristic of these animals, is a flexible, muscular tube. At the tip are two finger-like projections which can pluck leaves and grasses for transfer to its mouth. Used as a musical instrument, it produces loud trumpet-like sounds important in courtship and in communication. The large tusks, the much modified second pair of incisor teeth, grow throughout the life of the animal. These may reach a length of 3·18 m. (10 ft. 5 in.) and weigh 160 kg. (353 lb.). For such a huge creature it can move quite quickly, 9–13 km./hour (6–8 m.p.h.), and if enraged at 40 km./hour (25 m.p.h.) over a distance of 45 m. (150 ft.).

Elephants are social animals that live in herds. Each group is usually led by an old cow, though its nominal head is a large bull. Maturity is reached at about fourteen years. There is no fixed breeding season and the gestation period is long, perhaps twenty-one months. A single calf is born, covered with coarse black hair, and is about 90 cm. (3 ft.) tall at the shoulder. Suckled by its mother for two years, the calf remains with her for a further two years. A female African or bush elephant can live for sixty years; males for fifty years.

Elephant shrews are *insectivores and are so named because of their long, flexible snouts. The name is misleading as they are not true shrews, but form a distinct order, the Macroscelidae. They are usually active during the day and, when alarmed, can move quickly by hopping on their hind legs. Most of the fifteen species live in southern, central, and East Africa, with a few in the northwest. They inhabit grassland, forests, and rocky country.

Elephant-snout fishes are members of the family Mormyridae, such as the species *Mormyrus kannume*, which have their snout elongated into a trunk. They are African freshwater fishes which feed primarily on bottom-living insect larvae. Fishes of this family can produce an electric field by means of which they navigate in the rather murky waters of their river homes.

Elk *Moose.

Elms comprise about twenty species of trees of the north temperate zone extending southwards to the Himalayas, Malaysia, and Mexico. They all have leaves that are asymmetrical at the base and disc-shaped winged fruits with the seed set in the centre. The flowers usually appear before the leaves, though *Ulmus parvifolia* of eastern Asia produces them together. Some elms are small trees and dwarf forms are occasionally cultivated, but most are large, providing valuable timber used in construction and for making furniture.

The English elm, *U. procera*, is unusual in that it reproduces itself largely by suckers, rarely producing fertile seed. As a result, the majority of English elms are genetically similar. They have been greatly reduced in number through the depredation of Dutch elm disease, a fungus transmitted by beetles which burrow under the bark, causing the yellowing and eventual death of the foliage and finally of the tree itself. Many other species of elm are resistant and are being used in some areas to replace English elm.

Elver is the name given to young freshwater *eels, *Anguilla* species, when they are migrating up river following their migration as leptocephalus larvae from their spawning grounds in mid-Atlantic. In body form they are identical to the adult eel but are no thicker than a matchstick and about 8 cm. (3 in.) long. In Britain elvers run up rivers in huge numbers between February and April, the earliest runs being in western rivers.

Embryology is the study of the changes in shape, size, and composition of the growing embryo from the fertilized egg stage to birth.

Embryos are the early stages of animals before birth. After *fertilization, the egg divides into two similarly sized cells, each then dividing again to give a total of four cells. The process of division, or cleavage, moves through the eight, sixteen, and thirty-two cell stages, continuing until a ball of cells is formed. From this stage the precise details differ from animal group to group, but involve folding of cell layers. As the embryo develops, particular patches of cells begin to specialize, or differentiate, to form different types of tissue, such as skin, liver, or nerves. All the nutrient needs of an embryo are met either by stores provided for the egg, or in mammals through blood supplied by the mother via a *placenta. The developing embryo of many animals is protected from infection and mechanical damage by membranes, either inside the mother, or inside a tough 'egg shell'.

In humans the embryo is called a foetus after the first eight weeks of development. At this stage all the main organs have been formed. Development of embryos reflects the evolutionary origin of the particular animal. The human embryo goes through one stage at which it has gills, like a fish, and another where it has a tail. These

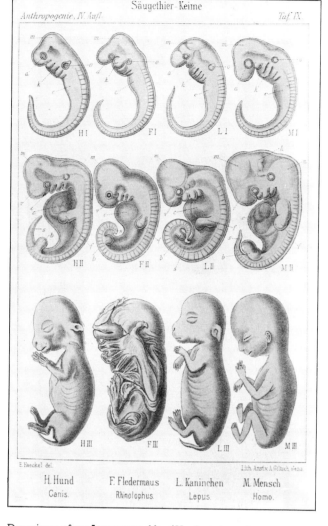

Drawings of **embryos** used by *Haekel show three stages in the development of (*left to right*) a dog, horseshoe bat, hare, and man. The similarities between the embryos are quite marked in the early stages but later they show the characteristics of their species.

features disappear as development proceeds, and the foetus comes to resemble a miniature human being.

Mammalian embryos, including those of humans, can be moved from one mother to another, or even implanted, after external fertilization of an egg. Their development can be monitored by ultrasound scanning and abnormalities noted before birth.

Emperor moths are among the largest moths in the world and include *atlas moths. Many have an eyespot or transparent window on each wing, and males have conspicuous comb-like antennae. Their caterpillars construct cocoons of silk, which in a few species is used commercially. Most are tropical, but the genus *Saturnia*, which includes some seven species, occurs widely in Europe, Asia, and North America. They include the European peacock moth, *S. pyri*, with a wing-span of 15 cm. (6 in.), and the North American *S. mendocino*, with a wing-span of up to 7 cm. (3 in.).

Emu: the only species of bird in the family Dromaiidae and is the Australian equivalent of the *ostrich. An adult

An **endangered species** hunted by bedouins for its meat, the Arabian oryx, *Oryx leucoryx*, had become extinct in the wild by 1972. Ten years later, under the organization of Operation Oryx, a captively bred herd was released into central Oman.

emu weighs some 35–40 kg. (80–90 lb.) and is a flightless, long-legged, long-necked bird, the female reaching 2 m. (6½ ft.) high. The smaller male incubates the eggs and cares for the chicks.

Endangered species are those plants and animals with population numbers so low that there is a danger of their becoming extinct. The cause is often habitat destruction or over-exploitation by man for commercial profit. Many plants and invertebrates are being threatened by clearance for agriculture. Over 350 species of amphibians, reptiles, birds, and mammals are considered to be in danger of extinction. Species such as the African mountain gorilla and the tropical green turtle are represented by only a few hundred individuals. *Conservation is not always successful and many species have already been lost.

Endive: a salad plant of Asian origin, belonging to the sunflower family, and introduced to Britain by the Romans. As its green leaves are bitter, they are blanched for a few weeks before harvesting.

The term 'endive' is sometimes applied to a lettuce with curly, branched leaves. Endives are closely related to true lettuce.

Endangered species

Some examples of presently endangered species:

	Species	Distribution	Main reason for endangerment
Mammals	Pygmy hog, *Sus salvanius* Kouprey, *Bos sauveli* Woolly spider monkey, *Brachyteles arachnoides* Mediterranean monk seal, *Monachus monachus*	Northern India Kampuchea & Laos & Thailand Southeastern Brazil Mediterranean coasts	Habitat loss Hunting; effects of wars Habitat loss; hunting Human disturbance; marine pollution
Birds	Mauritius kestrel, *Falco punctatus* Monkey-eating eagle, *Pithecophaga jefferyi* Kagu, *Rhynochetos jubatus* Lanyu scops owl, *Otus elegans botelensis*	Mauritius, Indian Ocean Philippines New Caledonia Lanyu (off S.E. Taiwan), Pacific Ocean	Habitat loss Habitat loss; some hunting Predation by introduced mammals; habitat loss Habitat loss
Reptiles & amphibians	Orinoco crocodile, *Crocodylus intermedius* Hawksbill turtle, *Eretmochelys imbricata* California legless lizard, *Anniella pulchra nigra*	Colombia & Venezuela Atlantic & Pacific & Indian Oceans Monterey Peninsula, USA	Hunting (extinct from Amazon & Orinoco) Commercial hunting Habitat loss
Fish	Ala balik, *Salmo platycephalus* Fiery redfin, *Barbus phlegethon* Alabama cavefish, *Speoplatyrhinus poulsoni* Totoba, *Cynoscion macdonaldi* Maryland darter, *Etheostoma sellare*	Zamanti River, Turkey Olifants River system, S. Africa Key Cave, Alabama, USA California, USA Maryland, USA	Uncontrolled fishing methods Predation by exotic fishes; habitat loss Pollution; over-collecting; habitat loss Severe over-fishing; disruption of habitat Pollution; disruption of habitat
Plants	Philip Island hibiscus, *Hibiscus insularis* Socotran pomegranate, *Punica protopunica* Neogomesia cactus, *Ariocarpus agavoides* Tarout cypress, *Cupressus dupreziana* Yeheb nut bush, *Cordeauxia edulis*	Philip Island, Pacific Ocean Socotra, Indian Ocean Mexico Algeria Ethiopia & Somali Republic	Grazing by feral goats, pigs, and rabbits Grazing by goats and cattle Horticultural trade (illegal export) Nomads and cattle damage; trees used for shelter Habitat loss; grazing by livestock

Some examples of recently extinct species:

	Species	Distribution	Year of extinction / Main reason for extinction
Mammals	Jamaican long-tailed bat, *Reithronycteris aphylla* Newfoundland white wolf, *Canis lupus beothucus* Bali tiger, *Panthera tigris balica* Caribbean monk seal, *Monachus tropicalis*	Jamaica Newfoundland Bali Caribbean Sea	c. 1900 / Habitat loss c. 1911 / Hunting (Government bounty) c. 1937 / Hunting c. 1952 / Commercial hunting
Birds	Tristan gallinule or Island hen, *Gallinula nesiotis* Arabian ostrich, *Struthio camelus syriacus* Madagascar serpent eagle, *Eutriorchis astur* Eskimo curlew, *Numenius borealis*	Tristan da Cunha, S. Atlantic Ocean Syria & Arabia Northeast Madagascar North & South America	c. 1890 / Predation by introduced rats c. 1941 / Hunting c. 1950 / Habitat loss c. 1970 / Hunting
Reptiles & amphibians	Barrington Island tortoise, *Geochelone* species Cape Verde giant skink, *Macroscincus coctei* Palestinian painted frog, *Discoglossus nigriventer* Jamaican tree snake, *Alsophis ater*	Galapagos Islands, Pacific Ocean Cape Verde Islands, E. Atlantic Ocean Hula Lake, Israel–Syria border West Indies	c. 1890 / Hunting c. 1940 / Mainly natural habitat loss c. 1956 / Habitat loss c. 1960 / Predation by introduced mongooses; hunting
Fish	Stumptooth minnow, *Stypodon signifer* Lake Titicaca orestias, *Orestias cuvieri* Shortnose sucker, *Chasmistes brevirostris* Blackfin & Deepwater cisco, *Coregonus* species	Chihuahua, Mexico Peru–Bolivia border Oregon, USA L. Michigan & Huron, USA/Canada	c. 1930 / Habitat loss c. 1950 / Disruption of habitat c. 1960 / Disruption of habitat c. 1960 / Fishing; predation by sea lampreys
Plants	Juan Fernandez sandlewood, *Santalum fernandezianum* Wine palm, *Pseudophoenix ekmanii* Roussillon forget-me-not, *Myosotis ruscinonensis* Interrupted brome, *Bromus interruptus*	Isla Robinson Crusoe, Pacific Ocean Dominican Republic S. France United Kingdom	c. 1916 / Commercial exploitation c. 1926 / Commercial exploitation c. 1960 / Habitat loss c. 1972 / Habitat loss

Endocrine glands are compact collections of secretory cells which discharge their secretions into the blood, instead of to a free surface through a duct. Present in all vertebrates and some invertebrates, they co-ordinate diverse cellular processes in different regions of the body. These processes may show daily, monthly, or seasonal rhythms.

There are five major ductless glands in mammals, whose sole function is the manufacture, storage, and release of one or more specific organic chemical compounds called *hormones. The five glands are the pituitary, the gonads, the adrenals, the thyroid, and the parathyroids. Other organs, such as the kidney, the stomach, and the pancreas, contain scattered patches of specialist cells with endocrine functions. Individual glands are able to modify or initiate specific chemical reactions in specific organs, often called target organs, elsewhere in the body. The discharge of their hormones is stimulated by signals received from the nervous system and endocrine systems, or other signals such as the blood levels of glucose or calcium salts. Target organs are tissues whose cell membranes are able to select and bind particular hormones from the mixture present in the blood.

Endocrine glands co-ordinate the long-term adjustments of cellular activity required for growth, reproduction, and many other functions. These actions of endocrine glands can be integrated with those of the brain by the hypothalamic control of the *pituitary gland.

Entomology is the study of insects. Like other branches of zoology, it began as a mainly descriptive subject, concerned with collecting and classifying specimens. Well over 1 million species of insect are known to science. The work of *Fabre in the nineteenth century did much to arouse interest in the life histories and behaviour of insects, and now entomology deals with all aspects of their biology.

Because insects are small and breed quickly, they are used as laboratory animals in many types of biological research. Some of these investigations are carried out because of the threat posed by disease-carrying insects or by those which eat crops.

Enzymes are types of *protein found in all living organisms. They allow all the chemical reactions of *metabolism to take place, regulate the speed at which they progress, and provide a means of controlling individual biochemical pathways. Enzymes owe their activity to the precise three-dimensional shape of their molecules. According to the 'lock-and-key' hypothesis, the substances upon which an enzyme acts (substrates) fit into a special 'slot' or space in the enzyme molecule, called the active site. A chemical reaction takes place at the active site and the appropriate products are released, leaving the enzyme unchanged and ready for re-use. This cycle can be repeated as often as 100,000 times per second.

Enzymes are very specific in relation to the substrates with which they work, and normally are effective only for one reaction or a group of closely related reactions. They function best in particular conditions of temperature and acidity (pH), called optimum conditions, and their action can be slowed or stopped by *inhibitors. Many enzymes need a *coenzyme in order to function. The human body contains at least 1,000 different enzymes.

Eohippus, or hyracotherium, was the first stage in the evolution of the modern horse, living about 55 million years ago in North America and Europe. There were several species of Eohippus, or 'dawn horse', all the size of a small dog and with small teeth which show that they browsed on soft vegetation. Unlike the modern horse, it had three hind-toes and four front toes which splayed out, presumably to accommodate walking on swampy ground. Eohippus became extinct some 49 million years ago and gave rise to a succession of horse-like mammals such as Orohippus, Mesohippus, and others.

Epidemics are sudden outbreaks of *infectious disease which spreads rapidly to affect large numbers of people. For an epidemic to occur there must be susceptible individuals, and a readily accessible source of disease, or toxin. A contaminated water supply can lead to outbreaks of contagious diseases like *typhoid or *cholera; their spread is particularly rapid in crowded conditions. Lack of sanitation, especially as once found in medieval cities of Europe, can result in epidemics which kill a large proportion of the population.

An epidemic may sometimes be due to a toxic rather than an infective cause, as in the Japanese outbreak of mercury poisoning caused by the consumption of fish contaminated by industrial effluent.

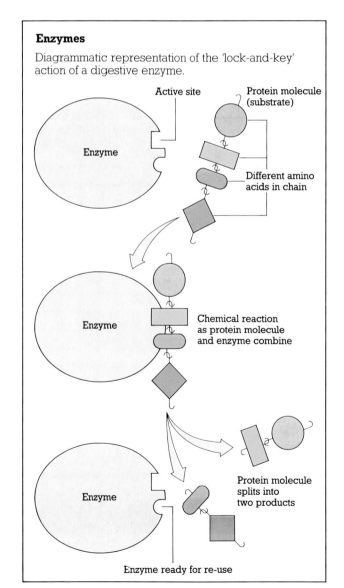

Enzymes

Diagrammatic representation of the 'lock-and-key' action of a digestive enzyme.

Active site

Protein molecule (substrate)

Enzyme

Different amino acids in chain

Enzyme

Chemical reaction as protein molecule and enzyme combine

Enzyme

Protein molecule splits into two products

Enzyme ready for re-use

Epilepsy is a condition in which recurrent interruptions of consciousnness are caused by abnormal bursts of electrical activity in the brain. It may take the form of 'absences' lasting only a few seconds, or prolonged and alarming rhythmic contractions of the whole body, often called fits. Between these extremes every shade of variant exists. Patients have no control over attacks. Brain damage due to accident or disease can lead to epilepsy. In such cases, the 'fit' may cause symptoms in localized parts of the body, depending upon the area of the brain damaged.

Epiphytes are plants which live attached to the outer surface of other plants, but are not necessarily parasitic. Some are climbers which may have some ground-roots, while others germinate on the host plant and have no contact with the ground. The latter are best represented in tropical rain forests, where orchids, *bromeliads, and ferns are the most common types. Some, such as the strangler fig, may grow to envelop the whole host and eventually kill it, the fig being left as a free-standing tree. In temperate regions the bulk of epiphytic plant species are mosses, liverworts, algae, and lichens. In the sea, algae may be epiphytic on other species.

Ergots are fungi which attack the flowering heads of grasses. They have two parts to their life-cycle; the first involves drum-stick shaped fruiting bodies (stromata) on the soil surface, and the second part is as a mass of fungal threads inside the grass seed. The grass flower becomes infected by spores released from the stromatal fungal stage and once inside the flower the spores grow into the developing grass ovary, eventually producing a distinctive hard, blackish-purple structure on the grass head called scelortia. These produce spores of their own which complete the fungus's life-cycle by growing into new stromata if dispersed to suitable germination sites.

The common rye ergot, *Claviceps purpurea*, produces sclerotia, which are commonly referred to as ergots, on the seed heads of rye. These ergots contain a mixture of *alkaloids, such as ergotamine and ergotinine, which if eaten can cause convulsions, or gangrene. The latter, which is first experienced as a burning sensation, gave rise to the popular name of St. Anthony's fire. This 'disease' was contracted by eating bread made from rye flour infected with ergot and caused thousands of deaths in medieval Europe.

Eschscholzias are plants of the poppy family containing about four variable species of annual and perennial herbs, all native to northwest America. The California poppy, *Eschscholzia californica*, is the best known and, as a garden plant, is one of the brightest of annuals.

Esparto grass is a large, tussock grass from North Africa, from which a valuable leaf fibre is obtained. Long-leaved varieties yield a product suitable for rope and twine manufacture, while other types are particularly valuable for paper-making. It has been cultivated in Spain for centuries and in Africa the natural stands are harvested by hand. This harvesting is deliberately controlled by limiting the leaf collection to certain months of the year, so that stocks do not become too depleted.

Ethology is the study of *animal behaviour in a natural environment. It is particularly concerned with the animal's interactions with others of the same species and the function of the behaviour. It is particularly directed towards how the evolution of behaviour has been influenced by *natural selection.

Eucalyptus is a genus of some 500 species of trees and suckering shrubs of Australasia, penetrating northwards into the Lesser Sunda Island of Indonesia. They are relatives of myrtles and cloves. They all have leathery leaves and an unusual lid, or operculum, over the flower bud, which is lost at flowering. They are typical of the Australian landscape and some, such as the coolabah tree, *Eucalyptus microtheca*, are found in arid country where few other trees will grow. They may be classified in groups according to the form of the bark: gums have smooth bark; blood-woods have scales; stringy-barks have fibrous bark; iron-barks are black when old, and so on. The suckering species are known as mallee.

The tropical *E. regnans* reaches some 97 m. (320 ft.) in height and 7·5 m. (25 ft.) in girth. Many species are of rapid growth, so that they are widely planted for timber. The blue gum, *E. saligna*, and other species yield oil of eucalyptus, used in medicine, and it and a few other species can be grown in cool, temperate regions.

They provide the staple diet of a number of animals, including koala bears. Their ecology is adapted to fire, for their oil is highly inflammable so that the trees can explode, and the stringy bark of some species encourages

These adult **euphausid shrimps**, *Euphausia superba*, are about 5 cm. (2 in.) long and are normally found on or near the surface of Antarctic waters. They feed on phytoplankton carried by southward currents.

flames to leap into the crown. The seeds of some species will germinate only after the fruits have been exposed to fire; the seedlings then develop in an environment where there is no competition, the parent trees having been burnt.

Eugenics is the scientific improvement of the hereditary properties of organisms, or the application of the science of *genetics to develop improved breeds or strains of domestic animals and plants. The main method is to encourage individuals with certain characteristics to breed and prevent others, which are weaker or lack the required characteristics, from doing so. Eugenics is responsible for the production of all improved agricultural crops and varieties. A eugenics movement was founded in the early part of the twentieth century by Sir Francis Galton, with a view to 'improving' the human race. His ideas are now generally unpopular.

Euglena is a genus of single-celled organisms found in fresh water. They are usually considered to be *algae as they contain the green pigment chlorophyll and can make food by *photosynthesis. If deprived of sunlight many species of *Euglena* can utilize ready-made foods from their surroundings, strictly speaking a feature of animals, not plants. All species have two long, whip-like flagellae with which they propel themselves through the water. They are often classified as part of the group Protozoa and called 'plant-like' flagellates.

Euphausid shrimps, or krill, are one of the most abundant orders of *crustaceans in the plankton of southern oceans, but with only ninety species. They are the staple diet of whales and large fishes, due to their convenient tendency to swarm. Primitive filter-feeding relatives of the true *shrimp, they are closely related to the crab, and lobsters. Most species of krill are luminescent, possibly as an adaptation to bring individuals together in swarms. Their density can reach 60,000 per cubic metre (1,700 per cubic foot), and a single blue whale may eat 4 tonnes (4·5 US tons) a day.

Euphorbias   *Spurges.

Eurypterids   *Sea scorpions.

Eutrophication occurs when there is an accumulation of minerals, particularly phosphates (from sewage and detergents) and nitrates (from agricultural drainage) in a river or lake. This nutrient enrichment accelerates the growth of all the plants but particularly that of aquatic *algae. Their rapid growth can block waterways and interfere with drinking-water supplies. In summer, vast population increases in the algae ('algal bloom') may create a dense green mat over the water. This may prevent light reaching other plants, which then die. The rapid growth of algae and the mass of decomposing plant material eventually use up all of the oxygen in the water. This leads to the eventual death of the algae themselves and proliferation of bacteria which do not require oxygen, and produce hydrogen sulphide, a toxic gas. This smells of 'bad eggs' and kills most animal life in the water.

Evening primroses, of which a common species is *Oenothera biennis*, are scented annual, biennial, or perennial plants with large, often yellow flowers. These open at dusk and are pollinated by nocturnal moths. They have fleshy roots which are sometimes used as a vegetable. The genus contains over 200 species, all native to North and South America and part of the same family as clarkia and fuchsias. Some are widely spread having escaped from gardens, and as such may be found throughout the Old World.

Evergreen oak   *Oaks.

Evergreens are woody plants which keep a cover of leaves throughout the year, even during adverse periods such as drought or winter. They include most *gymnosperms and a wide range of *angiosperms. Familiar examples in cold climates are pines, hollies, and ivy. The majority of tropical rain-forest trees are also evergreen. The leaves of all evergreens are usually leathery and more resistant to insect attack than those of deciduous plants. They may have a lifespan of several years but eventually fall as new leaves unfold.

Evolution is the natural process by which all organisms are constantly changing in shape, physiology, and behaviour. These changes are continual, albeit at a slow rate, and are a response to *natural selection. Evolution is one theory of the origin of species. Its main rival is the theory of separate creation, according to which different species have separate, not common, ancestry. The latter theory was widely accepted until 1859, but then Charles *Darwin assembled the facts of artificial selection, geographical variation, embryology, classification, and palaeontology, all of which suggested that evolution, rather than separate creation, was the correct theory of the origin of species. He also provided a theory, the theory of natural selection, which explained why evolution takes place. Other biologists before Darwin, such as *Lamarck, had suggested evolution, but Darwin was the first to bring about its wide acceptance. It is now supported by a mass of evidence, is accepted by virtually all biologists, and remains controversial only in some religious circles.

The course of evolution is now known in some detail. Life on earth originated about 3,500 million years ago and all life on this planet is probably descended from a single ancestral species. From the origin, increasingly complex but still single-celled organisms evolved until multicellular organisms arose about 700 million years ago. Most of the main kinds of animal, such as worms, arthropods, and molluscs, arose in a 'Cambrian explosion' about 600 million years ago. The first vertebrates evolved soon afterwards. Modern mammals and primates evolved only in the last 60 million years, after the dinosaurs became extinct.

Excretion is a term for the disposal of compounds which are surplus to requirements or harmful to the organism. Examples of common waste products include carbon dioxide, fatty acids, and nitrogen compounds such as ammonia and urea. Terrestrial animals, who have a particular need to conserve water, use the kidneys, lungs, alimentary tract, and skin as organs of excretion, although this is not their sole function. Of these, the lungs and kidneys are the most important. Excess carbon dioxide can be blown off rapidly from the lungs if the breathing rate is increased.

Exoskeleton refers to the characteristic tough outer skeleton or covering of some invertebrates, such as crabs and

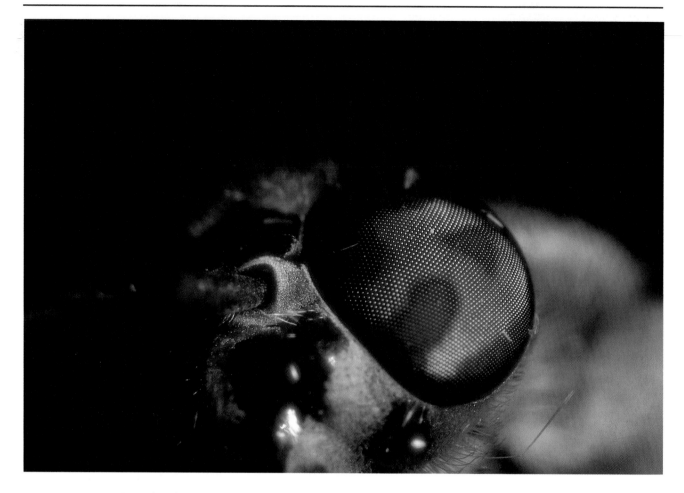

The highly developed compound **eye** of a horse fly, *Chrysops caecutiens*, consists of hundreds of individual facets. The colour bands, seen here, are characteristic of horse flies, and are thought to help the fly see with increased contrast by acting as filters.

insects. It does not grow, and so animals with an exo-skeleton can increase their size only by shedding or 'moul-ting' it at intervals. It is jointed, like armour, to allow movement, particularly by jointed limbs. It is made chiefly of a substance called chitin, which is exceedingly hard and durable.

Extrasensory perception is the supposed ability of some people to predict, or describe, events to which they can have had no direct access. Clairvoyance, precognition, and telepathy all claim to involve extrasensory perception.

Eyes are the organs of sight or light detection found in animals. In lower animals they consist of light-sensitive cells on the skin surface, usually in the fore-part of the body. In the course of evolution such cells have developed into complex organs. These vary from the simple eyes of planarians to the compound eyes of insects, or the eye of vertebrates. Flies have complex compound eyes, par-ticularly developed to distinguish movement and colour. Birds which hunt prey have eyes which focus in front, like those of man, while birds which are hunted have eyes on opposite sides of the head to cover a large field of vision. In man, apes, and some other higher species the eyes permit binocular vision and the ability to recognize and appreciate patterns and shapes. Colour vision is found to

some extent in virtually all animals. Some animals are, however, colour-blind and in some cave-dwelling or deep-sea animals the eyes have become *vestigial organs.

In mammals the eyes are globes lying in fluid-lubricated sockets in the skull. They are moved in all directions by muscles. The eye is protected by a lid lubricated with tears and is enclosed within a thick fibrous capsule modified in front as a transparent disc, the cornea. Beneath this is an adjustable ring of muscle called the iris, which can open in dull light and close in bright light. The lens of the eye lies behind the iris and is controlled by a muscle which can contract to make the lens thin, for looking at distant objects, or relax to make it thicker for close-up work. Inside the back of the eye is a lining of light-sensitive cells of two types. One type (cones) can detect coloured light, the other (rods) is responsible for black and white images. Between the lens and the retina is a transparent jelly. In man, light from outside passes through the lens and transparent jelly, to be focused on the retina. In twilight, colours cannot be seen because cones need a high light intensity to function correctly. In these conditions rods, which respond to lower levels of light, are used to dis-tinguish shapes and moving objects.

Eyeworms *Roundworms.

F

Fabre, Jean Henri (1823-1915), French entomologist, wrote *Souvenirs entomologiques* (10 vols., 1879-1908), which is notable for the great precision of its observations and wealth of patiently collected detail. He showed how large a part instinct plays in insect life, and rejected the theory of evolution as irrelevant to observed instinctive behaviour. His works have been extensively published as translated monographs, including *The Life of the Fly* (1913) and *The Life of the Weevil* (1919).

Fabricius, Hieronymus (1537?-1619), Italian anatomist and surgeon, published a description of the valves in veins (1603). This was of importance to his pupil, William *Harvey, who later described the circulation of blood in the body. Fabricius's chief work was on the anatomy and physiology of the foetus; he began the study of comparative embryology and described the formation of the chick in the egg.

Faeces are mixtures of dietary waste material, cells, bacteria, mineral salts, and secretory debris shed from the alimentary canal. In most terrestrial vertebrates they accumulate in the rectum to be discharged from the anus at irregular intervals. Their odour, colour, bulk, and consistency vary according to the animal and its food. Healthy humans excrete semi-solid faeces, the bulk of which depends on the amount of plant material eaten. Cellulose (roughage) passes unchanged into the faeces.

Fairy mosses are diminutive free-floating ferns, which cover the surface of pools and slow-moving streams. They are found mainly in tropical and subtropical regions but a few species reach temperate latitudes; of the five or six species in the genus *Azolla*, two are native to the USA and are naturalized in parts of Europe. Their stems bear a series of overlapping, paired leaves, the upper ones floating and green, the lower ones colourless. A few thread-like roots hang into the water and the upper leaves have cavities inhabited by a symbiotic *blue-green alga, *Anabaena* species, which 'fixes' nitrogen for its host. Fairy mosses reproduce sexually in the normal manner of *ferns, and can also reproduce vegetatively, parts of the plants being dispersed by water-birds.

Falcons form a large subfamily of diurnal birds of prey, typically with long, narrow wings, though the name includes broad-winged forest falcons and buzzard-like caracaras. True falcons include the tropical falconets, which are mostly small insect-feeding species, and the larger 'typical' falcons, fast-flying predators which attack other birds in flight. They range in size from the gyrfalcon, lanner, and peregrine to the much smaller hobbies and kestrels. Common characteristics are brown, or occasionally barred grey plumage, pointed wings, short tail, a short curved beak, and long clawed talons. With remarkable speed and agility they outfly other birds and strike them in flight, following them down to kill on the ground. Their spectacular hunting methods form the basis of the art of falconry. Falcons usually nest in isolation among crags and on cliffs, laying two to five eggs. There is a considerable trade in the eggs and young birds of the larger species, which are valued in falconry.

Fallopian (uterine) tubes in mammals carry the egg from ovary to uterus. The tube walls contain involuntary muscle and are covered internally by a membrane, lined with cilia, hair-like projections which help to propel the eggs. Fertilization takes place as the ovum passes along the tube.

Fallow deer: a species of *deer indigenous to southern Europe and Asia Minor, but now introduced to many parts of the world. The summer coat is reddish-brown with white or yellowish spots. Along its flank is a line of white hairs, while there is a line of black hairs down the spine and tail. In winter the coat becomes a uniform greyish-brown. The male or buck stands 90 cm. (3 ft.) at the shoulder; the doe is smaller. Only the male carries antlers; these take some six to seven years to develop fully. Usually one fawn is born; twins are rare. The fawn is notable for the large white spots on its coat. The fallow deer inhabits deciduous lowland woods, with thickets for cover and access to pasture land. It feeds on grasses, fruit, herbs, and shrubs, at dusk and dawn.

False scorpions *Pseudoscorpions.

False vampire bats are so called because it was once erroneously thought that they feed on blood, like true *vampire bats. A better name is yellow-winged bats. Most of the five species are carnivorous, eating small mammals and vertebrates, as well as insects and spiders. They occur in Africa, Southeast Asia, and Australia, and are characterized by very large, ears. The name is also given to some tropical American bats, such as the American false vampire, *Vampyrum spectrum*, a spear-nosed bat.

Families, within the system of *nomenclature in zoology and botany, rank below *orders and contain groups of *genera. One particular genus, which is considered to be typical of the whole family, gives its Latin name, with an appropriate ending (-idae for animals, -aceae for plants) to the family. For example, the cat family is called Felidae, as *Felis* is the most typical genus within that family, and the primrose family is Primulaceae, after the genus *Primula*, which includes the primrose. Large families may be divided into subfamilies by further grouping of similar genera.

Fantails are close relatives of the Old World *flycatchers. They are small long-tailed birds, 15-20 cm. (6-8 in.) long, that live primarily in Southeast Asia. They comprise their own subfamily of fantail flycatchers with some forty species in the genus *Rhipidura*; they are mainly brown and grey in colour, though many have white tips to the tail feathers. The Australian willie wagtail, *R. leucophrys*, is a glossy blue-black with white underparts and a white eye-stripe. It is found throughout Australia, except Tasmania. All species of *Rhipidura* are arboreal and feed on insects. They build finely woven nests in the forks of horizontal branches.

Fanworms are polychaete, *annelid worms, responsible for the protruding tubes found at low-tide levels on sandy beaches. A patient observer will see these tubes trans-

The spiral tentacles of the **fanworm**, *Spirobranchus giganteus*, extend from its burrow, which is excavated in living coral in a reef off the Cayman Islands.

formed when sea water covers the tube and the worm's head emerges, bearing beautiful fans of tentacles which filter detritus and plankton out of the water. The tentacles are also respiratory, and may bear simple eyes. Fanworms build their protective tubes out of sand-grains, chalk, or parchment-like mucus. The segmented part of the body remains concealed in the tube.

Fats are natural organic compounds consisting of a mixture of *lipids. They are solids at room temperature and each molecule of the most common biological fats consists of one molecule of *glycerol (a fatty alcohol) linked with three molecules of *fatty acids. They are an efficient energy store for any mobile organism, yielding more than twice as much energy per gram(ounce) as *carbohydrates. Fats are also used for thermal insulation (the subcutaneous fat layer) in birds and mammals and for the mechanical protection of delicate organs, such as kidneys. Fats are also present as energy storage molecules in plants.

Fatty acids are natural organic compounds which fall within the broad class of compounds called *lipids. They are used as subunits for making fats, oils, and waxes (lipids), and the phospholipid components of cell membranes. Each molecule consists of a carboxylic acid group (—COOH) attached to a long chain containing from thirteen to twenty-three carbon atoms with their associated hydrogen atoms. The chain may be 'saturated', containing single bonds only, or 'unsaturated', containing one or more double bonds. Fats normally contain saturated fatty acids, whereas oils contain unsaturated fatty acids; these differences are reflected in the physical state of such compounds at room temperature. Fatty acids in the diet are used as an energy source by the muscles, heart, and other body organs, and provide about 40 per cent of the total daily energy requirement in man. In hibernating animals and migrating birds they are the only important energy source.

Feathers are the most characteristic feature of birds. They are formed of keratin, the same substance as claws, hair, and nails. They vary from simple down feathers with very few 'hair-like' subdivisions (barbs), to the normal body (contour) feathers, which have barbs laid out in two rows on either side of the shaft (quill). Adjacent barbs are hooked to one another by hooks called barbules. The most elaborate feathers are those used for display, such as those of the peacock.

Feathers are very light, essential to a flying animal, and serve three main functions. Firstly they provide the basic overcoat which insulates the bird and prevents its losing heat too rapidly; in waterbirds the contour feathers, and a thick layer of down feathers, provide essential waterproofing. Secondly, they occur in a wide variety of colours, which are used by the bird for camouflage or display; and, thirdly, they form the large surfaces of the wings and tail, which are needed in flight.

Feather stars are fragile *echinoderms, closely related to *sea lilies, usually living well off shore on rocky outcrops or wrecks. The 550 or so species occur mainly in the Indo-Pacific seas. The feathery arms, commonly ten in number, form a food-catching funnel, and also beat alternately when the animal swims. Feather stars and their relatives (many of which are stalked and sessile) are the most ancient and primitive of echinoderms.

Feet (anatomy) are used by the majority of terrestrial animals as organs of propulsion, transferring the power of muscles into motion across the ground. They come in a wide range of shapes and sizes, depending upon the lifestyle of the animal. The feet of birds are modified with claws to grip a perch, or tear at prey. Those of frogs are webbed to facilitate swimming.

The human foot consists of three types of bones: tarsals, metatarsals, and phalanges. The foot is attached to the base of the leg bones through seven tarsal bones in the ankle. These form a flexible joint with the leg, and allow up and down movement of the foot. The tarsal bones are jointed to the slender metatarsal bones, five of them in each foot. The toes, which help man to grip the ground and give himself push, consist of phalangeal bones, the big toe having two and the others three. The human foot is unique in that the weight is supported by a flexible arch. Man walks by 'rocking' from the large ankle bone, called the calcaneum, to his toes. The flexible arch arrangement takes up any jolts which come from falls, or uneven ground.

Most vertebrates have this basic arrangement of bones in the foot, differing mainly in the size and position of each. The feet of horses, cows, camels, and other hooved animals represent a drastic modification, which enables them to run swiftly. They all walk literally on tip-toe. The horse has lost all of its toes except for the third. This is now elongated and, along with the metatarsals and tarsals, forms the lower part of the horse's leg. The hoof is not derived from bone, but is a pad of specialized skin.

Fennec foxes are the smallest species of *fox and yet the one with the largest ears. The length of its body from the tip of the nose to the base of the tail is 40 cm. (16 in.); the ears are 10 cm. (4 in.) long! These enormous ears are associated with the hot, dry climate of the Sahara Desert, the home of this fox and they help to disperse body heat.

The **fennec fox** is well adapted to desert life: its large ears serve to radiate heat from the body, the furred feet act as sand-shoes, and the overall light coloration blends in with the desert sand.

The fennec fox has a pale, sandy-coloured coat and feeds at night on rodents, lizards, birds, locusts, and fruit.

Fennel is a perennial member of the carrot family, native to southern Europe but naturalized in other places, particularly near the coast. It resembles dill in its feathery appearance but has a taste similar to that of anise. The seeds are used as a spice, while the swollen leaf-stems are treated as a vegetable.

Fenugreek is the common name for *Trigonella foenum-graecum*, an annual plant of the pea family. Although a native of the Mediterranean region, it is now widely grown in India, where the seeds are an ingredient of curries and the leaves are eaten as salad. Recent investigations suggest the plant may contain substances with contraceptive properties.

Fermentation, or anaerobic respiration, is the chemical breakdown of organic compounds carried out by living organisms in the absence of oxygen. It releases energy which is stored in the form of *adenosine triphosphate (ATP) for use elsewhere in the body. There are two main forms of fermentation. Alcoholic fermentation produces ethanol and carbon dioxide gas; it occurs in some yeasts and in bacteria and is exploited commercially in making beers, wines, and spirits. Lactic fermentation produces lactic acid only and occurs in the skeletal muscles of vertebrates. Their ability to respire without oxygen helps them to maintain muscular activity during severe exercise. Fermentation by bacteria is used in the manufacture of cheese and yoghurt.

Ferns are leafy plants of the class Filicinae, which are spread by *spores, not seeds. Like algae, mosses and liverworts, clubmoss and horsetails, they grow in two distinct forms that alternate during their life cycle (*see over*). For the main part of this cycle, a fern consists of a short stem (rhizome) from which roots grow down into the soil, and leaves (fronds) grow upwards. This is called the sporophyte generation because it produces spores inside spore-cases on the underside of the fronds. These capsules are shed into the air, and when they fall on moist soil they germinate to produce heart-shaped plants 2–3 cm. (about an inch) across, called prothalli. These bear the sex organs, which in ferns are cavities containing microscopic egg cells and sperms. The sperms swim to fertilize the egg when the prothallus is wet, so ferns are able to reproduce sexually only in damp habitats. After fertilization the egg cell develops into a new sporophyte plant.

Some ferns reproduce asexually by leaf buds or, in the case of bracken, by rhizomes. Many fossil ferns are known, some of which, unlike modern ferns, had seeds. Some modern ferns closely resemble fossil forms but others are considered to be more highly evolved. *Tree ferns are confined to the tropics, where they grow to 13 m. (40 ft.) high. Common temperate species include lady fern, male fern, polypody, and bracken.

Ferrets are domesticated *polecats. There are references to animals that were probably ferrets in the writings of Aristophanes and Aristotle, and domestication probably dates from prehistoric times. Their use then, as now, was to flush rabbits from their burrows. Later they were employed to kill rats. Ferrets spread into Europe, possibly with the rabbit, and historical records of them in England

Ferns: life–cycle

The life-cycle of a fern showing the alternation of generations between the asexual spore-producing fern plant, the sporophyte, and the much smaller sexual phase, the gametophyte. The spores are ejected from the sporangium by the straightening out of the annulus during dry weather.

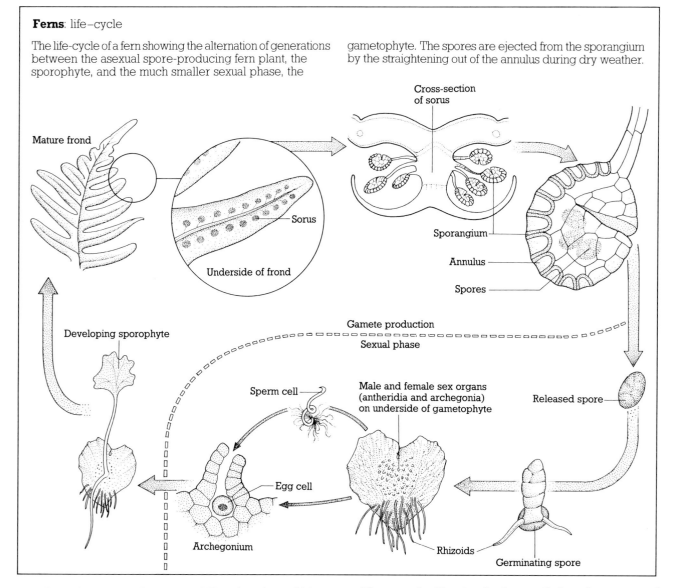

Cross-section of sorus

Mature frond

Sorus

Underside of frond

Sporangium

Annulus

Spores

Gamete production
Sexual phase

Developing sporophyte

Sperm cell

Male and female sex organs (antheridia and archegonia) on underside of gametophyte

Released spore

Egg cell

Archegonium

Rhizoids

Germinating spore

date from the thirteenth century. When hunting rabbits, the ferret is usually muzzled to prevent it killing and eating the rabbit in its burrow. Ferrets have also been used to draw cables through narrow underground pipes. Many are kept purely as pets or for show. There have been many escapes of ferrets into the wild and hybrids (polecat-ferrets) are common. In Britain there is a danger of the extinction of polecats through such hybridization.

Fertilization in animals and plants occurs during sexual *reproduction when a male *gamete (sex cell) fuses with a female gamete to form a zygote (which subsequently develops into a new organism). Self-fertilization is a process which is confined only to some plants and produces *clones.

Male and female gametes have become adapted to perform their different functions. The male gamete moves to fertilize the female and is characteristically smaller, to aid mobility, and more numerous to offset the risk of loss or damage. Female gametes usually contain food reserves for the developing zygote and are thus larger and less likely to move. Fertilization in animals may be external (fishes, amphibians, and some invertebrates) or internal (reptiles, birds, mammals, and some invertebrates).

The term fertilization is also applied to the process of adding manure or inorganic minerals to a soil in order to improve its nutrient status.

Fetus   *Embryos.

Fever, the elevation of body temperature above normal, is often a general reaction to infection or tissue damage. Temperature is stabilized by a 'thermostat' in the brain, which controls heat gain or loss. Heat can be produced by shivering, or prevented from leaving the body by cutting down blood-flow to the skin. Heat can be removed from the body by increasing skin blood flow, or by the evaporation of sweat. In fever, the 'thermostat' setting is raised, often with an exaggeration of the normal diurnal fluctuation. Fever probably has some beneficial effect in combating infection. Some fishes show behavioural 'fever' in that, when infected, they move into water at a higher temperature than they would normally choose.

Feverfew   *Chrysanthemums.

Fibre plants, used for the commercial production of vegetable fibres, have decreased in importance since the development of synthetic fibres. They have been essential to man since ancient times, for making textiles, ropes and twines, and clothing.

Three types of fibre occur in plants. The first includes *cotton and *kapok with the fibres associated with seeds. The second, exemplified by *flax and *jute, has fibres derived from stems. The third type of fibre derives from the leaves of plants such as *sisal.

Fieldfares are birds of the thrush family. This species breeds in wooded areas across much of the central and northern parts of Europe and western Asia, migrating to milder areas for the winter. They are mainly grey in colour, with a paler rump and speckled underparts. Like many thrushes they feed on berries and insects. The untidy, cup-shaped nest is usually made in a tree, and houses three or four eggs. Fieldfares may nest in loose colonies and attack any potential predator.

Fieldmice *Woodmice.

Field vole, or short-tailed vole: a species of rodent whose short tail is its most distinctive feature. Like all *voles, it has a rounded head and is strictly vegetarian. It prefers long, uncultivated grassland, but it may be found in most parts of northern Europe, provided there is some grass. It does not hibernate and may continue to breed into winter but, unlike the bank vole, is not restricted by the food supply.

Fighting fish: a species of Asiatic freshwater fish, *Betta splendens*, which has been kept in semi-domestication for centuries. Native to Thailand, it has now been distributed throughout the world. It occurs commonly in ponds, drainage ditches, and the backwaters of rivers, where it plays an important role in controlling malaria-carrying mosquito larvae. The water in which it lives is frequently deficient in oxygen and it breathes air by means of a respiratory organ above the gills. It lays its eggs in a nest of bubbles produced mainly by the male, and at first the young fishes live in the nest, which keeps them in

Male Siamese **fighting fish** confront each other with fins flared. Their intense coloration and silk-like fins have been developed by breeders over centuries. In contrast, the wild type is predominantly brown, although it does develop a greenish-blue sheen with coloured patches during the breeding season.

oxygen-rich water. The males were kept for their fighting qualities and have been bred to develop brightly coloured and aggressive forms.

Figs are large shrubs or small trees in the same family as hemp and mulberry. The most common of several species, producing the green, brown, or purple fruits, is *Ficus carica*. Originally native to western Asia, figs were known in Egypt as early as 4000 BC, and today are cultivated in the Mediterranean region and California. They have fleshy, fruit-shaped flowers and, in the widely cultivated Smyrna types, require cross-pollination by tiny fig wasps which parasitize a proportion of the flowers. When the plants are grown in a new area, the wasps must also be introduced, although the common or Adriatic fig does not require fertilization. Eaten fresh, dried, or preserved, figs have laxative properties. Many other species of fig are cultivated as ornamental plants, and one, known as the rubber plant (*Ficus elastica*), has become a common house-plant. This genus includes around 600 species.

Fig wasps, a family of minute, black *wasps related to *chalcid wasps, are essential to the maturation of the fruit of figs. The larvae develop in galls (growths) within figs. When the males, which are wingless, emerge, they eat into galls that contain females, and mate with them. These females enter the compound flowers of young figs and, if it is the right kind of fig, they lay their eggs as well as fertilizing the fig flower. If the fig is the wrong kind, they cannot lay, but nevertheless fertilize it. In Smyrna fig culture, suitable figs are hung in the trees to encourage the fig wasps.

Filmy fern *Ferns.

Finch is a term loosely applied to many small, stout-billed birds, but primarily to the Fringillidae, a group of about 125 species including canaries, serins, grosbeaks, and crossbills as well as finches. Most species occur in the Old World, but a small number are found in the Americas. Most are sparrow-sized, though many are stockily built and have very stout beaks for opening the seeds upon which they feed. The hawfinch can split open olive-stones, a feat requiring a pressure of up to 11 kg./cm² (159 lb./sq. in.), no mean effort for a bird weighing 55 g. (2 oz.).

Finfoots are three species of the bird family Heliornithidae. One species is found in tropical America, one in tropical Africa, and a third in Southeast Asia. Rather like small brown or black cormorants, they are up to 60 cm. (24 in.) long, but with large lobes on each toe, rather than webbed feet. They live on streams and in marshes, often within forests, and eat insects, small fish, and amphibia. They build nests of twigs or reeds in low bushes and lay four or five eggs. The young stay in the nest until quite well grown.

Fingers are that part of the hand made of the phalangeal bones. Many animals use fingers for grasping but man is unique among mammals, including apes, because of the ability to oppose thumbs to little fingers. This, among other finger and palm movements, allows humans to hold objects in their palms and to delicately manipulate tools. Fingers and thumbs are moved by numerous small muscles lying in the hands, assisted by the tendons of muscles in the forearm. The skin and muscles of the fingers are well supplied with nerves to permit the recognition of textures and shapes and to allow fine and delicate move-

A **fin whale** surfaces off the coast of Mexico to expel moisture-laden air from its blowhole. This species of whalebone whale has baleen plates designed to trap medium-sized plankton, such as krill and copepods, which form its staple diet.

ment. Each person has a unique set of fingerprint patterns, each finger often having a different pattern.

Fins are flattened appendages which are used in swimming, and are most highly evolved in fishes. The fins of fishes are essentially double flaps of skin with a skeletal support of bony elements in bony fishes and cartilage in sharks, skates, and rays. In bony fishes, especially, the fins can be erected and depressed and some can be moved by means of muscles attached to the bony rays. These rays are of two types: spines, which are strong and sharp and usually found in the anterior of the dorsal and anal fins; and soft or branched rays, which are segmented and finely divided to their tips.

Fins are named according to their position on the fish's body. The single or multiple dorsal fin run along the back; the anal fin projects from behind the vent or anus. The tail or caudal fin provides the chief propelling motion in swimming, while the paddle-like pectoral fins (each side behind the head) and pelvic fins (beneath the belly) keep the swimming fish in a level posture. The adipose fin, situated at the rear of the back, just forward of the caudal fin, is a 'false' fin in that it is made of fatty (or adipose) tissue. Some fins have a protective function, like the dorsal spines of the stickleback; those of the weever-fish and stone-fish are connected to poison glands.

The fin-like flippers of seals, whales, and dolphins have a very different structure. They are essentially modified limbs of the typical terrestrial mammal and have a bony skeleton.

Fin whale, or common *rorqual: a species of whalebone whale. It is the fastest of all the whales, swimming at speeds of 48 km./hour (30 m.p.h.), and is also capable of pulling a ship when harpooned; one individual towed a vessel at 19 km./hour (12 m.p.h.) for three hours while the ship's engine was running three-quarters speed astern. The fin whale is greyish in colour and white below, with a long, slender body, and can reach a length of 24 m. (80 ft.). Mating occurs in warmer seas during winter, before it moves north or south to the polar regions. Ten or twelve months later it returns to calve in tropical waters.

Fir is the name loosely used for a number of species of conifer, but it is more properly applied to the fifty or so species of *Abies*, a genus of the north temperate zone and Central America. All are evergreen, with needles borne directly on the stems and not on short side shoots like those of pines and cedars. The silver fir, *A. alba*, of the mountains of southern Europe, provides turpentine as well as good timber. The turpentine of *A. balsamea*, the balsam fir of eastern North America, is known as Canada balsam. Many species are cultivated for ornament as well as for timber.

Firebrats are larger relatives of *silverfish, coloured with white and black scales. They need a high temperature for survival and are found in bakehouses, large kitchens, and old fireplaces, where they live on residues of human food.

Firecrest: a bird species of the Old World *warbler family. Closely related to the goldcrest and the kinglets, it is a very small greenish bird, 9 cm. (3½ in.) long with a white eye-stripe. On the crown is a bold stripe of colour which is yellowish in females, reddish in males, and gives the species its common name. This stripe on the crown is often concealed by the other crown feathers and displayed only when the bird is excited. The firecrest is virtually restricted to Europe, living in forests and building its tiny nest high up in the canopy of coniferous trees.

Fireflies are soft-bodied, usually brown or black, nocturnal *beetles of the family Lampyridae, which are luminous as adults and sometimes as larvae. In some species, such as the glow-worms, *Lampyris* species, the females are wingless and look like larvae. The 1,700 species of fireflies and glow-worms are widely distributed, but particularly numerous in warm climates and damp places. The luminous organs are often borne on the last few segments of the abdomen and the colour, intensity, and intervals of flashing are particular to each species. The light production is often synchronized, and all members of the same species may flash together. Members of several other groups of beetles, such as the cucujos and other *click-beetles, are luminous and sometimes called fireflies.

Fireweed　　*Willow-herb.

Fisher, Sir Ronald Aylmer (1890–1962), English geneticist and statistician, wrote *Statistical Methods for Research Workers* (1925), which was extremely influential in biological research. It was followed by *The Genetical Theory of Natural Selection* (1930), putting forward his views on eugenics and contributing greatly to the study of population genetics.

Fishes are aquatic *vertebrates, which obtain their oxygen by means of gills. There are four major groups: the *cartilaginous fishes, *bony fishes, *lungfishes, and *jawless fishes.

Most fishes are predatory, and very few subsist wholly on plant material, although some eat both plants and animals. Most lay eggs, usually very numerous, which are typically shed in the water and are unprotected; but parental care of eggs is frequent, either in nests, by mouth brooding, or by brooding by the male. All fish are adapted to moving through the water using fins as the organs of propulsion. By far the best adapted group is that of the bony fishes, which contain the largest number of species of all the groups.

Fistulae are abnormal connections between two body cavities, or a cavity and the body surface. They may be caused by an *abscess bursting, or by the breakdown of a tumour, especially after treatment with X-rays. Non-tumorous fistulae tend to heal, provided the normal drainage of the cavities is unobstructed.

Flagellates are all *protozoan organisms which move using long whip-like appendages called flagella. They are confusingly intermediate between animals and plants. Some move freely but make their own food using *photosynthesis; some prey on other organisms; and some are parasites or symbionts. In fact the group has little in common *except* flagella. They are often divided into plant-like species (phytoflagellates) and animal-like species (zooflagellates).

Phytoflagellates include protozoans such as *Euglena*, *Chladymonas*, and *Volvox*, the latter species forming colonies of these single-celled organisms. Zooflagellates include the wood-digesting symbiont flagellates which are found in termite guts. Other examples live fully within animal tissues, and flagellates are responsible for the disease sleeping-sickness and leishmaniasis in man, both diseases being transmitted by the bites of insects.

Flag iris is a name used for colourful herbaceous *irises. One of the most familiar is the yellow flag, *Iris pseudacorus*, a wetland species from Europe and western Asia.

Flamboyant tree, or poinciana: a beautiful evergreen tree of the genus *Delonix*, and belonging to the pea family. It is native to Madagascar but very rare there. It is widely cultivated throughout the tropics for its large, bright orange-red flowers. A second species, rarely cultivated, is native to East Africa.

Flamingos are a small family of five species of water birds remarkable for their exceptionally long necks and legs and unique crooked bills, which are held upside down to filter algae or small animals from mud and water. They are white or rosy pink in colour, with vivid crimson and black wings, and the largest stand about 150 cm. (5 ft.) high. They inhabit remote salt-water, brackish, or alkaline lakes and lagoons in flocks which frequently number many

A greater **flamingo**, *Phoenicopterus ruber*, shields its newly hatched chick from the sun at Lake Elmenteita, Kenya. The chick's bill does not become fully formed and functional until some ten weeks after hatching.

thousands. Breeding in dense colonies, they build nests of mud in the shape of truncated cones and nurse their young in crèches.

Flash coloration is the term used to describe areas of bright coloration revealed when an animal is disturbed; they are intended to provide protection from predators. The phenomenon is commonest in insects, particularly grasshoppers and moths, which habitually rest in exposed sites and primarily rely upon *camouflage patterns. If disturbed by a predator such species will open their wings to reveal a brilliant patch of colour. Both the colour flashes and their sudden disappearance may deter or confuse a predator, allowing the insect to escape.

Flat-fishes comprise some 500 species of mostly marine fishes living in the Atlantic, Indian, and Pacific oceans, in shallow water down to depths of 1,000 m. (3,250 ft.). They are not very abundant in deeper water and very few are known from the polar seas of the Arctic. Some soles and tongue-soles live in fresh water, and some, such as the European flounder, do so temporarily.

Flat-fishes are distinguished by an asymmetric appearance, with both eyes on one side of the head, the mouth often twisted close to the eyes, and the dorsal and anal fins running along the edges of the body. The eyed side is coloured, the blind side white. Flat-fishes are totally adapted to life on the sea-bed. Upon hatching, the larva is bilaterally symmetrical and swims upright, like other fish larvae. Early in its development one eye migrates across the roof of the skull to lie next to the other eye, and the young fish then turns and begins to swim on one side. Whether it is the left or the right eye that migrates varies according to species or groups of species, giving rise to left-eyed and right-eyed categories. The changes which follow this movement of the eye may cause the jaws to become twisted, the pectoral fin on the blind side to atrophy, and the nostrils to be differently developed.

Flatworms, or planarians, can be found in almost any pond or stream, their tiny, flat bodies gliding along and their heads, often with two or more simple eyes, waving from side to side 'tasting' the water. These animals belong to the class Turbellaria, one of three in the phylum Platyhelminthes, the others being liver-flukes and tapeworms. Flatworms have no *coelom or blood vessels, so are always small, flattened, and very simply constructed to allow oxygen to diffuse into all parts of their body. Most free-living forms have a blind-ending gut, so they are unable to eat large items of food. They have a small, protrusible pharynx, used for feeding on detritus and tiny invertebrates. Some of the parasitic flatworms have no gut, and feed by absorption. Such flatworms are adapted for rapid reproduction.

Flax is a fibre plant that has been cultivated for thousands of years in the Mediterranean area and Europe generally. The best fibre is produced in cool, moist, temperate climates and from crops sown densely to induce slender, unbranched stems. The crop is not cut but pulled to maximize the length of the harvested stems, from which the fibres are extracted. Separation involves retting and scutching (beating) and the fibre may be woven into the fine, strong fabric—linen.

A faster-growing type, more adapted to the warmer climates of southern Asia and India, was selected for its

Flat-fishes

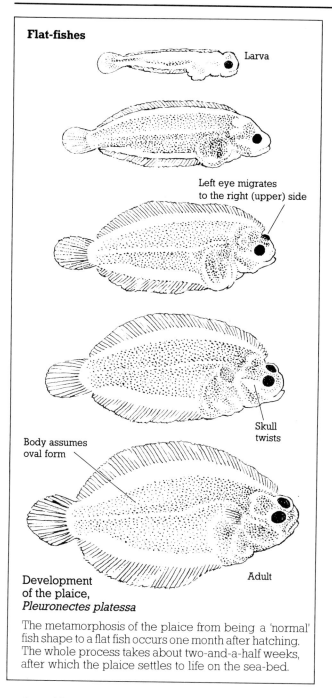

Larva

Left eye migrates
to the right (upper) side

Skull
twists

Body assumes
oval form

Adult

**Development
of the plaice,
*Pleuronectes platessa***

The metamorphosis of the plaice from being a 'normal'
fish shape to a flat fish occurs one month after hatching.
The whole process takes about two-and-a-half weeks,
after which the plaice settles to life on the sea-bed.

mals and birds. There are about 1,400 species, some sixty
of which occur in the British Isles. Being laterally compressed and hard, they move easily through fur and feathers; rows of backwardly pointing spines and bristles,
and strong claws, help them to cling to their host. Their
mouthparts are modified for piercing and injecting saliva
simultaneously with sucking. All fleas have powerful legs,
particularly the hind pair, and can jump more than 30
cm. (12 in.) high, two hundred times their body length,
thereby exposing their bodies to forces which would tear
a man apart. Females require a blood meal before laying
their eggs in the host's nest, or on the ground in its dwell-
ing. The elongate, pale larvae are scavengers, feeding on
detritus, and eventually they pupate in cocoons covered
by debris. Vibration, even after years of dormancy, stimulates emergence of the adults, which are attracted by
warmth. They can go several weeks before the first meal
and between meals.

The cosmopolitan human flea, *Pulex irritans*, became
associated with man when he lived in caves. Modern
homes are too clean and dry for larvae, and numbers have
declined, though they thrive on pigs. Female jiggers, or
chigoes, *Tunga penetrans*, found in many warm countries,
burrow under the skin of man and other animals, and
swell into pea-sized egg sacs. Rat fleas, *Xenopsylla cheopis*,
transmit the virus causing bubonic plague, while rabbit
fleas, *Spilopsyllus cuniculi*, are the main vectors of
myxomatosis.

Fleming, Sir Alexander (1881–1955), Scottish bacteriologist, in 1928 noticed the strong antibiotic effect of
a mould contaminating one of his cultures. The mould
produced a substance he called penicillin, which killed
organisms causing certain common infectious diseases, but
showed low toxicity to animals. The drug subsequently
proved effective against pneumonia, meningitis, and
venereal and other diseases.

Alexander **Fleming** at his laboratory bench in Queen
Mary's Hospital, London, where he discovered
penicillin. In 1945 Fleming shared the Nobel Prize for
medicine with Florey and Chain, who developed a
method of producing the drug in large quantities.

seeds and is known as linseed. This seed is rich in a drying
oil that is valuable as a solvent for paints and varnishes
and is also rich in protein. Used in small quantities, after
being boiled to remove poisonous compounds, the seed
may be fed to cattle to put a gloss on their coats.

Fleabanes (botanical) are perennial herbaceous plants
with showy, daisy-like flowers. This name is used for at
least three genera of the sunflower family. The common
fleabane of Europe is *Pulicaria dysenterica*, while the garden
'fleabanes' belong to the genus *Erigeron*. Others with this
as part of their name include members of the genus *Inula*.
The common fleabane was once used to treat skin disorders and dysentery.

Fleas are small, brown, wingless insects of the order
Siphonoptera, which feed by sucking blood from mam-

Flesh flies are greyish flies with a dark-striped thorax and mottled abdomen. Eggs hatch within the female's body, and larvae are deposited on decaying animal matter. This they liquefy with regurgitated *enzymes, and ingest while buoyed up by fleshy lobes surrounding the posterior breathing tubes. Pupation takes place in the soil. Flesh flies belong to the same family as *blowflies.

Flexner, Simon (1863–1946), American epidemiologist, was responsible for the discovery of a serum for treating cerebrospinal meningitis. He isolated the bacillus which can cause dysentery, and showed that poliomyelitis is caused by a virus.

Flickers   *Woodpeckers.

Flies are insects which belong to the order Diptera, which includes some 85,000 species worldwide. They are distinguished from other flying insects by having only one pair of wings, the hind-pair, present in other insects, having been modified into drumstick-shaped, gyroscopic organs called halteres. A few parasitic species, such as keds, are wingless. Most have large heads and eyes. Many other sorts of insect are popularly called 'flies', although they do not belong to the order Diptera. It is a diverse group that exploits a range of habitats and life-styles, and is successful by any criterion. Bot flies and others are pests of man or animals; bloodsuckers such as blackflies, and scavengers such as blowflies, are vectors of disease; and many, such as carrot-flies, are pests of cultivated plants.

Adults have sucking mouthparts variously modified for piercing (mosquitoes) or for lapping (houseflies), although they may be *vestigial. Most species eat nectar, but many are blood-sucking or predaceous. Some, such as horseflies, are stout, but the majority, such as midges, are small and soft-bodied.

*Metamorphosis is complete, and in many species the last larval skin forms a hard puparium, or 'pupal case'. The legless larvae (or maggots) have mandibles or sharp mouth hooks, and they feed in a variety of ways: some burrow in plant leaves, stems, or roots, or form galls; others are predatory in soil, water, or vegetation; but the majority feed on decaying organic matter. Many species are aquatic, and a few exploit such unlikely habitats as crude petroleum, salt lakes, and formalin-soaked carcasses.

Flight (vertebrate) is the mode of locomotion adapted by birds, bats, and insects. A few other animals have adaptations which enable them to glide, but true flight is unique to the above groups. The mechanisms of *insect flight differ from those of vertebrates such as birds or bats and are considered separately.

Bird wings have an aerofoil cross-section: the top surface of the wing is convex; the lower surface concave. As air moves over them they generate lift. Many gulls, terns, swifts, and albatrosses are able to glide long distances by heading into winds and upcurrents, thus gaining lift. Where there are no air currents, or where extra lift is required, the wings have to be flapped. The secondary *feathers connected to the 'forearm' give the wing its aerofoil section. The primary feathers, attached to the 'hand', can be rotated for extra control and on some birds may separate on the upstroke of the wing to save energy. The wings are raised by the elevator muscles situated along the breastbone and connected via long tendons. The powerful depressor muscles, also along the breastbone, then sweep the wings downwards and backwards to give a thrust that provides both lift and propulsion. The tail feathers provide braking, balance, and steering.

Bats, although they evolved separately from birds, have a very similar skeleton and mode of flight. The wings are supported by adapted 'forearms' and 'hands' and the power is provided by large muscles on the breastbone, the essential difference being that the aerofoil of the wing is provided by flaps of skin, which extend to the legs as well as to the 'arms'.

Florey, Howard Walter, Baron (1898–1968), Australian pathologist, with Sir Ernst Chain in Oxford purified penicillin and developed efficient means of producing this antibiotic. As a result of these advances, he became the main instigator of penicillin therapy. He tried it out on war wounds in North Africa in 1943 and shared the 1945 Nobel Prize for medicine with *Fleming and Chain. With the aid of six collaborators, he published *Antibiotics* (2 vols.) in 1949.

Flounders are *flat-fish living in coastal waters from the shoreline to 50 m. (162 ft.) depth. The European flounder, *Platichthys flesus*, is a right-eyed flat-fish which is unique in that the young fish lives in fresh water, mainly in rivers, which it ascends when about 2 cm. (1 in.) long, returning to the sea when adult. It lives on soft bottoms of both mud and sand, feeding on bottom-living crustaceans, worms, and molluscs, and is dull brown in colour, with the eyeless side opaque white. Many flat-fishes in North America are known as flounders, but only the starry flounder, *P. stellatus*, is a close relative of the European one.

Flowering onions   *Onions.

Flowerpeckers are agile birds which live in the tops of trees and bushes from Southeast Asia to Australia, comprising a family of some fifty-eight species. They are warbler- to sparrow-sized, mainly dullish brown birds, but a few species have strong colours. The scarlet-backed flowerpecker, *Dicaeum cruentatum*, has a flaming-red back. They probe into flowers for nectar (hence their name) and also eat many berries, especially those of mistletoe—some species are commonly known as mistletoe birds. Eight Australian species are called pardalotes.

Flowers are the parts of plants concerned with sexual reproduction. Each typically consists of an ovary surrounded by other structures. These usually include the male organs, or stamens, which produce pollen. Most flowers have leaf-like petals, which may be brightly coloured, and sepals, which often protect the flower in bud. Flowers may be solitary as in the tulip, or borne in clusters which are called inflorescences.

Flowers are often said to be unique to the *angiosperms, and they are referred to as 'flowering plants'. In a strict botanical sense *gymnosperms also have 'flowers', although these lack many of the structures of angiosperm flowers. In many trees and in some plants, the flowers are unisexual, bearing either ovaries, or stamens. When the sexual organs occur on different plants, the trees or plants are said to be dioecious (like most animals). When together on the same individual, either in the same inflorescence or not, they are said to be monoecious.

Flowers may be less than 1 mm. across, as in some trees,

Flowers

No two angiosperm flowers, whether monocotyle-donous or dicotyledonous, have exactly the same structure nor, necessarily, the same components. Nevertheless, some broad generalizations can be made and these diagrams show the structure of what may loosely be termed a 'typical' dicotyledonous flower and a more specialized monocotyledonous orchid with a precise insect pollinator. The function of most flowers is to secure successful fertilization of their ovules.

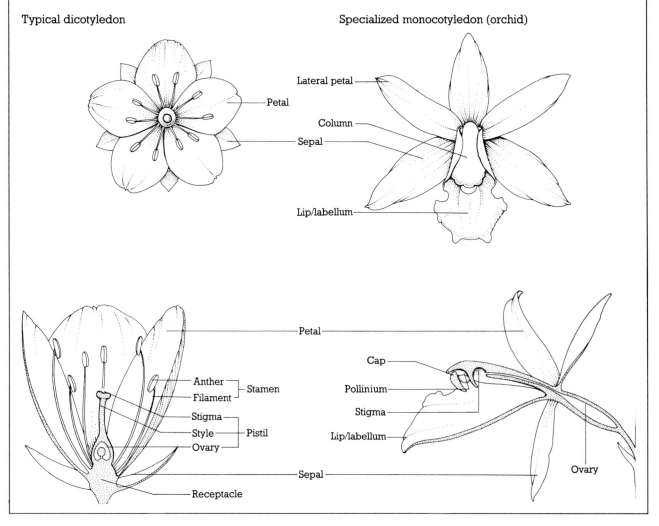

Typical dicotyledon

Petal

Specialized monocotyledon (orchid)

Lateral petal
Column
Sepal
Lip/labellum

Petal

Anther — Stamen
Filament
Stigma
Style — Pistil
Ovary
Receptacle

Cap
Pollinium
Stigma
Lip/labellum
Sepal
Ovary

or up to 1 m. (3 ft.) across, as in the parasitic *Rafflesia arnoldii*, a rare plant of Sumatra. Inflorescences may be several metres tall in some palms. What is commonly called a 'flower' on plants of the sunflower family and a few other groups, is in fact an inflorescence of many flowers.

The function of flowers is to present the *gametes for cross-fertilization, which is effected by the dispersal of *pollen grains. In plants where the pollen is carried by insects, the flowers are often large, brightly coloured, and fragrant to attract these pollinators. Most produce nectar on which the insects feed.

Flukes are *flatworms belonging to the class Trematoda. There are two main orders of fluke: the monogenean flukes which are usually ectoparasites on fishes, reptiles, or other cold-blooded animals, or the digenean flukes with complex life-cycles. Many of the 6,000 or so species of digenean flukes are parasites within the gut or internal organs of man and other vertebrates. They include *liver flukes and *blood flukes among many medically important species.

Fly agaric *Death cap.

Flycatchers belong to one of two distantly related families of birds, the New World (or tyrant) flycatchers, Tyrannidae, including about 360 species; and the Old World flycatchers, Muscicapidae, with 134 species. Tyrant flycatchers take their name from one of the first species described, the kingbird, *Tyrannus tyrannus*, which aggressively attacks birds of any size that come near its nest. Most members of both flycatcher families are insectivorous, many hunting by perching on branches and then sallying out after flying insects. Almost all are small, usually less than 23 cm. (9 in.) in size, and the large majority are dull brown in colour.

Flying fishes, of which there are about fifty species, are inhabitants of tropical and warm temperate oceans, moving into temperate seas with seasonal warming of the water. Members of the family Exocoetidae, they all have very large pectoral fins which are spread wide and used as 'wings' to glide once they break through the sea's surface.

To escape predators, **flying fish** surface over the Red Sea. Leaving the water at a speed of about 32 km./hour (20 m.p.h.), they open their pectoral fins, and skull the water with their tail to gain momentum. By opening their pelvic fins, they achieve the necessary lift to make themselves airborne; they may return to the skulling position for further glides.

Propulsion is provided by normal swimming under water and the fishes' 'wings' do not beat. They are highly adapted for life at the surface, being blue above and silvery white beneath.

Flying foxes are tropical, fruit-eating *bats of the sub-order Megachiroptera. The common name derives from the fox-like head for, unlike other bats, flying foxes lack the grotesque leaf-like folding on the nostrils necessary for echo-location. They do not need such navigational aids to find fruiting trees even though they are nocturnal and most species rely on their excellent eyesight. The name flying fox may be given to all 173 species of fruit bats, but it is usually restricted to the genus *Pteropus*, whose sixty-eight species range from Madagascar to Australia. They are of economic importance as pests of fruit crops and, in some countries, as human food. They include the largest bat, the Samoan flying fox, *Pteropus samoensis*, with a wing-span of 2 m. (6 ft. 6 in.).

Flying frogs, or gliding frogs, are members of one of the tree frog families, Rhacophoridae. They do not have a true flapping flight, but 'fly' by spreading the fingers and toes, increasing their surface area for gliding, and carefully controlling their descent from one tree to the next, using limb, finger, and toe movements. Their fingers and toes are unusually long and have fully developed webbing between them. There are also flaps of skin along the arms and legs. These tree frogs are found in Southeast Asia and have the added adaptation of sucker-tipped toes to help them grip leaves and tree trunks.

Flying gurnards are marine fishes occurring in the tropical and warm temperate oceans. There are about four species, all with blunt, heavily armoured heads and enormously long pectoral-fin rays. These are brightly coloured and are expanded suddenly in order to frighten predators, although they also act as 'wings' in short jumps out of the sea. These bottom-living fishes do not fly but walk along the sea-bed on their short pelvic fins.

Flying lemurs *Colugo.

Flying lizards are forest-dwelling *agamid lizards of the genus *Draco*, which occur in India, Southeast Asia, and Indonesia. They have the remarkable capacity of controlled gliding over distances of around 18 m. (59 ft.) from tree to tree. Their 'wings' are formed by flaps of skin on either side of the body; when the lizard is at rest they remain folded, but when it jumps into the air the flaps are extended by movable elongated ribs.

Flying phalangers are tree-dwelling *marsupials, which were once placed in the family Phalangidae, but the name now refers to species in three different families. The first is the tiny pygmy flying phalanger or feather-tailed glider, *Acrobates pygmaeus*, named after its long, hair-fringed tail and weighing only 15 g. ($\frac{1}{2}$ oz.). The second group are the gliders, with four species of *Petaurus*, weighing about 130 g. ($4\frac{1}{2}$ oz.). Their diet is not restricted to nectar but includes insects and small birds. Lastly, there is the greater flying phalanger, or great glider, *Schoinobates volans*, a vegetarian and much the largest at around 1·5 kg. (3 lb.) and capable of gliding for over 100 m. (33 ft.). All species are Australian, except for the sugar glider, *Petaurus breviceps*, of New Guinea.

Foam-nesting frogs, like the African species, *Chiromantis xerampelina*, and the Southeast Asian species, *Polypedates leucomystax*, make their nests from a liquid excreted from the female's vent. The mating pair beat the liquid into a foam, using their hind-limbs. Eggs are then laid in the foam, which prevents desiccation. The nests are sited in tree branches over water and when the rains come the developing tadpoles drop into the water where they complete their metamorphosis.

Foetus *Embryos.

Food chains, or webs, are a way of expressing the feeding relationships between organisms in a *community. They describe how energy, locked into food by plants and other *autotrophic organisms, passes from organism to organism. The most simple food chain, if such a thing existed, would consist of just three organisms: a plant, a herbivore, and a carnivore. The position each occupies in the food chain is called a trophic level and classifies them into producers (such as plants) and consumers (such as animals). The herbivore, which eats the plant, is called the primary consumer; the carnivore, which eats the herbivore, is called the secondary consumer.

Food webs describe situations closer to reality, where each organism may feed at several different trophic levels and produce a complex 'web' of feeding interactions. All food chains begin with a single producer organism, but a food web may involve several producers.

Footmen moths belong to the same family as tiger moths, but are slender and less colourful. The narrow fore-wings are folded tightly over and around the body at rest, the stiff, elongate appearance giving rise to the popular name. By day they are sluggish, and drop from vegetation if disturbed. The caterpillars, which are very hairy, mostly feed on lichens. These moths occur throughout the world but are predominantly tropical or north temperate species.

Foraminifera are related to *radiolarians, and like them are mainly marine members of the *protozoa. Un-

Copulating **foam-nesting frogs** cling to a branch over water in the Tunborati Game Park, South Africa. The foam hardens on the outside as a protection against the heat of the day and the eggs inside are kept moist.

like radiolarians, their shells are either calcareous or made of foreign particles and often built up of separate chambers added in sequence as the animals grow. Many species resemble strings of beads, because of their mode of growth, or have spiral shells like tiny snails. Foraminifera shells always have microscopic holes in the shell to allow extensions of the cell (pseudopodia) to protrude and be used for locomotion and feeding, rather like amoebae.

Forester moths   *Burnet moths.

Forests are areas of vegetation dominated by trees. They are the main plant cover over much of the earth but have been modified or removed by man from most of the temperate regions and are rapidly being felled in the tropics too. They are classified as *deciduous forests, coniferous forests, or tropical *rain forests, depending upon the climate and type of tree that makes up the forest. A forest may be just a few metres (yards) tall, as at the tops of tropical mountains, or may reach heights of 30 m. (100 ft.) or more. Tropical rain forests are the most species-rich of land vegetation, whereas, by contrast, some coniferous forests consist of a very small number of tree species. Their value lies not only in the production of timber and other materials such as drugs, fruits, and latex, but also in soil and water conservation.

Forget-me-nots are small, blue-flowered annual or perennial plants. They are all species of *Myosotis* and belong to the family Boraginaceae. This includes plants such as

borage, comfrey, and some 2,000 others. About forty species have been described, chiefly from Europe and Australia and New Zealand. Several are used as garden plants.

Forsythia is a genus of seven species of shrub, one native to southeast Europe, the rest to the Far East. From the latter have been derived the floriferous spring-flowering shrubs of gardens, notably the hybrid *Forsythia* × *intermedia*. Forsythias belong to the olive family, along with ash trees and lilacs. They all have yellow flowers and grow rapidly.

Fossil fuels are coal, oil, and natural gas, so called because they developed from the remains of living organisms. Oil and gas were formed from the slow decomposition and burial of *planktonic marine plants and animals which sank to the muds of the sea-floor. Coal is derived entirely from the accumulation of partially decayed land plants that grew in low-lying, swampy environments typical of the Carboniferous Period around 300 million years ago. As the material was buried, then subjected to heat and pressure, it went through a gradual transition from peat to coal and finally to anthracite. The ordinary bituminous household coal often has bright bands that originate from wood or bark material and dull bands that include plant spores and pollen.

Fossil hominids, which offer direct evidence about the origin of man, *Homo sapiens*, are very rare and usually consist only of fragments of the skeleton and, thus, many of the details of human evolution are unknown. Most of the important early fossil hominids have been found in Africa. The earliest man-like creature is called *Ramapithecus*, which lived about 10 million years ago and was fairly small, less than 1 m. (3 ft. 3 in.) high. It was very ape-like with a relatively small brain, but its teeth had developed a number of man-like features. Then, around 3·5 million years ago, a new species called *Australopithecus* appeared. It was rather larger, with a somewhat enlarged brain, and for the first time walked fully upright.

The fossil remains of the first member of the genus *Homo*, called *H. habilis*, is about two million years old. Although very similar to *Australopithecus*, from which it evolved, *H. habilis* did have a slightly larger brain. His scientific name, meaning 'handy man', derives from the fact that he used tools to kill and prepare food. The next stage to evolve was *H. erectus*, whose brain was rapidly approaching the size of that of modern man, and who lived from 1 million to 500,000 years ago. *H. erectus* was about 1·5 m. (4 ft. 10 in.) tall, and had characteristic, heavy brow ridges and a receding forehead. The fossil remains of this hominid have been found in many parts of the world, including Africa, China (Pekin man), Java (Java man), and Europe. Modern man's predecessor was Neanderthal man, the first true *H. sapiens*, who first appeared about 500,000 years ago and had a fully human-sized brain. Some palaeontologists refer to him as *H. neanderthalis*.

Fossils are the remains of once-living animals and plants preserved in rocks. The great majority of fossils consist of such hard parts as the internal skeletons of vertebrate animals and starfish, the shells of molluscs, lamp shells, and crustaceans, and the tough woody parts of plants. Occasionally, however, soft parts are preserved, such as whole insects trapped in amber (fossil resin) and certain dinosaurs that fell into natural tar pits. There are also *trace-fossils, which are signs, such as footprints or burrows, left by organisms.

Fossils are formed when an organism dies and its corpse is covered by sediments such as mud or sand before it is destroyed by scavengers or the weather. Over millions of years the sediments containing fossils are compressed and converted into rocks of the kind known as sedimentary rocks. Due to subsequent earth movements and erosion, the fossil-bearing rocks may become exposed on the surface of the land.

The chances of any particular organism finally turning up as a fossil are very small, and therefore only a small percentage of the species that have existed in the past are actually known. Nevertheless, those that have been found give a good general outline of the history of life on Earth and how it has changed as a result of *evolution. Fossils are very important in the dating of rocks in different places, samples of rock containing the same fossil species being of similar age.

Foxes are members of the dog family with worldwide distribution. Some twenty-one species occur throughout most parts of the world, with the exception of Australasia and Southeast Asia. They include species such as the Cape fox, *Vulpes chama*, Arctic fox, *Alopex lagopus*, crab-eating fox, *Dusicyon thous*, and the fennec fox. All are small dog-like carnivores which also eat fruits in season.

The European red, or common fox, *Vulpes vulpes*, is found from mountains to lowland areas and even around towns. It lives in a den and the surrounding area is often littered with bird, vole, rabbit, sheep, deer, mice, rat, and even insect remains.

The female fox, or vixen, bears four or five cubs which are weaned in thirty to forty days, at which time the vixen

begins to teach them to hunt and kill their prey. Initially small animals such as voles are taken, but their diet is gradually widened. The adult male reaches a length of 65 cm. (2 ft. 6 in.), has a tail 40 cm. (16 in.) long, and weighs 9 kg. (20 lb.); the vixen is smaller. Their lifespan is up to six years. The adult coat is yellowish or buff-coloured in the forms living in the desert or plains regions, and deep red in the woodland forms.

Foxgloves are biennial and perennial plants with erect spikes of purple, golden, or white flowers. All belong to the genus *Digitalis*, with around 100 species known from Europe, Asia, and northern Africa. They give their name to the foxglove family, Scrophulariaceae, which contains about 3,000 species, including antirrhinum, figwort, speedwell, and musk. This family is also known as the figwort or antirrhinum family. The common foxglove, *D. purpurea*, is cultivated for the extraction of compounds with heart-stimulant properties such as digitoxin and digitalin.

Fractures are breaks in bones, usually the result of considerable stress. A break in a bone previously weakened by disease is a pathological fracture, and may occur after a light blow. Fractures are usually complete, but in children an incomplete, so-called greenstick, fracture may occur. Fractures usually heal if the bone ends are in con-

A clean break or **fracture** in a tibia (shin-bone) shows up on an X-ray. The first phase of treatment consists of moving the broken parts of bone into their original positions; this is called reducing the fracture.

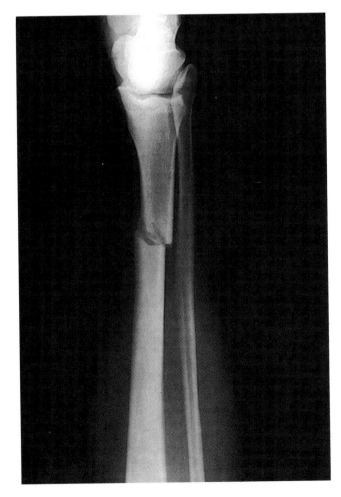

tact, and kept together, but they take several weeks to set, or consolidate. They are less of a problem nowadays as a result of the increased use of supporting metal plates and screws.

Frangipani is the common name of a sappy thick-stemmed shrub, *Plumeria alba*, which, like the other six species in the genus, is native to the warm Americas. As a member of the Apocynaceae family it is related to periwinkles and oleanders. It is much cultivated in the Old World tropics for its flowers, which are used in garlands in Hawaii and offered in Buddhist temples. A scent is prepared from the flowers.

Frankincense pine   *Loblolly.

Free-tailed bats are distinguished by the tail, which, unlike that in many other *bats, extends beyond the body, instead of being incorporated within the flying surfaces. This feature is common to all of the ninety-one species of bats in the family Molossidae. Their alternative name of mastiff bats stems from their massive, burly faces. They occur throughout subtropical and tropical regions, often forming large colonies in caves. Some species migrate distances of 1,000 km. (620 miles) to overwintering sites.

Freud, Sigmund (1856–1939), Austrian neurologist, initiated the use of psychoanalysis. Having discovered that a patient could be cured of hysteria by hypnotism, he urged other patients to talk freely, relying upon the patient's confidence in himself as practitioner, and a generally sympathetic atmosphere to treat mental disorders. The whole of a patient's mental history, often going back to early childhood, would be revealed. By drawing out memories from the subconscious mind, he would try to

Freud, seen here in the 1920s, proposed that adults can suffer mental illness because of events in childhood. The psychoanalyst following his methods probes the patient's past to unravel the origin of the disorder.

relieve the patient of inhibition and depression. The method was first called 'the talking cure'.

His insight into the workings of the mind told him much about the behaviour of healthy people too. Childhood impressions associated with such basic emotions as hatred, fear, and jealousy had, he found, a strong influence on many adults. To understand criminals, for example, might often require psychological insight into their behaviour.

Freud's ideas attracted a considerable following, though some psychologists, notably Jung and Adler, broke with him to develop systems of their own.

Frigate birds are five species of black tropical seabirds in the genus *Fregata*,; they all have with pointed wings, spanning 1·5–2 m. (5–7 ft.), and forked tails. They are called frigate birds because they attack other birds and force them to disgorge fish; they also catch flying fish as they leap out of the ocean. Frigate birds nest among rocks or low shrubs, raising single chicks. The male frigate birds have a large scarlet sac on their throats which is inflated during courtship displays.

Frilled lizard is the name of a highly spectacular species of mainly tree-dwelling, *agamid lizard from Australia and New Guinea. This species, *Chlamydosaurus knigii*, reaches a total length of about 1 m. (3 ft. 3 in.). It has a flap of skin (or frill) around its neck, which can be raised and extended to deter enemies. At the same time, the frilled-lizard opens its mouth in a wide gape to emphasize its aggressive display. On open ground, if alarmed, it runs quickly upon its hind legs.

Fritillaries (plants) are attractive plants of the lily family, with pendent bell-like flowers. About eighty species are known, chiefly from the temperate regions of the Old World and from North America. The snake's head, *Fritillaria meleagris*, grows in wet meadows, and is the best known of the European species. *F. imperialis* is the crown imperial. The bulbs of all species are very poisonous.

Fritillaries (zoology) are yellowish- or reddish-brown butterflies chequered with black, a pattern reminiscent of a dice-box (*fritillus* in Latin). The underside of the hindwing is usually washed or spotted with silvery or pearly white. It is a popular name for many similarly patterned butterflies in several genera of *brush-footed butterflies, occurring in Europe, temperate Asia, North Africa, and the Americas. Some genera, such as *Boloria*, are largely species of northern temperate regions, some reaching almost to the Arctic Circle. The Duke of Burgundy fritillary, *Hamearis lucina*, of Europe belongs to the *metalmark family. Most true fritillaries are fast and powerful fliers. Many of the European species feed as caterpillars on violets.

Frog-hoppers are squat, jumping Homopteran *bugs, which suck sap and occur worldwide. The nymphs of most species excrete a substance through which air is forced to form a blob of bubbles, often called cuckoo-spit. This may be produced on aerial parts of plants, as in the 800 species of the family Aphrophoridae, or on plant roots, as in the 1,400 species in the family Cercopidae. Some, such as sugar-cane frog-hoppers, are serious pests. (*see over*).

Frogmouths are a family of twelve species of bird related to the nightjars, which occur from Southeast Asia to Australia. Most species are about the size of crows, but the

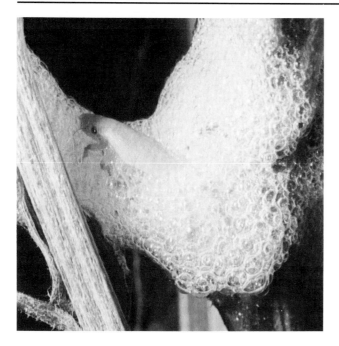

The **frog-hopper** nymph's bubbles form an environment suitable for its development. The bubbles are created by the nymph's posterior abdominal segments.

The brown-streaked plumage of a tawny **frogmouth**, *Podargus strigoides*, affords the bird perfect camouflage as it sleeps during the day in the fork of a tree in Queensland, Australia.

Papuan frogmouth, *Podargus papuensis*, is larger, being up to 55 cm. (22 in.) long. Frogmouths have a mottled brown and grey plumage, which acts as camouflage, a large head with large orange-yellow eyes and an enormous gape. They pounce on frogs and other small animals, mainly hunting at night. They make a loose nest of twigs on a tree branch and raise up to four young.

Frogs form part of one of the three major orders of *amphibian. Along with toads, they are distinguished in the adult stage by lack of a tail. The popular distinction between frogs and toads applies only in Britain, where frogs have slender bodies, long limbs, and a smooth, moist skin, and toads have stouter bodies, short limbs, and dry, warty skin. In other countries the distinction does not hold, for some families include species with smooth- *and* warty-skinned forms. The pectoral (or shoulder) girdle provides a good structural guide to the difference. Toads have a pair of overlapping cartilage elements in their pectoral girdle, while in frogs the same cartilages meet in the midline but do not overlap.

Members of the genus *Rana* have been called the 'true frogs', and are familiar to most people. Large numbers of the European common frog, *Rana temporaria*, and American leopard frog, *R. pipiens*, are used in school and college biology classes, and as experimental animals for research purposes. Larger members of the genus *Rana* are known as *bullfrogs.

Other frogs, to use the term in its more general sense, show a wide diversity of body shape. Probably the most remarkable frog is Darwin's frog, *Rhinoderma darwinii*, of Argentina. This is a mouth-breeding frog which comes in a dazzling variety of colour patterns; the male keeps the tadpoles inside his vocal sac until they are mature enough to fend for themselves.

Fruit bats   *Flying foxes.

Fruit flies are tiny, brownish or yellowish, red-eyed true *flies that are attracted to fermenting fruit and alcoholic drinks. Their flight is ponderous and slow, with a rather bulbous abdomen hanging down, so that they appear suspended in the air. They comprise a large, cosmopolitan family, and many species are common; the genus *Drosophila* contains some 1,000 species. They can be a minor pest in orchards or where fruit is stored, particularly in warmer countries. Most feed as larvae on or in decaying fruit and vegetation, and those found in fermenting fruit have been shown to be feeding on yeasts. Other species occur in fungi, on sap runs of trees, as leaf-miners, or in a few species as predators or parasites. Developmental stages of *Drosophila* species are adapted for life in a semi-fluid medium. The adaptations include floats to keep the eggs above the food and breathing 'siphons' in the larvae.

Drosophila species, particularly the vinegar fly, *D. melanogaster*, have been used extensively in genetics research because they have a short lifespan, are easy to culture, and have large *chromosomes in their salivary glands.

Flies of the family Tephritidae are also known as fruit flies or *gall-flies, and some species cause great damage to fruit crops. The Mediterranean fruit fly, *Ceratitis capita*, is found wherever fruit is grown.

Fruits are, in a botanical sense, the ripened ovaries of a seed plant and the surrounding tissues. They include many items such as acorns, pea pods, tomatoes, cucumbers, and wheat grains, which would not be thought of, in common usage, as fruits.

Around the developing seed is a tissue called the pericarp. This may be fleshy, juicy, leathery, oily, fibrous, or hard, and may be subdivided into further layers. The pericarp is involved in dispersal and protection of the seed, each type being adapted to a particular mode of dispersal. Fruits with a fleshy or juicy pericarp, such as *drupes or *berries, are designed to be eaten by animals and transported and subsequently deposited in the animal's droppings. Some fruits, such as burrs, are adapted to be distributed via the fur of animals—they have their outer pericarp layers formed into hooks. Wind dispersal is exploited by fruits with feathery plumes (dandelion), or membraneous 'wings' (sycamore). A few fruits, such as those of the tulip-tree, have fruits with a buoyant pericarp, adapted for water dispersal. The simplest form of dispersal is for the seed to be scattered as the pericarp springs open explosively, such as happens with gorse.

Fruits are classified according to whether they are dry or fleshy, and whether they split open when mature (dehiscent fruits), or remain intact (indehiscent fruits). Different types of fruit are given various names such as *berries, *drupes, *pods, *nuts, and *pomes. False fruits consist of true fruits with some accessory tissue in addition

to the pericarp. Some, like the 'berries' of juniper, are in fact modified cones.

Fuchsias are shrubs, sometimes with short-lived shoots, native to the Pacific region, especially tropical America. They belong to a family of world-wide distribution, which also includes the willow-herbs and evening primroses. The genus *Fuchsia* has some 100 species, although many familiar·garden fuchsias are hybrids with double flowers. Forms of *F. magellanica*, are widely used as hedging plants, notably in Ireland. The flowers of most fuchsias are visited by bees or honeybirds and the fruits are edible. In Australia, the term fuchsia is used for species of the unrelated genus *Correa*.

Fucus is a genus of brown *algae, which includes those species commonly called wracks or tangles, such as bladder and knotted wrack. They usually grow in the intertidal zone on the sea-shore, attached to rocks by a disc-shaped 'holdfast' and making a slippery mat at low tide. One function of the 'bladders', or air pockets, of some species is to give the fronds buoyancy, so that they are able to float on the water surface and receive light for photosynthesis.

Fulmars are large, heavily built birds of the same family as *petrels. The two species of the genus *Fulmarus*, the common fulmar, *F. glacialis*, and northern fulmar or silver-grey petrel, *F. glacialoides*, have narrow wings and gliding flight, and pick their food off the sea surface. They are native to Arctic and sub-Arctic regions.

Fumitories are a group of slender annual plants from the temperate regions of the Old World. They give their name to the family Fumariaceae, which contains 400 species of annual or perennial plants including bleeding heart, *Dicentra spectabilis*. The European fumitory, *Fumaria officinalis*, a weed of cultivation and waste places (and

Serrated wrack, a bladderless **fucus** of the middle-lower intertidal zone, is seen here growing on the coast of Devon, England. Its fronds are often covered by the growths of bryozoans, spiral worms, and hydroids.

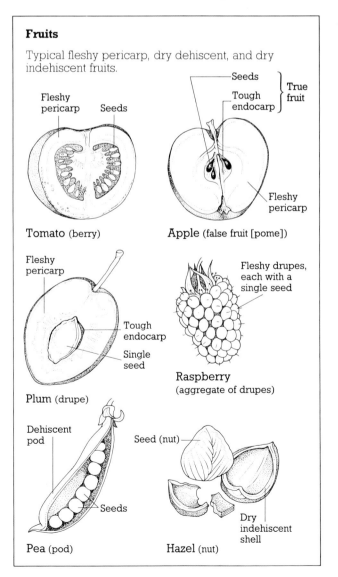

Fruits

Typical fleshy pericarp, dry dehiscent, and dry indehiscent fruits.

Tomato (berry)

Apple (false fruit [pome])

Plum (drupe)

Raspberry (aggregate of drupes)

Pea (pod)

Hazel (nut)

naturalized in the USA), was held in high esteem by
*Dioscorides for its purifying powers.

Fungi are classically regarded as non-flowering plants,
but are placed in the Protista kingdom by some biologists.
As they lack photosynthetic pigments they are dependent
on other organisms for their food. Most exist in the form
of microscopic threads called hyphae, which grow at their
tips, branching and forming a weft called a mycelium. All
fungi reproduce, survive, and spread mainly by means of
*spores, each microscopic spore potentially germinating
to produce new hyphae.

The fungi are divided into four major classes which are
based upon the way in which spores are produced within
fruiting bodies. The Phycomycetes are simple, thread-like
organisms such as water moulds, and include some 1,000
species. The class Ascomycetes includes some 15,000 species of fungi which produce spores in pod-shaped structures called asci. They include some *yeasts. The class
known as Fungi Imperfecti includes species which cannot
produce spores at all and reproduce by division. They
include 'moulds' such as *Penicillium*. The final class, the
Basidiomycetes, contains some 90,000 species of fungi
which bear spores on special structures called fruiting
bodies. They include mushrooms, toadstools, puffballs,
and many others.

Parasitic fungi cause diseases of plants and, less often,
of animals. They also form symbiotic associations with
plant roots called *mycorrhiza. Although they are essential to ecosystems in breaking down remains of dead
plants, their ability to cause decay also means that they
do much harm to man by spoiling food and rotting timber.

Fungus gnats are minute or small, delicate, true *flies
with long antennae and, in the females, external ovipositors. Their legless larvae live in damp places, feeding
on the hyphae of *fungi. Those which occur on cultivated
mushrooms can be serious pests. They include some 2,000
species distributed worldwide and resembling midges.

Fur *Hair.

Furniture beetles are small, brown *beetles of the family Anobiidae, and are related to the death watch beetle.
Their larvae, commonly called woodworms, are the most
important beetle pest of hardwood, even of wickerwork,
and naturally occur in dead wood in the wild. The head
of the adult beetle is under the thorax so as not to be
visible from above. It lays its eggs in cracks in unpainted
wood, and the larvae bore tunnels in the wood, filling
them with excreta as they feed. When nearly mature, they
move to just below the surface, where they pupate. The
adult leaves a circular hole when it eats its way out. There
are some 1,100 species of beetles in the Anobiidae, which
occur worldwide.

Gaboon viper: a very thick-bodied snake, sometimes
growing to 2·06 m. (6 ft. 9 in.) in length, and found in the
rain forests of equatorial Africa and south as far as Natal.
The fangs are notably long, over 5 cm. (2 in.) in large
individuals, and a bite can be very dangerous. When not
seeking ground-living animals as prey, the colourful, but
beautifully camouflaged, snake is often sluggish. It is not
a true viper but belongs to the same family as the *puff
adder.

Gadflies   *Bot flies, *Horse flies.

Galago   *Bush babies.

Galapagos finches, also known as Darwin's finches,
include fourteen species of the bunting family. They are
confined to the Galapagos Islands, except for one, the
Cocos Island finch. Most are sparrow-sized birds, the
males tending to be black, the females greyish. The main
variations are in bill shapes, which vary from very heavy,
finch-like beaks, through smaller bunting-like ones, to a
fine warbler-type. They are famous for their description
in Charles Darwin's theory of evolution. Darwin realized
that all the species were so similar that they must have
descended from a common ancestor. The present-day
range of species of Galapagos finches is a result of the
adaptation of different varieties to particular ways of life.
Some are seed-eaters, others insect-eaters, and both types
have bill shapes to match their way of life.

Galen, Claudius (129–199), Graeco-Roman physician,
was a surgeon to gladiators and became physician to the
emperor Marcus Aurelius. From experiments, including
the dissection of monkeys and other animals, he added
greatly to medical knowledge by describing functions of
the brain, spinal cord, ureter, and arteries (until then
believed to be full of air, not blood). He regarded the
body as the vehicle of the soul, and his works dominated
medicine throughout the Middle Ages.

Galingale is an ornamental perennial *sedge, *Cyperus
longus*, from lakes and stream-sides in Europe and North
America. The roots were at one time highly regarded as a
tonic or a cure for stomach disorders.

Gall bladder is the name for a muscle-walled sac attached closely to the liver. *Bile is manufactured in the
liver and passed to the gall bladder for storage and concentration. It passes its contents into the small intestine,
where the salts it contains lower surface tension and aid
digestion and absorption. The green pigments in bile are
waste products derived from the breakdown of red blood
cells; they cause the faeces to be coloured brown. Only
animals with a liver have a gall bladder.

Gall-flies are small flies with banded or spotted wings
of the family Tephritidae, often called *fruit flies. Their
larvae live and feed in plant tissues, inducing abnormal
growth and swellings known as galls. It is often possible

A **gaboon viper** , *Bitis gabonica*, on the forest floor, Angola. The remarkable camouflage of this snake, matching that of curled leaves and their shadows, enables it to ambush small animals for food.

to identify species by the host plant and by the site and shape of the gall. Many of the 1,500 species live in the flower-heads of Compositae, such as thistles and hawkweeds. Other members of this family have larvae that are leaf-miners, and feed within the tissue of a leaf. The larvae of several other insect families may also produce galls on plants.

Gallinules are several species of *rail in the genus *Gallinula*, which means 'little hen' in Latin and alludes to the birds' appearance and calls. Best known are the common moorhen or gallinule, *G. chloropus*, and the richly coloured purple gallinule, *G. martinica*, which breeds in the southeastern USA and which has deep purple and bronze-green plumage, a red and yellow bill, pale blue forehead, and

yellow legs. The latter name is sometimes used of the large purple moorhen or swamphen, *Porphyrio porphyrio*, a rail widespread in the Old World.

Gall midges are minute flies of the family Cecidomyidae, which have antennae resembling strings of beads, often covered with whorls of hairs. The habits of the larvae are varied, but many cause abnormal growths of tissue called galls on flowering plants and are consequently pests. The hessian fly, *Mayetiola destructor*, originally from Asia, is now a pest of cereals in Europe and North America. Among this large family of flies are species whose larvae feed on decaying vegetable matter, in addition to gall-forming types.

Gallstones consist of *cholesterol, *bile pigments, or a mixture of these and calcium salts. They form most commonly when cholesterol is precipitated from solution because it is present in excess or because of an infection. Gallstones are very common in elderly people, and oc-

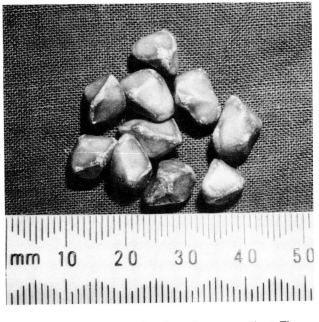

These **gallstones** were taken from just one patient. The stones may number several hundred, each the size of a small particle of gravel; at the other extreme, one stone may be as large as a goose's egg.

casionally cause attacks of pain, or obstruct the main bile duct and produce jaundice. Small stones may be dissolved, but larger ones may require surgical removal.

Gall wasps are a family (Cynipidae) of small, black *wasps which lay their eggs on selected parts of flowering plants, especially oaks and roses. The plants react to secretions of the larvae by producing characteristic growths called galls. Most species produce two gener-ations, one all-female, the other with both sexes. The plants produce different galls for the two generations. Oak-apple galls are an example of the galls caused by these wasps.

Galton, Sir Francis (1822–1911), was an English scien-tist whose interest in quantifying measurements of the human body led to biometric research and resulted in the classification of fingerprints. This, in turn, led to the adoption of fingerprint identifications of criminals. Galton was also an exponent of eugenics, the genetic control of human populations by selectively encouraged breeding.

Gametes are a special type of *cell used exclusively in sexual reproduction. Familiar types of gamete include the mammalian *ovum, or egg cell, and the male *sperm cell. Each is a typical cell apart from containing only half the normal amount of genetic information (chromosomes) of a body cell. Gametes from each of the sexes fuse together at fertilization to restore the full complement of chromo-somes before dividing to begin a new organism.

Gannets are large seabirds of the genus *Morus*. The three species, which are white and black, are native to tem-perate latitudes; but their family, Sulidae, also includes six pied or brown tropical species of *boobies. Gannets are long-winged, with webbed feet, a long conical beak, and wedge-shaped tail. They fly in groups low over the sea, catching fish by plunge-diving, after which their buoyancy allows a quick take-off. Their foods include shoaling fish and squid. They nest colonially on islands and isolated mainland coasts, in trees, on cliffs, or on flat ground.

A **gannet** colony of *Morus bassanus* on Bass Rock, Scotland. The chick is fed by both parents for about two months after hatching. By that time it is almost adult size, with a wing-span approaching 1·8 m. (6 ft.).

A pair of male Grant's **gazelles**, *Gazella granti*, lock horns in combat in the Serengeti National Park, Tanzania. The gazelles fight to defend their breeding territory and thereby mate with as many females as possible. They delineate the boundary of the territory by marking 'scent posts' with urine and then faeces.

Nests are rudimentary collections of sticks or seaweed and, although they lay up to three eggs, they seldom raise more than one chick successfully.

Gapers   *Clams.

Garden, Alexander (*c*.1730–91), was born in South Carolina, USA, of Scottish parentage, and became a physician; but he was noted as one of the most accomplished American botanists of his day. He sent many new plants to Europe, and the shrub gardenia was named after him. He also studied mineralogy and zoology and discovered some new animal species, notably certain species of eels.

Gardenias are a genus of Old World shrubs related to coffee. The species with evergreen leaves are widely cultivated for their heavily scented flowers, particularly in the semi-double white forms.

Garfish, or needle fish, of which there are nearly thirty species in the family Belonidae, are primarily marine surface-living fishes of tropical and temperate seas, although several species live in fresh water, principally in South America. All are slender-bodied fishes with very long beak-shaped jaws. Both mandibles of the jaw are lined with needle-like teeth, which they use when capturing a wide range of smaller fishes. Most species grow to about 1 m. (3 ft. 3 in.) in length, but some are larger. They leap out of the water very freely. They lay large eggs in the surface waters, each egg having long filaments which tangle with floating weed.

Garganey is the name for a small, mottled, brown and white dabbling *duck. The drake has a conspicuous curved white stripe from eye to nape. The single species is confined to the Palearctic region.

Garlic is a bulbous plant related to the onion, growing up to 1 m. (3 ft. 3 in.) in height, with flat leaves sheathing the lower stem, which ends in a cluster of small bulbs called cloves. Cultivated on a large scale in southern and eastern Europe, it has been popular at least since ancient Egyptian times. Used excessively it makes the breath ob-

noxious, which perhaps explains why it was valued in the Middle Ages for its alleged ability to keep vampires at a distance.

Garrod, Sir Archibald Edward (1857–1936), English physician and biochemist, described inborn errors of metabolism. His work on the changes in urinary pigments caused by disease led him to discover that the inability of the body to metabolize a specific food element was not due to an infectious agent, but was an inherited disease. His works included *Inborn Errors of Metabolism* (1909) and *The Inborn Factors of Disease* (1931).

Gastric ulcers   *Ulcers.

Gaur: a large species of Asian wild *cattle, sometimes called the Indian bison. The bull can attain a height of over 1·8 m. (6 ft.) at the shoulder, and is dark brown to black, like the yak, but with shorter hair; it has broad upturned horns tipped with black. The cow is smaller and reddish-brown in colour. The gaur, *Bos gaurus*, feeds on grasses and bamboo shoots and is vulnerable to the loss of its tropical forest habitats in Southeast Asia.

Gavial   *Gharial.

Gazelles are antelopes found in the hottest, driest parts of the stony plains or deserts of southern Asia and Africa. There are some eighteen species of gazelles, including the gerenuk, blackbuck, edmi, springbok, and Thomson's gazelle. Most are 90 cm. (3 ft.) or less in height at the shoulder. They usually have lyre-shaped, ringed horns that curve backwards and upwards, but these are generally smaller or absent in the female. Sandy-brown in colour, they may have black and white bands on the face and flanks. Gazelles are amongst the most graceful and fast-moving of animals, with an average speed of 48 km./hour (30 m.p.h.) and can easily pull ahead of a vehicle travelling at 64 km./hour (40 m.p.h.).

Gean *Cherry.

Geckos are a *lizard family of about 670 species, distributed in warm countries worldwide. Many species have rather flat bodies covered with granular scales. The majority are at least partly nocturnal and lack movable eyelids, the eye being protected by a transparent membrane. They differ from most other lizards in being vocal, the sounds produced being a series of chirping cheeps and clucks. Some house geckos can walk up vertical surfaces and even upside-down across ceilings; this is achieved by the action of specialized toe-pads that can cling to very small surface irregularities.

Geese are large, web-footed waterfowl, closely related to swans and ducks, with stout bodies and long necks. Although good swimmers, they are mainly terrestrial, spending much time grazing. They breed in high northern latitudes and migrate south in winter, when they form noisy flocks or gaggles on marshes, grassland, and arable land. They fly in skeins, sometimes in neat V-formation. The sexes are similar in colour in all fourteen species. There are nine so-called grey geese, such as the whitefront and the greylag, and five predominantly black species, which include the barnacle, brent, and Canada geese.

Genera (singular: genus), in the system of *nomenclature in zoology and botany, are groupings below *families and contain closely related *species, one of which is considered typical of the genus. Latin names of genera are written in italics, with an initial capital letter. Genera containing a large number of species may be divided into subgenera.

Genes are units of inheritance, consisting of short pieces of a *DNA molecule. They are passed from parents to offspring as part of a *chromosome. The word appeared early in the science of genetics, and as the science has progressed, its original meaning has been modified so much that it is a misleading term except when used informally. It started out as the hereditary factor that controlled a single character difference, such as blue or grey eyes in humans. In fact, very few characters, if any, have a one-to-one relation between a length of DNA and a character. A character is usually formed by the action of many genes. The currently accepted meaning is that a gene is a length of DNA that codes for a single *protein; this length of DNA is technically called a cistron. The word gene is still used to refer to the DNA that controls the development of a characteristic.

Genetic code is the language of the molecules of inheritance. *DNA consists of a sequence of molecules called nitrogenous bases and contains four kinds of such nitrogenous base: adenine (A), cytosine (C), guanine (G), and thymine (T). DNA influences the formation of characters in an organism by directing the production of *proteins. Proteins are made of *amino acids, and the properties of a protein are determined by its exact sequence of amino acids. The sequence in which amino acids are put together to form the protein is determined by the sequence of bases on the DNA molecule. The genetic code is based upon units of three nitrogenous bases (triplets); there are sixty-four possible triplets of the four nitrogenous bases. Each triplet encodes for a particular amino acid. The meaning of all sixty-four triplets has been determined by biochemists. The sequence GGC, for example, means 'insert a glycine amino acid at this point in the protein'. The genetic code was worked out in the 1950s and 1960s.

Genetics is that part of biology concerned with the structure, functioning, and evolution of hereditary factors called *genes. Its founding discovery, made by *Mendel, was that genetic factors are independent, in the sense that they do not 'blend'. Mendel crossed different strains of peas, which varied according to the height of the plants, texture of the 'seeds', or colour of the flowers. When he cross-fertilized dwarf and tall strains the progeny were always either dwarf or tall; they were never of medium height. He also inferred, from the pattern of inheritance of the independent factors, that each character organism, such as height of pea plant, must be controlled by two copies of each factor. Mendel's factors, which came to be called genes, could then be correlated with the *chromosomes, which also exist in two copies in normal body cells. Each of these factors, or genes, can exist in different forms called alleles. Each allele can exist as one of two basic types: a dominant allele which always produces its relevant characteristic, even if present as only one of a pair, and a recessive allele which is expressed only if both alleles of a pair are present. For any particular characteristic of an organism, if this is the result of a mixed pair of alleles (one dominant, one recessive), then the organism is said to be a carrier for that recessive allele. In such individuals the influence of the recessive allele is suppressed by the presence of a dominant allele. Recessive characters appear in organisms only if both alleles of the pair are recessive types. Examples of recessive characters include many congenital, or inborn, genetic diseases. An increasing knowledge of molecular genetics has made it possible to alter the genetic constitution of animals, or plants; this is genetic engineering, or biotechnology.

Genets are small carnivores of the mongoose family. They are closely related to civets and linsangs. There are eleven species, mainly African, although some extend into the Middle East and one, the common genet, *Genetta genetta*, occurs in southern Europe, where it is probably native. Genets feed on a variety of small birds, rodents, and insects. They also take fruit and a West African species, the forest genet, *G. maculata*, has been found to sip nectar, pollinating the flowers in the process. Genets are spotted and cat-like in appearance, but have elongated snouts with sharp, blade-like incisor teeth.

Gentians are mainly blue-flowered alpine plants from the mountains of Europe and Asia, though white and red-coloured species occur in New Zealand, and North and South America. They give their name to the gentian family which has 900 species. The bitter roots of the yellow-flowered *Gentiana lutea* from central and southern Europe have useful medicinal properties as a tonic and for jaundice.

Geraniums are, correctly speaking, hardy species of the plant genus *Geranium* which includes *cranesbills. The name is also used by amateur gardeners for varieties and species of the South African plant genus *Pelargonium*, of which there are several groups, including zonal, regal, and the scented- and ivy-leaved kinds. Most of them have bright, showy flowers and are often grown as pot plants. Oil is extracted from the scented-leaved kinds and may be used as a base for perfumes.

Gerbils are, strictly speaking, members of the rodent genus *Gerbillus* with thirty-four species, but the name is also used for the family Gerbillinae, with a total of eighty-one species (including the genus *Gerbillus*). All species are small, long-legged animals inhabiting desert regions of Africa and Asia Minor. They spend the day in sandy burrows, emerging at night to forage for seeds, roots, and insects. They often hop on their elongated hind legs like miniature kangaroos. They have many physiological adaptations to desert life, including a very efficient water-retrieval system in their hind gut.

Germination of a plant seed is the restarting of growth of an embryo after dormancy. It begins with the emergence of a root or shoot from the seed, once conditions of moisture, light, and temperature are suitable. Dormancy may be almost non-existent in some tropical trees, in which germination takes place before dispersal of the fruit. In other species, there is a period of 'after ripening' which can be broken only by exposure to certain wavelengths of light and plentiful water. Some seeds can remain dormant in the soil for up to a century.

Geums are perennial herbs with basal rosettes of leaves, and yellow, red, or white flowers. They are part of the rose family. The clove-scented roots of the herb bennet, *Geum urbanum*, were formerly used to flavour beer and as a medicament. Other species are popular, herbaceous garden plants.

Gharial, or gavial: a large *crocodile of the family Gavialidae, exceeding 6 m. (20 ft.) in length, and found in deep, fast-flowing rivers in parts of the northern Indian subcontinent. It has a very narrow snout, well adapted for seizing swimming fish. Adult males develop a tough, fleshy lump on the snout close to the nostrils; this bulbous structure probably aids recognition between sexes.

The false gharial, *Tomistoma schlegeli*, occurs in rivers of other parts of Southeast Asia. This belongs to the family Crocodylidae and not to that of the true gharial, *Gavialis gangeticus*.

Ghost pipefishes are five species in the family Solenostomidae, which are relatives of the *pipefishes. They have a long snout but the body is thickset, with large, bony plates and well-developed pelvic fins and two dorsal fins. They live in shallow water in the tropical Indo-Pacific. The females carry the eggs in a pouch formed by the pelvic fins.

Ghost swift moths *Swift moths.

Giant horsetail (fossil) *Calamites.

Giant salamanders are the largest of the living *amphibians, and may attain a length of 1·8 m. (6 ft.). They have squat, heavy bodies and broad, flattened heads. They are sluggish animals, living in cold, deep mountain streams. They undergo only partial metamorphosis for, although the adult lacks the larval gills, it never leaves the water. There are two species of giant salamander,

Megalobatrachus japonicus from Japan and *M. davidianus* from China.

Giant sloths are extinct mammals that lived in South America during the Pleistocene Epoch, about 1.5 million years ago. They reached a length of about 6 m. (19 ft. 6 in.) and, unlike the living sloths to which they are closely related, they were ground-dwelling and could rear up on their hind-feet to browse among the tree-tops.

Giant water-bugs are some 100 species of the family Belastomatidae of brownish, oval, flattened, aquatic *bugs. They are rapacious predators, and include fish and frogs in their diet. Some, such as *Lethocerus grandis*, are 11 cm. (4½ in.) long, and are among the largest insects. The hind legs are flattened and hair-fringed for swimming. They sometimes fly far from water and are attracted to lights at night.

Gibbons are the smallest of the *apes, lithe and slender with such long fore-limbs that when they stand the fingers of the arms touch the ground. No other ape or monkey can travel through the trees with the same speed. The nine species of gibbon live in the canopy of the forest in

The moloch, or silvery **gibbon**, *Hylobates moloch*, lives only on the western end of Java, and is photographed here in the reserve of Ujong Kulon. This species spends much time feeding on fruits, picked by using an opposable thumb and index finger.

Southeast Asia, swinging from branch to branch, or walking erect along horizontal limbs of trees, feeding on fruits, leaves, and young shoots, and occasionally on insects. A social animal, it travels in small groups, usually four individuals to a family, communicating with loud shrieks and cries, that can be heard over considerable distances. A single young is born, after a gestation period of seven months, and is carried by the mother for eighteen months while it slowly develops. Each species of gibbon occurs in a particular part of Southeast Asia.

Gila monster: a species of large, heavy-bodied lizard found in the southwestern USA in desert areas. Along with the beaded lizard, *Heloderma horridum*, it forms the family Helodermatidae. Amongst a total of 3,000 species of lizards, these two are the only poisonous ones. The gila monster, *H. suspectum*, has a short, thick tail which contains a fat-store, enabling it to survive considerable fasting periods. Its venom glands are situated in the lower jaw, the teeth are grooved, and venom flow is enhanced by chewing motions. The venom is remarkably toxic, comparable to that of the cobra, although few human fatalities have been recorded. It hunts by 'smelling' out prey such as birds' eggs in nests, or other reptiles in burrows. To this end, the gila monster has a special organ of taste, called Jacobsen's organ, on the roof of its mouth, which picks up scent particles off the tongue.

Gills are structures by which aquatic animals obtain oxygen from the surrounding water. They are extensions of the circulatory system which increase the area over which oxygen may be absorbed. In many invertebrates, gills are simply tuft-like projections, whereas in larger invertebrates, such as bivalve molluscs, they are elaborate structures. In vertebrates, such as fishes, they are hidden and protected within a body cavity called a gill chamber. To successfully function as a respiratory organ a gill must have a flow of water constantly passing over its surfaces. In fishes this is achieved by drawing water into the body through the mouth, and passing it out of the body across the gills. In fishes, gills are the feathery, blood-red structures beneath the gill-cover, arranged on the outer edge of a cartilaginous gill arch. The large, whitish projections on the throat side of the arch are gill rakers, which retain food. The gills are composed of slender filaments, each of which contains many small plates called gill lamellae, which are semicircular or leaf-like in form. Blood flows in fine capillaries through these very thin lamellae, and in this way the blood takes up oxygen from the water passing through the lamellae, as well as surrendering wastes, such as carbon dioxide, from the fish's body.

The circulation of water from the fish's mouth and across the gills is effected by two pumps, one in the floor of the mouth, which expands the throat as the mouth is opened, and the other in the gill chamber, which draws the water back and over the gills. Active fishes require more oxygen than inactive ones and their gills are correspondingly larger. Fishes under respiratory stress, caused by swimming fast, pump more water through the gills.

Ginger is a tropical, perennial species of monocotyledonous plant, which gives its name to the ginger family, Zingiberaceae, also including cardamom and turmeric. From swollen rhizomes, or roots, a spice is obtained, and appears on the market as the root, as powdered, dried ginger (particularly from West Africa, Jamaica, and

India), or as preserved ginger (chiefly from China). The latter is produced when the young rhizomes are boiled with sugar before being packed in syrup. The aromatic, highly pungent spice is used in cooking throughout the world and is one of the more important constituents of curry powder. Ginger is a native of tropical Asia, although the family as a whole occurs throughout the tropics. All species have branched, fleshy underground rhizomes, and have broad leaves growing from a central stem. The ginger family has some 1,300 species worldwide.

Ginkgo is a genus of *gymnosperms, comprising one living species, *Ginkgo biloba* (the maidenhair tree), and a number of species known only as fossils. The living tree is native to China, where it is grown around houses, probably for its edible seeds. It is also used as a street tree in other parts of the world. It provides a useful timber and an oil. The tree is deciduous with leaves shaped like those of the maidenhair fern. It is dioecious (bearing either male or female flowers) and the male gamete (sperm) is unusual in using hair-like cilia for locomotion.

Giraffe: the tallest species of animal in the world today, and quite unmistakable with its extremely long neck. The male is larger than the female and has a shoulder height of 3·5 m. (12 ft.), and a total height of 5·5 m. (18 ft.). There are nine subspecies, including the blotched giraffe, found over most of Africa south of the Sahara, and the larger, reticulated giraffe, found in East Africa.

The head bears short, straight horns which are bony outgrowths covered by skin. The large, brown eyes are protected by long eyelashes, and the tongue is prehensile and can be extended some 50 cm. (20 in.). A short mane runs along the neck and back. The long neck and tongue allow it to browse high in the trees. Special mechanisms prevent changes of blood flow when it raises its head from near the ground to its full height. The weight of the neck is balanced on the fore-limbs. In spite of its long neck, the giraffe is a graceful animal able to gallop at speeds of up to 50 km./hour (32 m.p.h.), the legs on the same side of the body moving in unison.

A single young is born after a gestation period of fourteen to fifteen months. Rather wobbly at first, the baby can move under its own power twenty minutes after birth.

Gizzard shad: a species of North American freshwater fish of wide distribution in eastern Canada and the USA. It is a member of the herring family, but is very deep-bodied with a bluntly rounded head. As an adult it feeds on minute algae, and is adapted to digest this food by having a thick-walled, gizzard-like stomach and an extremely long gut. The gizzard shad grows to a length of 30 cm. (12 in.).

Gladiolus comes from the Latin *gladius*, sword, describing the shape of the leaves. Gladioli are members of the *iris family, which includes the crocus. Their underground corms annually send up two or three leaves, followed by spikes of large, attractive flowers. Hybridization of the species, which occurs mainly in East and South Africa, has resulted in many colourful garden varieties, bearing spikes of trumpet-shaped flowers.

Glands are secreting organs, of varying structure and size, found in animals. They have important and diverse functions. Most of them, classified as exocrine, discharge

The splayed-leg posture of this drinking **giraffe** is necessary to get the head down to ground level. These reticulated giraffes at Samburu, Kenya, are one of nine recognized subspecies spread over the open forest and savannah of central and southern Africa.

their secretions through tubes or ducts opening onto a body surface or the gut. The remainder, the endocrine or ductless glands, discharge particular *hormones to the capillaries of the blood system. The preparation, storage, and expulsion of secretory material by glandular cells involve intricate, energy-using processes, controlled by the nervous and endocrine systems.

Exocrine glands are widely distributed in the skin of most terrestrial animals. Frogs, for example, possess two types of skin gland: one forms a mucous secretion which protects the skin from drying out; the other is a poison gland whose secretion discourages attacks by predators. Glands within the nostrils and eyes keep delicate surface membranes moist and free from obstruction. Sebaceous glands associated with hair follicles protect the skin of mammals. The sweat glands aid heat loss in mammals.

Exocrine glands of the digestive system provide saliva for the mouth, gastric juice for the stomach, pancreatic juice, and bile for the intestines. Mucous glands deliver a sticky coating for the lining membranes of the gut to protect them from the action of the digestive juices.

Glass snakes are not snakes but legless *lizards belonging to the genus *Ophisaurus*. There are several species related to the slow worm, found in North America, southeastern Europe, and parts of Asia. Their name refers to their fragility; when roughly handled or if struck by a stick, they may shed their long tails, a phenomenon which may lead an untutored observer to believe that the whole animal has broken to pieces.

Glassworts, or *Salicornia*, are plants with fleshy, jointed stems and no leaves. They are adapted to live in the desiccating environment of salt marshes and sea-shores throughout the world. They belong to the same family as beetroot and spinach. The ash glassworts have a high soda (sodium carbonate) content and were formerly used in the manufacture of glass and soap.

Glaucoma is raised pressure inside an eye as a result of poor drainage of the fluid within it. It is a common cause of loss of vision in elderly people. Peripheral vision is first affected, and may not be noticed at first, so that diagnosis is delayed. Treatment can slow down or arrest the damage.

Gliding frogs   *Flying frogs.

Globeflowers are members of the plant genus *Trollius*. About twenty species are known, mainly from Europe, Asia, and North America. They are close relatives of buttercups, but with larger, globe-shaped flowers, usually yellow or orange. All prefer moist habitats. (*See over*).

Glossopteris: a fossil tree that dominated the flora of the Southern Hemisphere in Permian times (286–245 million years ago). The leaves were shaped like long tongues with a prominent midrib and net-like veining. Seed-bearing reproductive organs were attached to the base of the leaves. The trunk grew 4–6 m. (13–19 ft.) in height and reached up to 40 cm. (15 in.) in diameter. Because the fossils are found in sedimentary rocks of the same age

The **globeflower**, *Trollius europaeus*, is found over much of Europe, usually in damp meadows. Up to 70 cm. (27 in.) in height, it has flowers 5 cm. (2 in.) in diameter, which appear from May to August and are pollinated by insects.

in India, South Africa, South America, Australia, and Antarctica, they provide good evidence that these continents were a single land mass (Gondwana) in the Permian Period.

Glow-worms *Fireflies.

Gloxinias are tuberous-rooted plants with rounded or heart-shaped leaves and colourful (often velvety) bell-like flowers. Members of the same family as African violets, they are a mainly tropical group and are popular greenhouse plants in temperate climates. Most of the florists' varieties are derived from the Brazilian species *Sinningia speciosa*.

Glucose is a simple sugar with the molecular formula $C_6H_{12}O_6$, belonging to the monosaccharide group of *carbohydrates. It is the most important fuel used for respiration in living organisms. Other carbohydrates in the human diet are converted into glucose, which circulates dissolved in the blood plasma until used by the body tissues. Once glucose has been consumed, it is stored as *glycogen in the liver and muscles, or may be converted into fat, being released later by the breakdown of these substances. Its concentration in the blood plasma is kept constant by the hormones insulin and glucagon.

Gluttons *Wolverines.

Glycogen is a natural organic compound belonging to the polysaccharide group of *carbohydrates. Each molecule comprises a branching network containing hundreds of *glucose molecules joined together. It can be very rap-

idly built up or degraded by enzyme action and is used as an energy storage material in bacteria, fungi, and most animals.

In vertebrates, the hormone insulin promotes glycogen formation in the liver and muscles, and other hormones, such as glucagon and adrenaline (epinephrine) cause its breakdown, thereby releasing glucose.

Glycolysis, or 'glucose splitting' is the first stage in the biochemical pathway of *respiration, carried out by virtually all living organisms. *Glucose ($C_6H_{12}O_6$) is progressively broken down to yield two molecules of pyruvic acid ($C_3H_4O_3$). Initially, *adenosine triphosphate (ATP) is used up in making the glucose molecule more reactive, but later steps release energy so that there is a net gain of two molecules of ATP for the process as a whole. Hydrogen atoms from the breakdown of glucose are taken up by *coenzymes and can be transferred to the *electron transport pathway, resulting in the production of more ATP. In *fermentation the pyruvic acid, produced by glycolysis, is transformed into lactic acid or ethanol. In aerobic respiration the pyruvic acid enters a series of reactions known as *Krebs' cycle.

Glyptodons are an extinct group of armadillo-like, large animals that lived in South America during the Pleistocene Epoch some 1·5 million years ago. They were up to 3 m. (10 ft.) in length, and the almost spherical body was covered by a complete shell of small, thick bones.

In most of the herbivorous **glyptodons**, the frontal dentition was absent. Limb bones were short and massive, and the claws of the forefeet well developed.

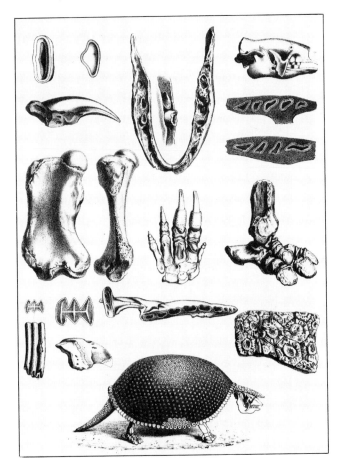

Herds of **gnus** may be migratory or sedentary, depending on the availability of green grass and water. Within those herds that migrate annually, large numbers may drown or be swept away at river crossings.

Some species of *Glyptodon* had a spiky armoured tip to their tail. They became extinct about 100,000 years ago.

Gnats   *Mosquitoes.

Gnus, or wildebeest, are species of grazing antelope, belonging to the same tribe (a taxonomic group below the subfamily) as the *hartebeest. The two species are found on the African veld, where they are perhaps the fastest of the animals. The brindled gnu, *Connochaetes taurinus*, travels in herds of a dozen to several hundred individuals, frequenting the open plains in search of grass and a regular supply of water. Brindled in colour, it has brown stripes on the neck and shoulders, and a black tail; the horns are smooth and spread outwards. A single calf is born after a gestation period of eight and a half months. The white-tailed gnu, *C. gnou*, was once common over much of the South African veld, but is now restricted to reserves.

Goannas   *Monitor lizards.

Goat moths are a family of large moths, some with a wing-span of up to 18 cm. (7 in.), whose large, wood-boring caterpillars smell of goats. These take three or four years to reach maturity and the caterpillars, sometimes known as carpenter worms, can inflict considerable dam-

age on trees. Their family is worldwide in distribution and in Australia includes the witchetty grubs, considered a delicacy by the Aborigines.

Goats are a type of antelope which make up the family Caprinae. The twenty-six species include the chamois, musk ox, and sheep. Three species are actually called goats: the mountain goat, *Oreamnos americanus*, the wild goat, *Capra aegagrus*, and the Spanish goat, *C. pyrenaica*. The wild goat is native to Asia Minor and parts of India, but has now been spread worldwide as a domestic breed. The horns are directed backwards and upwards, and are transversely ridged; males have a distinct beard and a gland beneath the short tail which gives off a pungent odour. The Spanish goat is similar in habits and appearance to the true wild goat, differing only in the shape of its horns. It is restricted to the Pyrenees. The mountain goat does not belong to the same genus as the wild and Spanish goats, but is more closely related to the chamois. It is found in mountainous and northern areas of North America.

Other species in the genus *Capra*, or true goats, include the ibex, *C. ibex*, markhor, *C. falconeri*, east Caucasian tur, *C. cylindricornis*, and west Caucasian tur, *C. caucasia*. These and other species have many, locally named, subspecies.

Goatsuckers   *Nightjars.

Gobies form the largest family of marine fishes, with more than 1,000 species. Mostly they live in shallow, in-shore waters from high-tide mark on the shore down to

200 m. (650 ft.), but a substantial number have invaded fresh water. They are small fishes, possibly averaging only 5 cm. (2 in.) in length, the largest being about 35 cm. (14 in.) and the smallest, a species from the Philippine Islands, grows to only 12 mm. (½ in.) long. This is one of the smallest known vertebrates. Gobies are blunt-headed fishes, with two dorsal fins, fully scaled bodies, and pelvic fins united to form a weak disc-like sucker. The family includes the *mud-skippers, which are abundant on tropical shores.

Godetias are annual plants native to western North America. They are closely related to the evening primroses, but differ in having rose-purple or reddish flowers. *Godetia amoena* and *G. grandiflora* are the parents of many colourful garden varieties.

Godwits are four species of large *wading birds of the Scolopacidae family, distributed worldwide. They resemble curlews but their long beaks are straight or slightly upturned. Like many waders they have two plumages: in winter they are dull grey-brown, but in summer their neck and underparts turn a bright chestnut colour.

Goitre is an enlargement of the thyroid gland which appears as a swelling of the neck. Simple goitre develops in animals and humans living in regions with very low levels of iodine in the soil. Iodine is an essential component of thyroid hormones, and if its supply is deficient the gland enlarges in an attempt to remedy the shortage. Iodine is now added to table salt before its sale, to prevent this type of goitre. Exophthalmic goitre, so called because protruding eyeballs are also a symptom, is a serious condition of thyroid enlargement and hyperactivity.

Goldcrest: a bird species of the genus *Regulus* in the family of Old World *warblers, and related to the firecrest and kinglets. It is a small greenish bird with a partially concealed black-edged crown stripe, yellowish in the female, golden in the male. At about 6 g. (less than ¼ oz.) in weight, it is the smallest British bird. Widespread in Europe and Asia, it builds its nest high up in trees, often in conifers or on the ivy-covered trunks of hardwoods.

Golden eagles are large, reddish-brown *eagles which are thinly but widely distributed over northern Europe, Asia, and throughout North America. Five subspecies are recognized. Northern populations tend to migrate south in winter. They feed mainly on small mammals and ground birds, which they catch in extensive home ranges, but are often scavengers in their mountain and moorland haunts. They nest among high crags, rearing one or two young.

Golden moles are a family of insectivores, only distantly related to true moles but of similar appearance and habits except that some of the eighteen species hunt on the surface. They have a flexible snout terminating in a stout nose-pad, which is used extensively in digging. Golden moles are found throughout Africa in suitable soils. They range in size from Grant's desert golden mole, *Eremitalpa granti*, which is 8 cm. (3 in.), to the giant golden mole, *Chrysospalax trevelyani*, which is 20 cm. (8 in.).

Golden orfe is the golden variety of the fish known as the ide, a member of the carp family which is widespread in central and eastern Europe. Because of its golden coloration it is kept as an ornamental fish in lakes and occasionally escapes from these into rivers. It has been found

A female **golden eagle** and chick on the nest in coniferous forest, Finland. Although two eggs are usually laid, in 80 per cent of cases only the first hatched young survives. The smaller chick is often attacked and killed by the larger one if food is short.

in several English rivers. Superficially like a *chub, it has moderately small scales. It spawns in late spring over water-plants in shallow water.

Goldfinch is the name for a number of bird species in the finch family, especially the American and the Eurasian goldfinches. The former, *Carduelis tristis*, is a small species in which the male is bright yellow, with black wings, tail, and cap. The female is dull olive-green without a black cap. It breeds over a wide area of northeastern North America. The Eurasian goldfinch, *C. carduelis*, is widespread over Europe and western Asia; both sexes have a brownish back and white belly with a striking red and white face. The wings are black, with a broad golden-yellow wing bar, which is conspicuous in flight and thus gives the species its name. Both these species live in open country close to woodland, and feed primarily on small weed seeds. The genus *Carduelis* has twenty-four species, some known as siskins or greenfinches.

Goldfish are best known as golden-coloured pond or aquarium fishes, often with elaborate fins. The tail fin may be double in some of these varieties. The variation is due to the selection of stock by fish-breeders to emphasize various colours, body forms, and fin shapes. Many goldfish are liberated into the wild and in warm temperate areas may form breeding populations, but generally they are seen and eaten by predators.

A native of Siberia, northern China, and Korea, the wild goldfish, *Carassius auratus*, lives in weed-filled lakes and back-waters of rivers. It spawns among vegetation in midsummer, the eggs adhering to the leaves. It is very resistant to extremes of temperature and low oxygen levels, two qualities which have made it a successful ornamental fish. In eastern Europe it is represented by a subspecies, called the giebel carp or Prussian carp.

Gold moths, or pollen-feeding moths, are tiny, metallic, day-flying moths of the family Micropterigidae: they are less than 12 mm. ($\frac{1}{2}$ in.) in wing-span. They comprise a unique suborder of Lepidoptera (moths and butterflies) which occurs worldwide, for, unlike other moths, they lack a proboscis but have jaws with which they feed on pollen. Their caterpillars probably feed on mosses, and pupae have movable legs and jaws.

Gold-tail moths *Tussock moths.

Goliath beetles *Hercules beetles.

Gonads are the male, or female, reproductive organs. They occur in pairs in most vertebrates but are reduced to a single organ in some birds, fishes, and reptiles. The male glands, or *testes, are found in the testicles of warm-blooded vertebrates, the female glands, or *ovaries, lie within the abdomen. The gonads have two functions: the production of reproductive cells called sperms (male) or ova (female); and the secretion of the sex hormones testosterone (male), and oestrogen and progesterone (female). The gonads are present at birth but are non-functional until the onset of sexual maturity (puberty), which occurs at different ages in different animals.

Gonorrhoea is a sexually transmitted infection caused by the bacterium *Neisseria gonorrhoeae*, which typically causes acute inflammation of the urethra or vagina. The incubation period is only three to five days and the first symptoms are pain on passing urine and a discharge of pus, although at least half the women infected have no symptoms. If it is promptly treated no permanent damage results, but if neglected complications, including sterility, may ensue. This infection was once easily treated with penicillin, but resistant strains are now becoming a problem.

Gooseberries are heavily thorned, hardy, deciduous shrubs, grown in cool temperate areas; they belong to the saxifrage family. Like their relatives, the currants, they produce edible fruits which vary widely in colour and size. Most are yellowish-green, some are distinctly yellow, and others reddish or purplish. Certain varieties produce berries sweet enough for eating raw, whereas others, containing more acid, are more suitable for cooking.

An unrelated species, *Physalis peruviana*, which belongs to the potato family, is known as the Cape gooseberry.

Gopher is a name used in North America for *ground squirrels of the genus *Spermophilus*, but the name is also applied to members of a quite distinct family of *rodents, Geomyidae. These are more accurately called pocket gophers because of their huge fur-lined external cheek pouches.

There are twenty-three species of *Spermophilus* in North America, with a further fourteen in the Old World. These ground squirrels are diurnal and mainly vegetarian, feeding on seeds, nuts, and roots, but they will also eat small birds and rodents if they can catch them. They usually live in burrows, hibernating in winter.

Pocket gophers are mole-like and live underground in extensive burrow systems, which they excavate with their huge, chisel-like, front teeth and powerful, clawed front feet. The thirty-four species, which are found only in North America, feed underground on roots and bulbs.

Goral: a species of goat-antelope, intermediate in appearance between these animals and part of the same subfamily as goats. It is found in mountains from the Himalayas to Korea at altitudes of 900–2,700 m. (3,000–9,000 ft.). When it is among tall grass it is difficult to see, as it stands only 70 cm. (28 in.) at the shoulder. It has a coarse coat of grey or brown. Both sexes have short horns that curve slightly to the rear. If alarmed it utters a penetrating hiss.

Gorgonians, or horny corals, are common octocorallian *corals of reef communities. They have a central skeletal axis, with *polyps arranged around this, often giving a branched tree-like appearance. They are usually brightly coloured, the precious *red coral, for example, being brilliantly coloured. The flexible branching skeleton is an ideal shape for trapping suspended food; the large surface area collects countless tiny animals as the coral sways with the current.

Gorillas are the largest of the great *apes. The male is heavily built and may stand 1·7 m. (5 ft. 6 in.) in height, with an arm-spread of 2·4 m. (8 ft.). It has a large head with a short neck, prominent mouth, thin lips, and small ears. The female is smaller than the male. This huge animal has formidable canine teeth, yet it feeds on fruit and vegetables and has no natural enemies. A group consists of about six individuals of the immediate members of the family—the adult male, one or two females, their babies and adolescent young. Most of the day is spent on the ground; they usually walk on all fours leaning on the knuckles of their hands, but occasionally upright. At night the group makes camp and each animal prepares a nest, the females and young in the trees, the adult males on the ground.

Three subspecies of gorillas are known: the huge western and eastern lowland gorillas, which live, respectively, in the rain forests along the west and east coasts of equatorial Africa, and the mountain gorilla of the montane forests around Lake Kivu and in the ranges that extend from the Congo east to the border of Uganda.

Gorse is the common name for plants of the genus, *Ulex*. They are spiny shrubs of the pea family, common along the Atlantic coasts of Europe on heathland. The spines are reduced shoots with minute leaves, though seedlings have more typical leaves. They all have yellow flowers with an unusual method of pollination. When a bee alights on the keel of the flower, its weight causes an 'explosion' within the flower which leads to the release of the stamens and forces pollen onto the bee's body. The seeds have two means of dispersal. Firstly, the pods split violently and scatter the seeds. Secondly, the seeds have a juicy area, attractive to insects, notably ants, which carry the seed away. The gorse is also known as furze or whin.

Gould, John (1804–81), British ornithologist and son of a gardener at Windsor Castle, became a taxidermist for the newly formed Zoological Society of London and travelled widely in Europe, Asia, and Australia. He published forty-one works on birds with 2,999 remarkably accurate illustrations by a team of artists, including his wife.

Gouramis are Asiatic fishes belonging to two families, the Belontiidae and the Osphronemidae. The former contains many small, brightly coloured, tropical freshwater fishes, including the *fighting fish. The Osphronemidae family includes the gourami, *Osphronemus goramy*, a 60 cm. (25 in.) long food-fish. Gouramis are found throughout Asia in ditches draining paddy-fields, swamps, streams, and ponds, and have a special breathing organ in the gills which permits them to breathe air. They make bubble nests at the water surface for their eggs.

Gourds are climbing, often perennial, plants belonging to the pumpkin family. This includes cucumbers, melons, and marrows. Several different species are given the name of gourd, although the name is used when referring mainly to *Cucurbita* species. Other species include the bottle gourd, *Lagenaria siceraria*, the bitter gourd, *Momordica charantia*, and the snake gourd, *Trichosanthes cucumerina*. Although some *Cucurbita* species have fruits which can be used as a vegetable, most are grown for their hard-skinned fruits which, although inedible, have a multitude of uses.

The flesh of the bottle gourd, usually too bitter to be eaten, can be scooped out, along with the seeds, and dried. Empty fruits are used as containers to store and carry solids and liquids. The bottle gourd has been known all over the world for at least 5,000 years. This wide distribution possibly resulted from the fruit's ability to float and survive in sea water for up to seven months. The fruits are variable in size, from 8 cm. to 1·5 m. (3 in. to 5 ft.) in length, and are flattened, round, club-, or bottle-shaped. As well as food containers they can be used as floats for fishing nets, and as pipes and other musical instruments.

The bitter gourd has an edible fruit used in Indian pickles and curries, and the snake gourd develops narrow, cylindrical fruits up to 1·5 m. (5 ft.) in length. This Asian fruit can be eaten when young, pickled or boiled. Small varieties of other gourds are used for ornamental purposes.

Gout *Arthritis.

Graaf, Regnier de (1641–73), Dutch physician and anatomist, was the author of works on the nature and function of pancreatic juice and the ovaries. He gave his name to Graafian follicles, small sacs in the ovary of a mammal in which the ova are matured prior to release and fertilization.

Grackles are the six species of New World *orioles or blackbirds, which comprise the genus *Quiscalus*. The males are mostly shiny black, with a bronze, blue, or green sheen, and have long tails which may reach 42 cm. (17 in.). The females are often browner and smaller. Grackles live in open, bushy country and out of the breeding season often gather in large flocks. They eat a wide range of foods, mainly insects in summer and seeds in winter and can be a serious pest to farmers.

Gram, Hans Christian Joachim (1853–1938), a Danish bacteriologist, devised a method of staining some bacteria with iodine and crystal violet, or similar dye, for microscopical examination. Those which retain the stain ('Gram-positive') are generally attacked successfully by penicillin, while 'Gram-negative' bacteria have to be treated with streptomycin.

Granadillas, or passion-fruits, are several species of *passion-vine, or *Passiflora*. The large granadilla, *P. quadrangularis*, has large yellowish fruits with white, juicy flesh; the young green fruits may be boiled and eaten as a vegetable. Granadilla is best suited to the hot, moist, tropical lowlands. Another tropical South American species, the sweet granadilla, *P. edulis*, is a popular crop in the mountains of Mexico and Central America, and has been naturalized in Hawaii. Its smaller, brownish fruits have more aromatic flesh than the large granadilla.

Grapefruit is a species of *Citrus* tree, grown in tropical and warm temperate regions of the world for its large yellow fruits, up to 12 cm. (5 in.) in diameter. It is probably so-named because its fruits are produced in clusters. About 70 per cent of the world's crop comes from the USA, but it probably originated in Barbados. Varieties exist which differ in their flesh colour (pale yellow or pink) and some are relatively seedless. Its West Indian name is pomelo, not to be confused with the pummelo or shaddock.

Grape hyacinths belong to the *Muscari* genus, with approximately fifty species of plants which grow from bulbs. They are native to the Mediterranean region and Asia Minor. They belong to the lily family and produce leaves and spikes of blue, or occasionally yellow or greenish, flowers in spring and early summer. The bulbs are dormant during the rest of the year.

Grapes, or grape-vines, are perennial, climbing plants of the genus *Vitis*. Along with Virginia creeper, they form the family Vitaceae, and are one of 700 species within this family. Many varieties of the European grape-vine are successfully grown in areas of the world with the dry, hot summers and cool, wet winters characteristic of Mediterranean areas. The fruits, or grapes, of many varieties of *V. vinifera* are used for wine production, and are juicy, seedy, and acidic. Grapes for eating tend to be less seedy or seedless, with up to 25 per cent sugar content.

The variety of grape-vine, soil, and climate all affect the chemical composition of grapes, creating the local characteristics of various wines. Red wines are made from the purplish to bluish-black grapes or from a mixture of purple and green grapes; white wines may be made from either purple or green grapes or from a mixture of the two. European vines are nearly all grafted onto the rootstock of an American species of *Vitis* that gives resistance to *Phylloxera*, a *scale-insect that was accidentally introduced to Europe, from North America, in 1860 and devastated the vineyards of France.

Italy, France, Germany, and Spain are the main wine producers, although California, Australia, and other regions have become more important in recent years. Raisins are the dried fruits of dessert varieties, sultanas are from smaller, seedless varieties, and currants are from small, black-fruited types.

Graptolites are small, colonial marine animals which are known only as fossils from the Palaeozoic Era. They first appear near the end of the Cambrian period (505 million years ago) and are among the most common Ordovician and Silurian fossils. They finally became extinct at the end of the Silurian Period, some 408 million years ago. They form branching colonies with stems bearing numerous little skeletal cups which protected the individuals. Graptolites are believed to be relatives of the living *acorn worms.

Grasses form one of the largest families of flowering plants, the Gramineae, with over 9,000 species, many of which are of major economic importance. They are *monocotyledons with fibrous roots and upright stems, and are distributed throughout the world, occurring in most kinds of habitat, from tropical rain forests and swamps to the hot, arid desert areas of the subtropics. They dominate such areas as the steppes of Asia, the prairies of North America, and the savannahs of Africa. This domination is largely a result of the grazing pressure of animals, which encourages grass at the expense of other herbaceous plants.

In an exceedingly diverse family, the only woody species are the *bamboos. Grasses differ from other plants in that the vegetative growth is made up mostly of leaf tissue. Tubular leaf sheaths support the parallel-veined leaf blades. The growing points, or buds, of grasses are near ground-level and consequently are not damaged by grazing; grasses form the main food of many grazing animals.

The characteristic branching pattern of grasses, known as tillering, occurs as a result of the buds developing into new side-shoots, and some species can spread by means of stolon or rhizome development. The flowers (florets) of grasses are insignificant; being wind-pollinated they have no need for showy petals. They are often clustered together in compact, elongated spikes, as in wheat. The 'seeds' (really *fruits) can be quite large and contain a large proportion of carbohydrate. The higher-yielding, easily collected, and cultivated species have been exploited as *cereals by man since agriculture began.

The presence and development of fossil **graptolites**, in this case *Didymograptus murchisoni*, provide a means of identifying specific zones in geological time.

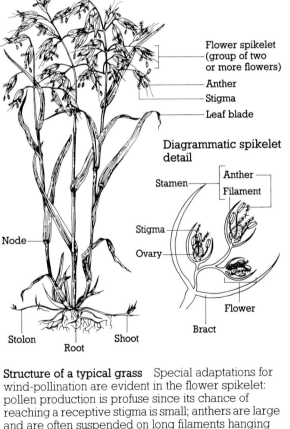

Grasses

Flower spikelet (group of two or more flowers)
Anther
Stigma
Leaf blade

Diagrammatic spikelet detail

Stamen
Anther
Filament
Stigma
Ovary
Flower
Bract

Node
Stolon
Root
Shoot

Structure of a typical grass   Special adaptations for wind-pollination are evident in the flower spikelet: pollen production is profuse since its chance of reaching a receptive stigma is small; anthers are large and are often suspended on long filaments hanging outside the flower so that wind movements readily shake pollen free; stigmas are spreading and feathery, snatching up floating pollen grains like a net; and flowers are borne on long slender stalks, exposed to the wind.

Grasshoppers are large to very large insects which, along with crickets, comprise the order Orthoptera, with some 17,000 species. All grasshoppers have strong, biting mouthparts and most species have leathery fore-wings; the hind ones, being broad and delicate, are folded fan-wise under the fore-wings when the insect is at rest. Some species have reduced wings or no wings at all. Almost all male grasshoppers produce sounds characteristic of the species, and are plant-feeders, although a few are carnivorous. They fall into two distinct groups; the *longhorn grasshoppers, and *shorthorn grasshoppers which include the locusts.

Grass moths are small moths with wing-spans of up to 3 cm. (approx. 1 in.) which belong to the subfamily Crambinae. They are abundant in grassy places, where they rest by day on stems, often head-down, with their straw-coloured wings tightly wrapped around their body and the prominent mouth-parts projecting forward like a horn. Their caterpillars, known as webworms, spin silken galleries low down among the grasses on which they feed.

Grass snake is a species of snake which occurs over much of Europe (including most of Britain) and parts of North Africa. It is a non-venomous snake, up to 2 m. (6 ft. 7 in.) in length, and is often found in moist habitats. Its diet includes frogs, newts, and occasionally fishes. Coloration is variable but the body is usually a greenish-grey and the neck often has a black-bordered, yellow, white, or orange collar. The grass snake usually deposits clutches of thirty to forty eggs in warm sites such as piles of decaying vegetation.

Gray, Asa (1810–88), American botanist, was author of many textbooks which greatly popularized botany. He wrote a *Flora of North America* (2 vols., 1838–43) with John Torrey, and was an enthusiastic leader in the discovery and classification of new species. Finding no conflict between evolution and his view of divine design in nature, he supported Darwin's theories at a time when they were anathema to many.

Grayling (fish): a species related to the salmon and trout but belonging to a separate family. It is distinguished by the adipose fin (a small dorsal fin near the tail) and high, many-rayed dorsal fin. It also has moderately large scales and is said to smell of thyme. Essentially a fish of fast-flowing, well-oxygenated, cool rivers, it is widely distributed throughout western Europe, although rare in some areas due to pollution. It is replaced in North America by the Arctic grayling, *Thymallus arcticus*. Both species feed on bottom-living insect larvae, crustaceans, and molluscs, and also take flies at the surface.

Greaved turtle, or Arrau turtle, is the largest of the *side-necked turtles, having a shell-length of up to 90 cm. (35 in.) and a weight of up to 90 kg. (200 lb.). It is found mainly in tributaries of the Amazon and Orinoco rivers of South America. Local people take adult turtles and eggs for food and oil, and the greaved turtle is now an endangered species.

Grebes are a family of some twenty species, related to bay ducks. They differ from other ducks in having lobed toes, pointed bills, slender erectly held necks, and practically no tails. They rarely come to land, even building their nests on water. The grebes have a world-wide distribution. The little grebe, *Tachybaptus ruficollis*, and a few other small grebes are known as *dabchicks.

Greenfinch: a sparrow-sized, greenish-yellow Eurasian bird of the genus *Carduelis*, with bright yellow patches in the wings and tail and a powerful beak. It is common in mixed woodland and open country over wide areas of Europe and western Asia. As a seed-eating species, it often comes to bird-tables in winter. In eastern Asia there is also a similar species, the oriental greenfinch.

Greenfly *Aphids.

Greengages are any one of several varieties of a tree regarded by some authorities as a subspecies of the European *plum, and by others as a separate species. Their small, round fruits differ from plums in having green or greenish-yellow skin, greenish-brown flesh, and a stone that does not separate easily from the flesh.

Greenshank are grey and white *sandpipers resembling redshank, but distinguished from them by having no wing-bar, green legs, and a longer and slightly upcurved beak. They breed in the USSR and Scandinavia, and winter as far south as southern Africa and Australia.

Greylag geese are typical Old World 'grey' *geese, looking like their farmyard descendants. Their plumage is greyish-brown barred with white, with a white stern. The western race has an orange beak and pink legs. In Britain grelag now breed naturally in northern Scotland and have been successfully reintroduced elsewhere.

Grey whale: a species of unique, bottom-feeding *whalebone whale. It was almost exterminated by whaling but is now protected; the two remaining populations are reaching reasonable numbers, although it is still threatened by hunting and pollution throughout its entire range. It migrates enormous distances each year, further than any other mammal on earth, with a round trip of 16,000 km. (10,000 miles) from the Arctic feeding grounds to the breeding lagoons of Baja, California. The female, which is slightly larger than the male, can reach 15 m. (50 ft.) in length.

Griffon vultures are large carrion-feeding *vultures with wings spanning almost 3 m. (10 ft.). There are four species widespread in Africa, southern Europe, and Asia. Birds of grassland and semi-desert, they soar high on thermals to spot fallen game, descending to feed in flocks. Griffons nest on crags and cliffs and rear a single chick which is fed by both parents on meat regurgitated from the crop.

Grosbeak is a name applied to several heavily beaked birds, especially some Northern Hemisphere members of the finch family, including relatives of the hawfinch. The pine grosbeak, *Pinicola enucleator*, is a relative of the crossbills and rosefinches. New World grosbeaks comprise six species in the bunting family related to the cardinals. All species primarily feed on seeds, using their large beaks to crush nuts in order to reach the seeds.

Gross, Samuel David (1805–84), American professor, was one of the most influential physicians of his day. His

System of Surgery (1859) was constantly revised in the light of his experience as a teacher of surgery, pathology, and anatomy.

Ground beetles are the typical 'black beetles' familiar to everyone; they make up the family Carabidae with some 25,000 species. Their main feature is a set of strong jaws pointing forward and a hard, black cuticle. They vary in size from small to large, are often metal-coloured, and almost all are ferocious predators, and thus help control pests. Most species live on the ground, although some live on trees. Several species have their wing-cases fused together and so cannot fly. Their larvae have long legs and are also carnivorous.

Ground cuckoos *Coucals.

Ground elder is a pernicious, perennial plant of the carrot family, native to Eurasia but naturalized in many parts of the world. It spreads by means of underground rhizomes, and is a particularly troublesome garden weed. In the past, it was often found in the vicinity of monasteries and, because of this and its reputation for curing gout, it is also called bishop's weed or goutweed.

Groundnut *Peanut.

Groundsel is the common name for several species of the plant genus *Senecio*. The north temperate common groundsel, *S. vulgaris*, is a rapid-maturing annual plant of waste places and cultivated ground. It belongs, along with other *Senecio* species such as ragworts and fleaworts, to the sunflower family. The seeds, which have long silky hairs, are easily distributed by wind.

Ground squirrels do not form a taxonomic grouping but the name refers to the many squirrel-like rodents,

When foraging for food, these **griffon vultures**, *Gyps fulvus*, may fly as far as 60 km. (37½ miles) away from their roosts. On sighting carrion, they usually alight some distance away and approach timidly; occasionally they will eat so much that they become too heavy to fly.

A thirteen-lined **ground squirrel**, *Citellus tridecem-lineatus*, emerges from hibernation and forages for food —nuts, fruit, and insects—in Michigan, USA.

especially those of the genus *Spermophilus*, in the family Sciuridae. They are most common in the tropics and are absent from Europe, but there are numerous species including some commonly called *gophers and *prairie dogs in North America, where they spend the winter in hibernation. All are predominantly vegetarian, but some will eat insects and other animal matter. Many species have prominent stripes down their sides and their common names often refer to these stripes or to their absence.

Groupers are members of the family Serranidae or sea basses, which comprises some 370 species. In general, the name grouper is applied to the larger members, such as the Queensland grouper, *Epinephelus lanceolatus*, which in Australian seas is known to grow to 270 kg. (680 lb.) in weight and 12 m. (37 ft.) in length. They are rather heavy-bodied, thickset fishes with big heads and wide mouths, with two dorsal fins, the first of which is spiny. All groupers are carnivorous, eating fishes, octopuses, and crustaceans. Many are hermaphrodites, some being male and female simultaneously, but most are female when first mature and male later in life. They are mostly tropical marine fishes but some are found in warm temperate seas.

Grouse is a name applied generally to the sixteen members of the subfamily Tetraoninae, a northern temperate region group including capercaillie, black grouse or blackcock, prairie chicken, and ptarmigan; it is more specifically applied to certain species of the family, such as the ruffed grouse, *Bonasa umbellus*, a common forest species in North America, and the red grouse, *Lagopus lagopus*, which is the British race of the Eurasian willow grouse, a moorland bird. Many species of this family, including ruffed and red grouse, are highly valued as sporting birds. All grouse nest on the ground, and lay between five and twelve eggs with very little nest material.

Growth in nature is a permanent increase in the mass of living material in an individual organism. A multicellular organism usually grows by mitotic division of its cells.

The spotted body pattern of **groupers**, such as this *Epinephelus* species from the Red Sea, helps to break up their body outline, a form of disruptive camouflage.

*Mitosis demands a source of energy and nutrients to allow replication of cell contents.

Plotting a graph of some measure of growth, such as weight or length, against time typically gives an S-shaped curve. In multicellular animals this is because they usually have a genetically determined maximum size. Such animals also show differentiation of their cells during growth to perform different functions and this, combined with differential growth rates within the organism, produces the final shape of the adult. By contrast, many plants are of indeterminate final size. Animals such as insects, crustaceans, and spiders show a stepped rather than a smooth growth curve. Only when their inelastic exoskeleton is shed can they grow rapidly in the brief period before the new, larger, outer shell hardens.

Overall pattern of growth is controlled by the genetic material in the *nucleus of the cells. Growth is modified by external and internal factors such as temperature, nutrition, and the availability of light for plants and hormones in animals.

Guacharo   *Oilbird.

Guan   *Curassow.

Guanaco is a species of llama and a New World relative of the camel. It is found in South America, where herds of about a hundred individuals range through the pampas and mountains. It stands about 1 m. (3 ft. 3 in.) at the shoulder and has a coat of soft woolly hair, which is pale yellowish-brown in colour on the body and ash-grey on the head. A single young is born after a gestation period of ten to eleven months. The guanaco is now raised on farms for its wool.

Guano is the name given to the nitrogen-rich droppings of fish-eating seabirds. It is valued as a fertilizer and was once commercially exploited in Peru.

Guavas are small tropical fruit trees, up to 10 m. (33 ft.) tall, with scaly reddish-brown bark and green to yellow berries. They belong to the myrtle family which also includes cloves and eucalyptus. One species in particular, *Psidium guajava*, originating in Central America, is widely grown there as well as in Africa, Asia, the Philippines, and the Pacific islands. The main commercial plantations are in India, Guyana, Brazil and Florida.

The large berries, up to 12 cm. (5 in.) in diameter, are extremely variable in flavour. The white, yellow, or even pink, juicy flesh is five times as rich in vitamin C as fresh orange juice. The seeds of the fruit are easily spread by birds and the plant has become almost a weed in some areas, notably Fiji. A smaller, bushier, closely-related tree with a grey–brown bark has dark-red fruits 3 cm. (1¼ in.) across, known as strawberry guavas. It is cultivated in the West Indies.

Gudgeon: a species of fish in the carp family which is native to Europe. Living close to the muddy or stony river- or lake-beds, it has a flattened belly and a pair of barbels at the corners of the mouth. It feeds on bottom-living insect larvae, crustaceans, and molluscs. It rarely grows longer than 20 cm. (8 in.).

Guelder rose: a deciduous European shrub, related to elder and honeysuckle. It belongs to the genus *Viburnum*,

The clusters of bitter red fruit of the **guelder rose** are produced only from its inner, fertile flowers. These fruit remain on the tree through the winter.

which includes about 120 species spread over much of Europe, Asia, and the Americas. The common guelder rose, *V. opulus*, has flattened heads of creamy-white flowers and bright red fruits. The outer flowers of the head have large petals but no sexual organs and act as visual attractants to insects.

Guillemots are small, fast-flying, seabirds of two northern genera. The common and thick-billed Brünnich's guillemots, *Uria aalge* and *U. lomvia*, are widespread on breeding cliffs of the northern Atlantic and Pacific and the Arctic Ocean. Black guillemots (tysties), *Cepphus grylle*, breed in the Arctic and north Atlantic areas only, and pigeon guillemots, *C. columba*, are north Pacific breeders. All are strongly gregarious, nesting in huge cliff or island colonies and feeding in groups at sea. Little nesting material is used as the normally solitary eggs and chicks are brooded on the feet of the adults. Guillemots feed mainly on plankton and small fishes, using their narrow wings to swim underwater.

Guinea–fowl are eight species of birds belonging to the family Numididae and related to other gamebirds. Most are the size of large domestic hens, usually greyish in colour, and speckled with paler markings. The vulturine guinea-fowl, *Acryllium vulturinum*, is unusual in having a bright blue neck and upper breast and long black and white hackles on the lower neck. All species are native to Africa south of the Sahara, but the helmeted guinea-fowl, *Numida meleagris*, has been widely introduced as a semi-domesticated bird and now breeds in a feral state in many parts of the. world. The nest is a scrape on the ground, usually with eight to twelve eggs. The downy young leave the nest soon after hatching.

Guinea-pig: a species of *cavy, related to the Patagonian hare, desert cavy, and rock cavy. It is known only as a domestic animal and attempts to identify the ancestral species among wild cavies have been unsuc-

A long-haired **guinea-pig**, one of the many, varied breeds kept as pets. The young are born with their eyes open and bodies fully furred. They are weaned within two to three weeks and can breed at six months.

cessful. The guinea-pig, *Cavia porcellus*, became known in Europe during the sixteenth century, following the Spanish conquest of South America, the home of present-day cavies. They had already been domesticated by the Incas, who kept them for food, allowing them the run of the house and surrounds to scavenge their food. The name guinea-pig is curious, for they do not come from Guinea and are certainly not pigs. It is quite likely that ships bringing guinea-pigs to Europe called at ports in Guinea and the animal was mistakenly associated with that country. There are many breeds of guinea-pig, which are now kept mainly as pets, although they continue to be used as food in South America.

Guinea-worms *Roundworms.

Guitar-fishes are *cartilaginous fishes, related to rays, which live in shallow water in tropical and warm temperate seas. They are flattened from above but the tail is well developed, as are the dorsal and tail fins. Guitar fish tend to be more active swimmers than rays or skates. Bottom-living, they feed mainly on crustaceans and molluscs and their small, closely packed teeth are well suited for eating hard-shelled invertebrates.

Gulls are seabirds which belong to the widespread subfamily Larinae which includes forty-six species. Generally white with fawn (juvenile), grey, or black backs, they are gregarious surface-feeders or scavengers, roosting in flocks and often nesting in large colonies. Most species live in the Northern Hemisphere. Gulls lay two or three cryptically coloured eggs in untidy grass nests; the chicks leave soon after hatching, but depend on parental feeding for several weeks. Many species take three to four years to reach sexual maturity and full adult plumage.

Gum-tree is a term commonly used for certain *Eucalyptus species in Australia, though it is also applied to a number of unrelated trees in other families, notably *Acacia species and *Nyssa* (*tupelo).

Gundis are one of the few families of *cavy-like rodents found outside South America. They are similar in size and appearance to guinea-pigs, except for their short tails and flat ears. There are four genera with some five species, occurring in desert or semi-desert regions of northern Africa. They shelter inside rock crevices but do not burrow. Diurnal and vegetarian, they are fond of sunbathing.

Gunnels *Blennies.

Gunneras are perennial herbaceous plants, chiefly from South America and New Zealand, which grow in moist habitats. The best known is the South American species *Gunnera manicata*, with giant leaves, 2 m. (6 ft. 6 in.) wide, on stalks up to 3 m. (9 ft. 9 in.) high. It is also known as hairy or giant rhubarb, although not in the same family as true rhubarb. It belongs to a small family of some 180 species, which includes water milfoil and other aquatic plants.

Guppy is a species of fish native to northeastern South America and Trinidad, but which has been widely introduced to warm temperate and tropical regions around the world. A popular aquarium fish, it has also been employed as a destroyer of mosquito larvae. It grows to 6 cm. (2½ in.) in length, the males being smaller. It bears live young and inhabits ditches, streams, and pools of fresh or brackish water.

Gurnards, of which there are about seventy species, are bottom-living fishes found in tropical and temperate seas and belong to the family Trigilidae. Known in America as sea-robins, possibly on account of their predominantly red coloration, they are well known for their ability to make noises by means of muscles attached to their very large swimbladders. They have hard, bony heads, frequently with long spines on the gill cover. The first of their two dorsal fins is composed of strong spines, and the lower rays of the pectoral fins are separate and mobile. Gurnards creep over the sea-bed searching for food with the sensitive finger-like rays of their pectoral fins.

The *flying gurnards are not closely related, and belong to the family Dactylopteridae.

Gymnosperms are one of the two major subdivisions of the seed-bearing plants. They are characterized by producing their ovules and seeds on the surface of a modified leaf, not enclosed within an ovary. Formerly more dominant, they have been ousted from much of the world's vegetation by *angiosperms, though the conifers, which are gymnosperms, dominate much of the cool, temperate forest in the Northern Hemisphere. The other groups of gymnosperms are probably not closely related, but merely represent a level of sophistication in the seed-bearing plants. They include *Ginkgo* and the *cycads. The gymnosperms comprise some 500 or so species.

Gypsophilas *Pinks.

Gypsy moth: a species of *tussock moth notorious for its accidental introduction to North America in 1869, which led to one of the worst insect plagues known. The caterpillars attack many species of deciduous and evergreen trees, often completely defoliating them. It is extinct in the British Isles, though common in the rest of Europe.

Gyrfalcons are the largest species of *falcon, with pointed wings and distinctive grey-brown, or even white plumage. In summer they range throughout the Arctic, feeding on ptarmigan, lemmings, and other small live prey. In winter they fly south to the sub-Arctic and warmer latitudes.

H

Habitat is the place in which an organism lives. It is defined by the food, space, micro-climate, other organisms, and physical and chemical conditions that it provides. An organism fitted to a particular set of conditions is said to show adaptation to that habitat. Most natural habitats are classified according to the physical features of their vegetation and include woodland, grassland, freshwater, or coastal habitats. Each broad category can be further subdivided; thus grassland comprises chalk grassland, hay meadow, pasture, and many other habitats.

Haddock: a member of the cod family which occurs in depths of 40–300 m. (130–980 ft.) on both the European and American coasts of the North Atlantic. It lives close to the sea-bed and feeds mainly on bottom-living animals. It is a valuable food-fish.

Haeckel, Ernst Heinrich (1834–1919), was a German biologist who was an enthusiastic supporter of the theories of Darwin. He applied these ideas in his zoological work and many popular books, most of which were translated into English. His work on embryology and the descent and relationship of various members of the animal kingdom, from protozoa to man, was influential in its time. Among the new words which he introduced to biology was that of *ecology.

Haemoglobin, a red-coloured protein containing iron, is the respiratory pigment of vertebrates. It is packed into red blood cells, thereby giving blood its characteristic colour. Its efficient oxygen-carrying ability enables about 200 ml of oxygen to be transported in each litre of blood. The haemoglobin molecule can carry oxygen or carbon dioxide, exchanging one for the other when passing through tissues with high concentrations of carbon dioxide. On its passage through the lungs, haemoglobin traps oxygen from the inhaled air, simultaneously releasing the carbon dioxide it carried to the lungs. However carbon monoxide, if inhaled, combines with haemoglobin to form a stable compound called carboxyhaemoglobin. In such cases the blood cannot transport sufficient oxygen to the tissues and death occurs rapidly.

Haemophilia is an inherited disorder characterized by failure of the blood to clot. This results from a deficiency of an essential substance, a coagulant known as Factor VIII. Abnormal bleeding can occur spontaneously or after the slightest injury. Beginning in early childhood, the end result of haemophilia may be severely damaged joints and muscles. It is inherited as a sex-linked *gene, which virtually restricts its occurrence to males. Females can 'carry' the haemophilia trait without showing any signs of the disease. In the past, severely affected children tended to die in adolescence, but now purified Factor VIII is given to stop bleeding.

Haemorrhages, or bleeding, occur when blood escapes from blood vessels. In healthy people they are commonly caused by injury. Effects depend on how much blood is lost and how quickly. Sudden loss of one-third of the blood may be fatal, but as much as two-thirds may be lost over twenty-four hours without death, provided the body can make replacement blood. Blood transfusion can be life-saving and was made safe by the discovery of the different blood groups. Certain diseases, such as *haemophilia, are particularly associated with haemorrhages.

Haemorrhoids, or piles, are small, irregular swellings caused by dilated veins inside the lower rectum and anus. Common causes include straining to open the bowels, a chronic cough, and other conditions which lead to pressure being placed on the veins of the rectum or anus. Internal haemorrhoids can bleed, cause mucous discharge with itching, and protrude after defecation. Occasionally blood clotting inside external haemorrhoids causes extremely painful swelling. Persistent protrusion or pain usually requires surgical treatment.

Hagfishes are one of the two orders of *jawless fishes (the other one consists of the lampreys), and are primitive fish-like vertebrates. In addition to lacking jaws they have no pelvic or pectoral fins and no true rays in the other fins. The mouth is a slit, with fleshy barbels, and they have one to sixteen gill openings along their sides. They live in all the major oceans, burrowing in the sea-bed and scavenging for food or catching worms and crustaceans. They have a rasp-like tongue which can be used to bore into dead or disabled fishes.

Hair is a general term for any long, threadlike structure, as found on plant stems and leaves. It is also applied specifically to the outer covering, or fur, of mammals, produced from a group of skin cells which form a hair follicle. The hair strand is continually produced from its expanded base within the follicle. This is the hair 'root' and comprises a mass of dividing cells. The main body of the hair strand is composed of a fibrous protein called keratin, also found in feathers, hooves, and horns. The core of the hair strand contains a pigment, giving the hair its colour. The centre may or may not be hollow. The strand is capable of being moved by small muscles attached to the hair follicle.

The lack of body hair in humans is a secondary feature, and thought to be associated with man's independence from his environment. Animal fur, or hair, is primarily used to insulate the body from heat loss. Some animals respond to the onset of cold winters by lengthening and thickening their fur. Others, such as the arctic hare, which rely upon their fur colour for camouflage, change colour by moulting. Colour and hair quality may be related to sex; the male lion's mane is a good example of hair used for sexual display. Human hair colour is controlled genetically but may lighten slightly in intense sunlight.

Hairstreaks, together with *blues and *coppers, are lycaenid butterflies. The name derives from distinct transverse lines on the otherwise uniformly coloured underside of the wings. As with many lycaenid butterflies, there is often an eye-spot on the underside of the hind-wings and one or more delicate 'tails'. These help the butterfly to confuse predators into thinking the eye-spot and 'tail' are the head, thus deflecting attacks away from the vulnerable body. The deception is most elaborate in tropical species, which include the majority of the 1,500 species.

Hakes are relatives of the cod family living near the continental shelf mainly in temperate areas of the Atlantic and Pacific oceans. They are long-bodied fishes with long-based dorsal and anal fins, large heads, and heavily toothed jaws. They feed on fishes, squids, and crustaceans, and usually live in moderate depths, down to 1,000 m. (3,280 ft.). They are valuable food-fishes; several of the ten or so species, including the European species, *Merluccius merluccius*, have been heavily over-fished. In North America the hakes are known as whitings, and the word hake is applied to distantly related *Urophycis* species.

Halibut: a slender-bodied but thickset species of *flat-fish with a large head, wide mouth, and powerful teeth. It lives in the deep waters of the cool North Atlantic. Its diet consists of fishes, squids, and crustaceans, and unlike most flat-fishes it actively hunts its prey in mid-water as well as on the sea-bed. A valuable food-fish, it is captured by long-lining as well as trawling.

Hamadryad   *King cobra.

Hammerheads are a family of *sharks containing nine species, distinguished by the remarkable head which is flattened and expanded in the shape of a hammer or spade. The eyes are placed at the extremity of these hammer-lobes and the nostrils are widely spaced on the front edge. The structure of the head may result in enhanced ability to trace scents in the water, but its function is not entirely certain. Living in all the tropical oceans, they are mainly coastal sharks but they do migrate far out to sea. They feed on a wide range of fishes and some of the larger species have been known to attack man. The great hammerhead, *Sphyrna mokarran*, of worldwide distribution, grows to 6 m. (19 ft.) in length.

Hammerkop, or hammerhead stork: a bird related to the storks and herons, but usually put in a family of its

Marine predators, such as this **hammerhead** shark, have far more operational freedom than those on land, being able to make twisting and turning scans in a three-dimensional space in search of vulnerable prey.

own. It lives in the warmer parts of Africa, stands about 60 cm. (2 ft.) high and is brown in colour, with a large crest spreading behind the head. It catches fish and amphibians with its powerful, heron-like beak. Its most unusual characteristic is that it builds an enormous, domed nest of sticks and mud among the branches of large trees. The entrance, or entrances, are in the side and the nest may be used for many seasons.

Hamsters are related to rats, mice, voles, and gerbils, with whom they share the family Muridae. Hamsters, of which there are twenty-four species, typically have soft fur, rounded faces, and large eyes. These features make them attractive as pets, of which the golden hamster, *Mesocricetus auratus*, is the most popular. All the captive stock of the golden hamster is descended from one female and twelve young which were collected in Syria in 1930. Its range includes much of the Middle East but the animal is little known in the wild. Another Middle Eastern species, the long-tailed hamster, *Cricetulus longicaudatus*, is rather rat-like in appearance. The common or European hamster, *Cricetus cricetus*, occurs in central Europe, the Caucasus, and western Siberia. In eastern Siberia, Manchuria, and northern China, there are three species known as dwarf hamsters, *Phodopus* species. All hamsters have huge cheek pouches in which they store food. They eat mainly seeds and grasses, and occasionally insects. They live communally in extensive burrow systems.

Hands have a very similar bone structure to that of feet, differing mainly in the names of the bones. The tarsals and metatarsals of the foot become carpals and metacarpals in the hand. They are best recognized in primates, including man, where the fore-limbs are used to collect and manipulate food and tools. The *fingers are the main agents of manipulation, although the wrist bones, the carpals, are more flexible than their equivalents in the foot. Human hands are distinguished from those of other primates by having wide tips, or balls, to the thumb. This, along with the ability to press each finger tip against the thumb ball, allows man to grip and manipulate objects with a much finer degree of control than can other primates. Races vary in their manual dexterity; the Asian races have

The forked horns of Coke's **hartebeest** and the difference in height between fore- and hindquarters give it an elegant appearance, although its movements are rather ungainly. The young are well camouflaged in grassland.

much more flexible finger joints than their Caucasian equivalents. The ridges and lines in the skin of the hand help prevent the palm from slipping when it is wet.

Harebells   *Campanulas.

Hares belong to one of the two lagomorph families of mammals, the other being the *pikas. They are distinguished from rabbits, with which they share the family Leporidae, by the production of young in an advanced stage, fully furred and with open eyes. Also, they never burrow like rabbits but rest during the day in 'forms', which are shallow depressions in the ground. There are some twenty-seven species of hare, including the American *jackrabbits, the African bushman hare, *Bunolagus monticularis*, and red rockhares, *Pronolagus* species. The common or European brown hare, *Lepus europaeus*, has a huge range from southern Finland to Iran. Hares have also been introduced into Australia, New Zealand, and South America. Some, including the blue or Arctic hare, *L. timidus*, change colour from brown to white in winter.

Harlequin bugs   *Shieldbugs.

Harpy eagles are one of the largest species of *eagle with wings spanning 2 m. (6 ft. 6 in.) They live in the canopy of South American rain forests, seldom appearing above or below it. Their prey includes monkeys, sloths, and birds.

Harriers (birds) are a cosmopolitan subfamily of ten species of *hawk with long, square-cut wings, and long legs. In flight they swoop low over the ground when hunting. They spot their prey from trees or posts, or by quartering systematically over likely ground. Small mammals, frogs, insects, and ground birds are typical prey. Many species of harrier are migratory, avoiding extreme cold. Harriers nest on the ground but courtship involves a spectacular aerial dance, during which the male passes food to the female.

Hartebeest is a species of grazing antelope found in large herds on the open plains and desert areas of Africa. It feeds during the day on grass and seems to need very little water. When they are grazing, a sentinel is posted on an ant-hill; this timid animal rarely attempts to defend itself even when wounded and is preyed upon by lions and other predators. It is amongst the fastest of the larger antelopes. There are many subspecies of hartebeest, including the korrigum, topi, senegal, western hartebeest, kongoni, and the bubal hartebeest. The coat is reddish or yellowish-brown, varying in the different subspecies. The face is long and thin, while the horns, present in both sexes, are short and stout.

Hart's tongue, unlike many other ferns, has undivided, strap-shaped fronds of a bright, shiny green. The spore-cases are in regular lines running from the central vein to the margin of the frond. It belongs to a group of relatively small ferns, many with fronds divided into tiny leaflets. Other species in this genus, *Asplenium*, include wall rue, *A. ruta-muraria*, rusty-back fern, *A. ceterach*, and several species of spleenworts.

Harvestmen, known as daddy-long-legs in America, are familiar spindly-legged *arachnids abundant in humus and on tree trunks. They resemble spiders, but lack a narrow waist. The legs of some species may be forty times the body length. A distinctive feature of many of the 3,200 worldwide species is that their eyes are often placed on a tubercule in the centre of the thorax, or carapace. Some harvestmen produce noxious secretions, and a few actually spray these at attackers. Defence can also involve deliberately shedding legs, if gripped by a predator, or feigning death for minutes at a time.

Harvest mice are one of the smallest species of *rodent with a body length of some 65 mm. (2½ in.). They have a wide distribution from Southeast Asia through Eurasia to the Arctic Circle. The young harvest mice, usually 3–8 in number, are reared in a nest of leaves which are left attached to the stems so that the nest hangs down between grass stalks. Contrary to popular legend, numbers of harvest mice have not declined drastically as a result of the introduction of combine harvesters since they breed commonly in hedges. The name harvest mouse is also given to nineteen species of very different mice of the genus *Reithodontomys*, from Central and South America.

Harvest-mites, or bracken-bugs, are the bright red parasitic larval stages of a very common subfamily of *mite, the Trombiculinae, which has free-living adults. The larvae bite any warm-blooded animals, especially rabbits, and can cause severe irritation to man when the knife-like mouthparts penetrate, especially around ankles and armpits. A salivary secretion is injected before tissue fluids are sucked out; fortunately these particular mites are not carriers of disease.

Harvey, William (1578–1657), was the English discoverer of the circulation of the blood, and physician to James I and Charles I. He studied at Padua, where he learned of the valves in the veins. Reflecting on the motion of the blood, he concluded that it was pumped by the heart, out through the arteries and back through the veins. The discovery was first made public in 1616 and was greeted with disbelief. Harvey's practice suffered as a consequence. Only later did the theory of blood circulation come to be regarded as a natural law, setting the ground for many subsequent discoveries in medicine.

Hawaiian geese, also called ne-ne from their calls, are related to Canada geese. With their habitat confined to Hawaii, their numbers had fallen to about forty before a remarkable rescue operation saved them from almost certain extinction. In 1950 three individuals were imported to a wildfowl reserve in Britain. They bred successfully and hundreds have since been raised there; some have been returned to their native Hawaii.

Hawfinches are three species of birds belonging to the genus *Coccothraustes*, within the finch family. The common hawfinch, *C. coccothraustes*, breeds over wide areas of Europe and Asia; the other two species, the masked hawfinch, *C. personatus*, and black-tailed hawfinch, *C. migratorius*, live in eastern Asia. All three species are greyish-brown with glossy blue-black wings and white-marked tails. They have very large beaks and break open large seeds to extract the kernels.

Hawk moths are powerful, fast flying moths with long, narrow fore-wings and stout bodies, occurring in most parts of the world, especially the tropics. They comprise the family Sphingidae, with some 1,000 species. Many species hover whilst feeding from flowers, using the proboscis to reach deep into the flower. The proboscis of the convolvulus hawk, *Agrius convolvuli*, is 15 cm. (6 in.) long, that is to say greater than its wing-span; and that of a Madagascan species exceeds 30 cm. (12 in.). Most feed by dusk or night on pale flowers, which they pollinate, but the hummingbird hawk, *Macroglossum stellatarum*, and bee hawks, *Hemaris* species, are diurnal. The caterpillars of many hawk moths have a horn at the rear, and in America are called hornworms; some, such as the *tobacco hornworm, are pests. The adult death's head hawk moth, *Manduca atropos*, will enter beehives to feed on honey.

Hawks are birds of the family Accipitridae, which includes the true hawks, buzzards, eagles, harrier-eagles, harriers, kites, fish eagles, bat-hawks, ospreys, and vultures. They are a large, highly successful family of diurnal birds of prey, ranging in size from tiny insect-feeding sparrow-hawks to eagles with a 2 m. (6 ft. 6 in.) wing-span. Common characteristics include dull, barred, brown or grey plumage, a powerful, strongly curved beak with nostrils mounted on a cere, long legs, and strongly taloned feet. Beak and feet are used jointly to tear open prey. Females are often larger than males. Flight patterns range from soaring and gliding (eagles, vultures) to high-speed precision swooping. Most take live prey exclusively, but some of the larger forms eat carrion as well, and vultures are almost exclusively carrion-feeding scavengers.

Eagles and other large hawks lay one or two eggs, the smaller species three to five. Their nests are often on crags and other high vantage-points, but some species breed on the ground. Many hawks migrate away from extreme winter conditions, some following migrating prey species. As 'top' predators in the *food chains, hawks are especially vulnerable to insecticides and other agricultural poisons. They are also destroyed as vermin, and the eggs and young of some species are in demand for falconry.

Hawthorns are species of the plant genus *Crataegus*, of which there are about 200 species in the north temperate zone. They are deciduous small trees and shrubs closely related to *Cotoneaster*. They hybridize readily with one another, especially in North America, which makes classification difficult; as many as 1,000 different 'species' have been named. Many are armed with spines which are really

The male catkins of the **hazel** appear before the leaves, and release ripe pollen. Note also the small female flower with its red styles. On windy days pollen will be blown to fertilize flowers on nearby hazels. The hard, woody-shelled hazel nuts, known as filberts or cob-nuts, are borne in clusters of up to four.

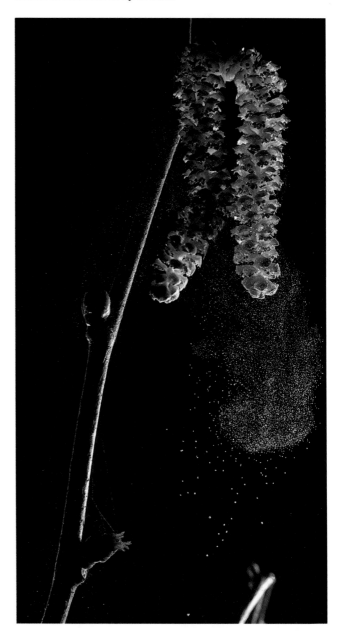

dwarf branches. Some American species are known as cockspur thorns. Hawthorn is widely used for hedging and the white or pink flowers are followed by red fruits (haws). The wood is used in the same way as *box. Hawthorns belong to the rose family and are thus related to apple.

Hay fever is an *allergic response of eye and nose membranes to pollen, causing profuse watery discharges and *common cold symptoms. It often occurs in several members of a family, and in association with asthma and eczema. Desensitization by injections of pollen extract will often prevent the attacks, but it needs to be started well in advance of the pollen season. Antihistamine drugs are moderately effective in treatment.

Hazels are deciduous shrubs, or more rarely small trees, of the genus *Corylus*. There are about fifteen species native to temperate regions of North America, Europe, and China. They belong to the same family as the *birches, alders, and hornbeams. The common hazel, *Corylus avellana*, is a deciduous hedgerow shrub of Europe valued for its nuts. In spring, the male flowers are conspicuous long catkins, in contrast to the unremarkable females. They are wind-pollinated. Formerly, hazel was an important constituent of woodlands where coppicing was practised. The slender, flexible shoots produced by coppicing were used for the hoops of barrels, thatching, fencing, and many other constructional uses.

Head, or skull: the front end of the body in all vertebrates, save humans and some apes, who have adopted an upright posture. It contains the brain, and the organs of special sense: the eyes, ears, nose, and tongue. The whole body is directed and co-ordinated by the head. Attached to the front of the head are the face, jaws, and mouth. In the course of evolution the head has assumed greater importance by reason of the enlargement of the brain. The reason for development of the head is that it is the part of the body which first meets new stimuli in the environment, on the assumption that most animals normally move in one direction.

Hearing is the ability to detect and interpret sound. Many animals have evolved this ability. It is usually based upon hair-like structures being directly or indirectly stimulated by mechanical vibration. These then send electrical messages along the auditory nerves to the brain. In vertebrates the sound receptor is frequently the *ear, but fish are sensitive to vibration along the length of their *lateral line.

Hearing has a number of functions. The most important are those of communication and location. Male mosquitoes locate females by sound. Grasshoppers and related species can transmit sounds which call mates, display aggression, and warn of danger. Many vertebrates, notably birds and frogs, use sound in a similar way. In man hearing is used mainly for complex auditory communication. Cetaceans, including dolphins, also communicate by sounds. In addition, the dolphin uses echo-location for navigation. Insectivorous bats use ultrasound echo-location to hunt insects. Some moths have evolved an ability to detect the high frequencies emitted by bats and thus avoid capture.

Heart is a general term for any muscular organ which pumps blood around the body. In many invertebrates it

---

Hearts

Diagrammatic comparison of the blood-flow system and muscular chambers in the heart of a bony fish, reptile, and mammal.

Bony fish

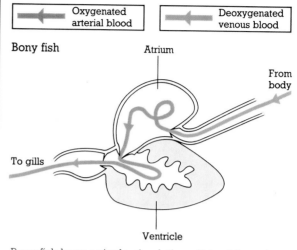

Bony fish have a single circulation of blood from the body to the gills.

Reptile

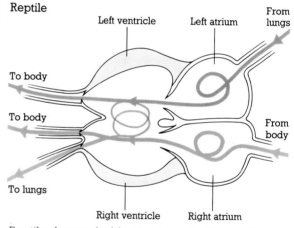

Reptiles have a double circulation, but incomplete separation of flow – only part of the blood being passed to the lungs.

Mammal

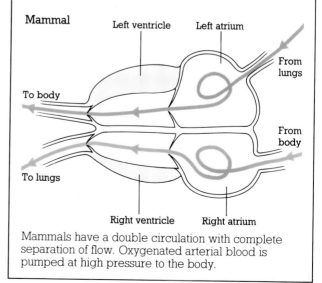

Mammals have a double circulation with complete separation of flow. Oxygenated arterial blood is pumped at high pressure to the body.

is a simple contracting tube which receives blood from one end and then pumps it to the other, creating an open flow along the body.

The vertebrate heart consists of muscular chambers, three in fishes and amphibians, four in others. Blood from the veins, at low pressure after passing through body tissues, is sucked into one of the upper chambers, called an atrium, usually the right in mammals, and then passes into the lower chamber, called the ventricle. In fish and amphibians there is a single ventricle, alternately shared by oxygenated and unoxygenated blood. Other vertebrates have at least a partial separating membrane, dividing the lower chamber into two ventricles. Complete separation in mammals leads to a much more efficient circulatory system, which avoids mixing oxygen-rich and unoxygenated blood. The ventricle contracts to force blood through the gills or lungs, where it is oxygenated before returning to the other atrium. This oxygenated blood is passed into the ventricle, which pumps it out of the heart through the body's largest artery, the aorta. From this, it flows around the body along arteries, then capillaries, before returning to the heart via veins.

The movement of blood through the heart is controlled by valves, allowing it to flow in one direction only. The heart muscle is of a special type, called cardiac muscle, and beats rhythmically throughout life, under the control of the *autonomic nervous system. In man the heart is cone-shaped, about the size of a clenched fist, and set in a cavity between the lungs. It lies within a double set of membranes which give it the freedom to beat without being squashed.

Heart attack   *Coronary thrombosis.

Heartburn is a burning pain in the centre of the chest, usually of short duration. It is associated with a movement of acid stomach contents into the gullet, and may be accompanied by their regurgitation into the mouth.

Heart urchins   *Sea-urchins.

Heath butterflies are yellowish-orange *brown butterflies which frequent heathland and grassland. There are some twenty species in Europe and temperate Asia, including a few in North America. The large heath, *Coenonympha tullia*, has conspicuous white-centred black spots on the under-sides of the wings, and is called the ringlet in North America.

Heather was originally the Scottish name for ling, *Calluna vulgaris*, though it is widely used nowadays for species of the closely-related *Erica*. The heather, or heath, family contains some 3,000 species which occur worldwide. The relatives of *Calluna* and *Erica* include rhododendron and bilberry. Virtually all species depend upon fungal *mycorrhiza for successful growth. Heathers are evergreen shrubs of acidic soil, especially moors and heaths.

Hebe   Speedwells.

Hedgehogs belong to the insectivore family Erinacidae, which they share with moonrats, a subfamily which lacks the spines of true hedgehogs. The western European hedgehog, *Erinaceus europaeus*, is replaced by a very similar species, *E. concolor*, in eastern Europe and Asia. The western European hedgehog has been introduced into New

Zealand, where it has become very common. Of the twelve species, four occur in Africa, and most resemble the European hedgehog, though three species of desert hedgehogs have distinctly larger ears and longer legs. Hedgehogs do not occur in the New World. They feed on a variety of invertebrates, as well as amphibians and the eggs or young of ground-nesting birds. All hedgehogs curl up when threatened so that the spines, which are present only on the back, protect the soft underparts.

Hedge sparrow, or dunnock: a member of the *accentor family of birds not closely related to the true sparrows. It breeds over much of Europe and western Asia. The hedge sparrow gets its name from its sparrow-like plumage, though it is a dull blue-grey below. It has a fine, sharply pointed beak with which it eats insects and small seeds. It lays three to five bright blue eggs in a nest concealed in thick vegetation.

Hellebores are plants of the genus *Helleborus*, of which there are about twenty species native to Europe and Asia Minor. They are herbaceous perennials, belonging to the buttercup family, with deeply divided, evergreen leaves and large white, green, or purplish flowers. The botanical name, from the Greek *elein*, injure, and *bora*, food, indicates the very poisonous nature of some species.

Hemlock is a tall perennial plant with fern-like leaves, related to the carrot or parsley and widely distributed in Europe, Asia, and North Africa. It is one of the most poisonous plants known to man and was used in Ancient Greece for the execution of criminals. *Conium* is both the botanical name and the name of the poison extracted from the plant. It was this poison that was taken by Socrates.

Hemlock spruces, are coniferous trees with some ten species in the genus *Tsuga*, spread over North America and the Far East. The name derives from the fact that the crushed foliage of some species smells like hemlock. The common, or Canadian, hemlock, *T. canadensis*, of North America is valued for its timber, its bark, and its pitch.

Hemp is a term properly used to describe the fibre extracted from the stems of the *cannabis plant, an annual relative of the hop, fig, and mulberry. Cultivated in Asia for thousands of years, it is now grown for its fibre wherever the climate is humid but temperate; the USSR, China, Japan, Italy, Yugoslavia and the USA are the major producers.

The cultivation of hemp is similar to that of the other temperate fibre crop, *flax, although the fibre itself is tougher and is used for coarse fabrics and ropes.

Several other plants are knows as 'hemp', but should not be confused with true hemp. Manila hemp is obtained from the banana-like *Musa textilis*; Maoris weave fine fabrics from New Zealand hemp, *Phormium tenax*, and sunn hemp is obtained from the leaves of the legume *Crotalaria juncea*, in India. Strong elastic white fibres from leaves of the bowstring hemp, a *Sansevieria* species, were used in African bows.

Henna, or Egyptian privet, is a shrub that has been cultivated as a dye plant for centuries in Egypt, Arabia, and India. The orange dye that can be extracted from its leaves was used by the women of ancient Egypt, and is still used by modern Muslims to colour their hands a reddish-brown. The dye is also used as a colouring in hair rinses and also for dyeing leather.

Hepatitis is an episode of acute damage to the cells of the liver, characterized by jaundice and caused by viruses, drugs, chemicals, or fungal toxins. One type of virus causes a form of hepatitis transmitted by contamination of food or water by sewage. This is known as infectious hepatitis. The second type of virus is not excreted, but is present in the blood. It is much less infectious, being spread either by sexual contact, or by blood transfusion, or from mother to unborn child. Severe infections of hepatitis can culminate in liver failure and consequent death.

Herbivores are animals that depend entirely upon plants for their food. Such a diet has a profound influence upon the teeth and digestive system. The cheek teeth are usually much developed, the premolars and molars having large and complex surfaces that function to divide the plant material into fine shreds as an aid to digestion. The front teeth have become variously modified and a gap (diastema) has developed between these and the cheek teeth. The diastema allows the passage of food backwards and forwards into the cheek pouches. The incisors of rodents, which are mostly herbivores, grow continuously and, having enamel on one side only, provide a sharp cutting edge.

The cell walls of plants are made of cellulose, a substance that is not readily broken down. Mammals lack the necessary enzymes for its digestion and so their alimentary canal is adapted for this task. In the horse, micro-organisms aid

Male **Hercules beetles**, *Dynastes hercules*, of tropical Colombia, have large 'horns'. These are used to wrestle with other males over a receptive female.

fermentation in the enlarged colon and large intestine. Food takes thirty to forty-five hours to pass through the gut of a horse. Ruminants regurgitate the cud and chew it again. They have a complex stomach consisting of four compartments; the passage of food takes seventy to one hundred hours in a cow. Rodents and rabbits pass food twice through the digestive tract in order to obtain essential vitamins not released on the initial passage of the plant material.

Herbs, in common parlance, are plants of use to man in medicine or cooking. In a botanical sense, however, they refer to non-woody seed-bearing plants, the aerial parts of which are ephemeral, hence the term herbaceous. Medicinal herbs are the source of a number of drugs still used in modern medicine. Culinary herbs are of value in flavouring foods, formerly often foods which would otherwise have been unpalatable. Certain herbs have so-called essential oils in their stems and leaves. For example, the leaves of mint, sage, thyme, and parsley, the stems of angelica and fennel, and the fruits of anise and caraway are rich in such oils. Often the last two are referred to as *spices.

Hercules beetles are among the largest of all *beetles, and may reach 16 cm. (6½ in.) in length. They typically have horns on the head and the front of the thorax, usually only in the males, but sometimes in both sexes as in the rhinoceros beetle, *Oryctes rhinoceros*. Most are tropical, and larval habits are varied; but some, such as the Goliath beetle, *Goliathus goliathus*, are pests of coconuts, sugar-cane, and rice in Africa. The same family of beetles (Scarabaeidae) also includes the South American elephant beetle, *Megasomas elephas*, and the Asian atlas beetle, *Chalasoma atlas*, both large bulky beetles.

Heredity is the transfer of properties of organisms from one generation to the next. There are two main kinds of heredity: *genetic and exo-genetic. Exo-genetic heredity is not caused by genes and is important only in humans, involving the transfer of property and cultural information by learning.

Only factors inherited from the parental generation are handed on by their progeny. Any acquired character, such as learned behaviour, cannot be genetically inherited. Genetic heredity is particulate and is carried by molecules called *DNA, on structures called *chromosomes.

Hermaphrodites are animals which possess both male and female reproductive organs. This is a common feature of many invertebrate phyla, such as annelids, molluscs, and related groups. Such animals rarely fertilize themselves, but still require two individuals to exchange sperm before their ova can be fertilized. This type of reproduction is an advantage in animals which may live at low densities in their habitat. They need only to meet another individual of their own species in order to produce viable offspring.

Some molluscs and a few species of fishes change sex as they mature. This is common in oysters, which begin as males and change to females, a phenomenon called protandric hermaphroditism.

The **hermit crab**, *Eupagurus bernhardus*, in symbiotic association with a sea anemone. This crab also often establishes a relationship with hydroid organisms from which it neither gains nor loses (*commensalism).

Hermit crabs differ from most other *crabs in having a soft, unprotected abdomen. They must seek security in abandoned mollusc shells, carefully choosing one to fit. The abdomen of these crabs is asymmetrically twisted, so it can be inserted snugly into snail shells, leaving only head and pincers protruding. Hermits may fight over a new shell when they have outgrown their present lodgings. Many carry 'passengers' on the shell, especially sea anemones, which can assist their defence, and they will carefully transfer their anemone onto the new shell. Some tropical hermits are land-dwelling, though they always require water nearby to dampen the body and for drinking; they have reduced gills, and breathe through modified lungs. Hermit crabs belong to two superfamilies of crustacean. Some species use bamboo or hollow mangrove roots as homes.

Hernia is the protrusion of some of the contents of a body cavity beyond its usual boundaries. A loop of gut may bulge through a weakness in the groin, forming an inguinal hernia. A hiatus hernia is the partial protrusion of the stomach through the diaphragm into the chest. Hernias can become painful, sometimes even gangrenous, or strangulated, and require surgical treatment.

Herons are stork-like birds, which, together with bitterns, belong to the family Ardeidae. They are typically slender, with long necks and legs. Some species of egret herons (*Egretta*) develop long plumes on their head, neck, and back. In flight the neck of most herons is retracted, the broad wings flap slowly, and the feet project beyond the tail. When hunting, a solitary heron will stand motionless

in the shallows, head on its shoulders, and snap up a fish, eel, or frog with a sudden lunge of its bill. Most of the forty-nine widely distributed species breed in colonies called heronries, building stick nests in trees.

Herpes is an infection with *Herpes simplex*, a virus causing repeated crops of small blisters around the mouth (cold sores) or in the genital area (genital herpes). The virus is present in the blisters, and between attacks is dormant in the nerves supplying the affected area. The virus occasionally causes serious infection of the brain.

The viruses which commonly affect the face and genital area differ slightly, being known as type I and type II. Both belong to the same group of viruses as *H. zoster* viruses, which cause *chicken pox and *shingles.

Herring: an abundant surface-living fish in the waters of the continental shelf on both sides of the North Atlantic. A subspecies of the same species occurs in the North Pacific. Both species are moderately slender, their bodies fully scaled and with a single dorsal fin. Their bodies are brilliantly blue on the back, silvery-white on the belly, with brassy tinges on the sides. Herring feed on planktonic animals, the major food item being copepod crustaceans of the genus *Calanus*, but any other planktonic creatures will also be eaten. They are themselves eaten by larger fishes such as cod, salmon, and tuna, by mammals such as dolphins and seals, and by seabirds, especially gannets. Young herring in the *whitebait stage are preyed upon by terns and puffins. Spawning occurs in spring and summer and each local population has a specific spawning ground, sometimes on shallow, stony banks in 15–40 m. (49–130 ft.) of water, or offshore as deep as 200 m. (650 ft.). The eggs are laid on the sea-bed, sometimes in thick layers.

The herring is an extremely important food fish, being caught in surface trawls, although for many centuries it was caught in drift nets. It is eaten fresh, and cured in various ways to be sold as kippers, bloaters, red herring, and roll-mops. In many areas it has been over-fished.

Herring gulls are the commonest northern *gull species, with grey mantle, black primary feathers, and pink legs. They breed throughout northern Europe, the northern USSR, North America, and the Arctic. Five or six subspecies are distinguished. Herring gulls integrate with the lesser black-backed gulls, *Larus fuscus*, in parts of their range.

Heterotrophic organisms are those which obtain organic food molecules, such as carbohydrates and proteins, ready-made from other organisms. All animals are heterotrophs, as are many fungi and bacteria. Many heterotrophs are herbivores, feeding directly on plants (*autotrophic organisms) which first manufacture the food by photosynthesis. In turn, carnivorous heterotrophs may feed on the herbivores. *Parasites are heterotrophs which derive their food from other organisms while they are still alive. *Saprophytes are heterotrophs that feed on the dead remains and waste products of plants and animals.

Hibernation permits animals to survive through winter when food supplies would otherwise be inadequate. Most hibernating animals allow their body temperature to fall. This saves energy by reducing the metabolic rate of the

The natural food of the **herring gull** is fish, small marine animals, shellfish, and the eggs of other birds. Rather than having to find food, however, they prefer to wait for scraps from trawlers or tourists.

organism, so that the body fat reserves last until food becomes available the following spring. There are two main strategies for hibernation: uncontrolled body temperature fall, or an overall lowering of body temperature to a pre-set level. Some species of bats use the first of these and allow their temperature to fall to that of their surroundings. Hence they are often found hibernating in caves or buildings where the risk of freezing is reduced. The main problem is that if the body temperature falls too much the animal might freeze to death. However, some species, such as the ground squirrel, will wake up if their body temperature falls dangerously low.

The other strategy is to control the hibernation body temperature. The dormouse maintains its temperature a few degrees above that of its surroundings. Bears often sleep through the winter, allowing their body temperature to fall to as little as 3–4 °C (37–39 °F). Most hibernators can restore their body temperature in minutes rather than hours and they may emerge during mild days in winter. A few birds have a hibernation-like state which lasts for short periods. Humming-birds, because of their high energy requirements and small size, allow their body temperature to fall overnight. There is no real distinction between hibernation and torpor, since both are physiological states where the animal's metabolic rate is lowered to conserve energy. Nevertheless some biologists classify the 'hibernation' of bats and bears as torpor.

Hibiscus is a genus of some 300 species of herbs, shrubs, and small trees of the tropics and subtropics. They belong to the same family as mallows and cotton. Widely cultivated are the shrubby *Hibiscus rosa-sinensis*, and the deciduous shrub *H. syriacus*, grown for hedging and naturalized in Europe. *H. esculentus* is *okra.

Hiccoughs, or hiccups, are violent, brief involuntary contractions of the muscles of the diaphragm in man, and can be very unpleasant for the sufferer if they persist for more than a few minutes. They can be caused by indigestion, alcohol, or other unknown factors.

Hickory is the common name, particularly in North America, for trees and timber of species of *Carya*, which are closely related to the walnuts. There are about twenty species in eastern North America and two species in China. Their edible fruits, like the related walnuts, are not true nuts. Those of *C. illinoensis* are the pecans of commerce. Their timber, which is very tough and elastic, is also valuable for tool handles and sports goods.

Hippocrates (*c.* 460–370 BC), the most celebrated physician of Greek antiquity, known as 'the Father of Medicine', was the first to separate medicine from philosophy and superstition. He taught that disease has natural causes, and his treatment was founded on the healing power of nature. The Hippocratic Oath, taken in some countries by newly qualified doctors as a pledge of ethical behaviour towards patients, cannot be ascribed to Hippocrates, although it does reflect his medical outlook.

Hippopotamuses, or hippos, are some of the largest living terrestrial mammals, weighing up to 3,200 kg. (7,050 lb.). They have an enormous head with a bulbous snout, and a large, barrel-like body, up to 3·5 m. (11 ft. 4 in.) long, set on short, stout legs. In spite of their size and weight, they are surprisingly agile and well able to overtake a running man. Most of their life is spent in water, in which they can float, or sink, and walk along the bottom. The sense organs, the nose, eyes, and ears, are all set on the upper surface of the head, and the nostrils can be closed.

Only two species are found today. The common hippopotamus, *Hippopotamus amphibius*, is 1·4 m. (4 ft. 6 in.) tall at the shoulders. Once it frequented most of the large lakes and rivers and even brackish waters of Africa, but it is now much less widespread. It feeds mainly on reeds and grasses in and around the water. At night a herd of twenty to thirty individuals will travel some 40 km. (25 miles) along a river gathering food. The hippo has the largest mouth of any mammal, with an awesome gape which exposes the long incisor and canine teeth, used in *display. The lower canines may reach a length of 60 cm. (24 in). A single calf is produced. It suckles while in water, can swim before it can walk, and will ride on its mother's back when she is in the water. The pygmy hippo, *Choeropsis liberiensis*, is about 90 cm. (3 ft.) tall at the shoulders. It is found in rivers in the forests of West Africa or in swamps. Its habits are similar to those of the common hippo.

Hispid bats, or slit-faced bats, belong to the family Nycteridae, with eleven species confined to Africa and the Middle East. The term 'slit-faced' refers to the longitudinal groove on the muzzle. There is a saucer-like depression between the eyes, which gives the group yet another name: hollow-faced bats. The name hispid is used in reference to the shaggy coat of these bats. They are insect-eaters and have a distinctive T-shaped tip to their tail.

Histology is the microscopic study of cells and tissues, either living or after they have been 'fixed' to preserve their structure. Stains or dyes are used to pick out specific structures within them. Thin, living tissues can be examined either under the phase-contrast microscope, or, after treatment with dyes, under the light microscope. Organs are examined after being cut into thin slices. Small samples can be harmlessly removed from living subjects to aid the diagnosis of disease. The electron microscope permits higher magnification, but requires pieces of tissue to be cut into ultra-thin slices and suitably stained. Tissue surfaces can also be examined at high magnification by using a scanning electron microscope, after coating the tissue with metals such as gold.

Hoatzin: an extraordinary bird, placed in a family of its own. It was once thought to be a distant relative of the gamebirds, but is now considered to be related to the cuckoos. About 60 cm. (2 ft.) in length, including a long tail, it is a dark brown bird, streaked with pale markings and with a loose, untidy crest. It lives in bushy vegetation along river banks in northern South America, eating leaves and fruit. It lays two to three eggs in a simple nest of twigs built in bushes. If disturbed, the young can jump from the nest into the water and, when danger has passed, clamber back into the nest using hooked claws on their still rudimentary wings.

Hobbies (birds) are a cosmopolitan group of three species of small, long-winged *falcon that hunt birds and insects on the wing. The European hobby, *Falco subbuteo*, grey-barred with white underparts, takes a higher proportion of birds than most. Widely distributed across Europe and Asia, it breeds in temperate latitudes and winters as far south as North Africa, Burma, and India. The others are the African hobby, *F. cuvierii*, and the Oriental hobby, *F. severus*.

Hodgkin, Thomas (1798–1866), Quaker physician and pathologist, was the first to describe Hodgkin's disease: a malignant disease of the lymphatic glands which may cause weight loss, fever, and itching, but is usually curable.

Hogs   *Pigs.

Holly is the common name for *Ilex aquifolium*, an evergreen tree of south and western Europe, valued for its attractive red berries borne through winter, and its white wood. Its most familiar use is in Christmas decorations, but it is also a plant surrounded with ancient superstition. The trees are dioecious, with male and female flowers on different trees. Hedgehog holly has prickles on the surface as well as the margins of the leaf. There are some 400 other species of *Ilex* throughout the world, not all evergreen. Other unrelated, prickly plants are sometimes called holly, such as *Olearia* species in New Zealand, and *Eryngium maritimum* (sea holly) in Europe.

Hollyhocks are tall and showy perennials related to mallows and cotton. The familiar garden varieties are derived from the Chinese species *Althaea rosea*. The flowers, which have medicinal uses, have also been used in dyeing.

Holm oak   *Oaks.

Homoeopathy is a system of treatment, devised by Dr Samuel Hahnemann (1755–1843), based on the doctrine that drugs given in very small doses have actions opposite to the effects of larger doses. It has the great virtue of never

The garden **hollyhock** has relatives with the same common name. The mountain hollyhock, *Iliamna rivularis*, shown here, is growing wild in the state of Idaho, USA.

harming the patient, since the remedies are prescribed in such minute quantities that they are incapable even of provoking an allergic response.

Homosexuality is the condition of emotional and sexual attraction between members of the same sex. Its expression is greatly influenced by social attitudes, which vary from acceptance to rejection. A significant number of people prefer homosexual relationships from the onset of sexual awareness in adolescence.

Honey-badgers   *Ratels.

Honey-bees are social bees of the genus *Apis*, which normally occur only in domestication. They may set up colonies in hollow trees but they do not usually survive long in the wild, having been bred for centuries in hives. Hives are kept throughout the world and there are a number of races which differ in productivity and temperament. A colony passes the winter with a single queen and a large number of workers, which keep close together to conserve heat. In the spring, the workers start to forage, bringing in pollen, called bee bread, and nectar, which mature into honey. They collect water to cool the hive and help to build and clean it. Wax is secreted from glands in between their abdominal segments, and this is used to build the vertical combs.

The queen bee controls the production of workers, drones (males) or new queens by laying two types of egg. Fertilized eggs develop into workers, which are technically non-reproductive females. Queens are produced by feeding some of these grubs with high protein foods. In late summer the queen lays unfertilized eggs which develop as drones. The new queens and drones leave the nest to mate. After a mating flight, the drones die, while the queens

return to their hive, collect workers, and swarm to set up new colonies. Honey-bees have complex methods of communication, particularly between the workers with their famous bee dances, which tell where nectar may be found. The queen produces substances which control the working of the hive, almost in the manner of some form of chemical slavery.

Honey buzzard is a small *kite widespread in Europe and Asia throughout the summer. Its plumage is dark reddish-brown, and its wing-span over 1·2 m. (4 ft.). Honey buzzards feed mainly on insects, notably wild bees and wasps, hunting both in the air and in trees, often destroying bee and wasp nests for the grubs inside. European populations move south after breeding to winter in Africa.

Honey-eaters make up the bird family Meliphagidae, with some 171 species inhabiting the forests of the Australasian region and some Pacific islands. Ranging from the size of a sparrow to that of a small crow, they are mostly brown and green in colour, but many also have conspicuous patches of bright yellow, white, or black. The beak is usually thin and sharply pointed, but is strongly curved in some species. They feed on a wide range of insects, fruit, and nectar (hence their name), and build open-cup nests in which they lay one to four eggs. In some species, several birds (in addition to the parents) may defend and raise the young. Some ornithologists consider the *sugarbirds to belong to this family.

Honey-fungus is an *agaric fungus, parasitic on a wide range of trees, shrubs and herbaceous plants, which it kills

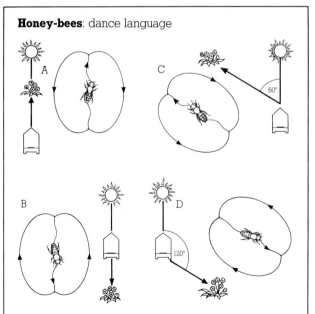

Honey-bees: dance language

The sophisticated 'waggle dance' performed by scout bees on the surface of the comb informs fellow workers of the precise whereabouts of a food source. Assuming that the top of the page represents the top of the hive, then the 'waggle' part of the figure-eight routine is danced in an upward direction (A), if the food source is to be found by flying towards the sun. Diagram B, C, and D show respectively the orientation of the dance if the food source lies directly away from the sun, at an angle of 60°, or 120° from it.

The fruiting bodies of the **honey-fungus**, which appear in summer and autumn, on a tree stump in the New Forest, England. In some species the young fruiting bodies are edible and are occasionally eaten by man.

and then lives on. A mass of fungal threads invades the wood and produces brown toadstools clustered at the base of the tree. It can spread to other parts of the tree, or neighbouring trees, by means of bootlace-like structures called rhizomorphs, hence its other name, bootlace fungus. It is faintly luminescent. Several other species of fungi of the genus *Armillaria* resemble tree honey-fungus, *A. mellea*.

Honey-guides are birds which are native to Africa, except for two species in India and Southeast Asia. The fourteen species make up a distinct family, and are sparrow-, or thrush-sized, with brownish or olive-green plumage. All of them appear to be parasitic, and to lay their eggs in the nests of other birds who then raise the young. They are insectivorous, in particular seeking out the nests of bees. Some species of honey-guide have bacteria in their digestive tract which enables them to digest the wax of bees' nests. They were given their name because the greater or black-throated honey-guide, *Indicator indicator*, has the habit of attracting the attention of people (and some animals, including the honey-badger or ratel) by calling and leading them to bees' nests. Once the person, or animal, has

broken open the nest to get at the honey the honey-guide can eat its share.

Honeysuckle, originally the name for red clover, is now the name of a number of trees, shrubs, and climbers which produce large amounts of nectar. In Europe and North America, these are generally species of *Lonicera*, which are members of the same family as snowberry, elder, and guelder rose. They are usually climbers (*lianas) with strongly scented flowers, visited by moths in the evening and known in Shakespeare's time as woodbine. In other countries, unrelated plants are called honeysuckle: in South Africa, for example, the name is used for *Tecomaria capensis* and species of *Halleria*, which have red, bird-pollinated flowers.

Hooded crow   *Carrion crow.

Hoofs   *Claws.

Hooker, Sir William Jackson and Sir Joseph Dalton, father and son, were both British botanists who became directors of the Botanic Gardens at Kew near London. William (1785-1865) revived the gardens and founded a museum there. Joseph (1817-1911) joined a voyage of exploration to the Antarctic and returned with an immense collection of plants from the Southern Hemisphere. Their distribution, he remarked, indicated an

A photograph of Sir Joseph Dalton **Hooker** taken in 1876. He contributed a considerable amount of botanical data to the book *On the Origin of Species*, written by his friend and ally Charles Darwin.

ancient linkage between the landmasses of Australia and South America, an idea later supported in the theory of continental drift. Later he spent three years in the northeast of India and sent home large numbers of rhododendrons, starting the fashion for them.

Hookworms are common parasites of man in tropical and subtropical regions. The parasites, two species of *roundworm about 1 cm. (½ in.) long, live in the upper gut, often in large numbers. One species, *Necator americanus*, occurs in the tropical regions of America, while the other, *Ancylostoma duodenale*, is an important parasite of the Old World regions. Each worm consumes a few drops of blood daily, and a heavy infestation will cause a severe anaemia. The eggs, passed in the faeces, develop in the soil into larvae which can penetrate the skin. They then travel by the bloodstream to the gut to complete their development. Modern treatment can eliminate the worms with little upset to the patient.

Hoopoe: a bird classified in a family by itself but related to bee-eaters and rollers. It occurs over wide areas of Africa, Europe, and Southeast Asia. The hoopoe is the size of a small crow, and has a salmon-pink body and large, erectile crest. Its wings and tail are strikingly patterned in black and white and its beak is long and curved. It lives mostly on the ground, catching large insects and lizards, and nests in a hole in a tree or cliff face, laying four to six white eggs. It has a characteristic undulating flight.

Hops are climbing plants belonging to the same family as figs, mulberries, and hemp. This is a large family, of some 3,000 species in tropical to temperate regions. The hop, *Humulus lupulus*, is cultivated as a flavouring for beer.

A number of bitter herbs have been used to flavour beer but hops have been used more than any other since the Middle Ages. The greenish, cone-like female flower-buds of this perennial are the only parts used in brewing. Hops are grown in localities scattered widely throughout the temperate areas of Europe, America, and Australasia.

Hormones are chemical messengers, usually proteins or steroids, formed in animals by *endocrine glands. They are discharged to the blood and circulated in low concentrations to all tissues. Individual hormones recognize their respective 'target' organs or cell systems by binding to specific chemical groups (receptors) on cell surface membranes. Once bound, they induce complicated, often quite slow, changes which lead to processes such as growth of the body, to sexual development, or to digestion. 'Target organs' are often some distance from their controlling glands. For example, gonadotrophins (pituitary hormones), released from the head, cause growth of the gonads in the pelvis.

Hornbeams *Birches.

Hornbills make up the bird family Bucerotidae with forty species native to Africa, Asia, and Australasia. They are medium to large birds 37·5–150 cm. (15–60 in.) long,

---

Hornbills

Cutaway view into the nest of a tree hornbill

The male hornbill on the outside is shown passing food to the female, who may remain holed in until the young are fully fledged. To keep the nest clean, faeces are ejected through the entrance and removed by the male from the nest area. To adapt to their cramped quarters young hornbills sit with their tails raised vertically.

and mostly brown, black, and white. Their most striking feature is the large, red or yellow, slightly curved beak, usually with a bony structure (casque) on top of the upper mandible. Many live on a diet of fruit, but some, such as the largest species, the great hornbill, *Buceros bicornis*, eat a wide range of animals, which they kill with their powerful beaks. Most species nest in large holes in trees: in many cases the female enters the hole before laying and then walls herself in with faeces and mud brought by the male; only a small crack is left through which the male can pass food. The female stays in the nest during egg-laying and incubation, until halfway through the nestling period, when she breaks out and helps the male collect food for the large young.

Horned toads get their name from the triangular, horn-like projections of skin which extend from their upper eyelids. The name is commonly applied to the South American horned toads, *Ceratophrys*, and the Asiatic horned toads, *Megophrys*, which are not closely related, but belong to different families of toads.

The name horned toad has also, confusingly, been used for a group of squat-bodied American iguanid lizards (also known as horned lizards). These spiny, scaled lizards are able to survive arid, even desert conditions by drinking the night dew, which collects around the scales surrounding their mouths.

Hornet is a name commonly used for any large species of social *wasp. They occur throughout north temperate regions and the Old World tropics and the queens in some species may have a wing-span of as much as 8 cm. (3 in.). The term is particularly applied to the European species,

The **horned toad**, or Texas horned lizard, *Phrynosoma cornutum*, can change its colour to match its surroundings. Fierce bony spines give extra protection.

Vespa crabro, which has a brown and yellow striped body, as opposed to the black and yellow of smaller wasps. They almost always nest in hollow trees and their sting is more unpleasant than that of wasps; that of some tropical species can be fatal.

Horns and horn-like structures (*antlers) are found especially among ungulate (hoofed) mammals. The term 'horn' is used to cover the entire structure, including the keratinized skin known as horn. A true horn is seen in members of the cattle family, and consists of a core of bone which is fused with the skull, encased within a hollow cone of the material called horn. The horny sheath grows throughout the individual's life and gradually thickens. Neither the core of bone nor the sheath of horn are shed except in the *prong-horn. Perhaps the most primitive type of horn, from which both horns and antlers may have developed, is that of the giraffe. Its horns are unbranched and covered with skin and hair; the bony core is separable in the young giraffe, becoming fused to the skull later. The horns of the rhinoceros consist of a mass of coarse keratinized strands which are a modified form of hair. They are not true horns as they lack a bony core.

Horntails *Sawflies.

Horse flies, or gadflies, are robust blood-sucking true *flies with colourful, iridescent eyes. They belong to the family Tabanidae (2,000 species), which also includes species of horse fly known as deer-flies and clegs. Only females feed on blood, first making a puncture with the strong, blade-shaped, sucking mouthparts, then continuing to lap up blood, which may flow for some time. The males are nectar-feeders. The usual victims are cattle and horses, but man is often bitten, leading to painful swellings at the site of the incision. Eggs are laid as compacted masses on plants growing in damp places, and, upon hatching, the

larvae drop to the ground and feed on worms and other small invertebrates. In some countries, horse flies are vectors of diseases, especially of a parasitic roundworm causing the disease lòa-lòa or loiasis.

Horsehair worms are a small phylum of worm-like animals with some 230 species. They may be up to 1 m. (3 ft.) long but are rarely even 1 mm. ($\frac{1}{25}$ in.) thick, so they are aptly named. Larval stages parasitize arthropods, especially aquatic insects, absorbing food across their body wall. The adults live freely in fresh water, or in soils, or in the sea, and do not feed.

Horse mackerel: a species of fish of the family Carangidae, which contains a number of species known as jacks in North America. Widely distributed in European seas, the horse mackerel, *Trachurus trachurus*, is distinguished by the series of broad scales along the sides, which towards the tail become heavy and sharply pointed. A schooling fish, it lives near the surface down to a depth of 100 m. (325 ft.) both inshore and in the open sea. Young horse mackerel shelter beneath the trailing tentacles of jellyfishes. Adults feed on small fishes, crustaceans, and squids. It is also known as scad.

Horseradish is a perennial plant, a member of the mustard family. It forms a stout, yellowish tap-root which can be used to make horseradish sauce. It probably originated in southeastern Europe but has been cultivated throughout that continent and western Asia since ancient times.

Horses belong to the same family as asses and zebras. All seven members of this family are closely related and can interbreed to produce sterile hybrids, such as *mules. All have only a single functional digit, the middle one, which terminates at its very tip in a hoof, on which the horse walks. The teeth are typical of *herbivores. The horse's ancestors include *eohippus and *mesohippus.

The wild horse is now represented by only one race, Przewalski's or the Mongolian wild horse, *Equus przewalskii*.

The **horseshoe crab**, *Limulus polyphemus*, most common along the northwestern Atlantic coast and down as far as the Gulf of Mexico, can often be found ashore in the intertidal zone along sandy bays and estuaries during the breeding season in spring.

It is found in the steppes and semi-deserts of southern Siberia, Mongolia, and western China and is a sturdy animal, with a rather large head, small ears, and heavy jaws. The mane is short and erect, black or brown in colour, edged with lighter hairs, and the tail is long-haired, black or brown in colour. In summer the body colour is reddish-brown with some white on the muzzle; in winter the hair is longer and paler. Przewalski's horses are now being bred in zoos and on a new semi-wild reserve with the result that numbers continue to increase.

Domesticated horses, *E. caballus*, can be divided into light and heavy. The light horses are for riding and drawing traps. Among the riding horses, the Arab is considered one of the most beautiful and has been crossed with other breeds to produce swift horses for racing. The heavy or draught horses are used for pulling carts and farm work.

Horseshoe bats are a family of *bats so called because of their horseshoe-shaped nose-leaf. There are about sixty-nine species spread throughout the Old World. These small, insectivorous animals are mainly tropical, but five species occur in Europe and two in Britain, including the endangered greater horseshoe bat, *Rhinolophus ferrumequinum*. Those species that occur in colder parts of the group's range practise *hibernation in caves during the winter.

Horseshoe crabs are not crabs but large primitive marine *arthropods, about 60 cm. (2 ft.) long, distantly linked to the arachnids. Only five species survive, in America and Asia; all their closest relatives are extinct, although they were common back to the Ordovician Period.

The horseshoe-shaped shell covers jaws, walking legs, and respiratory book-gills, while a long mobile tail-spine projects behind. Small invertebrates are gathered from the sea-bed by touch and smell; the tiny lateral eyes probably detect danger only from above. Horseshoe crabs come inshore in vast numbers for nocturnal mating. Males mount females, fertilizing 200–1,000 eggs as they are laid in a shallow sandy depression. Hatchlings are tail-less, and resemble fossil *trilobites.

Horsetails (plants) are primitive spore-bearing plants (*cryptogams), closely related to ferns, and forming the class Sphenopsida. They have whorls of green branches at regular intervals along the upright stem, and minute scale-like leaves. Like clubmosses, they were much larger in size and more diverse in the Carboniferous Period, 250 million years ago, when, as we know from fossils, forests of tree-sized horsetails, called *Calamites, existed. Today there are only twenty-five species and some, such as field, or common, horsetail, *Equisetum arvense*, are troublesome weeds on cultivated clay ground, and are poisonous to livestock.

Houseflies are true *flies of the family Muscidae that commonly enter houses. In particular, the name is used for the common housefly, *Musca domestica*, which has spread worldwide in association with man and is not only a nuis-

This **horsetail**, *Equisetum telmateia*, is native to Europe, Asia Minor, North Africa, and northwest America. Its 1·5 m. (5 ft.) sterile stems, shown here, appear after its shorter, spore-bearing fertile stems.

ance but a threat to health. It breeds in decaying organic matter, including excrement, and adult flies feed indiscriminately on this and on sweet substances. Contamination of food occurs through pathogens carried on the fly's feet, and also because, while feeding, they regurgitate saliva and partially digested food to liquefy sugar and other materials. This regurgitated liquid contains many micro-organisms harmful to man, which have been picked up from contaminated food by the fly. The liquid food is mopped up with the spongy proboscis or tongue. Fly-spots consist of regurgitated food or faeces. The lesser housefly, *Fannia canicularis*, has similar habits.

House martin: a species of bird in the swallow family, which breeds across most of Europe and Asia and spends the winter in Africa, India, or Southeast Asia. It is glossy blue-black above with a white rump and underparts. The house martin's nest of mud is usually built under the eaves of houses, hence its name. Like others in the swallow family, it feeds on flying insects.

Housemice have long lived in close proximity to man, but they are perfectly capable of surviving in the wild in hedgerows and on arable land. Indeed some mice alternate between houses in winter and the countryside in summer. The true housemouse is *Mus musculus* but other species of *Mus* enter houses in various parts of the world. The original range of *M. musculus* was probably southern Eurasia, including the Mediterranean region, but its distribution is now worldwide due to human introductions. It has been present in Britain since at least the Iron Age. It is a pest of stored foodstuffs, which it contaminates with its droppings as well as eats. It is potentially a health hazard in the spread of disease. The domesticated form is widely used as a laboratory animal as well as being popular as a pet.

Hoverflies are medium-sized true *flies of the family Syrphidae, many of which have pronounced black and yellow markings. The adults are often to be seen hovering with shimmering wings above flowers, only to dart off and reappear a short distance away. Adults eat nectar and pollen, but their larvae feed on plants, aphids, tree sap, decomposing organic matter, or debris in bee, wasp, or ant nests. The aphid-feeders are among the most useful insects in that they consume huge quantities of aphids which would otherwise eat crop-plants. The larvae of some hoverfly species are known as rat-tailed maggots, since they possess a telescopic breathing snorkel which enables them to live in stagnant mud.

Howard, Leland Ossian (1857–1950), American entomologist, profoundly influenced developments in his field by his studies of insect parasites; his published works include *The Insect Book* (1901), *The House Fly* (1911), and *The Insect Menace* (1931). He advocated quarantine controls to prevent insect pests from being imported from Europe, and also biological control methods for crop pests.

Howler monkeys include the largest species of the New World monkeys. They are found from Mexico through Central to South America in forests and more open woodland. They are thick-set, heavy-bodied animals, up to 75 cm. (2 ft. 6 in.) in body length with a prehensile tail up to 75 cm. (2 ft. 6 in.) in some species. Their fur is usually black in colour, but one species, the red howler, *Alouatta*

The **hoverfly**, *Rhingia campestris*, a common European species, has mouthparts adapted for feeding on pollen from long-stemmed flowers, such as this ragged robin.

seniculus, is bright golden red. The face is naked but a heavy beard hangs from the chin. A family group of five to twenty individuals, led by an old male, will occupy a territory. Most active in early morning and evening, the group will travel some 700 m. ($\frac{1}{2}$ mile) in a day through the tallest trees, choosing fruit to eat. Breeding takes place throughout the year.

Huckleberries are the fruits of two *Gaylussacia* species native to North America. The dwarf huckleberry or dangleberry has a large, round, bluish fruit. The black huckle-

berry has shining, black berries, and grows in woodlands and swamps of northeast America. Blueberries are sometimes wrongly called huckleberries.

Human skeletons are internal frameworks of numerous *bones. In the adult there is a total of 206 bones. Eighty bones form the head, face, neck, and trunk: the axial skeleton. In the upper limbs there are thirty-two bones on each side and there are thirty-one in each lower limb. Bones differ in form and texture from childhood to adulthood and from men to women.

The skeleton serves to control the body shape and supports the internal organs. Tendons attach muscles to the bones within the skeleton and power the system of levers which bones provide.

By the examination of skeletal remains, it is possible to determine the age at death and the sex of the individual. Moreover, carbon-dating enables the approximate period when the person died to be established. Such studies are of great value to forensic scientists, archaeologists, and palaeontologists. The study of bones in general — their shape, size, and variations in length from part to part — as well as dating, has advanced our knowledge of human evolution.

Hummingbirds form a large family containing about 320 species, and are confined to the New World, mostly to South and Central America. They include the vervain humming-bird, *Mellisuga minima*, the smallest of all birds at 6 cm. ($2\frac{1}{4}$ in.) long, and weighing 2 g. ($\frac{1}{4}$ oz.). The largest, the giant humming-bird, *Patagona gigas*, is only just over 20 cm. (8 in.) long. Most are dark green with a

The male broad-tailed **hummingbird,** *Selasphorus platycercus*, emits a loud and unique trilling sound from the beating action of its wings. This species is native to Utah, USA, where it has been observed to return to the same nesting site year after year.

bright sheen, but they may have bright iridescent patches of other colours, especially reds and blues. Many species are crested or have distinctively shaped tails. All have long, thin beaks and long tongues with which they can gather nectar from flowers as they hover in flight. Some also take insects and small spiders. They build neat, tiny nests and lay two pure white eggs; those of the vervain hummingbird weigh only 0·2 g. ($\frac{1}{40}$ oz.). Most stay in the same area all the year round, but the ruby-throated hummingbird, *Archilochus colubris*, which breeds as far north as Canada, migrates to Central and South America for the winter.

Humpbacked whale: a species of *whalebone whale, so named by whalers because it arches its back when rolling over to dive. These whales congregate every spring in Hawaiian waters, remaining there for several months, to give birth and to mate. One calf is born about a year after mating. Solitary males sing day and night for months on end, and a song can last for thirty-five minutes before being repeated. While following its regular migration routes, the humpbacked whale visits every ocean, spending the winter in the tropics, travelling in schools or alone. A large individual can reach a length of 15 m. (50 ft.) and weight up to 50,000 kg. (110,000 lb.).

Humus is formed by the decomposition of dead plants and animals in natural *ecosystems. It is a mixture of organic matter in various states of decay and the organisms that have caused its decomposition (largely bacteria and fungi). A modified form of humus may be produced naturally as *compost. Humus is essential to the structure and drainage properties of soil. It helps to retain nutrients in sandy soils and assist aeration of heavy, fine-particle, clay soils.

Hunger is a complex sensation, aroused by a combination of sensory stimuli, which drives animals to search for and consume food. Powerful contractions of the empty stomach (hunger pangs), low levels of sugar and fats in the blood, and alterations in the body's fat stores are all examples of changes that are perceived in the brain as hunger. In man, social and cultural influences, instead of hunger, tend to regulate feeding habits.

Hunter, William and John, brothers, were Scottish anatomists who were pioneers in several branches of medicine. William (1718–83) became London's foremost obstetrician. John (1728–93) studied under him, became a founder of scientific surgery, and made valuable investigations in pathology, physiology, and biology. He selected specimens to illustrate his ideas and exhibited them in a house which he built in London. Some of his exhibits were bought from the notorious 'resurrection men', who stole from graveyards to supply anatomists; he once paid £500 for the cadaver of an Irishman 2·5 m. (8 ft.) tall.

Hunting dog, or Cape hunting dog: a species of wild dog found in bush country south of the Sahara Desert. It is found in packs of fifteen to sixty individuals and preys mostly on small antelopes. The appearance of a pack is a signal for all other game to move out. As sick and aged animals lag behind they are caught and rapidly devoured by the dogs. The pack moves continually in search of food, communicating with a soft, clear, musical 'ho-ho'. At breeding time the pack breaks up and its members retire to find underground dens. Two to six young may be born and they may live for nine or ten years.

Hunting wasps are solitary *wasps whose females hunt, catching live prey to supply the nest cells in which their young develop. They sting the prey so that it is either killed or paralysed. If it is killed, they inject a chemical to prevent it from decomposing. The spider-hunting wasps (Pompilidae) catch only spiders, but others, including the *sand wasps, prey on various insects, especially caterpillars.

Nest cells are usually made in the earth, although in some cases they are constructed with shaped mud pellets. After each cell is filled with food, one egg is laid and the cell sealed. Many families of wasps contain species which can be described as hunting wasps.

Hutias are cavy-like *rodents and close relatives of the coypu. They differ in possessing harsh coats rather than the soft fur of coypus and in being terrestrial and not aquatic. They are confined to the Caribbean where twelve living species (nine restricted to Cuba) and four recently extinct species are known. The populations of those hutias that survive are generally in decline.

Huxley, Thomas Henry and Sir Julian, grandfather and grandson, were both English biologists. T.H. (1825–95) made his reputation as a marine biologist while ship's surgeon on a voyage to survey waters between Australia and the Great Barrier Reef. He studied in detail the surface life of the tropical seas and sent back many reports. Later, he studied fossils, especially of fishes and reptiles, and became a firm supporter of Darwin. He wrote *Man's Place in Nature* (1863) on the basis of a detailed study of anthropology, and considered himself an agnostic when challenged by men of religious orthodoxy. Sir Julian (1887–1975), son of Leonard Huxley, contributed to the early development of the study of animal behaviour, and applied his scientific knowledge to political and social problems. He was an outstanding interpreter of science to the public through writing and broadcasting, and became the first Director-General of UNESCO.

Hyacinths are perennial bulb-bearing plants, of which about thirty species are known, chiefly from the Mediterranean region. They are members of the lily family. The popular florists' varieties, with their densely packed spikes of scented flowers, are descended from *Hyacinthus orientalis* and its variety *provincialis*. The variation in flower colour, which occurs naturally in this species, has been exploited by plant breeders and a multitude of red, blue, creamy yellow, and white varieties are now available for indoor or outdoor garden cultivation.

Hybrids result from interbreeding between different species, varieties, strains, races, or populations. In nature, hybridization between *species is usually prevented by differences in courtship behaviour, physical incompatibility, or abortion of the hybrid embryo. This is important since, although any particular species is well adapted to its way of life, hybrids may not be so. When hybrids between species do arise, as in the case of the mule (offspring of a mare and a male ass) and the zebroid (from a horse or ass and a zebra), they are infertile.

In contrast, hybridization between individuals of races,

subspecies, strains or varieties of the same species often occurs. For example, all breeds of dog belong to the same species, and hybrids between breeds are easily produced. The same is true of many plants. The offspring of such mixed parentage frequently show hybrid vigour, growing much faster and to a greater size than either parent and often with the beneficial characteristics of both parents. This phenomenon has been used to advantage in the breeding of new varieties of crops such as maize and rice. The rice strain IR8, developed in the 1960s, shows high yield, low tendency to lodge (collapse), pest and disease resistance, and a favourable response to fertilizers. The superiority of hybrids has encouraged plant breeders in particular to attempt to overcome the barriers to inter-breeding between species, and many of the superior seeds on sale today are 'F$_1$ hybrids'.

Hydatids are a stage in the life cycle of the *tapeworm genus *Echinococcus*, which normally inhabits the gut of dogs. If ingested by humans, the eggs develop into cysts, several centimetres in diameter, which contain many daughter cysts. Surgical removal is difficult as rupture may lead to multiple recurrence.

Hydrangeas are shrubs of the genus *Hydrangea*, which belongs to the saxifrage family. This genus includes eighty species of deciduous shrubs, a few trees, and some climbers, native to an area stretching from the Himalayas to Japan and Southeast Asia, and to the temperate New World. Many are cultivated for their showy heads of flowers. In many species there are sterile, usually marginal, flowers which act as visual attractants to insect pollinators. There are also forms with globular heads of entirely sterile flowers, known as hortensias. The flowers may be white, pink, or blue, though some of the blue forms produce pink flowers if grown on lime-rich soils.

Hydras are cylindrical *polyps within the phylum Coelentrata and part of the class Hydrozoa along with colonial hydroids such as the Portuguese man-of-war. Hydras may be found amongst almost any pondweed; each polyp will extend to about 1 cm. ($\frac{2}{5}$ in.) if undisturbed. The tentacles sway gently, until contacted by *Daphnia* or other aquatic organisms, whereupon stinging cells discharge and the prey is drawn into the mouth, its empty skin being ejected hours later. Hydras can move by cartwheeling on their tentacles to reattach their base to some plant or rock. They can reproduce by budding from their own stalk releasing the new hydras once they are large enough. Most species exist as separate males or females; males release sperm into the water around females to fertilize the egg. The egg is produced as a resistant shell which overwinters.

Hyenas are a family of four Old World species including the *aardwolf, found on the plains of Africa, Arabia, and India. The coat of coarse fur is tawny grey or black with either irregular blotches of black or brown, or black stripes on the body and legs. The head is broad with large, rounded ears. They may be 80 cm. (2 ft. 7 in.) tall at the shoulders and, as the fore-limbs are longer than the hind, the body slopes down to the long, 30 cm. (1 ft.) tail of coarse hair. Their gait is rolling since the fore- and hind-limb on each side move forward simultaneously.

Although usually considered scavengers, with powerful jaws and teeth, all species will kill animals for food. The spotted hyena, *Crocuta crocuta*, will hunt in packs to kill

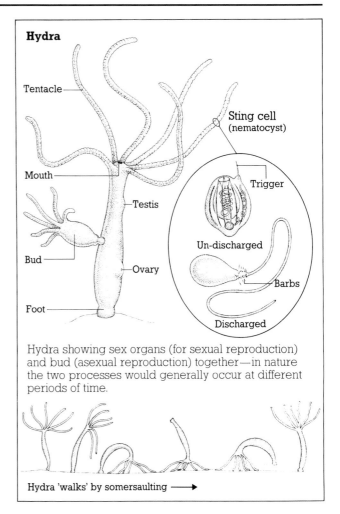

Hydra showing sex organs (for sexual reproduction) and bud (asexual reproduction) together—in nature the two processes would generally occur at different periods of time.

Hydra 'walks' by somersaulting ⟶

prey as large as a zebra. The hyena can detect a carcass several kilometres (miles) away either by scenting its smell, or from vultures overhead. They cope with bones by re-gurgitating pellets of the indigestible parts of their food. The striped and brown hyenas, *Hyaena hyaena* and *H. brunnea*, are primarily scavengers, often living a solitary life.

Hypertension is raised arterial blood pressure. The increased strain on the heart and blood vessels can cause various problems, such as heart failure and stroke. Treatment, which nearly always requires drugs, can ease the pressure and largely avoid these complications.

Hypnosis is a condition which refers to two slightly different phenomena, one in humans and one in animals. Human hypnosis is a trance-like state, induced by a skilled therapist. The treatment opens the subject's mind to suggestion and enables past events to be recalled in detail. It is used to treat addictions, phobias, and other psychological disorders and also for the induction of anaesthesia, especially for dentistry. Animal hypnosis is a natural phenomenon whereby the animal becomes immobile, or paralysed, as a reaction to fear. This is also known as tonic immobility, similar to catatonic trance in humans, which is often caused by fear too. Hypnosis can be experimentally induced in all animals with a fear of man by physical restraint. It is a last resort when an animal is attacked; it serves to reduce its chances of being killed and eaten; most predators respond to the movement of prey and will not attack what is apparently dead.

Hypothalamus: a region of the brain in vertebrates. Its main function is in co-ordinating nerve systems which control the viscera, or internal organs. It also performs other important functions. In reptiles, birds, and mammals it is the site of the 'thermostat' that regulates body temperature. In mammals, it controls heartbeat, respiration rate, and blood pressure. It also integrates the *nervous system and *endocrine system.

Hypothermia is reduced body temperature. It is a common cause of death following immersion in the sea, or exposure. It is not uncommon in elderly people, living alone and immobilized by illness, or in newborn babies, whose temperature-regulation systems are immature.

Hyraxes are about the size of a large domestic rabbit and look like guinea-pigs. They are, however, very different creatures, being primitive ungulates (hoofed mammals), distantly related to elephants and sea-cows. The eleven species occur in Africa and the Middle East. Their resemblance to elephants is seen in the skeleton, the tusk-like incisors, and in the reproductive system, particularly the testes, which do not descend into a scrotum but remain within the abdomen. There are three main groups: rock hyraxes, bush hyraxes, and tree hyraxes. All species are herbivorous.

Hyssop is a bushy herb of the Mediterranean, sometimes evergreen and formerly much used in medicine. It was also used in cooking, like its relative, mint. There are about fourteen other species within its genus, *Hyssopus*, extending from the Mediterranean to central Asia; the hyssop of the Bible was probably capers, an unrelated plant.

Hyraxes are social animals and even different species may live together. Here a group of Bruce's yellow-spotted hyraxes, *Heterohyrax brucei*, emerges from a rocky outcrop at the Masai Mara Game Reserve, Kenya.

Ibex: a species of wild *goat found in Asia and Europe. It lives in the vicinity of precipitous cliffs and mountain crags at high altitudes, often above the snowline. It is a typical bearded goat with enormous horns, up to 1·2 m. (4 ft.) long, which rise close together and sweep back in a wide arc. The coat is usually yellowish-brown in colour. One or two young are born in May or June.

Ibises are medium-sized, long-legged birds of the same family as spoonbills, with long, down-curved bills. The glossy ibis, *Plegadis falcinellus*, is widely distributed throughout the world, but most other species are tropical or subtropical. The sacred ibis, *Threskiornis aethiopicus*, was treated as holy by the ancient Egyptians and often mummified.

Icefishes are mainly Antarctic fishes of the family Channichthyidae which have large, rather spiny heads, and long bodies which lack scales. Their most notable feature is their pale, almost colourless appearance, even the gills being pallid, due to the lack of the red blood pigment haemoglobin, which normally transports oxygen within the body. In icefishes sufficient oxygen to meet their needs is carried in solution in the blood plasma, for they have a sedentary, inactive life-style. In recent years they have been caught in numbers by trawling.

Ichneumon flies, as they are popularly called, are not flies but small parasitic *wasps. They comprise the largest suborder within the Hymenoptera (bees, ants, wasps, and sawflies) with an estimated 30,000 species. Females search out characteristic micro-habitats, such as rotting vegetation, and lay eggs in, on, or near to any suitable host they find. Each species of ichneumon may have a specific host which it parasitizes, or may attack a broad range of insects. Their larvae feed on the larvae or pupae of other insects and spiders, and on spider eggs. Since the host is gradually consumed and eventually dies, they are more like predators than parasites and are now termed parasitoids. Adults are wide-ranging, some feeding on nectar or honeydew, but many do not feed at all. The females characteristically have a long thread-like ovipositor with which she places her eggs into hosts. Some species do not have wings and can reproduce by parthenogenesis.

Ichthyology is the study of *fishes, including all aspects of their biology. Marine fishes are an important food source for mankind and may become even more so in future. Much research has been done on the life histories and migration patterns of economically important species, so that careless over-fishing or the netting of young fish does not destroy the fishery, as has happened with, for example, the European herring. Success in fish conservation depends on international political co-operation as much as increased ichthyological knowledge. Research has also been applied to improve methods of fish farming.

Ichthyosaurus: an extinct marine reptile which lived in the Mesozoic Era (245–66·4 million years ago). It had evolved a long, slender snout armed with teeth for catching fish. The body was very fish-like in shape, complete with dorsal and tail fins, while the four reptilian limbs were reduced to no more than fin-like steering organs. It clearly lived permanently in the seas, and is known to have produced live young, rather than having to emerge onto land to lay eggs as other reptiles must.

Id or 'the it' is a term used in psychoanalysis to describe the subconscious, or impersonal part of the mind. This part contains the basic instincts of birth, life, and death and is distinct from the ego, or conscious state of mind.

Iguanas are a *lizard family comprising over 630 species which occur mainly in the New World, but also with representatives in Madagascar and Polynesia. Like their Old World counterparts, the agamid lizards, iguanas are a diverse family. They are active during the day, and terrestrial, tree-dwelling, and partly aquatic species occur. Most species eat insects but some larger species eat vegetable matter.

The common, or green, iguana, *Iguana iguana*, may attain a total length of about 2 m. (6 ft. 6 in.), the tail being much longer than the body. It lives mainly in trees but it is also an excellent swimmer. Other species include the *anoles, *basilisks, and *sea lizards.

Iguanodons formed one of the groups of bird-hipped *dinosaurs which existed 120 – 65 million years ago. They were large, up to about 11 m. (36 ft.) long, and were

A common **iguana** rests in the branches of a tree, over a tributary of the River Amazon. This is a favoured position from which iguanas can easily escape into the water if disturbed by predators.

capable of walking bipedally. They had numerous small blade-like teeth suitable for a diet of vegetation. When worn down, the teeth could be replaced by new ones growing from below. The 'thumb' of the iguanodon terminated in a sharp, horny spine which may have been used for defence. Their fossil remains have been found in western and central Europe.

Ilex *Holly, *Oak.

Immunity (biological) is the state of being highly resistant to disease, especially infections. It depends on the body's ability to recognize and react to harmful foreign organisms or substances called antigens. The body does so by the formation of *antibodies which react against these antigens. Immunity is usually actively acquired through natural infection, but can also be given by inoculation, the administration of a small quantity of live or killed organisms. Both confer long-lasting immunity. Specific antibodies to a potential infection can be given, but passive immunity thus acquired is short-lasting. Baby mammals acquire passive immunity, conferred by antibodies from the maternal blood late in the pregnancy, and from colostrum, the 'first milk' secreted by the mammary glands after birth.

Immunological deficiency, except for that produced by drugs, is a rare condition. When hereditary defects in either the hormonal or cellular components of the *immunity system occur, they render their victims liable to repeated infections. Drugs used in the treatment of cancer all impair the patient's immunity as well as killing cancer cells. Deliberate depression of immunity is essential for the survival of organ transplants (except between identical twins), and is also used in conditions where an aberration

of an immune mechanism is damaging the patient's own tissues (auto-immune disease). Patients with impaired immunity are subject not only to increased risk of common infections, but are also liable to opportunistic infections by organisms incapable of attacking normal subjects. A recent type is acquired immune deficiency syndrome (*AIDS).

Immunology is the study of the body's responses to foreign substances or organisms. Initially, the immune system appeared solely to confer resistance to infection. Now, its known functions include development of hypersensitivity (allergy), immunity to cancer, reaction to a transplanted organ, and involvement in disorders damaging the body's own tissues (auto-immune diseases). All these depend on the body's ability to distinguish 'self' from 'not self'.

The immune response to a foreign substance or antigen has two components. One is the formation of *antibodies, which react immediately with the corresponding antigen. The second is the 'programming' of specialized cells, lymphocytes, to react against a specific antigen and produce substances which amplify the immune response. The result is recognition and memory of foreign substances leading to specific responses and clearance.

Indian bean tree   *Catalpas.

Indian fig   *Prickly pear.

Indri: the largest species of *lemurs, *Indri indri*, which is up to 70 cm. (2 ft. 3 in.) long. It is found only in the northern part of the east-coast rain forests of Madagascar. Grey and black in colour, it has very long back legs with which it can make enormous vertical leaps from one tree to another. It lives in small social groups of two to four adults and communicates with eerie howls which carry over great distances. A single young is born and is carried by its mother. It is a much-threatened species because of the destruction of its habitat.

Infection is the invasion of the body by harmful or pathogenic micro-organisms, which results in the activation and possible breakdown of the body's defences. Common causes include bacteria, viruses, fungi, and parasites. Many of the effects of infection are due not only to the pathogenic organism, but also to the host's immune response. Antimicrobial drugs have revolutionized the outlook in many infections, for example penicillin for bacterial infection. Equally important are measures to control infectious agents in the environment, by good hygiene, treatment of water, and sanitation. Conferring immunity by immunization protects individuals and has eradicated at least one infection, namely smallpox.

Infectious diseases are those that can be transmitted from one person to another either by direct contact, or indirectly, for example via food. Many diseases caused by micro-organisms are infectious. Many virus infections, such as measles, are so easily transmitted by exhaled droplets that any susceptible person near the patient is likely to become infected. For this reason most people catch such diseases in childhood and are thereafter immune. At the other extreme is leprosy, which is not very infectious, so that the risk for those working in leper colonies is very small. Most *epidemics are due to infectious disease.

Inflammation is the response of tissues to injury, the signs of which are redness, heat, swelling, and pain. These result from increases in local blood-flow and distension of the smallest blood vessels. This in turn leads to leakage of fluid and protein from the blood, and migration of white blood cells into the damaged area. Inflammation, triggered by chemical agents, is beneficial, ensuring an excess of blood cells and proteins required for protection against infection and for tissue repair.

Influenza is a contagious disease caused by one of several strains of the influenza *virus. The strain responsible for *epidemics has a remarkable capacity for mutation. This means that exposure to the virus in a previous epidemic gives no immunity against a subsequent one. A fresh vaccine must be prepared for each epidemic strain. The infection damages the lining of the bronchial tubes, and opens the way for bacterial invasion. This was responsible for most of the deaths in the disastrous European epidemics of 1918-19. Antibacterial drugs have greatly reduced the death rate.

Inhibitors slow down or prevent the catalytic action of *enzymes. Competitive inhibitor substances resemble the normal substances upon which enzymes work and fit into the active site. Non-competitive inhibitors, including poisons such as cyanide, bind elsewhere on the enzyme molecule, changing its shape so that its activity is destroyed. Naturally occurring inhibitors control the *metabolism of living cells by switching off enzymes that are no longer needed.

Insanity, or madness, is used to describe a state of mind involving loss of reason or control. The main use of the term is in law: people declared medically insane are not responsible for their actions and thus cannot enter into legal contracts.

Insect flight is achieved by beating one or, more usually, two pairs of wings up and down. These wings consist of two thin sheets of *cuticle, strengthened by tubes, or veins, and enclosing an air-filled cavity. Insect wings can be moved either by the indirect action of muscles attached to the thorax (to which the wing bases are attached), or by a combination of this and the direct action of muscles attached to the wing bases. The thorax of insects has remarkable elastic properties, due partly to the mechanical elasticity of cuticle, and partly to a natural rubber called resilin, which forms the hinges at the base of each wing. In species with a slow, flapping flight, the wings are raised by muscles pulling down the 'roof' of the thorax, the return stroke is caused by the thorax regaining its shape (elastic power) and by the contraction of muscles attached to the wing bases. Species with a very high frequency of wingbeat, such as the mosquito, which is capable of up to 600 beats per second, possess a unique type of muscle (asynchronous muscle). This distorts the thorax by rapid contractions which are initiated by a single nerve impulse—all other muscles work on the basis of one impulse, one contraction.

The intricate twisting of the wings is created by muscle movements at the wing base, or by the complex manner in which the wing is anchored to the thorax. The frequency of wingbeat can vary from 8 beats per second in some butterflies to over 1,000 beats per second in some midge species. In flight, insects, such as hawk moths, can reach

Using stroboscopic flash photography, **insect flight** can be captured. This lacewing, *Chrysopa* species, has two pairs of wings which act independently. Here, once clear of the leaf, the insect performs a backward somersault before levelling off into stable flight.

speeds of 15 m./sec. (34 m.p.h.), but most 'flies' move at around 4 m./sec. (9 m.p.h.).

Insectivores comprise an order of 345 species of mammals related to the primates. So close is the relationship that one group, the tree shrews, has alternated between the orders according to the changing ideas of those studying them. Insectivores are certainly primitive and the simpler forms probably resemble the ancestral mammals. Primitive features include the simple teeth, five digits on each limb, and flat feet. Insectivores are generally small, though some, such as hedgehogs, reach moderate sizes. The longest is the giant otter shrew, *Potamogale velox*, which

The giant otter shrew, *Potamogale velox*, is an **insectivore** of West and central Africa. It spends most of its active hours in the water, scouring stream banks for crabs and vegetable matter, as well as insects.

is up to 60 cm. (2 ft.) in length. The body covering is usually soft fur, but the hedgehog and some tenrecs are spiny. Excluding the tree shrews, the eight living families are the tenrecs, otter shrews, solenodons, golden moles, hedgehogs, elephant shrews, shrews, and moles. A further family, the West Indian shrews, became extinct very recently, so recently that their remains were found in fresh owl pellets in 1930. As the name suggests, insectivores feed on insects but most take earthworms, slugs, and snails, and even small vertebrates such as frogs.

Insect pests are those which can kill or damage man himself, his domestic animals, his food crops, or other plant products used by him. They also include many insects which are more of a nuisance than a serious threat. Those which cause physical damage to man are of outstanding importance in warm and tropical climates, and of these the *mosquitoes are the most harmful. They carry the pathogens of the various forms of malaria, as well as those of yellow fever and other dangerous diseases. *Fleas carry bubonic plague from rats to man.

Among pests of domestic animals are tsetse flies, bot-flies, keds, stable flies, and lice. Every plant which is of use to man has its own insect pests, which eat the whole plant or parts of it. Those feeding on roots below ground include chafers, leather-jackets, and wireworms. Of those feeding above ground on whole plants, aphids, scale-insects, and *locusts are probably the most important, although cutworms and other caterpillars do much harm.

Examples of those which are of nuisance value to man include the ones that bite in the summer, such as mosquitoes, biting midges, and wasps. Others are household pests: silverfish, cockroaches, clothes moths, and bedbugs, none of which are dangerous, but all of which are considered a threat to man.

Insects are a class of *arthropods, their bodies being divided into three parts; head, thorax and abdomen. The head bears many structures, including one pair of antennae, two compound eyes, and the mandibles. The thorax consists of three segments, each of which bears a pair of legs, although these may be modified as grasping or swimming organs in some species. In most insects the thorax also bears two pairs of wings on the second and third segments. The wings are strengthened with veins and are usually transparent, although they may be covered with scales or hairs. The front pair of wings may be thickened for protecting the hind ones, as in beetles and earwigs. Some insects, such as silverfish and springtails, are wingless. The insect abdomen has six to thirteen segments and may bear various appendages, especially those for mating and egg-laying. The insect body receives oxygen through fine tubes, the tracheae, ramifying through the body and opening into spiracles on the sides of the body. They have a simple digestive system, and waste products are removed from the open circulatory system by organs called Malpighian tubules. These extract nitrogenous wastes and excrete them into the hind gut as crystals of uric acid.

Insects range greatly in adult size, from less than half a millimetre to over 150 mm. (6 in.). The number of insect species, roughly estimated at over 750,000, easily outnumbers all other species of plants and animals combined. They are divided into some thirty orders. Of these, nine have a life history involving complete *metamorphosis, and the rest have incomplete metamorphosis.

Instinct is the mechanism by which animals can perform complex behaviour patterns without learning or conscious effort. Instinctive behaviours are inherited and have evolved to be adaptive, fitting the organism to its particular role. Such behaviours are thus specific to a species and common to all its members. Instinct is of particular importance in *animal behaviours such as courtship, mating, and other reproductive activities. More general behaviours such as feeding and defence may have an instinctive base. Many birds, some grasshoppers, frogs, and a number of other animals have song or call patterns that attract mates and are based upon instinct. The courtship of sticklebacks is typical of such behaviours in that it demands correct responses from each of the participants in a complex sequence.

Instinctive behaviours often require a stimulus or releaser to initiate them. The herring-gull chick pecks the red spot on the adult's bill, releasing its instinctive feeding behaviour. A releaser will operate only if conditions, both internal and external to the organism, are suitable. Thus the bright plumage of many species of bird disappears outside the breeding season. Instinct may be modified by learning, particularly in vertebrates.

Intelligence is the mental faculty for understanding and solving problems. It is measured by means of intelligence tests, devised to test skill at verbal and non-verbal problems. The result is expressed as the Intelligence Quotient (IQ) in which the average score is 100. Scores below 50 imply severe mental handicap and IQ scores of 50–70 are regarded as evidence of mild mental handicap, or 'sub-normality'. The main value of IQ tests lies in defining those groups who do badly so that special educational provision needs to be made.

Animal behaviour is often considered to be purely a matter of *instinct, rather than intelligence. It is difficult to demonstrate reasoning, or the ability to solve new problems, in animals. Monkeys show intelligence when presented with a wide variety of visual problems, while rats, with poor eyesight, respond intelligently to problems with smells as cues. The use of tools by birds, such as the woodpecker finch, and by mammals, such as the *sea otter or apes, is an example of intelligent behaviour in animals.

Interferon is a soluble protein produced by animal cells which have engulfed living or dead *virus particles. It is released rapidly to provide partial protection from viral infections, before effective concentrations of *antibodies specific to the virus can be built up in the body.

Intestine is that part of the *alimentary canal that connects the stomach to the anus. It ranges from a single tube in invertebrates to a complex system of segments devoted to different tasks in mammals. Its purpose, in all animals, is to aid the breakdown of food into substances such as sugars, amino acids, or fats, which can then be absorbed by the intestinal cells. This is achieved through the secretion of *enzymes by the intestinal cells, and by keeping the food in the body long enough for them to act.

The intestine of humans is divided into two main regions, the small intestine, and the large intestine. Most of the breakdown and absorption of food occurs in the former, while the latter, which includes the *colon and *rectum, is largely involved in extracting water from the food remains. Other names for the intestine include the gut, and bowel. The intestinal contents are moved along by wave-like contractions of the muscular intestinal wall, called peristalsis. Absorption of substances is helped by small finger-like extensions of the intestine wall, called villi, which greatly increase the surface area of the intestine.

Invertebrates are all the animals without backbones and include about 97 per cent of all living animals. They are much more diverse than the vertebrates, and much less well studied, despite their enormous diversity. They range from simple one-celled protozoa to such complex animals as the octopus, lobster, or bumble bee. Some are built with a circular plan of radial symmetry, like jellyfish and sea-urchins; others have long, thin bodies, which are bilaterally symmetrical, either soft and sinuous like the worms, or with a stiff outer covering and jointed legs, like the arthropods. All the features familiar in vertebrates are foreshadowed in the invertebrates, from which the former are thought to have evolved.

Invertebrates live in all possible habitats, from deep seas to deserts and the polar regions. On land they are usually small, seeking protective humid crevices unless they are provided with a built-in shell or cuticle; but in the sea they can be very large and dominate their environments, either as individuals like the giant squid, or in colonies such as corals. Invertebrates are also good at inhabiting the bodies of other organisms, as parasites, and this may be the commonest of all their ways of life.

Flowering between May and August, the yellow flag, *Iris pseudacorus*, is a common **iris** of marshes, and pond and river edges. It is found throughout Europe, the Caucasus, and North and West Africa.

Irises, of which there are approximately 200 species, are plants found in temperate regions of the Northern Hemisphere. Most have flattened or, less often, rounded leaves, which arise from a basal rootstock called a *rhizome or from a *corm below the ground. The flowers, which are often large and showy, have equal numbers of upright and pendulous petals, termed standards and falls. The bearded irises (with hairy filaments on the falls) are a group of irises which have been extensively hybridized by man as garden plants. *Iris florentina* is the source of orris root, used as a base for perfumed powders.

Irish elks were giant deer that lived in the Pleistocene Epoch, 1·5 million years ago, in Europe and Asia. Their antlers were enormous, having a span of up to 4 m. (13 ft.). They probably died out only a few thousand years ago and their partially fossilized remains are found in peat bogs.

Ivy is the common name of the genus *Hedera*, belonging to the Araliaceae family. One of the few *lianas of temperate Europe, it is an evergreen climber which attaches itself by means of its roots. Ivy leaves have three to five triangular lobes in juvenile foliage but are unlobed and narrowly pointed in mature, sexual shoots. These shoots produce no roots and bear green flowers in the autumn, when they are visited by many insects including moths and wasps. The black fruits survive the winter and are a valuable source of food for birds at this time. Ivy has a bad reputation as a strangler of trees, although in healthy trees this is unfounded. Other plants of this family are also known as ivy.

J

Jacamars are birds which live in the tropical forests of Central and South America. They are sparrow- to thrush-sized birds, with long tails and long, straight beaks. They are iridescent green or greenish-black above and most have chestnut underparts and white throats. They perch on branches and dart out after passing insects. They nest in holes in banks and lay one to four white eggs.

Jacanas, or 'lily-trotters', are seven species of tropical and subtropical long-legged *wading birds. They have elongated toes and hind-claws which spread their weight and enable them to walk on floating vegetation. They frequent sheltered inland waters including rivers.

Jacaranda is a genus of some fifty species of trees of the tropical Americas, related to catalpas. The one most widely seen is *Jacaranda acutifolia*, a fast-growing tree from Brazil, with feathery leaves and large heads of pale violet flowers.

Jackals are the common wild dogs of the warmer parts of the Old World. The four species are usually found in open or lightly wooded parts of Africa and Asia, sometimes even in outlying districts of towns and villages, where they search for food amongst garbage. They are scavengers who serve a useful purpose in removing carrion, and are also excellent mouse and rat catchers. Stealthy and secretive, they spend most nights hunting in family groups or packs of three to six individuals.

The female gives birth, in a cave or burrow, to a litter of usually four to nine pups. Only when jackals are about

If startled, the Australian **jacana** or lotus bird, *Irediparra gallinacea*, scuttles into the air or dives underwater, and its red comb may turn yellow.

to produce a family, or are caring for the pups, are they really sociable. In captivity they can survive for thirteen years, but their lifespan is probably shorter in the wild. The four species of jackals are the golden, *Canis aureus*, silver or black-backed, *C. mesomelas*, Simien or Ethiopian, *C. simensis*, and sidestriped, *C. adjustus*.

Jackdaw: a species of bird belonging to the crow family which occurs over much of Europe and western Asia. It is a smallish bird, 32 cm. (13 in.) long, black above except for the nape and the area over the ears, and underparts, all of which are greyish-black. It has a striking greyish-white eye. In eastern Asia there is a form which has a white collar and white underparts; this is called the Daurian jackdaw. The jackdaw lives in open, wooded country, in city parks, or on cliffs, nesting in holes in trees or among rocks. It also commonly nests in holes in tall buildings such as church towers. It has a reputation for stealing money and other shiny objects.

Jackrabbits are not rabbits but true hares of the same genus (*Lepus*) as European hares. They have larger legs and ears than other hares but otherwise they are similar in appearance and behaviour. They are particularly noted for their leaping habits. Jackrabbits are confined to western North America, where they inhabit arid rangeland. Six species are recognized, including the white-tailed jackrabbit, *L. townsendi*, which inhabits the slopes of the northwestern Rocky Mountains, and the black-tailed jackrabbit, *L. californicus*. The latter is called the varying hare because it often turns white in winter.

Jaegers are *skuas of the genus *Stercorarius*, which are sometimes separated from the genus *Catharacta*. They include three slender, fast-flying hunters of the northern tundra: the Arctic, pomarine, and long-tailed jaegers, all of which feed on small mammals and birds. They nest on the ground or on cliffs, laying two or three eggs. In autumn they fly south and feed mainly at sea, penetrating far into the Southern Hemisphere.

Jaguar: the biggest *cat species found in the New World, from Patagonia through South and Central America as far north as Texas, New Mexico, and Arizona in the USA. One of the most handsome of the cats, it may be as much as 2·43 m. (8 ft.) long. It has a rich yellow or tawny coat marked with a chain of black spots down the back, bordered by five rows of black rosettes running lengthwise along each side. The tail, limbs, and head are heavily spotted and lined with black. Black jaguars may be found in the valley of the River Amazon. Among its prey are capybara, alligators, turtles, peccaries, and even man. Mating may take place at any time and gestation lasts just over three months. Two to four kittens are born; they are more heavily spotted than the adults. The jaguar's lifespan is twenty years.

Japonica is the common European name for *Chaenomeles speciosa*, known in North America as the Japanese quince. It is a deciduous Chinese shrub, related to the quince, and produces bright red or pink flowers early in spring before the leaves expand. The quince-like fruit can be used to produce jelly.

Jasmine is the common name of a number of species of *Jasminum*, a genus of some 300 species, related to the forsythias. They are mostly climbers of the tropics and subtropics except in North America. The most frequently seen are the white-flowered *J. officinale* of southern Europe, the oil of which is used in perfumery, and winter jasmine, *J. nudiflorum*, a Chinese species with yellow petals.

Jaundice is a yellow staining of the skin and other tissues, caused by a bile pigment in the blood. This is a product of the breakdown of *haemoglobin in red blood cells. The pigment is taken up by the liver, and excreted in the *bile. Jaundice may therefore be due either to a large increase in red cell breakdown, or to disorder of the liver, or to obstruction to bile-flow. Except for severe jaundice in new-born babies, it is in itself harmless, its importance depending on its cause.

Jawless fishes, or agnathans, are very primitive fishes which have not evolved the paired jaws of other fish. They form one of the three classes of fishes and include the very earliest fossil vertebrates from the Palaeozoic Era, approximately 465 million years ago, such as **Jaymoytius* and *Cephalaspis*. Agnathans are primitive in other ways too, most of the fossil species being covered in heavy dermal armour and with poorly developed fins. The jawed fish must have evolved from forms such as these, although their exact ancestor is not known. The living jawless fishes are subdivided into two orders: *lampreys and *hagfishes. These are rather specialized, indirect descendants of the extinct forms and are considered to be the most primitive living vertebrates.

Jaws are the opposing sets of bones which form the framework of the mouth and masticating apparatus in vertebrates. The upper jaw is fixed to the skull. The lower jaw is freely movable, being slung by strong muscles and fibrous cords from joints at the base of the skull just in front of the ear. It can be moved upwards and from side to side in mammals. In contrast to the powerful closing muscles, the muscles which open it are few and weak, opening depending on gravity in man and many other animals. The lower jaw of some snakes can be dislocated downwards to allow large prey to be swallowed whole.

Jaymoytius: one of the oldest known fossil fishes. It lived about 400 million years ago and its remains have been found in rocks of this age in Scotland. It was only about 3–4 cm. ($1\frac{1}{4}$–$1\frac{3}{4}$ in.) long, and is a member of the *jawless fish group, lacking the true jaws of more advanced fish, the mouth being no more than a simple slit. A row of small, round gill-openings supported by cartilage runs along the side of the head. Neither scales nor paired fins are present and it is thought that *Jaymoytius* may be an indirect ancestor of the living lampreys and hagfishes.

Jays are large birds belonging to the crow family and their thirty-seven species are found in many countries. About 35 cm. (14 in.) in length, they are sturdily built birds with powerful bills. Many are quite brightly coloured, often having blue in their plumage; and a number of species are crested. They live in woodland and eat insects, small birds, fruits, and seeds. Several species live all the year round in tight-knit family groups, which together may even care for the offspring of a single nest.

Jellyfish are marine invertebrate animals of the phylum Coelentrata. Like the related corals and sea-anemones, they subdue prey with special stinging-cells on their tentacles which hang below their floating body. Occasionally these stings can even harm humans. Jellyfish have only a thin layer of cells on their inner and outer surfaces, most of the body being non-living, jelly-like material forming the familiar 'bell'. Simple muscles pull this bell in and out, expelling water from the mouth, so the animal jets into the upper water and parachutes gently down again, trapping prey as it descends. They have a complex life-cycle involving a tiny larval (planula) stage and a *polyp stage, which buds off small immature jellyfish, which may develop directly into an adult jellyfish or medusa.

Jenner, Edward (1749–1823), English naturalist and physician, discovered vaccination against smallpox. He helped in the preparation and arrangement of specimens brought back to England by Joseph Banks in 1771. After two years as a pupil of John Hunter in London, he returned to practise medicine in the country and to continue his observations of animal life. He described the behaviour of the baby cuckoo and showed the value of worms in arable soil.

Jenner observed that infection with the mild disease called cowpox made people immune to smallpox. In 1796 he performed the first vaccination by inserting cowpox matter into two scratches made on the arm of a healthy eight-year-old boy. A few months later the boy was inoculated with smallpox, and the disease refused to take. The discovery made Jenner famous. The practice was not without opposition, but before his death smallpox vaccination was practised throughout the world.

Jerbils   *Gerbils.

Jerboas are desert *rodents distinguished by extremely long hind legs. The range of the thirty species extends from North Africa across central Asia to northern China.

The Egyptian **jerboa**, *Jaculus jaculus*, has tufts of hair on the undersides of its feet to help it move over soft sand. It uses its long tail as a balancing organ when jumping, and as a prop when standing upright.

Although only mouse-sized, they are capable of leaps of up to 3 m. (10 ft.) when fleeing from predators. While foraging, they either hop on all fours like rabbits or walk on their hind legs. Their food consists of plants and insects.

Jew's ear fungus commonly grows mainly on branches of elder but also on elm, beech, walnut, and pine. It forms soft brown, ear-shaped fruit bodies, which are gelatinous when wet and bear the spores. The jew's ear fungus belongs to an order of *fungi which also includes witch's butter, *Tremella lutescens*, and sticky coral fungus, *Calocera viscosa*.

Jiggers   *Fleas.

John dory: a deep-bodied, flattened fish with a large head and expansible mouth. It has a very distinct black blotch with an encircling yellow ring on each side of its body. Living in shallow coastal waters of Europe, the John dory, *Zeus faber*, has a close relative which lives in North American Atlantic coastal waters, called the American John dory, *Zenopsis acellata*. They swim slowly, either singly or in a small school, often sheltering among algae or under moored boats, and feed on fishes, which are captured by stealth. Dories comprise several species within the family Zeidae and are regarded as a food-fish.

Joints (biological) are the sites where movement can occur between two or more parts of a skeleton or rigid tissues. In vertebrates there are three main types of joint: the freely movable or synovial joints, which include ball-and-socket joints of the hip; pivot joints, as at the base of the neck; and hinge joints, as in the elbow or knee. Slightly movable joints, in which bones have the potential to move over one another to a small extent, occur along the spine. Immovable joints include the interdigitating sutures of the skull bones, mobile only in babies.

Jujube is the common name of a number of *Zizyphus* species, which are relatives of the *buckthorns. The Chinese jujube, *Z. jujuba*, is one of the five major fruits of China, where some 300 varieties of it are grown. It is a deciduous tree or shrub bearing oblong *drupe, fruits. The name is also used for lozenges or rubbery sweets originally prepared from jujubes but today usually made with synthetic materials.

Jumping plant lice are small *bugs of the family Psyllidae, which contains some 1,300 species, with transparent wings and strong hind legs for jumping. In temperate regions most species overwinter as eggs; the young hatch in the spring, and become adult in the summer, so that there is only one generation a year. The adults are about the size of an aphid and resemble tiny cicadas. The larvae can do considerable damage to fruit trees, and also produce copious amounts of sweet-tasting honeydew which spreads over leaves.

Jung, Carl Gustav (1875–1961), Swiss psychologist, broke with *Freud to found a school of analytical psychology. It differed from psychoanalysis in its use of concepts of the 'collective unconscious' (a kind of race memory) and the libido (psychic drive or energy). He advocated a highly complex classification of types of personality, and introduced the terms 'extrovert' and 'introvert'.

Joints

Joints occur wherever two or more bones meet. The type of movement possible at freely moveable joints depends on the shape of the bones at the point where they rub together.

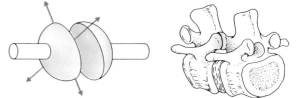

In a sliding or gliding joint, two flat surfaces rub together providing twisting and bending movement. An example is the joint between the projections on adjacent vertebrae of the spine.

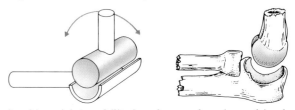

In a hinge joint, mobility in only one plane is combined with strength, as in the elbow joint.

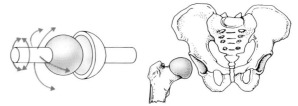

In a ball-and-socket joint a rounded head on one bone fits into a cup-shaped cavity in another. A wide range of play is possible, as in the hip joint.

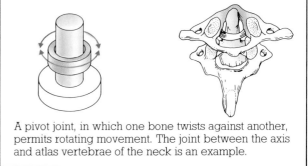

A pivot joint, in which one bone twists against another, permits rotating movement. The joint between the axis and atlas vertebrae of the neck is an example.

Junglefowl are four species of the genus *Gallus* which is part of the pheasant family. All live in Southeast Asia in wooded country. The males are shiny greenish-black on the wings and tail, and many of the tail feathers are elongated and curved; the feathers of the neck and back are orange-chestnut and the underparts are black. The females are brown. All species have a distinctive red comb and wattles. The red junglefowl, *G. gallus*, is the ancestor of the domestic chicken.

Junipers are evergreen trees and shrubs of the conifer genus *Juniperus*, which comprises some sixty species of the northern temperate zone, extending to the tropics in the Kenya highlands. The common juniper, *J. communis*, is the only tree species native to the northern temperate regions

Both species of **jute** are annual plants. Sown in spring, they are cut in early autumn and dried in the sun, as here, near Dacca, Bangladesh, before being steeped in water and the fibres teased out.

of both the Old and New Worlds. This and other species have needle-like leaves throughout life, while some other species of *Juniperus* have small ones pressed close to the stems, as in *J. sabina*, the savin of Eurasia. The cone is modified to form a fleshy mass surrounding the seeds and is eaten by birds. The 'berries' of the common juniper are used in the production of gin, to which it has given its name. The timber of some species is valuable, that of the American red pencil cedar, *J. virginiana*, being used in pencil manufacture.

Jussieu, Antoine Laurent de (1748–1836) was one of a family of French botanists whose home was a centre for plant collection and research. In *Genera Plantarum* (1789), he developed the system on which modern plant classification is based. From extensive observation, he grouped plants into families on the basis of common features.

Jute is a fibre plant that became important relatively recently as a replacement for hemp when a method of spinning the new material was developed in Scotland in the mid-nineteenth century. There are two species of plant of the genus. *Corchorus* from which jute fibre is extracted. They belong to the lime-tree family, the Tiliaceae. *C. capsularis* is grown in India while *C. olitorius* is cultivated for jute fibre in part of Africa.

　　The plant stems are cut by hand and, after drying, are immersed in water for up to five weeks. The extracted fibres can be up to 3 m. (10 ft.) long but are weaker and less durable than flax or hemp. The spun yarn is woven to produce hessian, usually used for sacking and bag manufacture.

Kagu: a bird in a family of its own, found only in the forests of the Pacific island New Caledonia. It is related to the cranes and rails, and resembles a squat heron, standing about 60 cm. (2 ft.) high with long legs and beak. It is a pale grey in colour, speckled with brown, and has a loose crest. The kagu eats worms, snails, and other small animals, and builds a simple nest of sticks on the ground in which it lays a single egg.

Kale　*Cabbages.

Kangaroo rats are twenty-two species of desert *rodents of North America, where they fill a niche occupied by the jerboas in the Old World. Like them, they live in burrows during the day and emerge at night to feed on plant material and insects. They hop on their very long hind legs after the manner of kangaroos. They are part of the *pocket mice family.

Kangaroos are probably the best-known animals of the Australian fauna. The name is usually reserved for the fifty species of the family of big-footed *marsupials which also includes the *wallabies. The red kangaroos, *Macropus rufus*, and the grey kangaroos, *M. giganteus*, may reach a

The red **kangaroo**, *Macropus rufus*, lives chiefly in the highlands of southern and eastern Australia. Here, a joey protrudes from its mother's pouch; it will be mature at two years of age and live for about fifteen.

weight of 90 kg. (200 lb.) and a height of 1·8 m. (6 ft.) or more. The large, heavily built, muscular hindquarters and tail serve to balance the small, lightly built, mobile fore-quarters. The fore-limbs are held clear of the ground during bipedal locomotion, and when the animal is collecting and eating food. For such large animals they can move apparently effortlessly with leaps of 8 m. (26 ft.) in length. The teeth and digestive system are adapted for a herbivorous diet. Although it does not chew the cud, as cattle and true ruminants do, the food is regurgitated and swallowed again.

Like all marsupials, the kangaroo has a pouch into which the minute young, or joey, crawls immediately after it is born. The joey remains in the pouch for some 190 days before it leaves for short periods. A second offspring may be born while the larger young is still suckling; and the mother produces two milks vastly different in composition to satisfy their different needs.

Kapok is a fine fluffy material which surrounds the seeds of various tropical trees. The main commercial source is the trees *Ceiba pentandra* and *Bombax ceiba*, both members of the tropical rain-forest family, Bombacaceae, which includes balsa, baobab, and durian trees. The main centres for commercial plantation are Asian countries such as Thailand. The fruit capsules of these trees are full of wax-coated fibres up to 3 cm. (1¼ in.) long and these are believed to aid seed or fruit dispersal by water. This floss is very buoyant and light (eight times as light as cotton) and is used to fill lifejackets and cushions, and for insulation. The fibres are naturally smooth but may be spun after chemical treatment has roughened them.

Katydids   *Longhorn grasshoppers.

Kauri pines are broad-leaved, tropical conifers, yielding valuable timber and resin. They are found only in Southeast Asia and Australasia and, in spite of their common name, are not true pines. They are related to the monkey-puzzle tree and are conifers well adapted to the tropical climate. The New Zealand kauri, *Agathis australis*, rivals the redwoods in size and age, and the klinki pine, *A. hunsteinii*, is a valuable source of plywood.

Kea: a species of bird in the parrot family confined to the South Island of New Zealand. It is the size of a large crow, bronze-green all over with blue and red in the wing and a red rump. It has a powerful beak, more curved in the male than in the female. Keas eat a wide range of fruits and insects and will also feed on dead sheep, an action that has led to the erroneous belief that they are sheep-killers. They nest in holes among rocks or in hollow trees.

Keds   *Louse-flies.

Kelps are exclusively marine, brown *algae. They include the largest known alga, *Macrocystis*, which may be 60 m. (197 ft.) long. *Laminaria* (oarweed) is common off European coasts and grows below the low-water mark. They have long, strap-like fronds attached to rocks by a root-like holdfast. Some parts of coasts have extensive beds of kelp, and are referred to as kelp forests.

Kenaf is a tropical, annual plant, related to cotton and mallow, with tall, slightly prickly straight stems up to 4 m. (13 ft.) tall. The species *Hibiscus cannabinus* is cultivated in Africa and Asia for the fibre obtained from its stem. Similar to jute but somewhat coarser, it is used mainly in the manufacture of sacking. Its seeds contain about 20 per cent of oil similar to cotton-seed oil.

Kestrels are small *falcons of the genus *Falco*, with broad, pointed wings and a long square-cut tail, characteristically seen hovering 5–6 m. (16–19 ft.) above ground while searching for prey. The European kestrel or windhover, *F. tinnunculus*, is heavily barred in pale and dark brown, and is widespread over Europe, Africa, and Asia. Its counterpart in the USA is the sparrow hawk, *F. sparverius*, and about a dozen other closely related species occur more locally in Africa, Australia, and on oceanic islands. Their prey is insects, small rodents, reptiles, and occasionally small ground-living birds.

Kiang: the largest and most handsome subspecies of the Asiatic wild *ass. It inhabits the high mountain plateau of Tibet up to 5,000 m. (16,000 ft.). The dung of this ass is gathered for fuel.

Kidneys are normally present as a pair of organs in the abdomen of vertebrates. They function mainly as organs of excretion, but also play a part in the control of blood salt concentration. The general pattern for all vertebrates is that blood passes continually through the kidney where it is 'cleaned' of all waste compounds, or excesses of useful salts and substances. Inside the kidney are special tubes, called nephrons, which filter out water and soluble substances. They also pass small molecules from the blood. The nephron tubules then actively absorb the useful compounds, and some water, back into the blood, carefully controlling their level in the blood. The waste products and water are passed out of the kidney as urine, and collect in the *bladder.

During the breakdown of proteins the body discards large amounts of nitrogen, usually in the form of the toxic but soluble gas, ammonia. This is carried away from cells by the bloodstream. In aquatic freshwater animals, water loss is not a problem and they rid themselves of ammonia by producing a large quantity of dilute urine. Terrestrial animals cannot afford to do so; instead they convert the toxic ammonia into the less toxic form *urea. This is then filtered out by the kidneys and excreted in a more concentrated form. In birds and reptiles most of the ammonia is converted into an insoluble compound called uric acid. This packs more nitrogen into a smaller space and is excreted as a white paste, equivalent to urine.

Killifish, or toothcarps, are a family of at least 300 species of fresh- and brackish-water fishes, found in tropical and warm temperate regions of the world. Mostly small, with the largest species reaching 15 cm. (6 in.), they have a fully scaled body and a single dorsal fin, with rather flattened heads and fine teeth in the jaws. Killifish lay eggs, and one group, the rivulins (particularly common in tropical Africa and South America) lay eggs before the onset of the dry season which hatch after the next rains. As a result the fish never live longer than one year. They are poorly represented in southern Europe, being found only on the Mediterranean coastlands, but are common enough elsewhere.

Kingbirds belong to the tyrant *flycatcher group and are restricted to the New World, where they live in open, wooded country. They are the size of a small thrush, blackish or greenish with paler underparts. They get their name from their aggressive domination of any other species of bird which comes close to their nests.

King cobra, or Hamadryad, of Southeast Asia, is the world's largest species of venomous snake, with reliably recorded lengths of up to 5·72 m. (18 ft. 9 in.). The female builds a nest out of vegetation, into which she deposits her eggs, and remains coiled above them until they hatch. The king cobra feeds mainly upon other snakes.

King crabs are large relatives of the stone crabs, found in the northern Pacific. They have long, slender legs, with a span over 1 m. (3 ft.), and both the legs and body have a heavily sculptured exoskeleton. These crabs are trapped in large numbers off Alaska, as food.

Confusingly, the *horseshoe crabs, which are not crustaceans at all, are also sometimes known as king crabs.

Kingcup, or marsh marigold: a species of plant, *Caltha palustris*, with large yellow flowers. It is a member of the buttercup family and is widely distributed over much of Europe and Asia, being found chiefly in water meadows and by the sides of streams.

Kingfishers are a cosmopolitan family, the Alcedinidae, with ninety-one species of small to medium-sized birds

The common **kingfisher** swoops with deadly accuracy into the water, rarely missing its carefully sought victim. Returning to an observation point, this remarkable bird then swallows the fish whole, always head first.

The **kinkajou**, *Potos flavus*, though mostly nocturnal, may lazily emerge from its nest in a tree hollow on hot and humid days. Superbly agile, this creature advances stealthily in unfamiliar territory.

with heavy-looking beaks and heads, short legs, rounded wings, and brilliant plumage. Many species plunge into water to catch fish, but the majority live on insects and other small animals. They nest in holes in banks or tree trunks. The common kingfisher, *Alcedo atthis* has subspecies which span much of the Eurasian landmass as well as many Pacific islands down to New Guinea.

Kinkajou is a species of carnivore of the racoon family. Although mainly a fruit-eater, it also takes insects, small mammals, and birds' eggs. In appearance it is rather like a monkey, with its flat face, round head, and prehensile tail. It is a stocky animal, sometimes reaching 57 cm. (2 ft.) in length, without the 55 cm. (22 in.) tail, and weighing from 1.5 to 3 kg. (3–7 lb.). It lives in tropical forests of Central and northern South America and spends its time entirely in trees. It is nocturnal and usually solitary. Its soft, dense fur is used locally.

Kites (birds) are a subfamily of *hawks with slender bodies and wings and long tails. True kites live in open country, partly as hunters but mostly as scavengers. Red kites, *Milvus milvus*, once scavenged London streets but are now rural over most of Europe. Black kites, *M. migrans*, still scavenge widely in Africa, Australia, and Eurasia. The name applies also to fork-tailed kites (honey-buzzard) and white-tailed kites, a subfamily of small kites that hover like kestrels and feed on insects and small mammals.

Kitten moths   *Puss moths.

Kittiwakes are two species of cliff-nesting *gulls: *Rissa tridactyla*, common on most Arctic and cool temperate coasts, and *R. brevirostris*, confined to the Bering Sea area. They breed on sea cliffs, raising one to three chicks on narrow, unprotected ledge nests. They feed communally at sea on fish and large plankton, and are seldom seen inland.

Kiwi fruit is the fruit of a climbing plant, *Actinidia chinensis*, native to southern China. The name kiwi fruit comes from the national bird of New Zealand, since that country exports the fruit to Europe and the USA. The fruits are also known as Chinese gooseberries because they resemble large brown, hairy gooseberries. Their pale green flesh is extremely rich in vitamin C.

Kiwis are three species of birds that live in New Zealand. They are about the size of a large chicken with a large head and a long, slightly curved beak. They are flightless, with very powerful legs and long, loose, brownish or greyish feathers which look rather like hair. The females are considerably larger than the males. They are mainly nocturnal, and eat a wide variety of food, including berries and worms, some of which they trace by smell. They nest in a hole in the ground, the male doing most of the incubating of the usually single egg.

Klipspringer: a species of antelopes, found on rocky slopes in Africa from the Cape of Good Hope to Ethiopia. It is a dainty animal only 50 cm. (20 in.) high at the shoulder. The bucks have horns that rise almost straight from the head, bending only slightly forwards. Its coat is yellowish-brown in colour, speckled with yellow, the hair being long and brittle. It is an agile creature that can obtain a foothold on a rocky projection 2–3 cm. (1 in.) in diameter. Family groups feed together on leaves, shrub roots, and grass. If alarmed the klipspringer can utter a shrill whistle.

Knotweed, or knotgrass, is the common name for many species of the plant genus *Polygonum*, related to rhubarb and sorrels. This genus includes 150 species of annual and perennial plants with pink or white flowers, which are found throughout Eurasia. The creeping stems of the common knotweed, *P. aviculare*, have distinctive swollen joints, narrow leaves, and very small flowers; it is a plant of cultivated land and waste-ground. Other species of *Polygonum* include the tall, perennial Himalayan knotweed, *P. polystachyum*. The closely related Japanese knotweed, *Reynoutria japonica*, is now a common naturalized plant in Europe.

Koala: a species of herbivorous *marsupial found from Queensland to Victoria, Australia. It lives in eucalyptus and other trees, and feeds on their foliage. Superficially it resembles a small bear, having a wide head and a thickset body, and weighs about 9 kg. (20 lb.). Like all marsupials, it has a brood-pouch with a pair of teats; a single young is born about thirty-five days after mating, and it makes its way into the pouch immediately after birth. Not until it is four to five months of age will it venture out of the pouch to travel on its mother's back, returning to the pouch at intervals until it is six months old. It may live for up to twenty years. Considered an endangered species, it is today increasing in numbers, saved by an active programme of conservation.

Koch, Robert (1843–1910), German country doctor and bacteriologist, studied an outbreak of the disease called anthrax in cattle. He found that the sick animals contained a bacillus which could be cultivated outside the body, and by the use of dyes was able to distinguish it. Using this technique, he discovered first the tuberculosis bacillus and then the bacterium which causes cholera. He also studied typhoid and malaria and formulated rules relating to conditions which must be satisfied before a disease can be ascribed to a specific micro-organism.

Kohlrabi is a plant derived from the same species as cabbage, but it has a very different shape. The edible base of the single stem is swollen, and can develop to a large size, resembling a knobbly green turnip. Its value as a vegetable has been known since Roman times.

Komodo dragon, or ora: the largest living species of lizard, attaining lengths in excess of 3 m. (10 ft.). These *monitors are restricted to a few of the Lesser Sunda Islands, Indonesia. An adult Komodo dragon is a formidable predator, quite capable of killing and eating small deer, which it brings down by grasping a leg or the throat with its sharp, curved teeth.

The powerful **Komodo dragon** is now a threatened species, as local people have hunted it for its palatable flesh. Only some 6,000 of these reptiles remain.

Kookaburras are very large birds, up to 45 cm. (18 in.) in length, which are members of the kingfisher family. They live in Australia, New Guinea, and adjacent islands in open, wooded country, feeding on large insects and lizards and nesting in holes in trees. The most familiar species is the laughing kookaburra, *Dacelo novaeguineae*, which is common in many parts of Australia, especially down the eastern side of the continent. Its very loud, almost hysterical, peals of laughter are one of the most conspicuous features of the Australian dawn.

Kraits comprise a dozen species of venomous *snakes, related to cobras, that occur in Southeast Asia and western Indonesia. They often have a colour pattern of black and white (or yellow) bands. They are mainly nocturnal and rather docile and, when threatened, are inclined merely to coil up loosely and make jerky body movements. The banded Krait, *Bungarus fasciatus*, grows to 1·2 m. (4 ft.) in length. Their venom is very toxic and bites are extremely dangerous when they occur. Most species feed on other snakes, though some are fish-eaters.

Krebs' cycle, also known as citric acid cycle or tricarboxylic acid cycle, is the second stage in the biochemical pathway of *respiration. Unlike the first stage, *glycolysis, it takes place when oxygen is available and can be carried out only by aerobic organisms. Krebs' cycle is called a cycle because it involves a series of chemical reactions, each controlled by an enzyme, which forms an interconnected cycle of events. At a specific point in the cycle, a modified form of pyruvic acid (produced in glycolysis) combines with one of the compounds in the cycle. By way of eight separate reactions the original pyruvic acid is broken down into carbon dioxide and hydrogen atoms ($CH_3COOH + 2H_2O \rightarrow 2CO_2 + 8H$). *Adenosine triphosphate (ATP) is produced directly at one point, and the hydrogen atoms resulting from the process are transferred to the *electron transport chain, which produces more ATP.

Krill *Euphausid shrimps.

Kudu are large, lightly striped, African antelopes. The greater kudu, *Tragelaphus strepsiceros*, may reach a height of 1·3 m. (4 ft. 5 in.) at the shoulders, with horns 1 m. (3 ft. 3 in.) long with an open spiral of at least two and a half turns. The female is smaller and lacks horns. A rich fawn in colour, it has a few thin stripes of white on the back. This beautiful, shy, and wary animal is found in East, central, and southern Africa, mostly in hilly or broken country with thorn bushes or tall grass. A browsing animal, it lives in small groups and feeds at dawn and dusk. There is an elaborate courtship and a single calf is born. The lesser kudu, *T. imberbis*, is similar in appearance and habitats, but occurs in East and parts of northeast Africa.

Kwashiorkor is a serious nutritional disease of children fed solely on a diet of maize, prevalent in early infancy in tropical regions. A lack of animal protein causes delayed growth, severe loss of weight, and retention of tissue fluids. Affected children develop distended bellies and swollen legs as well as other symptoms. The disease can be fatal if not treated promptly.

Laburnums are poisonous deciduous trees of the pea family, with pendent inflorescences of yellow flowers. Of the three species in the genus *Laburnum*, the most familiar is *L. anagyroides* of southern Europe.

Labyrinth fishes are a group of related fishes which can breathe air through a convoluted organ in the upper part of the gill chamber. Air is taken in through the mouth and passed over the labyrinth, where oxygen is absorbed by blood in the capillaries. As a result these fishes, which include gouramis, fighting fish, and snake-heads, can live in water which is frequently stagnant, and hence deficient in dissolved oxygen. Several families are represented in fresh waters of southern Asia and Africa.

Lacebark *Bark.

Lacebugs are small, plant-feeding *bugs of the family Tingidae, which have a network of fine sculpturing on the whitish wings and thorax. There are 800 species, found mainly in Mediterranean regions. Their eggs are frequently inserted into plant tissues, and covered with a brown secretion that hardens to a scab. Nymphs are often spiny and lack the adult patterning.

Lacewings are delicate green or brown insects, with two pairs of large, net-veined transparent wings and, often, bright golden eyes. They comprise the order Neuroptera, with around 5,000 known species, including the *ant lions and some species with mantis-like forelegs. Lacewings lay their eggs at the tips of silk stalks on plants, and both adults and larvae are active predators of plant-eating insects and important destroyers of aphids.

Lackey moths *Tent caterpillars.

Ladybirds are *beetles of the family Coccinellidae, with over 5,000 species; they are roundish in outline and have strongly convex wing-covers. They are red, yellow, or black, coloured often in spots or blotches. Both adults and larvae are active predators of plant-eating insects and important destroyers of aphids.

Lady fern *Ferns.

Lamarck, J.-B. P. A. de Monet, Chevalier de (1744–1829), French biologist, among others anticipated *Darwin in conceiving the idea of organic evolution, but accounted for it by the theory that all living organisms are continually trying to improve themselves. Like Darwin, he wrongly believed that characteristics acquired in an individual's lifetime are passed on to its offspring. This was central to Lamarck's theory, but played only a minor part in Darwin's ideas about natural selection.

Lammergeyer: a species of large Old-World *vulture of eagle-like appearance, with a wingspan of 3 m. (10 ft.), fully feathered neck, and dark 'beard' on either side of the beak. Soaring birds of remote mountain regions of Africa,

An adult river **lamprey**, *Lampetra* species, attached to a rainbow trout by means of its wide, sucker-like mouth. Lampreys have an eel-like form and a slimy, scaleless skin. Their eyes are well developed.

Asia, and southern Europe, they are rare everywhere and persecuted by man. They breed on high crags, and are mainly carrion feeders, though suspected of attacking farm stock as well.

Lampreys comprise one of the two classes of *jawless fishes (the others are hagfishes). They are primitive eel-like vertebrates which have seven separate gill-openings each side of their body, and lack pectoral and paired fins. Lampreys are mainly freshwater fishes. Some species migrate to the sea to feed but return to breed in fresh water. They occur in the temperate zones of both Northern and Southern Hemispheres. Many lampreys are parasitic when adult, sucking blood from fishes. The sea lamprey, *Petromyzon marinus*, greatly affected the fish populations of the American Great Lakes once it had become established there, but others do not feed as adults. All have a larval life of several years when they lie buried in river mud. During this time the larvae are called ammocetes.

Lamp shells, or brachiopods, resemble *bivalve molluscs yet are actually related to *bryozoans and form the phylum Brachiopoda, with 280 living species. Their two-valved, chalky shell is not symmetrical, having a large ventral valve and smaller dorsal valve rather than matching left and right halves as in bivalves. Within this shell is a complex, ciliated, filter-feeding organ, quite unlike the flat, sheet-like gills of a bivalve. Most species of lamp shells live attached to rock or to other organisms in the sea, and a few burrow in sand. They are very common as fossils, with 30,000 named fossil 'species', and have considerably declined in number since the Mesozoic Era.

Lancelets, also known as amphioxus, are small, slender, fish-like animals growing up to 9 cm. (4 in.) in length. Their position in animal classification has always been a matter of debate as they have features of both vertebrates and invertebrates and have been claimed to be a link between them. There are about twenty-five species known, of worldwide distribution in shallow seas. They burrow in clean sand where the tidal current is moderate and lie with the head end exposed to catch floating food particles in the fine tentacles around the mouth. These animals are considered to be a subphylum, equivalent to that of the vertebrate subphylum. Along with two other more obscure subphyla, they belong to the phylum Chordata.

Lanner: one of the largest of the true *falcons, widespread throughout the Mediterranean area, Africa, and Arabia. Narrow-winged and speedy, it is a desert predator, feeding on spiny lizards and other ground prey, and also on birds, especially migrants on passage.

Lantanas   *Teak.

Lantern fishes are small marine fishes of the family Myctophidae, living in the open ocean in all areas including polar seas; some 220 species are known. They are distinguished by their blunt snout, large eyes, usually a small dorsal fin near the tail (adipose fin), and the presence of small light organs on the head and body. The arrangement of these light organs is unique for each species, and there is no doubt that they play an important role in species recognition. Fishes of the deep sea, they ascend towards the surface at night although in daylight they will be caught at depths of 700–1,000 m. (2,000–3,250 ft.).

Lantern flies are medium-sized to large, often brightly coloured, tropical *bugs of the family Fulgoridae. In many of the 600 or so species the front of the head is enormously enlarged and inflated which, despite the insects name, does not glow in the dark. They feed by sucking the sap from trees.

Lappet moths   *Eggar moths.

Lapwings are birds of the genus *Vanellus*, and members of the *plover family. Many of the twenty-three species have crests, like the one popularly known in Britain as the peewit, or green plover, *V. vanellus*. Others have wattles on the face, and sharp spurs on the wings, as in the African spur-winged plover, *V. spinosus*. Lapwings are fierce defenders of their breeding territories.

Larches are deciduous conifers of the genus *Larix*, which comprises ten to twelve species native to the north temperate zone. Belonging to the pine family, and closely allied to cedars, they are unusual among temperate region conifers in being deciduous. The European larch, *L. decidua*, produces valuable timber, used for stakes, telegraph poles, and planking. The North American tamarack larch, *L. laricina*, is also grown widely for its timber.

Larks are birds of the family Alaudidae, which includes some seventy-five species. Most are greyish-brown and buff above, usually streaked with black, and pale below with little streaking. The tail is often edged with white.

This elaborately coloured **lantern fly** from Central America, *Fulgora servillei*, has acquired two apt names: 'alligator bug' from its tooth- and jaw-like markings, and 'peanut bug' because of the shape of its head. The genus is named after the goddess of lightning.

The beak, used for feeding on insects or small seeds, is usually longish and sharply pointed. Most species of lark are less than 22 cm. (8 in.) long. They are found in many parts of the world, but are poorly represented in South America. They live in open country, from grassland to desert. Many, such as the skylark, *Alauda arvensis*, have beautiful songs given in flight. They nest on the ground, usually under cover of the edge of a tuft of grass or a stone.

Larkspur   *Delphiniums.

Larvae (singular: larva) are immature stages of any animal, which differ from the adult in appearance and often in habits as well. In land animals, the larva is usually a stage of growth, like a caterpillar, and the transition to the adult form may occur through a resting stage, the pupa. In marine animals, the larva is more often a stage by which the organism is distributed and the adult form is reached before much growth in size has taken place. In some animals, particularly insects, there may be several larval stages, each differing in appearance, before the adult form appears.

Larynx, or voice box, consists of two large pieces of elastic *cartilage, the thyroid and cricoid cartilages. It occurs in the throat of most amphibians and other vertebrates. The thyroid and cricoid cartilages are jointed and can move in relation to one another. Two very small cartilages are jointed to the back of the cricoid and from each an elastic membrane, the vocal cord, passes forward to join onto the thyroid cartilage. A complex series of tiny muscles allows the cords to be voluntarily opened, closed, shortened, and lengthened. In man the larynx is capable of adjustment to form a greater range of sounds than any other animal.

Lateral line: a conspicuous line visible along the sides of a fish. This marks the position of a series of sensory organs which act as vibration receptors, and are used by the fish to detect other individuals from the vibrations they set up in water.

Latex is a fluid produced by some *angiosperms, often confined to a system of long, branched tubes in the phloem. It is a milky liquid containing substances in suspension which, if separated out, have a rubbery quality. Their function is unknown, but in some plants, such as *spurges, the latex is poisonous. Several plant species produce latex which can be used to produce rubber. Opium is obtained from the latex of the poppy, *Papaver somniferum*, and most natural rubber from that of *Hevea brasiliensis*.

Laurel is the name used for several quite distinct but superficially similar shrubs, though strictly it should be applied only to the *bay laurel, *Laurus nobilis*. The laurels in northern Europe are usually species of the cherry genus, *Prunus*, notably cherry laurel, *P. laurocerasus*, and Portugal laurel, *P. lusitanica*.

Lavender is one of about twenty-eight species of *Lavandula*, a genus of shrubs belonging to the mint family. They are found from the Atlantic islands through the Mediterranean to India. *L. vera* is widely cultivated for its flowers and for its oil which is used in scent and soap.

Laver is an edible, pinkish-purple seaweed of the genus *Porphyra*, common world-wide on rocks near the edge of the sea. Its thin, wavy fronds are used as food in Wales, parts of southwestern England, and Japan.

Lawes, Sir John Bennet (1814–1900), English agriculturalist, experimented on his Hertfordshire estate with animal feeds and fertilizers. He studied soil fertility after many years of continuous cropping with a variety of plants and manures, and patented a mineral superphosphate fertilizer (1842). He founded the now famous research station at Rothamsted.

Leaf beetles, or chrysomelids, are a large family of *beetles, with over 20,000 species. The adult beetles are often convex, metallic-coloured species with large 'feet' (tarsi) and bead-like antennae. Their larvae, and often the adults themselves, feed on plants and the family includes many pest species such as the *Colorado beetles. Among their number are the flea beetles, tortoise beetles, and bloody-nosed beetles.

Leaf-chinned bats make up the family Mormoopidae with eight species. All have a plate-like extension of the lower lip, which presumably functions like a nose-leaf in echo-location, since the latter is absent. They occur in Central and tropical South America, as well as in the West Indies. Their preferred habitat is forest or riverine forest, where they feed on insects.

Leaf-cutter bees are small, solitary *bees of the genus *Megachile*. They build their nest cells, often in rotten wood, out of pieces cut from plant leaves; the shape of the piece varies according to the part of the cell for which it is to be used. They are world-wide in distribution.

Leaf-hoppers are *bugs of the family Cicadellidae, with almost 8,500 species distributed worldwide. Both the adults, which can reach 1 cm. (½ in.) in length, and

The wing-covers of this **leaf insect**, *Phyllium pulchrifolium*, from New Guinea, mimic the veins and colour of real leaves. This deception is further enhanced by the twig-like appearance of its broad legs.

Leaves

Leaf shapes and arrangements are many and varied, often being used as one of the key features in the identification and classification of plants. Special terms describe the overall shape of the whole leaf, the shape of its margins, tip, or base, or the arrangement of individual leaflets.

Acuminate	Acute	Aristate	Attenuate	Bipinnate	Ciliate	Cleft	Cordate	Crenate
Cuneate	Cuspidate	Deltoid	Dentate	Elliptic	Emarginate	Entire	Filiform	Hastate
Incised	Lanceolate	Linear	Lobed	Mucronate	Oblanceolate	Oblong	Obovate	Obtuse
Orbicular	Oval	Ovate	Palmate	Pectinate	Peltate	Perfoliate	Pinnate	Reniform
Retuse	Rhomboidal	Sagittate	Serrate	Spathulate	Subulate	Trifoliate	Truncate	Undulate

nymphs suck the sap of plants. They are one of the commonest groups of bugs and are very agile jumpers.

Leaf insects are large, green relatives of the *stick insects with which they make up 2,500 or so species of the order Phasmida. The body is flattened and bears leaf-like extensions which closely mimic the plants on which they feed. They are confined to the Old World tropics and include *Phyllium crucifollium*.

Leaf-nosed bats, of the family Hipposideridae, have a well-developed nose-leaf, which forms part of the echolocation system. Of the sixty-one species, the nose-leaf is most elaborately developed in the flower-faced bat, *Anthops ornatus*. Leaf-nosed bats are confined to the tropics and warmer parts of the Old World. They are typical insect-eating bats and often very common. They roost in caves and houses in groups of several hundreds. The name leaf-nosed bats is also given to some species of New World tropical fruit bats and the false vampires.

Leather-jackets *Crane flies.

Leaves are outgrowths from the stem of a plant which are usually green and flattened. They form the foliage of a plant and act to trap sunlight to be used in *photosynthesis. A leaf consists typically of a petiole or stalk and a thin sheet of cells strengthened and served by veins. Its surface, particularly that away from the light, is covered with openings, called stomata, which regulate the movement of gases such as carbon dioxide, oxygen, and water vapour, in or out of the plant.

Leaves can be simple, with a single leaf blade, or compound, with several leaflets. Their surface can be naked, hairy, or waxy depending upon the habitat in which a plant is adapted to grow. Leaves can be modified to form protective scale leaves, or may be lost. Many desert cacti lack leaves, but have green stems through which they photosynthesize.

Leeches are mostly found in fresh water, where they seek other animals from which they extract a meal of blood. They are one of the three main classes of *annelids, but lack clear segmentation and have a sucker at either end. They are usually flattened rather than cylindrical, though this may not be true when they are bloated after a large meal. Most leeches prefer invertebrates or cold-blooded hosts such as fishes. Relatively few will take blood from mammals, but the famous medicinal leech, *Hirudo medicinalis*, does so, and has a powerful anticoagulant to keep the blood flowing freely. It was used extensively in the past for bleeding patients. Leeches have razor-like jaws inside the front sucker and are believed to anaesthetize the host's skin before biting.

Leeks belong to the same family as the *onion and originate in Near Eastern and Mediterranean regions. They have been cultivated as a vegetable since ancient times for their solid, white leaf-bases which form elongated, cylindrical bulbs.

Leeuwenhoek, Anton van (1632–1723), Dutch naturalist, made a small, single-lens microscope for each specimen that he examined, and was the first to observe bacteria, protozoa, and yeast. He accurately described red blood corpuscles, capillaries, striated muscle fibres, spermatozoa, and the crystalline lens of the eye.

Legionnaires' disease is an acute feverish illness caused by a small bacterium that can grow in water-tanks. Outbreaks have occurred from contamination of air-conditioning plants or the water-supply in large buildings. The most commonly diagnosed type of illness is a serious lung infection. The condition takes its name from the first recognized outbreak, at a convention of the American Legion at Philadelphia in 1976 when 182 cases were reported.

Legs are limbs modified for walking. Any slender appendages, usually jointed, used by animals in locomotion are loosely referred to as legs. The simplest form of leg occurs in the *arthropods, whose name literally means 'jointed legs'. Those of insects, for example, are made of five sections which, starting with the basal section, are called the coxa, trochanter, femur, tibia, and tarsus.

The legs of vertebrates follow the basic pattern of having two main sections to the leg, jointed in the middle at the knee. The larger section is formed of the femur, or thigh-bone, which is jointed to the hips. The lower section is composed of two bones, a thick tibia, or shin-bone, and a thinner bone of the outside part of the leg, called the fibula. The bases of the lower leg bones are jointed to the foot. In mammals, including man, there is a small bone called the patella, or kneecap, at the joint between the upper and lower halves of the leg. The leg is pulled forward by powerful thigh muscles, and back by the hamstring muscles. The 'hamstrings' are tendons from the upper muscles which pass beneath the knee to the lower part of the leg. The Achilles tendon extends from the calf muscles in the lower leg to the foot.

Legume is a name commonly applied to plants of the *pea family that are used as food crops. Legumes are rich in protein; the pods, seeds, or leaves of many species have always been an essential part of man's diet in virtually every part of the world. Those which have edible seeds, such as beans, chick-peas, and soya beans, are called grain legumes, or pulses. The proteins of these pulses contain a different range of amino acids to those contained in cereal proteins, and to a large extent complement such true grains. The staple diet of many different peoples consists of a mixture of cereal and pulses. Other legumes such as alfalfa and clover are economically important fodder crops, or are used to improve pasture quality.

All legumes have the ability to convert nitrogen from air into compounds suitable for uptake by the plant, and ultimately protein synthesis. This ability is unique among flowering plants and conferred by their possession of the bacterium *Rhizobium* in root nodules. This micro-organism is able to 'fix' nitrogen from a gaseous form into nutrients suitable for uptake into the plant. This example of *symbiosis is the reason why legumes are protein rich and why alfalfa and clover enrich soils depleted of nitrates.

Leidy, Joseph (1823–91) was the pre-eminent American anatomist of his day. His monographs on the fossil fauna collected by the geologist F. V. Hayden in the far west of the USA are landmarks in palaeontology. He was the first to identify extinct species of the horse, tiger, rhinoceros, camel, and sloth in the USA. He also broke new ground in parasitology with the publication of *Flora and Fauna within Living Animals* (1853).

Lemmings are part of a large family of 110 species, which includes ninety-nine species of *vole and mole-vole. Of the eleven species of lemming the Norway lemming, *Lemmus lemmus*, is the best known. Other species occur throughout the Arctic in both the New and Old Worlds. They are about the size of a stocky rat, with dense fur and small ears. They live in steppe regions, where they construct deep burrows, foraging for vegetable matter in the winter along tunnels in the snow. Lemmings are most famous for their so-called migrations, which are, in fact, mass movements following build-ups of population. The lemmings peak in numbers about every three or four years, probably in relation to over-exploitation of the food resources. Migrating lemmings sometimes meet the sea, and attempt to swim across what, for all they know, is a narrow channel. Mass drownings may occur, giving rise to the myth of deliberate suicide.

Lemon: one of sixty species of *Citrus* tree which produces distinctively shaped fruit with nipple-shaped ends. The juice of this fruit contains about four times the level of citric acid found in the sweet orange, making it unsuitable to be eaten raw. As a crop, it is intolerant of temperature extremes and is grown mainly in regions with a Mediterranean-type climate. It is commercially exploited as a major source of citric acid.

Lemon sole: a species of *flat-fish which is not a sole but is related to the plaice and flounder. It lives on the Atlantic coast of northern Europe and in the North Sea, on mud, sand, and gravel bottoms, at depths of 40–200 m. (130–650 ft.). It has a small head and mouth which limits the size of its prey; mostly it feeds on worms and small crustaceans. It is a moderately valuable commercial fish, usually caught in trawls.

Lemurs are primates found only in the forests of Madagascar. The destruction of these forests threatens many species with extinction. Lemurs vary in size from small, mouse-like creatures to animals the size of a domestic dog. Some of the twenty-two species can stand erect and walk or hop when they descend to the ground; others walk quadrupedally. They have large, dark eyes that look forward over the small, slightly pointed nose. All are timid creatures that feed on insects and fruit. Some species feed at night, while others are active during the day. They breed once a year and produce a single offspring; the mother carries it spreadeagled on her breast or on her back. They are the living representatives of an ancient group of primates called prosimians, and have found a niche in Madagascar due to the absence of monkeys.

Typical lemurs include the ruffed lemur, *Varecia variegata*, the lesser mouse lemur, *Microcebus murinus*, and the fat-tailed lemur, *Cheirogaleus medius*. All of the lemurs are tree-dwellers except the ring-tailed lemur, *Lemur catta*, which lives in a dry, rocky habitat almost destitute of trees. The *aye-aye and *indri are also lemurs.

Lentils are annual *legumes which originated in the Middle East and were among the first domesticated crops. They are now widely grown in Europe, Asia, and America for their protein-rich seeds. The bright orange split lentils sometimes seen are prepared by removing the seed coat and dividing the seed into two.

Leopard: one of the smaller and most widespread of the big *cats, found throughout Africa, Asia Minor, India, China, Korea, and Siberia. It can reach a length of 2 m. (6 ft. 6 in.), 90 cm. (3 ft.) of this being tail. The coat is a dark shade of yellow marked on the back and flanks with black rosettes, the centres of which are dark yellow. The texture of the fur varies from thick and rich to short and coarse, depending upon the climate in which the animal

Ring-tailed **lemurs**, *Lemur catta*, are gregarious, living in groups of between five and thirty, with both male- and female-dominated hierarchies. The tail is used for balance and, in different positions, serves as a strong visual symbol to other troop members.

For the **leopard**, *Panthera pardus*, a tree is not only a strategically safe place to relax and shelter from the heat after exertion or during digestion, it is also a convenient larder. Skilled hunters, leopards have been known to jump from trees on to prey.

lives and the subspecies. It preys upon deer, monkeys, dogs, pigs, porcupines, and other such animals. Leopards are solitary except in the breeding season. After a gestation period of about three months, two to five cubs are born, covered with fur but blind for the first ten days of life. The family remains together for about two years until the cubs are nearly mature. One individual has lived in captivity for twenty-three years, but the lifespan of the leopard is shorter in the wild.

Leopard moths, which belong to the same family as goat moths, occur worldwide. The adults are stout-bodied and fluffy, with coal-black spots on the white wings and thorax. The leopard moth caterpillar has a wedge-shaped head and burrows through the wood of various trees, taking two or three years to mature. It is occasionally a pest of fruit trees in Europe and North America.

Leopard snake: a species of harmless *colubrid snake, up to about 1 m. (3 ft. 3 in.) in length, found in parts of southern Europe, and Asia Minor. It has a distinctive pattern of reddish-brown blotches or stripes and is widely regarded as the most beautiful of the European snake species. It feeds mainly on rodents and is usually encountered in rather rocky habitats.

Lepidodendron is an extinct genus of *pteridophyte plant which grew as a giant tree in the coal-bearing swamps around 300 million years ago. Its tall, unbranching trunk reached over 30 m. (98 ft.) in height and 2 m. (6 ft. 6 in.) in diameter at the base. At the top it bore a crown of branches carrying true leaves and reproductive, spore-bearing cones.

Leprosy is a very slowly progressive infection caused by the bacterium *Mycobacterium leprae*. The organism grows in the skin and the lining of the nose. It causes disfiguring swellings, and affects nerves close to the surface of the skin, producing loss of sensation, and damage to the extremities. The condition is infectious, but the risk to contacts is low. It used to be common in Europe, and the reasons for its disappearance are obscure. Treatment can halt, but seldom cure, leprosy.

Lesser panda   *Red panda.

Lettuce is probably the best-known of all green salad vegetables, and has been cultivated for so long that its origin is uncertain. It belongs to the sunflower family and varieties have been derived from the species *Lactuca sativa*. These can be divided into two major groups. Cabbage lettuces have roundish heads, and their leaves vary in texture from soft to crisp. Cos lettuces have more elongated, upright heads.

Leukaemia is a cancer of the white *blood cells. There are several types of leukaemia, each affecting a different type of white blood cell. For example, that affecting the cells which produce lymphocytes is called lymphatic leukaemia. Different types tend to affect different age groups and progress at different rates. The one which particularly affects children responds in a substantial proportion of cases to prolonged treatment with drugs and radiation. Damage occurs because the cancerous white blood cells are produced at the expense of normal red blood cells, platelets, and normal types of white blood cell. This leads to an increased chance of infection and *anaemia.

Liana is a name used for any woody climbing or scrambling plant which supports itself on bushes or trees. Lianas are most conspicuous in tropical rain forests. They include *rattans, which scramble upwards using hooks. Others

This brownish-orange **lichen**, often the initial colonizer of bare substrata, here encrusts rocks near the Antarctic shore. The green growth is a species of moss.

may have twisting stems or leaf-stalks, sucker-pads, or spines. Of the small number of temperate lianas, clematis, ivy, and honeysuckle are best known.

Lice is an imprecise name for any small, flattened, generally wingless insects which are external parasites of birds and mammals. There are three distinct orders of insect which include species that are given the name of louse. They are the *booklice, *biting-lice, and *sucking lice. The name is usually associated with the sucking lice species, *Pediculus humanus*, which infest man.

Lichens consist of two different organisms in partnership (*symbiosis): a fungus which forms the main body, and an alga whose cells are contained in a layer within the fungus. By *photosynthesis, the green or blue-green alga makes sugars which are used by the fungus for food. The combined organism grows more slowly and is longer lived than either the fungus or the alga by itself. Some lichens make chemical compounds that neither partner can produce alone, including some which are used as dyes.

Lichens grow on exposed surfaces such as tree trunks, rocks, and walls; they are sensitive to atmospheric pollution by sulphur dioxide gas. Some grow as crusts on surfaces but others form spreading branches. The fungal partner produces spores, and the lichen partnership is reproduced and spread by algal and fungal cells in small dry clumps which blow away from the parent lichen. They grow exceedingly slowly and are one of the few organisms to have colonized Antarctica.

Life as we know it probably first arose on this planet about 3 to 4 thousand million years ago. Many explanations of how this occurred have been proposed, although the following sequence of events is most widely accepted at present.

At that time, the Earth's atmosphere was composed largely of a mixture of ammonia, methane, hydrogen, and water vapour. It has been shown that simple organic molecules can be synthesized if mixtures of these gases are exposed to gamma radiation, electrical discharges, or ultraviolet light under various physical conditions. All of these sources of energy are likely to have occurred in the atmosphere. Having been formed, these molecules could have fallen to the ground in rain to accumulate in the rivers and seas.

The organization of these molecules into cells capable of obtaining energy and reproducing themselves is the most difficult step to explain. In some way, suitable combinations of molecules came together, became surrounded by a type of membrane, and acquired integration. These early organisms must have required energy to grow and multiply. They probably took in organic molecules from their watery surroundings and then broke them down by anaerobic *respiration (since the atmosphere probably lacked free oxygen) to release energy. The first *reproduction was presumably simple asexual division.

The subsequent development of a huge variety of life forms from these beginnings is explained by *evolution.

Life cycles describe the phases of an organism's life, and the manner in which it produces the next generation. This includes the production of *gametes and their fusing during *fertilization and the transition from an *embryo stage to the final mature adult stage.

Most organisms have a life cycle involving a single adult stage preceded by egg and juvenile, or larval, stages. The stages within each life cycle are adaptations for optimal survival of offspring. Most vertebrates have young which resemble their parents in most respects except sexual maturity. Many invertebrates have highly specialized immature stages, such as the caterpillar stage of some insects, or the floating planktonic larvae of crabs and marine molluscs.

In plants, including mosses, ferns, and some algae, alternation of generations may be observed—they have two types of 'adult' plant. In mosses, these consist of a leafy gamete-producing plant called a gametophyte, and a spore-producing plant called a sporophyte. In ferns, the dominant plant is the sporophyte, while the other type, the gametophyte, is a small fleshy plant.

Aphids also show alternation of life cycle with regard to their mode of reproduction and structure of the adult.

Lignum vitae is the wood of six species of the tree genus *Guaiacum*, native to tropical America. All are evergreen trees and the common name of lignum vitae is often applied to *G. officinale*. This has a greenish-brown, hard, and heavy wood from which medicinal guaiacum is obtained.

Lilacs are small trees, of the genus *Syringa*, with some thirty species, all with scented flowers. They are members of the olive family. Common lilac, *Syringa vulgaris*, has white or purple flowers in conical, drooping bunches. Persian lilac, *Syringa × persica*, is a hybrid with common lilac as one parent. This name is also used for an Indian tree, *Melia azedarach*, belonging to the mahogany family.

This tree is known by several names including China tree, pride of India, bead tree, and Indian lilac tree.

Lily is the common name for many members of the family Liliaceae, whose 3,500 worldwide species make it one of the largest families of monocotyledonous flowering plants. This family includes many popular garden plants such as tulips, *Tulipa* species, hyacinths, *Hyacintha* species, and the true lilies, *Lilium* species. It also includes vegetables such as onions, *Allium* species, leeks, *A. porrum*, and garlic, *A. sativum*. Most members of the family arise from bulbs, corms, or fleshy roots.

The true lilies, with some eighty species and many hybrids, include the Madonna lily, *L. candidum*, the martagon or turk's cap lily, *L. martagon*, and the tiger lily, *L. tigrinum*. Other species within the lily family, which are given the common name of lily, include the African lily, *Agapanthus* species, the glory lily, *Gloriosa rothschildiana*, and the daylilies, *Hemerocallis* species. The name is also applied to unrelated plants such as *water-lilies and the Guernsey lily, *Nerine sarniensis*

Limes (citrus fruit) are trees belonging to the genus *Citrus* which produce a thick-skinned greenish fruit. These are the smallest of the commercial citrus fruits, replacing the lemon in tropical areas. The juice is very sour, containing up to 8 per cent citric acid, which may be extracted commercially. The fruit dries out too rapidly to be transported far and very little is exported. All citrus fruit juices are rich in vitamin C and lime juice was once drunk by sailors in the British navy to prevent scurvy—hence their nickname of 'Limeys'.

Lime trees, or linden trees, belong to the genus *Tilia*, of which there are fifty or so species, mostly confined to the north temperate zone. They have heart-shaped leaves and sweetly scented flowers which produce large amounts of nectar to attract insect pollinators. Their soft, pale wood is used for plywood and brush handles, that of *T. americana* being the basswood of commerce. Limes form part of the family Tiliaceae, which also includes many species whose bark is a source of fibre or jute, for rope making and matting.

The lime fruit is produced by unrelated trees of the citrus family.

Limpets are conical gastropod *molluscs common on rocky shores. They cling tightly to the rock using a broad,

The underside of *Cellana tramoserica*, a **limpet** of southeastern Australian shores, showing its muscular foot, which is used to cling to the rocky substrate.

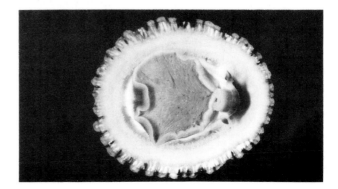

muscular foot, resisting wave-action and predators. When the tide covers them, they move over the rocks to browse on algae and always return to the same precise home-site. The trail left by their rasping 'tongue' is a familiar sight on otherwise algae-coated rocks. There are many species around the world.

Limpkin: an American water-bird, with a single species in the Aramidae family. Glossy olive-brown flecked with white, it inhabits wooded marshes, feeds on snails (leaving the shells undamaged), and utters loud wailing cries, mainly at night.

Linacre, Thomas (1460?-1524), English physician and classical scholar, has been called the 'restorer of learning' in England. He wrote textbooks on Latin grammar, and his students of Greek included Thomas More and probably Erasmus. Linacre's translations into Latin of Galen's Greek works on medicine and philosophy brought about the revival of interest in anatomy, botany, and clinical medicine in Britain. He founded medical lectureships in Oxford and Cambridge, and was the first President and one of the main founders of the Royal College of Physicians.

Ling   *Heather.

Ling (fish): an elongate fish of the cod family, with two dorsal fins, a single anal fin, and the typical chin barbel. It is a moderately deep-water fish living on rocky grounds

A portrait of Carl **Linnaeus** by Alexander Roslin, 1775, hanging in the Palace of Versailles. Renowned for his high fees, on this occasion Roslin was honoured to paint the great naturalist for 'love'.

and especially on wrecks, and growing to a length of 2 m. (6 ft. 6 in.). It feeds almost exclusively on fishes, ranging from northern Norway and Iceland, southwards to the Bay of Biscay.

Linnaeus, Carl (1707-78), Swedish naturalist, was the founder of modern systematic botany and zoology. Interest in the variety of stamens in flowers led him to devise a system of classifications of plants, dividing them into twenty-four classes on this basis. Endowed with a strong sense of order, he described over 7,000 plants, giving them binomial Latin names. His classification system was based upon the concept of a species being the smallest unit. The species names given by Linnaeus usually describe some feature or the origin of the plant. Species which shared sufficient features in common were placed in a genus. The use of the generic name and the specific name, together, gave a particular organism a unique binomial name. It was a practical classification and the use of Latin made it internationally acceptable. This was the basis of modern plant nomenclature, although his classification was later superseded by that of Jussieu.

In the tenth edition of *Systema Naturae* (1758), he recognized the importance of mammary glands for suckling young, and formed a group, 'mammalia', which included man (called *Homo sapiens*). His classification of animals on the whole was less satisfactory than that of plants as he paid little attention to internal anatomy. Even so, his influence was enormous.

He travelled and corresponded widely and inspired his students to go on collecting expeditions over much of the world. He was an astute observer of nature and his experiments in controlling insect pests using other insects led him to the idea of a 'balance in nature' where there is a continual 'war of all against all'.

Linnets are common *finches found in much of Europe and western Asia, where they occur in open, bushy country. The female is rather sparrow-like, but the male has a brown back, a greyish head with a red crown and a red breast. The species *Acanthis cannabina* is commonly kept as a cage-bird because of its pleasing, melodious song.

Lions, the largest of the big *cats, are now restricted to Africa, except for a few Asiatic ones surviving in a reserve. The African lions live in prides of three to thirty individuals on the veld and in scrub country, each pride with its own territory. Prides can cover 48 km. (30 miles) in a night. When chasing, lions can reach a speed of 48 km./hour (30 m.p.h.) over a short distance. Their prey includes zebra, waterbuck, and antelope, but they will eat anything from carrion to fish.

The male lion stands about 90 cm. (3 ft.) at the shoulders; the lioness is a little smaller. The mane on the shoulders and head of the male makes him seem even larger than he is. After a gestation of about 105 days, two or three cubs are born, each about 30 cm. (1 ft.) long.

Lipids are a major group of natural organic compounds comprising fats, oils, waxes, phospholipids, and steroids. The molecules of all these substances contain a high proportion of CH_2 groups and therefore are not very soluble in water, but are readily soluble in organic solvents such as ethanol and chloroform. Fats and oils are built up from *fatty acids and glycerol and are commonly used for energy storage. Waxes are formed from fatty acids and

From the age of about five months, **lion** cubs start to explore without their mother. Here the adult male falls prey to the play-acting of a future killer, though cubs will be one year old before being taken on their first real hunt by the lioness.

complex alcohols and provide a waterproof surface layer on the leaves of plants and the exoskeletons of insects; bees use wax for building their honeycombs. Phospholipids have a similar structure to fats, but one of the fatty acids is replaced by a phosphate group which links the molecule to a complex alcohol. This creates a molecule one end of which dissolves easily in water (hydrophilic), while the other end is insoluble in water (hydrophobic).

Liquorice is a perennial herbaceous plant, *Glycyrrhiza glaba*, which is unusual among legumes in that it is cultivated for the strongly-flavoured extract from its roots, as opposed to using the seed or pods. Since ancient times it has been used to mask unpleasant tastes in foods. More recently, its flavour has been used in sweets, and also in various drinks and cough mixtures. It is grown mainly in the Mediterranean region, where the dried roots are often chewed.

Lister, Joseph, first Baron (1827–1912), English inventor of antiseptic surgery, was influenced by Pasteur's discovery that infection is due to micro-organisms, and at first chose neat carbolic acid as an antiseptic in his operating room. When he found that this caused burns and irritation, he diluted it to produce an effective disinfectant and advocated the aseptic routines employed in hospitals today.

Litchi, or lychee: an evergreen tree about 10 m. (33 ft.) tall, and one of 2,000 species making up a family of tropical and subtropical trees, shrubs, and lianas. The species, *Litchi chinensis*, is a native of southern China and produces a round, thin-skinned, spiny fruit about 3 cm. (1¼ in.) in diameter. The fruits may also be dried, when they are known as litchi nuts. Cultivation throughout the tropics has been only partially successful as the tree requires a cool, dry season to fruit well and does not prosper in moist, lowland regions. The rambutan has a very similar fruit with a prickly red skin. This is sometimes known as the hairy litchi.

Little owls are small grey-brown *owls of the genus *Athene*. One of the three species, the little owl, *A. noctua*, is widely distributed over Europe, Asia, and Africa, and has been successfully introduced into New Zealand. Evening and early-morning hunters of open woodlands and plains, they have taken readily to parkland, towns, and farms, where they hunt small rodents and insects. Their calls are

An operation, dated 1882, in which **Lister's** antiseptic spray was used. Summing up his work, the *British Medical Journal* said that he: 'saved more lives . . . than all the wars of the nineteenth century together had sacrificed'.

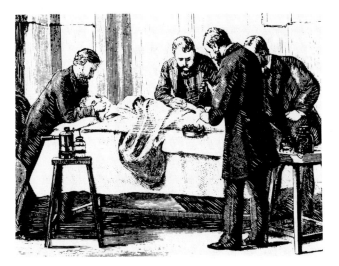

shrill whistles and shrieks and they nest in trees, buildings or on cliffs, laying four or five eggs.

Liver: both the largest and most important gland in vertebrates. It is the main organ of *metabolism. It lies in the abdominal cavity, near the stomach, and receives blood from the intestines. This blood carries the products of digestion, either to be stored or broken down into products readily available to tissues. It also receives blood from the *spleen with broken-down products from the digestion of red blood cells. These, together with similar products formed locally, are discharged into the intestine as *bile pigments. Bile salts formed in the liver are also discharged into the intestine with the pigments, and aid digestion. In man, excess alcohol may indirectly lead to irreversible liver destruction and early death.

Liver flukes are *flukes which are entirely parasitic. There are two species commonly called liver flukes: the Chinese liver fluke, *Opisthorchis sinensis*, and the sheep liver fluke, *Fasciola hepatica*. Both are leaf-like as adults, without obvious heads and a sucker at one end. They live within liver tissues of vertebrates, hanging on with the suckers. The Chinese fluke has two intermediate hosts before it reaches man. Eggs released from the gut of man hatch to infect a secondary carrier, usually a snail. The larvae multiply, then a free-swimming stage leaves the snail to seek a new vertebrate host such as a fish. Man can therefore pick up liver flukes by eating undercooked meat or fish. The sheep liver fluke uses a snail as a single intermediate host for the primary host, the sheep, is infected by eating grass with liver fluke larvae encysted on it.

Liverworts are plants related to mosses which occur in moist terrestrial, or aquatic habitats and may be *epiphytes. They are *cryptogams in which the gametophyte (sex-organ bearing stage) is predominant and it may be either leafy, like a delicate moss plant, or simply flat and spreading. Most grow on the surfaces of soil, trees, or rocks, but a few grow in water. The spore-bearing stage (sporophyte) is shorter-lived than that of mosses, lasting only a day or two. Its capsule releases the spores by splitting open into four segments, and not by scattering them from one end as most mosses do.

Livingstone daisy   *Mesembryanthemums.

Lizards, of which there are about 3,000 living species, comprise a highly diverse array of *reptiles distributed throughout the world except for some cold regions. They have a wide range of life-styles: many are terrestrial, whereas others are adapted for tree-dwelling, aquatic, or burrowing existences. Most have well-developed limbs, although these are reduced or lost in some snake-like species such as glass snakes, the slow worm, some species of skinks, and pygopodids. Their tails are usually quite long and many species, if caught, are able to shed their tails, a mechanism called autotomy. This act often distracts the predator for long enough to allow the lizard to escape.

Some of the main lizard families are: agamids, chameleons, monitors, pygopodids, iguanas, skinks, geckos, and slow-worms.

Llama: a New World species of the camel family, domesticated like the alpaca and related to the guanaco and vicuña. Found at altitudes of 2,300–4,000 m. (7,600–13,000 ft.), it is the principal beast of burden in the Andes of Peru, Bolivia, Argentina, and Chile though only stallions are used to carry loads, of up to 80 kg. (176 lb.) for up to 30 km. (19 miles) per day. The llama provides meat, wool, fat for candles, and valuable hide. It is about 1·2 m. (4 ft.) tall at the shoulders and has a thick fleece, which varies in colour from white to black. The neck and head are long, and the eyes and ears quite large. A foal is born after a gestation period of eleven months, and is suckled for four months.

Loaches are a family of small, eel-like, freshwater fishes, related to the carp family. Loaches are most abundant in tropical Asia, but their range extends across Europe and temperate Asia and part of northeast Africa. They are adapted for life on the river-bed, either hidden under stones or burrowing in mud, and having numerous barbels around their mouth. There are about 150 species.

Lobelias are a genus of annual or perennial plant species, both herbs and shrubs, which have a worldwide distribution. The flowers may be red, yellow, or white, though the predominant colour is blue. In the mountains of central Africa and South America are species which grow to 2 m. (6 ft. 6 in.) or more in height. In contrast *Lobelia erinus*, the South African parent of numerous garden varieties, is a dwarf species only 10 cm. (4 in.) in height. Lobelias form a family of some 1,200 species which includes other genera with a wide range of small growth forms. Some are epiphytes, others resemble palms, and yet others have developed needle-like leaves.

Loblolly, or Frankincense pine: one of the yellow pines of North America, native to the southeast USA, though widely planted elsewhere, especially in Australia. It may reach some 30 m. (110 ft.) in height and its needles are arranged in bunches of three. The cones are egg-shaped and the seeds are four-angled and winged. The wood is used for many purposes in the USA and a turpentine is also extracted from the tree.

Lobster moth: a species of large moth belonging to the same family as the puss moth. It has a bizarre, reddish-brown, lobster-like caterpillar, with pointed humps on its back. When alarmed, the caterpillar throws back the head, raising the unusually long, red, thoracic legs, and throws forward the swollen hind-end, which bears a pair of stiff filaments.

Lobsters can weigh up to 23 kg. (50 lb.), which makes them the most massive of all living *arthropods. They belong to the *crustacean order called decapods (crabs, prawns, crayfish), and have eight walking legs and two large pincers. Unlike crabs, lobsters and *crayfish have an elongate body; lobsters also have unequal pincers, the larger one being used to crush shellfish.

Lobsters usually lead a scavenging or carnivorous existence and they can retreat rapidly using an emergency flicking action of the abdomen when threatened. They live in crevices in rocky sea-beds and corals, which explains their willingness to enter lobster-pots, probably explored as potential homes. A lobster takes several years and many skin changes to reach maturity, and the females then breed at two-year intervals. They can produce at least 100,000 eggs, which they carry around beneath the abdomen for many months before releasing them as tiny larvae.

Desert **locusts** darken the sky in western Ethiopia during the plague of 1958. This particular swarm, one of the biggest ever seen in East Africa, covered an area of 1,000 sq. km. (386 sq. miles) and contained an estimated 40,000 million individuals.

Locusts are *shorthorn grasshoppers which exist in two phases, the solitary and the migratory, each differing in appearance and behaviour. In the solitary phase, the adults and the young tend to be green and fat and individuals lead almost solitary lives. The migratory-phase individuals are darker and gather in groups when quite young. These hopper bands eventually mature and swarms are formed. Migratory swarms left uncontrolled will totally destroy all vegetation, but eventually, after bad breeding seasons, they revert to the solitary phase. The best-known locusts are the migratory locust, *Locusta migratoria*, which occurs from the Mediterranean region to New Zealand, usually in areas of good rainfall, and the desert locust, *Schistocerea gregaria*, which occurs in dry areas, especially of North Africa and the Near East; this is the locust of biblical plagues. The red locust, *Nomadacris septemfasciata*, of South Africa, can also be destructive, as can species in North and South America.

Locust tree   *Robinia.

Logan, James (1674–1751), Irish-born scholar, emigrated to America as an assistant to William Penn in Philadelphia. Later, while Governor of Pennsylvania, he performed botanical experiments which showed con-clusively that pollen was essential for fertilization in corn. He also observed that pollination was effected by wind-blown pollen from one maize plant to another. Linnaeus named the plant genus *Logania* in his honour.

Loganberries are believed to have originated in about 1890 in California, in the garden of Judge J. H. Logan, as a natural hybrid, probably between the raspberry and the dewberry.

Longhorn beetles are characterized by the length of the antennae, which are as long as, or longer than, the rest of the animal and generally longer in the males than in the females. The body is elongate, often colourful, and varies in size from small to very large. Indeed, they include the longest known beetles such as the South American species *Titanus giganteus*, which can reach 20 cm. (8 in.) in body length. The adults, if they feed at all, eat leaves or pollen, but the larvae feed internally on plants, often in wood, and their lifespan may be three or more years. They occur throughout the world and many of the 20,000 species are injurious to timber.

Longhorn grasshoppers, also known as bush crickets, and katydids in America, have antennae which are longer than their bodies. The front wings of the males are modi-fied for sound production, while the females have promi-nent scimitar-like ovipositors. Their stridulation (song) is loud and frequent, much more so than that of *shorthorn grasshoppers. They are almost all plant-feeders, often on trees or shrubs, but also in marshy country, and even in

This **longhorn grasshopper** of Costa Rica is excellently camouflaged to blend with its environment. The American name katydid for species such as this derives from the sound of the male's nocturnal stridulations.

caves. They insert their eggs into slits cut into plants. They comprise some 5,000 species in the family Tettigoniidae and are found worldwide.

Longhorn moths have long, thread-like antennae, often many times the length of the body, especially in males. They belong to the family of moths known as Incurvariidae. They are small, often brilliantly metallic coloured, and fly by day. Their tiny caterpillars feed by mining within leaves or flower heads; older ones make cases of leaf fragments.

Long-tailed ducks, known as old squaws in North America, are a species of small black and white sea duck, abundant throughout the Arctic. They dive to remarkable depths and fly very fast. The drake has long needle-like tail feathers. Both sexes have different seasonal plumages.

Loofah: a member of the pumpkin family along with cucumbers and marrows. The young fruits of this Asian climbing plant are eaten fresh or cooked. The ripe fruit has a complex network of fibres which are exploited in the familiar sponge-like bathroom loofah. It is also used in industrial filters, like those of diesel engines. The loofah material also has shock- and sound-absorbing properties, useful, for example, in military helmet linings.

Loons *Divers.

Loopers are caterpillars of the moth family Geometridae, which includes the carpet moths. The name geometer means ground-measurer and describes how they move by arching or looping the body, bringing the hind claspers or prolegs up to the thoracic, or true, legs and then moving these forward. Many geometer caterpillars rest with the body attached to a twig by the claspers and held erect, in imitation of a twig.

Loquats are small evergreen trees from China. They belong to the genus *Eriobotrya*, which is included within the rose family, like their close relatives the apples. They are cultivated in the Mediterranean region, Japan, and India, in addition to their native country. They produce smallish, yellow, pear-shaped fruits.

Lorises are two species of small, nocturnal primates, with enormous, forward-facing eyes and well-developed, grasping hands. They are omnivorous, eating insects, fruit, leaves, birds' eggs, small lizards, and mammals. The adults sleep in the hollow of a tree or clasped to the fork of a branch. They are solitary or live in pairs. A single offspring is born after a gestation period of three to five months. The baby is carried by the mother until it becomes independent. The slender loris, *Loris tardigradus*, found in southern India and Sri Lanka, is 23 cm. (9 in.) long. The slow or grey loris, *Nycticelous coucang*, dwells in the forests of southern Asia and the East Indies region. It moves more slowly than the slender loris. Dark brown in colour, it is 33 cm. (13 in.) long.

Lotuses are two species of *water-lilies of the genus *Nelumbo*. The pink-flowered sacred lotus, *N. nucifera*, is native to eastern Asia, though it has been introduced into other parts of the tropics; the yellow-flowered American lotus, *N. lutea*, is native to the southern USA. The flowers and leaves of both species are held well above the water and their fruits are carried in a swollen receptacle which, after ripening, breaks away from the stem and floats; when this decays, the fruits are released and sink to the bottom. The fruits and rhizomes are used as food in some parts of the world.

Louse-flies are highly modified true *flies of the family Hippoboscidae, which are external blood-sucking parasites of birds and large mammals. They are flattened and leathery, with long, curved claws on the feet. Some species are winged; others, such as the sheep-ked, *Melophagus ovinus*, are wingless and tick-like. Females produce only one fully developed larva at a time, and this immediately pupates.

Louseworts are small semi-parasitic plants usually found on grasses growing in wet meadows. The generic name *Pedicularis* is from the Latin *pediculus*, a louse, because it was assumed they were responsible for lice in sheep. The

500 or so species are widely distributed, chiefly in the northern temperate regions of the world. Together with figworts, antirrhinums, and others, they belong to the foxglove family.

Love-birds comprise nine species of the parrot family. All are African mainland species, except one which lives in Madagascar. All are small, up to 16 cm. (6 in.) long, and mainly green, though many have white, red, or dark brown on their heads. The females of several species carry grass and other materials for nesting by tucking it in among their feathers rather than carrying it in their beaks. They were given their name because they live in pairs and spend a great deal of time preening each other.

Love-lies-bleeding is a plant typical of the family Amaranthaceae, which also includes cockscombs (*Celosia* species). This family of over 900 species has a mainly tropical distribution, but provides many popular garden species for temperate regions. Love-lies-bleeding, which is native to India, may reach a height of 1·5 m. (5 ft.) or more. The red flowers are borne in long pendent chains; hence the common name.

Lucerne (plant)   *Alfalfa.

Lugworms are the familiar polychaete *annelids dug by fishermen everywhere as bait, from sandy beaches where their worm-casts betray them. They live in U-shaped burrows and feed by swallowing sand and detritus. They void the sand, as the cast at the tail-end of the burrow, roughly once every forty minutes. Lugworms are related to bristle worms, but lack the lateral paddles; instead they bear frilly, red gills, to facilitate breathing in stagnant mud or sand.

Lumbago   *Discs.

Lumpsucker: a species of almost spherical fish with rows of large, coarsely spined plates on the sides and back. A North Atlantic fish, it is found from the shoreline to a depth of 200 m. (650 ft.). It has a round sucker-disc on the belly, by means of which it attaches itself to rocks, especially in late winter, when it lays its eggs in clumps which are guarded by the male. It is known as lumpfish in North America.

Lungfishes, one of the four main orders of *bony fishes, comprise two families: the Australian lungfishes, and the South American and African lungfishes. The first family contains a single species, *Neoceratodus forsteri*, which has a single 'lung' and functional gills. The South American lungfish, *Lepidosiren paradoxa*, and the four African species of *Protopterus* have a pair of sacs to their 'lung' and have non-functional gills. There is an extensive fossil record of the group since the Triassic Period, 225–180 million years ago, and these six species can be seen as living fossils, as well as evidence for continental drift. The African and South American lungfishes live inside a burrow in the bed of the lake during the dry season, and can survive drought. The African species form a 'cocoon' in the mud and can survive up to two years of drought. The Australian lungfish lives in waters which do not annually dry out and uses its lung when the water becomes stagnant. Lungfishes are relatively large freshwater fishes, some attaining 1·5 m. (5 ft.) in length.

Both sexes of the Tanzanian yellow-collared **love-bird**, *Agapornis personata*, are very similar. They often frequent acacia bushes, feeding mainly on their seeds, though they are also partial to cultivated millet.

Lungless salamanders are the largest family of *salamanders with approximately 215 species. A few species fail to metamorphose and remain as permanent, gilled larvae, breeding in the larval state, but the majority undergo complete *metamorphosis. However, unlike other salamanders, they do not develop lungs to replace the larval gills, depending instead upon respiration through the thin, moist skin of the body and pharynx (throat). They vary widely in size, from 4 to 30 cm. (1½ to 12 in.), and may occupy a variety of aquatic, terrestrial, and arboreal habitats.

Lungs are a pair of organs used to draw in air and bring it into contact with the blood in most vertebrates. They

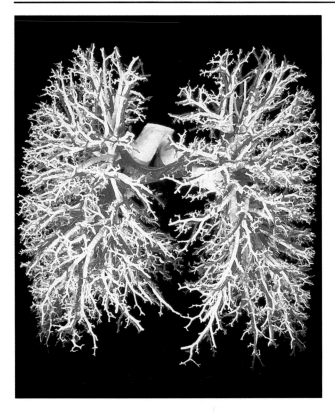

A resin cast of human **lungs** outlines the bronchial passages. The pulmonary arteries (the only arteries in which deoxygenated blood flows) are shown in red. Blood in the arteries absorbs oxygen from inhaled air and transfers carbon dioxide to be breathed out.

lie in cavities in the chest, on either side of the heart. In adult amphibians the lungs are simple sac-like structures which are inflated by forcing air into them using the tongue and mouth cavity. In virtually all other vertebrates the lungs have an elastic tissue framework in which lie numerous thin-walled air-containing sacs richly served by capillaries (small blood vessels). Air entering the lungs passes down rigid-walled tubes which give way to muscular-walled tubes then on to the air sacs. Air is sucked into the lungs when a muscular membrane, called the diaphragm, descends and the rib-cage elevates. Air is expelled due to the elastic recoil of the lungs when the muscles relax.

Lungs also exist in some snails and scorpions, but do not have a regular ventilated system of air movement. They rely simply on diffusion of oxygen across their walls.

Lungworms are parasites of many large mammals, including cattle, deer, pigs, and sheep, as well as amphibians such as frogs. They are *roundworms, especially of the family Metastrongylidae, living within the lung; the host coughs up eggs and swallows them. The young larvae are released in faeces, up to five million per day. New hosts then pick them up when grazing. Lungworms are white and threadlike, about 5 cm. (2 in.) long, and severe infestations cause coughing, fever, and even death in livestock.

Lupins are annual or perennial species of the genus *Lupinus*, some of which are popular herbaceous garden plants. The long spikes of pea-like flowers, in a wide variety of colour combinations, are a classic example of selective breeding. They are members of the pea family and related to clovers, beans, and many other familiar plants.

Lychee *Litchi.

Lymph glands should properly be described as nodes since they contain no glandular tissue. They are compact structures, often the size and shape of a small bean, distributed in groups throughout the body along vessels of the lymphatic system. Like blood vessels, these are placed so that fluid (lymph) draining from tissues passes into them. Valves in the vessels entering and leaving them direct the lymph towards collecting ducts which convey it back to the blood.

Nodes, which receive a rich blood supply, consist of dense collections of white blood cells called lymphocytes, and a series of channels, organized as a fine mesh, which act to filter out all large particles before they can enter the blood system. Scavenger cells, or macrophages, process the trapped matter and stimulate nearby lymphocytes to produce antibodies specific to any foreign protein molecules (antigens) released.

Lymphocytes recirculate from the blood to the nodes and back every few hours, thus ensuring that their exposure to antigenic material, and therefore the body's immunity to foreign molecules, is kept up to date. Infected nodes enlarge and can become very painful.

Lynx: a species of the *cat family, which occurs in the north of the Northern Hemisphere from Siberia through Europe and North America. There are several subspecies including the Canadian lynx, which is found from Alaska to the northern part of the USA. It is some 120 cm. (4 ft.) long and 60 cm. (2 ft.) high at the shoulder. The coat is a mixture of black, dark brown, and tawny yellow, and the underparts are cinnamon, the cheek ruffs black and white, and the tail has a black tip. The European and Spanish lynxes are smaller but have similar colouring. The former ranges through woods of Europe and Asia, and the latter survives in Portugal, the Pyrenees, and in a reserve in Spain.

Lynxes are solitary, hunting at night by sight and smell, and their prey includes small deer, badgers, foxes, hares, rabbits, small rodents, and even beetles. Mating occurs in March and two to four young are born two months later in a den. The kittens are blind; the eyes open ten days later. They begin hunting with their mother at two months of age.

Lyre-birds have very restricted ranges in eastern Australia, where the two species live in damp, eucalyptus forests. They resemble long-legged brown chickens except for the striking, long tail-feathers of the male, the outer two of which are white, banded with brown and curved outwards at the end. When the tail is raised the shape resembles a lyre. The displaying male is very noisy and mimics a wide range of other birds and even dogs and cats. The female builds a large, domed nest of twigs and leaves, often on the ground, in which she lays a single egg.

M

Macaques form a genus of some fifteen species of Old World monkeys that possess cheek pouches in which they temporarily store leaves and fruit. The hardiest of all the monkeys, they can live in quite cold climates as well as in the hottest parts of India and Burma. The majority are found in Asia and the East Indies region. About the size of a collie dog, they have dark brown or yellowish hair. The head is large with a naked face, pointed ears, prominent eyebrows, and large eyes; the buttocks are bare and often a bright colour; the tail is usually long. They congregate in troops of anything from a few up to a hundred or so individuals, spending their time almost equally between the ground and the trees. They are most active in early morning and evening, searching for food. Fully adult when four to five years old, a female bears a single young. At first the baby clings to its mother, but is soon in-dependent. The macaques include the rhesus monkey, *Macaca fuscata*, well known because of the human blood factor named after it. Other species include the Barbary ape and the pig-tailed, lion-tailed, long-tailed, and toque macaques.

Macaws are large members of the parrot family, all of which live in South or Central America or the Caribbean islands. Although some are smaller, the best-known ones are about the size of large crows with long tails. They are brilliantly coloured, mostly with blues, reds, yellows, and greens, and have very powerful bills with which they can break open large nuts; they also eat many fruits. They nest in holes in trees and lay white eggs. Several species have very restricted ranges and are in danger of extinction; one, the Cuban macaw, was exterminated in about 1864.

Mace (spice)   *Nutmeg.

Mackerel are part of a family of torpedo-shaped fishes, often with a blue-green back, irregular black curving lines, and silvery sides. A series of finlets above and beneath the tail shows their relationship with the tunas. The common mackerel, *Scomber scombrus*, occurs on both sides of the North Atlantic and in the Mediterranean, living in huge schools near the surface. They are found both inshore and offshore in summer, but lie inactive near the sea-bed in

Macaws are gregarious birds: here scarlet- and green-winged macaws, *Ara macao* and *A. chloroptera*, dig out and eat mineral-rich soil from a rock-face in Peru.

winter. Their food comprises all kinds of surface-living small fishes and crustaceans; and they in turn are eaten by larger fishes, sharks, and dolphins. Large quantities are caught for human consumption.

Madonna lilies *Lily.

Maggots, or gentles, are the larvae of species of true *flies, in which the head and the mouthparts are replaced by hardened structures developed in the thorax. The best known are the white maggots of blowflies, flesh flies, and houseflies. When fully grown, maggots form *pupae within the last larval skin, which hardens and darkens. Their feeding habits are varied: those of blowflies feed on rotting flesh, and it is these which are most often used for fishing bait.

Magnolia is a genus of some eight species of trees or shrubs native to North and Central America, eastern Asia, and the Himalayas. They have rather large leaves and conspicuous flowers, often with large, conspicuous sepals which are indistinguishable from the petals. The seeds are red and dangle on threads from the fruits when these split open. Some are evergreen, like *Magnolia grandifolia* of the southeastern USA, and others are deciduous.

Magpie is the common name for about nineteen species of the *crow family with a wide distribution in both Old and New Worlds, with the exception of Africa. They have smallish thrush-sized bodies, but long tails. Most are strikingly coloured, often blue and white or green and white. The European or American common magpie, *Pica pica*, is shiny blue-black and white and one of the few birds to be common in both regions. They build domed nests of sticks in bushes or trees, and have a reputation for stealing almost anything that shines, thus collecting a wide range of objects, hence the metaphorical use of their name.

Magpie lark: an Australian bird that lives in open woodlands and is common in gardens. It is a striking black and white thrush-sized bird with a staring white eye; it eats mainly insects. The magpie lark builds a nest of plant fibres which are held together with mud; and usually placed on the horizontal branch of a tree.

Magpie moths are mottled or spotted, usually black on white, and most belong to the family Geometridae. The magpie, or currant moth, *Abraxas grossulariata*, is strikingly black-spotted on white with touches of orange, and is abundant and widespread throughout Eurasia. Their similarly patterned looper caterpillars are often a pest on currant and gooseberry bushes.

Mahoganies are trees of tropical and subtropical regions which form the family Meliaceae. This contains about 550 species, among which the genera *Swietenia* and *Khaya*, each with eight species, produce the highly decorative reddish timber called mahogany. The *Swietenia* species are native to the Americas, while *Khaya* is restricted to Africa. The original mahogany, brought to Europe by sixteenth-century Spanish explorers, came from *S. mahagoni* from the Caribbean area. Most of the wild trees of this species are now poor specimens, the best having been removed long ago. The impoverishment of the stock is a classic example of genetic erosion. Most American mahogany now comes from *S. macrophylla*, native to South and Cen-

tral American rain forests. The paler timber of American mahogany comes from three different species of *Khaya* native to West African rain forests.

Many other trees of this and other families, produce dark timber which may be sold as mahogany. Such timber does not necessarily have the same qualities as true mahogany.

Maidenhair ferns are a mainly Mediterranean group of *fern of the genus *Adiantum*. All are deciduous, with unmistakable fronds in the form of wedge-shaped leaflets along branching, wiry, blackish-brown stalks. The name maidenhair is used for many similar species.

Maidenhair tree *Ginkgo.

Mail cheek fishes are species in which the bone beneath the eye is strengthened and enlarged to form a bony ridge from the cheek to the gill-cover. They are included in the order of fishes which contains the *scorpion fishes, gurnards, and bullheads or sculpins. A similar ridge is also very prominent in the scorpion fishes.

Maize is an annual grass, popularly known in the North America as corn. It grows up to 4.5 m. (15 ft.) tall and is the only cereal of American origin. It was first cultivated over 5,000 years ago in Central America and is now grown throughout the tropical, subtropical, and warm temperate regions of the world. The plant is unusual among the grasses in that its male and female flowers are separate and occur on the same plant. Most of the maize varieties cultivated today are hybrid forms. These give high yields but may be susceptible to disease.

The maize grain is rich in starch, but contains up to 15 per cent protein and 5 per cent oil. The ground grain provides cornstarch (or cornflour) which can be converted with dilute acids to make corn syrup. This is unique among plant sugars as it consists mainly of the easily digestible sugar glucose, making it particularly useful for baby-food formulations. Corn whiskey is made by fermenting the grain.

There are several types of maize, differing principally in the characteristics of the grain borne on the cobs. 'Flourcorn' has grains with soft starch that can be stirred into a paste, whereas 'Flintcorn' cannot, as the starch is quite hard. The grain of another type, known as *sweet corn, contains mostly sugars rather than starch and is eaten as a vegetable. Yet another kind of maize has a high water content in its centre; when it is heated the steam produced causes it to explode and turn inside out to produce the familiar popcorn.

Malaria is an infectious disease caused by certain parasitic protozoa belonging to the genus *Plasmodium*. These micro-organisms have a developmental stage within the red blood cells of man and other animals. Attacks involve proliferation of the parasite within the blood cells, eventually causing the cell to burst. The destruction of millions of blood cells leads to a characteristic periodic fever, often with a cold stage, hot stage, and finally a sweating stage. The loss of red blood cells also leads to *anaemia.

The disease is dependent on, and transmitted by, *Anopheles* *mosquitoes, the female of which picks up the parasites while feeding on mammalian blood. The parasites multiply within the cells of the mosquito's stomach, and are eventually re-injected into another host when the mosquito feeds. Malaria is a very important cause of infant mortality in some regions. Different species of *Plasmodium*, produce different types of malaria. Control is mostly effected by destruction of the breeding sites of the mosquito. Several drugs are effective in preventing or controlling the disease. Other names for malaria include ague, marsh fever, and periodic fever.

Male fern: the 'fern' of popular speech, with mid-green, divided fronds which grow to about 1 m. (3 ft. 3 in.) high

from a stocky, upright rootstock. They have chaffy brown scales on their stalks and round brown spore-cases on their undersides. There are several other species of *Dryopteris*, the genus to which the male fern belongs, most with male fern as part of their common name. The common male fern, *D. filix-mas* was once thought to be the male sex of the lady fern, *Athyrium filix-femina*.

Mallard: a widely distributed, familiar species of dabbling *duck, from which most farmyard breeds are descended. The drake's green head, white collar, and purple-brown breast are distinctive. Females are mottled brown. Mallards inhabit all types of inland water, even ornamental ponds in towns, and usually spend the winter on estuaries and coasts.

Mallee fowl are ten species making up a family which is related to other gamebirds. They are mostly dullish brown or greyish birds and rather partridge- or pheasant-like in size and appearance. They occur in Southeast Asia, Australia, New Guinea, and some Pacific islands. Their most extraordinary feature is their nesting habits, because instead of incubating their eggs they bury them in sand warmed by the sun, warm volcanic ash, or rotting veg-

Widely distributed in the Northern Hemisphere, the **male fern**, seen here in a damp woodland glade, is also native to South America. Its fronds, which come from a large clump of crowns, die back in the autumn.

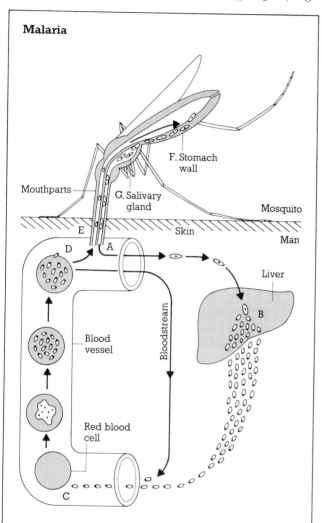

Malaria

Life cycle of the malarial parasite The parasites are injected into the human bloodstream along with saliva through the piercing mouthparts of an infected *Anopheles* mosquito (A), and are carried to the liver. Here they multiply asexually (B) and some attack red blood cells (C), feeding on their contents and further multiplying asexually within the cells. Infected blood cells subsequently burst (D), generally at regular 48 or 72 hour intervals, causing attacks of severe fever. After several generations of re-infestation of red blood cells, some of the parasites develop into a sexual form which will only develop further once sucked back (E) into a mosquito's stomach (F). Here eggs and sperm are released and fertilization occurs. The zygotes then penetrate the stomach wall, rapidly reproduce asexually, and migrate to the mosquito's salivary glands (G), completing the cycle ready for re-injection into the human bloodstream.

etation. The heat of these substances hatches the eggs. On hatching, the young fight their way to the surface and can fly within a few hours; they receive no parental care. The mallee fowl, *Leipoa ocellata*, of central and southern Australia, builds a special mound, which it keeps opening and adjusting so as to keep the temperature constant. Their family name of Megapodes refers to their large feet, which are adapted to digging 'incubation' holes for the eggs.

Mallows are a group of annual, biennial or perennial plants from Europe, Asia, and North Africa. They belong to the mallow family, which also includes cotton, hibiscus, and hollyhock. They have lobed leaves and purple, rose, or, occasionally, white flowers. Most species are in the genus *Malva*, though the name is also applied to the perennial marsh mallow, *Althaea officinalis*, a plant much valued since early times as a vegetable and for its medicinal properties.

Malpighi, Marcello (1628–94), Italian microscopist, investigated the structures of the kidney, skin, and blood vessels, confirming Harvey's views on blood circulation. He also traced the development of the chick embryo, studied the anatomy of the silkworm, discovered the breathing system of insects, and examined plant cells. The Malpighian body, part of the nephron, the Malpighian layer of the skin, and Malpighian tubules in insects were named after him.

Mambas are some of the world's most venomous *snakes. They are related to cobras and range through tropical Africa south to Natal and South West Africa. There are four species: three arboreal green mambas, *Dendroaspis* species, and the black mamba, *D. polylepis*, which is essentially terrestrial. The black mamba is a large, fast-moving snake, which chases its prey (usually mammals or young birds). Its bite is very dangerous; the venom contains a powerful nerve toxin which has the effect of inhibiting breathing. The green mambas, *D. angusticeps*, *D. viridis*, and *D. jamesoni*, although venomous, are much less aggressive and feed on small birds, lizards, and eggs.

Mammal-like reptiles (synapsids) are a fossil group which formed the first really successful development of reptiles, preceding the dinosaurs. The earliest ones were the pelycosaurs, best known from North America, or species such as *Dimetrodon incisivus*, distinctive for the large 'sail' along its spine, which appeared about 300 million years ago. These were superseded by the more advanced therapsids, which consisted of carnivorous and herbivorous species. These gradually became more and more like mammals, evolving, for example, complex teeth, a single bone in the jaw, and a more upright gait. The first known mammals evolved from the therapsids about 190 million years ago as small rodent-like animals, such as *Morganucodon*. At this point in evolution, the mammal-like reptiles, or synapsids, became extinct.

Mammals are vertebrate animals which bear their young alive and feed them with milk, a secretion of specialized skin glands called mammary glands. They are also warm-blooded animals with large brains, sweat glands in their skin, and body hair. The toes and fingers terminate in claws, nails, or hoofs. The dentition includes incisor, canine, premolar, and molar teeth, the number and shape varying greatly with the mode of feeding. There is only one replacement of teeth and the dentition and digestive system ensure efficient food utilization.

There are three subclasses of mammals: the egg-laying or monotreme mammals, the *marsupials, and the placental or eutherian mammals. There are around 4,250 species of mammals which have a wide distribution on land, in water, and in the air. In size they range from the pygmy white-toothed or Etruscan shrew, *Suncus etruscus*, only 80 mm. (3¼ in.) in length including its tail, to the largest animal that has ever lived, the blue whale, 33·27 m. (109 ft. 4 in.) in length.

The majority of mammals are quadrupeds; only a few have given up this gait. The exceptions include jumping mammals, such as kangaroo and jumping mice, and species which have adapted to an aquatic life. A few groups of mammals have modified their body plan for flight, as in bats, or gliding, as in phalangers. The diversity of forms and their ability to occupy all available habitats is a measure of their success in adaptation.

Mammary glands are characteristic of female mammals which produce milk for suckling their young. The supply of milk after birth depends on its regular removal by suckling or milking. The glands lie beneath the skin of the abdominal wall and open into separate teats, whose number depends, in general, upon the size of the litter. Pigs and dogs have numerous pairs whereas there is usually only a single pair in humans, although supernumerary glands and nipples are sometimes found in line along the chest and abdominal wall.

In primates, including humans, both sexes have mammary glands and teats or nipples, but the breasts enlarge only in females, on reaching sexual maturity, due to fat deposits around the gland. The gland becomes active and greatly enlarged during pregnancy and after birth due to hormonal stimulation.

Mammoths were large elephants which occurred in Africa and the Northern Hemisphere from the Pliocene Epoch onwards. They reached a height of 4·5 m. (15 ft.) at the shoulder and had a pair of long, curved tusks up to 5 m. (16 ft.) long. A number of frozen carcasses have been found in Siberia, from which it is known that they had long, dense hair and also a thick layer of blubber. They disappeared about 12,000 years ago, probably because of the warming of the climate. Cave paintings of mammoths by Stone Age man suggest that they constituted an important source of food, clothing, and ivory for tools.

Man is the sole member of the family Hominidae and as such is part of the Primate order. The chief specializations of man, which distinguish him from other primates, include elaboration of the brain and behaviour, communication by speech, erect posture, and long postnatal development. The brain of adult man is at least twice as large as that of any adult living ape.

The difference in behaviour between man and apes is considerable, but perhaps the most striking is related to communication by speech, along with the power of abstract thought. Man is also able to store information in written records, and more recently by other means, and to pass it on from one generation to another. The gait of man is totally bipedal, the body being fully and continuously balanced on the two hind-limbs.

After a gestation period of nine months one baby (but occasionally larger numbers of non-identical or identical

Siberian permafrost preserved this baby **mammoth** in excellent condition until its discovery in 1978. Nicknamed 'Dima' (a pet form of Dmitri) by the Soviet press, the mammoth was later exhibited across Europe.

offspring) is born which suckles for periods ranging up to several years before weaning, depending upon cultural practices. The rate of development is slow and the period of childhood prolonged, which has a great influence on the family and social organization of man.

Manakins are a family of fifty-five species of birds that occur only in rain forests in South and Central America, feeding mainly on small fruits. They are cousins of the cotingas but are only sparrow-sized. The males of many species are brightly coloured, often with black bodies and white, yellow, or red heads. Some have blue bodies, and one or two have elongated tail feathers, though most have short tails. The males of most species display in leks, each bird having a separate court, though some have communal displays. The females look after the young by themselves.

Manatees are large, entirely aquatic, herbivorous mammals belonging to one of the genera of *sea-cows. The three species are found on tropical and subtropical Atlantic coasts, estuaries, and great rivers, where they feed upon marine or freshwater plants. The West Indian manatee, *Trichechus manatus*, can reach a length of 4·6 m. (15 ft.) and is torpedo-shaped, with the digits of the fore-limbs modified as paddles; the hind-limbs are absent. The tail is powerful, flattened horizontally, and provides the main propulsive force. The flattened, rather pig-like, bristly snout has a large, muscular upper lip, the corners of which pluck the vegetation. The nostrils are high on the snout and have valves, the eyes are round, and the ears minute. The skin is tough, almost hairless and grey-brown in colour. A single calf is born after a gestation period of about five months; the mother takes great care of it and suckles it when she lies on her back. Exploitation of these animals has occurred for their meat. The other species are the West African manatee, *T. senegalensis*, and the Amazonian manatee, *T. inunguis*.

Mandarins (oranges), along with tangerines and satsumas, are all varieties of the *Citrus* tree, *C. reticulata*. They all produce fruit which is smaller than an orange and is sometimes slightly flattened in shape. A range of types exists varying particularly in skin colour, but all easily peeled. Cultivated in Japan and China for centuries, they are now grown in southern Europe as well as the southern USA. The mandarin is a parent of several hybrid citrus fruits, including the clementine and tanget.

Mandrake, or devil's apple: a plant known to botanists as *Mandragora officinarum*, a small perennial of the potato family. This strange plant is native to southern Europe and the Mediterranean region, and since ancient times has been a source of mystery and superstition. The fleshy root is said to shriek when pulled from the ground. This root was used in ancient Greece to alleviate pain and as an aphrodisiac. The white or purple flowers are produced at ground level. They are followed by poisonous golf-ball-sized fruits. The American mandrake, or may apple, *Podophyllum peltatum*, is a quite unrelated plant of damp woodlands and flood-meadows. Other species in the genus *Podophyllum* may be referred to as mandrake, but are more closely related to barberry.

Male **mandrills** look formidable with their bright facial markings and large head-crest. These features are used to intimidate other males and predators in the defence of their troop.

Mandrill: the largest and perhaps the most colourful species of *baboon. The mandrill, *Papio sphinx*, is dark brown with white cheek fringes, a yellow beard, and a tuft of hair on the crown. The face of the adult male is brilliantly coloured, unlike that of its close relative the drill, *P. leucophaeus*, which is black. The colours are present but much duller in the female. The buttocks are also brightly coloured, blue and white.

It is found mainly near the coast of West Africa, where it lives on the forest floor but will also climb. It eats fruit, berries, bark, and roots. Troops of up to sixty individuals may travel together.

Mangel-wurzels (mangolds)   *Beetroot.

Mangoes are large, spreading, evergreen trees widely distributed throughout the tropics but native to India. They belong to a family of some 600 species of trees, shrubs, and climbers which includes cashew and poison ivy. A well-defined dry season is vital for successful fruit development but, even then, good crops usually occur only in alternate years. The yellow, green, or red fruits have an unpalatable skin but the orange-yellow flesh of some varieties is edible.

Mangosteen: a small conical tree, native to the humid equatorial countries of Malaysia and Indonesia. It belongs to the family Guttiferae of around 1,000 species of tree and shrub, found all over the world but particularly common in the tropics. Its close relatives include the mammey apple and St. John's wort. It has globular reddish-purple berries, the size of a small orange, with a thick skin surrounding several ivory-white segments.

Mangroves are a family of tropical trees which grow on muddy coastlines and in estuaries. These trees, especially species of *Rhizophora*, develop extensive roots which trap mud and silt. At high tide, the roots are inundated with salt water, a physiological stress which would kill any other tree. Though mangroves are adapted to this, it is not a requirement for their successful growth as they are easily raised in botanic gardens away from the coast. They cope with salt in a variety of ways. Some species have 'filters' on their roots to exclude it; others allow the salt in but excrete it from the twigs, or let it accumulate in leaves which are then dropped. Many species have seeds which germinate in the fruit and produce a long, sharp-pointed root which embeds into the mud when the fruit is dropped. Others have salt-resistant floating seeds.

Manta rays, cartilaginous fishes which belong to the family Mobulidae, are giant *rays which live in tropical seas. About ten species are known of which the largest, the Atlantic manta, *Manta birostris*, may be 6 m. (19 ft. 6 in.) wide. They swim near the surface, feeding on small fishes and planktonic animals which are funnelled into the large mouth between the fleshy flaps ('horns') on the head and are then filtered from the water by the gill rakers. These horns are responsible for the alternative name of devil ray.

Mantis shrimps are an order of flattened, shrimp-like *crustaceans, with one pair of legs enormously expanded and so resembling those of the praying mantis. They wait in burrows, shooting out these barbed limbs to grasp and spear passing shrimps or fish. When mantis shrimps are in captivity, blows from these powerful limbs can crack aquarium glass. They are mainly tropical, up to 36 cm. (14 in.) long, and brightly coloured. Temperate species such as *Squilla empusa*, though smaller and less colourful, show the same remarkable habits.

Manure is the most commonly used natural or organic fertilizer. The faeces of almost any animal will provide a rich source of the mineral nutrients essential for healthy and productive plant growth. Manure has several advantages over inorganic fertilizers. It is cheap, readily available in most agricultural communities, and gives a slow, steady release of nutrients, which cannot be washed out of the soil in a single rainstorm.

Maples are trees of the genus *Acer*, comprising some 200 species of the north temperate zone. They are deciduous and have winged fruits. Some of the North American species, like the sugar maple, *A. saccharum*, have sap which is the source of maple syrup. A number of Japanese species

Marmosets

Each species of marmoset has a distinct facial appearance, though as a group they resemble each other closely. They are usually diurnal and live in family groups.

Saddle-backed marmoset, *Saguinus fuscicollis*

Pied marmoset, *Saguinus bicolor*

Emperor marmoset, *Saguinus imperator*

Cotton-top marmoset, *Saguinus oedipus*

White-lipped marmoset, *Saguinus nigricollis*

are widely grown for the spectacular red or yellow colour of their leaves. Timber and charcoal are obtained from many species. One of the most widespread maples in Europe is the sycamore, *A. pseudoplatanus*.

Marabou: a large species of *stork, reaching 1·5 m. (5 ft.) in height, and found throughout most of the warmer areas of Africa. It is mainly greyish above with paler underparts, and has a whitish ruff at the base of the neck and a long pink pouch. It is best known as a scavenger and is commonly seen gathering with vultures at kills. These storks will also take small frogs and other small animals such as locusts.

Marigolds, of which there are three kinds, French, African, or pot, are all members of the sunflower family, and as such are related to the daisy. Strangely, the names do not indicate the countries of origin, for both the French

The **marabou**, *Leptoptilos crumeniferus*, has a powerful bill with which it can drive away even vultures from carrion. When duelling for a mate early in the season, as these two males are, lasting injury is common.

and African marigolds are derived from Mexican species, *Tagetes patula* and *T. erecta* respectively. The pot marigold, *Calendula officinalis*, is an annual plant native to southern Europe. It has long been cultivated for its attractive orange flowers and as a kitchen and medicinal herb.

Marine iguanas *Sea lizards.

Marjoram is the name of several species of plant of the genus *Origanum*. They belong to the mint family, and like mint are used as herbs. Wild marjoram, *O. vulgare*, is an erect perennial, growing up to 70 cm. (2 ft. 4 in.) tall; it is widely distributed over Europe. Sweet marjoram, *O. majorana*, is more popular as a culinary herb. Like the smaller pot marjoram, *O. onites*, it comes from the Mediterranean region. Wild marjoram from warmer countries, which has a stronger flavour, is dried and sold as the herb oregano.

Markhor: a species of wild goat found from Kashmir through Afghanistan to the USSR, living in the open or in forests on steep barren slopes and precipitous crags. It rarely goes above the snowline. Males may reach a shoulder height of 1·04 m. (3 ft. 5 in.); females are smaller. The compressed, spiral horns vary in size and shape, the longest being 1·65 m. (5 ft. 5 in.). The hair is long and silky and, in the male, grey-brown in colour, becoming white in old age; the females are dark fawn. One or two kids are born after a gestation period of six months.

Marlins are *spearfishes with a spear-like bill which is round in cross-section. The dozen or so species are all streamlined fishes with a powerful body, a long dorsal fin, and high lobes to the tail fin. They are probably the fastest swimming fishes in the world, living near the surface of the sea in all tropical and warm temperate oceans. The larger species include the blue marlin, *Makaira nigricans*, which grows to 4·6 m. (14 ft. 8 in.) in length and around 450 kg. (992 lb.) in weight. They are popular sporting fishes in certain tropical areas and also commercially important as a food-fish. Members of the genus *Tetrapturus* are rather smaller, but include the white marlin, *T. albidus*, which occurs as far north as Nova Scotia.

Marmosets are a curious group of *primates found in South America. The eight species are all less than 30 cm.

(1 ft.) in length, and have extremely long tails which are not prehensile. The digits on their hands and feet have hooked claws, except for the big toe which has a nail. They are tree-dwellers in equatorial forests, a life to which they are well adapted. They show a range of coloration from drab to quite vivid reds, browns, and greys. Many species have distinctive crests, whiskers, and beards.

Marmots are closely related to the *ground squirrels but are rather stockier in build, weighing up to 7·6 kg. (16½ lb.) and have a relatively shorter tail. Thirteen species occur throughout North America, much of temperate Asia, and in parts of Europe. They construct large burrows; some species, such as the European alpine marmot, *Marmota marmota*, form colonies, but others are more solitary. They are vegetarian and can become pests of crops. Marmots such as the hoary marmot, *M. caligata*, are famous for their prolonged hibernation, which usually occupies six months of the year and in the northern parts of their range may extend over eight months.

Marram grass is a species of grass found in arid habitats, such as the sandy coasts and dunes of western Europe. It is a perennial with tightly rolled leaves, very resistant to wind and salt spray, and has a creeping rhizome which effectively binds otherwise loose sands. It has been usefully employed for stabilizing sands subject to erosion by wind and tides.

Marrow, or vegetable marrow, is the name used in Europe when referring to varieties of the plant species *Cucurbita pepo*, which produces long, cylindrical fruits. These are usually green but may also be yellow, white, or striped. Like their relatives melons and cucumbers, they are climbing or trailing annuals, with large, prickly leaves and stems. In North America these fruits are known, along with other varieties, as summer *squashes and include the American pumpkin. They originated in tropical America and are now grown all over the world as a vegetable. Other varieties of *C. pepo* include ornamental gourds, and those with fruit used when young as courgettes, which are also known as zucchini.

Marsh, Othniel Charles (1831–99), American palaeontologist, collected fossil vertebrates from Nebraska to California, discovering 500 new species, 225 new genera, 64 new families, and 19 new orders. His ideas on links between fossil groups, such as the extinct toothed birds evolving from reptiles, raised Darwin's prestige; and his reconstruction of the evolution of the horse from *Eohippus, its ancestor, is also notable.

Marsh deer: a swamp-loving animal, the largest species of the South American *deer. It stands 1·10 m. (3 ft. 8 in.) at the shoulders. The coat is of a rich, deep, chestnut-red colour in summer, becoming browner in winter. The males usually have eight points to their antlers. It occurs in marshes across central South America.

Marsh harriers are slender *hawks with long wings and tail, and are widely distributed in Britain, Sweden, northern Asia, and eastern Mediterranean countries, though they are nowhere numerous. Characteristically, they fly low over moorland and marshes, hunting for the small mammals and insects that are their main prey. In contrast to their solitary breeding pairs they form greg-

arious groups on their annual migrations to Africa and southern Asia.

Marsh marigolds *Kingcups.

Marsupial moles are similar in appearance and size to true moles but rather different in habits as they do not construct permanent burrows; those they dig collapse behind them. Notable features include a nose-shield, powerful front claws for digging, and silky white to golden fur. The eyes are vestigial and the, unusually for a marsupial, pouch opens to the rear. They are the only Australian mammal to have adapted to a burrowing life.

Marsupials are a subclass of mammals now mainly confined to Australasia, but represented in the New World by the opossums. Marsupials differ from other mammals primarily in their reproductive processes, in that the young are not nourished in the womb through a placenta but are born in a very undeveloped state. Most growth takes place in the pouch, which is a fold of skin, usually facing forwards, enclosing the nipples. The pouch is absent from most American opossums. There is little evidence supporting the supposed inferiority of the reproductive system, although it is true that many marsupials seem unable to compete with introduced placental mammals and several species have become rare or extinct as a result. Their abundance in Australia is due to the separation of that continent from other land masses before placental mammals could become established there.

An interesting feature of marsupial evolution is that it shows a parallel with the evolution of mammals elsewhere. Thus there are equivalents to wolves, bears, badgers, cats, ant-eaters, rats, mice, and moles. Marsupial 'cats' are really comparable to genets. Bandicoots are rabbit-like in appearance but insectivorous while others look like true insectivores. The *flying phalangers parallel flying squirrels and there are even otter-like water opossums. Australian 'opossums' resemble monkeys. Kangaroos, can be considered as equivalent to antelopes as they occupy similar ecological niches.

Martagon lilies *Lily.

Martens are carnivores of the family Mustelidae, which includes the weasels, stoats, otters, and badgers. Martens look like large versions of the related stoats, except that the tail is bushier. There are eight species which occur throughout temperate regions of the Northern Hemisphere. All are carnivorous, hunting in trees, with squirrels as their favourite prey, although they will take any other small mammal or bird including chickens. There are two species in America: the fisher, *Martes pennanti*, and the American marten, *M. americana*. The former belies its name in being more of a terrestrial carnivore than a fisherman. The Old World martens are sometimes named after their preferred habitat such as pine marten, *M. martes*, and beech, or stone marten, *M. foina*. These two species extend south to the Mediterranean region, but the *sable, *M. zibellina*, is confined to the coniferous forests of Siberia. Its range once extended into Europe but demand for its beautiful fur has exterminated it over wide areas.

Martins comprise twenty different species of bird that are not particularly closely related within the *swallow family, Hirundinidae. They vary from about 10 to 20 cm.

(4 to 9 in.) in length and most have glossy black plumage with a purple or green sheen. Very agile on the wing, they feed on flying insects. Martins occur in most areas of the world, those that breed at high latitudes migrating long distances to warmer climates for the winter.

Mason bees are related to the leaf-cutter *bees. Like them, they make their cells in rotten wood, or in holes in masonry, but instead of using leaves they make them of small blobs of clay moulded together.

Mason wasps, or potter wasps, are solitary *wasps of the family Eumenidae. Their females build piles of mud cells, placed one on top of another, to form cup-shaped nest cells. Each cell is provisioned with insects before an egg is laid and one cell is capped before the next is started. Similar species also use mud to build nest cells in the cracks of walls.

Mastication is the act of chewing. Movements of the jaws enable the molar teeth to break up solid lumps of food into small particles, which are mixed thoroughly with the saliva during the process. Thus the food eaten is converted to a moistened paste which can be easily swallowed and subsequently digested.

Mastitis is inflammation of the breast. In women the term is often applied to non-infectious, benign cysts. Their presence makes the recognition of cancer more difficult. Infectious mastitis is an important disease of cows, which, by contaminating the milk, can cause disease in humans. It can also occur in women who are breast-feeding but responds to antibiotic treatment.

A **matamata** from Brazil. Its strange appearance with warts and bumps, together with flaps of loose skin on the head, neck, and limbs, serve as effective camouflage as it lies in the mud awaiting passing prey.

Mastodons are extinct elephants, which were very abundant during the Tertiary Era (66 to 1·6 million years ago), throughout the world. In addition to the curved upper pair of tusks, in the male mastodon there was a shorter pair of lower tusks which protruded forwards. Like the mammoths, mastodons had long, hairy coats.

Matamata: a species of large *snake-necked turtle, *Chelus fimbriata*, with a shell-length up to 40 cm. (16 in.), from northern South America and Trinidad. It has a very bizarre appearance with its very flattened head being characterized by fringes of skin, tiny eyes, elongate snout, wide mouth, and rough shell. It feeds by suddenly opening its jaws and expanding its neck, thereby creating an inrush of water to carry a fish into its mouth.

Mather, Cotton (1663–1728), son of the American Puritan pastor, Increase Mather, became his father's colleague and was involved in the Salem witchcraft trial. As a naturalist, he is famous for a description of variety crossing in Indian corn (maize), the earliest authenticated account of an artificial plant hybrid. An interest in medicine led him to insist on the smallpox inoculations by *Boylston in the Boston epidemic of 1721.

May (blossom)   *Hawthorn.

Mayflies are delicate, weak-flying insects of the order Ephemoptera, with two or three filamentous 'tails', found near fresh water. The 1,500 worldwide species have short antennae but large eyes, and at rest the wings are closed vertically. They have two pairs of gauze-like wings, the front pair much larger than the hind pair. They are well known to fishermen, because the aquatic, herbivorous, three-tailed nymphs are a major fish food. The flying, hair-covered, sub-adult, or dun, which moults into the adult proper (spinner), is used as a fishing fly. Adults often live less than a day, and do not feed, but may emerge in

spectacular numbers. They mate, lay eggs, and then die. The duns may live for longer than a single day and are often assumed to be adults.

Meadow pipit: a species of *wagtail or pipit that is a streaky, olive-green bird with white outer tail feathers. It is the size of a small sparrow with a thin, sharply pointed beak, and breeds in all but the southernmost areas of Europe and western Asia. This species lives on rough, open, grassy ground and nests on the ground near the side of a clump of grass.

Meadow saffron   *Autumn crocus.

Meadowsweet is a perennial plant belonging to the rose family. Its normal habitats are flood-meadows, marshes, and the margins of wet woodlands distributed throughout Europe, the Mediterranean, and Asia Minor. It also occurs as a garden escape in the eastern USA. The creamy-

The fruit of the **medlar**, *Mespilus germanica*, which follows solitary white flowers with red anthers, is about 3 cm. (1¼ in.) across. Cultivated forms of the tree are usually thornless and their fruits are larger.

white, scented flower-heads were made into a popular herbal drink in medieval times.

Mealy bug   *Scale-insect.

Measles is a very infectious virus disease, mainly affecting children. Its characteristic rash appears first on the forehead and neck, then spreads to the trunk, then to the limbs. The first symptoms include a fever and a cough about four days before the appearance of a rash. Serious complications, due mainly to bacteria, are now much less frequent because of available treatment. However, the disease can be devastating in populations not previously exposed to it. Very rarely, the virus causes a slowly progressive disease of the brain.

Measuring-worms   *Loopers.

Medicine in the broad sense has two aspects: firstly, the study of all types of human abnormality and disease; secondly, the attempt to cure illness and prevent premature death. The term medicine is also used when referring to the way in which illness and disease are treated by non-surgical methods. The drugs or substances prescribed for such treatments are colloquially referred to as medicine. This applies especially to preparations which are taken by mouth in the form of a solution.

Medlar: a species of deciduous tree of the genus *Mespilus*. Like its relative, the apple, it is part of the rose family. Native to southeastern Europe and Asia Minor, it is grown throughout Europe as much for its ornamental appearance as for the edible fruits, which are small, orange-brown, and plum-shaped. The fruits are unusual in that they have a large opening in the bottom through which the seed vessels are visible.

Meerkats are South African *mongooses. The name literally means lake cat, but they are not particularly associated with water. There are two species, the red meerkat, *Cynictis penicillata*, and the grey meerkat or suricate, *Suricata suricatta*. Meerkats are highly sociable and forage by day for insects and small vertebrates. Their habit of sitting upright on their haunches is very characteristic.

Megalosaurus, one of the reptile-hipped *dinosaurs, so far has been found, in fossil form, only in Europe. It was a fairly large biped, standing about 3 m. (10 ft.) high, and possessed well-developed, sharp teeth for eating its prey. It was a close relative of Tyrannosaurus.

Meganeura: an extinct gigantic *dragon-fly, very much like the present-day species, that lived in swamps during the Carboniferous Period around 300 million years ago. The wing-span could reach 70 cm. (27 in.) and it has been calculated that it flew at 69 km./hour (43 m.p.h.). It is the largest insect known to have lived.

Megapodes   *Mallee fowl.

Megatherium   *Giant sloth.

Meiosis is a special type of cell division (compare with *mitosis) that produces *gametes for sexual reproduction. A cell normally contains two copies of every chromosome but a gamete contains only one of each different chromosome. Meiosis involves halving the number of chromosomes. The process of gamete formation involves two division stages. In the first, the chromosomes come together as pairs and exchange genetic material; this is called recombination or crossing over. Each chromosome of a pair (now consisting of two chromatids) migrates to opposite ends of the cell. In the second division stage, the chromosomes undergo a mitotic type of division to produce four gamete cells, each with only one copy of each chromosome from the parent cell.

Melons are members of the very variable species of annual, sprawling plant, *Cucumis melo*. Originating in tropical Africa, they are closely related to the cucumber and members of the pumpkin family. The various types are distinguished mainly by characteristics of their fruit.

Membrane (biology) is a term denoting any thin layer of tissue that encloses or covers part of a living organism. This includes multicellular membranes such as the peritoneum, which encloses the organs on the mammalian abdomen, and the mucous membrane, which lines the intestine and respiratory tract. Membrane which forms the boundary of a cell is called plasma membrane. It is present in both plant and animal cells. Each cell membrane is thought to consist of a double layer of fat and protein, and it serves to maintain the cell's integrity.

Memory is the capacity to record and later recall information, events, and experiences. It takes place in special areas of the brain and understanding how it operates is one of the great challenges in science. A better understanding of the memories of animals and computers may shed light on the disorders of human memory.

Menarche *Menstruation.

Mendel, Gregor Johann (1822–84), was Abbot of Brünn in Moravia and founder of the science of *genetics. His experiments in the breeding of garden peas led to the formulation of 'Mendel's laws' of heredity, published in 1866. By carefully crossing pea plants with contrasting characters, such as smooth and wrinkled seeds, or red and white flowers, he was able to show that their offspring were not a simple blend of their parents' characteristics. After careful analysis he arrived at the concept of dominant and recessive genes. After publication, his results lay unnoticed in the journal of a local society until they were rediscovered in 1900 as three botanists, working independently, came to the same conclusions as Mendel. The experiments of Mendel were too original for many scientists of his day, yet the science he founded is now one of the most important fields in modern biology.

Meningitis is inflammation of the membrane covering the nervous system. Outbreaks are usually due to bacteria, such as those causing pneumonia or tuberculosis, which grow in the throat or lungs, and spread readily in crowded communities. It can also be caused by viruses. The symptoms of severe meningitis are high fever, headache, and vomiting. The muscles of the neck become stiff and the patient may lose consciousness. Most types of meningitis can be effectively treated if caught early.

Menopause is the time in a woman's life when menstruation becomes irregular and finally ceases. It happens

Gregor **Mendel**, photographed in about 1870, two years after he had become Abbot of Brünn (Brno). His study of hereditary factors was based not only on work with peas but also on surveys of other plants and of bees.

normally between the ages of 45 and 55. A decline in activity of the ovaries, which fail to respond to the pituitary gland's gonadotrophic hormones, leads to reduction in the output of oestrogen, a female sex hormone. During this period physical symptoms such as hot flushes (sensations of warmth spreading up to the face), sweating, obesity, and some wasting away of breasts, uterus, and vagina may occur. Emotional problems and depressions can trouble some women. Severe menopausal symptoms due to oestrogen deficiency may be helped by a course of oestrogen treatment.

Menstruation is a phase of the menstrual cycle of sexually mature women and other female primates. A flow of blood, cell debris, and mucus from the inner linings of the uterus escapes from the vagina for about five days. It occurs once a month on average from the first menstrual cycle, or menarche, until the *menopause. The menstrual cycle is under hormonal control. Gonadotrophins from the pituitary, acting on the ovary, induce an output of oestrogen, the shedding of an ovum about two weeks after menstruation begins, and release of progesterone. Oestrogen and progesterone cause renewal of the linings of the uterus cast away during menstruation. Should the ovum, on its passage to the uterus, become fertilized, it implants there. Implantation induces hormonal adjustments which prevent the appearance of menstruation, an early indication of pregnancy.

Mental illness is a disturbance of thinking and behaviour not attributable to brain damage. Acute illnesses of this sort can usually be classified as *neurosis or *psychosis. Treatment with psychiatric counselling and drugs may be effective.

Mental retardation, also called subnormality or mental handicap, means reduced capacity for reasoning and understanding, arising in childhood. It may be caused by genetic defects like *Down's syndrome, or by acquired diseases of the nervous system. Affected children are slow to achieve the usual 'milestones' in development like sitting, walking, and talking. Subsequently, formal testing of *intelligence confirms the problem. Severely retarded children never catch up and as adults are rarely able to live independently.

Mergansers, or sawbills, are bay ducks which chase fish under water. Unlike other ducks, they have slender, tapering beaks, each mandible of which is edged with saw-like teeth for grasping slippery prey. The six species

A male red-breasted **merganser**, *Mergus serrator*, on a favoured rest site, a grassy bank just beyond the water. Mergansers have a northern distribution and frequent sheltered, shallow bays and inlets.

include the red-breasted merganser, goosander, and smew. Most of them have crests.

Merlins are small, swift *falcons, widespread in northern Europe, Asia, and North America as eight or nine subspecies. Their swift, erratic flight enables them to strike at birds and insects on the wing. In winter they fly south to warm temperate and tropical areas such as North Africa.

Mesembryanthemums, or Livingstone daisies, are part of a large family of succulent plants found mainly in South Africa. The leaves are swollen and fleshy, and most species are adapted to grow under arid, desert-like conditions. In some, only two leaves are produced on a very short stem and the relatively large flower is produced from a slit between the leaves. The flowers are, for the most part, large and colourful as in the popular garden annual *Mesembryanthemum criniflorum*.

Mesites are three little-known species of birds confined to Madagascar. Related to the rails but looking rather like ground-living pigeons, they are often said to be related to the *babblers. These unusual birds are about 25 cm. (10 in.) long, and greyish, greenish, or brownish in colour. They feed on a wide range of snails and insects and build nests of twigs low down in bushes.

Mesohippus: an ancestor of modern horses, intermediate between *Eohippus* and *Equus*. It lived between 32 and 26 million years ago in North America. In size it resembled a large dog, standing about 60 cm. (2 ft.) at the shoulder, and like *Eohippus* was a forest-dwelling browser. Three toes were present in the feet as the limbs became adapted for swift running and trotting.

Mesquite: a deciduous tree of tropical America with a taproot which reaches down 15 m. (50 ft.) or more. This characteristic makes it valuable in reafforestation of devastated dry country. A member of the pea family, it produces protein-rich 'beans' which are used as fodder. There are some forty species of *Prosopis*, and all those in cultivation are also called mesquite or algaroba.

Metabolism is the collective term for all the chemical reactions taking place in an organism. It is subdivided into catabolism, whereby complex substances are broken down into simpler ones, and anabolism, whereby new compounds are built up. The most important catabolic reactions are those of *respiration, which releases energy. Almost 60 per cent of the energy released in catabolism is in the form of heat. The rate of heat loss from an organism's body is called its metabolic rate, and provides an indication of the rates of all these chemical reactions; it is normally measured in kilojoules of heat released per square metre of body surface per hour ($kJ/m^2/h$). The rate increases during times of growth, muscular activity, or illness, and decreases during inactivity. Basal metabolic rate (BMR) is the minimum rate of energy release needed to keep the organism alive. For an adult man, the BMR is about 165 $kJ/m^2/h$, while for a woman about 150 $kJ/m^2/h$ is normal; children have a much higher BMR. Anabolism uses basic organic compounds which are originally produced by *photosynthesis to make proteins, lipids, and other complex molecules. Anabolic processes use energy in the form of *adenosine triphosphate, which has been produced by catabolic processes. The body of an organism normally maintains a delicate balance between catabolic and anabolic processes.

Metal marks, or judies, are butterflies found almost exclusively in the New World. The 1,000 or so species of small, colourful butterflies comprise the family Riodinidae, which also includes the snouts (Libythinae). Most species range from 2 cm. (¾ in.) to 6·5 cm. (2½ in.) in wing-span. In many ways they resemble the blue butterflies, some species having metallic colours. The sole European representative, the Duke of Burgundy fritillary, *Hamearis lucina*, is not a true fritillary, but resembles one in colour and pattern.

Metamorphosis is a change of form, and often of physiology as well, during the life history of an animal. Such changes are distinct from changes in mere size and those associated with sexual maturity. Metamorphosis is described as either complete or incomplete and it may be abrupt or gradual. In a complete metamorphosis, as between a caterpillar and a butterfly, an intermediate stage called the *pupa is found. In incomplete metamorphosis, as in a grasshopper, the young resemble the adults except that they have wingbuds in place of wings, and the feeding habits are the same. The most abrupt changes are also found in insects: a legless crawling maggot may turn into a housefly in the course of a few days. The transformation from tadpole to frog is an example of gradual metamorphosis, with the development of legs, and loss of gills and tail, taking place over a considerable period. Metamorphosis does not imply an increase in complexity; many parasitic adult animals are much less structurally complex than the free-living larvae from which they come.

Metasequoia is a genus of deciduous conifers with a single living species. This is a primitive group of trees which flourished many millions of years ago and is related to the *redwoods. The dawn redwood, *Metasequoia glyptostroboides*, was discovered in China in 1944 and is really a 'living fossil' as all other members of the genus are known only from fossils. It is readily recognized by the soft, pale green leaves arranged as flat 'fronds' in opposite pairs. It was confined to a single valley and in danger of extinction, but is now widely planted as an ornamental tree throughout temperate regions.

Mexican burrowing toads are extraordinary in appearance. They have small heads, small eyes, round distended bodies, and short limbs, which are partially hidden in the body skin. They are well adapted to burrowing, using large spade-like prominences on the feet. They feed almost exclusively on ants and termites, which they lick up using a rod-like tongue, apparently a unique method of feeding in frogs and toads.

Meyer, Adolf (1866–1950), was a Swiss-born American psychiatrist whose system of psychobiology established the principle that the patient's problems must be viewed in terms of total personality. His *Commonsense Psychiatry* (1948) expounds the principle.

Mice are considered by some authorities to be members only of the genus *Mus*, but the name is applied to many other small rodents. Members of the family Muridae can quite properly be described as mice but other small rodents are also called mice because of their appearance.

Examples of the latter include pocket mice, kangaroo mice, New World mice, dormice, and jumping mice. Within the murid family there is no clear distinction between rats and mice, but large murids tend to be called rats and small ones mice. To the purist, only rodents of the genus *Rattus* are true rats.

There are many murid mice. Some typical examples are the temperate *harvest mice and *woodmice, the African grass and striped mice, the Australian native mice, and the Asiatic tree and spiny mice. The most familiar mouse is the *housemouse, which can be a health risk. Some other mice are economically significant as pests of growing crops or stored foodstuffs, particularly in the tropics. Most mice are omnivorous, although predominantly vegetarian, taking insects only incidentally. Mice include some of the smallest of mammals, the pygmy mouse weighing only about 7 g. (¼ oz.).

Michaelmas daisies   *Daisies.

Microbiology is the study of micro-organisms such as bacteria, viruses, algae, yeasts, and moulds. This field of science started in the seventeenth century with the invention of the microscope, and since then the techniques

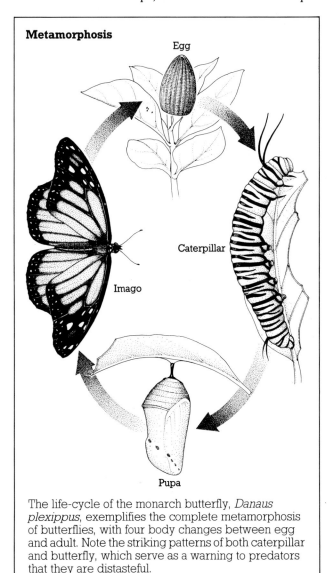

Metamorphosis

Egg

Caterpillar

Imago

Pupa

The life-cycle of the monarch butterfly, *Danaus plexippus*, exemplifies the complete metamorphosis of butterflies, with four body changes between egg and adult. Note the striking patterns of both caterpillar and butterfly, which serve as a warning to predators that they are distasteful.

have improved to the state where micro-organisms are now being studied at a molecular level. The advances made possible by the invention of the electron microscope have led to new scientific disciplines such as genetic engineering and virology.

The techniques of microbiology have had great impact on the medical sciences, enabling pathogenic organisms to be identified and isolated, and a cure diagnosed. Applications in biotechnology help to produce antibiotics, vitamins, metals, and proteins industrially.

Midges are, popularly, small *flies which swarm, particularly at dusk and near water, in such numbers as to constitute a nuisance. The biting midges, of the family Ceratopogonidae, are known in North America as punkies or no-see-ums. They are tiny flies whose females suck blood from insects, birds, or mammals, including man. Their bites are out of all proportion to their size. Their scavenging larvae are aquatic or live in damp places.

Non-biting midges, of the family Chironomidae, resemble mosquitoes, and the males have feathery antennae. The larvae of most are aquatic, feeding on decaying plant material, and include the bloodworms of stagnant water and sewage works, These are red in colour because they contain the respiratory pigment *haemoglobin, which helps them to obtain oxygen in stagnant water.

Midwife toads are species of the genus *Alytes*, which are so called because of the unique mode of parental care exhibited by the male. He carries a rosary-like string of eggs entwined around the hind-limbs. Initially the eggs are yellowish but, as development proceeds, they darken

The male European **midwife toad**, *Alytes obstetricans*, carries a string of eggs until the tadpoles hatch after several weeks (depending on the temperature). The young swim off when he next enters the water.

and are taken to, or deposited near, the water by the male. The tadpoles reach maturity in two to three months in the summer; late brood tadpoles usually overwinter. Three species of midwife toads are known, one from each of western Europe, Majorca, and Morocco. All are nocturnal and have a whistling call likened to the sound of chiming bells—hence the alternative name of bell toads. In the daytime they may dig burrows or hide away in the spaces in drystone walls.

Migraine is a common type of irregularly recurrent, severe headache. An attack usually starts with a flickering or blurring of vision and this may progress to coloured zigzag lines and partial loss of vision. At the same time, the headache begins, usually on one side of the head, and the sufferer may be sick. Attacks last 2–72 hours. The cause is poorly understood, although dietary factors are suspected.

Migration is the movement by animals from one place to another to find suitable breeding conditions or food. The distance travelled may be large or small. The arctic tern travels up to 40,000 km. (25,000 miles) each year. The daily vertical migration of zooplankton in the ocean is measured in metres. Not all migrations are return journeys. Mass migrations of locusts occur when they have stripped one site of all vegetation. Reindeer move on when food supplies are depleted, as do wildebeest in Africa and the red kangaroo of Australia. The period between migrations may be from a few hours to several years.

Animals frequently migrate to suitable breeding grounds. The wandering albatross, the fulmar, and other oceanic birds return to a particular site to breed, as do sea turtles, such as the green turtle. Frogs and other amphibians must migrate to suitable water for mating and reproduction. Many whales feed in Arctic or Antarctic waters but travel to the tropics to breed. Some bats travel

Milkweed's pollen is attached to clasps in slits in the ring of the male organs (above the reflexed petals). These clasps catch the legs of a visiting insect, dusting them with pollen, which is then carried off by the insect and then deposited on the female organ of another milkweed.

several hundred kilometres to hibernate. Monarch butterflies in Canada fly south to overwinter in vast numbers. The mechanisms by which animals navigate along their migration paths are not fully understood. Learning, olfactory senses, and magnetic orientation may all be involved. Birds are known to set a compass course by the sun, stars, and an inborn magnetic sense.

Mildew can mean any mouldy fungal growth, for example on damp stored materials. However, there are also two kinds of plant disease known as downy and powdery mildews, caused by fungi of the family Peronosporaceae and of the order Erysiphales, respectively. Powdery mildew of cereals, caused by *Erysiphe graminis*, is economically serious, and is controlled by breeding resistant crops or by the application of fungicide.

Milk is a fluid of high nutritional value manufactured by female mammals to nourish their young. Its principal constituents are water, milk sugar (lactose), fat, and protein. It is rich in calcium, poor in iron, and contains vitamins. Most substances dissolved in the maternal blood appear in the milk, including drugs, alcohol, and antibiotics. Milk is produced by a complicated sequence of

chemical and physical reactions by mammary glands. The glands of mammals enlarge during pregnancy under the influence of hormones. At the same time the secreting tissues and ducts are prepared for the production and ejection of milk after the birth of the young.

Although all animal milks are very nourishing, each species secretes milk of a different composition, particularly suited at any point in time to the growth patterns and age of its young. Rabbits' milk is rich in protein so that baby rabbits double their birthweight within a week. Human milk contains less protein than other milks, and babies take about six months to double their birthweight. About two days after birth, milk flows from the ducts of the mammary glands in response to suckling of the teats or nipples. The 'first milk' secreted, a yellowish fluid called colostrum, gives the young some protection against infections.

Milkfish: a species of large, silvery fish which grows to 1·8 m. (6 ft.) in length, and looks superficially like a herring. An Indo-Pacific fish, it has transparent, needle-like fry which are collected from estuaries and tidal pools when about 1 cm. ($\frac{1}{2}$ in.) long, and reared in freshwater pools until they reach a marketable size.

Milkweed butterfies   *Monarch butterflies.

Milkweeds are some 200 species of perennial plants belonging to the genus *Asclepias*, which consists of species native to America and Africa. Their brightly coloured flowers, often reds and yellows, have reflexed petals and are carried in umbels. The milky sap of many species contains poisonous *alkaloids and this property is used by insects such as the *monarch butterfly, whose caterpillars feed on milkweeds and obtain their toxins from them.

Miller's thumb   *Bullheads.

Millet is the name of a particular group of grasses characterized by their very small seeds and used as a cereal since ancient times. With the exception of bulrush millet, they are smaller plants than sorghum or maize. They are important cereals of the tropics and warm, temperate regions, tolerating drought and intense heat, and growing rapidly and maturing, even on poor soil. They keep well, the finger millet, *Elleusine coracana*, of the semi-arid areas of southern India and Africa being particularly useful as a famine reserve, since heads of it can be stored for five years.

Tef, *Eragrostis tef*, is the staple millet of Ethiopia, where it was first cultivated thousands of years ago. Common millet, *Panicum miliaceum*, is widespread as a food crop for man and livestock in the USSR, China, Japan, India, and southern Europe; in North America it is known as hog millet. Foxtail millet, *Setaria italica*, is known in Britain and North America only as bird-seed, but elsewhere in the warm temperate regions it is grown for human food.

A **millipede**, *Opisthospermorpha* species, crawls across the rain forest floor in southeast Brazil. Millipedes feed on decaying plant material which, when fragmented by them, is more easily broken down by micro-organisms.

Millipedes are many-legged, herbivorous *arthropods, living in thick humus, burrows, or even caves. They considerably outnumber their relatives, the carnivorous centipedes, especially in the tropics, with over 7,500 species. They can be distinguished by shorter legs, arranged with two pairs per segment; and the body is usually more cylindrical than in the flatter, more agile centipedes. They protect themselves with unpleasant secretions such as phenols or hydrogen cyanide, or by rolling into a ball when threatened.

Mimicry in nature is a form of defence, using colours or shapes, to confuse potential predators. There are two main types of mimicry, Batesian and Müllerian, named after those who first described them. In Batesian mimicry a harmless individual, the mimic, resembles a poisonous or dangerous species, the model. A predator which has tried unsuccessfully to eat the model will avoid the mimic. The American monarch, or milkweed butterfly, *Danaus plexippus*, normally contains poisons derived from its larval food plants. Several other species of non-toxic butterflies mimic the monarch butterfly in colour and pattern. After a predator tastes the unpleasant butterfly, other adults of similar pattern are likely to be avoided. Batesian mimicry is most effective where the model is common and the mimic rare; otherwise the predator may not learn to associate a particular pattern with distaste.

In Müllerian mimicry a group of distasteful or well-defended organisms have developed a common colour pattern. Many wasps and bees have the same barred black and yellow coloration. The advantage to these species is that a predator will sample a few unpleasant individuals

A slow-growing plant, this fruiting **mistletoe**, *Viscum album*, on an apple tree host, is likely to be at least ten years old.

of any one species and then generalize the experience to avoid all similar species. Fewer individuals from each species are sacrificed.

Some plants, such as the bee orchid, mimic an insect. When a bee attempts to copulate with it the orchid transfers its pollen to the bee. This is later deposited on another flower when the bee is fooled again, and thus pollination is achieved.

Mimosa   *Acacias.

Miner's cat   *Ring-tailed cat.

Mink are carnivores related to weasels and otters. There are two species, the North American mink, *Mustela vison*, and the much rarer European mink, *M. lutreola*, which is now confined to the USSR east of the Urals. Mink are prized for their fur and many are bred on fur farms. Escapes have led to the American mink becoming established as a wild species in Scandinavia, Iceland, Germany, USSR, and most of mainland Britain. Mink are generalized carnivores and feral mink occasionally become pests.

Mints are aromatic plants belonging to the genus *Mentha*. The most familiar of several species is the common or spear mint (*M. spicata*), a perennial, lilac-flowered plant growing up to 90 cm. (3 ft.) tall. Indigenous to southern Europe, this species is now widely grown as a garden herb. Other mints, including water mint, *M. aquatica*, and corn mint, *M. arvensis*, are true species, but some such as peppermint and French mint are hybrids. The leaves of mints are covered with glandular hairs and, when dried, are used as culinary herbs.

The Labiatae family, which is typified by mints, contains some 3,000 species of herbs and shrubs. They are found worldwide and include thyme, marjoram, *Salvia*, and *Coleus*.

Mirid bugs, sometimes called capsids, occur on vegetation almost everywhere, from sea-shore to mountaintop. Most species suck plant juices, but a few are carnivorous, and they vary widely in appearance and habits, constituting the largest family of *bugs with over 6,000 species. Few exceed 1 cm. ($\frac{1}{2}$ in.) in length, most are elongate and soft-bodied, and many are green or brown. The tarnished plant bug, *Lygus pabulinus*, and apple capsid, *Plesiocoris rugicollis*, are pests wherever fruit trees are grown. They feed on buds, young shoots, and developing fruit. Some species are good mimics of ants, and probably feed on their larvae.

Missel-thrush, or mistle-thrush, is a largish greyish *thrush, 25 cm. (10 in.) long, with a rather upright stance and a harsh chattering call. It is a species found across most of Europe, and on the higher ground of eastern Asia. It lives in open country with scattered clumps of trees, and feeds on many small animals, fruits, and berries. The name, mistle-thrush, is derived from its habit of eating mistletoe berries.

Mistletoes are a family of parasitic plants with green leaves and roots which penetrate their host plant. They include some 1,300 species with a mainly tropical distribution, but extending into temperate regions. Most species absorb nutrients from their host and some tropical species are a problem in citrus plantations. They produce small flowers which develop into small fruits with sticky flesh which adheres to the beaks of birds. The mistletoe seeds are distributed by birds which feed on their berries, often being placed on new host plants as the birds clean their beaks.

They are used as Christmas decoration in Europe and North America. The common Eurasion mistletoe, *Viscum album*, is a parasite of trees such as oak, hawthorn, and apple. The equivalent species in North America is *Phoradendron flavescens*.

Mites are tiny and often beautifully coloured *arachnids. Most are less than 1 mm. ($\frac{1}{25}$ in.) long and so able to live in minute habitats unavailable to other arthropods, usually on land but sometimes under water. They comprise the order Acarina, which also includes *ticks among the 35,000 species, some of them a threat to human health or to commercial products. In mites the front and back parts of the typical arachnid body are fused to give a simple ovoid shape with eight short legs. Males and females are similar, although the female may be larger, and mating involves the transfer of sperm packages by the

male, using his jaws. The young stages have only six legs.

Mites are very diverse in their habits. Some eat green plants, and the spider mites are pests of cotton and fruit trees. Many scavenge on dry products, like the house-dust mite, *Dermatophagoides*, which is notorious for causing human allergies. 'Chiggers', *Trombicula*, are attracted by warm carbon dioxide in mammals' breath. Their larvae are parasites, feeding off skin and producing severe itching in man. Mites can transmit disease in the tropics, above all mange and scabies. Many species hitch a lift on insects or birds to transfer between food sources.

Mitochondria are minute, cigar-shaped structures found in the cytoplasm of all *cells except those of *bacteria and *blue-green algae. About 0.003 mm. (0.0001 in.) long, they are the site of a vital process called *respiration, in which energy is released for the cell's needs. The series of chemical reactions which occurs in respiration is accelerated by *enzymes, positioned on the folded inner membrane of mitochondria. The outer membrane forms a simple barrier between the mitochondrion and the rest of the cell.

Mitosis is the process by which one *cell divides to produce two genetically identical cells, each called a

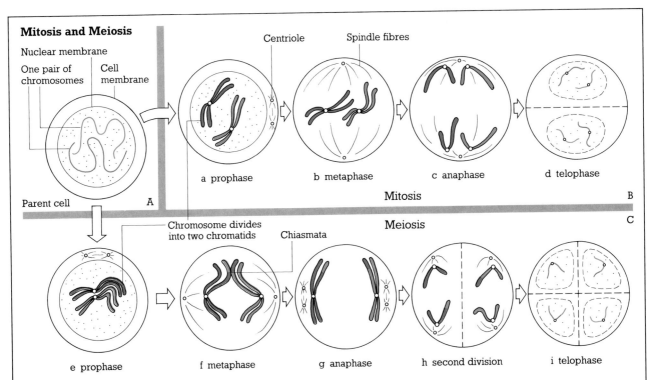

Mitosis and Meiosis

Nuclear membrane
One pair of chromosomes — Cell membrane — Centriole — Spindle fibres

a prophase   b metaphase   c anaphase   d telophase

Parent cell   A   Mitosis   B

C   Meiosis

Chromosome divides into two chromatids — Chiasmata

e prophase   f metaphase   g anaphase   h second division   i telophase

Cells can divide in one of two ways: mitosis (B) produces exact copies of the parent cell; meiosis (C) produces cells (gametes) which contain only one chromosome from each parental pair. The cell shown here (A) contains only one pair of chromosomes, each depicted in a different colour to simplify events in cell division. The prophase (a) of mitosis involves a shortening and thickening of the chromosomes, and their division into two chromatids (each equivalent to the original chromosome). In metaphase (b) a cell organelle called a centriole migrates to each pole of the cell and produces a web of spindle fibres. In anaphase (c) these fibres are believed to pull the chromatids apart from the equator of the cell. After

separation the chromatids, now true chromosomes, gather inside a new nucleus before the cytoplasm divides (d) and two new cells are formed. In meiosis the chromatid pairs come to lie alongside each other in prophase (e) and become attached at points called chiasmata. As these chromatid pairs are pulled apart in metaphase (f) some sections break off and rejoin the 'wrong' partner. This serves to exchange genetic material present in chromosomes. In anaphase of meiosis (g) each pair of chromatids is still associated and is not separated until the subsequent division stage (h) which is absent in mitosis. This division (i) produces four daughter cells, each with only a single chromosome from each cell pair.

A European **mole** peers into the sunshine, its eyes able to register only changes in light level. Its comparatively huge front claws are well adapted for digging.

daughter cell. In plants and animals, all growth occurs through mitotic divisions. Mitosis first involves the division of the nucleus and *chromosomes, then the division of the cytoplasm to form two discrete cells. There are four phases recognized in mitosis (prophase, metaphase, anaphase, and telophase) and these are preceded by the cell enlarging and the chromosomes becoming apparent as thread-like strands in the nucleus. These strands shorten and become more prominent during prophase. There are normally two copies of every chromosome in each cell, and during prophase each chromosome reproduces itself. This gives each chromosome a double appearance as each now consists of two halves or chromatids. During metaphase and anaphase, these 'double' chromosomes are pulled apart, each chromatid (now called a true chromosome) migrates to opposite ends of the cell where it is surrounded by a new nuclear membrane; the cell then separates to form two new cells.

Moas are extinct flightless birds related to the living kiwis and cassowaries. Some species grew as tall as 3 m. (10 ft.) with long legs for running and very long necks. They were confined to New Zealand, and one species, *Dinocras maximus*, was still present when man reached those islands. The last known specimens were seen in the late seventeenth century.

Mockingbirds are members of a family confined to the New World and including also catbirds and thrushes. In appearance a little like long-tailed thrushes, several of the thirty-one species are greyish or brownish above with dark-spotted, pale underparts. The Galapagos mockingbird, *Nesomimus trifasciatus*, is striking for having developed several different forms on different islands, though these have not diverged as far as the Galapagos finches. They were so named because of the remarkable ability of the eastern mockingbird, *Mimus polyglottos*, to mimic the songs of a wide range of other birds.

Mock oranges are deciduous shrubs of the genus *Philadelphus*, native to the northern temperate zone, particularly the Far East. They are widely cultivated for their white flowers, which have a heavy scent, similar to that of orange blossom. Sometimes they are known as syringa, causing confusion with the unrelated plants of the genus *Syringa*—the lilacs.

Mole crickets are large *crickets which burrow in soil. Their front pair of legs is much modified for digging and they are unable to jump. The males trill loudly underground. They fly freely at night and may damage root crops. The common mole cricket, *Gryllotalpa gryllotalpa*, one of fifty species, has been introduced to North America from continental Europe.

Mole rats belong to two distinct *rodent families, whose similarity is due to convergent evolution. The family Bathyergidae, or African mole rats, includes some nine species, while the Muridae, or rat and mouse family, includes sixteen species such as the blind mole rats, East African mole or root rats, and Central Asiatic mole rats or zokors. All dig extensive burrows and feed mainly on roots. Unlike moles, they burrow, not with their feet, but with their huge front teeth. The most unusual is the African naked mole rat, *Heterocephalus glaber*, which is the nearest mammalian equivalent to social insects, with reproductive and worker 'castes'.

Moles are insectivores belonging to the family Talpidae. In addition to the European mole, *Talpa europaea*, the family includes twenty-eight other species, such as the very similar American moles, mole-like shrew moles, and desmans. Most moles construct a system of underground

A bird that tastes a toxic North American **monarch butterfly** generally avoids all further contact. Colonial roosting reinforces the warning implicit in the coloration and is thus safer for all the members of the colony.

burrows in grassland, throwing up the excavated soil at intervals to form molehills. Once a burrow system is constructed, active tunnelling ceases. The mole uses its burrows as a trap to catch earthworms that fall through into the tunnels. Surplus earthworms are immobilized by chewing their anterior ends and the still-living worms are stored in underground chambers for later consumption.

The desmans, of which there are two species, the Russian, *Desmana moschata*, and Pyrenean, *Galemys pyrenaicus*, are well adapted to an aquatic life. Their fur is two-layered, with a dense, waterproof underlayer, and their toes are partly webbed. The star-nosed mole, *Condylura cristata*, of North America has a sensitive snout which extends into pinkish, fleshy tentacles; these are used to 'feel' for food.

Mole salamanders are a small family of approximately thirty-three species of small to moderately large salamanders, up to 40 cm. (16 in.) long. The majority are stoutly built, terrestrial animals with broad heads and sturdy limbs, and have deep, vertical grooves along the body (marking the positions of the ribs). An aquatic stage is always present and persists in some, including the well-known *axolotls. Typically, adult male salamanders are terrestrial and capable burrowers.

Molluscs rank as the second largest of animal phyla, with 120,000 species; far behind the arthropods but at least twice as numerous as vertebrates. They perhaps rank highest of all in terms of diversity of appearance: snails, oysters, sea slugs, and octopods are all molluscs. They live in every possible ecological niche, from deep sea to deserts, and have evolved forms as diverse as mobile carnivores, sessile filterers of planktonic food, or rapacious consumers of garden vegetables.

The basic molluscan body-plan is fairly constant. There is a soft body, often with a creeping, slug-like foot and a covering of hard chalky shell, made of crystals of calcium carbonate embedded in a special protein, giving great strength. The head and foot can be withdrawn into the shell for safety in species with a one-piece coiled shell, such as snails and winkles. In those species with a two-piece shell, the bivalves, all the soft parts are enclosed and even the foot rarely protrudes. In the third main class, the cephalopods, the shell is often *vestigial and the soft body is adapted for high-speed jet propulsion instead.

Molluscs are abundant animals, especially in shallow seas and on rocky shores, where their protective body-plan is especially appropriate. They form a vital food source for fish, birds, and man. They are one of very few invertebrate groups successful on land; a special ribbon of teeth, the radula, enables them to eat tough green vegetation, and the shell or thick mucus secretions give considerable protection against desiccation.

Monarch butterflies, or milkweed butterflies, are large, tough, toxic, butterflies, strikingly patterned with some combination of orange, black, and white. They include some 200 species with a mainly tropical distribution, which are sometimes known as tiger butterflies. Most species incorporate poisons from their larval food plants and serve as models for palatable mimics. Males have scent patches on the hind-wings, and clustered hairs at the tip of the abdomen for disseminating the scent. Most are found in the Asian tropics, but they also occur in Africa, Australia, and America. The North American monarch,

Danaus plexippus, migrates south in winter to enormous communal roosts in southern California, Florida, and Mexico. Successive summer broods move as far north as Canada, and occasional vagrants reach Britain. This same species is also common over most of the Pacific region, including Australia.

Mongolism *Down's syndrome.

Mongoose is not a precise scientific name but is applied to a family of carnivores which includes the genets and civets but may be applied specifically to the subfamilies Herpestinae and Galidiinae. The latter includes four species of Madagascan mongooses, some of which are striped in colour. The African and Asian mongooses (subfamily Herpestinae) include twenty-seven unpatterned species. They feed on a variety of prey (including snakes), some being purely insectivorous, but most take rodents, eggs, and any other small vertebrate they can catch. Most mongooses are solitary but some, such as the banded mongoose, *Mungos mungo*, and meerkats, are social and forage in bands. Mongooses are roughly the size of an otter but the smallest, the dwarf mongoose, *Helogale parvula*, weighs only about 0·7 kg. (1½ lb.). Some Indian species have been introduced to the West Indies to control rats in sugar plantations but usually they become very destructive to native mammals and birds.

Monitor lizards are a family which includes the Komodo dragon, the largest living lizard. Roughly half of the thirty known species are found in Australia, where they are commonly referred to as goannas; the remainder occur in Africa through Asia to the East Indies. Most are strikingly elongate and of alert appearance. They flick out their long, forked tongue to examine their environment, rather in the manner of snakes. They can defend themselves with their strong jaws or heavy claws and by thrashing around their powerful tails. Monitor lizards are all very similar in form, being long and slender, with a long neck, head, and tail. Most are heat-loving, terrestrial or arboreal species, but at least one species, the Nile monitor, *Varanus niloticus*, is at home in water.

Monkey flower *Musk.

Monkey-puzzle tree: a native of Chile and an example of a subtropical conifer. Also known as Chile pine, though not a true pine, it is one of eighteen species of the genus *Araucaria* found in New Guinea, eastern Australia, the islands of the Pacific, and parts of South America. The monkey-puzzle has a regular dome-shaped crown of downward-pointing branches with close-set, leathery, pointed leaves. These stay on the tree for up to twenty years, so that even the old branches are clothed in leaves. It is the nearest living example to the trees of the Carboniferous Period, which gave rise to our main coal deposits. The seeds of the monkey-puzzle are edible. Other species of *Araucaria* include the Norfolk Island pine, *A. heterophylla*, and parana pine, *A. angustifolia*.

Monkeys are higher *primates and have relatively large brains, acute vision, and hands developed as useful grasping organs. Two distinct groups have arisen, apparently independently, from prosimian ancestors. One group, the New World monkeys, is found in South America. These are called platyrrhines (flat-noses) and have prehensile (grasping) tails. This group is represented by a variety of forms, including capuchins and marmosets. The second group is the Old World monkeys, called catarrhines (drop-noses), of Africa and Asia. They include the colobus monkeys, mangabeys, baboons, and the apes. The Old World monkeys are not markedly different in general habits and organization from those of the New World and it would appear that the two groups have made many changes in parallel.

Sometimes growing to over 30 m. (100 ft.) in its native Chile, the **monkey-puzzle tree**, *Araucaria araucana*, attains a height of about 24 m. (80 ft.), when planted in Britain, like this specimen. Male and female flowers are usually borne on separate trees and the cones of the latter, which take two seasons to mature, are about 15 cm. (6 in.) across.

Monkfish: a *cartilaginous fish of the family Squatinidae, also known as angel sharks. The body is flattened from above and the pectoral and pelvic fins are expanded, though the former are not joined to the head as in the rays. It is an ugly fish, with gill-openings on the side of the head and a terminal mouth, both of which are shark features. It is found in shallow water in the warm temperate parts of the Atlantic and Pacific, and one species is a summer visitor to southern Britain. Its food comprises mostly bottom-living fishes, especially flat-fishes and rays. It gives birth to litters of between nine and sixteen young in summer. It is a food-fish with very firm flesh.

Monkshood   *Aconites.

Monocotyledons are a subgroup of *angiosperms. They are characterized by having a single *cotyledon in the seed, leaves with parallel veins, and flowers with parts arranged in threes or multiples thereof. Those, such as palms and bamboos, which have woody stems, develop in a different way from *dicotyledons. There are many exceptions to these general features, for example yams have a network of leaf veins like dicotyledons. Monocotyledons include palms, grasses and bamboos, orchids, irises, and lilies. Although there are fewer families and species than in the dicotyledons, they dominate the grasslands of the world and are economically important as food crops. *Cereals are the most important monocotyledons to man, but others include onions, pineapples, bananas, and many fibre or spice plants.

Monsoon forests   *Forests.

Montbretia: a hybrid species of plant derived from South African plants with swollen rootstocks (corms) and belonging to the iris family. The leaves are sword-shaped and the spikes of trumpet-shaped flowers may be yellow, orange, or red. The popular garden montbretia is a cross between *Crocosmia* species. Other true species of *Crocosmia* are sometimes also called montbretias.

Moonfish   *Opah.

Moor was originally any tract of open land. Today the word is commonly used for all vegetation types where peat accumulates more quickly than it is broken down. Land like this is found in cold and wet areas where soil micro-organisms cannot act fast enough to decompose dead vegetation, and where basic mineral ions which usually promote breakdown are in short supply. The dominant vegetation of moors often includes heathers or pines, which can tolerate acid conditions created in bog peat. Fens are a type of moor, dominated by plants which can tolerate neutral to basic peats with low nutrient levels.

Moorhens   *Gallinule.

Moose, known in Europe as the elk: the largest living member of the *deer family. A bull stands 1·8 m. (6 ft.) at the shoulder and has huge antlers up to 1·8 m. (70 in.) across with numerous points. The head is long, the ears large, and the nose bulbous. A growth of skin and long hair hanging from the throat is known as the bell. The moose can reach high into willow trees to browse on leaves and twigs. In the rutting season in autumn, the bulls fight for possession of the cows. One or two young, or on rare

Moose, like other deer, shed the fur-covered skin of their antlers (the velvet) each autumn. To do so the antlers are thrashed against vegetation, as this mature bull moose in Canada has done.

occasions three, are born. The calf is like its mother in colour, unlike the young of other deer which have stripes or spots.

Moray eels are a family of some eighty species of *eel, mainly found in tropical and subtropical waters. These eels, which reach 1·5 m. (5 ft.) in length, are voracious predators with sharp teeth, normally eating other fish but also striking at divers who approach too closely.

Morels belong to the group of cup fungi. The common morel, *Morchella esculenta*, has a fruiting body which is club-shaped with a pale stalk and darker head covered with a network of ridges; in all it is about 15 cm. (6 in.) high. It is edible and appears in spring. Several other species of the genus *Morchella* have morel as part of their common name.

Morphology is the study of the form and structural arrangements exhibited by living things. Humans and other mammalian embryos are morphologically indistinguishable until they reach the foetal stage. By contrast morphological differences do exist between races, and between individuals of the same race. These differences have a genetic origin, but some morphological changes can be brought about by environmental influence. For example, in man, teeth can be straightened in the jaw by use of metal bands. The study of morphology strongly supports the theory of evolution and is the foundation of *palaeontology.

Morphos are large, South American, forest butterflies of the sub-family Morphinae. In most of the 200 or so species, the male is a vivid metallic blue on the upper-side. The iridescent blue colour is a result of fine structures on the wing scales splitting up white light, rather than a pigment in the scale. Their wings have been used abundantly in decorative work, particularly jewellery. In some countries they are now protected by law. Females and males of some species are brown with yellow, white, or blue markings, and one species is silvery-white. Some fly slowly over the ground, others soar around tree-tops. The gregarious caterpillars are usually vividly coloured, often bristly, with yellow and red stripes; they feed on forest trees or bamboos.

Mosquitoes are delicately built true *flies with long legs and narrow wings bearing scales and hairs. Their wing-span ranges from less than 3 mm. up to 15 mm. ($\frac{1}{8}$ to $\frac{1}{2}$ in.). Their antennae are long, with whorls of hairs which are more developed in the males. The mouthparts are adapted for piercing and sucking and most feed on the blood of mammals or birds. Their larvae are always aquatic, in fresh, brackish, or sometimes fully salt water. Some freshwater species may live in old cans and tree holes, while others use lakes, ponds and reservoirs. The buoyant eggs are laid either in floating rafts, or singly. The larvae (wrigglers) have brush-like mouthparts and feed by filtering microscopic plankton or detritus from water. They breathe through modified organs on the tail, some lying with the body hanging down below the water, others with it parallel to the surface. Some gain air by piercing plant stems. The pupae are active with well-developed muscles in the tail and they breathe through trumpet-shaped organs on the thorax. Mosquitoes are by far the most important insect vectors of the causative agents of human disease, the various forms of *malaria being foremost, although yellow fever and filariasis are also frequently spread by mosquitoes. More than 1,600 species are known throughout the world.

Moss animals *Bryozoans.

Mosses are small green plants, simply constructed compared with other land plants. Like liverworts, ferns, horsetails, and clubmoss, they exist in two different forms during their life-cycle. The first form is the leafy plant, popularly called 'moss', which is the larger and lasts longer. This carries the male and female organs and, after fertilization, produces the second form of moss 'plant'. This is the spore-bearing capsule which is raised on a leafless stalk above the 'moss' plant. The spores released from the capsule blow away and germinate in damp places to produce new leafy plants.

  Mosses have no roots, and their leaves are very simple in construction. Because they are not able to keep their internal tissues supplied with water from the soil, they can grow only in habitats that are permanently or frequently wet, forming either transverse wefts or cushions of vertically growing shoots. They grow on soil, rocks, and tree trunks and a few species grow under water. *Sphagnum moss can cover large areas of wet, acid uplands.

Moths, together with butterflies, constitute the order Lepidoptera, containing more than 100,000 species. Moths are characterized by two pairs of wings covered with microscopic, overlapping scales. The scales are coloured, and some are specialized as scent-gland outlets. Moths tend to be stouter-bodied and duller in colour than butterflies and comprise 80 per cent of the total number of species of Lepidoptera. Most species fly at night; some, however, such as burnet moths, are brightly coloured and day-flying. Unlike butterflies, most rest with the wings overlapping the body, and fore- and hind-wings are held together with a hook-and-eye arrangement.

  They range in size from large atlas moths to the tiny so-called microlepidoptera, many of which have a wing-span of only 2 mm. ($\frac{1}{12}$ in.) or so. Characteristically, moths use their proboscis to feed on nectar; but gold moths have jaws and eat pollen, and some adults, such as emperor moths, never feed. The antennae of many males are

This **moss**, *Polytrichum juniperinum*, is often abundant on scorched areas, as here on a heath in Surrey, England. Its spore-carrying capsules are in the background.

feathery or comb-like, and are covered with microscopic scent receptors to detect the female *pheromones.

The caterpillars of most species of moths are plant-feeders, many, like the owlet moths, being catholic in choice of food plant. Moth caterpillars also include species which bore into timber, eat hair, wax, or even drugs. The variety of life-styles is enormous, ranging from tiny, concealed leaf-miners to the gaudy black and yellow caterpillars of the cinnabar moth, *Tyria jacobaeae*. As in butterflies, *metamorphosis is complete, and a pupa, often enclosed in a cocoon spun of silk, is formed in protected crevices, or in the soil.

Motion sickness, or travel sickness, is caused by overstimulation of the organs of balance in the inner ear. People differ greatly in their tolerance of motion, and most will become more tolerant on repeated exposure to such movement. Common symptoms include headache, nausea, or vomiting. There are drugs which are moderately effective in preventing the condition.

Motmots are a small family of eight species of birds allied to the kingfisher, all restricted to warm areas of the New World. They vary in size from that of a small thrush to that of a crow and are mainly green, though the head is often marked with blue and black. The tail is long, especially the two central tail feathers which are bare of vanes for a short distance before the end, so giving a racquet-like appearance to the tip. They hunt by flying out from perches to catch flying insects and nest in burrows which they dig in banks.

Mouflon: the smallest of the wild *sheep, native to Corsica and Sardinia and related to the argalis. It inhabits the mountains, living in small flocks, in rough and often precipitous terrain. The coat of the mature male is reddish-brown with a whitish saddle. Both sexes usually have horns, although those of the female are often very small. After a gestation period of about six months, one to three lambs are born.

Moulds are *fungi—they may be *saprophytes or parasites—that grow as a mass of microscopic threads (mycelium) with small lollipop-like, spore-producing bodies. Mould spores are abundant in the air, so they can easily infect and decay damp organic materials, or cause plant or animal diseases in the case of parasitic fungi. Moulds produce both useful and poisonous chemicals, for example the antibiotic penicillin and the poison aflatoxin. Blue mould is caused by saprophytic species, such as *Penicillium roquefortii*, and gives certain cheeses their flavour. Grey mould on fruit is caused by the parasite *Botrytis cinerea* and causes serious damage to crops, but is welcomed as '*pourriture noble*' by wine-growers as it aids the production of sweet wines. Fungal moulds are not to be confused with the *slime moulds.

Mountain ash   *Rowan.

Mountain beaver: the most primitive *rodent alive today, occurring in western North America. It is neither a beaver nor a mountain animal: a more appropriate name of this species is sewellel, which comes from a Chinook word for the cloak made from the animal's skin. It does not even look much like a beaver as it has only a short tail, and is up to 1.5 kg. (3⅓ lb.) in weight and 40

cm. (16 in.) in body length. The animal spends much of its time in its extensive burrow. It is a vegetarian and is usually found near water.

Mousebirds make up a small family of six species, all of which are confined to Africa. Although only a little heavier than sparrows, they look larger because of their greatly elongated tail. They are all pale grey or brown, with marked crests; some have red, blue, or white on the head. They live in small flocks and roost closely together at night.

Mouse deer   *Chevrotains.

Mouse-tailed bats form a family of *bats which occur in the Middle East and southern Asia. The three species are characterized by a very long tail, which at around 6 cm. (2¼ in.) is roughly equal to the length of the body. They roost in large numbers, often in man-made structures such as the Egyptian pyramids. They are insectivorous and have a seasonal state of torpor when insects become scarce.

Movement, or the ability to change position, is one of the fundamental characteristics of living organisms. This capacity is most highly evolved in members of the animal kingdom in response to their need to seek or catch food.

Some micro-organisms move by simply drifting in air or water. More advanced protozoans can extend pseudopodia (finger-like projections filled with cytoplasm) or propel themselves by means of flagella. Bundles of contractile fibres, known as muscles, occur in most multicellular animals. By shortening in length (contracting) and acting against non-contractile structures (skeletons), these fibres cause parts of animals or entire organisms to change their position relative to the surroundings. Pseudopodia, cilia, and muscles may also produce movement within animals, for example the movement of food through the alimentary canal (peristalsis).

Plants generally lack specialized structures for generating movement. Most higher plants, however, exhibit tropic movements (tropisms), such as the growth curvatures of stems and roots, towards or away from stimuli such as light, gravity, water, or touch. Unlike animal movements, these growth responses are largely irreversible and occur slowly. Plants also show nastic movements, responses to non-directional stimuli, for example the opening of flowers during daylight.

Mud eels   *Sirens, *Congo eels.

Mud puppy is the common name for *Necturus maculosus*, a species of *salamander found in the eastern USA. It does not completely transform into an adult salamander, remaining as a permanent larva. The mud puppy belongs to the family Proteidae, which includes the olm. It is stout-bodied, and 20–45 cm. (8–18 in.) in length, including the tail. It has bright red gills, small eyes, and four toes on both fore- and hind-limbs. An inhabitant of rivers, streams, and lakes, it lays 30–200 eggs, which develop normally into tadpoles that may take up to six years to reach maturity.

Mud-skippers are a group of *gobies belonging mainly to the genus *Periophthalmus* and found in inshore waters, especially mangrove swamps and muddy intertidal flats,

A **mud-skipper** astride a mangrove root erects its dorsal fin, warning others to keep their distance. This semi-aquatic fish is about 14 cm. (5 in.) long, with mottled brown coloration.

of the Indian and west Pacific Oceans. They frequently leave the water altogether and lie on the mud, or climb the exposed mangrove roots, using their pectoral fins as levers and the pelvic fin sucker to stay in place. Their eyes are placed high on the head and give good all-round vision. They are very alert to predators, skipping away across the mud to hide in the burrows of crabs or returning to water at the first sign of danger.

Mulberries are fruit-trees of the genus *Morus*. The black mulberry, a tall, deciduous tree, growing up to 10 m. (33 ft.), has been grown in Europe since ancient times although it is native to western Asia. The delicious purplish berries resemble loganberries. The white mulberry is a small tree native to China. Its pinkish-white fruits are edible, and the bark has been used for making paper, but it is grown particularly for its leaves, which are fed to silkworms. Mulberries belong to the same family as figs and hemp.

Mules are hybrids between a *horse and an ass, more especially the offspring of a male ass and a mare. The hybrid from a she-ass and a stallion is often called a hinny. Mules have the shape and size of a horse, and the large head, long ears, and small hoofs of an ass. They are usually sterile, although occasionally a hinny will produce a foal. Mules have great endurance and are sure-footed, making excellent pack or draught animals.

Mullein is the common name for a large group of chiefly biennial herbs related to antirrhinums, foxgloves, and figworts; their generic name is *Verbascum*. They are natives of Europe, Asia, and North Africa, but are also widespread as introduced weeds in other parts of the world. The basal rosettes of large, often grey and woolly, leaves, are pro-duced in the first season and the handsome spikes of yellow or reddish-purple flowers, growing up to 2 m. (6 ft. 6 in.) tall, develop in the second year.

Mullets belong to two quite unrelated families of fishes, namely the grey mullets of the family Mugilidae, and the red mullets of the family Mullidae. The former are found in all tropical and temperate seas; some species occur as freshwater fish in parts of the tropics. Greyish in colour, they are broad-headed, torpedo-shaped fishes with two dorsal fins, the first containing only four strong spines. Living in schools, they are most common in shallow water and can be seen browsing on the fine green algae on rocks and harbour walls, and also sucking up mud from the sea-bed. Internally, they have a thick-walled stomach and a very long intestine which allows them to digest algae.

Red mullets, known in North America as goat fishes, are usually red in colour, although many become paler at night. They have two long barbels on the chin, which are used as feelers to detect worms, crustaceans, and molluscs buried in the mud and sand. Worldwide in range, they are most abundant in tropical seas.

Multiple sclerosis is a disease affecting the *nervous system. It is characterized by repeated attacks, each of which clears up to a considerable extent but which contributes to an increasing state of muscular weakness. The condition may be arrested at any time, and the course and final outcome are unpredictable. The disability is caused by progressive damage to the membranes covering nerves. The cause of the condition is a mystery in spite of a number of clues. There are great geographical differences in incidence. For example, it is rare in South Africa, except among first-generation immigrants from northern Europe.

Mumps is a specific virus infection, causing fever, headache, vomiting, and eventually swelling of a pair of glands that produce saliva. In contrast to highly infectious diseases, such as measles or chicken pox, it affects relatively few children, so that many adults are susceptible. Once caught, it is mildly infectious for up to ten days after symptoms appear. The infection may spread to other salivary glands and to the pancreas, brain, and testicles. In adult men, mumps can cause sterility.

Muntjac are a subfamily of small *deer, containing six species widely distributed in the wild in southern and Southeastern Asia. The antlers are simple and supported by long, skin-covered pedicels which continue down the forehead as convergent ridges marked by lines of dark hair. Their coat is deep chestnut in colour, and both sexes have tusk-like canine teeth in the upper jaw that are used for fighting. Reeves's muntjac, *Muntiacus reevesi*, has a loud dog-like bark which is repeated at regular intervals when it is alarmed or disturbed—hence it is commonly called the barking deer. This species has been kept as an ornamental park animal in parts of Europe, and is now feral in many countries there.

Muscles can be discrete bodies or sheets of tissue, both capable of shortening in length, or contracting. They are used to convert stored chemical energy into mechanical movement. The muscle consists of bundles of fibres, each made from strands, or fibrils, of two types of protein molecules, actin and myosin. These substances are attached to one another in such a way that a change in the chemical conditions surrounding them can alter their structure and make one slide over the other. This involves the use of energy, supplied by a substance called *adenosine triphosphate.

There are three basic types of muscle, each distinguished by the appearance of its fibres when magnified. Striated muscle, which looks banded under the microscope, is the familiar skeletal muscle which we see as 'meat'. It is used to control movements of the skeletal bones and is usually under voluntary control. Smooth muscle, lacking bands, is found in many internal organs such as the stomach wall. It is usually under the control of the *autonomic nervous system. The last type is cardiac muscle, found only in the walls of the heart.

The energy needs of muscle mean that they are richly innervated with blood vessels, hence their rich, red colour. The blood also removes waste products of the contraction process, particularly a substance called lactic acid. All muscle contractions are initiated by an impulse from a nerve ending. This releases a chemical which makes the membranes of the muscle fibres permeable to certain chemicals. These in turn create the conditions under which contraction occurs. The relaxation of muscles happens when these contraction-causing chemicals are taken back across the membrane.

Mushrooms are spore-bearing, umbrella-like fruiting bodies of certain fungi belonging to the group Agaricales. Like toadstools they have a stalk and a whitish cap with brown or pinkish gills underneath which produce the spores. Fungal mycelium (the vegetative part) may grow for many years in soil, producing mushrooms at intervals, usually at the same time each year. There is no biological difference between mushrooms and toadstools.

Musk deer are three species of small, solitary *deer, formerly widespread in the forests and scrubland of East Asia. Devoid of antlers, they have instead long upper canine teeth. These tusks, used in fights, are 7 cm. (3 in.) long in the male, and shorter in the female. The thick coat of dark brown is slightly mottled with grey; the hair is distinctive because of air-filled compartments enhancing insulation. They feed on lichens and mosses, which are

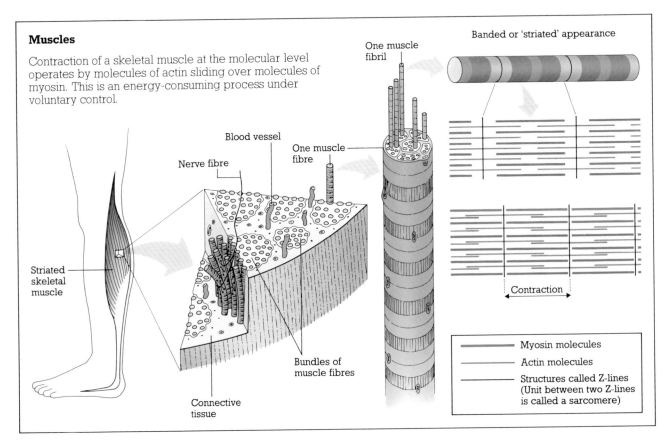

Muscles

Contraction of a skeletal muscle at the molecular level operates by molecules of actin sliding over molecules of myosin. This is an energy-consuming process under voluntary control.

Banded or 'striated' appearance

One muscle fibril

Blood vessel

Nerve fibre

One muscle fibre

Striated skeletal muscle

Bundles of muscle fibres

Connective tissue

Contraction

Myosin molecules
Actin molecules
Structures called Z-lines
(Unit between two Z-lines is called a sarcomere)

chiselled from rock and tree surfaces by spatula-shaped lower incisor teeth. Hunted for centuries for the brown musk contained in a gland near the tail, and used in medicine and perfume, they are now a subject for conservation.

Muskellunge: the largest fish species of the *pike family, growing occasionally to a length of 2·4 m. (7 ft. 9 in.) and a weight of 45 kg. (112 lb.). Restricted in its distribution to the Great Lakes of North America and the St. Lawrence and Ohio rivers, it is typically a fish of large waters, especially where aquatic plants grow densely. It is a predatory species, feeding on smaller fishes, with occasional water-birds and mammals featuring in its diet. It is a favoured angling fish but has become rather uncommon.

Musk ox: a species of goat-antelope which lives in the Arctic region of North America. It is more closely related to sheep and goats than to oxen. It stands 1·5 m. (5 ft.) at the shoulders; the head is long and bears horns curving outwards and downwards and then upwards. The coat is of long black fur and hangs like a fringe around the stocky legs, which in contrast, have white fur. For most of the year it lives in herds of four to a hundred individuals. The bulls are extremely aggressive in the rutting season and their musk gland exudes a strong smell. A single calf is born and the mother protects it fiercely. Suckling continues for nine months, and the female calves only once every two years.

Muskrat: a species of aquatic *rodents that look like very large water voles. They are little modified for life in the water, apart from the laterally flattened tail. The muskrat, *Ondatra zibethicus*, is native to North America. Its soft, dense fur is highly regarded in the trade under the name musquash and more of these skins are sold than of any other species. The animals are kept on fur farms and escapes have resulted in the species becoming well established in Europe, where they have become serious pests by undermining river banks with their burrows. The name refers to musky secretions of the male.

Mussels are perhaps the commonest of all *bivalves, and may occur in dense beds several miles long on coast lines and estuaries, wherever there is plentiful suspended food. Like most other bivalves, they filter out this food with their large internal sheet-like gills, and then carry it to their mouths using beating hairs (cilia). They secrete tough byssus threads by which they anchor themselves to the sea-bed. Both horse mussels, *Modiolus modiolus*, and common mussels, *Mytilus edulis*, are extremely abundant between high and low tide-marks. Often young, adult specimens are transported to favourable sites to provide supplies for the seafood industry.

Mustard is the name for a number of European, yellow-flowered, annual plants. They give their name to the mustard family, or Cruciferae, a large family with over 3,000 species world wide. This includes many economically important species, such as cabbage and oilseed rape, and many ornamental garden species. White mustard, *Sinapis alba*, is grown as a salad plant, and is eaten as a seedling. The seeds of white mustard, brown mustard, *Brassica juncea*, and black mustard, *B. nigra*, are ground to yield mustard flour. When mixed with water, or vinegar, this produces the condiment mustard.

A common **mussel**, *Mytilus edulis*, releases a jet of spawn into the waters of an estuary in Devon, England. The spawn stimulates other mussels in the bed to breed.

Mutations are mistakes in heredity, which cause the imperfect transfer of a characteristic from one generation to the next. Organisms carrying a mutation are known as mutants. Dwarfism and albinism are examples of mutations in man. At the molecular level, they are changes in the *DNA. A change in the type of nitrogenous base at a particular position on a DNA strand can affect the properties of a protein, which may then affect the functioning of the organism.

Most mutations are deleterious, that is they produce an organism less adapted to its environment than the normal one. Typically, natural mutations occur very rarely, less than one in a million generations. The frequency of mutations can be greatly increased by certain chemicals or radiation.

In addition to mutation of the molecular sequence of DNA, a type called chromosomal mutation can occur when chromosomes fail to separate or fragment during *meiosis. The occurrence of beneficial mutations is the main mechanism which enables evolution to occur. A harmless mutation may arise and remain at low level in a population, but a mutation which improves a character can quickly become common in the population, through *natural selection.

Mute swans are the familiar Old World *swans with gracefully curved necks, whose wingbeats produce a musical note. Their orange beaks have black basal knobs. They have been semi-domesticated for centuries, and young birds are still tagged at 'swan upping' ceremonies in Britain.

Mutton bird is the name given to two different species of petrel. In New Zealand they are the sooty shearwaters, *Procellaria grisea*, in southern Australia they are the short-tailed shearwaters. *P. tenuirostris*. Fledglings of both are taken from their cavity nests, cooked, and preserved in their own fat as part of the winter diet of the Maoris or Aborigines. Both species of mutton bird nest underground in abandoned burrows on offshore islands. The adults are pelagic-feeding.

Mutualism   *Symbiosis.

Mycorrhizas are the *symbiotic associations of fungi with the roots of plants, in which the plant supplies the fungus with organic compounds and the fungus improves the efficiency of nutrient take-up by the plant roots. Many plants, such as heathers and orchids, cannot grow without their mycorrhizal fungus. There are many different kinds of mycorrhizal association. In the ectomycorrhiza of trees,

Various features of an internal **mycorrhiza** are shown in this longitudinal section (× 85) of the cells in a leek root. The fungus, *Glomus mosseae*, in this association has hyphae (thread-like filaments) with prominent swellings, or vesicles. The function of the latter, which contain fats, is not yet known.

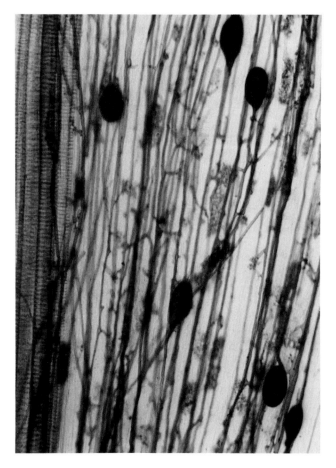

the rootlets are surrounded by a sheath of fungal tissue which is sometimes also connected to a fungal fruit-body such as the fly agaric. Endomycorrhizal associations are those in which the bulk of the mycorrhiza is present inside the plant roots. This type of association is thought to be most common. Mycorrhiza occurs in most types of land plant, including cryptogams.

Mynah bird is a name used for a number of species belonging to the starling family. They originated in Asia, but some, particularly the common mynah, *Acridotheres tristis*, have been introduced to a wide variety of warmer parts of the world by man. Thrush-sized, blackish or brownish birds, often with bright yellow legs and beak and with conspicuous pale patches in the wings, mynahs are perhaps best known for their ability to mimic the calls of other birds. The common mynah is a popular cage-bird and can be taught to talk.

Myocardial infarction   *Coronary thrombosis.

Myriapods comprise some 10,500 animal species, commonly known as *centipedes and *millipedes. They are a major class of *arthropods, characterized by having legs of similar shape on virtually every segment. Unlike insects, their *cuticle is not waxy and waterproof, and their fine breathing tubes cannot be closed by valves. This makes them very prone to water-loss; hence they are most frequently found living in dark and humid places beneath stones and humus. They rely on scents and vibrations to explore their twilight world, as they lack the elaborate compound eyes of insects. Most types of myriapod are herbivores, but true centipedes are predatory.

Myrtles are part of, and give their name to, the large myrtle family with some 3,000 species of small shrubs and trees. The family is mainly centred on tropical and subtropical regions and includes *Eucalyptus* and cloves. The European myrtle, *Myrtus communis*, is an evergreen shrub of southern Europe and western Asia.

Myxophyceae   *Blue-green algae.

N

Nails (anatomy) are the horny coverings of the tips of fingers and toes. Hooves or claws are their equivalent in some vertebrates. They develop in the same way as skin but advance from specialized regions of dividing cells, nail roots, which are mostly covered by skin. Nails rest on thin skin but are tucked beneath finer skin along their sides. The half-moon mark, or lunular, on the base of the nail is the limit of firm body attachment to nail roots. In man, nails grow at a fairly regular rate of half a millimetre ($\frac{1}{50}$ in.) a day. Claws are used for grooming and scratching as well as the retention of prey. Nails and claws of wild animals are worn away naturally in the course of physical activity.

Narcissus is the name given to those varieties of the bulbous plant genus *Narcissus*, in which the petals are united to form a small cup, as opposed to the trumpet shape of daffodils. The flowers may be borne singly on the flower stem or in clusters. Hundreds of garden varieties have been produced, in which the size and colour of the cup vary considerably.

Narcotics are pain-relieving drugs, particularly morphine and other drugs derived from the *opium poppy. They react with parts of the nervous system responsible for the sensation of pain, and relieve the feeling of severe pain. Most also produce a sense of detached well-being, medically valuable, but creating a danger of drug addiction.

Many narcotics need to be given in increasingly large doses as the body builds up a tolerance to them after regular use. Physical dependence on a narcotic can make its withdrawal difficult. Cocaine is probably the most addictive of narcotic drugs but causes no physical dependence; its addiction acts in a psychological way.

Narwhal: a species of *toothed whale that lives almost exclusively in Arctic and North Atlantic waters. The bull is unmistakable because of the very long, straight, spirally fluted, single tooth, or tusk, in the upper jaw. The narwhal itself may be as much as 6 m. (20 ft.) long and its tooth may project a further 2·7 m. (8 ft. 10 in.). It does not appear to use its tusk as a weapon, although it may serve in battles for possession of females. It is greyish in colour with darker mottling along the sides and back, and it feeds on fishes, squids, and crustacea.

Nasturtiums are varieties of the plant species *Tropaeolum majus*, an annual climbing plant with rounded leaves and large orange flowers, found wild in Peru. Together with perennial species such as canary creepers, *T. peregrinum*, they form the family Tropaeolaceae, containing about ninety species from the mountainous parts of South and Central America. They are now very popular summer-flowering garden plants. The young leaves were at one time used in salads and the green fruits may be used as a substitute for capers. They are not to be confused with the genus *Nasturtium*, which includes watercress, from the mustard family.

Natterjacks, the smaller and rarer of the two native British *toads, occur throughout western and central Europe. They are usually greenish above with darker markings and a central yellow stripe (occasionally absent) on the back. Male natterjacks have a loud ratchet-like call, the sound carrying for considerable distances. They are short-limbed, and run rather than hop, leaving characteristic 'tank-tracks' in sand. They are typically found in sandy habitats, often with midwife toads.

Natural selection is the process of differential survival and reproduction that enables *evolution to take place. The concept was invented by *Darwin, and is the accepted explanation both of evolution and of the fact that organisms are well designed for living in their natural surroundings.

The argument is as follows. Organisms produce more offspring than can survive. Thus, there will be competition for survival. Only those organisms that are well designed for survival will live to reproduce. If the properties that enabled them to survive are inherited, then the organisms of the next generation will resemble the successful members of the previous generation. Thus, the organisms best fitted for survival are selected by nature for reproduction. As environments change, so also do the factors that make one individual better 'fitted' than another to reproduce. The constant selection, or survival of the 'fittest', is the eternal force that drives evolution.

Natural selection has been confirmed by observation, and studied by experiment. In a population of peppered moths, *Biston betularia*, for example, of which some were better camouflaged than others, birds have been seen to eat more of the poorly camouflaged type. This results in evolution towards an increase in the proportion of better-camouflaged moths. The agent of natural selection, in this example, is visual hunting by birds. Even a very slight advantage of one individual over another is sufficient to cause changes in the population.

Nature reserves are natural areas that may be managed so as to maintain their original state or to avoid a *succession. They are usually set up to protect a type of habitat, such as heathland, and the species normally present. Commercial gain is usually unimportant. Visiting may be restricted. A warden is often present to ensure minimum disturbance. The Royal Society for the Protection of Birds, Nature Conservancy Council, and county naturalists' trusts have many nature reserves in Britain, often managed by volunteers. The US National Park Service administers national seashores, lake shores, wild and scenic rivers, scenic trails, and national preserves (wildlife parks).

Nausea is an unpleasant and distressing sensation felt in the throat or the stomach. It is often, though not always, associated with vomiting, which it precedes. Distension or stretching of the walls of the stomach, duodenum, or oesophagus can activate sensory nerve fibres, causing the sinking feeling of the abdomen often felt in a rapidly moving lift. Greater stretching of these parts induces nausea, which may be accompanied by bouts of sweating and pallor. Nausea may be experienced in *motion sickness and certain disorders of the brain.

Nautiluses (invertebrates)   *Paper nautiluses, *Pearly nautiluses.

The condition known as **neoteny** is exhibited by the Mexican axolotl species, *Ambystoma mexicanum*, of which there are two known types, both shown here: a black form and a white, or albinistic, pink-gilled strain.

Navel, or belly-button: the scar formed in mammals, soon after birth, by healing at the site of severence of the umbilical cord. In the womb this carries blood between the floating foetus and the *placenta.

Nectarines are a variety of the peach. These fruit-trees, *Prunus nectarina*, produce distinctive smooth-skinned fruits, which are smaller and more brightly coloured than those of the peach tree, *P. persica*, with white or yellow flesh. The trees are less hardy than peaches and are grown in protected positions in cooler temperate areas.

Needle fish   *Garfish.

Nematodes   *Roundworms.

Nemesias are multicoloured varieties of South African annual plants of the genus *Nemesia*. There are about fifty species, some of which are low-growing perennial shrubs; others are popular garden plants of temperate regions.

Ne-ne   *Hawaiian geese.

Neoteny is an unusual condition that occurs in some animals. The adult retains a larval form but is capable of reproduction. The best-known example is of the Mexican *axolotl, a salamander which develops as a large larva with feathery gills. Under normal circumstances it never becomes a true adult, but injection of thyroid-gland extract will cause metamorphosis to the adult form.

Nerines are a genus of eighteen species of South African plants of the lily family. Their bulbs have a cycle of growth and rest which corresponds to the wet and dry seasons, with attractive and brightly coloured flowers appearing at the onset of the rains.

Nerves are bundles of nerve fibres, each capable of carrying unidirectional signals to the brain from all parts of the body. They are present in virtually all multicellular animals. Each fibre is the elongated portion of a nerve cell (neurone). The cell body of a neurone is located within the central *nervous system. Each fibre is surrounded by a sheath of insulating material (myelin), produced by a special type of cell, to prevent electrical interference between adjacent fibres in the nerve. Signals carried in the fibre are called impulses because they travel as discrete bursts of electrical activity. Each burst is followed by a short period when no further impulses can travel along the nerve. These 'spaces' enable information to be coded into bursts, as in morse code. Information from the receptor end of the nerve fibre is thus coded into burst patterns, eventually to be translated into appropriate action by the brain. The junction between two nerve fibres is called a synapse, and the transfer of impulses is mediated by the release of chemicals called neurotransmitters across the gap of the synapse. Amplification, or modification, of the information can occur at this point. Most fibres split into several smaller fibres at their ends, enabling each neurone to communicate with many other fibres.

Nervous system: a network of nerve cells specialized to carry signals between parts of the body. The central nervous system, including the brain and spinal cord, relates the input and output conveyed through the peripheral nervous system. In higher animals, some *reflexes are situated within the spinal cord, reflecting the segmental organization of the invertebrate nervous system. Other reflexes pass through the brain, where, in addition, more complex processing is possible. The *autonomic nervous system also has central and peripheral components. Its activity is entirely reflex, but it can be influenced by other parts of the brain.

Nettles are of two kinds, 'stinging' and 'dead'; the stingers belong to the genus *Urtica*, which has about fifty species armed with single-celled stinging hairs. The

Nervous system

A typical reflex action　Signals from a sense organ, in this case a pain receptor in the human finger tip, travel via a sensory nerve fibre into a dorsal root of the spinal cord. Here they are passed to a relay neurone before passing out of the spinal cord and along a motor nerve fibre to the muscle which effects a response to the pain stimulus – arm muscles pull the finger away.

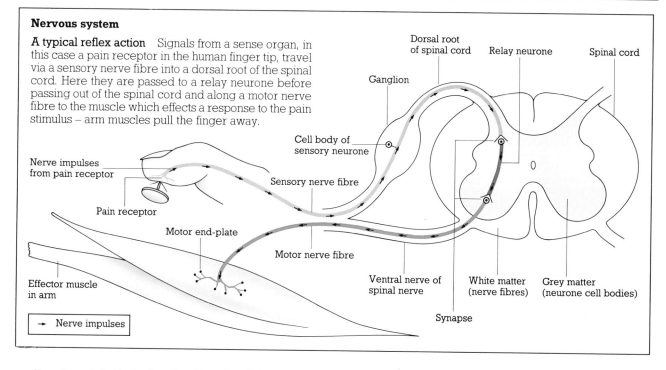

Dorsal root of spinal cord
Relay neurone
Spinal cord
Ganglion
Cell body of sensory neurone
Nerve impulses from pain receptor
Sensory nerve fibre
Pain receptor
Motor end-plate
Motor nerve fibre
Effector muscle in arm
Ventral nerve of spinal nerve
White matter (nerve fibres)
Grey matter (neurone cell bodies)
Synapse

→ Nerve impulses

swollen tips of the hairs break off at the slightest touch, releasing into the wound an irritant mixture of formic acid and enzymes. They are part of the family Urticaceae which is found worldwide. Dead-nettles (*Lamium* species) belong to the mint family; they have no stinging properties but have hairy, nettle-shaped leaves.

Neurology and neurosurgery are the specialities dealing respectively with the medical and surgical aspects of diseases of the *nervous system.

Neurosis is a mental illness in which excessive anxiety and fearfulness are the main symptoms. In contrast to

This close-up of a stinging hair of the **nettle**, *Urtica diocia*, shows the hollow tube along which the irritant fluid flows when the hair is broken.

*psychosis, contact with reality is fully maintained in neurosis. Depression, phobias, obsessions, and hypochondria are also types of neurosis. Such symptoms are often the result of overstress, especially when it interacts with a particularly vulnerable type of personality. Hypnosis and psychoanalysis were developed to try to understand and treat neurosis. Treatments are based on *behaviourism, drugs, or psychotherapy.

Newts are members of the family Salamandridae, which also includes the true salamanders. European newts of the genus *Triturus* form a distinct group of species. They are terrestrial for part of the year, but enter the water, usually a pond or lake, to breed in the spring. At this time, the males become generally brighter in colour, and some species, such as the crested newt, *Triturus cristatus*, develop a crest of skin along the back and tail. Male alpine newts, *T. alpinus*, are strikingly coloured, having plain red-orange bellies, bordered with sky blue, and a low cream and black spotted crest. Courtship in newts is elaborate, the male displaying to the female. Eggs are laid singly, attached to aquatic plant leaves.

In North America there are two groups of newts, the chiefly aquatic eastern newts, and the more terrestrial western newts. Two species of eastern newt have a tadpole stage, which is followed by a distinct terrestrial, sexually immature, or eft, stage. This lasts for up to three years, after which the newt returns to water to become a sexually mature adult.

Nicotiana, or tobacco plant, is a mainly tropical American plant genus with about forty-five known species. They are mostly annual or perennial herbs with large rosettes of sticky leaves and tall spikes of tubular flowers. *Nicotiana tabacum* is the parent species, from which the many varieties of *tobacco have been developed. All the species contain the poisonous substance nicotine, which has useful properties as an insecticide. Most have fragrant flowers which open at night, but colourful, day-flowering varieties of *N. alata* have been bred.

Nightingales belong to either of two species of the genus *Erithacus* of the thrush family. They are smallish and rather nondescript brown birds, 15 cm. (6 in.) in length with a reddish-chestnut tail. They breed in much of Europe and western Asia, mostly in wooded country, but spend the winter in Africa. Seeking a breeding site on or close to the woodland floor, they make a small nest of leaves and lay 4 or 5 dark olive-brown eggs. Their food consists of insects and small fruits. The European nightingale, *E. megarhynchos*, is particularly famous for its beautiful song, frequently heard at night as well as in the day, and the inspiration of much poetry.

Nightjars are birds which have a worldwide distribution with the exception of the cold polar areas, although they are only summer visitors to high-latitude temperate areas. Most of the sixty-nine species are the size of thrushes, although two species have elongated tail feathers of 65 cm. (26 in.) or more. Their plumage is mottled with soft greys, blacks, buffs, and white, making them extremely well camouflaged as they sit on the ground. They feed on flying insects at night or dusk, aided by their large eyes and enormous gapes, which enable them, to open their mouths very wide to snap up their prey. They have distinctive, often very repetitive calls which carry long distances. Several species are named by the sound of their calls, for example whip-poor-will and chuck-will's-widow. The female lays either one or two well-camouflaged eggs on the ground, without any nesting material. The European nightjar, *Caprimulgus europaeus*, was popularly called the goatsucker from the erroneous belief that it sucked milk from goats in the night.

Night monkey: a New World monkey with large eyes; it is also known as the owl-faced monkey or douroucouli. The only nocturnal monkey, it seems to be blinded by daylight. In the day it remains hidden in hollow trees, coming out at sunset to prowl in search of insects, eggs, birds, and fruit. It has a long tail which is not prehensile, and is grey-brown in colour with three black stripes along the head. It ranges from Guyana to Brazil and Peru.

Nightshade is the name usually applied to some species within the potato family. The deadly nightshade, *Atropa belladonna*, is a tall perennial plant with black cherry-like fruits. Although cultivated for the drugs hyoscyamine and atropine, the whole plant is extremely poisonous, the roots particularly so. Woody nightshade, *Solanum dulcamara*, is a perennial climber with clusters of purple flowers followed by red fruits. The black nightshade, *Solanum nigrum*, is an annual plant with small black fruits, found as a weed of cultivated land the world over. Enchanter's nightshade, *Circaea lutetiana*, is an unrelated, small perennial plant related to the willow-herb.

Nilgai: a species of large antelope found on the lightly wooded hills of India. The bull has short, smooth, keeled horns, and an iron-grey coat. The female is smaller, without horns, and tawny in colour; the calves are also tawny. Small groups of four to ten cows, calves, and young bulls are found together; adult bulls are solitary or live in bachelor groups.

Nipa palm: a *palm of mangrove swamps and estuaries from southern India to the Pacific. It has leaves several metres long arising from a creeping stem which forks regu-

The shiny black berries of deadly **nightshade** appear in August to October, following drooping, bell-shaped violet-green flowers. Usually found in waste places, the plant is a native of Europe, west Asia, and North Africa.

larly like a seaweed. The fruits are different from those of most other palms in that they float and the plant is thereby dispersed by ocean currents. Nipa palms are the oldest palms known from the fossil record and one of the seven earliest *angiosperm plants. They were much more widely dispersed in past geological eras.

Nitrogen cycle is a term for a *biogeochemical cycle involving the element nitrogen. Nitrogen gas forms 78 per cent of the atmosphere but is extremely unreactive and can be used, or 'fixed', by only a few living organisms, namely the blue-green algae and certain types of bacteria. These organisms are able to convert nitrogen gas into ammonia (NH_3), which they use to make amino acids, proteins, nucleic acids, and other nitrogenous compounds. Leguminous plants, such as peas, beans, and clover, form a *symbiotic association with *Rhizobium* bacteria, producing root nodules in which these nitrogen-fixing bacteria can live. Nitrogen fixation also occurs during thunderstorms when lightning generates nitrous oxide (NO) and nitrogen dioxide (NO_2) gases in the air; the gases dissolve in rainwater to form nitrates (NO_3^-) in the soil.

All plants take up nitrates from which they produce all their essential nitrogenous compounds. Some of this nitrogen is lost when leaves, seeds, and fruits are shed, but most remains trapped until the plants die or are eaten. Animals obtain their nitrogenous substances directly or indirectly from plants and excrete excess nitrogen in their faeces and urine. These substances and the dead bodies of all organisms are acted on by decomposing and nitrifying bacteria, which convert ammonia into nitrate, ultimately resulting in the return of nitrates to the soil. Denitrifying bacteria, which live in waterlogged soils, obtain oxygen for respiration by breaking down nitrates and producing nitrogen gas, which is released to the atmosphere to complete the cycle. Modern agriculture upsets the natural balance by adding nitrate fertilizer to the soil, most of which is washed out into rivers.

Nival plants are those growing in or near snow. They include the 'red snows' of unicellular algae which live on the surface of the snow in Arctic regions and a certain number of angiosperms which flower once the snow above

Nitrogen cycle

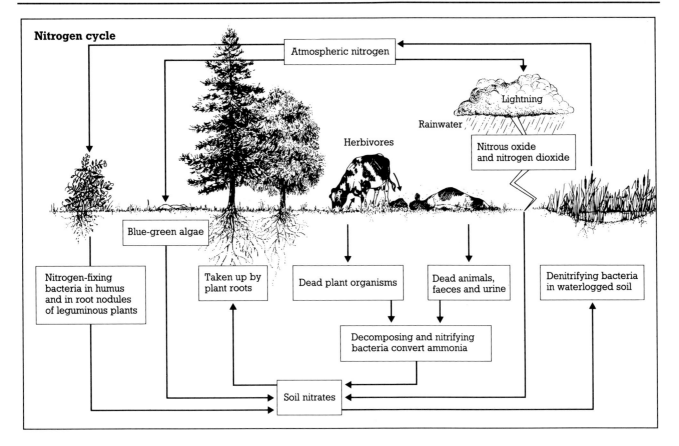

them has melted. Many alpine species can be considered as nival plants. The snow protects these from the excesses of the winter climate.

Noddies are *terns of the genera *Anous*, *Procelsterna*, and *Gygis*, widely distributed in warm seas. Common, or brown noddies, *Anous stolidus*, have dark plumage in striking contrast to the white tern, *G. alba*, and the blue-grey noddy, *P. cerulea*. All lay single eggs, usually in small nests on cliffs or in trees.

Nomenclature in zoology is used to provide a system whereby any kind of animal has one correct name, by which it is internationally known. The system stems from the work of the Swedish naturalist Carl Linnaeus (1707–78). The starting-point for all naming of organisms is 1758, the date of the tenth edition of his *Systema naturae*, all names before that date being ignored. Today the *International Code of Zoological Nomenclature* has developed from it. This code is produced by an International Commission, which modifies the rules as needs be and adjudicates in cases that are doubtful.

The language of zoological nomenclature is Latin, the language of science in Linnaeus's day, and the correct name of a *species is the basis of the system. This name, called the scientific name, is composed of two words, the generic name, printed with an upper-case first letter, and the specific name, with a lower-case first letter. The whole scientific name, which is always printed in italic, may be followed by the name of the first user and the date of first use, although these are not obligatory. If one species has received two or more names (synonyms) then the earliest is correct, and if two different species have received the same name, called a homonym, then the earlier has the right to it.

The rules also cover the names of subspecies, which are written as trinomials, a third word being added to the name. Higher groupings, up to *family names, add -idae to the root of the genus name. Subfamilies add -inae and superfamilies -oidea. Higher groupings, from *order to *phylum, are not considered in the code, but are usually understood by consensus of opinion.

Norway lobsters *Dublin Bay prawns.

Nose: an important organ in air-breathing vertebrates, particularly mammals. It serves two distinct purposes, firstly to clean and warm inhaled air and, secondly, to provide the sense of smell. In its roof there is a moist membrane in which lie nerve endings. These are receptive to different chemical substances which dissolve in the fluid on the membrane, and signal olfactory sensations to the brain. Compared to most mammals, the human sense of smell is very poorly developed. The nose is used by most vertebrates to find food, to scent danger from predators, and to recognize other individuals. Mammals deprived of the sense of smell die.

Most mammals warm incoming air by passing it over moist scroll bones inside the nose. These are kept warm by a rich blood supply. In many mammals, especially those of arid regions, the nasal membranes act as water conservation devices by trapping condensation from exhaled air.

No-see-ums *Midges.

Nostoc is a genus of microscopic *blue-green algae consisting of a chain of attached cells including colourless cells that can use nitrogen from the atmosphere to produce food. Free-living species of *Nostoc*, along with other blue-

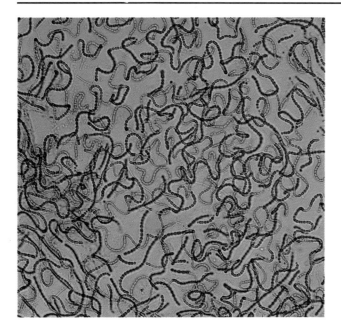

This species of **nostoc** occurs everywhere in fresh water and moist places. Cells are 0·01 mm. across but a colony, of which this is part, may reach 1 cm. ($\frac{3}{8}$ in.).

green algae, are thought to maintain fertility in some tropical topsoils.

Notornis, or takahe: a large chicken-sized flightless bird, resembling a moorhen, which belongs to the rail family. It is mainly bright blue in colour with a wide range of iridescent sheens; its heavy beak, the frontal shield above it, and the legs are bright red. It is found only on the South Island of New Zealand, where it lives in hilly grasslands. It is very rare, possibly due to competition for grasses (its staple food) with the introduced red deer. It makes a simple nest on the ground and lays one or two cream-coloured eggs.

Nucleic acids include *DNA and *RNA; they are found in the cells of all living organisms and are responsible for protein synthesis and for the cell's hereditary characteristics. Both substances consist of long chains of nucleotide subunits; RNA molecules are single-stranded but DNA is double-stranded. In advanced organisms such as vertebrates, plants, and the majority of invertebrates, DNA is confined to the nucleus of the cell, with small amounts in the mitochondria and chloroplasts; RNA is found mainly in the cytoplasm.

Nucleotides are the subunits of nucleic acids; each nucleotide consists of a five-carbon sugar linked to a phosphate group and an organic, nitrogenous *base. In *RNA the sugar is ribose, whereas in *DNA it is deoxyribose. The nitrogenous base may be adenine, cytosine, guanine, or thymine in DNA; adenine, cytosine, guanine, or uracil in RNA. When RNA is transcribed from DNA, the nitrogenous base uracil is substituted for thymine.

Nucleus (biological): the structure in higher (eukaryotic) *cells that contains the *chromosomes. It is surrounded by a double *membrane, and pores which allow the transfer of substances between the *cytoplasm and the interior of the nucleus. In dividing cells, the individual chromosomes are condensed and easily visible, but normally they are uncoiled and form a substance called chromatin, which fills the nucleus. Some cells, such as bacteria, blue-green algae, and red blood cells, lack a nucleus.

Nudibranchs *Sea slugs.

Nurse shark *Carpet sharks.

Nuthatches are birds of the family Sittidae, which occur in Europe, Asia, and North America, mostly in forest. The twenty-five species range in size from that of a warbler to that of a large sparrow, and are grey or greenish-blue above and paler below, often with dark caps, eye-stripes, or other facial patterns. They have very short tails and longish, straight beaks. They run up and down tree trunks and along branches, probing behind bark for insects. In autumn some species store seeds behind the bark of trees for use as food during the winter. They nest in holes in trees, often plastering the hole with mud so as to reduce the size of the entrance to prevent larger birds and mammals preying upon their eggs or young.

Nutmeg comes from the tree *Myristica fragrans*, 20 m. (75 ft.) or more in height, which is native to the Molucca Islands of Indonesia. It belongs to a family of tropical rain-forest trees with 380 species spread throughout the tropics. The fleshy fruits resemble large apricots. The nutmeg is the large brown seed which, after drying, is marketed either whole or in a powdered form and used as a flavouring. The trees are either male or female and in plantations about ten per cent male plants must be planted to ensure pollination of the female flowers. The seed is surrounded by a red-coloured, fleshy network called the mace, another source of flavouring.

Nutrias *Coypus.

Nutrients are those inorganic and organic chemicals essential as nourishing substances for the maintenance of life.

When ripe, the fruit of the **nutmeg** tree splits to reveal the mace surrounding the single seed. The nutmeg itself, enclosed in a kernel, is extracted after being heated for several days over a gentle fire.

Both *autotrophic and *heterotrophic organisms require inorganic chemicals, or mineral salts. The basic elements of all nutrients are ultimately recycled in an *ecosystem.

Plant nutrients, usually synonymous with mineral salts, are categorized as macro- or micro-nutrients. Macro-nutrients are required in large quantities and include nitrogen, phosphorus, and potassium. Micro-nutrients, such as copper, zinc, and molybdenum, are essential but may be toxic if present in large amounts. Heterotrophs vary in their requirements, but vertebrates require a balanced diet incorporating carbohydrates, fats, proteins, and small quantities of vitamins.

Nutrient deficiency can cause poor growth, deformity, malfunctioning, and sterility. A range of characteristic *deficiency diseases is recognized in man.

Nutrition is broadly concerned with the study of the types and amounts of food materials that promote the growth, health, and reproduction of plant and animal life.

The adequate nourishment of animals requires an intake of certain raw materials at intervals controlled by hunger, appetite, or thirst. Foods supply the energy needed to drive chemical reactions in the body as well as all nutrients needed to build, repair, and support the body. In a majority of animals, essential nutrients are obtained by active and selective hunting for appropriate foodstuffs and water, initiated by sequences of intricate reactions in the brain.

The specific nutritional needs of animals are known for relatively few species, apart from man and domesticated animals. They differ widely between species. Warm-blooded animals require more food for energy production than do cold-blooded animals. Within a species, requirements vary according to size, genetic make-up, activity, sex, and reproductive state.

Nuts are, in the botanical sense, a particular type of *fruit in which the wall enclosing the seed (pericarp) is a hard, woody shell, and does not split open when the seed is ripe. However, the term nut is used for any firm, oil-rich kernel, usually surrounded by a hard shell. Of those commonly eaten in temperate regions, only the hazel nut, filbert, and sweet chestnut are, strictly speaking, nuts. Acorns and beechmast are also true nuts. Others, such as Brazil nuts, cashew nuts, pistachios, and macadamia nuts, are really seeds with a hard shell derived from the seed coat (the testa), not from the pericarp. They mostly develop inside true fruits, an exception being the cashew. Peanuts are seeds enclosed within a pod, this being a form of pericarp. Walnuts, pecans, almonds, and coconuts are all seeds of *drupe fruits. Betel nuts, widely chewed in India, are the seeds of a palm, and contain a mild narcotic. Pine nuts are the seeds of pine trees, notably those of the stone pine, native to the Mediterranean region.

Nuttall, Thomas (1786–1859), English emigrant, became a pioneer in American palaeontology and botany. He recorded his expeditions across the continent to the Pacific coast in *The Genera of North American Plants* (1818) and *A Journal of Travels into the Arkansas Territory* (1821).

O

Oak-apple galls (oak-apples) are large, soft, spherical growths on the terminal buds of oak trees (*Quercus* species). They are growths which develop in the summer as a response to the presence of the larvae of a species of *gall wasp, *Biorhiza pallida*. Winged males and wingless females emerge in the autumn from the oak-apple gall. The eggs of this generation are laid on oak roots and these larvae cause a different type of gall. From these, wingless females, without males, emerge in the spring to climb the tree and lay eggs in terminal buds, and their larvae stimulate the formation of oak-apples again. Oak-apples are often confused with marble galls, which are spherical and produced by a different species of gall wasp.

Oaks are trees and shrubs of the genus *Quercus*, and belong to the same family as beeches and sweet chestnuts. This genus contains some 450 species which occur principally in northern temperate zones, though a few are native to mountains in the tropics. Cool temperate species are usually deciduous and have leaves with lobed edges. Almost half of the total number of *Quercus* species live in warmer temperate regions and are usually evergreen, often with unlobed leaves. Oak trees have their male catkins and inconspicuous female flowers on the same tree and produce their fruit in the form of a nut called an acorn. The most common species of northern Europe is the common, or pedunculate, oak, *Q. robur*, while the durmast, or sessile oak, *Q. petraea*, is found on upland acidic soils. Southern European species include the turkey oak, *Q. cerris*, holm, or evergreen oak, *Q. ilex*, and cork oak, *Q. suber*. Among the many North American species are the red oak, *Q. rubra*, white oak, *Q. alba*, and pin oak, *Q. palustris*. Many species, such as common and white oaks, are valuable timber trees, long used for the construction of houses and once famous for their use in English ships. In addition to a hard, durable timber, some species, such as the cork oak, also provide cork.

Oar-fish: a species of fish distributed worldwide in the open sea, living mainly in mid-water at depths of 300–600 m. (975–1950 ft.). Its long, compressed, silvery body, up to 7 m. (23 ft.) in length, and deep red fins may have given rise to some reports of sea-serpents, when the fish was near the surface. It feeds mainly on small, mid-water shrimps called euphausids. (*See over.*)

Oarweeds *Kelps.

Oats are an important cereal crop, particularly in cool, moist regions of North America and northern Europe. It can tolerate poor or acid soils, but not drought. It is grown mainly as a livestock food, and may be harvested unripe and fed fresh, or as silage. It is more usually fed to livestock as crushed or rolled ripe grain. Relatively little is now used as human food, the grain having a high proportion of husk that is not easily removed. Some is used to produce oatmeal and porridge oats.

Like rye, oats were probably selected from wild varieties that were carried with wheat and barley as their cul-

The biology of the **oar-fish**, *Regalecus glesne*, is largely unknown, most studies having been made of specimens washed ashore. Unlike their stronger swimming parents, larvae such as this one may be trapped in trawls.

tivation spread northwards throughout Europe. Selection was for a type that did not shed its seeds and gave better yields. The wild oat, *Avena fatua*, remains an unwelcome weed in wheat and barley crops. A close relative, the black or bristle oat, is even more tolerant of acid, infertile conditions and was a useful crop in areas of northern Europe before the introduction of artificial fertilizers.

Obesity, an unwanted amount of body fat, is a problem of appetite rather than of metabolism. Most people under middle age unconsciously adjust their food intake to balance varying energy expenditure, and thus effortlessly maintain a steady weight. It is not known how this is achieved. Those who lack this ability must, if they wish to avoid gaining weight, substitute conscious control of their food intake.

Octopuses are *cephalopods which have entirely lost their ancestral molluscan shell, leaving a rounded bag-like body suited to a cave-dwelling life. They can use jet propulsion, but more often crawl slowly over surfaces using their eight suckered tentacles. They feed mostly on crabs and other shellfish, grasping them with their tentacles,

biting with strong beak-like jaws, and injecting a poison. Paralysed prey is taken back to the den for feeding. To assist hunting, the octopus can learn complex behaviours, retaining new information in its elaborate brain, which rivals that of many vertebrates in respect of its capacity. Mating behaviour is also complicated, with ritual colour changes. Females brood the large eggs, and generally die thereafter.

Octopuses usually prefer warmer seas, but even there never gain the sizes credited to them in legend; the body rarely exceeds 40 cm. (16 in.) in width, though the arms may be up to 5 m. (16 ft.) long.

Oesophagus, or gullet, in mammals is a tube stretching from the pharynx, behind the nose and mouth, to the stomach. Its muscular wall is lined by a layer of mucus-covered cells. Food is retained in the lower end of the gullet by a sphincter (muscle ring) which opens at intervals for the gullet to squeeze its contents into the stomach. In *ruminants the lining of the gullet is especially tough so that their coarse diet does not damage its wall. In these animals food is returned from the stomach via the gullet to the mouth for the chewing of the cud.

Oil beetles are soft-bodied, black *beetles with no hind-wings and the front wing-covers reduced. They exude an oily, evil-smelling fluid from their joints when alarmed. Their larvae are parasitic on solitary bees or grasshoppers and pass through several different stages. Along with *blister beetles, they comprise the family Meloidae, with some 2,000 species found worldwide.

Oilbird, also known as the guacharo or diablotin, an aberrant relative of the nightjars, placed in a family of its own. It inhabits tropical forests of northern South America. It is the only nocturnal fruit-eating bird and probably finds its food by using its sense of smell. It nests in colonies in the near-total darkness of caves and finds its way by echo-location. This involves emitting a series of clicks and timing how long the echoes take to return, rather after the manner of bats. It lays two to four eggs and the young remain in the nest for up to four months.

Octopuses display a quite remarkable degree of colour change. Intensifying its coloration as it leaves a rocky shelter, this lesser octopus, *Eledone cirrhosa*, may be issuing a warning to potential predators.

The dark cave habitat of the **oilbird**, *Steatornis caripensis*, affords total protection from predators. Its weird cries and red eyes, which glow when lit by torchlight, have given rise to its local name of 'little devil'.

Oil exploration fossils are the minute plants and animals retrieved from sea-floor sediments which aid the search for petroleum. These microfossils are so small that they can be recovered intact from the drill chippings or cores. These organisms make good time-markers for relating one piece of geological strata to another, hence giving clues about possible oil-bearing rock formations. Among the most valuable are *foraminifera, *radiolarians, ostracods, diatoms, *spores, and *pollen. Ostracods (sea shrimps) are minute crustaceans, which, along with foraminifera and radiolarians, have contributed to marine sediments since the Palaeozoic Era (570 million years ago). Diatoms are unicellular plants whose cell walls contain silica. Formations of the rock diatomite under the sea-bed can indicate the presence of oil and gas. Spores from marine sediments can show by their colour the temperature at which a rock was formed, and indicate whether oil is likely to have formed.

Oil plants are those which store oils or fats as food reserves, usually in their seeds but sometimes (like the oil palm and olive) in the surrounding fruit wall. Most plants do this to a limited extent, but the thirty or so species exploited commercially contain between 40 and 65 per cent of oil in their fruits or seeds. About 90 per cent of the world's production comes from twelve species. Soya bean, cotton (seed), groundnut, and sunflower are the main sources, plus oilseed rape, coconut (copra), oil palm, linseed, sesame, olive, castor, and safflower, in that order of importance. Invariably the oil is extracted by some means of milling or pressing.

The range of oil plants is very wide, from annuals to shrubs and trees, with representatives of the sunflower, pea, mustard and palm families, among others. The oil produced from different plant families has very different characteristics. There is an ever-increasing demand for vegetable oils, principally for the production of salad and cooking oils and for the manufacture of soaps. Some species, like castor-oil plants and rape, yield high-grade lubricating oils, and others, like linseed, yield a type of oil that dries on exposure to air, making it useful as a base for paints and varnish and for the manufacture of linoleum.

The palm species yield fats rather than oils, their products being solid at normal air temperatures. Cocoa butter is another fat, extracted from the seeds of cacao and used as a base for chocolate.

Okapi: a species in the giraffe family, discovered in 1900. It lives in deep forests of the upper Congo in the equatorial

The behaviour of the **okapi**, *Okapia johnstoni*, in the wild remains a mystery. Certain courtship phases in females and indications of a territorial imperative in males suggest a sedentary, rather than nomadic, form of society.

zone of Africa. The neck is much shorter than that of the giraffe. It is 1·5 m. (5 ft.) tall at the shoulders. Only the male has horns, covered for most of their length with hair. The neck and body are a rich dark brown, the head is cream-coloured, and the limbs and hindquarters are unique with slantwise stripes of black and white, while the lower part of the limbs is white. The tail, up to 45 cm. (18 in.) long, ends in a tuft of hair. Like the giraffe, it has a long, prehensile tongue with which it browses on the leaves of the forest trees. The young are born singly, with the same colouring as the adults.

Okra is an erect annual plant up to 2 m. (6 ft. 6 in.) tall, native to tropical Africa, though it is now grown throughout the lowland tropics. A relative of cotton, hollyhocks, and mallows, it belongs to the genus *Hibiscus*. It is grown for its fruits, long, ridged pods, also known as lady's fingers or quambo. These grow up to 30 cm. (12 in.) in length and are harvested while still green. The plant has showy yellow flowers with some red patterning.

Old squaw   *Long-tailed duck.

Oleanders, or rosebays, are plants of the genus *Nerium*, which is part of a family of about 1,500 mainly tropical trees, shrubs, and lianas, including *frangipani and *periwinkle. Oleanders are atypical in that they are native to waterside habitats in the Mediterranean area. The common oleander, *Nerium oleander*, is a shrub with narrow leaves, typical of stream-side plants (rheophytes). It has large white or pink flowers and is often cultivated, but all parts of it are poisonous.

Olives are narrow-leaved, greyish, evergreen trees of the genus *Olea*, with over thirty-five species in warm temperate and tropical regions. They belong to the same family as lilac, jasmine, and ash trees. The common olive, *O. europaea*, is native to the Mediterranean area and is also cultivated in the dry subtropics; it grows up to 20 m. (65 ft.) in height, and produces oil-rich black fruits, or olives. The tree is tolerant of arid, stony, infertile soils, and fruiting is favoured by dry summers and cool winters, particularly near the sea, where the crop is protected from extremes of temperature. Italy is the largest producer of olives, followed by the USA (principally California), Spain, and France.

When ripe, the black fruits contain up to 60 per cent of oil, which is extracted by crushing and pressing and is valued as cooking and salad oil. About 10 per cent of the crop is harvested unripe or green and pickled in brine. Both ripe and unripe olives have an inherent bitterness which is removed by soaking in an alkaline solution.

Olm: a single species of aquatic, cave-dwelling *salamander found in underground lakes and streams in northeastern Italy and Yugoslavia. The adult, which is a permanent 'larva', has a slender, cylindrical body up to 30 cm. (12 in.) long, including a crested tail. It has large, feathery, pinkish gills, poorly developed eyes, only three toes on the fore-limbs and two toes on the hind-limbs, and is pale, usually whitish in colour. It often lives deep underground, sometimes living for as long as twenty-five years, and lays between twelve and seventy eggs. When only a few eggs are produced these may be retained in the body and live young are produced. The olm is a member of the same family as the American mud puppy.

The common **oleander**, photographed in southern Spain. Its flowers are followed by long pods containing a large number of hairy seeds, which are dispersed by wind.

Onions probably originated in central Asia and were cultivated in ancient Egypt, India, and China. They are now widespread in all temperate areas, and the USA, Japan, and Europe are the greatest centres of production. All varieties of onion are derived from a single species, *Allium cepa*, and are naturally perennial plants. They belong to the monocotyledonous family, the Liliaceae, which includes lilies, tulips, leeks, and asparagus. The onion plant has a very short stem from which a number of sheathing leaves arise. Towards the end of the first season food is stored in the leaf-bases to form a swollen bulb. The pungently flavoured bulbs are harvested when dormant and keep well over a long period. The crops may be grown direct from seed, or from small bulbs or 'sets'. Young plants, known as spring onions or scallions, are eaten as a salad vegetable.

The Welsh or Japanese bunching onion, *A. fistulosum*, is a close relative, principally grown in gardens of China and Japan; it forms a number of somewhat elongated, smaller bulbs, after the manner of shallots.

The 500 or so species of the genus *Allium* are distributed mainly in north temperate regions. Their flowers, in a wide range of colours, are borne in umbels, as much as 20 cm. (8 in.) across in some species, though in others they are replaced entirely or partially by small bulbs called bulbils.

Opah, or moonfish: a deep-bodied, little-known, oceanic fish, which is brilliantly coloured, deep blue on the back, shading to pinkish on the belly, with rounded milk-white blotches on the body and blood-red fins. It grows to over 1·5 m. (5 ft.) in length and can weigh up to 50 kg. (110 lb.). Despite its stout appearance, it is an exceptionally fast swimmer and feeds on squids and mid-water fishes. It probably lives in mid-water at depths of 100–400 m. (325–1,300 ft.), but most of the specimens studied have been either accidentally captured by commercial trawlers or stranded on the beach.

Opium poppy: a species of widely distributed, summer-flowering annual plant of the poppy family. It is grown in hot, dry climates, including parts of Europe and east Asia, for the drug opium, which is extracted from its sap. The raw material is refined to produce medically important pain-killing drugs, including laudanine, morphine, and codeine. Misuse of opium drugs, including heroine, has addictive and tragic effects and has led to drug trafficking and even wars.

Opossums are New World *marsupials of the family Didelphidae, but some Australian *phalangers are loosely called opossums, as are members of the separate family of South American *rat opossums. There are some seventy-five species of didelphid opossums, most of which live in Central or South America. Some show convergence with placental mammals in resembling mice or shrews and there is an 'otter' in the form of the yapok, or water opossum, *Chironectes minimus*. Most opossums are small or medium-sized mammals, with the Virginia, or common opossum, *Didelphis virginiana*, being the largest at up to 5·5 kg. (12 lb.). It is noted for its habit of feigning death, or playing possum, when attacked. Most opossums have prehensile tails and live in trees. Their diet is varied but predominantly insectivorous or carnivorous.

Opuntia   *Prickly pear.

Ora   *Komodo dragon.

Oraches are members of the plant genus *Atriplex*, of which there are several species, and part of the beetroot family. They are tall-growing plants or small shrubs of the sea coast, often covered with a powdery meal and bearing small flowers. The leaves of some species have been used as a substitute for spinach and as a cure for gout.

Oranges account for 75 per cent of the world's output of citrus fruits. The sweet orange, *Citrus sinensis*, cultivated for thousands of years in China, is now grown wherever a suitable Mediterranean-type climate exists. Its juice contains up to 12 per cent sugar and 1 per cent citric acid (the least acid of all citrus) and is rich in vitamin C. The sour, or Seville, orange, *C. aurantium*, from Southeast Asia, is too bitter to be eaten fresh and has a citric acid content of about 3 per cent. It is valued for marmalade production and is an important crop of the Seville area of Spain.

The characteristic colour, as in all citrus fruits, develops only when the ambient temperature is less than 13 °C (55 °F) at the end of the ripening period. Fruits grown in the humid tropics remain green even when fully ripe.

Orange-tips   *Whites (butterflies).

Orang-utan: a species of large *ape related to chimpanzees, gorillas, and man. It is the largest fruit-eater in the world and is known to eat at least 200 species of fruit. It lives in lowland, tropical rain forests of Borneo and Sumatra, usually within river boundaries or mountain ranges, where it is an endangered species.

The prominent flanges of fat at the side of the face are very striking in large adult males. The jaws project forward from the face, the eyebrow ridge is only slightly pronounced, and the eyes and ears are small. The coat is coarse, long, and shaggy, especially over the shoulders and arms, where it may grow to 45 cm. (18 in.) in length. The fur varies in colour from shades of orange to purplish-blackish-brown, becoming darker with age. The skin of the face is brownish-black in mature adults, and the arms and hands are extremely long. When extended the arms may span 2·4 m. (7 ft. 10 in.) and their length is exaggerated because the standing height of the animal is only 1·4 m. (4 ft. 6 in.). There is marked sexual dimorphism: the female is smaller, standing only 1·06 m. (3 ft. 6 in.) in

Young **orang-utans**, *Pongo pygmaeus*, in Sabah, north Borneo. This ape's long arms and hook-shaped hands and feet are especially useful when it moves through the trees. Adults live for about thirty-five years.

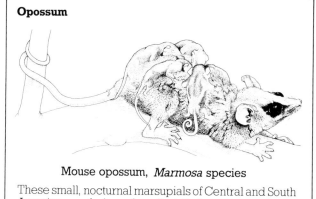

Opossum

Mouse opossum, *Marmosa* species

These small, nocturnal marsupials of Central and South America use their prehensile tails for support when moving through the trees. They breed throughout the year and have a short gestation period of seventeen days. The young are carried until they are about three months old.

height and weighing only half as much as the male. Sexual maturity is reached at about ten years of age and, after an elaborate courtship, mating occurs in the trees. The baby is born after a gestation period of eight months and it is nursed for two to three years.

Orchids are members of the plant family Orchidaceae, of which there are possibly 30,000 species, making it one of the world's largest flowering-plant families. They are found on all continents, with the exception of Antarctica, and occur in habitats ranging from the leafy canopies of tropical forests, through forest and woodland floors, to grassland, heaths, semi-deserts, and marshes.

They vary greatly in form and habit. Some have swollen stems (pseudobulbs), others, such as *Vanilla*, have climbing stems, while those that live on the branches of tropical trees are known as *epiphytes. Most species have green leaves or stems (containing chlorophyll), but certain species, such as the European bird's-nest orchid, *Neottia*, have no chlorophyll, obtaining their nutrients from soil humus. The flowers of orchids, which vary considerably in size, shape, and colour, are adapted for pollination by a great variety of insects. These intricate plant–insect relationships have been studied by many eminent biologists, including Darwin. The seeds, which are dust-like, are produced in large numbers and are wind-dispersed. In the wild, they require the association of a certain kind of fungus in order to germinate.

Orchids have long been cultivated by gardeners, and many thousands of artificial hybrids have been produced.

Orchis is one of the 600 or so genera in the plant family Orchidaceae. About eight species are known, of which two are North American and the remainder mainly from Asia, Europe, and North Africa. The leaves and flower stems are produced annually from an underground, tuberous rootstock and the flowers, which are relatively small, are far less exotic than the tropical epiphytic orchids.

Orders, in the classification of animals and plants, rank below *classes, and are built up of *families. Many orders, particularly if they include two or more distinctive clusters of families, may be divided into suborders. The members of each order or suborder have characteristics indicating common evolution by descent, and generally have more in common than all members of larger groups. The Latin names of plant orders are usually indicated by the ending -ales, as in Fucales, an order of brown seaweeds.

Organ-pipe coral has a skeleton of parallel tubes, each containing one feeding *polyp, so it looks very like a set of organ-pipes, often bright orange or red. This *coral has a fleshy covering when alive, and eight tentacles per polyp, a typical feature of octocorals. It is common in shallow Indo-Pacific reefs.

Orioles are the twenty-eight species of birds belonging to the family Oriolidae, which occurs in most of the warmer areas of the Old World. They are thrush-sized birds, and the males of many species are yellow or rich brown with black wings and head markings. The females are darker, often olive-green. Orioles live in wooded country and nest high up in trees. Many have very beautiful, flute-like calls.

This name is also often used for the species of the New World family Icteridae, which includes the troupials, caciques, New World orioles, and blackbirds. Most are thrush- to crow-sized, and predominantly black (especially the males; the females may be brown), often with striking patches of red or yellow in the wings and tail. Most live in wooded country though some, such as the bobolink, dickcissel, and the meadowlarks, live in open country. They are insectivorous, fruit- or seed-eating birds, some being serious pests of soft fruit and rice. Many of those breeding in North America migrate to warmer areas for the winter. The cowbirds are parasitic, laying their eggs in the nests of other birds. Many of these 'New World' orioles breed in colonies and some, such as the oropenolas, *Psarocolius* species, weave long, tube-like nests hanging from branches of trees.

Ormer *Abalone.

Ornithology is the study of *birds, and includes investigation of behaviour, biology, and biogeography of birds. The physics of bird *flight has been studied in detail, and birds have been used extensively in investigations of animal behaviour. Bird *migration has always intrigued ornithologists and much information about their travels has been obtained by banding or ringing birds, and by tracking them on radar.

Birds are conspicuous and popular animals, which are threatened if their habitats are destroyed by man. The concern of ornithologists has done much to draw attention to problems of conservation in general.

Orthopaedics is a branch of surgery dealing with deformities of bones and muscles. The term originally applied to children but now refers to bone and muscle deformities in persons of all ages. A sound knowledge of anatomy forms a necessary basis for corrective surgery, plastering, or manipulation.

Ortolan: a member of the *bunting family, and is a small bird which breeds over much of Europe and western Asia, favouring open country with trees. It migrates southwards for the winter and had long been prized as a delicacy in some countries.

Oryxes are three species of antelopes with almost straight, slender horns placed immediately behind the eyes. The oryx or gemsbok, *Oryx gazella*, has a cream-coloured coat and chocolate 'stockings'. The face has dark eye- and nose-stripes, and often a dark forehead patch. This pattern accentuates the length of the head. The animal can be up to 1.2 m. (4 ft.) tall at the shoulder, with horns up to 1 m. (3 ft. 4 in.) long.

The scimitar-horned oryx, *O. dammah*, is one of the most threatened antelopes. Its survival would seem to depend upon the establishment and maintenance of reserves with adequate protection. The majority of the wild population is found in Niger, and to some extent in Chad, and it is extinct north of the Sahara. It breeds well in captivity, and captive-bred stock could be returned to herds in reserves. The Arabian oryx, *O. leucoryx*, once roamed the deserts of Arabia, Syria, and Mesopotamia in large numbers; today it is extinct in the wild. Fortunately, a herd was established in 1962 to prevent its complete extinction and has been reintroduced into Oman. An inhabitant of wild, arid areas, it is one of the few large mammals that can exist for very long periods without water.

Osler, Sir William (1849-1919), a Canadian-born physician, revolutionized medical education in North America. He introduced bedside teaching for medical students in hospitals and close cooperation between wards and laboratories. Professor of Medicine at the University of Pennsylvania, at Johns Hopkins University, and then at Oxford, he wrote a classic text, *Principles and Practice of Medicine* (1891).

Osmosis is the diffusion of water or another solvent through a semi-permeable *membrane, such as a cell wall. This membrane is like a sieve, allowing solvent molecules, which are small, to pass through it, but preventing larger molecules dissolved in the solvent from passing through. If a semi-permeable membrane separates two solutions of different concentrations then solvent molecules pass through the membrane from the side of lesser concentration to the side of greater concentration. This tends to equalize the concentrations on either side of the membrane.

Osmosis is of great importance in many processes which occur in animals and plants. If red blood cells, which are normally found in blood serum, are placed in pure water, then the cell walls allow the water to diffuse inwards, and the cells swell and eventually burst. Osmosis also controls the intake of water by the hairs on plant roots, and the production of urine in the kidney. The turgidity of plant cells is due to their maintaining a difference in pressure across their cell walls. Water enters their concentrated cell contents and effectively 'inflates' the cell.

Osprey, or fish-hawk, is a species of bird which is widespread throughout the world on lakes and sea coasts, breeding on all the continents except Antarctica and South America. Eagle-like in size and disposition, it specializes in catching fish close to the surface, spotting these prey from a height, diving with wings folded, and entering the water with clawed feet forward ready to grasp. The pads and claws are rough to facilitate holding; surprisingly large fish can be lifted and carried to the nest. For a time extinct in Britain, ospreys are now returning slowly; individual nests are carefully guarded each season against egg-hunters. The single species of osprey is placed in a unique family, the Pandionidae.

Ostrich: a flightless species of bird which is the sole member of its family. It is the largest living bird and occurs only in Africa. The male stands up to 2·5 m. (8 ft.) tall and weighs up to 120 kg. (270 lb.); it has a long, bare, pinkish neck and long, bare legs, with pink thighs. The feathers are loosely structured, and black in colour, except for white feathers in the rudimentary wings and in the tail. The female is a little smaller and has brown instead of black plumage. She lays her eggs in simple scrapes in the ground, each egg being up to 20 cm. (8 in.) long and weighing up to 1·3 kg. (3 lb.). The tail and wing plumes have long been valued for head-dresses and the bird is now kept on farms for this purpose.

Otters are some twelve species of carnivores in the weasel family, with worldwide distribution except in polar regions and oceanic islands. Long-bodied and streamlined with short limbs and webbed paws, they have a wide, flexible tail and rich brown, waterproof fur. They inhabit rivers and freshwater lakes, and are some of the fastest aquatic mammals, cruising along at 10 km./hour (6 m.p.h.). When swimming, they can shut their nose and remain submerged for four minutes or more without coming up for air. Otters feed mostly on fish, usually the rather sluggish species. Male otters are larger than females and may be about 1·4 m. (4 ft. 7 in.) long. There are usually two or three cubs in a litter, but occasionally as many as five may be produced. They are blind and toothless at birth, and the mother cares for them. Unlike the otters of the genus *Lutra*, which includes the Eurasian otter, *L. lutra*, the Oriental short-clawed otter, *Aonyx cinerea*, and Cape clawless otter, *A. capensis*, they catch their food with their front limbs, not their jaws. One species, the sea otter of North America, *Enhydra lutris*, is marine, and feeds on shellfish, which it opens by using pebbles as tools.

Ounce   *Snow leopard.

Ovaries are the female reproductive organs of animals which produce the egg cell, or *ovum, or they can be the lower part of a flower which contains the ovules. In mammals they produce single, or multiple, egg cells at regular intervals. In humans this involves the release of a single egg from one of a pair of ovaries after twenty-eight days. After another twenty-eight days the other ovary in its turn releases an egg, and so on. Occasionally more than one egg will be released at once. The release of eggs in all mammalian ovaries is controlled by *hormones, some of which are produced by the ovary itself. The ovaries of invertebrates can consist of several parts, each capable of continuously producing eggs.

The ovary of a plant is that part which encloses the ovules, destined to become seeds. It is usually associated with the style and stigma; these organs, situated inside the flower, trap and encourage pollen to grow down into the ovary, where fertilization occurs.

Ovenbirds are members of a South and Central American family which contains some 217 species. Varying in size from that of a warbler to that of a large thrush, most

A rufous **ovenbird** sits atop its mud nest reinforced with root fibres, on the Argentinian pampas. The fortress-like nest has evolved in response to the lack of naturally protective nest sites in grasslands.

are uniform brown in colour, with pale underparts. Their wings are short and rounded, and their beaks slender, and occasionally curved. They live in a wide variety of habitats from forest to semi-desert. Their nests are usually domed, built of sticks or mud. Some, such as the red, or rufous ovenbird, *Furnarius rufus*, build a round nest with a side entrance; this sets very hard and looks rather like a simple clay oven, hence their name.

Ovum: the female egg cell in animals. Before development it usually has to undergo *fertilization by a male sex cell, or *sperm. Unlike normal body cells, the nucleus of an ovum contains only half of the genetic material, or *chromosomes, required for full development. Some animals can produce individuals without fertilization of the ovum, but such offspring are sterile. Mammalian-fertilized ova develop within the mother, whereas birds and reptiles lay fertilized ova which develop outside the body. In fishes and amphibians ova are deposited outside the body and fertilized externally by the male.

Owlet moths are mostly small to medium-sized, night-flying moths, often dull brown in colour and well camouflaged, matching the stereotyped idea of a moth. Some, however, are day-flying and brightly coloured. They are the largest family of moths or butterflies, known as the Noctuidae, and include some 20,000 species. The majority have a wing-span of 2–4 cm. ($\frac{3}{4}$–$1\frac{1}{2}$ in.), but one of the largest of all moths, the giant owl moth, *Thysania agrippina*, with a wing-span of 30 cm. (12 in.), is a noctuid. Other common species of noctuid include the angle shades moth, *Phlogophora meticulosa*, and the large yellow underwing, *Noctua pronuba*.

Owls comprise an order of nocturnal birds of prey which falls into two distinct families united by many common features. All have soft, cryptically coloured plumage, rounded wings, and a short tail. Their large heads have forward-facing wide eyes usually set in a prominent facial disc of feathers. The beak is small and curved, and the legs are often feathered. Like all birds of prey, their feet are strongly clawed for grasping and lifting. Owls fly remarkably silently, characteristically striking at prey on the ground. Accuracy is achieved by very good nocturnal vision and directional hearing through asymmetrically set ears. Owls scream, hoot, whistle, and click in conversation. Their courtship is often noisy, conducted over their large territories by night. Nests are untidy collections of sticks, often taken over from other species. Most owls lay two to five white eggs, though the number is variable and may exceed a dozen. The eggs hatch asynchronously, giving the best opportunities of survival to the earliest hatchlings.

One of the two families includes about ten species of barn owls, distinguishable by their heart-shaped mask. The other family includes the remaining 120 or so species, ranging widely in size, habits, and geographical distribution. From the smallest pygmy owls to the largest eagle owls, all are unmistakably owls. Though seldom seen by day, they are common birds in every kind of habitat from polar tundra to tropical rain forest, taking rodents, insects, and small birds as their main prey.

Oxen is an imprecise term for members of the family Bovidae which includes gazelles, elands, impalas, cattle, yaks, bison, buffaloes, goats, and sheep. The name of ox is used in reference to species such as the musk ox and the Kouprey, or Cambodian forest ox, *Bos sauveli*. In some parts of the world, large, heavy builds of cattle used as draught animals are termed oxen.

Oxlips *Primrose.

Oxygen cycle is a term for a *biogeochemical cycle in which the element oxygen circulates between living organisms and the non-living environment. The primeval atmosphere of the Earth contained very little oxygen, but with the evolution of plants it become an important part of the atmosphere. Over many millions of years its concentration has gradually increased to the present level of 21 per cent by volume. Virtually all of this oxygen was formed by *photosynthesis carried out by *blue-green algae, and later by green plants. Oxygen is removed from the atmosphere by the aerobic respiration of living organisms, by the burning of fossil fuels, and by the formation of oxides. Respiration and the burning of fossil fuels produce carbon dioxide gas, which can be used again for photosynthesis, which in turn releases oxygen into the atmosphere to complete the cycle. The oxygen cycle parallels the *carbon cycle in most respects.

Oystercatchers, also known as sea-pies or kleepers, are large, portly, pied or all-black waders with long, blunt, scarlet bills. These bills are laterally flattened like knife-blades and are used by the birds to hack at and lever open bivalves (though not oysters in Britain). The loud 'kleep' call, often uttered in flight, is distinctive. Generally seen on the sea-shore, they are sometimes found in large flocks outside the breeding season. They constitute a small family of only about seven species, but the common oystercatcher, *Haematopus ostralegus*, in its many subspecies breeds in places as far apart as Iceland and New Zealand.

Oysters are *bivalve molluscs encountered in vast natural or cultivated beds in coastal shallows throughout the world. Prized everywhere as food, some forms are also valued for pearl production.

Oysters live with their flattened, left shell valve underneath, whereas the upper valve assumes rough and often bizarre shapes, according to the currents and sediments where they grow, unlike the beautifully sculptured symmetry of the valves of cockles or scallops.

Tiny pearls can occur in all oysters, but are rarely of value in temperate species, being made of irregular, chalky shell material rather than the special smooth nacre with which true pearl oysters coat intrusions. Most Western oyster farmers therefore concentrate on raising large meaty specimens, keeping at bay predators and other competing bivalves which encroach on the beds.

P

Pacas are large South American *rodents, weighing up to 10 kg. (22 lb.), which are closely related to the *agouti. The two species are pig-like in appearance although with a rounded muzzle. They prefer tropical forests and lie up during the day in burrows, which they leave in the evening to forage for roots, grasses, and fallen fruits. They take readily to water and can swim well.

Paddle-fishes are species of fishes so-named because of their long, blunt and flattened snout, which looks like a paddle or lolly-stick. The American paddle-fish, *Polydon spathula*, and a relative, *Psephurus gladius*, in China, are the only members of their family, and are related to the sturgeons. Living in the Mississippi river system, the American paddle-fish grows to 2 m. (6 ft. 6 in.) and feeds on plankton, principally small or larval crustaceans which it catches by swimming with its huge mouth open. At one time caught for food, it is now considerably reduced in numbers.

Paddle worms   *Bristle worms.

Pain is a sensation which can range from the mildly unpleasant to the excruciating. It is a primitive protective sense which is difficult to quantify in man and extremely hard to assess in animals. Networks of naked nerve endings, which respond only to extreme stimulations, exist in many tissues. Two distinct spinal pathways connect these endings with the brain, where processing determines the quality of the pain. Even extreme pain may be disregarded during conditions of wound shock or intense excitement. Pain relief may be obtained from drugs (analgesics) which either inactivate the nerve endings or interfere with neural processing.

Painted-lady butterflies are members of the *brush-footed butterfly family and occur worldwide. Many are migratory; the species occurring in Britain in summer, the painted lady, *Cynthia cardui*, is pinky-orange with black and white markings, and flies fast and powerfully. Every year this species spreads throughout Europe from its winter base in the Mediterranean region and North Africa, and migrates south in autumn. There are nine species of painted-lady butterfly; seven are found in America, and one in Australia, whereas *C. cardui* is worldwide in distribution.

Painted snipes are a family of *snipes comprising two species of tropical and subtropical wading birds. They look like true snipes but have shorter beaks and fly weakly with legs dangling, like rails. The females are bigger and more brightly coloured than the males and are the dominant partners.

Pair-bonding is found in species where a single male and female mate and remain together for the duration of the breeding season or longer. It is common in birds but is also found in many other animals. Extended courtship and other ritualized behaviour patterns maintain a stable family unit that gives the maximum chance of survival to the young.

The **paca**, *Agouti paca*, is well camouflaged in the forest by its spotted coat. Timid by nature, it sticks to well-tried paths in undergrowth or on riverbanks.

Palaeontology is the study of past life on the earth, based on the evidence of *fossils. As well as being concerned with individual organisms, it provides information for evolutionary studies and palaeoecology (the study of ancient ecosystems). In addition, fossils can be clues to other aspects of the earth's history, such as geography and climate.

Palaeontology is indispensable to the geologist because it helps to indicate the relative age of rocks. Fossil invertebrates are commonly used for this purpose, and different groups of these are used in the dating of different geological periods. For example, subdivisions of the Cambrian Period (570–505 million years ago) are dated by means of *trilobites. In the twentieth century the study of microscopic fossils, especially *foraminifera, has become important, because they can be used to identify sediments potentially containing petroleum.

Fossils are commonly found only in rocks deposited after the beginning of the Cambrian Period. The reason for this is not known, although it has been suggested that earlier organisms lacked the hard skeletons suitable for fossilization. In recent years much older microscopic fossils have been reported, including organisms resembling *blue-green algae that are over 3,000 million years old. Future evidence from this very early period may give us information about the origin of life itself.

Palm-chat: a sparrow-sized bird found only on the West Indian islands of Hispaniola and Gonave, and placed in its own family. It is greenish and brown above, with pale, heavily streaked underparts, and has a broad and heavy beak for feeding on berries. Several pairs build a communal nest of twigs, though each pair has a separate entrance.

Palm civets are a subfamily with some seven species of carnivores, intermediate between genets and mongooses but more closely related to the former. There is an African species, *Nandinia binotata*, but most occur in tropical Asia. They are arboreal and, although they certainly occur in palm trees, they are by no means confined to them. Their diet consists of small mammals, birds, and insects but they do not take much vegetable matter apart from some fruit. Another closely related subfamily consists of the five species of banded palm civets, which are similar in all respects to the palm civets.

Palmetto, originally palmito, is the Spanish name for the dwarf fan palm, *Chamaerops humilis*. The name is also used for other palms, notably *Sabel palmetto*, a fan palm of the southern USA which reaches 25 m. (80 ft.). The dwarf fan palm is the only palm native to the European mainland, being found in the western Mediterranean region, where it is typical of the short leathery-leaved vegetation known as *garigue*. It is usually stemless but may have a short trunk.

Palms (trees) are a family of some 2,500 species of woody *monocotyledons. They occur mainly in the tropics, with a few extending into the Mediterranean region, subtropical China, and New Zealand. Most are unbranched with a terminal head of pinnate (divided) leaves, like the coconut, or fan-shaped leaves like the talipot. Other palms sucker from the base, like the sago palms, or are branched, like the doum palm of Africa or the *nipa palm, whose stem creeps along the ground. The rattans are climbing plants, or *lianas. Some palms, like the talipot, flower once and die, others flower several times from lateral shoots. *Lodoicea maldivica* of the Seychelles has the largest seed of any plant and one of the largest of all known leaves. All palms produce *drupes, juicy like the date, or fibrous like the coconut.

Some human societies, notably in the Pacific, have an economy completely based on palms, which provide their shelter, clothing, food, drink, utensils, and matting. In New Guinea, sago palms provide the staple food of many people and throughout the Asiatic tropics, the principal narcotic is derived from the betel nut, a palm fruit. Palms provide not only thatch and building materials, but also margarine and vegetable oil, toddy and other drinks, vegetable ivory, raffia, and a host of lesser products. Unlike dicotyledonous trees, such as oak or maples, palms grow only from the top of their 'stem' and do not expand their trunk as they grow.

Palynology is the study of *pollen grains and spores. The shape and structure of pollen grains can be used in plant classification. It also provides a tool for understanding vegetation of past ages, for the grains are preserved in peat bogs and other places. Their recognition and their relative abundance provide a basis for discovering what plants grew in a site and thence the local climate at the time they grew.

Pampas (singular: pampa) are a type of natural grassland in South America, the name being originally restricted to that of Argentina. They are usually dominated by the pampas grass, *Cortaderia selloana*, which may reach some 2 m. (6 ft. 6 in.) when in flower.

Pampas grass, of which there are six species in the grass genus *Cortaderia*, is native to tropical and temperate South America. The plants are large perennials with long, narrow leaves and tall silky-plumed flower-heads. *C. selloana* is the common kind frequently cultivated as an ornamental garden plant. The sexes are separate, and female plants produce larger flower spikes than the males.

Pancreas, or sweetbread: a large gland which lies in the mammalian abdomen. It contains two separate types of cell. One group manufactures digestive juices which enter the small intestine via ducts. The other type consists of isolated patches of cells, called islets, scattered among the digestive juice-secreting cells and their ducts. The islets together form an *endocrine gland and secrete their products into the bloodstream. These endocrine cells are responsible, among other things, for the manufacture of insulin, used to treat diabetes.

Panda, or giant panda, is the biggest and most familiar member of the racoon family, its relative the *red panda being much smaller. The giant panda has been adopted as the symbol of the World Wildlife Fund and, with its black body, white head, black ears, and black 'spectacles', it has a most endearing appearance. It lives high in the mountains of southwestern China and eastern Tibet, where it moves overland with a lumbering gait on its flat, bear-like feet. Perhaps surprisingly, it is quite agile and climbs trees. It lives in dense forests of bamboo on which it feeds almost exclusively and of which it must ingest a prodigious amount to obtain its nourishment.

Before producing her young, the mother will select a large hollow tree, often a fir, to use as a den. The helpless

The bleak **pampas** of Bolivia are in the high plateau region between the Andean ranges. They are dominated by tussock grasses, evenly spaced because of root competition due to the area's scanty rainfall.

infant remains there, cared for by the mother, until three months of age. This animal is endangered; ten reserves have been created in China and a breeding station is being established. Leopards are the main predator of young pandas.

Pangolins are found in the tropical regions of the Old World. They are edentate mammals and are the only mammals with scales; those of the great ground, or giant pangolin, *Manis gigantea*, are 7·5 cm. (3 in.) long and 13 cm. (5 in.) wide. The seven species of pangolin, four in Africa, and three in Asia, have a small head, a long, tapering body, and broad, heavy tail. The head and body of the giant pangolin is 85 cm. (33 in.) long and the tail is up to 80 cm. (31 in.). It moves slowly, dragging the heavy tail; if threatened, it curls into a ball and so is protected by the sharp-edged scales. It feeds on insects, usually ants or termites, and after tearing open a nest will mop them up with its long, thin, sticky tongue.

Pansy   *Violets.

Panthers   *Leopards.

Papaya, or pawpaw: a rapid-growing, soft- or hollow-stemmed small tree, *Carica papaya*, with a crown of large, segmented evergreen leaves. Of Central American origin, it is widespread as a garden plant throughout the tropics but seldom cultivated on a larger scale. It gives its name to the small family Caricaceae, comprising thirty species of tropical trees; other *Carica* species also produce edible fruit. The orange or yellow fruits of the papaya are eaten fresh, canned, or made into drinks. Immature fruits may be cooked as vegetables. The latex of the tree contains the useful enzyme, papain, which breaks down protein, and

is used to prevent cloudiness in beer, to shrink-proof wool and silk, or as a meat tenderizer.

Paper nautiluses are small floating *octopuses, of the genus *Argonauta*, which have no true molluscan shell. Yet mature females secrete a beautiful, thin-walled, chalky egg-case using the expanded tips of two tentacles. This is used as a brood-chamber and carried by the female once eggs are laid. It is usually large enough to be used as a shelter by the female, who can retreat fully inside; the tiny males also occasionally cohabit.

Paracelsus, Philippus Aureolus (1493–1541), Swiss physician whose real name was Theophrastus Bombast von Hohenheim, was a controversial figure, who lectured in German rather than the accepted Latin. He condemned all current medical teaching not based on observation and experience. His study of diseases suffered by miners was one of the first accounts of an industrial disease. He taught that man was part of nature, and he believed in nature's healing power. He introduced mineral baths and chemicals such as sulphur, iron, lead, arsenic, and mercury as medicinal treatments.

Parakeet is a name used mainly for three separate groups of the parrot family. The first includes birds of the genus *Cyanoramphus* which occur in New Zealand and a number of other Pacific islands. The second group belongs to the genus *Psittacula*, with thirteen species, found mainly on Indian Ocean islands, and in Southeast Asia. The third group belongs to the genera *Brotogeris*, *Forpus*, and *Bolborhynchus*, with nineteen species from Central and South America. All are smallish, though some, with the exception of the New World species, have long tails. The rose-ringed parakeet, *P. krameri*, is a common cage-bird which has escaped and become established in the wild in a number of countries, including England.

Paralysis is a loss of muscular action caused by damage to *nerves regulating voluntary movements, or by interference with the cellular processes of muscle contraction by cold, drugs, or poisons. Damage which crushes or severs nerves within the spinal cord may paralyse part of the body below the damaged point. *Poliomyelitis, fortunately now rare because of immunization, is a virus disease which destroys spinal nerves associated with muscle action and can leave individual muscles permanently useless and wasted. Paralysis to one side of the body, for example of the facial muscles, arm, and hand, as in a stroke, may develop after small blood vessels, which supply the nerve centres in the opposite side of the brain, bleed or clot.

Certain species of scorpions and spiders, and the krait, inject paralysing nerve poisons to immobilize their prey. Certain arrow poisons, such as curare, bind selectively to the surfaces of vertebrate muscles at their nerve terminals, blocking the transmission of impulses.

Paramecium is a genus of freshwater, ciliated *protozoans, and part of the class Ciliatea. They are the most complex of all unicellular animals. These protozoans are slipper-shaped, about 0·15 mm. ($\frac{1}{200}$ in.) long, and swim actively, engulfing small food particles via mouth and gullet. The surface of all species of *Paramecium* is covered with beating cilia, all interconnected and co-ordinated, and with trichocysts, tiny toxic barbs which are discharged into prospective predators or used for anchorage. All ciliates, including *Paramecium*, can reproduce by asexual division, when they simply divide into two halves. Most species can also undergo sexual reproduction by a process called conjugation; two sexually compatible types come together and fuse, eventually dividing into new daughter cells once genetic material has been exchanged.

Paraplegia is paralysis of both legs due to damage of the spinal cord. It involves paralysis of some of the body muscles, loss of sensation in the lower body, and often loss of control of bladder and bowels. The paralysed muscles become stiff and often move involuntarily because of reflex activity in the spinal cord below the damage.

Parasites are *heterotrophic organisms living in or on another animal or plant (the host), from which they obtain food and, frequently, shelter. Many animal groups include some parasites. Parasitic plants are much rarer and often can photosynthesize so that they are not wholly dependent on their host for food. Occasionally the host may be killed but a 'true' parasite is so well adapted to its host that this is unusual. Ectoparasites live on the outside of their host. Endoparasites may live in the cells of their host (intracellular) or within its body cavities and fluids (extracellular).

Many parasitic organisms, such as tapeworms, flukes, nematodes, and protozoa, protect themselves by having a tough outer skin or by producing substances which deceive the host's chemical defences. Insect parasites, or parasitoids, which attack other species of insect, almost always kill their host. Also among the insects is the phenomenon of hyperparasitism whereby parasitic species are themselves attacked by yet smaller parasites.

Reproduction presents great problems for endoparasites. A secondary host may be used as a vector to carry the parasite from one primary host to the next. The vector itself may be a parasite. The mosquito *Anopheles* carries the parasitic protozoan *Plasmodium*, which causes malaria in man. Most endoparasites need to produce thousands of eggs to ensure that some of their offspring are passed on to other hosts. The endoparasitic threadworm *Ascaris lumbricoides* can produce 250,000 eggs a day. Some endoparasites are self-fertilizing hermaphrodites, or simply reproduce parthenogenetically, as in aphids or parasitic wasps (ichneumons).

Pardolotes   *Flowerpeckers.

Parkinson, James (?–1824), English surgeon and palaeontologist, was author of *An Essay on the Shaking Palsy* (1817), describing Parkinson's disease, which is a disorder of ageing people characterized by trembling. His work on fossils culminated in the publication of *Organic Remains of a Former World* (3 vols., 1804–11).

Parrot-fishes are brightly coloured fishes of the family Scaridae, which live on reefs in shallow tropical and subtropical seas. Their teeth are fused to form beak-like tooth plates joined at the midline, while in the back of the throat there are an upper pair of pharyngeal teeth and a single lower set of teeth. These grind up the algae and soft coral rock, which the fish bites off with its powerful front 'beak'. Parrot-fishes are responsible for the erosion of many coral reefs. About seventy species are recognized, mostly small and less than 45 cm. (18 in.) in length. However, since the two sexes are often differently coloured (the males usually being brighter), and in many species females change sex as they age, there has been much confusion about exactly how many species there are.

Parrots belong to a very large family containing about 330 species, which occur in most of the warmer areas of the world. The family includes parrots, budgerigars, lovebirds, parakeets, lorikeets, cockatoos, and macaws. The majority are forest-dwelling species. They vary in size from the tiny buff-faced pygmy parrot, *Micropsitta pusio*, 8·5 cm. (3½ in.) long, from the lowland forest of New Guinea, to the hyacinth macaw, *Anodorhynchus hyacinthinus*, 1 m. (3 ft. 3 in.) long, from tropical South America. Many parrots are very brightly coloured, although a few are black or white. The commonest colour is green with yellow, blue, or red markings. They have short beaks with the upper mandible hooked downwards over the shorter

A **parrot-fish**, *Scarus* species, asleep at night wedged into a crevice of a reef, off the Maldives. The fish wraps itself in a mucous cocoon, which is thought to provide protection against predators such as moray eels.

lower one. They feed primarily on fruits and nuts though some of the smaller ones have brush-like tips to their tongues and drink nectar. Most nest in holes in trees and lay round, white eggs. A number of species are restricted to very small areas, such as a single island, and are highly vulnerable to habitat destruction. Some species attract high prices as cage-birds.

Parsley is a biennial herb, *Petroselinum crispum*, which probably originated in southern Europe, but it has been cultivated for so long that this is uncertain. It is a member of the carrot family, or Umbelliferae. The wild form has a plain, deeply segmented leaf, but the cultivated form has curled and crisped segments. The leaves are used whole or finely chopped as a garnish in a variety of dishes and are rich in vitamins.

Parsnips are vegetables derived from a wild biennial species, *Pastinaca sativa*, of the Umbelliferae, or carrot family that occurs throughout Europe and western Asia. This plant has been cultivated for its uniquely flavoured roots since ancient times and has been introduced throughout the world. The tapered yellowish roots contain both sugar and starch, hence their sweet flavour.

Parthenogenesis is a mode of reproduction which involves the development of an unfertilized egg. Animals or plants using this method of reproduction produce offspring which have the usual two pairs of chromosomes per cell but are genetically identical to the parent. Parthenogenetic reproduction may alternate with normal sexual reproduction in plants such as dandelions and in animals such as *aphids. Some species, such as the Indian stick insect, *Dixippus morosus*, very rarely produce males and depend almost entirely upon parthenogenesis.

Partridges are a number of species of birds within the pheasant family, not all closely related, but all medium-sized and short-tailed. They live on the ground in open country and some, such as the Himalayan snow partridge, *Lerwa lerwa*, live high in mountains. All have short, stubby beaks and eat seeds, other vegetation, and also some insects. The European, Hungarian, or grey partridge, *Perdix perdix*, is an important gamebird and has been widely introduced in North America.

Pasque-flowers are so called because they flower at Easter. They are perennial plants with finely divided leaves and purplish cup-shaped flowers. They belong to the genus *Pulsatilla* and are thus related to buttercups and columbines. These medicinally useful plants usually grow in chalk grassland throughout Europe and western Asia. Larger-flowered varieties with violet and reddish petals have been developed as garden plants.

Passenger pigeon is now an extinct species of dove, although it was formerly one of the most abundant of all bird species, living in huge flocks in North America. Due to a combination of over-hunting and destruction of its woodland habitat, it had become extinct by about 1900.

Passion-fruit *Passion vines.

Passion vines are part of the family Passifloraceae, which includes 600 species of vines, trees, shrubs, and herbaceous plants, found in the tropics and subtropics.

Although a climber, the **passion vine**, *Passiflora vitifolia*, carries its brilliant red blooms near the ground—the feeding level of the hermit hummingbird which, in gathering nectar, pollinates the flower.

The genus *Passiflora*, with 400–500 species, includes many that are woody climbers with edible fruit, such as the *granadillas. The purple passion-flower, *P. edulis*, and its cultivars are grown in many parts of the world for their round or egg-shaped, aromatic fruit. Most species of *Passiflora* have very distinctive flowers with conspicuous central filaments forming a corona. The flower was once regarded as a symbol of Christ's Passion — hence the name. Several species of *Passiflora* are cultivated as ornamental climbers.

Pasteur, Louis (1822–95), French chemist and biologist, founded the science of bacteriology. He did pioneer work in a form of biochemistry called stereochemistry. It was when he was appointed to promote science in the brewing industry that he made his greatest discovery: fermentation was not a purely chemical reaction as had been thought, but was caused by living micro-organisms (yeasts). Similarly, he postulated that putrefaction must be caused by *bacteria. These could be killed by heat, and he devised the process now known as pasteurization. During investigation of the problems of the French silk industry, he isolated *bacilli attacking silkworms and found methods both of detecting diseased stock and of preventing the disease from spreading. This saved the silk industry of France and other silk-producing countries. He then turned his attentions to the fatal cattle disease of anthrax and to the problem of chicken cholera. In both diseases, he was able to isolate the bacillus responsible and cultivate less virulent forms which he then used to vaccinate healthy animals. His last triumph was against hydrophobia, or rabies, making a vaccine which not only protected dogs from contracting the disease, but also cured humans already bitten by rabid animals.

Pastures are grasslands maintained for the grazing of animals or for haymaking. Some have never been ploughed up, whereas others are ploughed and replanted

Dr **Pavlov** lived and worked at St. Petersburg (now Leningrad). Professor of Pharmacology and a Director at the Institute of Experimental Medicine there, he won a Nobel Prize in 1904, when this photo was taken.

(often after crops have been grown on the site) using a mixture of legumes, and grasses. In Europe two common species used for replanting are white clover, *Trifolium repens*, and perennial rye-grass, *Lolium perenne*, although in America a wider range of clovers and grasses is used, reflecting the wider range of conditions under which pastures are established.

Patchouli oil is extracted from a number of species of *Pogostemon*, woody herbs of the Asiatic tropics which belong to the mint family. The oil is insecticidal and a leech-repellent. It has some medicinal uses but is most familiar as a perfume much favoured in India.

Patella, or kneecap: a bone in the tendon at the front of the knee. The tendon extends from the kneecap to be inserted into the upper end of the shin-bone in the leg. The kneecap protects the lower end of the thigh-bone, with which it articulates.

Pathology, the study of diseased organs and tissues, is best understood historically. Advances in the classification and understanding of disease were made in the early nineteenth century by increased use of post-mortem examinations. These were usually performed by the physician who had treated the patient. A further great advance followed the routine microscopic examination of the tissues (histopathology: see also *histology). This tended to be carried out by a specialist who also conducted the post-mortems. Pathology is still taken to mean this combination of post-mortem and histopathology. The pathologist also undertakes the microbiological, chemical, and immunological analysis of specimens of blood and tissue, taken from the living patient. These aspects, still considered part of pathology in the widest sense, have now developed into the disciplines of medical microbiology, chemical pathology, and immunology.

Pavlov, Ivan Petrovich (1849-1936), Russian physiologist, was noted for his research on the secretions of glands, especially in the digestive system. He also studied the activity of the brain and the functioning of reflexes. Psychology has been influenced by his work on reflexes in dogs. Salivation in dogs normally occurs in response to the smell of food. Pavlov trained his dogs to salivate in

A newly emerged adult **peacock butterfly**, *Inachis io*, hangs limply from its pupal case. Gradually, as blood is pumped into the veins, its wings will expand and, after a period of drying, the insect will fly away.

response to the sound of a bell even when food was no longer present. This is called a conditioned reflex.

Pawpaw   *Papaya.

Peaches are small, deciduous trees that originated in China as *Prunus persica*. This species has been cultivated throughout the warm, temperate zone of the Old World since ancient times and has given rise to many varieties. Closely related to the almond and cherry, the peach belongs to the rose family along with apples, pears, and many other shrubs, herbs, and trees. The velvety skin of the peach fruit is greenish-white, or yellow, often flushed with red. The flesh also varies from white to yellow, according to variety. Some varieties of the peach tree are grown for their ornamental white, pink, or red flowers, but all require warmth for satisfactory culture. The *nectarine is one of the most distinct varieties of peach.

Peacock butterfly: a species of *brush-footed butterfly, *Inachis io*, with large spectacular eye-spots; it occurs throughout Eurasia and is common almost everywhere that its larval food plant (nettles) thrives. If attacked by a bird, the peacock butterflies orientate the eye-markings towards the attacker and make a hissing noise by rubbing the wings together.

Other butterflies may also be locally known as peacock butterflies because of eye-spot patterns.

Peacock moths   *Emperor moths.

Peacocks, or peafowl, belong to one or other of three species of the pheasant family. The male common peacock, *Pavo cristatus*, comes originally from India and Sri Lanka and is famous for its beautiful train of multicoloured feathers, which it fans during courtship. It has been widely domesticated for so long that nothing is known of the history of domestication (although Solomon may have imported peacocks into Palestine—1 Kings 10: 22) or of the origins of a number of colour varieties, the most striking of which is an all-white form.

There are also six species of the genus *Polyplectron*, which are known as peacock-pheasants.

Peacock worms   *Fanworms.

Peanut, also known as groundnut and monkey-nut: an annual *legume of South American origin, now widely cultivated throughout the tropical and subtropical areas of the world. The familiar wrinkled fibrous pods contain two to four red or brown seeds with white flesh and are harvested from below soil level. After fertilization, the flowering stalks grow downwards into the soil where the seed pod develops, perhaps safe from predators. This habit has earned the plant the alternative name of groundnut.

The seeds are extremely nutritious, containing about 30 per cent protein and up to 50 per cent oil, and are rich in vitamins B and E. The whole seeds (for they are not true nuts) may be cooked, as in tropical Africa, or eaten raw or roasted, with the skins removed, as in the USA and Europe. The edible oil is becoming of increasing importance in world trade and is widely used for cooking, for margarine, and for soaps and lubricants. The protein-rich seed residue is widely used in animal feeding.

Peanut worms   *Sipunculids.

Pearl mussels are elongate, slightly kidney-shaped, freshwater *bivalves which were cultivated in Roman times because they often produce tiny pearls in the body wall in response to an irritating parasite. Although not comparable in size with oyster pearls, these are of good quality, and the Romans established considerable pearl fisheries in some parts of Europe.

Pearl oysters, occurring in warmer Pacific seas, produce the finest natural pearls by coating any foreign object intruding within their shell valves with the same iridescent mother-of-pearl material (nacre) that lines their shells. All *bivalves do this, but in most types the resultant pearl will be very irregular, or fuses to the shell itself, so becoming commercially useless. Pearl oysters, such as *Pinctada margaritifera* and *P. mertensi*, can be 'seeded' artificially, producing valuable pearls within three years.

The soil around this **peanut** plant, *Arachis hypogea*, has been scraped away to reveal the pods below the surface. Its yellow blooms, typical of the Leguminosae family to which it belongs, are pea-flower shaped.

A **pearly nautilus**, *Nautilus macromphalus*, from the southwest Pacific, lies on its side showing the marked shell, fleshy hood, and tentacles. The animal is active at night, swimming backwards by forcing water out of its siphon. During the day it rests, secured by its tentacles, in a rock crevice on the sea-bed.

Pearly nautiluses are the only *cephalopods with external shells, which are secreted as a spiralled series of gas-filled buoyancy chambers. They live in the tropical Pacific from the surface down to 500 m. (1,600 ft.). The animal occupies the most recently made shell chamber, with its thirty-eight tentacles projecting. During its nocturnal feeding activities, the nautilus usually swims backwards.

Pears have been grown throughout Europe and western Asia since the earliest historical times. There are several wild species, but cultivated pears are all descended from *Pyrus communis*, a large, deciduous, long-lived tree, growing up to 15 m. (49 ft.) in height, which is slow to bear fruit. They belong to the rose family and are related to the apple, cherry, and peach. Nowadays, pears are grafted on to quince root-stocks so that they may fruit earlier. The green, globe-shaped fruits of pears are classified as dessert types, culinary kinds, or perry pears, which are crushed and used to produce a type of cider.

Pearson, Karl (1857–1936), English biologist and mathematician, wrote *The Grammar of Science* (1892), which stimulated a generation of young scientists. He applied statistical measurement to the Darwinian theory of evolution by natural selection. Later in his life, his energy was a prime factor in the establishment of statistical laboratories throughout the USA and Europe.

Peas are climbing, annual plants with terminal leaflets modified into tendrils. They give their name to the pea family or Leguminosae (*legumes), one of the largest plant families, with over 17,000 species of trees, shrubs, climbers, and herbaceous plants. This worldwide family includes *beans, *peanuts, *vetches, and *acacias. Their characteristic features include divided leaves, a single carpel on the flower, root nodules containing nitrogen-fixing bacteria, and seeds that are held inside pods.

Several species of pea have been cultivated since prehistoric times but only *Pisum sativum* is widely grown today. It probably originated in the Middle East. The pods contain numerous seeds, which vary in size and form according to variety. The small-seeded types have the best flavour, while the so-called sugar or mangetout peas have tender pods which can be eaten whole if they are picked before the seeds swell. The more usual pea is grown to be picked before the seeds ripen. Two major forms exist, differing in their seed shape, round or wrinkled. The former are more hardy but the latter are preferred for quick-freezing and canning. Field peas are a variety grown for feeding livestock and were once commonly grown with cereals such as oats and barley.

Some other legumes are called peas, for example chick-peas, but none belongs to the true pea genus, *Pisum*.

Pecan: a large tree which grows up to 60 m. (195 ft.) tall. It is found along streams in the arid areas of Mexico and southern-central USA. In North America it is now widely cultivated in orchards, where it is kept pruned to 15 m. (48 ft.) A relative of the walnut, the fruit of the pecan is a nut with a smooth red shell and a sweet flesh.

Peccaries are members of the hippopotamus family and are the New World equivalent of the *pigs. There are three species found in South and Central America. They are forest animals that travel in groups of a few to 300 individuals. They dig for tubers and roots, and also eat fruit and small animals, including snakes. The collared peccary, *Tayassu tajacu*, occurs from sea-level up to altitudes of 5,500 m. (18,000 ft.). It has thick, coarse, bristly

Collared **peccaries** are known to have at least six different vocalizations, most being specific to alarm or anger. Territorial and gregarious, they can also often be boisterous and quarrelsome with each other.

Emperor **penguin** chicks, once they are independent of their parents, are kept together in crèches protected by adult guardians. When the long winter passes and their moult is complete, they all migrate to the ice-floes.

hair, grizzled in colour, with a light-coloured 'collar' almost encircling the body at the shoulders. The peccary reaches a height of 50 cm. (20 in.) at the shoulders. The white-lipped, *T. pecari*, and Chacoan, *Catagonus wagneri*, peccaries are larger, and live in tropical and thorny forests respectively.

Pedipalps are important structures in the *arachnids, situated just behind the jaws. In spiders they are used by the male for sperm transfer, and are enlarged and knob-like, resembling boxing gloves. The space within the knob is filled with sperm by dipping it into a globule of semen previously deposited on a tiny silken mat, then the filled pedipalp is inserted into the female.

In scorpions and false scorpions the pedipalps are modified as the pincers. In mites and ticks they are reduced in size and used as additional mouthparts or lost altogether.

Pekin robin   *Babblers.

Pelargoniums are perennial plants or small shrubs native to South Africa, often with scented leaves and large, colourful flowers. They are often called *geraniums, although they do not belong to the genus of this name. The genus *Pelargonium*, along with true geraniums (*Geranium* species), forms a major part of the family Geraniaceae, with some 750 species of subtropical and temperate-region plants. Most of the varieties popular as garden and glass-house plants are of hybrid origin, produced over many years by breeding and selection. The four main groups recognized by horticulturists are: scented-leaved, ivy-leaved, zonal, and regal or show. The wild species, of which there are around 250, are South African in origin and include many succulent kinds.

Pelicans are a family of water-birds allied to boobies and cormorants, and well known for their very large beaks and extraordinary throat pouches. They appear ponderous creatures, but are majestic and spectacular both in flight and on the water. The brown pelican of North America, *Pelecanus occidentalis*, catches fish by plunging into the water from above like a booby, but most species are surface-feeders, dipping their heads under water or scooping prey up in their throat pouches. Pelicans are gregarious birds; they often hunt in parties, roost communally, and usually nest in colonies. The eight species are widely distributed in warmer latitudes.

Pellitories are slender, perennial herbaceous plants which belong to the stinging-nettle family. The common European pellitory of the wall, *Parietaria judaica*, is a plant used medicinally for complaints of the bladder. As its name suggests, it typically grows on walls.

Pelvis: a term applied to the lower part of the abdomen in most vertebrates, or more specifically the bony structure formed by the hip bones and lower part of the spine. It is shaped differently in male and female. In the female, the pelvis is broader and the central space rounder, wider, and with smoother walls, for the easier exit of the foetus. Measurement of skeletal pelvic remains can reveal the sex of the individual. The pelvic contents consist of male or female sex organs, the bladder, and the rectum.

Penguins are flightless seabirds with eighteen living species. Confined entirely to the Southern Hemisphere, they range from the Galapagos penguin, *Spheniscus men-*

The fleshy, single-seeded fruit of a species of **pepper,** growing in the rain forest of the Solomon Islands. Its leaves have drip-tips that facilitate the flow of water off the plant.

diculus, to those of Antarctica (emperor and Adélie). Emperors, *Aptenodytes forsteri,* are the largest, standing 1 m. (3 ft.) tall, and little blue or fairy penguins, *Eudyptula minor,* are the smallest at 20 cm. (8 in.). Their wings are modified as flippers to propel them through the water and their webbed feet and stiff tail feathers act as a rudder. Food includes fish and swimming or floating crustaceans; larger species dive deeply for bottom-feeding fish and squid. Most species nest colonially, some in huge colonies. Tropical species nest under cover, some in burrows and caves.

Penis: the male copulatory organ present in mammals. The term is also used to describe a similar organ in some invertebrates, including slugs and snails. In mammals it is essentially a tube, called the urethra, surrounded by spongy, erectile tissue. In some, including man, the tip is expanded into a cone-shaped structure called a glans, which is covered by the prepuce, an extension of skin lined by mucous membrane which keeps it moist. The urethra connects to the bladder and is used to void urine. Before use in the transfer of sperm to the female vagina the organ is erected. Erection of the penis is caused by blood entering the cavernous spaces within membrane-bound tissue but being prevented from leaving by contraction of all draining veins. With the exception of man, most primates, rodents, and several other animal groups, have a bone, called a baculum, within the erectile tissue. In animals erection is most usually caused by olfactory stimuli coming from a receptive female. In man, erection is under voluntary as well as emotional control. The glans is liberally supplied with sensory nerves which when stimulated by friction within the vagina lead to the forceful ejection of semen.

Penis worms are small, sluggish seashore creatures of which only eight species are known. Despite this they are placed in a separate phylum, Priapulida, because of their unique body structure. Their body is cucumber-shaped, with a protrusible proboscis at the front and, in *Priapulus bicaudatus,* two frilly 'gills' at the back. They burrow in soft sand, feeding on passing invertebrates, such as polychaete worms, using spines around the tip of their proboscis. They are simply built worm-like animals, lacking segmentation, blood vessels, or eyes, and have a primitive nervous system. They have been found off the coasts of North America and Siberia, and in the Baltic Sea and Antarctic waters. There is still controversy about which other groups of animals may be their nearest relatives.

Pennyroyal   *Mint.

Pentstemons, or penstemons, are a genus of perennial plants mostly native to North and Central America. They belong to the foxglove family. The name refers to the flower's five stamens (four fertile and one abortive stamen). The stem bases are often woody and this gives a low shrubby habit to some species. The tubular flowers, which are often large and colourful, make them popular as garden plants.

Peonies are perennial, tuberous-rooted, herbaceous plants or small woody shrubs. They form the family Paeoniaceae, which contains only the genus *Paeonia* with thirty-three species. They are widely distributed in Europe, Asia, China, and northwest America. The large and colourful flowers of most species make them popular as garden plants. Many ornamental varieties have been bred from the herbaceous species *P. lactiflora* and *P. officinalis,* while the shrubby tree peonies are derived from *P. suffruticosa,* the moutan.

Pepper, or true pepper (*Piper nigrum*): a perennial, woody, climbing shrub native to southwest India and now cultivated as a crop in most parts of Southeast Asia, Brazil, and Madagascar. It belongs to a family of small tropical trees, shrubs, and climbers with around 2,000 species. Historically it has been the most important of spices, known even to the ancient Greeks and Romans. It is still universally used as a condiment and flavouring in all kinds of savoury dishes.

Flourishing only in hot, wet, tropical climates with a long rainy season, the pepper plant produces red fruits borne in long clusters. Black pepper is made from the dried, whole fruits which become black and wrinkled and are known as peppercorns. White pepper, a less pungent form, is produced if the fleshy red outer skin is removed before drying. The pungency is caused by various resins and a yellow alkaloid, piperine. Pepper is an important crop for countries like India and the USA. Sweet and chilli peppers are obtained from a completely different group of plants, the *capsicums.

Peppered moth: a species of Geometrid moth which rests by day on tree trunks, where the usual black-mottled white colour form acts as camouflage among lichens. In industrial areas, where tree trunks are blackened and bare, an all-black form is more frequent, a phenomenon known as industrial melanism. This moth is one of the few documented cases of 'evolution' in response to a change in its environment.

Peppermint is a hybrid between water mint and spearmint, widely cultivated in Europe, North Africa, and America. Unlike other mints, it is prized for its oil, which is distilled from fresh flowering plants and used in cordials and confectionery. The oil is mildly antiseptic and is also used to treat indigestion.

The adult European **peppered moth**, *Biston betularia*, with a range as far as temperate Asia and Scandinavia, appears in late spring and summer. The normal form (*right*) is well camouflaged here against lichens on a birch trunk, as compared to the black form (*left*).

Pepper shrikes belong to a family containing only two species, both of which occur in Central and South America. Around 15–18 cm. (6–7 in.) long, they are olive-green above and yellow or white below, with grey cheeks and orange and brown eyestripes. They have stout, hooked beaks with which they eat insects and fruit.

Peptic ulcers   *Ulcers.

Perch is the name for two species of freshwater fishes. The European perch, *Perca fluviatilis*, is found across the Eurasian land mass from Ireland to Siberia. A similar, but distinct, species, the yellow perch, *P. flavescens*, ranges across northern North America. Both are very distinctive fishes with a large, spiny dorsal fin, bold dark bars across the body and blood-red pelvic and anal fins. They are typical fishes of lowland rivers and lakes, often swimming in small schools, although large specimens tend to be solitary. They feed on invertebrates when young, but eat larger prey as they grow, including fishes and often young perch. Breeding takes place in late spring, the eggs being shed in long strings and wound among plants, tree roots, or branches lying in the water. They are popular fishes with anglers, and on the European continent a common food-fish.

Peregrines are a species of large *falcon with some nineteen subspecies recognized throughout the world. They are noted for their swift, versatile flight and spectacular hunting of birds on the wing, characteristics which endear them to falconers. They live on open ground, sea coasts, and other areas where birds are plentiful, northern populations following their prey southward in winter. Two to four eggs are laid, but egg survival has recently been threatened by insecticides, which lead to weakening of egg-shells and so reduces viability.

Perennials are plants that live for more than two years (as opposed to annuals and biennials). In a garden sense, they are usually herbaceous plants with perennial root-stocks which annually produce new flowering shoots above the ground. Woody kinds retain a permanent framework of branches while the climbing kinds may be herbaceous or woody.

Peritonitis is inflammation of the peritoneum, the membrane lining the abdominal cavity. It is caused by bacteria, either through their movement in the bloodstream from an infected site to the peritoneum, or through their release when an organ bursts. The latter cause is the reason why appendicitis and gall-bladder disorders are potentially dangerous. It is a serious condition, but one whose risks have been much reduced by modern treatment.

Periwinkles (botanical) are species of *Vinca* and *Catharanthus*, both members of the same family as oleander. The first is a genus of shrubs and climbers of Eurasia, some of which are grown in gardens; the second comprises tropical shrubby plants, mostly native to Madagascar. The Madagascar periwinkle, *C. roseus*, is one of the commonest of tropical weeds. *Alkaloids found in the plant have been used in the treatment of leukaemia in children.

Periwinkles (molluscs)   *Winkles.

Persimmons are deciduous trees which grow up to 13 m. (43 ft.) tall and are native to warm temperate regions. They belong to the genus *Diospyros*, which includes some 200 species, mostly tropical and evergreen, and are in the same family as ebony. The Japanese persimmon, *D. kaki*, is native to the warm, temperate regions of China and Japan where it is an important crop on account of its edible fruits. It is also popular in the southern USA, the south of France, and Australia, where it is principally grown as a garden tree. Its fruits are yellowish-red, tomato-like, and are also known as the Chinese date-plum. The native American persimmon, *D. virginiana*, has smaller, darker red fruits which, although they are often picked from the wild, are not exploited commercially, although the tree itself is used as a rootstock for the Japanese persimmon. The Eurasian date-plum, *D. lotus*, has fruit which may be yellow or purple.

Personality is the distillation of the most characteristic features of how a person thinks, experiences emotion, and behaves. There have been many efforts to describe it scientifically, none of them entirely successful. Traits such as impulsiveness, passivity, dependence on others, coldness in personal relationships, and obsessional behaviour are of particular interest in psychiatry. One classification distinguishes two major personality variables: sociability (extroversion/introversion) and worrying (neuroticism). Patients with pathological personalities differ only in degree from the general population.

Perspiration, or sweat, is a clear watery fluid, produced in the sweat glands and discharged onto the surface of the skin of mammals. It is released in response to nervous signals from the brain, which are initiated if the body temperature rises above normal. The evaporation of water cools the skin and the blood passing through it. In hot conditions sweat may be produced at up to four litres per hour. The body then loses significant quantities of water and salt.

Pests are of great economic importance. Broadly speaking, they include organisms that cause disease in man, or that cause damage or disease to crops or livestock. Agriculture has favoured pests by concentrating their foods in specific areas, and emphasizing yield rather than disease-resistance in the crops grown. Some pests, like the locust, reproduce and disperse rapidly when conditions are favourable. Fungi, such as rusts and smuts, attack crops, and reduce yields. Some viruses, like the tobacco mosaic virus, affect plants. Foot-and-mouth disease, fatal in domestic animals, is caused by a virus. Parasites, such as threadworms, can be a problem in livestock.

Pest damage may be reduced by crop rotation, or by intercropping where two or three crops are grown among each other. Vaccination can protect livestock. *Biological control of pests, the use of resistant crop species, and other non-chemical pest controls are generally preferable to chemical spraying. For many pests become resistant to particular chemicals and those toxic substances may enter ecosystems, possibly damaging wildlife and eventually reaching man.

Petrels are seabirds which belong to one of four families: the Diomedeidae (albatrosses), Procellariidae (true petrels and shearwaters), Hydrobatidae (storm petrels), and Pelecanoididae (diving petrels). All are oceanic seabirds, with characteristic tubular nostrils which help to drain the salt glands which are positioned close to the eyes. About a hundred species are known, over half of them prevalent in temperate or cold latitudes of the Southern Hemisphere. Most species are highly social, nesting colonially and feeding together in huge floating 'rafts'. All species lay a single white egg, tending it for the unusually long incubation period of six to seven weeks; chick growth is correspondingly slow. Petrels range in size from tiny storm petrels and diving petrels, 15 cm. (6 in.) long, to majestic, wandering albatrosses with wings spanning 3.5 m. (11 ft.). Modes of flight range from the fluttering and whirring of the smaller members to the soaring and gliding of fulmars, shearwaters, and albatrosses. Several of the smaller species pursue prey under water but most feed at, or close to, the surface.

Petunias are colourful, large-flowered annual plants, related to the tobacco plant, *Nicotiana*, and the potato. The genus *Petunia* contains over thirty species of annuals or perennials, all from South America. The garden petunias are chiefly hybrids between *P. integrifolia* and *P. nyctaginiflora*.

Peyote: a small, round cactus, native to the Rio Grande Valley, Mexico, and central Texas, which is the source of the hallucinogenic drug mescaline. The botanical name of the species is *Lophophora williamsii*, and unlike most cacti it does not possess spines. It has religious significance to the Peyote Indians.

Phalangers, or cuscuses, are *marsupials belonging to the family Phalangeridae. They are often called opossums, but this name is used specifically for the very different South American opossums. The family Phalangeridae does however contain four species known as brush-tail possums. Phalangers have a wide distribution throughout

Two male golden **pheasants**, *Chrysolophus pictus*, square up to each other at the beginning of the mating season in May. Females are considerably less decorative, being generally brown with glossy black markings.

Australasia and include some ten species, the majority of which are lemur-like, such as the black-spotted cuscus, *Phalanger rufoniger*. Most species of cuscus and brush-tail possums are vegetarian, although some species also eat insects. The common brush-tail possum, *Trichosurus vulpecula*, was introduced as a potential fur species but has done much damage to native trees by its feeding. It also carries tuberculosis which may infect cattle.

Phalaropes are three species of small *wading birds with lobed feet and needle-like beaks. Habitual swimmers, they sit high in the water and spin like tops to stir up food in shallows. The males, smaller and less colourful than the females, incubate the eggs and care for the young.

Phanerogams, or spermatophytes, are seed-bearing plants. They consist of two main subgroups: the *angiosperms and the *gymnosperms. Their common feature is that they reproduce, and are dispersed, by means of seed as opposed to the spores of *cryptogams. The seed contains the developing plant embryo and a store of food for the initial growth of the germinating plant. The whole seed is enclosed within a hard protective coat, or testa.

The phanerogams arose around 120 million years ago from diverse groups of fern-like plants and the two subdivisions are not closely related to each other.

Pheasants are gamebirds of the family Phasianidae, which contains about 183 species found all over the world, with the exception of southern South America, areas of high latitude, and the dry areas of North Africa and the Middle East. There is a great range of size from the smallest quails, 13 cm. (5 in.) long, to the long-tailed peacocks, 2 m. (6 ft. 6 in.) long, including their tails. The females of most species are mottled brown in colour, which serves to camouflage them when they incubate their eggs on the ground. The males are often brilliantly coloured, with long, elegant tails. The family also includes francolins, wood-quails, partridges, and jungle-fowl.

The name of 'pheasant' is applied more particularly to about twenty-five species within this family. These are mostly chicken-sized birds, closely related to the jungle-fowl, and with long tails. The males of some are among the most beautiful of all birds and have been highly prized by collectors for thousands of years. They include the simply coloured but very striking silver pheasant, *Lophura nycthemera*, black below and silvery white above with a black crown and red wattled face, and the golden pheasant, *Chrysolophus pictus*, brilliant red below, with bright blue wings, a yellow back, and a golden head. Most species come from the Himalayas and mountain ranges of China, where they live in clearings in forests. The ring-necked pheasant, *Phasianus colchicus*, originally probably from eastern Asia, has been so widely introduced to other areas that its natural distribution is now obscure. It is now an important sporting bird in most parts of Europe and North America.

Pheromones are highly volatile substances secreted by an animal in order to influence the behaviour of another individual of the same species. They work at very low concentrations and can be released from special glands, or simply discharged from the body in perspiration, urine, or faeces. They function as a chemical signalling system, of particular importance in the life of insects. Pheromones play a very important part in the sexual behaviour of all animals. A virgin silkmoth releases a pheromone from

Photosynthesis

This process consists of two stages, one occurring only in sunlight (light reaction), the other at any time (dark reaction). The former uses chlorophyll pigments to convert light energy into chemical energy which in turn is used in the dark reaction to convert carbon dioxide and hydrogen into sugars such as glucose.

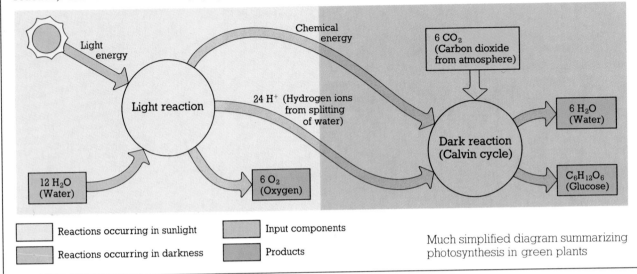

Light energy → Light reaction

Chemical energy

$24 H^+$ (Hydrogen ions from splitting of water)

$12 H_2O$ (Water)

$6 O_2$ (Oxygen)

$6 CO_2$ (Carbon dioxide from atmosphere)

Dark reaction (Calvin cycle)

$6 H_2O$ (Water)

$C_6H_{12}O_6$ (Glucose)

Reactions occurring in sunlight
Reactions occurring in darkness
Input components
Products

Much simplified diagram summarizing photosynthesis in green plants

special glands which may attract males from several kilometres away. Female mammals produce a scent when 'on heat' to advertise their readiness to mate. Other uses of pheromones include territory marking, trail marking, and alarm signalling. Similar scents designed to have effects on other species of animal are called allomones.

Phloem vessels are plant tissues which are involved in the movement, or *translocation, of food substances around the plant. They do so by using chemical energy to transport substances up or down the plant, from the leaves to the roots, or vice versa. In woody plants the phloem vessels constitute the innermost layer of the bark.

Phloxes, from the Greek word for flame (in allusion to the colour of the flowers of these plants), are mostly perennial plants of temperate America and Eurasia. They give their name to the phlox family, Polemoniaceae, containing about 300 species. The many garden varieties are derived from the North American *Phlox paniculata*. They have perennial rootstocks and stems about 1 m. (3 ft. 3 in.) in height, terminating in large clusters of brightly coloured, perfumed flowers. Other popular garden phloxes include the trailing or mat-forming species, with tiny leaves and relatively large flowers, and the annual phlox, which is a hybrid between *P. drummondii* and other species from Texas and New Mexico.

Phobias are types of *neurosis in which severe anxiety and fear are generated by particular objects or places. The sufferer may go to inconvenient lengths to avoid these and may be restricted as a consequence. Common types of phobia include the fear of spiders, needles, crowded shops, or open spaces. Behaviour therapy is sometimes effective.

Photosynthesis is the utilization of light energy from the sun to produce food in the form of *carbohydrates. It is a process carried out by all plants, using special pigments, called *chlorophylls, which can absorb the radiant energy of the sun and convert it to chemical energy. This chemical energy is used to make sugars from the gas carbon dioxide and water. In this sense 'food' is simply stored energy, capable of being released either within the plant at a later date, or utilized by animals.

In higher plants, the photosynthetic pigments are green chlorophylls and are contained in small 'packets', called chloroplasts, in the surface layers of leaves and young stems. Some algae and bacteria have red, blue, or brown photosynthetic pigments called phycobilins, which function in a manner similar to that of chlorophyll.

Phylum: one of the first group of divisions into which the animal kingdom is classified. In general, phyla are groups which show no evolutionary relationship to others except in that all have the basic characters of animals. All phyla contain *classes, although some large ones, notably chordates, arthropods, and echinoderms, are divided first into subphyla. There are no rules for the naming of phyla, except that they are in Latin form, but they are universally recognized. Among plants, groups at the level of phyla are called divisions, and their Latin names are distinguished by the ending -phyta.

Physiotherapy is the treatment of muscles and joints by physical means. It promotes healing more quickly than does rest alone, with minimal restriction of movements. For example, massage increases blood-supply, reduces swelling due to oedema (water retention), and relieves pain. It is very valuable following sports injuries. In older people it relieves pain resulting from everyday injuries and helps alleviate joint stiffness and arthritis. It can also be used to relieve the pain from muscular tension.

Phytogeography *Biogeography.

Pickerel are small members of the *pike family found only in eastern North America. Two species are known, the chain pickerel, *Esox niger*, and the redfin or grass pickerel, *E. americanus*, the latter being divided into two subspecies. They are similar to the pike in many ways but not in size, the chain pickerel growing to 75 cm. (30 in.) and the grass pickerel to 38 cm. (15 in.). Both feed on small crustaceans and insect larvae, and eat fishes once they are longer than 10 cm. (4 in.). They live in slow-flowing, heavily weeded streams and lakes.

Piddocks are *bivalves which gain protection against predators by burrowing into rocks, clay, or even masonry foundations of man-made structures like piers. The shell valves gape open towards the front and back of the animal, allowing the foot to protrude for burrowing and the siphons to be extruded for irrigating the burrow with clean water. The foot holds the piddock in its burrow while the shell valves rock back and forth to scrape rock away.

Pigeons are birds of the *dove family with an almost worldwide distribution. This family contains about 280 species which range in size from about 15 cm. (6 in.) long for the smallest species, to large-crowned pigeons from New Guinea, *Goura cristata*, which are almost the size of a small turkey, 80 cm. (32 in.) long. Pigeons have a wide range of plumage colours, especially greys and greyish-browns, though the twenty or so species of fruit pigeons are predominantly green. The typical pigeon beak is of medium length, but thinnish. Most species are monogamous and live in pairs during the breeding season, though many aggregate into flocks, sometimes of considerable size, at other times of the year. The extinct passenger pigeon of North America, *Ectopistes migratorius*, once migrated in flocks of perhaps hundreds of millions and also bred in enormous colonies, a factor in its extinction, since people were able to harvest the birds in large numbers.

All species are vegetarian, eating leaves, seeds, or fruits. Pigeons build flimsy nests of twigs, usually in the branches of trees, but sometimes in holes in trees, on rock ledges, or on the ground. Most species lay one or two white eggs. The young are raised on pigeon's milk, a product of the cell walls of the parents' crop, for the first few days, followed by increasing amounts of vegetable material.

Some species, such as the Barbary dove, *Streptopelia risoria*, have been kept in captivity by man for many centuries. Others, such as homing or racing pigeons, derived from the rock dove, *Columba livia*, have been kept for delivering messages. In medieval times the dovecot, a nesting aggregation of birds in semi-captivity, provided an important source of food.

Pigmy moths are tiny moths of the family Nepticulidae, with wing-spans from 3 mm. ($\frac{1}{8}$ in.) to 10 mm. ($\frac{1}{2}$ in.). They occur worldwide. The wings are hairy, and the fore-wings usually dark and metallic coloured. The caterpillars are leaf-miners, most constructing a mine with a characteristic shape.

Pigs are a family of nine species, widely distributed in the Old World. They are members of the large order of Artiodactyla (even-toed ungulates) but, unlike other members of this order (with the exception of the hippopotamus), they are non-*ruminants. The pig family includes wild boars, bush pigs, babirusas, and wart hogs. All are terrestrial and their modest-sized body is covered with bristles. All have an elongated head which terminates in a snout. This is a tough, mobile, disc-like plate of cartilage, reinforced with a small bone to withstand the stresses exerted when it is used for pushing, lifting, digging,

and breaking through tangled brush. Their canine teeth are tusk-like and grow upwards and outwards. They are omnivores, feeding on roots, tubers, fruit, berries, and, occasionally, small animals.

The gestation period is three to five months and the female has many teats to accommodate the large litters produced. They are prolific breeders and a female may have more than two litters per year. Pigs were domesticated thousands of years ago in both Europe and Southeast Asia, but they were half-wild and sometimes dangerous. During the nineteenth century these were crossed with more docile varieties from Asia to produce the breeds of today. The equivalent family in the New World is the *peccary family.

Pikas are mammals of the order Lagomorpha which otherwise includes only rabbits and hares. They resemble rabbits in their dentition and in the habit of refection, in which the night-time droppings are eaten. Unlike the rabbits, however, they have short ears, hind legs that are not much longer than the front ones, and they lack a tail. They are unique among mammals in making hay in the late summer for use in the winter, for they do not hibernate. Grasses and other vegetation are carried to rocks exposed to the sun and, when dry, stored in piles at the entrance of their burrows or under rock overhangs. The fourteen species range from the Urals to Japan in temperate Asia and there are two North American species, the collared pika, *Ochotona collaris*, and the North American pika, *O. princeps*.

Pike are slender freshwater fish with dorsal and anal fins placed close to the tail fin, and a pointed head. They occur as five species in the family Esocidae across northern Europe, Asia, and North America in slow-flowing rivers and lakes. Pike are predatory fishes, eating mainly aquatic insect larvae and crustaceans when young but later eating other fishes with the occasional duck or mammal. They typically lie in wait for prey close to vegetation, camouflaged by their mottled coloration, and make sudden charges, the huge teeth in the lower jaw stabbing the prey.

Pike-perch: a European fish *Stizostedion lucioperca*, now more commonly known as the zander. Its original range was in central and eastern Europe, but it has been introduced in France, the Netherlands, and England. A slender-bodied relative of the perch, it has a long-based, spiny dorsal fin. Feeding mostly at dawn and dusk, it is an efficient predator of smaller fishes and competes successfully with both perch and pike.

Pilchards are members of the herring family, and in the case of the European species, *Sardina pilchardus*, are called pilchards when adult and sardines when immature. This species occurs throughout European seas (except the Baltic) although it is uncommon north of Britain. It is a schooling fish living from the surface to a depth of 55 m. (180 ft.) and feeding on plankton. Other species of pilchard are abundant off southern Africa and Australia, and these belong to the genus *Sardinops*, as does the Californian or Pacific sardine, *S. caeruleus*. The pilchard was an immensely valuable food-fish in the first half of the twentieth century but the industry collapsed, possibly due to over-fishing or oceanic circulation changes.

Piles *Haemorrhoids.

Pill beetles are small, drab, and almost spherical beetles of the family Byrrhidae, with some 270 species distributed worldwide. They live at the roots of grass or moss and under stones. When disturbed, pill beetles draw their antennae and legs close to the body and feign death; in this state they are difficult to detect.

Pill bugs   *Woodlice.

Pilot-fish: a species of oceanic fish, worldwide in distribution in tropical and subtropical seas, which occasionally occurs in cool temperate seas. It belongs to the same family as the horse mackerel, and like it is streamlined in body shape, but has five to seven broad, dark blue bars on its sides; it grows to 70 cm. (28 in.) in length. Pilot-fishes have the habit of swimming close to larger animals, such as sharks and turtles, and often accompany sailing ships, usually keeping slightly below but close to the front of the animal or ship. As juveniles they swim close to floating weeds and jellyfishes.

Pimento is also known as allspice because the dried, unripe fruits of this small tropical tree, belonging to the myrtle family, yield a spice that combines the flavours of cinnamon, cloves, and nutmeg. It is native to the West Indies and tropical Central America and is used to flavour foods. An oil is extracted from the berries, and another from the leaves. These are used in toilet preparations, soap, and hair tonics, and bay oil is obtained from a closely related species. The name pimento is occasionally used for *sweet pepper.

Pimpernels are a group of plants which belong to the genus *Anagallis*, of which the best-known European species is the scarlet pimpernel, *A. arvensis*. This tiny annual with bright red, pink, or even blue flowers, is also called poor man's weather-glass because the flowers open only in fine weather. The genus contains twenty-eight species, all of which are restricted to northern temperate regions. They belong to the primrose family. The Mediterranean species, *A. linifolia*, is a perennial in its native habitat but is often grown in gardens as an attractive annual with several colour forms.

Pinchot, Gifford (1865–1946), the first professional American forester, believed that the nation's forest reserves and natural resources should be managed scientifically as a matter of government policy. A prolific writer and speaker, his far-sighted work in conservation covered many areas of ecology and had the backing of President F. D. Roosevelt.

Pineal gland, or pineal body: a small ball of tissue attached to the brain of vertebrates. In some vertebrates, such as frogs and sharks, this organ contains sensory cells which respond to light and in these animals acts like a third 'eye' under the surface of the skin. In mammals it is apparently devoid of nerve cells, but consists of secretory cells which are activated by fibres of the *autonomic nervous system. The cells contain serotonin, which is converted to a hormone, melatonin. The amounts of melatonin in the gland, brain fluids, and *plasma fluctuate in a twenty-four-hour rhythm, peaking during darkness. In some animals, melatonin suppresses the growth and activity of the *gonads, by an action on the hypothalamus or pituitary. Its function in man is unknown.

The feet of these **pink-footed geese** are indeed pink, but those of immature birds may be light grey or even tinged with yellow. Like many other types of geese, they roost mainly on the coast or in estuaries.

Pineapple: a plant of South American origin cultivated particularly in Hawaii, Malaysia, Thailand, and other tropical areas. It belongs to a family of monocotyledonous plants, which include the *bromeliads, Spanish moss, and many ornamental species. The short-stemmed perennial plants have a spiral of large, sword-shaped leaves, topped by a swollen stem carrying between 100 and 200 flowers. Their fruits join together to produce the large, pine-cone shaped, sweetly flavoured fruit, which contains about 15 per cent sugar and is rich in vitamins A, B, and C.

Pine marten   *Martens.

Pines are evergreen conifers of the genus *Pinus*, of which there are some seventy to a hundred species in the north temperate zones and mountains of the tropics. Together with silver firs, spruces, hemlock spruces, Douglas firs, cedars, and larches, they make up the main family of conifers. All pines produce resin and have female cones which take two to three years to ripen after pollination. The wind-blown pollen is produced in such quantities that it can turn rain yellow and leave a scum on standing water. Pines are important constituents of northern temperate forests and are widely planted in plantations throughout the world. In some areas, such as the Cape region of southern Africa, they can be a menace, taking over the native vegetation. In Britain, where the Scots pine has been introduced by man to southern England, it often invades heathland.

Soft pines are those with little resin and include the Weymouth pine, *P. strobus*, of eastern North America. Hard pines have harder timbers, often resiny, as in the pitch pine, *P. rigida*, and the Corsican pine, *P. nigra* var. *maritima*. Pines can be classified according to the numbers of needles on each short shoot.

Most have winged seeds, dispersed by wind, but the stone pine, *P. pinea*, of the Mediterranean has large, highly nutritious seeds (pignons or pine nuts) which are dispersed by animals. Many other conifers are called pines but are not members of the pine family.

Pink-footed goose: a species of goose that can be distinguished from the larger bean and greylag geese by the darker head and neck, black and pink beak, and pink legs. They breed in Greenland, Iceland, and Spitzbergen and winter in western Europe, including Britain, where they are the most numerous visiting goose.

Pinks (plants) belong to one of the largest genera, *Dianthus*, which is a member of the family Caryophyllaceae. This large family of over 2,000 species has a worldwide distribution and includes many popular garden plants, including *Gypsophila* and campions. Pinks are species of *Dianthus* occurring naturally or as hybrids. The carnation is a form of *D. caryophyllus*, while the 'garden pink' is derived from *D. plumarius*. Both parent species are native to Europe and have been selectively bred for over 300 years to produce the garden pinks and carnations. Since they are hybrids some are sterile, and others may not breed true from seeds.

Pintail: a species of slender, long-necked dabbling *duck, widespread in the Northern Hemisphere. The drake (male) is mainly grey, with a chocolate and white head, neck, and breast. Both sexes have pointed tails, but the drake's is greatly elongated, and needle-like.

Pinworms are a family of *roundworms which are parasites in the gut of vertebrates and invertebrates. The human pinworms, *Enterobius vermicularis*, are a common, but relatively harmless, parasite of children throughout the world. They pass from child to child as eggs derived from foods or by hands contaminated with the faeces of an infected individual.

Pipefishes, of which there are about 150 species, are relatives of the sea-horses, and like them are slow swimmers. They are found in marine and brackish water in all but the polar seas, although they occur most abundantly in the tropics. A few species, principally in Southeast Asia, live in fresh water. Mostly they are found in shallow seas down to a depth of 90 m. (295 ft.), but a few live at the surface of the open sea, often in association with floating seaweed. The body is long and thin and encased in hard bony rings; the snout is also long with a small mouth. Their fins are small and weak. The males carry the eggs on the under surface of the body or tail, or in some species in an open pouch.

Pipe snakes *Cylinder snakes.

Pipistrelles are small to medium-sized insectivorous *bats of the family Vespertilionidae, with some forty-seven species in the genus *Pipistrellus*, found in most parts of the world except South America. The common pipistrelle, *Pipistrellus pipistrellus*, is the commonest bat in Britain and is the one most often found roosting in attics and seen flying just before dusk. The name derives from pipistrello, the Italian word for bat.

Pipits include about thirty-four species of birds belonging to the *wagtail family. All are small, 15–18 cm. (6–7 in.) long, greyish or greenish above, and paler below; many of them are heavily streaked. Most have white outer tail feathers. They have a worldwide distribution, and live in open country or on the edge of woods, usually nesting on the ground.

The teeth of most **piranhas** actually interlock and can slice through flesh like a scalpel. All have toothed scales under the belly. The size of adult fishes varies from about 30 cm. (12 in.) to about 60 cm. (2 ft.) in length.

Piranhas are predatory fishes in the same family as the characins and tetras, but are confined to the tropical fresh waters of South America. They are deep-bodied, disc-shaped fishes of the genera *Serrasalmus*, *Rooseveltiella*, and *Pygocentrus*, with a blunt head, a moderately long dorsal fin, and an adipose fin near the tail. Their most notable features are the teeth and jaws. The teeth are triangular in the lower jaw, but have several cusps in the upper; all are pointed and razor-sharp, while the jaws are strong with powerful muscles. Piranhas live in shoals and eat fishes, and can quickly strip all the flesh from a large, disabled fish, leaving only the bones. They also attack mammals, like the capybara, which live at the Amazon water's edge, but the stories of their attacks on man have been greatly exaggerated.

Pirarucu *Arapaima.

Pistachio nuts come from a small evergreen tree, related to the mango, and native to the Near East and western Asia. It has been cultivated there and in the Mediterranean area for up to 4,000 years. Male and female flowers are on separate trees and from the latter are obtained the pleasant but mild nuts (in fact a seed, not a true nut) with green kernels.

The pitchers of this species of **pitcher-plant**, *Nepenthes ampullaria*, growing in Borneo, are mostly at ground level. About 10 cm. (4 in.) across, and with small lids, they form a rosette produced from a stem buried in leaves and humus. The plant climbs by means of leaf tendrils, developing further pitchers as it does so.

Pitcher-plants are *carnivorous plants with leaves or leaf-parts modified as pit-fall traps, into which insects are attracted, ensnared, and utilized as a source of food.

This way of life has evolved in three different families of plant. The genus *Nepenthes*, with some seventy species, forms its own family of jungle plants of India, Australasia, and Madagascar. Their pitchers may be up to 30 cm. (12 in.) deep, and contain up to 2 lit. (4·25 US pints) of water. The family Sarraceniaceae contains seventeen species, divided into three genera, of North or South American species. Their narrow pitchers may reach 15 cm. (6 in.) in some species, and are attractively coloured. The Australian flycatcher plant, *Cephalotus follicularis*, is the sole representative of the exclusively Australian family Cephalotaceae. Its pitchers, which reach 5 cm. (2 in.) in depth, have lids, or hoods.

All species absorb nutrients from dead invertebrates, the last two families using enzymes, or bacteria, to speed up the decay process.

Pittas are forest-dwelling birds which inhabit warm areas of the Old World including Africa, Southeast Asia, and Australia. In shape and size they are rather like short-tailed, long-legged thrushes. Most species are strikingly coloured, with patches of blue, red, green, or black, many having turquoise blue on the wing or bright red under the tail. They hop around the forest floor catching a wide variety of small animals. They build a domed nest, made of twigs, on or close to the ground.

Pituitary gland, or hypophysis: the most important of the vertebrate *endocrine glands, despite its small size. It lies in a bony pocket at the base of the skull. The anterior lobe consists of six specialized cell groups, each producing a different hormone. Outputs of these hormones can be adjusted individually by specific releasing factors, which are then transported in blood leaving the hypothalamus. The endocrine functions of sex glands, thyroid glands, and adrenal glands are regulated in this manner. Other hormones secreted by the pituitary include somatrophin (growth hormone), which controls bone and muscle growth, and prolactin, which controls milk production by mammary glands. The posterior lobe is attached to the brain by a short stalk containing neuro-secretory fibres from the hypothalamus. Some of these release vasopressin to act rapidly on the reabsorption of water by the kidney; others release oxytocin, a hormone which causes milk ejection when teats are pulled.

The close association between hypothalamus and pituitary, common to all vertebrates, enables the brain to co-ordinate a wide range of endocrine activities with seasonal variations in daylight, temperature, and weather. The pituitary also affects skin coloration in some mammals and amphibia. It is sometimes known as the master gland.

Pit vipers are venomous snakes with folding fangs. There are some sixty species living throughout the world, although they are most diverse in the New World. They differ from other *vipers in their ability to detect their prey (such as mammals) by using heat-sensitive pit organs that are situated on the head between the eye and the

Pit viper

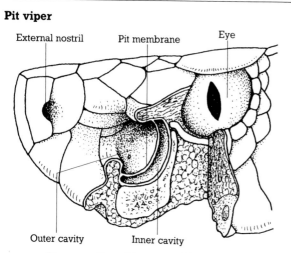

Cross-section of heat-sensitive organ

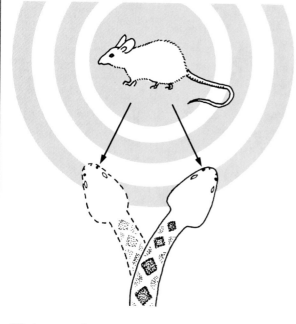

Pit vipers can locate warm-blooded prey in total darkness. Between nostril and eye there is a deep pit divided into two chambers by a heat-sensitive membrane. When radiant heat from an animal falls on the membrane, the snake pinpoints its exact location by moving its head from side to side, allowing for lack of stereo sensing at a distance. Accuracy is achieved by receiving signals from different angles and constantly revising the sensory information as it gets nearer.

nostril. Some of the most feared snakes of the tropical Americas, such as the fer-de-lance, jararaca, and bushmaster, belong to this group. Other pit vipers include the rattlesnakes, the copperheads, and cottonmouths or water moccasins. Many of the Asiatic pit vipers are treedwelling. Pit vipers belong to the family Crotalidae and give birth to live young.

Placenta: an organ within the *uterus which provides the developing embryo with nutrients. It is connected to the embryo via the *umbilical cord. The mass of cells and blood vessels which form the placenta permit food and

oxygen to be transferred from the mother to the embryo through a cell barrier. Waste-products from the embryo pass across the placenta into the mother's bloodstream for excretion.

Placoderms are a group of ancient, extinct fishes, living from around 400 to 200 million years ago. They were among the first fishes to evolve true biting jaws and are often known as the ancient jawed fishes. The head and front part of the body were covered in heavy bony plates for protection, and they had a pair of pelvic fins similarly encased in bone. Some of the earliest placoderms were found in sediments laid down in freshwater lakes or seas.

Plaice: a species of *flat-fish with both eyes on the right side of the head and orange-red spots on the coloured side. It is a European species which is common in shallow coastal water from the tideline down to 50 m. (162 ft.); it prefers sandy bottoms but will live on mud or gravel. It eats a wide range of bottom-living animals, especially molluscs. The plaice has stronger teeth on the underside of the jaws than above. Spawning takes place in late winter; the eggs float near the surface at first, hatching in ten to twenty days.

Planaria   *Flatworms.

Plane trees belong to the genus *Platanus* and include some ten species of deciduous trees of which eight are native to the southwest USA and Mexico. The most widely cultivated species are the oriental plane, *P. orientalis*, from southeastern Europe, and the buttonwood, or American plane, *P. occidentalis*, of eastern North America. The London plane is believed to be a hybrid between these two species. Because it is one of the most pollution-tolerant of all trees, it is widely planted in towns. It has a thin, flaky bark and fruits which resemble drumstick heads on short ropes. In North America some species are commonly called sycamores, a name reserved for a type of maple in Europe.

Plankton is the assemblage of microscopic organisms, both plants and animals, that are suspended in the upper

A **plankton** sample from the Great Barrier Reef, Australia, contains two kinds of protozoans. The rounded bodies are dinoflagellates and the spiked ones radiolarians. The silica skeletons of radiolarians contribute to the ooze of sediments on the ocean floor.

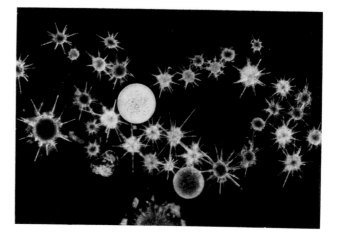

levels of seas and fresh water. Their location is mainly dependent on currents and water clarity, as the photosynthesizing plants require sunlight. The diatoms, tiny algae, and small animals drift freely, while larger animals swim independently. In marine plankton, these animals include representatives of virtually every marine group; some are totally planktonic, others only during their larval stages. Plankton is the basis of all aquatic food-chains. The plants *photosynthesize, are eaten by planktonic animals, and then fall prey to larger invertebrates, fishes, and even whales. Dead plankton sinks, and becomes food for detrital feeders.

Plantains are a family of perennial or occasionally annual plants of which 250 out of a total for the family of 253 species belong to the genus *Plantago*. They have low-growing rosettes of leaves and slender spikes of inconspicuous flowers. They are found in all temperate regions, and on mountains in the tropics. Some species are troublesome weeds of lawns.

The name is also used incorrectly for culinary or cooking varieties of *banana.

Plant classification is necessary (as is *nomenclature in zoology) to put order into the vast assemblage of species: the flowering plants (*angiosperms) alone encompass some 250,000 species. Within the plant kingdom the most basic division is between those which produce seeds, the phanerogams or spermatophytes, and those which reproduce by spores, the cryptogams. The simplest cryptogams with unicellular sex organs (thallophytes) are the fungi and algae. Those with multicellular sex organs, but no clear differentiation of vegetative organs into roots, stems, and leaves, are mosses and liverworts (bryophytes), whereas those with such a clear differentiation are ferns, clubmosses, and horsetails (pteridophytes). Phanerogams are split into gymnosperms, where the seeds are not enclosed in a protecting carpel, and angiosperms, where they almost always are. This description of major plant groupings is often said to be 'artificial' and does not represent the real relationships between plants. Nevertheless, the general system, variously modified to give different ranks to the major groupings, is widely accepted and is convenient to use, two major requirements of a good classification.

The major groupings are in turn variously subdivided into more natural subgroupings, the most important of which are *orders. Orders are built up of *families and these consist of *genera. The hop, for example, is a member of the genus *Humulus* and its specific name is *Humulus lupulus*; the genus belongs to the family Cannabaceae (which also includes the genus *Cannabis*) in the order Urticales (which also includes the nettle, fig, and elm families) in the subclass Dicotyledonae of the angiosperms.

Plant-cutters are three species of birds in a family of their own; they are found in Peru, Chile, Bolivia, and northwestern Argentina, where they inhabit open, bushy country. They are sparrow-sized, brownish or greyish with dark streaks above, the underparts and the top of the head being a dull red. The beak is finch-like, short and stout, and they feed mainly on fruit and seeds.

Plant defences have been evolved to protect them from being eaten by animals. This can be done by mechanical deterrents such as spines, prickles, or thick layers of hairs, or by chemical deterrents, which include a wide range of compounds. One of the main groups of plant deterrents, or poisons, are the *alkaloids which affect the metabolism of animals through the nervous system. Many plant poisons are effective only if eaten by an animal, and usually act to deter rather than kill. Others, like that of poison ivy, inflict damage by simple contact. A combination of physical and chemical attack is provided by the stinging hairs of the *nettle.

Plant fossils are usually found within sedimentary rock and may have undergone one of four main types of preservation. Firstly, the rapid burial and partial decay of plants forms rocks such as coal. Secondly, fossil impressions may form when all organic matter has decayed. These are similar to plant fossils which have undergone the third type of preservation, namely the formation of casts, formed when soft sediments infill hollow cavities created by the decay of soft plant material. Fourthly, petrifactions (like fossil forests) occur when mineral solutions enter plant tissues and replace or modify the softer parts.

The fossil plant record is poorly known because, even before death, many plants break up and their parts become separated. Also, the places where plants grow are remote from the sedimentary basins, such as seas or lakes where future rocks are being formed. However, we know that the earliest plant fossils are microscopic, marine algae around 3,000 million years old. The first land plant fossils evolved in the late Silurian Period, about 410 million years ago. In the Carboniferous Period, 300 million years ago, there were great swamp forests in the Northern Hemisphere, dominated by *Lepidodendron* and *Calamites*. By the end of the Permian Period, 245 million years ago, drier climates prevailed worldwide and many swamp plants became extinct. The early Mesozoic Era, 193 million years ago, was dominated by *gymnosperms. In the early Cretaceous Period, 118 million years ago, flowering plants evolved and rapidly became the dominant land vegetation.

Plants are living organisms, composed of many cells, which produce their own food from water, carbon dioxide, mineral salts, and sunlight. They do this by the process of *photosynthesis and are often called autotrophs, meaning 'self-feeding'. This term also covers algae and some bacteria, which also contain pigments capable of trapping radiant energy from the sun. These simple organisms are sometimes called plants or classified along with protozoa and fungi as a group called Protista.

Groups normally considered as plants include liverworts, mosses, ferns, and seed plants. They are often called primary producers as they are the first link in the food chain. Plants form the basis of life for all animals. They differ from animals in that their cell walls are rigid, being reinforced by cellulose, and connected one to another by pores.

Although many plants try to deter animals from eating them, the *angiosperms in particular rely upon insects or other animals for pollination or seed dispersal. There are parasitic forms, some of them *epiphytes, and some which derive energy from the breakdown of dead animal and plant tissue (*saprophytes).

Plasma (of blood) is a water-based solution of substances such as glucose, amino acids, salts, antibodies and vitamins. It occurs as the fluid part of the blood of a wide range of vertebrate and invertebrate animals. The substances dissolved in it form a reservoir of raw materials which can be tapped by tissue cells according to their needs. The plasma fraction of blood also carries waste products such as urea until they are removed by the excretory organs. The main function of blood is to carry oxygen to all parts of the body using respiratory pigments, such as *haemoglobin. These pigments are carried in the plasma, either as large molecules, or in special blood cells. In man, plasma forms about 55 per cent of blood volume, the remainder being mostly blood cells.

Platy: a group of freshwater or brackish-water tropical aquarium fishes. The name derives from the genus in which it was formerly placed, *Platypoecilia*. The male has the first rays of the anal fin modified to transfer sperm to the female, who gives birth to live young. They originate in the Atlantic coastlands of Central America, where they live in swamps, pools, and slowly flowing rivers.

Platypuses, along with the echidnas, make up the mammalian subclass Monotremata, distinguished by their egg-laying habits. They are furry and secrete milk but show many reptilian features in the skeleton, so they may represent an intermediate stage between reptiles and mammals. The eggs are incubated by the female in an underground nest and the young are nourished by milk that exudes from diffuse mammary glands. There are no teats. Specialized features include the leathery and very sensitive duck-like bill, the flattened tail, and webbed feet. The webbing on the front feet extends beyond the claws, but it is folded back when the animal is on land. The male has a poison spur on each hind-foot, but it is not known how or when the poison is used. Platypuses are aquatic and feed on freshwater crustaceans, insects, fish, and amphibians. There is only one species, found in eastern Australia and Tasmania.

Pleistocene megafauna is a term used in reference to a number of extinct giant species of mammals, many of which are closely related to more modest-sized modern forms. They occurred in the Pleistocene Epoch, which began about 2 million years ago, and ended with the close of the last Ice Age, about 10,000 years ago. Among the giants were mammoths, megatheres (giant sloths), glyptodonts, the Irish elk, and giant lemurs, bears, rhinos, sabre-toothed tigers, kangaroos, and others. The reason why they all died out around the close of the Pleistocene Epoch is unknown. Perhaps it was due to climatic changes, or even to man's activities.

Plesiosaurus: an example of a group of extinct marine reptiles of the Mesozoic Era (245–66 million years ago). Its most characteristic feature was a long, flexible neck to help it catch fish. The four legs were modified as powerful, flat paddles which beat up and down to propel the body through the water.

Plovers are a large family of fifty-six species of *wading birds with a cosmopolitan but mainly tropical range. They differ from sandpipers in having short, straight beaks, thick necks, and boldly patterned plumages, frequently with a dark band across the breast and/or tail. Typically, they inhabit open country near water, and lay their eggs in a shallow scrape in the ground, protecting them and their young by feigning injury. The main genera

The lapwing, *Vanellus vanellus*, is a wading **plover**, yet inland flocks of this species are also often found on moist grasslands or arable fields in winter. It feeds almost entirely on animal matter.

are: the lapwings; the golden plovers, remarkable for the long migrations made by the American species; and the sand plovers, which include small shore-birds like the ringed plover. The crab plover, *Dromas ardeola*, is the only member of the family Dromadidae, and breeds in parts of the Middle East.

Plumbago (plant) is a genus of some twelve woody herbs and scramblers of warm regions, which are part of the same family as sea lavender and thrift. They have glandular hairs on the outer surface of the fruit, which are thought to aid their dispersal by animals. Many members of this genus, and general family, are grown in gardens as climbers or small shrubs.

Plume moths are moths of the family Pterophoridae; they usually have the wings divided into several hair-fringed lobes and are long-legged, giving an extremely fragile appearance. Most are greyish, yellow, or brown, and they fly feebly at dusk. The twenty-plume moth, *Orneodes hexadactyla*, with each wing divided into six lobes, belongs to another family, the Orneodidae.

Plums are members of the cherry genus, *Prunus*, which occur wild throughout the Northern Hemisphere. A number of different species of *Prunus* have contributed to the development of the modern kinds. They are small, deciduous trees, often grown as bushes, with small, white flowers and fleshy, short-stalked fruits containing a single, hard-shelled seed. The European plums were developed from crosses between the native sloe, or blackthorn, and the cherry plum, *P. cerasifera*, of western Asia. They vary in skin colour (greenish-yellow, yellow, red to purple) and season of ripening (late July to late September). Cooking plums are more acidic than dessert types. Certain varieties, more suited to the warmer temperate areas, are grown to produce prunes, the purple-skinned fruit being dried on

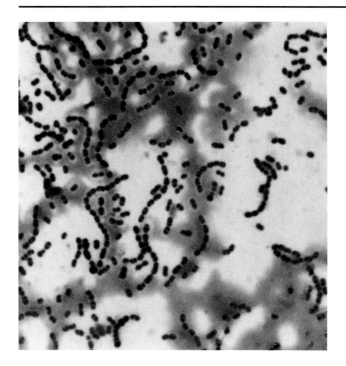

Most forms of **pneumonia** are caused by bacteria such as *Diplococcus pneumoniae*, above. But viruses, fungal infection, some parasites, and even certain irritants inhaled may also cause the disease. The symptoms are initially similar to those of influenza.

the tree or artificially. Other important species are the American plum and the Japanese plum. The greengage and damson are both types of plum.

Pneumonia is inflammation of the lung, and may be caused by one of several species of bacteria. The commonest cause is the bacterium *Streptococcus pneumoniae*, which used to be an important cause of death from middle age onward. Now, this infection is effectively treated by antibacterial drugs. Pneumonia may occur in a previously undamaged lung, but sometimes it is associated with, and draws attention to, previously unsuspected abnormalities of the lung.

Poa grass is the name used for several species of grass in the genus *Poa*. They are identified by a prefix such as bulbous, alpine, wood, and annual. *P. annua* (annual poa) is possibly the commonest grass weed of cultivation and is found throughout the world. Meadow species such as smooth meadow-grass, *P. pratensis*, are useful for grazing stock and hay production, while others with narrow leaf-blades are ingredients of lawns.

Pochards are a tribe of mainly freshwater bay ducks that collect food from the shallow river- or lake-bed, or from underwater vegetation. Like other diving ducks they patter along the surface of the water before taking off. The sixteen species of bay ducks include several called pochard and tufted ducks. The American term for all of them is bay ducks.

Pocket gophers *Gophers.

Pocket mice are a family of sixty-five species of *rodents, restricted to the New World. They can be divided into

two habitat groups: the kangaroo rats (*Dipodomys* species), kangaroo mice (*Microdipodops* species), and pocket mice (*Perognathus* species) live in hot, dry habitats of North America; the spiny pocket mice (*Heteromys* and *Liomys* species) live in tropical rain forests of Central and northern South America. They all eat seeds and other vegetable material and are usually active at night.

Pods are the fruits of beans, peas, and their allies in the Leguminosae, or pea family, though in common parlance other fruit types may be included. They usually split down one side to expose a row of seeds, though some, like the peanut, are indehiscent (do not burst open).

Poinsettias *Euphorbia.

Poisoning is the condition caused by accidental or deliberate ingestion, inhalation, or absorption of a poison. Poisons are substances capable, in small quantities, of causing structural tissue damage or functional disturbance in the body through their chemical action. Poisons are widespread in nature, where they may be used to subdue or kill prey, as in spiders and some snakes, or as a defence, as in many insects, fishes and plants. A social problem is self-poisoning with alcohol and by drug overdosage, a common means of suicide. Accidental drug overdosage can occur after a prescribing error, or when children mistake tablets for sweets. Inhalation of poisonous gas, such as carbon monoxide, is often accidental, but solvents are increasingly inhaled as a form of drug abuse.

Poison ivy is the common name for *Rhus radicans*, a climbing shrub with aerial roots, native to North America. Although it resembles ivy, it is a species of *sumac and is related to cashew and mango rather than to true ivy. Like many other members of its family, when touched it provokes a severe skin reaction in many people, giving them painful blisters and sores. Its ally, poison oak, *R. toxicodendron*, has similar properties.

Poisonous animals may use the poisons they produce offensively, to catch their food, or defensively, against predators. The term is often used for those species which contain poisons as a deterrent; those which use stings or bites are called venomous animals. The type of poison and the delivery mechanism varies.

Many coelenterates, including some sea anemones, jellyfish, and corals, have stinging cells (nematocysts) which, when triggered, inject paralysing toxins into fish or smaller prey. Species of the marine snail *Conus* have developed a poisonous tooth which they use to stab their victims. Some insects, like many bees and wasps, have the ovipositors in the tip of their abdomen modified to form stings which inject poison. These may be used in defence or, in some species, to immobilize prey. Scorpions can use their venom for defensive purposes and some produce neurotoxins that can have unpleasant effects on man. All spiders use toxins to subdue their prey and some, such as the black widow spider, produce venoms which may sometimes prove fatal to man. A number of snakes, including the adders, rattlesnakes, and cobras, have hollow fangs through which poison is injected as they bite their prey. Some *poisonous fishes have defensive spines on their bodies and a number of toads and salamanders, such as arrow-poison frogs, secrete poisonous substances over their skin surface. Many poisonous animals, including those that use poisons in defence, have *warning coloration.

Poisonous fishes are found in several unrelated orders over a wide geographical range, although most are tropical. Three types can be distinguished. Firstly, there are those which are poisonous from natural toxins within the body. Most notable are the puffer-fishes, whose liver, gonads, and blood are extremely poisonous to man. In spite of this, they are commonly eaten in Japan, where they are known as fuju. The eggs of a number of fishes are poisonous, like the cabezon of California, the barbel, and the pike.

Secondly, there are fishes which secrete natural toxins which are poisonous to a greater or less extent. The slime from the soap-fishes, *Rypticus* species, is bitter-tasting to man and fishes quickly reject them. Trunk-fishes, *Lactophrys trigonus*, also give off toxic secretions and if they are kept in captivity with other fishes the latter will die.

Thirdly, venom glands associated with fin spines or gill-cover spines are common and are found in the Indo-Pacific stonefish, the European weevers, and the toadfishes and star-gazers of the North American and other coasts. Sting-rays also have venom glands on the dagger-like spine on the tail, and these have been known to cause death in man. The sting-ray raises its tail above

A **polar bear** on an ice-floe near Moffen Island, north of Spitsbergen, Norway. Polar bears tend to be solitary and roam the pack ice over large areas. They are most numerous where the ice is kept in motion by wind and currents and open water provides access to seals, which serve as their main prey.

its back and stabs forwards and upwards, often causing fearful wounds in its victim. Despite this, poison and venom glands have a protective function; they are never used offensively.

Poisonous plants   *Plant defences.

Pokeweeds, also called pokeberries or poke roots, belong to the plant genus *Phytolacca*. This forms part of the family Phytolaccaceae, which includes some 125 species of trees, shrubs, woody climbers, and herbaceous plants of mainly tropical and subtropical regions of the New World. The best-known species is the Virginian pokeweed or red ink plant, *P. americana*, which occurs naturally in Mexico and Florida, but has been introduced elsewhere. This tall plant, growing up to 3·5 m. (12 ft.), is a perennial, which produces spikes of small white or lilac flowers and bears red berries from which a red dye is extracted. Other members of the family also yield red dyes and the roots of some species contain medically valuable drugs.

Polar bear: a single species of bear, *Thalarctos maritimus*, which is one of the largest carnivorous animals in the world. A mature male can reach a length of 2·75 m. (9 ft.) and a height of 1·5 m. (5 ft.) at the shoulders. The female is smaller but otherwise similar in appearance. The polar bear lives amongst the ice-floes of the Arctic and has even travelled to Iceland and Greenland. It is quite at home in the water. The coat of dense, long fur is white tinged with yellow. The feet are heavily haired to provide insulation and give a grip on the cold, slippery ice of its habitat. It has better vision than most bears and a good

sense of smell, both essential in searching for prey in such a bleak region. Its favourite foods are seals and walrus cubs, but caribou, foxes, birds, shellfish, and any other form of animal life is eaten.

After mating, the female will find a sheltered bank and dig a cavern in the snow. She then curls up to sleep for the coldest months of the year. In January, nine months after mating, the cubs are born—usually two. Only about 25 cm. (10 in.) long at birth, they are practically hairless and their eyes are closed for their first six weeks. By March they emerge to learn to hunt and fend for themselves, but remain with their mother until sixteen or seventeen months of age. Mating occurs only every two years because of the dependence of the cubs on the female. Polar bears can live for up to twenty-five years.

Polecats are carnivores of the subfamily Mustelinae, which also includes the weasels; indeed two out of four species belong to the same genus as the weasels and stoats. They are the European polecat, *Mustela putoris*, and the steppe polecat, *M. eversmanni*, of eastern Europe. The others are the marbled polecat, *Vormela peregusna*, and the African polecat or zorilla, *Ictonyx striatus*. The American *skunk is sometimes called a polecat. The European polecat is found in forests throughout Europe, with the exception of Scandinavia. It is a nocturnal predator, preying on hares, small rodents, birds, and invertebrates. It is noted for the foul-smelling secretions from its anal glands. The steppe polecat is the probable ancestor of the ferret.

Poliomyelitis, or polio, is an infection by a virus which specifically damages those nerve cells in the spinal cord which initiate the contraction of muscles. Infection results in a variable degree of weakness and wasting. Because formerly this often occurred early in life it was known as infantile paralysis. The virus appears in the faeces, and its spread is favoured by poor hygiene. With improvement

Pollen grains of two species: *below left*, the stinking hellebore, *Helleborus foetidus*; *below right*, the black alder, *Alnus glutinosa*. Magnified about 3,000 times, they show how different the exterior patterns can be. In the former species the pollen tubes emerge from the furrows, while in the latter they exit from the apical pores.

of hygiene, the disease became prevalent in young adults rather than children and, in consequence, caused more extensive paralysis. Vaccination has now virtually eliminated it.

Pollack: a member of the cod family distinguished by its brownish-green back shading to yellowish-green on the sides, and by its lack of a chin barbel. It occurs off the Atlantic coast of Europe and in the western Mediterranean, swimming in schools near rocky reefs and wrecks in depths of 5–200 m. (19–650 ft.). It feeds on fishes and large quantities of crustaceans. Spawning takes place in early spring in deep water; both the eggs and larvae float at the surface and drift inshore. Young fishes are common in shallow water in midsummer.

Pollen consists of grains containing minute male 'plants' encased in a tough wall. This is often characteristically shaped or ornamented so that in many cases the pollen grain's parent plant can be identified. The wall's function is to protect the male *gametes from desiccation or other injury during their passage to the female gametes in *pollination. Pollen grains are produced by structures called stamens in angiosperms.

In late spring and early summer, pollen, particularly from wind-pollinating plants like conifers and grasses, may be so copiously produced as to cause scums on ponds. Some people are sensitive to high levels of airborne pollen and have an allergic reaction known as *hay fever.

Pollen-feeding moths *Gold moths.

Pollination is the process prior to fertilization in plants whereby pollen grains are brought into contact with ovules in *gymnosperms, or stigmas in *angiosperms. The effective agents may be wind, as in many trees and in grasses, water, as in some aquatics, or animals, particularly insects such as bees. Most plants promote the transfer of pollen from one plant to another. This is called cross-pollination and produces offspring with a genetic make-up in between that of their parents. The pollen-producing structures often mature well before the ovule, or stigma, is ready to receive pollen; this prevents plants from becoming self-pollinated. There are many exceptions

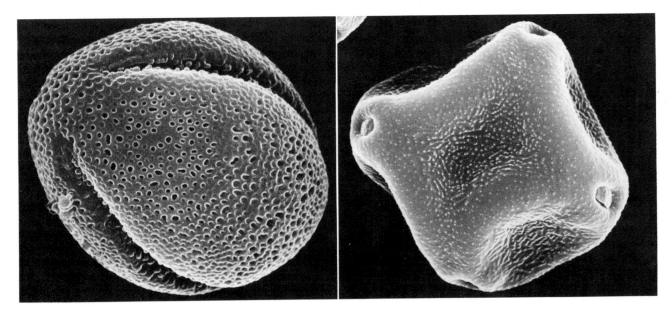

to this rule, however; some plants use only self-pollination, while others can use both types.

Pollution occurs when a harmful substance is released into the environment in sufficient quantities to damage living things. Air pollutants include sulphur dioxide and smoke, which harm animals and plants. The London smog of 1952 killed over 4,000 people.

Water and land pollution may be caused by waste materials from agriculture, industry, or towns. In water, these may cause eutrophication, or deoxygenation, or simply kill by poisoning organisms. On land, heavy metals such as lead may build up in animals through the food consumed. Many pollutants may be neutralized by the environment over a long period.

Polyanthus   *Primrose.

Polyp (invertebrate) is a general name for any sessile animal with radial symmetry, living singly or in a colony, and catching food in its waving tentacles. Polyps include such creatures as sea anemones, hydras, and corals, all of which are rather plant-like in appearance. The polyp phase of these animals can be a larval or an adult stage. In some jellyfish the polyp is the sedentary larval stage, developing into a free-swimming 'umbrella' with tentacles beneath. Such forms are called medusae and are the dispersive stage of polyp-bearing animals.

All polyp-type animals bear stinging-cells on the tentacles which are used to immobilize small prey. All polyps have a simple construction, with a single opening, or mouth, and only two cell layers making up their body wall. Animals with a polyp-stage are among the most primitive of animal phyla.

Polypodies are ferns of the genus *Polypodium*. Rather variable in appearance, the fronds of *Polypodium* can reach 40 cm. (16 in.) long depending on species; they are herring-bone shaped and bear rows of round, orange-brown spore-cases. The fronds remain green through the winter and only die back when new ones appear in spring.

Pomegranates are two species of deciduous bush, or small tree, which form their own plant family of just one genus, *Punica*. One species, *P. protopunica*, is native to an island near South Yemen, while the other, *P. granatum*, is native to the area between the Balkans and the Himalayas. This is the species now widely cultivated wherever there are hot, dry summers and cool winters, as in California and the Mediterranean area. The fruit has a leathery, brown to red skin and contains many seeds in a pink, juicy, somewhat acid pulp. Although it is eaten as fruit, it is more conveniently exploited as juice. The plant is also grown for its ornamental white, orange, or red flowers, particularly in temperate areas.

Pomelo   *Grapefruit.

Pomes are false *fruits consisting of a central 'true' fruit surrounded by swollen tissue derived from a flower-stalk. Apples, pears, and quinces are examples, the 'core' being the true fruit.

Pompanos are fishes belonging to the same family as the horse mackerel. Most are placed in the genus *Trachinotus*, and the name is in wide use in North America.

Pond snails popular with aquarists include the great ramshorn snail, *Planorbis corneus*, whose shell grows to over 2·5 cm. (1 in.) in diameter. This specimen is shown laying its eggs on the glass side of a tank. Each batch is contained in a protective gelatinous capsule.

They are deep-bodied, rather blunt-snouted fishes with long-based dorsal and anal fins. They are especially abundant in shallow water and moderate depths near reefs, and swim in schools, feeding on crustaceans and small fishes. They are particularly graceful and fast swimmers. Most species grow to a length of 50 cm. (20 in.). Pompanos are well known as a good-tasting food-fish.

Pond-skaters are elongate, blackish *bugs which live on the surface film of ponds and streams. There are over 200 species of pond-skater in the family Gerridae. They lay their eggs in batches on water plants below the surface and both young and adults are carnivorous, usually feeding on other insects trapped by the surface-tension of the water. A few, such as *Halobates*, belong to that small number of insect species to live on the sea. These water-skaters are found in tropical and subtropical seas, often hundreds of miles from land.

Pond snails occur in almost every freshwater habitat, and range from tiny spire-shaped species, such as *Hydrobia*, to the giant ram's-horn snails, *Planorbis*. While most breathe by gills and are related to winkles and whelks (prosobranchs), some are amphibious lung-breathers (pulmonate snails), having adapted from land-dwelling ancestors and reverted to freshwater life. These often act as hosts to parasites of vertebrates, such as liver flukes. A few pond snails give birth to live young, but most species lay gelatinous egg-masses on aquatic vegetation.

Pondweed may be of a great variety of species, though the plants most usually designated under this common name belong to the monocotyledonous family Potamogetonaceae. This family includes the genus *Potamogeton*, of which about fifty species are known with a worldwide distribution. Most are submerged aquatics, with slender stems and pale green to brownish, almost translucent, leaves. In keeping with many plants adapted to live below water-level, the tissues are not strengthened by the substance lignin, and collapse on removal from the water. They are to be found in still and running water. Some species spread with great rapidity and may eventually prove troublesome in lakes, filtration plants, and drainage systems. Some species have two kinds of leaves, broad, floating ones and narrower, submerged ones.

Poplars are trees of the genus *Populus*, allied to the willows and native to northern temperate regions. There are thirty-five species, all deciduous, with simple leaves and flowers in pendulous catkins. The female catkins produce seeds covered in down, hence their name of cottonwood in North America. The seeds are dispersed by wind. The trees are fast-growing and relatively short-lived, the timber being much used for pulp, matches, and other small objects. In plantations, the most commonly cultivated poplars are hybrids between the European black poplar, *P. nigra*, and the American eastern cottonwood, *P. deltoides*. The spire-shaped Lombardy poplar, *P. nigra* var. 'Italica', is a popular ornamental species. The aspen, *P. tremula*, a tree native to Europe, has leaves distinctive for their 'trembling' in the slightest breeze.

Poppies are members of the plant family Papaveraceae, with over 400 species of mainly temperate distribution. The chief genera are *Papaver* and *Meconopsis*, each having about forty species. To the former belong the corn or field poppy, and the oriental, opium, and Iceland poppies. They have divided leaves, a milky sap (latex) and large, colourful flowers. Among other genera for which this name is used are *Eschscholtzia* (California poppy), *Glaucium* (horned poppy), *Argemone* (Mexican poppy), and *Romneya* (tree poppy).

Population (ecology) is the collective term for all the representatives of one species which live and breed at a particular site. This distinguishes it from a flock, or swarm, which is usually part of a much larger interbreeding population. The maximum size of the population will be limited by the *carrying capacity of the habitat. Populations are rarely stable, especially where *succession is occurring; most fluctuate within certain numerical limits in response to food, space, and disease. Species, particularly those involved in direct predator–prey relationships, may show great changes in population numbers as the populations of the two species interact. Factors which act on populations irrespective of their abundance are called density-independent factors. Climate factors are usually quoted as the best examples and can cause severe fluctuations on an irregular basis in any population.

Porbeagle: a large, heavily built shark which ranges across the whole of the North Atlantic region and possibly also the Southern Hemisphere. It lives close to the surface, migrating northwards in summer but rarely coming close inshore. It is a very powerful swimmer, with a broad, deep tail and a body temperature higher than the surrounding sea. It feeds mainly on fishes and squids. Despite its size of up to 3 m. (9 ft. 9 in.) long, it is not dangerous to man.

Porcupines are large *rodents characterized by quills, or spines, which are modified hairs, though they are not well developed in all species. There are two distinct

The prehensile-tailed **porcupine** of Central American forests, *Coendou villosus*, is an agile tree climber.

families, both with eleven species, the Old World porcupines (Hystricidae) and those of the New World (Erethizontidae). The largest species, and the one with the best-developed spines, is the crested or African porcupine, *Hystrix cristata*, which weighs up to 27 kg. (60 lb.). It often defends itself by rushing backwards towards its attacker. The other African genus of porcupine, the brush-tailed, is much smaller at about 3 kg. (7 lb.). The Asian porcupines, such as the Asiatic brush-tailed porcupine, *Atherurus macrourus*, are more like typical rodents in appearance, and the Bornean long-tailed porcupine, *Trichys lipura*, has only rudimentary spines. The New World porcupines are markedly more arboreal than those from the Old World and have less well-developed spines. Two species have a prehensile tail (*Coendou* species). There is only one North American species, *Erethizon dorsatum*, which extends as far north as the Arctic Ocean.

Porgies are North American sea breams belonging to the family Sparidae. Ten species of porgy are recognized on the Atlantic coast and one on the Pacific coast, most of them deep-bodied, with a single dorsal fin. They have well-developed teeth in the jaws, those in the sides being blunt, crushing teeth. They are bottom-dwelling fishes, living close to reefs as well as on open sandy and muddy bottoms. Most are carnivores, eating sea-urchins, molluscs, and crustaceans, but some also eat plants as well. They are good food species.

Porpoises are the smallest marine mammals, with some six species found in the north temperate zones, the west Indo-Pacific, and the coastal waters of southern South America. They are often found near the coast and in estuaries, and sometimes swim up-river for many miles. They may reach a length of up to 2·25 m. (7 ft. 4 in.). They are black above and white below, with a low dorsal fin and small flippers. Superb swimmers, they will pursue shoals of herrings, sardines, and mackerel. Mating takes place in the summer and gestation lasts until the following June or July. The calf is large, being about half the length of its mother; it is suckled with extremely rich milk. The harbour or common porpoise, *Phocoena phocoena*, is found around the coasts of Europe and in the North Atlantic and Pacific oceans.

Port Jackson shark: a member of a small family of rather primitive sharks, often called bullhead sharks, which are found only in shallow water in the Indo-Pacific region. Living off southern Australia, it is similar to the Californian horn shark, *Heterodontus francisci*, and grows to a length of 1·5 m. (5 ft.). Its body is deep and heavy, with spines in front of the dorsal fins, and its teeth are distinctive, being sharp in the centre of the jaws and rounded, for crushing, at the sides of the jaws. Their teeth are well adapted to their diet of sea-urchins, molluscs, and crustaceans, all hard-shelled prey. These sharks lay their eggs in spirally twisted egg-cases which have a tendril at each corner; these tangle in seaweeds.

Portuguese man-of-war is the name given to a colony of the animal *Physalia physalis*, which produces different types of individual which, by co-operating together, resemble a large jellyfish. One modified individual secretes a gas-filled float, and the other individuals produce feeding tentacles which hang from this, the whole structure reaching 30 cm. (12 in.) across. Each individual is either

A **Portuguese man-of-war** clearly shows the division of the body into the gas-filled float, the long, coiled fishing tentacles armed with stinging cells, and the stout feeding tentacles used to digest its prey.

a modified *polyp or a medusa, and those specialized for feeding have tentacles reaching several metres down. These kill fish on contact and draw their dead bodies upwards to the feeding polyps. The tentacles are capable of inflicting painful stings on bathers. Other individuals are adapted for reproduction. A colony, blown by winds, can sometimes appear on coasts far from its native subtropical areas.

Possums are Australian *marsupials of the suborder Diprodonta and include species in several families. The pygmy possums of the family Burramyidae comprise seven species, including the pygmy glider, *Acrobates pygmaeus*. The sixteen species of ringtail possums are grouped in the family Pseudocheiridae and, like most possums, are nocturnal tree-dwellers. Like many other species of possum, members of this family are leaf or sap eaters, unlike the striped possums (*Dactylopsila*) of the family Petauridae (gliders), which are insectivorous. A specialized possum in a unique family is the honey possum, *Tarsipes rostratus*. This small shrew-like marsupial, up to 8 cm. (3 in.) in body length, feeds exclusively on nectar and pollen. Other possums include the brush-tail possums, which are relatives of the phalangers.

Potatoes are one of the most important food crops of the family Solanaceae and are thus related to the tomato. They are grown throughout the world, particularly in cool, temperate regions, for their swollen tubers which act as storage organs for carbohydrate food reserves. A perennial plant, the potato is cultivated as an annual, with fresh tubers (or seed potatoes) being planted each spring. Many varieties exist, differing in their season of use, skin colour, flesh colour, flesh texture after cooking, and tuber shape.

The potato was first cultivated in South America, being selected and bred from native species, particularly *Solanum tuberosum*, with bitter, poisonous tubers. These grew on the high Andean plateau around Lake Titicaca, where the cold, damp climate had produced plants capable of growing in northern temperate regions. The potatoes brought to Europe from the Americas by sixteenth-century explorers had relatively small, irregularly shaped tubers. The modern potato varieties have been selected to have shallow 'eyes' (the sites of the buds), a uniform shape, and smooth, blemish-free skin. They are eaten in large quantities in North America, Europe, and the USSR. In addition to being an energy source, potatoes provide up to 5 per cent of our protein, 7–10 per cent of iron and riboflavin, and 25 per cent of our vitamin C requirements.

Potoos are birds of five species in the family Nyctibiidae, which inhabit the northern half of South America, parts of Central America, and the Caribbean region. They are about the size of a crow, and beautifully camouflaged in browns, buffs, greys, and blacks. Their beaks are small, but, like the nightjar, the gape is enormous and used for catching insects at night. They lay their single egg on a broken stump of a tree, incubating it by sitting vertically above it, looking like an extension of the stump.

Pott, Percivall (1714–88), English anatomist and surgeon, was renowned for his brilliant and successful operations. He introduced many improvements to make surgery more humane and less painful, and his many excellent books revolutionized the practice of surgery in Britain. His description of an occupational cancer, that of chimney sweeps which he attributed to soot, resulted in him being regarded as the founder of modern oncology (the study and treatment of tumours).

Potter wasps   *Mason wasps.

Pottos are two species of primate belonging to the same family as the lorises and bush babies. They live in trees in the rain forests of West and central Africa. The short limbs and grasping hands and feet are used to cling tightly to branches. The potto, *Perodicticus potto*, and the golden potto, or angwantibo, *Arctocebus calabarensis*, are nocturnal animals up to 32 cm. (13 in.) in body length, and with a tail 1–5 cm. ($\frac{1}{2}$–2 in.) long. Both have thick fur, brownish-grey to rufous-brown on the back, paler on the underside. The gestation period lasts for about five months and there is usually a single offspring. The baby crawls immediately on to the mother, who carries it until it is self-sufficient. Pottos eat insects, leaves, birds' eggs, and other small animals and have large forward-facing eyes, well adapted to seek prey at night.

Pouched mice is the name loosely given to mouse-like *marsupials in Australia, but the genus usually intended is the genus *Smithopsis*, which includes the fat-tailed mouse, or dunnart, *S. crassicaudata*. Other mouse-like marsupials often called pouched mice include the kultarr, *Antechinomys laniger*, a jerboa-like animal, and the mulgara, *Dasycerus cristicauda*. All are members of the family Dasyuridae and tend to be insectivorous and carnivorous.

Poultry is a collective term used to describe domesticated birds, particularly those farmed for their eggs or meat, such as chickens, turkeys, ducks, and geese. All domestic breeds of duck are descendants of the mallard, including even the rather striking breeds such as the Indian runner, with its upright stance, and the crested duck, with its conspicuous topknot of feathers. In some areas Muscovy ducks are kept in a semi-domestic state. The domestic goose has been derived primarily from the wild greylag goose, though in eastern Asia the so-called Chinese goose (not a wild species) is a domestic derivative of the swan-goose. The Egyptian goose was widely domesticated in ancient Egypt but has now been replaced by other domestic strains. Other main species which have been domesticated for their eggs or flesh are guineafowl, pigeons, and peafowl. Others are kept for a variety of reasons and include the ostrich (for plumes), the cormorant (by the Chinese for fish-catching), and birds of prey (for falconry).

Prairie-chicken, or prairie-hen, are two species of birds from the genus *Tympanuchus*, part of the grouse family. Both lesser and greater prairie-chickens live in North America in tall grass prairies. About the size of chickens, they are heavily barred, brown birds. The males display communally, by inflating large air-sacs (red in the lesser, orange in the greater) in the side of the neck. The females tend the eggs and chicks.

Prairie dogs, in spite of their name, are really *ground squirrels. The five species are all small, fat rodents of the open plains and Rocky Mountain region of North America. They are social animals and were once famous for their 'towns', systems of tunnels and dens that stretched over large areas beneath the surface. Such a township would have been inhabited by hordes of prairie dogs. Today a few rather small towns exist in remote places. They feed upon green vegetation or, if that is scarce, upon roots, and will also eat insects when these reach plague proportions. From four to six young are born in May in an underground nest, and at four weeks they emerge to begin feeding on plants. The black-tailed prairie dog, *Cynomys ludovicianus*, is about 30 cm. (1 ft.) long with a tail of 8 cm. (3 in.) and is buff or cinnamon in colour apart from its tail. The white-tailed prairie dog, *C. leucurus*, is smaller and, apart from the white tip to its tail, resembles its relative.

Prairies are grasslands on the great level tracts of land in central North America, formerly occupied by herds of buffalo and now mostly converted to the cultivation of cereals. The term has now been extended to grasslands elsewhere.

Pratincoles are eight species of birds which belong to the family of *wading birds. They are widespread in warm areas of the Old World, living in open country, often near to water. They are brown, short-legged, long-winged

267

birds, resembling very large swallows as they catch their insect prey on the wing. They lay two to four eggs in a shallow scrape on the bare ground.

Prawn is a name, sometimes used as a synonym for shrimp, but which more usually refers to the larger, inshore, decapod *crustaceans, especially those with a long projection between the eyes. Prawns tend to live below tide-level in winter, but are found in rock-pools in the summer, and some species move up estuaries, burrowing in the sand. They have a sharply bent abdomen and usually lack enlarged pincers, as they feed only on small pieces of vegetation and debris.

Prawns are particularly renowned for their colour-changes, as they adapt to match their background, using special, pigmented cells just below the cuticle. The chameleon prawns, *Praunus* species, are particularly good at this, being red, green, or brown by day, and transparent and blue-tinged at night.

Praying mantids are large insects in which the first segment of the thorax is elongate and the front legs are modified for seizing prey. Mantids sit on plants in a 'praying' position until prey appears. The females sometimes eat the males during mating. They are one of two major divisions of the order Dictyoptera, the other being the cockroaches. There are around 1,300 species of mantids, found mainly in the tropics and subtropics.

A **praying mantis**, *Acanthops* species, (*below*), from the rain forest of Costa Rica, uses its uncanny resemblance to dead leaves to ambush insect prey.

The pale-colour phase of the hooded **prawn**, *Anthanas nitiscens*, (*above*), caused by contracted pigment cells. This species, about 2 cm. ($\frac{4}{5}$ in.) long, ranges from the Mediterranean to the Baltic, from extreme low tide down to 70 m. (230 ft.).

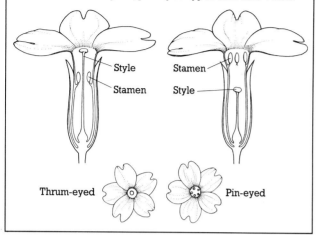

Primroses

Certain *Primula* species produce two types of flowers to reduce self-pollination. Thrum-eyed, long-styled flowers are receptive only to pollen carried from the anthers of short-styled, pin-eyed types and vice versa.

Style

Stamen

Stamen

Style

Thrum-eyed

Pin-eyed

Pregnancy is the presence within the body of a developing *embryo or foetus. It covers the time from fertilization of a female ovum by a male spermatozoon until birth, in the human female normally about 266 days or, more conveniently, forty weeks from the last menstrual period.

Pricklebacks *Blennies.

Prickly pear is the common name for the 300 or so species of *cactus in the genus *Opuntia*, which includes the Indian fig, *O. ficus-indica*. Most species bear large pear-shaped, prickly fruits, from which their name is derived. They are native to northern South and Central America. but many species have been introduced to other parts of the world. Several species are now well established in South Africa and were once a serious weed in Australia.

Primates are an order of eutherian mammals with arboreal habits, omnivorous diet, and comparatively unspecialized teeth. They have dextrous, grasping hands and feet with the first and second digits usually being opposable. Instead of claws, their fingers and toes are equipped with nails, with sensitive pads on the tips of their digits. They have forward-looking large eyes with stereoscopic vision, and a large brain. Most primates are specialized for an arboreal life, and the few that live mainly on the ground are descended from tree-living forms. The need to use co-ordination between hand and brain in tree-life has had an important influence on the evolution of primates, including man. Furthermore, as tree-living creatures they depend upon quick reactions to keep safe. Primates have been most successful in tropical and subtropical areas, where most of the 181 species occur. There are two main suborders: the Prosimians, which include lemurs, bush babies, pottos, and tarsiers; and the Anthropoids, which include monkeys, apes, marmosets, and tamarins.

Primitive moths are tiny, brilliantly coloured, day-flying moths of the family Eriocraniidae, less than 12 mm. ($\frac{1}{2}$ in.) in wing-span. The proboscis is short, and unlike

most moths they have rudimentary jaws. Females of the European species lay eggs in slits in leaves, usually birch, and the legless caterpillars are leaf-miners. They pupate in the ground.

Primrose is the common name for several of the 500 or so species in the plant genus *Primula*, which includes cowslips, *P. veris*, true oxlips, *P. elatior*, and other species or hybrids known as primulas. The common primrose, *P. vulgaris*, is a spring-flowering perennial, native to western Europe and Asia Minor. Like most other *Primula* species, it produces a rosette of leaves at ground level and flowers held on stalks. The common primrose hybridizes naturally with the cowslip to produce the false, or common oxlip, *P. vulgaris × veris*. Similar hybrids between these parents also give rise to the garden varieties of polyanthus and auricula.

The primrose gives its name to the family Primulaceae, comprising some 1,000 species of mainly northern temperate region plants. Relatives of the primroses include the shooting stars, *Dodecatheon* species, *pimpernels, *Anagallis* species, and *cyclamens, *Cyclamen* species.

Privets are a number of species of *Ligustrum*, an Old World genus of fifty species in all, belonging to the same family as lilacs, olives, and ashes. The wild privet of Europe, *L. vulgare*, was formerly much used for hedging, but it has been displaced by the Japanese *L. ovalifolium*, a plant which is tolerant of pollution.

Proboscis monkey: a species of Old World monkey that is confined to Borneo, where it dwells in forest regions close to water. It will sit for long periods of time on a branch or travel in small troops and can swim well. Its

A male **proboscis monkey** sits among mangrove leaves, Brunei, northeast Borneo. At over 23 kg. (50 lb.), the adult male is about twice the size of the female.

call is a drawn-out, resonant honk. A vegetarian, it feeds on leaves and fruit. About 75 cm. (2 ft. 6 in.) long, it has a reddish coat with some grey hair on the limbs and lower back. It is remarkable for its extremely long nose in the male, which can be 75 mm. (3 in.) in length and reach below the chin.

Proboscis worms *Ribbon-worms.

Prominent moths, together with kitten moths, the puss moth, and the lobster moth, belong to the family Notodontidae, and are world-wide in distribution. When the insect is at rest, projecting tufts of scales at the middle of the hind margin of each fore-wing are brought together, forming a prominence above the centre of the sloping wings. Prominent moths are stout-bodied, white, greyish or brownish, with pale hind-wings and long, narrow fore-wings, camouflaged like bark or wood. The caterpillars characteristically have one or more humps on their backs, and tend to rest with the hind-part, and occasionally also the fore-part, somewhat elevated. They feed on the leaves of a variety of trees.

Prong-horn, or prong-buck: a species of even-toed ungulate (hoofed mammals) found in semi-arid regions of western North America. It is the only surviving representative of a family that was once widespread in that continent. The prong-horn is the only horned animal that sheds the horn sheath each year, and the only one with branched horns (as opposed to antlers). The male can reach a height of up to 1 m. (3 ft. 3 in.) at the shoulders; the female is smaller. The coat is brown, and the underside of the body and neck and part-way up each side is white. The neck is banded and the cheeks are white. The rump has a patch of long, white hairs that are erectile and act as a warning signal to others. The females give birth to twin fawns after a gestation period of eight months.

Proteas are some 130 species of evergreen shrubs in the genus *Protea*, mostly from the Cape region of South Africa. They form part of the family Proteaceae, which contains over 1,000 species of shrubs and trees distributed throughout the Southern Hemisphere. *Protea* species have inflorescences that consist of many flowers enclosed in coloured bracts. One of the most spectacular is the king, or giant protea, *P. cynaroides*, with flower-heads almost 30 cm. (12 in.) across. Such robust flowers are ideally suited for pollination by sunbirds, which feed on the nectar. Other species of the family Proteaceae, such a those of the Australian genus *Banksia*, are pollinated by small nocturnal animals (by marsupials in Australia).

Protected species are those plants and animals protected by law from collection or interference by man. The protection may be temporary, perhaps lasting only for the breeding season, as for many gamebirds and freshwater fish, or it may be permanent, where a particularly valuable or endangered species is given local, national, or international protection throughout the year. This protection may be worldwide, as for Steller's albatross, or limited to certain countries in regions where a threat exists. Several whales and fishes have been protected, as an aid to their conservation, by internationally agreed, annual catch quotas. Many species are indirectly protected through living in *nature reserves or wildlife parks.

Protective coloration is found in many animals. It may be used for *mimicry, *camouflage, warning, or surprise. The conspicuous black and yellow markings of the bee and the wasp warn of their painful sting and they are left alone; but so too are the harmless hoverflies which mimic their black and yellow colouring. The eye-spots of many moths and butterflies can be revealed suddenly to frighten off predators. Many ground-nesting birds, such as nightjars, are coloured to blend perfectly with their surroundings.

Proteins are molecules composed of long chains of *amino acids linked by peptide bonds; they make up about half of the dry weight of the bodies of living organisms. The 'primary' structure of a protein is the order, or sequence, of amino acids in its polypeptide molecule. Proteins differ in the sequence of amino acids

The densely packed flowers of **proteas** are surrounded by rows of coloured bracts to make, as in *Protea magnifica*, a large flower-head which, with its copious supply of nectar, is particularly attractive to sunbirds.

in their polypeptide chain; the precise sequence determines the final three-dimensional shape of the protein molecule. The stages in the formation of a three-dimensional shape from a straight polypeptide chain are called secondary, tertiary, and quaternary levels. The 'secondary' structure is the regular three-dimensional folding of the polypeptide chain formed by hydrogen-bonding between the peptide bond regions of the chain. The commonest secondary structure is a right-handed spiral shape, called an α-helix; this occurs in proteins such as keratin from hair, horns, nails, and wool, and helps to explain the physical properties of these structures. Globular proteins have rounded, compact, soluble molecules produced by an additional folding process involving chemical interaction between different amino acids in an already regularly folded, 'secondary' level structure. This gives each molecule its 'tertiary' structure. Globular proteins include enzymes, antibodies, and many other types. Some proteins, such as haemoglobin, show 'quaternary' structure, in which complete, 'tertiary' level molecules are fitted together to make a larger protein assembly.

Other proteins include muscle, snake venom, hormones, and blood-clotting agents. Proteins in the human diet are needed to repair or replace existing molecules, or for growth, but may also be exploited as an energy source.

Protista: one of the three major kingdoms of living organisms, the others being the plants and animals. They usually include the algae, bacteria, fungi, and protozoa, some of which may also be considered to be 'plants'.

Protoplasm   *Cytoplasm.

Protozoa literally means 'first animals', and it was probably from simple unicellular life-forms like these that the animal kingdom evolved. Most living protozoans, a vast, diverse assemblage of one-celled non-photosynthetic organisms, are very far from simple, being themselves the product of lengthy evolution. There are at least 50,000 species, which together with unicellular algal plants make up a separate kingdom, the Protista.

The cell of a protozoan contains all the minute organelles and chemical apparatus needed to live, together with locomotory structures (cilia, flagella, or pseudopodia) and sometimes an exoskeleton. Each can also reproduce itself, sometimes sexually. Yet most forms still measure less than 0·1 mm. ($\frac{1}{200}$ in.), very few being visible to the naked eye. Protozoan cells are more complex and versatile than any single cell found in a higher animal.

Protozoans are incredibly diverse in habits. *Flagellates, *radiolarians, and *foraminifera are abundant in plankton, while *amoebas and *ciliates like Paramecium occur in fresh water. Many protozoans are parasitic (causing dysentery, sleeping sickness, and malaria), or symbiotic. In fact, examples of protozoa can be found eating almost anything, which reflects their amazing adaptability.

Protura: an order of minute insects, comprising some 200 species, each not more than 2 mm. ($\frac{1}{8}$ in.) long, and found in humus-rich soils throughout most of the world. They have piercing mouthparts and no eyes or antennae. The first of three pairs of legs are held forward to serve as sensory organs. On hatching, they have only nine abdominal segments and add one more at each of the first three moults. They probably feed by sucking nutrients from the hyphae of fungi. They are usually classified with the insects, but some biologists place them in a class on their own.

Pseudoscorpions, or false scorpions, are tiny *arachnids, most less than 4 mm. ($\frac{1}{7}$ in.) long, living among humus, bark, and soil worldwide. The 2,000 or so species are rarely seen, though they are quite common. They resemble true scorpions but lack the long abdomen and sting, and are much smaller. The *pedipalps are modified as pincers, bearing poison glands; hence the false scorpions carry their sting not in their tails but in their arms. They build silk nests for over-wintering and for protection when moulting.

Psoriasis is a common skin disorder characterized by patches of reddened skin covered with silvery scales. Although the name implies itching, this is not usually prominent, and for most patients the condition is merely a nuisance and embarrassment. The cause is quite unknown. It is possible, with ointments, to make the patches disappear, but recurrence is the rule. The condition is sometimes associated with a type of arthritis.

Psyche: the mind or soul, as distinct in Western thinking from the body. It includes all actions of the mind in thinking, feeling, and sensing, together with associated memories, fears, and wishes. The relationship between the psyche and the brain is the central problem in philosophy and psychology.

Psychiatry is the medical speciality dealing with mental illness. Most patients are seen on a voluntary basis, but particularly severely disturbed people with *psychoses sometimes have to be treated against their will.

Psychoanalysis is the method of understanding and treating *neurosis devised by Sigmund *Freud. The recommended procedure is for an analyst to see a patient for one hour, five times a week for several years. The patient relaxes on a couch and describes what comes into his head; the analyst says little but tries to interpret what the patient says in the light of the evolving relationship between them. Understanding of this is believed to benefit the patient's morale and his relationships with others.

Psychology is the scientific study of the functioning and properties of the behaviour of man and animals. One major branch is experimental psychology, which can be applied to man and to animals, and is largely concerned with the measurement of behaviour in response to different stimuli. Social psychology and clinical psychology use scientific quantification to evaluate normal and pathological human thinking and behaviour. Other studies termed 'psychological' are purely descriptive, and often subjective, and therefore not scientific.

Psychoses are severe mental illnesses. Sufferers have abnormal beliefs, called delusions, or false perceptions, called hallucinations. Delusions are illogical, held against all argument, and tend to colour the patient's interpretation of everyday life. Hallucinations are usually heard or seen, and may seem real to the patient, although they have no substance in reality. The onset of psychosis may be sudden, making admission to a psychiatric hospital necessary for a short time.

Psychoses can be broadly divided into a so-called 'functional' group and an 'organic' group. The first includes *schizophrenia and manic-depressive states. The second includes those conditions caused by physical factors such as drugs and brain diseases.

Psychosomatic disorders are physical illnesses believed to be caused, or to have their severity determined, by psychological or emotional factors. It is thought that physical illnesses such as asthma, eczema, and peptic ulcers develop partly as a response to stress or anxiety. They are usually treated as physical illnesses but the symptoms may be reduced by the removal of stress.

Ptarmigan: a species of bird belonging to the grouse family. It lives in northern temperate areas of both Old and New Worlds, usually on treeless moorland or on mountains. All subspecies of ptarmigan moult into a white plumage in winter so as to be well camouflaged in the snowy areas in which they live. They may shelter from extremely cold nights by digging burrows in the snow. They are vegetarians, living on shoots and leaves.

Pteraspis: a primitive *jawless fish which lived in fresh or brackish water in early Devonian times (around 390 million years ago). The front of the body was covered in thick plates of dermal bone which tapered in the front to form a long snout. The eyes were large and placed on either side of the head shield. *Pteraspis* had no fins but bony spines were used as substitutes. The tail was covered in diamond-shaped scales and was well developed and flexible. No fossils of the internal skeleton have been found but it is assumed that it was made of cartilage.

Pteridophytes are *ferns. Some botanical classifications also include in this group all the horsetails and clubmosses. They were grouped together because they all have true stems, leaves, and roots, and have an organized system of xylem. As a division of the plant kingdom the Pteridophytes were very important in terms of number of

A male white-tailed **ptarmigan**, *Lagopus leucorus*, in winter plumage. It feeds mainly on dwarf willow on high, windswept ridges, and is found from New Mexico as far north as central Alaska.

species and physical size in the Carboniferous Period. Present-day species are concentrated in the tropics with relatively few temperate species.

Pterodactyls, or pterosaurs, are a group of extinct flying reptiles which lived in the Mesozoic Era (245–66·4 million years ago), at the same time as the dinosaurs. The fourth finger of each hand is greatly elongated and supported a flight membrane. As a further adaptation to 'flight', the skeleton was extremely lightly built. They probably flew mainly by gliding.

Puberty marks the stage in children when the output of hormones from the *gonads increases sufficiently for reproduction to be possible. Its onset is influenced by the brain and occurs at a variable age, which in girls ranges from 10 to 16 years and in boys from 14 to 18 years. It is associated with a period of rapid bodily growth and the gradual appearance of the secondary sexual characteristics. In boys the scrotum, testes, and penis enlarge, hairs sprout on face and body, and the voice breaks. In girls the breasts enlarge, the hips broaden, the distribution of body fat changes, and *menstruation begins.

Puff adders are stoutly built *vipers, up to about 1·2 m. (4 ft.) in length, from Africa and western Arabia. There are eight species which live in a wide range of habitats but avoid extreme desert conditions and rain forests. Although nocturnal, the puff adders often bask during the day. If disturbed, they makes loud hissing (puffing) noises, hence their common name. In southern Africa, species such as the common puff adder, *Bitis arietans*, or gaboon viper, *B. gabonica* (not a true viper, but a type of puff adder), account for the greatest number of human deaths due to snake-bite.

Puffballs are the spore-bearing structures of a group of fungi called Lycoperdales. The powdery mass of dry spores develops inside a hollow ball of fungal tissue which is white or buff. When ripe this has a hole at the top from which clouds of spores are puffed when the puffball is struck. The giant puffball, up to 100 cm. (3 ft.) across, is one of the largest fungal fruiting bodies known.

Puffbirds are birds found in Central and South America. They make up a family of some thirty-two species and are found mostly in forests, where they sit upright on perches and fly out after insects. An exception is the swallow-winged puffbird, *Chelidoptera tenebrosa*, a longer-winged species which spends more time in the air than the others. In general, puffbirds are thick-set, black, brown, grey, and white birds with long, strong beaks (often yellow or red). They nest in holes in banks or in the ground.

Puffer-fishes are fishes well known for their ability to inflate their stomachs with water or air (in which case they float helpless at the surface). There are about 130 species known in the family Tetraodontidae, mostly tropical and subtropical, coastal sea fishes, with a few living in the open ocean and others living in fresh water in Africa and Southeast Asia. They are all rather plump, with small mouths, and four flattened teeth, forming a parrot-like beak. Their fins are small and rounded, and are used as paddles in swimming. All have spines in the skin, especially in the belly, and these become more

Normal and inflated spiny **puffer-fishes**, *Diodon holocanthus*, from the Caribbean. These fishes inflate themselves when disturbed to avoid attacks by predators. At the same time the eyes take on a glaring attitude, while the black spots on their back also contribute to their threatening appearance.

prominent when the fish is inflated. The internal organs and blood of many species are highly toxic to humans, but despite this the rest of the body is used as food, especially in Japan, where they are known as fugu.

Puffins are a tribe of *auks with coloured beaks and occasionally head plumes. They nest colonially in burrows dug with feet and bill, and feed by swimming and diving. Atlantic puffins, *Fratercula arctica*, breed on North Atlantic and Arctic coasts, while three other species inhabit the northern Pacific.

Pug moths are small moths of the family Geometridae which rest with the wings spread away from the body and pressed flat. Their *looper caterpillars mostly feed inside flowers or seed pods, although some eat leaves, and a Hawaiian species has recently been found to catch and eat flies. The pupae of most pug moths are brightly coloured and are formed in silken cocoons in the ground.

Puma: a species in the *cat family which ranges over the whole of North and South America. It lives in a wide variety of habitats from sea-level to altitudes of up to 4,000 m. (13,000 ft.). It resembles a slender lioness and may reach a length of 2·9 m. (9 ft. 6 in.), about a third of which is tail. In colour the short, close fur varies from yellowish-brown to red, the underparts are white, the ridge of the back and the tail are usually marked by a darker line, and the tail is black-tipped. A solitary animal, which avoids humans, it can ·travel 48–80 km. (30–50 miles) when hunting. It feeds on a wide variety of prey from deer to slugs and snails. Breeding occurs only every two to three years, when one to six kittens are born in a den after a gestation period of three months. The kittens

are weaned in one to three months, but are not fully independent until about two years of age. They have a lifespan of about eighteen years. The puma, *Felis concolor*, is also known as the cougar, mountain lion, or panther.

Pumpkins are typical members of the family Cucurbitaceae, which includes cucumbers, melons, marrows, and others. The family contains around 700 species, all typically perennial, climbing or trailing plants which produce rotund fruits with a leathery skin. The pumpkin can be one of two species; the name is used in Europe when referring to a variety of *Cucurbita maxima*, and in North America when referring to a variety of *C. pepo*. Both are orange in colour and only the European pumpkin is invariably round.

Punkies   *Midges.

Pupae (singular pupa) are the resting stage in the *metamorphosis of many insects, which comes between the larva and adult. Pupae show the outlines of the appendages of the adults, but only some abdominal muscles are functional. These are enough for some, like mosquito pupae, to swim, but most do not move about. Pupae fall into two groups: those in which the appendages are closely stuck to the body, like butterflies, and those in which they are free, like beetles. Most pupae are contained in silken cocoons or earthen cells, but those of two-winged flies live inside the hardened and darkened last larval skin, or puparium. The function of the pupa is to give the insect a resting period in which to reorganize its tissues. This is necessary because of the extreme changes, both structural and physiological, needed to bridge the gap between a larva and an adult.

Purple emperor butterflies are dark brown with a white band and spots, but males have a rich purple or blue iridescence. There are several species within the *brush-footed butterfly family, which occur in Europe and Asia, with one British species. They are powerful fliers, generally soaring high above the tree-tops, but they are attracted down by carrion. The adults never visit flowers, feeding instead on honeydew.

Pus is a yellow fluid, containing lipids, proteins, cell debris, and bacteria. This is formed from the decomposition of dead and damaged cells by white cells of the blood, which migrate to infected tissues.

Puss moths are fat-bodied, furry moths, typical of the family Notodontidae, with their white fore-wings delicately etched with grey. The caterpillars of the European puss moth, *Cerura vinula*, are green with a white-edged, purplish-brown saddle and a forked tail; when irritated, they rear up, exposing a red false face, protrude a pair of lashing threads from the tail, and squirt formic acid. The family to which they belong also includes the kitten moth, *Furcula* species, with similar caterpillars, and the *prominent moths and *lobster moths.

Pygopodids are a family of legless *lizards found in Australia. These snake-like creatures do not burrow into the soil like many similar families of legless lizards, but hunt other species of lizards for food. The widespread Burton's legless lizard, *Lialis burtonis*, grows to 30 cm. (12 in.) in length.

A European **puss moth** caterpillar disturbed from feeding on the leaves of willow. It assumes its grotesque defence posture by tucking the head into the thorax and raising both front and rear ends. This exposes the brightly coloured false head and the extended whiplash processes of the 'tail'.

Pyrethrum is the name formerly used for a group of perennial, herbaceous plants now included in the *Chrysanthemum* genus. The garden forms are derived from *C. coccineum*, a variable species from Iran and the Caucasus, and have divided, fern-like leaves and large daisy-like flowers in red, pink, or white. The insecticide pyrethrum is produced from the closely related *C. cinerariifolium*.

Pythons, of which there are about twenty species, are snakes which occur in warm regions from Africa to Australia and are absent from the Americas and Madagascar. They resemble boas in retaining claw-like vestigial hind-limbs but differ in being egg-layers (boas bear live young). They are found in a wide range of habitats: some species live close to water, many climb trees and bushes. They feed mainly on small mammals which they kill by constriction. Their jaws are well endowed with backward-pointing teeth which, although not venom-producing, will hold prey firmly. Most have heat-sensitive pits in their upper lips, similar in function to the facial pit of pit vipers, which assist the snake to locate its prey. The reticulated python, *Python reticulatus*, from Asia, is the largest snake of the Old World, reaching a length of 10 m. (33 ft.).

Quagga: a zebra that once lived in southern Africa. It has been exterminated by man, the last survivor having died in a Dutch zoo in 1883. It was much less prominently striped than other zebras, with the hind region almost plain coloured.

Quahogs   *Clams.

Quails are small birds belonging to the pheasant family. They are short-tailed, not more than about 25 cm. (10 in.) in length. The best known of the New World species are the bobwhite, *Colinus virginianus*, and the California quail, *Lophortyx californica*; both are brownish above, mottled and barred below, with striking black and white face patterns, the California quail having a black plume standing vertically on the top of its head and curving slightly forwards. The best-known Old World species is the common quail, *Coturnix coturnix*, a species with a wide range over Europe, Asia, and much of southern Africa. The European birds migrate over the Mediterranean to Africa for the winter.

Quassia is a genus of forty species native to the tropics, particularly the Americas. They are evergreen trees, the

A brace of common **quails**, in which the male can be distinguished by the pronounced neck-band. Quails rarely fly, except during their annual migration.

most important being quassia wood or bitter quassia, *Q. amara*, a tree some 6 m. (20 ft.) tall with divided leaves. Its intensely bitter wood and root was formerly much used in medicine.

Quetzals belong to the trogon family of birds. They live in tropical forests of Central and South America. The size of a small crow, they are brilliant green above and red below. By far the most striking is the resplendent quetzal, *Pharomachrus mocinno*, in which the male's upper tail-coverts are brilliant iridescent green, and extend for some 60 cm. (2 ft.) beyond the tail. They sit motionless on branches just below the canopy of the forest and sally forth to catch passing insects and to pluck fruits off trees. They nest in holes in trees, laying two to four pale eggs.

Quinces are trees or shrubs native to southern Europe and Asia, closely related to *japonica and other plants of the rose family. The common quince, *Cydonia oblonga*, has been grown in Europe since Roman times for its golden-yellow, pear-shaped, aromatic fruits. Somewhat hard, and too acidic to be a dessert fruit, they are used to make jam or jelly. The quince is also used as a rootstock for pears. Ornamental quinces belong to the genus *Chaenomeles*, and are small hardy shrubs which produce attractive pink, red, or white flowers.

A resplendent **quetzal** at his nest-hole entrance. When incubating eggs he sits in the nest facing outwards, his tail-coverts protruding from the tree over his head. He leaps backwards into flight from a perch.

Rabbit-eared bandicoots, or bilbies, are three species of Australian *marsupials so called because of their long rabbit-like ears and hopping gait. They are about the size of rabbits and live in burrows but otherwise are not similar, with their pointed snouts, long tails, and insectivorous diet.

Rabbit-fishes belong to the family Siganidae, which is widespread in the Indian and Pacific oceans, and are so-called because of the blunt snout and rabbit-like appearance of the jaws. About ten species are known, all shallow-water reef fishes which feed on algae. The spines on the dorsal, anal, and pelvic fins (each pelvic fin has two spines) have a toxic mucus, which can cause painful wounds.

Rabbits are members of the mammalian order Lagomorpha, which also contains hares and pikas. The names hare and rabbit are used rather loosely, but some sixteen species of rabbits are distinguished by their young, which are born hairless and blind in fur-lined nests, usually underground. Young hares or leverets, on the other hand, are born above ground, are fully furred, and are able to run soon after birth. Rabbits tend to be smaller than hares and to have proportionately shorter legs and ears. There are also differences in the skull-bones.

There are two principal groups: the Old and New World rabbits. Of three species of Old World rabbit the most familiar is the European rabbit, *Oryctolagus cuniculus*, from which domestic rabbits are descended. Its original range was around the Mediterranean region but, through introductions, its distribution is now worldwide except for the tropics. Another Old World species is the Bunyoro rabbit, *Poelagus marjorita*, of central Africa. It is frequently a pest of crops. New World rabbits, which include seven species of cottontails among a total of fourteen species, occur southwards from southern Canada. They are all rather similar to each other and to the European rabbit. The volcano rabbit of Mexico, *Romerolagus diazi*, is restricted to two mountain ranges and is one of the smallest species at up to 36 cm. (14 in.) long. Most rabbits burrow extensively in the ground.

Rabies is a virus infection of the brain, which affects many different animals. Man is infected by a bite by a rabid animal, usually a dog. The condition causes convulsive spasms, which are typically provoked by swallowing. The popular name of hydrophobia (fear of water) refers to the connection between the victim drinking water and the severe spasms caused by its swallowing. Established human infection is always fatal, but the interval of several weeks between infection and onset allows the effective use of vaccination. The six-months' quarantine on imported animals, which has eliminated the disease in Britain, is necessary to establish whether or not they harbour rabies.

Race (biology) refers to a group, population, breed, or variety within a *species. The term is rarely used scientifically because of difficulties in its exact definition.

A common **racoon** forages at night in woodland, New York State. Racoons are probably polygamous, with one male visiting two or three females during the breeding season, normally in late January and early February.

Subgroups within a species are thought to evolve from a series of biological changes through many generations. These groups reflect local differences in the distribution of *genes, arising from the isolation of breeding populations and subsequent loss of interbreeding. In human populations, social customs, and religious and language differences can lead to the formation of distinct races.

Racoons are carnivores of the genus *Procyon*, which are related to coatis and kinkajou. They give their name to the family Procyonidae. Racoons range from southern Canada to northern South America. There are six species, of which the most widespread in the USA is the common racoon, *Procyon lotor*. Affectionately known as 'coons', these animals were often adopted as pets by early settlers. Their intense inquisitiveness can create havoc in a household, with jars and bottles being opened by their highly manipulative fore-paws, which are as adept as hands. They are heavily furred, with a long bushy tail, and their pelts are valued in trade. Racoons are found near water and take much of their food from streams, particularly frogs, crayfish, and fish. They also eat insects, small mammals, and a certain amount of vegetable matter such as seeds and nuts.

Radiolarians, among the largest and most beautiful *protozoa, are related to amoebae and are found in the marine plankton. Many species are visible without microscopes and all have an elaborate silica skeleton, usually with radiating rods forming a spiny glass-like sphere. Between the spines fine extensions of the cell protrude, to act as a filter for food particles, which are drawn to the interior

of the body. These protrusions also assist locomotion in some species, although many radiolarians just float passively, often aided by tiny oil droplets which act as 'floats'. Radiolarians can be so numerous that after death their shells sink to the ocean floor and form a distinctive type of sediment, called radiolarian ooze.

Radishes are a popular salad vegetable, various forms of which are cultivated in most parts of the world. The small red or white, pungent, swollen roots vary in shape from round to cylindrical. They are relatives of cabbage, turnip, and other members of the mustard family.

Raffias are tropical *palms of the genus *Raphia*, which yield the material used for tying up plants, or for handicrafts. The fibre is stripped from the upper surface of the youngest leaflets. There are several species, all with gigantic leaves. In parts of Africa they provide the principal material used in local house-building, the stalks providing the framework and the leaves the roofing.

Ragfish: a single species, *Icosteus aenigmaticus*, which is the only representative of its family; it lives in the North Pacific, on both the Asian and North American coasts. Although it grows to a length of at least 2 m. (6 ft. 6 in.), its bones contain very little calcium, and so are somewhat flexible. This gives it its peculiarly floppy (rag-like) appearance out of water. Young fishes live in inshore waters near the surface and are brownish in colour; the adults live in mid-water in the open ocean.

Ragworms   *Bristle worms.

Ragworts belong to the plant genus *Senecio*, itself part of the sunflower family. Common ragwort, *S. jacobaea*, is a tall perennial with yellow flowers, a weed of waste places and pastures which, because of its poisonous nature, is avoided by grazing animals. The so-called Oxford ragwort, *S. squalidus*, is one of the most rapidly spreading alien weeds in Europe. It is an annual, branching plant from Sicily, with yellow daisy-like flowers, and seeds which are supported by silky hairs, to facilitate dispersal by wind. They lodge in any convenient place and are able to germinate and grow with a minimum of moisture. In Britain, it is one of the first colonizers of waste ground.

Rails are a large family of birds which includes many called rails, crakes, gallinules, and coots. A typical 'rail' is a small or medium-sized, chicken-like, marsh bird with long legs which dangle below it in flight, short rounded wings, and a stubby tail which it jerks as it walks. Its beak can be long, short, or conical. Many rails, such as the corncrake, *Crex crex*, live on the ground in thick vegetation and, because of their skulking habits and reluctance to fly, are rarely seen. Although absent from the Arctic and Antarctic regions, rails are widely distributed. Familiar European species are the moorhen, coot, and water rail.

Rainbow trout: a species of fish native to the rivers and lakes of the western coast of North America from northern Mexico to Alaska, but they have been introduced to the rest of North America and other temperate regions of the world. In the north of its native range, it is migratory and, although breeding in fresh water, spends part of its life in the sea (when it is known as a steelhead trout). Heavily spotted on back and fins, the rainbow trout has an iridescent streak along its side. It feeds on insects, their larvae, and crustaceans, and large specimens eat other fishes. Preferring clean, cool water it is nevertheless tolerant of low oxygen levels and this has made it suitable for intensive culture in fish farms as a food species.

Rain forest occurs mostly in tropical and subtropical regions with adequate rainfall and an absence of any marked seasonality. Rain forests also occur in the temperate, northwest coastal areas of the United States. The trees in such forests are evergreen and often very tall.

Tropical rain forest is probably the richest of all vegetation types in terms of the number of plant and animal species involved and is notable for the high proportion of *lianas and *epiphytes. It represents a huge resource in terms of timber, drugs, and other produce, but is rapidly disappearing through overexploitation of a few species. The remaining forest species are neglected or burnt and many have undoubtedly become extinct without even being discovered. Once destroyed, tropical rain forest cannot be re-established due to the severe soil erosion which occurs after the plant-cover is removed. The tropical rain forests are also very important in stabilizing the climate of the areas in which they exist. The sheer mass of vegetation acts like a sponge, holding and slowly releasing water into rivers.

Rain frogs are a group of small African frogs within the family Microlylidae, with almost spherical bodies, short limbs, and a very short snout. They live in burrows for most of the year, emerging to feed (almost exclusively on termites) and to breed when the first rains fall in southern

Fields of **rape** in Sussex, England. The pollen of this plant, using static electricity, can jump the gap between it and a visiting insect. When carried to another flower, it can then leap from the insect to the stigma.

Africa. The males are considerably smaller than the females. During mating the male becomes glued to the female's back by his throat, chest, and fore-arms. If threatened, the rain frog will inflate its body and can exude sticky droplets which deter most predators.

Ramazzini, Bernadino (1633–1717), Italian physician, was recognized with *Paracelsus as a founder of occupational medicine. He made meticulous observations of the diseases of workers in a large number of different professions, in order to find relationships between disease and occupation. He recognized the dangers inherent in the use of certain elements: for example, metals used by gilders; coloured glass containing antimony, used by glass-workers; mercurial preparations used by surgeons; and certain elements used by chemists in preparing medicines. His investigations resulted in an overall picture of the incidence of many diseases, as well as detailed descriptions of the pathological changes seen in patients.

Rambutan *Litchi.

Ramshorn snails are *molluscs, with a flat, spiralled shell unlike the asymmetric coil of more typical snails. They are found in fresh water, browsing among dense weeds. The many species of ramshorn snails breath via a 'lung' within the mantle cavity. They also have blood which contains the respiratory pigment haemoglobin; this enables them to live in stagnant waters.

Ranunculus *Buttercups.

Rape (botany) is a plant with several varieties belonging to the species *Brassica napus*. It belongs to the mustard family and is related to the cabbage. It is an annual plant, grown either to be grazed by cattle or, increasingly in recent years, for its oil-rich seeds, hence the name oil-seed

Rain forest

A profile of a South American rain forest shows the various layers of vegetation and some typical animal inhabitants. The emergent trees may be gigantic, often reaching up to 60 m. (200 ft.) and usually having an umbrella-like crown. The gaps between the scattered emergent trees are filled by an almost continuous tree-top canopy. Over 50 per cent of forest animal species live in this stratum, which is rich in fruit and flowers. The middle stratum or understorey has the densest plant growth, containing numerous shade-loving palms and climbers as well as philodendrons, monsteras, and epiphytic orchids. The smaller trees typically have conical crowns above an unbranched axis. Between the understorey and the forest floor is a rather sparse shrub layer. The forest floor itself is rich in smaller organisms, including the bacteria and fungi which rapidly break down the rotting vegetation. The decomposition rate is high — leaf litter can be broken down in six weeks compared with a year in a temperate forest.

Morpho butterfly

Harpy eagle

Emergents

Parrot

Toucan

Tree frog

Hummingbird

Capuchin monkey

Canopy

Spider monkey

Vine snake

Sloth

Understorey

Silky ant-eater

Emerald tree boa

Ocelot

Shrub layer

Giant armadillo

Tapir

Tarantula

Forest floor

rape. Autumn or spring sown, the 1 m. (3 ft. 3 in.) tall, erect, leafy plant produces masses of vivid yellow flowers. Up to 40 per cent of the seed is oil, and is used in the manufacture of margarine and cooking oils, though certain varieties of rape yield high-grade lubricating oils suitable for use in jet engines.

Raspberries are prickly, deciduous bushes of the genus *Rubus*, related to the blackberry and other fruit bushes within the rose family. They are cultivated commercially in parts of Europe, and western North America. The fruit is made up of many single-seeded, round fruitlets, resembling a small loganberry but characteristically separating from the conical white core when it is picked. These fruits are usually red, but yellow and black-fruited types also occur. Raspberries are perennial plants with upright, slightly prickly canes, and the fruit is usually borne on two-year-old canes.

Ratel, or honey badger: a badger-like carnivore, closely related to true *badgers, that occurs throughout most of Africa and southern Asia as far east as the Bay of Bengal. The single species, *Mellivora capensis*, is also known as the honey badger on account of its habit of breaking open the nests of wild bees. It is extremely aggressive and, although only about 12 kg. (26 lb.) in weight, it has been known to drive lions away from their kill.

Ratfish   *Chimaeras.

Rat kangaroos are small *marsupials belonging to the same family as the large kangaroos. They are called rat kangaroos mainly because of their small size, for they do not look much like rats except for one species from Queensland. Most of the ten species have kangaroo-like hind-limbs, which, although relatively short, are used for hopping.

Rat opossums, or shrew opossums, belong to the family Caenolestidae, a group of five species of small South American *marsupials confined to the Andes. They have long tails like rats but their pointed snouts give them a closer resemblance to shrews. Like shrews, they are insectivorous. The females lack the marsupial pouch.

Rats are a widespread group of *rodents belonging to the family Muridae, which also includes mice, but the name is also applied to several other subfamilies, including the Cricetomyinae, Otomyinae, and Rhizomyinae. Many other rodents are also called rats if their size and appearance justifies it, and the name is applied to unrelated rat-like animals such as rat opossums. The name is generally used, however, with reference to members of the genus *Rattus*, of which there are sixty-three species with a worldwide distribution.

Two species are of particular economic importance because of their destructive feeding habits and their role in the transmission of disease. These are the black rat, *Rattus rattus*, and the brown rat, *R. norvegicus*, but as the colour varies in either species, the alternative names of ship rat and common rat are preferred. Both are commensals of man. The ship rat requires the warmth of human buildings to survive but the common rat can live independently in the countryside. The former was distributed throughout the world by ships. It arrived in Britain during the eleventh century and was the species implicated in the Black Death of the Middle Ages. The common rat, a native of China, arrived in Europe around 1729, rapidly displacing the ship rat.

Diseases spread by rats include plague, typhus, leptospirosis, toxoplasmosis, and food poisoning.

Rat-tails (fishes), of which there are around 250 species, are deep-water marine fishes found in all oceans. They are blunt-headed and have an extremely long tail, a short, high dorsal fin, and a many-rayed but low anal fin. They make up their own family, the Macrouridae, and are distantly related to cod. They are near-bottom feeders, and possibly the most abundant fishes in the deep ocean. Many species make drumming sounds by means of muscles in the swimbladder.

Rattans are climbing *palms, of the genus *Calamus*, with about 200 species, particularly common in the forests of tropical Asia. They include some of the longest known organisms, some exceeding 100 m. (330 ft.). They climb upwards as a host tree grows, and collapse as it decays; then, the growing tip climbs up again as a new host grows into the canopy, and so on. Many are fiercely armed with thorns. Their major importance is in the manufacture of cane furniture, an important forest export of countries like Indonesia and Malaysia.

Rattlesnakes, venomous snakes of the *pit viper group, are confined to the Americas, where their distribution extends from southern Canada to northern Argentina and Uruguay. Their most remarkable feature is the rattle at the end of the tail. The rattle consists of horny interlocking segments and makes a buzzing sound when vibrated rapidly. New segments are added to the rattle each time the snake's skin is shed; barring damage, the older the specimen the longer the rattle. The purpose of the rattle is to warn other animals of the snake's presence and prevent if from being accidentally trampled on. Sound production in snakes has no part to play in communication between individuals, as snakes are deaf.

Large rattlesnakes feed mainly on small mammals; smaller forms tend to prefer lizards. All species bear their young alive. There are two main subgroups of rattlesnakes: the pygmy rattlesnakes of the genus *Sistrurus*, which grow to 60 cm. (2 ft.) and have small rattles, and the larger rattlesnakes of the genus *Crotalus*, which can reach 2·4 m. (8 ft.) in length.

Ravens are ten species of the genus *Corvus*, part of the crow family. The best-known member of the group is the common raven, *C. corax*, a glossy blue-black bird that may grow to 65 cm. (26 in.) in length. Its hoarse croak has given it the reputation of being a bird of ill omen. Ravens are found in virtually every county of the world, with the largest number of species in Australia.

Rays are a moderately large group of *cartilaginous fishes containing some 300 species. They are related to the sharks but are distinguished from them by having the five or six paired gill-openings on the underside of the head, and the eyes and spiracles on top of the head, and particularly by the enormous development of the pectoral fins, which join onto the head. These features make them well suited to life on the sea-bed; in particular, the dorsal spiracles mean they can pump water over the gills without raising themselves from the sea-bed. However, several

rays, notably the manta rays and eagle rays, have adopted a mid-water to surface life-style.

Rays have evolved in several ways. Most numerous are those of the family Rajidae (known as skates in North America), which lay eggs in rectangular capsules. The electric ray family, which includes *torpedoes, is heavy-bodied with large dorsal and tail fins, and powerful electric organs in the head and back region (skates also produce weak electric currents). There are several small families of sting-ray in which the fins are reduced, but all of which have a long, serrated, dagger-like spine on the tail, with venom organs attached. These include the South American river sting-rays, which are the only *cartilaginous fishes to live in fresh water. Sawfishes are a family of ray and are unmistakable with their long, serrated snout.

Razorbills are *auks of the north Atlantic and Arctic oceans, breeding from France to Greenland. Their deep bills, upright stance, and diving habit make them the closest northern equivalent to penguins. Their head and back are black and their underparts white. They nest colonially on cliffs, raising a single chick which leaves the nest before its wings are fully feathered. Razorbills feed mainly on small fishes and plankton.

Razor-shells are long rectangular *bivalves shaped like old-fashioned, cut-throat razors. They are specialized for burrowing in sand of coastal waters and can escape from predators at speed by a very efficient digging action. Normally only the feeding siphons, used to filter out detritus for food, are exposed at the surface of sandy beaches; if the animal senses vibrations in the sand it stops feeding and rapidly digs downwards until danger passes.

Rectum: the terminal part of the large intestine of animals, in the posterior part of the pelvic cavity. It functions as a store for faeces before they are voided through a ring of voluntary muscle known as the anal sphincter.

Red admiral butterflies, originally called 'red admirables', are velvety brown *brush-footed butterflies with red bands on the fore- and hind-wings; the fore-wing apex is white-spotted black. Their caterpillars usually feed on plants of the nettle family. The common red admiral,

Razorbills, *Alca torda*, spend much of their time sitting like penguins on coastal rocks, or flying low over the sea in groups. The black and white areas of their upper plumage fade slightly in winter.

Vanessa atalanta, occurs throughout Europe, Asia, North Africa, and North America. Other species are more restricted in their distribution, such as the blue admiral, *V. canace*, of Southeast Asia.

Red-backed voles *Bank vole.

Red bugs are *bugs of the family Phrrhocoridae, contrastingly marked with red or orange and black. Most are flattened, plant-feeding bugs, although some resemble ants, and at least one is carnivorous. They include the cotton stainers, *Dysdercus* species, widely distributed in warm countries, which pierce cotton bolls to feed on the seeds, thereby admitting a fungus which stains the fibres. They are serious pests in India and North America.

Red coral, found mainly in the Mediterranean Sea and off the coasts of Japan, is a *gorgonian type of coral greatly valued for making jewellery. The colonial animal, with a central stalk of bright red and fused chalky spicules, is easily spotted and eagerly sought by divers.

Red currant: a European native shrub, *Ribes rubrum*, which like its close relative, the black currant, belongs to the saxifrage family. Cultivated types may be either derived directly from *R. rubrum* or from crosses with other *Ribes* species. These crosses have produced a variety of fruit colours ranging from red to white (the white currant). The latter are often preferred as the fruit is less acid. In contrast to black currant, the fruit is produced on the older wood.

Red deer: a species of deer widely distributed in North America (where it is known as the wapiti), Europe, and Asia. In summer the coat is a rich red-brown colour, becoming a greyish-brown in winter. The stag may have a shoulder height of up to 1·4 m. (4 ft. 7 in.) with antlers up to 1 m. (3 ft. 3 in.). Only the stags bear antlers; these are cast each year and become progressively more complex, reaching full development when the stag is six years old.

For most of the year the sexes are segregated. In the rutting season, each stag collects a harem of hinds and guards them and his territory against all challengers. About eight months after mating, the hind leaves the herd and gives birth to a single calf. Some hinds bear each year, others only in alternate years. The young is left to lie alone for much of the first few days of life, the mother returning two or three times a day to suckle it. Once able to stand, the calf follows the hind. The calf is suckled for twelve months.

Redfish are members of the genus *Sebastes* and so belong to the scorpion-fish family. The name is used for the three North Atlantic species, especially the ocean perch, or redfish, *Sebastes marinus*. In the North Pacific, where there are more than sixty species of *Sebastes*, they are known as rockfishes, although similar in all other respects to the redfish. Most species are thickset, with large heads, and spiny dorsal and anal fins, and are red in colour. Different species live at different depths from the shoreline down to 1,000 m. (3,250 ft.).

Red grouse: the British and Irish subspecies of the willow grouse or willow ptarmigan, *Lagopus lagopus*, which is a member of the grouse family. In all other subspecies the wings are white and much of the body plumage turns white in winter, but in red grouse the wings and body plumage stay a dark reddish-brown all the year round. The red grouse lives on open moorland and eats mainly heather shoots, though the growing chicks take many insects.

Red panda: a species of the racoon family related to the giant *panda. It is found in an area from the Himalayas to south China. It has long, luxuriant fur of a deep, rich chestnut colour, except for the face, which is white with a narrow, dark stripe running from the eyes to the corners of the mouth. The tail is ringed with bands alternately light and dark. In many ways this animal resembles a racoon. At night it forages for bamboo shoots, grass roots, acorns, and fruit. It usually rests in trees during the day, with its long bushy tail curled round its body.

Red sea-bream are fishes distributed widely in European and North African coastal waters, although they are mainly summer migrants in British seas. Deep-bodied, with spiny dorsal and anal fins, they are coloured rosy-brown to grey on the back and pink on the sides, with a black patch behind the head. Young fish live close to rocky shores and wrecks but the adults, growing to 50 cm. (20 in.), live in deeper water. Sea-breams are members of the same family as porgies, and several other species of sea-bream occur in European seas.

Redshank are two Old World species of *sandpiper with scarlet legs. The white rump and hind border of the wing are conspicuous in flight. The typical call of the common redshank, *Tringa totanus*, is a musical 'tew-hew-hew', falling in pitch. When on the ground, they frequently bob their heads up and down in characteristic sandpiper fashion. The spotted redshank, *T. erythropus*, occurs in northern Europe and winters in Africa and China.

Red spider mites are one of the commonest orchard and greenhouse pests. They are herbivorous *mites of the family Tetranychidae, with needle-like mouthparts with

The **red panda**, *Ailurus fulgens*, is native to mountainous regions among bamboo or rhododendron forests. It grows to 1·1 m. (3 ft. 6 in.) in total length, and its tail can be three-quarters as long again as its body.

which they pierce plant cells and suck out juices. Their tiny resistant eggs are attached to plants by silk. There may be several generations per year depending upon temperature. Adults move between crops by 'parachuting' on silken threads.

Redstarts are two groups of small birds, one belonging to the New World *warbler family and the other to the Old World *chats. The first group includes eleven species

The common European **redstart**, *Phoenicurus phoenicurus*, flits between trees in search of insects or berries. Nests may be in trees, within walls, or in tunnels in undergrowth. Both parents feed the young.

of the genus *Myioborus* and the American redstart, *Setophaga ruticilla*. The *Myioborus* species are found mainly in Central and South America, though the painted redstart, *M. pictus*, occurs in North America. All are brilliant red and black, often with white wing and tail markings. The male of the American redstart is black with white underparts and orange-red patches in wings and tail. Several small relations of the Old World chats are also called redstarts, particularly those belonging to the genus *Phoenicurus*. Only slightly longer than warblers, these birds have orange-red tails and black, grey, or red bodies. Although widespread in Eurasia, most occur in the mountain ranges of Asia.

Redwing: a species of bird, *Turdus iliacus*, in the thrush family. It is widely distributed across the cooler parts of Europe and Asia. It is about 20 cm. (8 in.) long, brown above, with a whitish eye-stripe, and pale with dark streaks below. There are orange-red patches on its flanks and under the wings, which are clearly visible when it flies. It feeds largely on insects and worms in summer, but takes many berries and fruits in winter. The same name is sometimes used for the red-winged blackbird, *Agelaius phoeniceus*, one of the New World orioles.

Redwoods are tall North American conifers of the genera *Sequoia* and *Sequoiadendron*, each with one species. The Sierra redwood, *Sequoiadendron giganteum*, is also known as the mammoth tree, wellingtonia, giant sequoia, or big tree. It reaches some 96 m. (over 300 ft.) with a trunk diameter of up to 10·5 m. (35 ft.), and is native to the Sierra Nevada hills of California. The coast redwood, *Sequoia sempervirens*, is one of the tallest known trees, reaching 120 m. (400 ft.) in height. It is native to southern California and produces a valuable soft, fine-grained timber, much in demand for building construction and carpentry. This has led to the loss of many of these trees, with the consequence that they are now conserved in forest refuge areas. Both species of redwood are very long-lived; some Sierra redwoods are considered to be 3,000 years old. They are relic species of a once much more numerous group of conifers; many genera contain fossil 'species'.

The dawn redwood, *Metasequoia glyptostroboides*, is considered by some to be closely related to the true redwoods.

Reed, Walter (1851–1902), was the American bacteriologist and army surgeon after whom the hospital in Washington is named. He helped prove that typhoid fever among troops can be transmitted by flies, but more important was his work in Cuba in identifying the species of mosquito responsible for carrying yellow fever. Control of the mosquito resulted in eradication of the disease there. Application of similar measures in Panama made possible the construction of the Panama Canal.

Reedfish   *Bichirs.

Reeds (plants) are a group of moisture-loving grasses which belong mainly to the genus *Phragmites*. Three species of *Phragmites* are known, two of which are tropical, from Argentine and Asia, while the third, *P. communis*, the common kind, is distributed widely throughout the temperate regions of the world. All colonize ditches, streams, and pond and lake margins by means of perennial rhizomes from which arise leafy stems up to 4 m. (13 ft.) in height. In Europe, the common reed has long been used for

The giant sequoia is the rarer of the two **redwoods** which inhabit California, USA. This particular specimen, called the Grizzly Giant, is in a protected grove in Yosemite National Park in the Sierra Nevada Mountains. Some idea of its immense size can be gained by comparing it with the figure standing at its base.

thatching the roofs of houses, fencing, and furniture-making. The name reed is also used to describe species within the monocotyledonous families Cyperaceae, Typhaceae, and Sparganiaceae.

Reed warblers are a group of about twenty-five species of the genus *Acrocephalus* belonging to the Old World *warbler family. Varying in size from 13 to 20 cm. (5 to 8 in.), they are mostly dull olive-green birds, though a few are buff and streaked. They live in thick vegetation, such as reed-beds and scrub, and often make their presence known by characteristic, powerful songs. The marsh warbler, *A. palustris*, is a great mimic of other birds' songs.

Reflexes are the built-in involuntary responses of the *nervous system to a variety of external or internal stimuli. The simplest involves two *nerve cells: a sensory nerve cell with a receptor, and a motor nerve cell which is connected to a muscle or gland. The sensory nerve cell transmits a signal directly to the motor nerve cell, through a connection within the brain, or spinal cord. This is known as a reflex arc. Other reflex actions, controlled within the *autonomic nervous system, include diversion of blood flow between organs, and the regulation of blood pressure and the digestive system. Reflexes also help to control walking and limb movements.

Reindeer, or caribou: a species of deer found in the northern parts of Eurasia and in equivalent regions of North America and Greenland. It is the only deer in which both sexes have antlers; even fawns of two months have small spiky antlers. The adult is large, standing up to 1·5 m. (5 ft.) at the shoulder, and is usually brown to grey with a paler mane. It is the most migratory and sociable of the deer, travelling in herds of thousands of individuals over long distances. Many subspecies exist, especially in North America, where it is known as the caribou.

The reindeer is the only domesticated member of the deer family and is used in Scandinavia for riding, and pulling sledges. It also provides meat and milk and other dairy products. Its hide is used for clothing and shoes, the sinew for thread, and the hair for stuffing mattresses. It lives further north than any other hoofed animal, even entering the Arctic Circle.

Reindeer moss is the common name for *Cladonia rangiferina*. It is not a moss, but a lichen, which grows as clumps of branched, grey stems, up to about 8 cm. (3 in.) tall. It is abundant in the Arctic regions and is grazed by reindeer.

Remoras, of which there are about eight species, are members of the family Echeneidae, marine fishes found in all tropical oceans. They have an elongate, streamlined body with a flattened head, on the top of which is a large sucking disc (formed by the modified first dorsal fin, the rays of which cross the head like slats of a Venetian blind). This disc is used to attach the fish to large sharks, manta rays, bony fishes, turtles, and even whales, on which they ride. Some remora species are found only on certain hosts, others, like the sharksucker, *Echeneis naucrates*, attach to several kinds of host. Most remoras eat small fishes which they catch by swimming free of their host; a few eat the skin parasites of their host.

Reproduction in nature ensures that species will continue even though individuals eventually die. The aim of all organisms is to pass on their *genes to the next generations. There are two mechanisms of reproduction, asexual and sexual.

In asexual reproduction there is only one parent. One or several cells develop and detach from the parent to give a new individual which is a close copy of the parent. Such individuals may be called *clones. Single-celled organisms such as amoebae often simply divide into two similar individuals by splitting into equal halves. Asexual reproduction is largely confined to micro-organisms and to plants. This type of reproduction in plants is called vegetative reproduction and includes such things as runners, rhizomes, and root suckers which will grow into new plants if removed.

Sexual reproduction usually involves two parent organisms. They produce male or female reproductive cells (*gametes) in their sex organs by *meiosis. Each gamete contains half of the parent's genetic information. During fertilization, male and female gametes fuse to form a zygote. This develops into a new individual possessing some of the characteristics of each parent. Hermaphrodite animals like the earthworm have both male and female organs in their body but normally cannot fertilize themselves. Plants may be uni- or bi-sexual and some are able to fertilize themselves, but others have mechanisms to prevent self-fertilization. Many plants and invertebrate animals can reproduce both sexually and asexually. Sexual reproduction gives more variety in the offspring than asexual and makes adaptation to new or changing environments easier.

Reptiles are a group of vertebrates, generally with dry, scaly skins, whose living representatives include lizards, snakes, amphistaenians, crocodiles, turtles, and the tuatara. It is thought that, during the Mesozoic Era (245–66 million years ago), reptiles arose from the same source as dinosaurs, but are not direct descendants of them. Reptiles are extraordinarily diverse, and share with mammals and birds the ability to produce eggs, which contain amniotic membranes that enclose the developing embryo in a bag of fluid. Reptiles, birds, and mammals are collectively called amniotes, in reference to this shared characteristic. The amniote egg represents an important difference from the reproductive patterns which occur in most amphibians, in that hatchling reptiles are essentially miniature

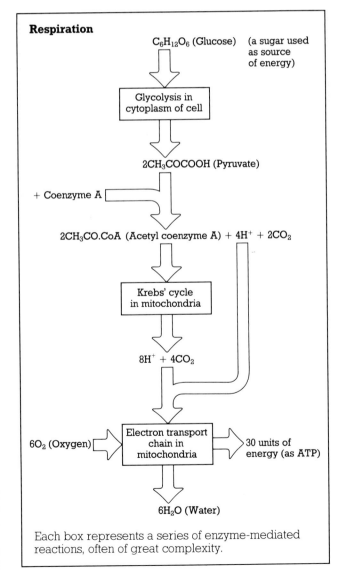

Respiration

$C_6H_{12}O_6$ (Glucose) (a sugar used as source of energy)

⬇

Glycolysis in cytoplasm of cell

⬇

$2CH_3COCOOH$ (Pyruvate)

+ Coenzyme A

⬇

$2CH_3CO.CoA$ (Acetyl coenzyme A) + $4H^+$ + $2CO_2$

⬇

Krebs' cycle in mitochondria

⬇

$8H^+$ + $4CO_2$

$6O_2$ (Oxygen) ➡ Electron transport chain in mitochondria ➡ 30 units of energy (as ATP)

⬇

$6H_2O$ (Water)

Each box represents a series of enzyme-mediated reactions, often of great complexity.

The common **rhea**, *Rhea americana*, has a mainly vegetarian diet, although it may also eat some insects and even small vertebrates. Rheas therefore prefer well-vegetated countryside to open grassland.

replicas of their parents. Thus reptiles have successfully eliminated the aquatic, larval (tadpole) stage. Reptiles, however, resemble amphibians in being cold-blooded, that is, their body temperature primarily depends on absorbing heat from the environment. Birds and mammals, in contrast, are warm-blooded.

There appears to be no one structural characteristic that is possessed uniquely by all reptiles. The name reptile is really a convenient label for a grade of biological organization intermediate between amphibians and birds and mammals. Although not as independent of their environment as birds and mammals, in so far as they are cold-blooded, reptiles have the advantage that their dietary requirements are comparatively low. They are thus well suited for survival in conditions of irregular food supply.

Respiration refers collectively to the physical and chemical processes in living organisms which result in energy liberation. Since every cell requires energy to carry out its living functions, all organisms respire. The energy released by respiration is derived from energy trapped in food molecules.

Some single-celled organisms use a form of respiration which achieves only partial breakdown of food molecules and yields relatively little energy. This is known as anaerobic respiration, since oxygen is not required. Many organisms and tissues, for example vertebrate muscle during exertion, revert to this alternative when oxygen is scarce. The presence of oxygen allows organisms to use the more efficient 'aerobic' respiration. In a sequence of cellular biochemical reactions, food is completely broken down into simple inorganic molecules like carbon dioxide and water, releasing far more energy than that gained through anaerobic respiration.

Efficient aerobic respiration depends upon a system of supplying oxygen and removing waste products such as carbon dioxide. A large respiratory surface, such as lungs or gills, is needed for this, and an efficient circulatory system to carry the oxygen to tissues and waste products away from them. The process of breathing and getting oxygen to cells is called external respiration, and the energy-releasing biochemical reactions within *mitochondria of the cell are called internal or cellular respiration.

Retinitis, or retinopathy, is a term applied to a range of conditions affecting the light-sensitive layers at the back of the eye. These often seriously impair vision. An important hereditary form is retinitis pigmentosa, in which irregular black spots can be observed on the back of the eye with an ophthalmoscope. Retinopathy is one of the most serious complications of diabetes, arising from damage to the blood vessels supplying the eye.

Rheas are two species of large, flightless birds confined to South America. Somewhat similar to ostriches, they are long-legged, long-necked birds with a loose covering of feathers on the body, standing up to 1·5 m. (5 ft.) tall and weighing up to 25 kg. (55 lb.). They escape from their enemies by running rapidly. The males are polygamous, up to six females laying their eggs in the male's nest. He then incubates them and looks after the young.

Rheumatism is a word without a precise medical meaning. Rheumatic fever is an abnormal reaction to infection with a specific bacterium, and is an important cause of heart disease in many parts of the world. Rheumatoid arthritis is a disorder of joints, having nothing to do with the preceding condition. Polymyalgia rheumatica is another distinct condition, while muscular rheumatism is a wide variety of aches and pains, which hardly ever arise in the muscles. A rheumatologist is a specialist particularly concerned with the varieties of arthritis.

Rhinoceroses, or rhinos, are huge, ungainly creatures, equalling hippopotamuses as the second largest land mammals. The largest of five species, the white or square-lipped rhinoceros, *Ceratotherium simum*, stands 2 m. (6 ft. 6 in.) at the shoulders, and its horn may reach a length of 1·5 m. (5 ft.). The head is massive and armed with one or two horns, these being formed from a consolidated mass of hair. The huge body is covered with scantily haired hide, the legs are thick, and the three-toed feet have a horny sole.

The black, or hook-lipped rhino, *Diceros bicornis*, once common in Africa south of the Sahara, is today found only in East Africa. It browses, using its prehensile upper lip to grasp and draw twigs and leaves into its mouth. It can reach a speed of 56 km./hour (35 m.p.h.) in spite of its bulk and relatively short legs. About eighteen months after mating, one calf is born, and a few hours later it is mobile enough to follow its mother. The lifespan is probably about twenty-five years, but one lived for forty-seven years in captivity. The white rhino is today found in central Africa. The Indian or greater one-horned rhino, *Rhinocerus unicornis*, is found in Asia, and its thick, almost hairless, hide is folded into large plates or shields. The Javan or lesser one-horned rhino, *R. sondaicus*, is similar to the Indian but smaller, being only 1·4 m. (4 ft. 6 in.) in height. The Sumatran or Asiatic two-horned rhino, *Dicerorhinus sumatrensis*, is similar in size to the Javan rhino. The newborn young is covered with thick, brown hair which it loses as it grows.

Rhizomes are swollen underground *stems bearing roots and some scale-leaves as well as the aerial parts of the plant. There is no clear distinction between rhizomes and *corms or stem *tubers, and some, like those of *Iris*, grow at ground-level. They act as storage organs during

An adult white **rhinoceros**, *Ceratotherium simum*, may weigh up to 2,100 kg. (2·3 US tons), though it gives birth to relatively small calves. This species, along with all rhinos, has been hunted by man in the Orient for the medicinal value of its horns.

adverse periods, such as winter in temperate regions or dry seasons in the tropics.

Rhododendrons are evergreen or deciduous shrubs and trees of the genus *Rhododendron*, which comprises some 500–600 species in the north temperate zone, the mountains of tropical Asia, and the Arctic. They have glossy, dark green leaves, and large flowers varying from white and yellow to red and pink. Like most of their relatives in the heather family, they are pollinated by bees and birds. Some are very stout trees with massive stems and big buds and leaves, while others are low creeping shrubs, and some are *epiphytes. Azaleas are species, mostly deciduous, of *Rhododendron* and are widely cultivated, often as hybrid forms, for their flowers.

Rhubarb, although often referred to as a fruit, is really a vegetable since it is the long, fleshy, leaf stalks that are eaten. It is a perennial plant with very large leaves arising from a substantial rootstock, producing a supply of leaf stalks over a long period of the year. Forced rhubarb is produced from rootstocks lifted in the winter and placed in the dark. The fresh leaves must not be eaten as they contain a high level of oxalic acid, a poison to humans. Rhubarb is related to sorrel and docks.

Rhynia: a fossil plant, dating from about 400 million years ago, which grew in marshes or bogs. It consisted of a creeping rhizome with erect stems that carried spore-bearing bodies at the tips, holding them out of the water to distribute the spores. An extinct type of *pteridophyte, *Rhynia* may be the ancestor of all the later land plants to evolve.

Ribbon-fishes belong to the family Trachipteridae and are found in all oceans. They are elongate and very compressed, mid-water inhabitants of the open ocean, living in depths of 300–700 m. (1,975–2,275 ft.). They are usually bright silver in colour with reddish fins, but their biology is little known. There are about seven species, of which the North Atlantic dealfish, *Trachipterus arcticus*, is one of the largest, growing to 2·5 m. (8 ft.).

Ribbon-worms, or proboscis worms, comprise some 650 species of worm-like animals of the phylum Nemertinea. They are often very long, and sometimes brightly coloured, and can be found in tangled knots under sea-shore rocks, or in shallow water of temperate seas throughout the world. A few species live in fresh water, or are terrestrial in the tropics. These animals have no *coelom, and are unsegmented, and move by slow undulations of the body, or glide like *flatworms. All have a long proboscis, sometimes barbed, which can be shot out to catch prey; some use it to grasp nearby objects and pull their bodies along after it.

Ribosomes are small spherical organelles within the cytoplasm of cells. They are the site of *protein synthesis and consist of a type of *RNA and a protein.

Ribs are flat curved bones jointed to *vertebrae behind, and directly or indirectly to the breast-bone in front. Collectively they form the rib-cage. The front ends of the ribs end in springy cartilages which join them to the sternum. This allows the chest to be expanded and compressed by muscle action. In the cavities of the ribs there is red,

Rice, seen growing here in northern Colombia, South America, is the most productive of the world's cereal crops. Plants reach a height of 1·8 m. (6 ft.) and are generally harvested by hand. Each drooping head of seed can be up to 50 cm. (1 ft. 6 in.) long.

blood-forming marrow. The ribs, within the structure of the rib-cage, are interconnected by muscles. Some pull the ribs together during breathing out, others pull the ribs apart during inhaling.

Rice is a tropical annual grass, *Oryza sativa*, which provides the staple cereal diet of half the world's population and is second only to wheat in terms of total output. An Asian species, it is unique among cereals as it is usually cultivated in standing water (paddy fields), the rice stems being adapted to allow oxygen to pass downwards to supply the waterlogged root system. It is an invaluable crop in the high-rainfall regions of the tropics and subtropics and its cultivation has spread from Asia to Central and South America. Good control of water-levels is needed as, like all cereals, rice needs a dry ripening period. Upland rice, or dry rice, is grown on the hills, where the paddy-field system is impossible but rainfall is sufficiently high. Rice is always established from seed, and the best yields are obtained by transplanting nursery-raised seedlings, a process carried out by hand in most parts of the world.

The grain is usually white, but there are red-, brown-, or black-skinned types. This skin, plus the layer beneath and the embryo, are rich in oil, minerals, vitamins, and protein. These are removed during the milling and polishing processes. In brown rice only the husk is removed, making it superior in food value. Polished white rice is deficient in vitamin B_1 and if used as the basis of a diet may cause beri-beri. To offset this deficiency, vitamin and mineral extracts are often added during preparation of the grains for sale. To reduce losses of nutrients the grain may be parboiled before milling. African rice, *O. glaberrima*, is grown in the flood plains of West Africa.

Right whales are a family of three species of *whalebone whales, distinguished from all other whales by the absence of a dorsal fin, and by grooves on the throat and chest. Right whales were so called by whalers as they were slow-moving, easily captured, and floated when dead. Included in this family is the common right whale, *Balaena glacialis*, the bowhead whale, *B. mysticetus*, and the pygmy right whale, *Caperea marginata*. This last is the smallest, being only about 6 m. (20 ft.) long; it is very rare and found only in Australian and New Zealand waters. Colour varies from black to slate-grey according to species. Right whales have very large heads and are rich in whalebone, with some 500 plates of baleen hanging from the roof of the mouth, the longest plate being about 2 m. (7 ft.). The body is rather chunky with a velvety appearance and in *B. glacialis* and *B. mysticetus* is up to 15 m. (50 ft.) long. When the right whale surfaces to expire, the spout is double and is directed forwards and upwards about 4·5 m. (15 ft.) in the air. The common right whale has several subspecies, each given common names according to their location.

Ring-dove *Woodpigeon.

Ringhals, or rinkals: a species of venomous snake, *Haemachatus haemachatus*, mainly restricted to eastern parts of southern Africa. It is nocturnal in its habits and eats mammals, birds, and toads. The rhingals is a species of spitting *cobra, which, when threatened, as a defensive reaction may spray venom for distances of up to 3 m. (9 ft. 9 in.) into the eyes of an intruder, causing at least temporary blindness.

Ringlet butterflies are *browns (Satyridae) of several genera, usually with prominent white-centred black spots near the edge of the otherwise brown wings, on both upper- and under-sides. They occur in many countries, and their caterpillars feed on grasses. Some species have conspicuous yellow-ringed eye-spots near the apex of the wings, but others, such as the genus *Erebia*, have only small black spots. The spots to which the name ringlet refers act as deflection marks during attacks by birds.

Ring-ouzel: a species of bird of the thrush family which breeds in northern parts of Europe and winters further south often on moorland, or in open forest. The male is greyish-black with a white crescent across its breast; the female is browner with a smaller crescent.

Ring-tailed cat, or miner's cat, is a species of mammal related to the racoons. It is found in Mexico and the southwestern USA as far as Alabama and Oregon, where it lives in dry, rocky habitats. It is racoon-like in appearance with a tail banded with grey and dark brown rings, and feeds on small mammals, insects, and fruit.

The cacomistle, *Bassaricus sumichrasti*, is a similar species which occurs in Central America.

Ringworm is an infection of the skin with one of a variety of *fungi. It causes a mild inflammation, with redness and scaling, which spreads outwards, leaving a more normal central area, and thus forms a ring. It was once a common infection of the scalp in children and difficult to treat, but modern drugs have largely removed the problem. When it occurs between the toes, ringworm infection is called athlete's foot, or tinea pedis.

River blindness, or onchocerciasis, is an infestation with a species of parasitic *roundworm, *Onchocerca volvulus*. It causes inflammation of the skin, since the body responds to the worms' presence by forming fibrous tumours around them. Migration of their larvae into the eye causes blindness. It derives its name from the fact that the worms are transmitted by mosquitoes which keep close to the rivers in which they breed. It is a serious cause of blindness in some parts of Africa. Other forms also occur in Central and South America.

River dolphins are five species of *toothed whales, found in fresh or brackish waters of rivers of southern Asia and tropical South America. There are river dolphins called Ganges, *Platanista gangetica*, Indus, *P. minor*, whitefin, *Lipotes vexillifer*, Amazon, *Inia geoffrensis*, and La Plata, *Pontoporia blainvillei*. All have a long, narrow beak and jaws studded with teeth, are blind or nearly so, and small, being 1·5–3 m. (5–10 ft.) long. The La Plata river dolphin travels and breeds in small schools in the estuaries of large rivers along the coast of South America. The whitefin dolphin is found only in the Yangtze and lower Fuchunjian rivers of China.

RNA (ribonucleic acid) is a *nucleic acid molecule consisting of a long chain of *nucleotide subunits. The subunits are joined in a sequence which is precisely determined by *DNA in the nucleus of the cell. Three different types of RNA are distinguished; messenger RNA, or mRNA, transfer RNA, or tRNA, and ribosomal RNA, or rRNA. Messenger RNA contains up to 3,000 nucleotides and each molecule carries the coded instructions for making one protein, copied or 'transcribed' from the DNA. It passes into the cytoplasm and becomes attached to ribosomes, cell components within which protein synthesis takes place. Transfer RNA molecules contain seventy-five to ninety nucleotides and occur in the cytoplasm of the cell. They are responsible for moving amino acids into the correct position as proteins are assembled. Each type of tRNA 'recognizes' a particular group of nitrogenous bases, called a codon, on the mRNA, to which it becomes attached. The other end of the tRNA carries a specific amino acid which thus becomes incorporated in the correct location in the growing protein chain: the process of reading the code is called translation. Ribosomal RNA contains 1,000 or more nucleotides and plays an as yet unexplained role inside the ribosomes.

Roach (fish): a species in the carp family, widely distributed in western and central Europe. Because of its popularity as an anglers' fish it has been introduced to many areas outside its natural range. Living in lowland rivers and lakes, it is relatively undemanding of water quality and adaptable to different foods. The roach thrives in large ponds, reservoirs, and even slightly polluted water. It eats insects, crustaceans, snails, and plants. It spawns in late spring, shedding its yellow eggs on plants where they hatch in nine to twelve days. The roach is distinguished by its moderately deep body, bluish or greeny-brown back, silvery sides, and orange to red pelvic and anal fins.

A **road-runner**, *Geococcyx californianus*, catching a lizard in northern Mexico. Road-runners kill their prey by first pounding it on the ground. They can run at a speed of over 37 km./hour (23 m.p.h.).

Roaches   *Cockroaches.

Road-runners, belonging to the cuckoo family, are two species of bird found in semi-desert areas of North and Central America. They are heavily streaked, brown birds with shaggy crests, long tails, and powerful legs; they usually run instead of flying. They feed on lizards and other small animals. Their nest of twigs is usually built low down in a bush or cactus and in it they lay three to five eggs.

Robber crabs   *Coconut-crabs.

Robber flies are a family of some 4,000 species of medium-sized to large flies, some robust and hairy, most with a long, narrow abdomen, which prey on other insects. The legs are long and strong and, on close examination, the head is grooved between the eyes, and the face bearded. Some catch their food in flight using the legs as a scoop; others lie in wait and pounce. The proboscis is horny and is used to pierce the prey and suck its juices. Eggs are laid in damp soil, rotting wood, or similar places, and larvae eat decaying plant material.

Robin is a name applied originally to the European robin redbreast, *Erithacus rubecula*, a small bird of the thrush family, widely distributed in Europe. It is brown with an orange-red breast. Tame and confiding, this bird is especially well known in Britain, where it is regarded with great affection and is commonly portrayed on Christmas cards. Pioneer settlers in many other parts of the world tended to call any local small bird with a red breast a robin and hence the name is used for other birds in most parts of the world, especially the American robin, *Turdus migratorius*, the African scrub robins of the genus *Cercotrichas*, and the Australian robins, especially those of the genus *Petroica*.

Robinia is a genus of some twenty species of deciduous trees or shrubs native to eastern North America and Mexico. Along with acacias and laburnum, they are part of the pea family. The locust tree, or false acacia, *Robinia pseudacacia*, is widely planted in the Old World for timber, but it readily naturalizes and has become a weed in many places. This and other species are cultivated for their feathery foliage and pretty pea-flowers.

Rockfishes *Redfish.

Rocklings are about twenty species of fishes belonging to the cod family, similar to the ling in body shape, but much smaller, growing up to 50 cm. (20 in.) in length. They have two, three, or four barbels round the mouth and another on the chin. Most species live close to the sea-bed or near the shore among rocks.

Rock lobster *Crawfish.

Rock roses are low-growing, slender-stemmed shrubs of spreading or mat-forming habit. The plant genera to which they belong, *Helianthemum* and *Cistus*, are both part of the rock rose family of around 165 species. Of these, about seventy species of *Helianthemum* and twenty of *Cistus* are known from Europe, Asia, and North America. The perennial garden varieties with an extensive range of colours have mainly been produced from the European *H. nummularium* and other closely related species.

Rocky Mountain spotted fever, or tick fever, is one of a group of infections resembling *typhus. It is caused by *Rickettsia rickettsii*, a micro-organism infecting rodents and other small mammals. In spite of its name, it occurs in South as well as North America and is transmitted to man by *ticks. The onset is sudden with headache, chills, and fever which persist for two to three weeks. A rash usually appears on the fourth day. In severe cases there is delirium, shock, and kidney failure. The fever responds to antibiotic treatment.

Rodents comprise a large order of mammals, with around 1,700 species. Few are bigger than a squirrel, although the largest, the capybara, is the size of a sheep. Rodents are primitive mammals showing few departures from the body plan of the first mammals. They are found in all parts of the world except Antarctica and are mainly terrestrial, although a few, such as the beaver, are aquatic. Many of them are of great economic importance either because they are pests or because they provide a desirable product such as fur. Some are used extensively as human food, particularly in the tropics.

Rodents are recognized by the single pair of front teeth in each jaw. The teeth are chisel-like and the cutting edge is maintained by the upper and lower pairs working against each other. The cheeks can be drawn in behind the front teeth so that material being gnawed falls out of the mouth and is not swallowed. (The name rodent comes from the Latin *rodere*: to gnaw.)

The classification of the rodents is complex and imperfect, but three main types are recognized, distinguished by the structure of the skull. The first are the squirrel-like rodents. Representatives, besides the squirrels, are marmots, prairie dogs, gophers, beavers, and the South African spring hare. Next are the mouse-like rodents, which include rats, mice, hamsters, jerboas, voles, and lemmings. Finally, there are the cavy-like rodents, including the guinea-pigs, viscachas, agoutis, and capybara. Some authorities put these South American rodents in a separate group, the Hysterimorpha.

Roe deer: a species of small *deer with a wide distribution in Europe, the Middle East, and northern Asia. It inhabits the edges of forests, open wood, or scrubland, which provides thick cover during the day. It is active mainly at night, when it browses on leaves, and eats berries, fungi, and grass. The smallest of the European deer, the buck is usually about 73 cm. (29 in.) at the shoulders, and its antlers are 22 cm. (9 in.) long. Only the male has antlers and these are shed at the beginning of winter. Shortly afterwards the new antlers form rapidly, their size and complexity increasing until the buck is about four years old. Twins are usual and are born in the spring. The fawn has a pale brown coat flecked with white, providing it with excellent camouflage. This is essential, as it will lie motionless while the mother is away feeding.

Rollers (birds) belong to a family of birds which is widespread in warmer parts of the Old World. The eleven species of rollers vary in size from that of a large thrush to that of a crow. They are brightly coloured, many with dark blue in the wings and pinkish brown or turquoise blue on the bodies. A few have elongated outer tail feathers. They are birds of open country, sitting conspicuously on branches and flying off to catch large insects, lizards, and other small animals with their powerful, brownish, slightly hooked beaks. They nest in holes in trees or banks and get their name from their tumbling display flight. (There is also a roller *canary.)

Rook: a species of bird in the *crow family. It is about 50 cm. (18–19 in.) long and has a black plumage with an iridescent sheen. The adult has a bare, whitish area on the face. Rooks breed throughout Europe and Asia and nest in colonies (called rookeries) in the tops of trees.

Roots are the organs through which a plant absorbs water and mineral nutrients. They also serve to anchor the plant in the soil. Generally they are underground, though some are aerial, and they lack chlorophyll. Where there is one major swollen root with small offshoots, it is known as a tap-root. Where there are several swollen ones, as in the dahlia, they are known as root tubers. Roots are usually covered at their growing points with fine hairs, called root

A lilac-breasted **roller**, *Coracias caudata*, perches on a thorn bush in the Okavango Basin of Botswana.

hairs. These are the site of water and mineral uptake; the rest of the root is covered with a tough 'skin' equivalent to the bark in woody plants.

Rorquals are a family of *whalebone whales, with a dorsal fin and longitudinal grooves below the throat and chest. There are six species, the four most important being the blue whale or Sibbald's rorqual, *Balaenoptera musculus*, the fin whale or common rorqual, the sei whale or Rudolph's rorqual, *B. borealis*, and the humpbacked whale. The majority feed on small crustaceans and krill, but some will also take small fishes. Females give birth once every two to three years to a single calf after a gestation period of ten to eleven months. Most young at birth are 3·9-4·5 m. (13-15 ft.) long, although the calf of the blue whale is larger. For six months the young feeds on the mother's milk alone; this is extremely rich and growth is rapid.

Rorschach, Hermann (1884-1922), Swiss psychiatrist, devised a test to measure personality traits. It assesses the subject's interpretation of ten ink-blots of varying shape and colour.

Rosebay willow-herb   *Willow-herb.

Rosechafers are white-flecked, metallic-green *chafers of the genus *Cetonia*, slightly smaller than cockchafers. Their larvae live in the soil feeding on roots and their adults feed in the hearts of roses. Some tropical species are very brightly coloured.

Rose-finches are grouped in the genus *Carpodacus* of the finch family. Although most of the twenty or so species occur in Eurasia, a few, such as the common house finch, *C. mexicanus*, are widespread in the New World. As their name suggests, the males are usually largely or partly pink in colour; the females tend to be greyish-brown above, paler and heavily streaked below. All rose-finches feed primarily on seeds for most of the year, but bring insects to their young.

Rosemary is a densely leaved, evergreen shrub belonging to the mint family. It is native to dry, Mediterranean scrub regions, where it grows up to 2 m. (6 ft. 6 in.) tall. It has been cultivated in Europe for several centuries as a culinary herb. Commercially it is valuable as a source of an oil used in perfumes, shampoos, and soaps. Several varieties, developed from the common rosemary, *Rosmarinus officinalis*, are also used as ornamental plants.

Rose of Jericho: a small, annual plant of the mustard family, known to botanists as *Anastatica hierochuntica*. If dried out by drought, the shrivelled plant can resume its form when wetted. The inward curling of the branches of the dried, mature plant allows it to be blown across the deserts of its native Syria, thus dispersing its seeds.

Roses are scrambling shrubs of the genus *Rosa*, consisting of some 250 species native to the north temperate zone and mountains of the tropics. Most have rather pithy stems with thorns. The flowers are followed by brightly coloured false fruits called hips, which are fleshy receptacles formed from the stalk beneath the flower. The true fruits are the 'seeds' within.

The rose family is one of the largest families of woody

or herbaceous species with over 3,370 species distributed worldwide. It includes many large and showy garden plants such as rowans, cotoneasters, and japonicas, as well as many fruit-bearing species such as apples, plums, cherries, and strawberries. Virtually all species have simple flowers which produce a lot of pollen to attract insect pollinators.

Cultivated roses have a complicated history involving a number of wild species, which include the original China rose, *R. chinensis*. They have been bred for increased flower size and continuous flowering as well as a dwarf habit. Flower colour varies from intense red and almost black to yellow and white; breeders are still trying to produce a blue rose. Roses have been cultivated on a large scale for their petals, which are used in scent manufacture, notably for attar (otto) of roses, derived from *R. damascena*. The term 'rose' is used for a number of other plants with superficially similar flowers.

Ross, Sir Ronald (1857-1932), British medical officer, discovered, while serving in India, the final link which brought malaria under control. His proof that it was transmitted by mosquitoes helped to show how yellow fever, sleeping sickness, typhus, plague, and other epidemic diseases are spread.

Rotifers are a phylum of around 1,500 species of tiny invertebrates and get their name from a ciliated crown which can appear to rotate. This 'crown' can be used to

Female freshwater **rotifers**, *Floscularia ringens*, live in a tube fixed to plants, and grow up to 1·6 mm. ($\frac{1}{20}$ in.) long. The four-lobed crown of cilia on top of the body filters particles from the water, some of which are used in the construction of the tube.

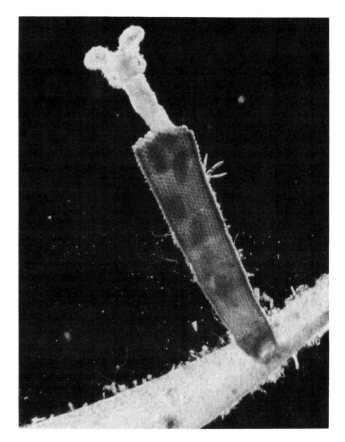

suck planktonic organisms into the mouth of sessile roti-fers, or as a locomotory organ in free-living species. They are among the tiniest of multicellular animals, often no bigger than protozoa, yet they contain numerous tissues, with complex sensory organs and eyes, and show elaborate behaviour and locomotory patterns. Some have very beautiful skeletal coverings.

Most rotifers feed on suspended particles, but some (despite their small size) are rapacious carnivores. All can reproduce sexually, with separate males and females. But by far the majority are female, since many generations arise by parthenogenesis (reproduction without fertilization).

Roundworms, or nematodes, are the largest phylum of worm-like animals, with over 10,000 species. Many species are so tiny that they pass unnoticed, although they may be extraordinarily abundant in soils or within other organisms. They include free-living species, usually less than 1 mm. ($\frac{1}{25}$ in.) in length, which occur in soil, sea water and fresh water, but by far the most common are the parasitic species. All have a cylindrical body, bounded by a single muscle layer and a tough *cuticle. They feed on other smaller animals or bacteria, or may simply swallow food from tissues of their host.

Many serious parasites or pests are roundworms: *eel-worms in plants, and *hookworms, *lungworms, Guinea-worms, and elephantiasis-producing filaria mammals. Virtually all groups of animals and plants have at least one roundworm parasite associated with them. Many, such as *pinworms, are relatively harmless and wide-spread. Others, like the large *Ascaris*, which grows up to 30 cm. (12 in.), live in the gut of vertebrates such as pigs, where they may weaken the host if present in large numbers.

Rove beetles are a very large family of *beetles with an estimated 27,000 species throughout the world. They are small to medium-sized, elongate, and with short wing-covers; the large hind-wings are elaborately folded be-neath them. The whole of the abdomen is thus exposed. Many species fly readily. Both adults and larvae are active carnivores and are an important part of the soil fauna. They include the familiar devil's coach-horse beetles.

Rowans are trees of the genus *Sorbus*, which includes whitebeams, *S. aria*, and the wild service tree, *S. domestica*. Most rowans have divided leaves with small, toothed lea-flets. Their flowers are produced as clusters and are fol-lowed by red, yellow, or white berries, depending upon species. The European rowan, *S. aucuparia*, also known as the mountain ash, is a tree some 18 m. (60 ft.) in height and is native to Europe and western Asia. Species of *Sorbus* hybridize readily and can produce a large number of intermediate varieties.

Royal fern   *Fern.

Royal water-lily, or Victoria water-lily: a freshwater aquatic plant of the family Nymphaceae. This is one of three tropical South American species in the genus *Victoria*, named in honour of Queen Victoria. It has very large tray-like, floating leaves up to 2 m. (6 ft. 6 in.) in diameter. The large white flowers, some 30 cm. (12 in.) across, remain open for only two days and the petals turn pinkish-red on the second day.

Rubber is contained as particles in the latex of several plants, but 90 per cent of the world's production comes from plantations of the Brazilian tree *Hevea brasiliensis* in Southeast Asia, particularly Malaya. This is a tall tree growing to 25 m. (81 ft.) in height, belonging to the *spurge family and related to the castor-oil plant. The bark of this native of the Brazilian Amazon basin is cut carefully to collect the latex running from the wound. The rubber particles, comprising about 30 per cent of the latex, are coagulated with acid and then separated. The raw rubber has to be further processed before it can be used to produce tyres and other products.

Natural rubber is also obtained from several other plants. Panama rubber comes from *Castilloa elastica*, a tall, Central American forest tree, related to the fig and mul-berry, and Ceara rubber from *Manihot glaziovii*, a small Brazilian tree related to cassava. *Parthenium argentatum*, a low-growing desert shrub of the same family as the daisy, from the southwestern USA and Mexico, gives better yields than other substitute plants and was cultivated dur-ing the Second World War to produce Guayule rubber.

The **rubber** tree has been planted in several parts of the world where the climate is suitably tropical. This tapped tree is in a plantation in the Ivory Coast, West Africa. The shallow, half-spiral cuts, made every two or three days, are started when the trees are between five and seven years old, and can be continued for about thirty years. Subsequent cuts are made below the existing one to provide more crops of latex.

Before the Brazilian rubber tree was introduced elsewhere, *Ficus elastica*, the rubber plant, was grown in Asia, producing India rubber.

Rudbeckia   *Coneflowers.

Rudd: a species of fish closely related to the roach but distinguished by its steeply angled mouth, bronzy-yellow sides, and deep, blood-red pelvic and anal fins. It is widely distributed in western and central Europe, but it less adaptable than the roach, and feeds mainly on surface-living insects and crustaceans. It lives in lowland lakes and back-waters of rivers, often among dense vegetation, and grows to 40 cm. (16 in.).

Ruff: a species of bird within the Old World sandpiper family. It is notable for the extraordinary, variably coloured ruffs and ear-tufts developed by the males in the breeding season. These can be erected and feature prominently in courtship display, which takes place in special arenas, where the males congregate. The smaller females are traditionally called 'reeves'.

Ruffin, Edmund (1794–1865), pioneer in soil chemistry, lived in Virginia, USA. By reviving worn-out tobacco lands with the application of marl, a form of powdered rock containing lime, he showed the value of a natural and readily available fertilizer. He founded *The Farmer's Register* (1833), a journal which encouraged a scientific approach to farming.

Ruminants are *herbivorous mammals which feed on plants and regurgitate a food bolus for further mastication, known as chewing the cud. The ability to ruminate not only provides for better utilization of the food but has the added advantage that the time needed for grazing, in a dangerous environment with little cover, is reduced to a minimum. The chewing of the meal can be carried out later in a safe place. The characteristic feature of these animals is the huge rumen, or storage pouch, which is part of the stomach. The rumen is the living quarters of symbiotic micro-organisms which break down the cell walls of the plants before the cud or food bolus is returned to the mouth to be chewed again. Among the ruminants

In spring, adult **ruffs**, *Philomachus pugnax*, compete for the attention of the reeves in arenas within breeding grounds known as 'hills'. Most contests are in display only, although some encounters are fiercely physical.

are camels, lamas, peccaries, chevrotains, deer, giraffes, okapis, pronghorns, sheep, goats, bison, and antelopes.

Rushes are monocotyledonous plants of the family Juncaceae. They occur worldwide, with the exception of the tropics, and form a family of some 400 species. Most are perennial, with rhizome-like rootstocks bearing grass-like or tubular leaves. They are plants of bog or marshy habitats and, though they have little value as ornamental plants, they were frequently used at one time as floor-covering and for basket- and mat-weaving. The name is also used, with suitable prefixes, for several other un-

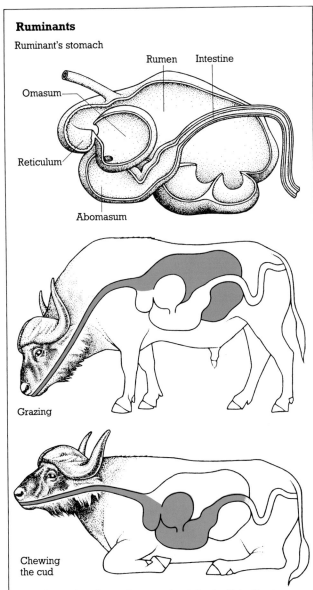

Ruminants

Ruminant's stomach

Rumen   Intestine
Omasum
Reticulum
Abomasum

Grazing

Chewing the cud

The four-chambered stomach of the buffalo (*top*) enables it to feed rapidly and digest even the coarsest grass. When the animal is feeding (*centre*), the grass passes into the rumen and reticulum, the storage stomachs, where it is partially broken down by protozoa and bacteria. Later the grass is regurgitated as the animal lies down to chew the cud (*bottom*). The grass is chewed once more and, when reswallowed, it passes through the omasum, where water is absorbed, and then to the abomasum and intestine, where digestion is completed.

related kinds of plant, such as flowering rush, *Butomus umbellatus*, which is the sole member of its family, and bulrush, *Typha* species, a genus in the family Typhaceae.

Rust fungi are parasitic fungi which attack plants, causing disease. Rust-red spores form in patches on the leaves and stems of diseased plants, and when blown to disease-free plants they will infect these too. Cereal rusts, such as wheat rust, *Puccinea graminis*, have serious economic consequences; they reduce crop yields, and the fungal infection can spread rapidly when a crop of a susceptible species is grown over a large area. Clouds of spores are then produced which travel hundreds of miles on the wind. Rust fungi are very host-specific: that is, each species of fungus infects only one or two species of plant.

Rutabaga   *Swede.

Rye is a grass which has been cultivated in central Europe since Roman times. Like oats, it is thought to have been selected from grass seeds that were carried unintentionally with wheat as its cultivation spread from western Asia. It was more successful than wheat in central Europe, being more tolerant of the cooler conditions and, in particular, able to survive the cold winters. It is valuable on relatively infertile, acidic, sandy soils, and is more resistant than wheat is to drought.

Rye grain is similar to, but longer than, wheat grain and its flour is used to make the very dark, strong-flavoured black bread of eastern Europe and the USSR. Rye flour is also mixed with wheat flour to make a lighter bread. A limited quantity is grown to produce rye-based crispbreads. The grain is also used to produce the characteristic whiskey of the USA, as well as gin (Netherlands) and beer (USSR). In addition, rye is valuable as a livestock food, grown particularly as an autumn-sown crop to produce early spring grazing.

It is a very close relative of wheat and a successful cross has been made between them to produce a hybrid known as triticale, which combines the higher yield and quality of wheat with the winter-hardiness of rye. A fungus, *ergot, can attack rye grains and, if eaten by man, can cause hallucinations, abortion, or gangrene.

Sable: a species of *marten that inhabits the coniferous forests of central Siberia. It hunts in the trees for squirrels, birds, and eggs but also eats berries and seeds. It is noted for its fine, dark brown to black fur, which has always fetched very high prices in the fur trade. Sable is also a name given to the sable antelope, *Hippotragus niger*, from southern Africa. It is so named because the bulls turn black with age, sable being the word for black in heraldry.

Sabre-toothed tigers are extinct members of the cat family that lived a few million years ago during the late Tertiary Era. Species such as *Smilodon* evolved a pair of huge upper canine teeth while reducing the lower canines, and the lower jaw could open extremely wide. The action of the teeth was a powerful, rapid stabbing movement as the head was brought down over the prey. Their main food was probably large, tough-skinned animals such as elephants and rhinos. Several species of sabre-tooth tigers are known from the Oligocene (36 million years ago) until the Pleistocene (1 million years ago).

Safflowers, annual or perennial plants or shrubs with spiny leaves and yellow flowers, are native to the Mediterranean region. They belong to the genus *Carthamus*, part of the sunflower family. *C. tinctorius* is known as the saffron thistle because of its spines, and its flowers which yield yellow or red dyes.

Saffron is obtained from the orange-red stigmas of an autumn-flowering Asian crocus, *Crocus sativus* (not to be

The **saffron** crocus, seen here, has three long, toothed stigmas drooping between two petals. Grown commercially in Old World countries, its cultivated forms are sterile and are increased vegetatively by means of cormlets produced around the old corms.

confused with the poisonous autumn crocus, *Colchicum autumnale*). It is very expensive to produce, as 150,000 flowers are needed to produce 1 kg. (2 lb. 3 oz.) of saffron, but a mere pinch gives an intriguing taste and a golden colour to rice or saffron cake. As well as being a flavouring, saffron has been used for thousands of years as a dye for cloth.

Sage is a low-growing shrub of the genus *Salvia*, belonging to the mint family, and with a woody stem. It is native to arid areas of southern Europe, though cultivated as a culinary herb in other, cooler regions. Its greyish, velvety leaves are very strongly and distinctly flavoured. Variegated forms of sage are also used as ornamental plants.

Sago is a starch product obtained from the trunks of two species of sago palm, and sometimes from other monocotyledons, including certain cycads. The trees are cut down just before flowering and the trunks split to remove the pulp. These palms occur in the lowland swamps of Papua New Guinea and parts of Indonesia, providing a staple food for the Papuans and a useful substitute for rice in other areas. A flour is made from the pounded and washed pulp, which does not keep well. Pearl sago, produced when a paste of the flour is sieved onto a hot plate, is the form marketed in Europe for use in puddings.

Sailfish: a species of fish of the Atlantic, Indian, and Pacific oceans, which migrates with seasonal warming of the sea. It is easily recognized by its long, pointed snout, which is rounded in cross-section, and its very high, sail-like dorsal fin. It is dark blue on the back and the dorsal fin is cobalt blue and ornamented with rounded dark spots. It grows to a length of 3.5 m. (11 ft. 6 in.), but despite this length is not especially heavy, attaining a maximum of 125 kg. (275 lb.), on account of its slender build. Sailfishes are popular with game fishermen because of the spectacular leaps they make out of the water. They live near the surface, occasionally coming close inshore when they follow schools of smaller fishes and squids.

Sainfoin is a perennial, leafy *legume with a deep tap-root, widely grown in North America and Europe. It is a valuable crop for hay, particularly for racehorses.

St. John's wort is a plant belonging to the genus *Hypericum*, which contains about 300 species of mainly perennial plants and shrubs widely distributed in Europe, Asia Minor, China, Japan, and North America. The shoots terminate in heads of yellow flowers with a central cluster of long stamens. The large-flowered kinds such as the spreading *H. calycinum* and the shrubby *H. patulum* are used as garden plants. *H. perforatum*, from Europe and Asia, has medicinal uses and is also used to make oil of St. John's wort. They form part of the family Guttiferaceae, which includes the mangosteen and mammy apple.

Saithe pollock   *Coalfish.

Saker falcons are large *falcons of eastern Europe and Asia, which winter in the eastern Mediterranean region, Arabia, and northeast Africa. Dark-plumaged, they are otherwise similar in most respects to the peregrine, being fierce and skilled hunters of birds on the wing. Unlike the peregrine they are adapted to life in dry, hot desert conditions.

Salamanders are animals belonging to the order Caudata, or tailed *amphibians. The name is popularly used in a broad sense for all tailed amphibians, including the true salamanders, newts, Asiatic land salamanders, sirens, giant salamanders, mole salamanders, lungless salamanders, the olm, and mud puppies. The true salamanders, belonging to the family Salamandridae, include the newts and species such as the fire salamander, *Salamandra salamandra*.

Typically, true salamanders are lizard-like, although the skin is smooth or rough, but never scaly. Most species have a tadpole, or larval, stage, though some species, such as the European fire salamander, give birth to fully metamorphosed (transformed) young after retaining the eggs inside the female's body until they are just about to hatch. Some species of 'false' salamander, like the olm and the blind salamanders from Texas and Georgia, USA, remain as permanent larvae which are capable of breeding in the larval state. Salamanders are principally a northern temperate group of amphibians, found in North America, Europe, North Africa, and Asia.

Saliva is a juice containing an enzyme (ptyalin), salts, and mucus. It is formed by the salivary glands and lubricates the lining membranes and contents of the mouth so that swallowing becomes possible. It dissolves substances such as sugar, which increases sensations of taste. Ptyalin breaks down starches into sugars.

Salmon, of which there are seven species, are migratory fishes which breed in fresh water and migrate to the sea to feed and mature, returning to spawn between one and four years later. The Atlantic salmon, *Salmo salar*, breeds in rivers in Europe and North America, and many travel to the seas around Greenland for their feeding migration. In the north Pacific region there are six species of salmon, all members of the genus *Oncorhynchus*, which are distinguished from the Atlantic species by the greater number of rays in the anal fin. Many of these, like the sockeye, *O. nerka*, and the king or chinook salmon, *O. tshawytscha*, are important commercial fishes for canning, but all salmon are valuable food-fishes. They feed on crustaceans and other fishes while in the sea and grow rapidly. All the Pacific salmon die after spawning in the gravelly shallows of rivers, as do many Atlantic salmon, but some of the latter survive to spawn two or three times.

Salp   *Sea squirts.

Salpiglossis is a small genus of plants from Chile, related to the potato but far more ornamental. *Salpiglossis sinuata*, a large-flowered annual species with a great deal of colour variation, is grown as a pot plant in greenhouses or as an outdoor border plant in temperate regions.

Salsify is a member of the sunflower family with large, attractive, purplish flower-heads resembling those of a dandelion. Sometimes grown as an ornamental, it flowers in its second year. A native of southern Europe, it is more commonly grown for its carrot-like, white roots which may be eaten boiled, baked, or in soups and are said to resemble oysters in taste. The young leaves of this 'oyster plant' are sometimes used in salads.

Salvias form a large genus of annual, biennial, or perennial plants or small shrubs with about 500 species. They are distributed throughout temperate and sub-tropical regions and belong to the mint family. Many species are used as decorative plants in the greenhouse and garden, while some such as *sage, *Salvia officinalis*, are herbs. Varieties of *S. splendens*, with compact spikes of scarlet flowers, are treated as half-hardy annuals in temperate regions and are firm favourites for bedding schemes; this species, however, is a perennial sub-shrub in its native Brazil.

Sand boas comprise ten species of non-venomous burrowing snakes of the *boa family, which occur in southern Europe, North Africa, and central and southwestern Asia. They are generally small, rarely more than 90 cm. (3 ft.) in length; their heads are rather pointed and they have short, blunt tails. They generally occur in rather dry habitats and feed mainly on lizards or small rodents, killing by constriction.

Sand-dollars are flat, disc-shaped *sea-urchins which have lost the regular spherical symmetry of their relatives. They are often biscuit-like, only the five-rayed marking on the surface revealing their *echinoderm affinities. Some have slits or notches in their internal skeleton, or test, and the spines are usually very short to allow efficient burrowing by the many specialized hydraulic tube-feet.

Sand eels, or sand lances, are small fishes of the family Ammodytidae, and twelve species are known. The family is best represented in the northern Atlantic, where there are eight species. They are long, slender fishes with a pointed head, protuberant lower jaw, long dorsal and anal fins, and a forked tail. The largest is the European greater sand eel, *Ammodytes*, growing to 32 cm. (13 in.). It burrows

The **sand-dollar**, *Leodia sexiesperforata*, from Barbados, shows the holes, or lunules, formed originally from indentations of the rim. The petaloid area at the centre of the outer skeleton, or test, marks the position of the tube-feet and shows its radial symmetry.

in clean sand but feeds on planktonic organisms. All sand eels are eaten by fishes and seabirds.

Sanderling: a species of small *sandpiper, which has chestnut and white plumage while breeding in the Arctic, and white and grey plumage for the winter migration to the southern continents. They are exceedingly active birds, following the tideline to feed upon shrimps and molluscs stranded by the receding waves.

Sand-flies are small blood-sucking flies which inflict painful bites. They are found in warm countries, and are notorious as vectors for various disease organisms, including those that cause kala-azar and oriental sores in South America, northern Africa, and southern Asia. They belong to the same family (Psychodidae) as the tiny, hairy, owl midges or moth-flies of northern temperate areas, which are harmless to man.

Sand-grouse live in arid or semi-arid areas of Africa, Europe, and western Asia. The sixteen species make up a separate family and look rather like ground-living pigeons. Their upper parts are usually well camouflaged in mottled greys and browns, but their undersides may have bolder markings such as yellow, reddish-brown, or black. They often fly in very large flocks, and regularly move long distances at dusk or dawn in order to reach water. They lay two or three eggs in a simple scrape in the ground. When they are rearing young, the males transport water back to the chicks by walking into the water and soaking their belly feathers. These act like sponges and enable the chicks to drink without leaving their desert homes. Sand-grouse feed mainly on fruits, berries, and seeds.

Sandhoppers, or beach-fleas, are terrestrial *amphipods that live among driftwood or other debris along the strandline of the sea-shore. They dig short burrows amongst the dead seaweed, and scavenge. They scull rapidly along the sand, and can jump using a rapid flick of the abdomen. One species, *Talorchestia*, only 2 cm. ($\frac{4}{5}$in.) long, can leap a distance of 1 m. (3 ft.). Sandhoppers have well-developed eyes, and they use the angle of the sun and polarization of its light to locate their normal zone on the beach.

Sand lances   *Sand eels.

Sand lizard: a species of *lizard that is found over much of Europe, extending to central Asia. It is a stout species, growing up to 19 cm. ($7\frac{1}{2}$ in.) in length, and a representative of the family Lacertidae. This family includes many familiar European species, such as the viviparous, or common, lizard, *Lacerta vivipara*, and the green lizard, *Lacerta viridis*. In England it is a protected species found on sandy heathland and dunes, but in other parts of its range it also occurs on embankments, and in hedgerows, and fields.

Sand martins are four species of martin in the genus *Riparia*, and are members of the swallow family. The most common sand martin, *R. riparia*, is known in the New World as the bank swallow. It breeds over wide areas of Europe, Asia, and North America, migrating southwards to the tropics in winter. It is a small bird, 12 cm. (5 in.) long, brown above and whitish below with a brown breast-band. Sand martins are insectivorous and get their name from building nests (often in colonies of considerable size) in the banks of rivers, sand quarries, and the like.

Sandpipers are *wading birds of the family Scolopacidae, allied to godwits, curlews and snipe, and comprising some seventy-five species. They are generally small, brown and white birds, 15–30 cm. (6–12 in.) long; some species are slender with long beaks and legs, like the redshank, others are dumpy and short-legged like the dunlin. The common sandpiper and some others prefer inland habitats, but the majority are birds of the estuary and sea-shore. Unlike their relatives in the plover family, which are largely tropical, the sandpipers are mainly northern in distribution. Some species are known as peeps in North America.

Sand wasps are solitary *wasps which dig burrows in sand. They belong to the family Sphecidae. Some species provision each burrow with a single large caterpillar before laying an egg and then sealing the burrow. Others go on providing food as the larva grows, showing early traces of social behaviour. Some adult sand wasps are quite large insects with a very thin 'stalk' joining their abdomen to the thorax.

Sap is the product of *photosynthesis in plants. It flows in the *phloem of seed-bearing plants and is rich in sugars such as glucose. It may be tapped and eaten fresh, for example as with maple syrup, or fermented into alcoholic drinks, like pulque from *Agave* species in Mexico or various toddies from palms throughout the tropics.

Sapajous *Capuchin monkey.

Sapodilla: a 20 m. (65 ft.) tall, evergreen tree, *Archras zapota*, native to Mexico and Central America but now widespread in the tropics. It is part of a family of about 800 species of trees which yield timber, latex, or edible fruits. The yellow-brown flesh of the ripe apple-sized fruit (sapodilla plum) is prized for its sugary flavour, but the tree is better known for the chicle gum obtained by collecting a milky latex from cuts in the trunks. It is the basis of natural chewing-gum and was used as such by the Aztecs.

Saprophytes are organisms that need organic compounds as food, and do not produce their own through photosynthesis. Saprophytic fungi and bacteria are very important in *food chains as they help to decompose dead plants, or animals and recycle nutrients. Only 'plants' or single-celled organisms are called saprophytes.

Sapsuckers *Woodpeckers.

Sardines *Pilchards.

Sarsaparilla is the name given to the dried underground parts of at least eight South American monocotyledonous plant species of *Smilax*, a genus of some 350 species of woody or herbaceous climbers with hooks or tendrils. This genus provides many extracts considered of importance in medicine.

Sassafras trees include three species of deciduous trees of the genus *Sassafras*, relatives of the avocado, bay laurels,

and cinnamon. The name is also used for an oil derived from roots of the American Sassafras, *Sassafras albidum*, a tree with simple leaves, native to North America, and reaching a height of some 30 m. (100 ft.) in the wild. It was formerly used in medicine or to kill lice and treat insect-stings. It is still important in scent-making.

Satsumas *Mandarins.

Savannah, or savanna, is the name given to level areas of grassland interspersed with isolated trees. It is a kind of open woodland, though originally the term was applied to large treeless areas in America, and has even been applied to prairies. It occupies a belt between tropical forests and grassland and is a particularly common vegetation type in southern and East Africa.

Savory is the name of two herbs of the genus *Satureija*, members of the mint family. Summer savory, *S. hortensis*, is a fragrant annual herb well known in Roman times, whose leaves and young shoots are used in combination with other culinary herbs. Winter savory, *S. montana*, is also of Mediterranean origin, with similar culinary uses, but it differs in being a small, woody, perennial, bushy plant. Having a stronger, more pungent flavour, it is rarely used on its own, but is included in mixed herbs for its hot spiciness.

Sawfishes are *cartilaginous fishes of the genus *Pristis*, within the *ray order. There are about six species and all have a very elongate snout, along the sides of which are large teeth, all more or less uniform in size. Although usually marine inshore fishes of tropical and subtropical oceans, they are often caught in estuaries and even in fresh water. The 'saw' is used mainly for stirring up the bottom mud in search of buried crustaceans, worms, and molluscs. The greater, or small-toothed, sawfish, *P. pectinata*, is found in all tropical oceans and can grow up to 7.7 m. (25 ft.) in length.

Sawflies belong to the order Hymenoptera, which includes ants, bees, and wasps, but they are distinguished from them by the absence of a narrow 'waist' between thorax and abdomen, and the consequent tubular or flattened appearance. Many of the 2,000 or so species have a saw-edged ovipositor through which eggs are inserted into plant tissues. Most are black, or black and yellow. Some are fast-flying, others feeble. The adults feed mainly on pollen and nectar and many species reproduce by parthenogenesis.

The larvae of most, such as the abundant gooseberry sawfly, *Nematus ribesii*, feed on leaves and resemble caterpillars, from which they differ in having more than five pairs of abdominal legs. The larvae of large, black and yellow wood wasps or horntails, *Uroceras gigas*, which have conspicuous ovipositors, burrow in wood; other species bore in stems or fruit or are leaf-miners; and a few are parasitic on wood-boring beetles. Some species are gall-forming, such as those that cause bean-shaped red galls on willow leaves (*Pontia* species); these swell in reaction to egg-laying, and the larva eats the tissue inside the gall.

Saxifrages are a large group of mainly alpine plants from the mountains of the Northern Hemisphere, but extending into the Andes of South America. They belong to the family Saxifragaceae which includes currants and

The purple **saxifrage**, Saxifraga oppositifolia, growing on stony, lichen-encrusted ground in Greenland. Flowering between May and August, this species inhabits arctic and subarctic areas, and mountains in much of Europe. It grows on rocks and screes, and in crevices, forming cushions or mats sheltered from cold winds.

hydrangeas. This large and widespread family contains some 1,250 species. They mostly form rosettes of leaves, which may be hairy or succulent. The flowers vary considerably in colour from white through yellow, pink, and red; some are solitary, others appear in long, arching panicles. They form one of the most popular groups of plants for rock garden cultivation and many varieties and hybrids have been bred.

Scabies is an infestation of the skin by an itch *mite, Sarcoptes scabiei. The female mite burrows into the skin to lay her eggs, and her feeding causes small itchy spots. It is spread by close bodily contact, usually in bed, when the mite is active. It is easily cured by skin creams, or lotions, which kill the mites.

Scabious are annual or perennial plants of the genus Scabiosa, which forms part of the family Dipsacaceae. The name is derived from the Latin scabies, itch, alluding to the reputation of the plants for curing this condition. Over fifty species are known within the genus but others commonly called scabious, belong to other genera in the same family. The Dipsacaceae, which also includes teasel, has some 350 species in Europe, Asia, and Africa. Some species of Scabious, such as the annual S. atropurpurea and the perennial S. caucasica, are popular as garden plants.

Scale-insects are a superfamily (Coccoidea) of plant-feeding *bugs with 'normal' insect-like winged males and wingless, squat, scale-like females. The minute, first stage larvae are active in both sexes, but at the first moult those destined to be females become inactive, or sessile, often losing all appendages and secreting a coating of wax and a scale which may entirely cover them—hence their name. Under this scale the female remains until fully mature and ready to produce eggs. The females are thus fixed to their food plants for life, and look like an excrescence of it. The male-producing larvae usually have a resting stage before the emergence of the adult males. They have one pair of delicate wings, legs, and long antennae.

Scale-insects are among the most damaging of all insect pests, although a few species, such as the lac insect, Laccifer lacca, of India, which secretes a resinous substance (shellac) used as varnish, produce substances that are valuable to man.

Scales are bony, or horny plates which cover the outer surfaces of reptiles and fishes. The wings of some insects, such as butterflies or moths, are covered in 'scales' which are really modified, flattened hairs.

There are four major types of scale in living fishes. Placoid scales (or dermal denticles) have a disc-like basal plate of dentine and an enamel cap, similar to that on human teeth; they occur on sharks, skates, and rays. Cosmoid scales have similar dentine on their surface, but it does not form a hollow, tooth-like structure, and the scale has two layers of different structure beneath. They are found only on the *coelacanth. Ganoid scales occur only on some primitive fishes, such as sturgeon; they are shiny and hard-surfaced, covered with an enamel-like substance called ganoine, beneath which are two other layers. All bony fishes with scales have so-called bony-ridge scales which are thin and translucent, and their surface is marked with ridges alternating with shallow depressions. They grow on both surfaces and the ridges represent new growth. By counting the groups of ridges it is possible to ascertain the age of the fish.

In reptiles, scales can be of two types: horny plates derived from the skin, or bony plates derived from bone-tissue. The majority of lizards and snakes have horny scales covering their skin. Crocodiles and a few other groups retain the primitive bony armour-like scales, which are still covered by a horny 'skin'.

Scale worms are polychaete *annelids belonging to the families Polynoidae and Sigalionidae. They have plate-like extensions of their upper surface which overlap like tiles on a roof. Most species occur in shallow waters around sea coasts and many are *commensalistic with other animals.

Scallops are *bivalves which are highly prized as food, and live on sandy sea-beds throughout most of the world's seas. The two shell valves have wavy or 'scalloped' edges, covering many tiny eyes and feelers on the edge of the scallop's mouth (body folds). When these detect predators

A **scallop** of the genus *Chlamys* from Puget Sound, Washington, USA. Long sensory tentacles and eyes line the shell edges, while the small, dark guard tentacles within control the water intake of the body cavity.

The black-striped, green coloration of this juniper **scarab**, *Plusiotis gloriosa*, provides good camouflage among juniper foliage in Arizona, USA. It is a leaf-eater and its white grubs feed on roots.

like starfish the valves are clapped together to jet water out, and the scallop 'jumps' away to safety. Like all bivalves, scallops are filter-feeders.

Scaly-tailed squirrels are a family (Anomaluridae) of gliding *rodents unrelated to the flying squirrels or other true squirrels. They live in the forests of central and West Africa and all but one of the seven species has a web of skin between the limbs. The scaly underside of the tail helps them to obtain a grip on tree trunks.

Scampi   *Dublin Bay prawns.

Scarabs are members of a large family (Scarabaeidae) of heavy and often colourful beetles which includes *chafers, *Hercules beetles, and the famous sacred scarab, *Scarabeus sacer*, worshipped by the ancient Egyptians. Adult sacred scarabs make and roll balls of dung, sometimes coated with clay, for food. Their eggs are laid in buried cones of dung. There are some 2,000 species of scarab beetle throughout the world.

Scarlet fever is an infectious disease with a characteristic rash, caused by the bacteria of the genus *Streptococcus*. Previously a severe and dangerous illness, it became less virulent during the first thirty years of the twentieth century, and is no longer a problem. This change occurred before most antibacterial drugs were available, and has never been explained.

Scaup are three species of *pochards. The New Zealand scaup, *Aythya novaeseelandiae*, is purplish black while the other two species look like tufted ducks, except that the drakes have grey backs and no tuft. The greater scaup, *A. marila*, is marine, with a circumpolar breeding range, whereas the lesser scaup, *A. affinis*, is a North American bird which prefers inland habitats.

Schizanthus, or butterfly flower, is a genus of colourful and showy annual plants. The fifteen species of *Schizanthus* are mainly native to parts of South America, and are related to the potato and tobacco plant. *S. pinnatus* and its varieties are popular garden plants.

Schizophrenia is a severe mental illness, one of the major *psychoses. A patient with an acute attack may lose contact with reality, express peculiar ideas, and imagine that he hears voices. In addition sufferers may believe that their thinking is interfered with and that their body and mind are controlled by external forces. Some patients recover completely. The causes of schizophrenia are not yet understood but may include genetic factors and stress. It does not mean split personality.

Sciatica *Discs.

Scilla is a genus of bulbous plants of the lily family. Over 100 species have been recorded, chiefly from Europe, Asia, and the Mediterranean region, but extending into southern Africa. They are mainly spring-flowering, with white, blue, or greenish flowers. *Scilla siberica* and *S. bifolia* can be naturalized in the garden to produce carpets of blue flowers. *S. non-scripta* is the English bluebell.

Scorpion fishes are distributed mainly in tropical and warm temperate seas. They are bottom-dwelling fishes, often found in rocky areas, and are well camouflaged to match their surroundings by colouring as well as by the presence of numerous flaps of skin on the head and body. Many species have venom-secreting glands at the bases of the sharp fin spines and can inflict very painful wounds. In general they are stout-bodied with a large head which has a strong bony ridge on the cheek; the head is often spiny. They eat a wide range of fishes and crustaceans, and rarely grow larger than 50 cm. (20 in.).

Scorpion-flies belong to the order Mecoptera, and derive their name from the upturned, bulbous end to the abdomen which characterizes the males of some of the 400 or so species. All have a head which extends downwards into a beak bearing the biting jaws. It is a small but ancient group, differing little from fossils some 250 million years old. Some feed on dead or dying insects and occasionally fruit, and the minute, wingless, snow flea, *Boreus hyemalis*, eats the mosses among which it lives. Another species catches insects with its hind legs while hanging by the front pair. The larvae are scavengers in soil or leaf litter.

Scorpions are the most primitive *arachnids, and were perhaps the first arthropods to adapt to life on land. About 800 species are known, being commoner in tropical areas. Most are secretive and nocturnal. Most species are 3–10 cm. (1–4 in.) long, but fossil species grew to almost 90 cm. (35 in.).

The oval, front part of the body of a scorpion bears eight legs, pincer-like *pedipalps, and jaws, while the mobile abdomen bears the barbed sting. The venom of most species is merely irritant to man, although several species, notably *Androctonus australis* of the Sahara region, have venom powerful enough to kill a man. The poison is mostly used on insects sought as food. To avoid mistakes during mating, scorpions use elaborate courtship dances to avoid each other's sting. After hatching, youngsters ride on the mother's back for safety.

Scorzonera is a perennial plant of the sunflower family with yellow, dandelion-like flowers, native to central and southern Europe. The black-skinned, fleshy, tap-roots of *Scorzonera hispanica* may be boiled and eaten as a vegetable, like those of its close relative, salsify. Like chicory, it is the source of a coffee substitute and the fresh young leaves may be used in salads. *S. humilis*, a species not used as a vegetable, is commonly known as viper's grass.

Scoters are three Holarctic species of bulky, blackish, sea ducks allied to the eiders. Like them, they are birds of the far north. Most of the common scoters, *Melanitta nigra*, and all the velvet scoters, *M. fusca*, which are seen in Britain, are winter visitors.

Scots pine: a conifer native to Europe, reaching 68° N. and as far south as Spain. Natural Scots pine forests still persist in parts of Scotland, though most of the trees seen in the British Isles have been introduced by man from European continental stocks. Sometimes it is referred to as Scots fir, though it is a conifer of the pine genus. Its timber is valued for carpentry and general building construction.

Screamers belong to a family of three species of birds confined to South America. They are large and slightly turkey-like, mainly black or grey above and light grey or white below. They carry a sharp spur on the leading edge of the wing. Although they do not resemble ducks, they are thought to be related to them. Screamers live in marshes and have rather long toes to support their weight on the soft ground. They eat vegetable matter and build a shallow nest of reeds in which they lay three to six white eggs. The young leave the nest soon after hatching. Screamers have a powerful, far-carrying call which is the origin of their name.

Screw-worm flies *Blowflies.

Scrub-wrens *Warblers.

Scud *Horse mackerel.

Sculpins *Bullheads.

Sea anemones are related to jellyfish, hydras, and corals, and as members of the phylum Coelenterata all have radial bodies bearing tentacles. They are solitary *polyps, usually less than 5 cm. (2 in.) long, though some reef-dwelling forms reach 1 m. (3 ft.) across. They can be spectacularly coloured, especially in tropical waters.

Anemones catch small invertebrates, or fishes, after paralysing them with stinging cells (nematocysts) on their tentacles. The prey is then pushed into the large digestive cavity inside the sea anemone. While most animals succumb to the anemone poisons (some even being painful to man), a few creatures, such as the fish *Amphiprion*, have special defences which enable them to live within the mass

The closed mouth of the stalked **sea anemone**, *Sagartia elegans*, is surrounded by rings of stinging tentacles to catch prey. This species is widely distributed around the British and Irish coasts.

of tentacles, cleaning the anemone in return for protection and bits of food.

Sea bass *Bass.

Sea-bear *Sea-lion.

Seabirds are, generally speaking, birds that swim and take a substantial part of their food from the sea, though the term excludes many kinds of birds that spend part of their lives close to the sea. Species groups which are referred to as seabirds include the penguins, petrels, auks, tropic-birds, gannets, frigate birds, terns, gulls, cormorants, and many species of pelican that feed mainly at sea. In addition several kinds of diving ducks are largely marine. Other phalaropes, some divers and grebes, herons, and birds of prey spend part of their year at sea.

Except for Emperor penguins, which nest on sea-ice, all seabirds breed ashore. Species feeding close to the shore, such as gulls, skuas, and cormorants, often nest singly or in small colonies; those feeding farther out on the open sea, such as penguins, auks, or petrels, may be numerous enough to form colonies of many hundreds of thousands. Most true seabirds lay one or two eggs; those with two often rear only one chick. Incubation and the care of chicks take longer than for land-birds of similar size, for growth appears to be geared to irregular and often inadequate food-supplies. Many seabirds do not breed until several years after reaching maturity, and may live into their second or third decades.

Most seabirds feed communally, following each other's movements, plunging, diving, dabbling, or chasing active prey under water, and sharing finds of food.

Sea-bream belong to the family Sparidae, which also includes the North American porgies and Australian snappers. They are marine fishes, especially abundant in trop-

ical and warm temperate seas but also occurring in temperate regions. They are deep-bodied, fully scaled on head and body, with spiny dorsal and anal fins. Their teeth are specialized according to their diet and can be either strong and sharply pointed, blunt and rounded, or flat. The *red sea-bream is a typical example of this group.

Sea-cows, or Sirenia, are an order of mammals whose members are largely extinct. There are four living species, three of *manatee of the Atlantic and the *dugong of the Pacific and Indian oceans. A fifth, the enormous Steller's sea-cow, was discovered in 1741 in the Bering Sea, but became extinct twenty-seven years later. The population was isolated and virtually imprisoned on Bering Island, as any attempt to leave would have led to their death from large marine predators.

Sea cucumbers are worm-like leathery creatures, which are actually a class of *echinoderms. Their internal skeleton is reduced to small, calcareous ossicles or spicules, and the tube-feet are concentrated around the mouth for feeding, or in rows below for walking. Most sea cucumbers live within crevices or burrows. When they are disturbed they eject entangling gummy strands (hence their other name of 'cotton-spinners'), or even their own guts. In some countries these animals are dried as 'trepang', valued as a food. Most common species are between 10 and 30 cm. (4 and 12 in.) in length, but some Pacific Ocean species can reach 1 m. (3 ft.) in length and 60 cm. (2 ft.) in diameter.

Sea ducks are *bay ducks which spend much of their life at sea or on coastal waters. They find their food, which

The fine fishing tentacles of the **sea gooseberry**, *Pleurobrachia* species, from the Great Barrier Reef, Australia, form a wide net to catch planktonic animals, such as small crustaceans.

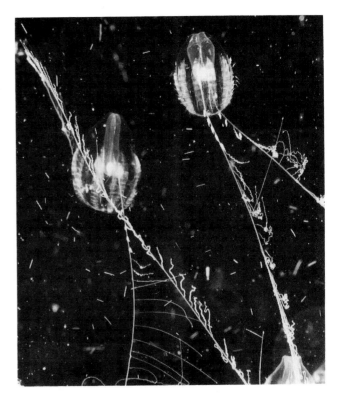

is principally molluscs, on the sea-bottom. Typical sea ducks include the eiders, scoters, and the long-tailed duck.

Sea eagles are the eight species of *Haliaeetus* which occur thinly across northern Europe, Asia, and North America. Among the largest of the eagles, they are typically found on islands or mainland close to the coast, where fishes are readily obtainable. Arctic populations migrate south in winter when the sea freezes.

Sea-fans are *gorgonian corals, with a regular flattened branching pattern which results in a fan-like structure, often 50 cm. (20 in.) across. Each *polyp within the colony catches small suspended animals from the sea water, and usually the fans span the prevailing current. They occur mostly in warmer seas, and are familiar in marine aquaria. Wherever they are found, they tend to harbour a variety of encrusting organisms, which obscure their beautiful latticed outlines.

Sea gooseberries are strange, floating marine animals, also known as comb jellies, forming the phylum Ctenophora, with some 90 worldwide species. They get these odd names from their ovoid jelly-like shape, and the regular rows of cilia which provide propulsion. They also have a pair of trailing tentacles, which bear adhesive cells to trap plankton. This simple structure suggests links with the sea anemones and jellyfish. Sea gooseberries are often luminescent, and sometimes become stranded in shimmering masses on coastlines.

Sea holly is the common name for two totally unrelated plants. The sea holly of Europe and Mediterranean coasts is *Eryngium maritimum*, a member of the carrot family. This is a glaucous, spiny-leaved, perennial with attractive grey-blue stems and leaves, and blue flowers. The leaf colour is the result of a waxy secretion which enables the plant to withstand wind and salt spray. The name is also used for the species of *Eryngium* cultivated as garden plants.

The Asian sea holly, *Acanthus ebracteatus*, belongs to the family Acanthaceae. Its leaves are used to make a cough medicine popular in Malaysia.

Sea-horses are fishes which are closely related to the pipefishes, but their tails are curled forwards and their heads are inclined forwards. They swim upright, propelled by an undulating dorsal fin, and can anchor themselves by means of the tail. The male carries the eggs in an abdominal pouch. The Australian leafy sea-horse, *Phyllopteryx foliatus*, grows up to 30 cm. (12 in.) long and is unusual in having fleshy growths on its body which mimic seaweed.

Seakale, although not a variety of kale, is a member of the same mustard family. It is a native of the sea-cliffs and shores of western Europe. The leaf stalks, if blanched, may be boiled and, although somewhat bitter, have a nutty flavour. Seakale is rarely cultivated nowadays.

Sea-lilies, or crinoids, are stalked relatives of the *feather stars. They are now largely extinct with only about eighty species remaining, all in deep waters. The sea lilies comprise part of a class of *echinoderms which flourished in Palaeozoic seas, their feathery arms filtering out abundant planktonic food. Some of the fossil sea lilies had stalks 25 m. (80 ft.) long. Such forms gave rise to a very diverse

One of the most decorative **sea-horses** is the leafy seadragon, *Phycodurus eques*. Like its fellow Australian, the leafy sea-horse, it is festooned with loose flaps of skin, which presumably act as camouflage when it hides among the seaweed fronds.

echinoderm fauna, but only sea-urchins, starfish, sea cucumbers, and brittle stars now remain.

Sea-lions, with seals and walruses, are marine mammals which belong to the order Pinnipedia, or fin-feet, living both on land and in the water. There are fourteen species of sea-lions, or eared seals, found in the North and South Pacific, the South Atlantic, and other southern waters. The Californian sea-lion, *Zalophus californianus*, and the northern fur seal, or sea-bear, *Callorhinus ursinus*, are found in the Northern Hemisphere. The remaining species live in the Southern Hemisphere. All sea-lions have a streamlined body with paddle-like limbs, the digits of which are joined by a web. The adults are entirely at home in the sea but the young have to learn how to swim. The largest male may reach a length of 3 m. (10 ft.); the females are smaller. They range in colour from the dark chocolate-brown of the Californian to the greyish-yellow of the Australian sea-lion, *Neophoca cinerea*. Considering their size, sea-lions are quite mobile on land, as they can rotate their hind-limbs forward to support their body and can even climb cliffs. Herd-living, they are even more sociable than the seals. In the breeding season harems are formed and ter-

A **sea otter** cracks a clam with a stone, while floating on its back in the water. This appealing animal is protected along much of the Californian coast but conflicts have arisen due to its predation on clams, and abalones which are also harvested as seafood.

The brilliant electric colours of the **sea slug**, *Chromodoris quadricolor*, outshine surrounding corals and sea-squirts on Australia's Great Barrier Reef.

ritories defended. The pups are suckled for at least three months after birth.

Sea lizards are large marine *iguanas which bask, sometimes in large aggregations, among lava rocks on the shores of the Galapagos Islands. They are essentially herbivorous, feeding on seaweeds for which they often have to dive. Submergence time may vary from about three minutes in shallow water to about one hour in deeper water. They are large, stocky lizards growing to 1·2 m. (4 ft.) in length, and are unmistakable on account of a crest running from the neck to the tail. They lay 2 or 3 eggs in tunnels dug in sandy beaches.

Sea mat   *Bryozoans.

Sea mice are beautifully iridescent marine animals of the genus *Aphrodite*, which are actually polychaete *annelids, and certainly not 'mice'. Seen from below, the segmentation and the stout bristly paddles on which they walk reveal their true worm identity. But from above they are covered by scales and long, glistening hairs, and when moving along half-buried in sand they do resemble rather sluggish mice, with very colourful fringes. They are related to scale worms.

Sea moths, of which there are at most six species, are small fishes found only in the tropical Indian and west Pacific oceans. They belong to the family Pegasidae, and grow up to about 14 cm. (6 in.) long. The body is oddly shaped, broad and flat at the front, then tapering toward the tail. The body appears to be encased in rings of bony plates. The snout is almost duck-billed, while the pectoral

fins are large and like moths' wings. Usually brown in colour, they live mainly on sandy bottoms.

Sea otter: a species of *otter that lives along the shores of the west coast of America, swimming and floating among beds of seaweed. A single pup is born in a nest of floating kelp, and lives entirely at sea. It is able to swim at 16 km./hour (10 m.p.h.) and may dive to a depth of 30 m. (100 ft.) or more in search of food. It feeds on sea urchins, crustaceans, cuttlefish, mussels, clams, abalone, and other shellfish. One of the few tool-using mammals, it will rest a stone on its chest and crush clams and sea urchins against it.

Sea perch is a name used for several species of fishes in the family Embiotocidae, or surf perches, with about twenty-three species native to the North Pacific, and one species found in fresh water in California. Most are around 25 cm. (10 in.) in length and are deep-bodied fishes with large scales, and a spiny dorsal fin. They live in shallow coastal water, often among the kelp beds. The male has a fleshy modified anal fin, which acts as an intromittent organ (for transferring sperm), for these fishes give birth to live young.

Sea potato   *Sea-urchins.

Sea robins   *Gurnards.

Sea scorpions, or eurypterids, are extinct, aquatic, freshwater arthropods, thought to be true *arachnids. They were common in fresh water during the Ordovician to Permian eras (505–245 million years ago), and were rapacious carnivores. Some species achieved a very large size indeed, up to 3 m. (10 ft.) long, and their main prey must have included contemporary species of fishes.

Seasickness   *Motion sickness.

Sea slugs, or nudibranchs, are spectacular gastropod *molluscs, related to snails but usually with a reduced or absent shell. The exposed flattened body is often brilliantly coloured, signalling to predators the possession of distasteful or toxic secretions and so conferring protection despite shell loss. A few species even eat sea anemones and acquire their prey's stinging cells, which are then stored in projections on the back to be used for the sea slug's own defence. Most species breathe through filamentous projections, which are often brightly coloured.

Sea snails are well-known seashore *molluscs, including winkles and whelks. Most are asymmetrically coiled, as their body twists during the larval stage, and the shell grows spirally. Many of the thousands of species scrape up algae for food, though a few use their gills to filter out plankton, and some are predatory, drilling through other molluscs' shells and sucking out their victims' tissues. A few are free-swimming, with keeled shells.

Sea snakes mainly occur in the shallow tropical seas of Southeast Asia and northern Australia, although some of the fifty or so species are more wide-ranging. They are poisonous, their venom apparatus being similar to that of *cobras. The majority are totally aquatic and give birth to live young, but some less specialized species lay eggs on land. Their marine adaptations include paddle-shaped tails and valvular nostrils (often on top of the snout); glands under the tongue are modified for the elimination of excess salt. Sea snakes belong to the family Hydrophidae.

Sea spiders, though often overlooked, are quite common marine animals, especially around rocks and pilings. They resemble spiders, having eight to twelve legs, and form the arthropod class Pycnogonida, with some 600 species; they are remotely related to the arachnids. Curiously, they have reproductive organs in their legs, and males have special egg-bearing limbs. The male gathers eggs as the female lays them, cementing them together and carrying the spherical mass until hatching-time.

Sea-squirts are found in shallow seas and resemble irregular bags of jelly, with two openings where water is

Sea snails

Sea butterflies, *Limacina* species, are sea snails that live in plankton and swim by graceful movements of the foot, which is divided into a pair of muscular flaps. At each stroke the flaps make a figure of eight. The foot also has a feeding function as it collects plankton on extensive ciliary tracts.

Blue and perforated sac-like **sea-squirts**, seen here clustered off a reef at a depth of 20 m. (65 ft.) off the Komodo Islands, Indonesia. Sea-squirts are usually found in waters that are rich in suspended organic matter.

drawn in and out to be filtered. Despite this primitive (and sometimes colonial) way of life, sea-squirts are quite advanced animals, sharing with *acorn worms and *lancelets some features found in vertebrates. These features are most obvious in the larvae, which are tadpole-like and swim actively, using propulsive muscular tails.

Sea-urchins, though related to the five-armed starfish and brittle stars, are the only *echinoderms to have a complete rigid test (an internal shell), the five arms having joined to give a spherical shell. From this, spines and protective tiny claws (sometimes poisonous) protrude upwards, while the tube-feet mostly emerge below to allow walking. The sea-urchin's mouth is armed with five elaborate rasping jaws to scrape up algae.

Most sea-urchins are regular in shape, ranging from 1 to 40 cm. ($\frac{2}{5}$–16 in.) across; but a few asymmetrical burrowing forms occur, like sand-dollars, heart urchins, and sea potatoes. Some species commonly burrow beneath sandy beaches.

Sea walnut   *Sea gooseberries.

Seaweeds are marine *algae, ranging in size from a few centimetres to many metres long. They may be green, red, or brown and the different coloured, photosynthetic pigments are accompanied by differences in life cycle, construction, and habitat. The largest are the brown algae, or brown seaweeds, including the *kelps and wracks, which are anchored to rocks on the sea-bed by a

strongly adhesive disc, the holdfast, while their fronds move in the water currents. Green seaweeds such as the edible sea lettuce, *Ulva lactuca*, often grow in brackish water. The purple red seaweeds, such as the laver, *Porphyra umbiliacus*, and the red dulse, *Rhodymenia palmata*, are edible, while others are used as fertilizers, as grazing for sheep in some areas, and as sources of iodine and alginates.

Secretary bird: the sole member of the family Sagittariidae, found in open country over much of Africa south of the Sahara. It stands up to 1·5 m. (5 ft.) high and has pale grey plumage except for black wings and thighs. The feathers of the loose crest look rather like quills stuck behind the ear, and may be the origin of the name. It is long legged and runs about on the ground, catching snakes and lizards. It is related to the birds of prey and, like many of them, builds nests of branches in the tops of trees and has two or three young at a time.

Sedges are a large group of perennial monocotyledonous plants with rhizomatous or creeping rootstocks, solid triangular stems, and grass-like leaves. They belong to the genus *Carex*, with over 1,000 species which are very widely distributed throughout the temperate, Arctic, and Antarctic regions of the world. The family Cyperaceae, to which they belong, contains over 4,000 species, some of which are referred to as reeds or even 'grasses'. They have little ornamental or economic value, though *C. arenaria*, from Europe and Siberia, is useful for stabilizing loose sands and grows mostly on sea-shores. Sweet sedge, *Acorus calamus*, is a member of the arum family and is botanically unrelated to the true sedges.

Sedge warbler: a species of bird in the family of Old World *warblers. It breeds in damp areas and in wet ditches over wide areas of Europe and western Asia and spends the winter in Africa. It is a small brownish, heavily streaked bird, 12·5 cm. (5 in.) long, and has a conspicuous whitish eye-stripe.

Seeds are the products of fertilization in plants and consist of a plant embryo surrounded by a protective seedcoat. In seeds of *angiosperms the embryo is embedded in fleshy tissue known as endosperm, which provides a store of food for the developing embryo and the seedling after germination. In many plants, the food reserves are stored in the cotyledons, which are the first leaves of the seedling.

There may be up to four layers in the seed's protective coat, but more frequently there are two, known as the testa (outer) and tegmen (inner). A third layer, the aril, is sometimes present as a fleshy layer enveloping the whole or part of the seed. The testa and tegmen may be toughened by different layers of tissue, giving protection to the seed from desiccation and crushing.

Seeds vary in size from dust-like ones, such as those produced in millions in the capsules of orchids, to those of the palm, *Lodoicea maldivica*, which weigh several kilograms each. Seeds are often part of a *fruit and this helps their dispersal.

Seer *Spanish mackerel.

Semen is the watery fluid containing sperms that is ejaculated from the penis of vertebrates. The fluid is a mixture of secretions produced by the prostate and other accessory

It is not only food that the **secretary bird**, *Sagittarius serpentarius*, brings to its chicks. Both parents also bring water, as this one just has. On reaching or leaving the nest, each adult growls to calm the chicks.

glands of the male reproductive tract. It acts as a supporting medium for sperms, providing them with the sugar, fructose, which is converted to energy by the sperms. It also includes other compounds which act on the womb in a manner not fully understood.

Sempervivum *Stonecrops.

Senecio is a plant genus of the sunflower family, to which ragworts belong. It is the largest plant genus with about 2,000 species with a worldwide distribution. In habit they vary considerably and include small annuals such as *Senecio vulgaris* (groundsel), desert succulents, climbers, and the giant *S. grandifolius* from Mexico, which grows to over 5 m. (18 ft.) in height. In the mountains of East Africa are found strange tree-like forms which have adapted to survive the extremes of temperature between night and day. Several shrubby species such as *S. laxifolius* are valuable garden shrubs, while the Canary Island *S. cruentus* is the parent of the pot plant called cineraria.

Senescence, or senility, is the progressive failure of any biological system with time. Senescence of individual cells may sometimes be due to failure of DNA repair, a recognized factor in some rare human conditions causing premature ageing. Many other endogenous and exogenous processes contribute to human ageing. Senescence is not, of itself, an explanation for age-related diseases such as *dementia.

Sensitive plant: a small shrubby perennial from tropical America belonging to the pea family. It is part of the genus *Mimosa*, which includes between 500 and 3,000 species of tropical trees and shrubs. The leaves are divided into many leaflets which quickly fold together when the plant is touched. This sensitivity and movement of plant leaves also occur in other plant species within the pea family, such as *Neptunia*, or in other families such as the genus *Biophytum* in the family Oxalidaceae.

Septicaemia is the presence of disease-causing bacteria in the blood. Often referred to as blood poisoning, it may complicate a local infection such as pneumonia. The bacteria can enter the bloodstream by a trivial wound such as a needle prick. Almost any bacterium capable of infecting man may be a cause of septicaemia, and may lead to widespread tissue damage through its spread, or release of toxins. The condition is always serious, but prompt treatment can usually cure it.

Sequoia *Redwoods.

Seriemas are two species of birds related to the cranes, and confined to grassland areas of central South America. They are tall, thin birds standing 75–90 cm. (30–36 in.) tall, and are greyish above, paler below. They usually run rather than fly. They eat a wide range of insects and other small animals as well as some fruits. They build simple nests of twigs in a bush and rear two or three young at the same time.

Serow are two species of goat-antelopes. The mainland serow, *Capricornis sumatrensis*, frequents the forest lands on the mountain slopes of northern India, Burma, and China, at heights of up to 4,000 m. (13,000 ft.). It is an elusive, agile animal, and if discovered can leap 4·5–6 m. (15–20

A young mainland **serow** browses in the forests of northeastern India, its long, tasseled ears alert for danger. Although secretive in disposition, serows are territorial and can be aggressive.

ft.) at one bound. As it makes off through the thickets, it utters a loud, angry, shrill, whistling snort. In the rutting season rams often 'horn' the trees, making the surface of the horns smooth. It will take readily to water and can swim well. The Japanese serow, *C. crispus*, is similar in appearance but is adapted to the lower temperatures of Japan and Taiwan.

Serpent star   *Brittle star.

Service tree   *Rowan.

Sesame is an ancient oil plant of African origin, principally grown in the drier areas of India, China, Mexico, Burma, and the Sudan. True sesame, *Sesamum indicum*, is one of fifty species in the sesame family, and the most important as a commercial crop. It is an erect, hairy annual, 2 m. (6 ft. 6 in.) tall, which develops white to purplish, bell-shaped flowers that produce capsules containing small white, brown, or black seeds. The edible oil that forms 50 per cent of the seeds is used for cooking, and in the manufacture of margarine. Husked seeds are traditionally used for garnishing cakes, confectionery, and bread, particularly in Mediterranean countries.

Sewellel   *Mountain beaver.

Sexually transmitted disease is a term that has largely replaced venereal disease. This is a group of diseases, caused by infective micro-organisms, which are mainly dependent upon sexual intercourse for their transmission. It includes *gonorrhoea and *syphilis. Once important causes of illness, and eventually death, such diseases are now easily treated by drugs.

Seychelles frogs are a group of three of the five species of frogs found on the Seychelles islands in the Indian

Ocean. *Sooglossus gardineri* and *S. sechellensis* are small, about 1·5 cm. (⅔ in.) and 2·5 cm. (1 in.) long respectively, and frog-like; *Nesomantis thomasseti* is larger, 4·5 cm. (2 in.), and more toad-like. All three species are found in the moss forests on Mahé and Praslin islands. Their relationship to other frogs is uncertain. They have been grouped at different times with either the true frogs, the spadefoot toads, or the Asiatic horned toads, but are now placed in a family on their own.

Shad are species of fishes of the herring family which mostly belong to the genus *Alosa*. In general they resemble large, deep-bodied herrings, growing sometimes to 60 cm. (25 in.) in length, and with a notch in the upper jaw. Many species, like the European twaite shad, *A. fallax*, are migratory and spawn in rivers after migrating from the sea. Most feed on plankton but the North American hickory shad, *A. mediocris*, eats small fishes. The gizzard shads are not members of this genus, but are close relatives.

Shaddock: the largest *citrus fruit, up to 30 cm. (12 in.) in diameter. It is also known as pummelo (not to be confused with pomelo, a West Indian name for grapefruit). It is coarse-skinned and rather bitter and, although not grown extensively anywhere, it is grown throughout Asia as a dessert fruit.

Shag is an alternative name for *cormorants widely used in New Zealand and Australia. In Britain the name is used more specifically for the green cormorant, *Phalacrocorax aristotelis*, to distinguish it from the common cormorant, *P. carbo*.

Shaggy ink cap: a *fungus whose fruiting bodies, or 'toadstools', commonly appear in autumn among grass on rich soil. White at first, the gills underneath the cap later

Often found in large groups, the **shaggy ink cap**, *Coprinus comatus*, is common in Europe, North America, Australia, New Zealand, and South Africa. Up to 12 cm. (10 in.) tall and about 3 cm. (1 in.) across, specimens are edible if picked before the cap has expanded.

The sand **shark**, *Odontaspis taurus*, is common in the tropical and subtropical waters of the Atlantic. A voracious fish-eater, it sometimes rips open fishermen's nets in order to get an easy meal.

break down into a fluid that looks like ink because it is full of black spores. Many other *agarics of the genus *Coprinus* are called inkcaps and the breakdown of their fruiting body helps to disperse the spores. The majority of species grow in rich soils, often favouring dung heaps or compost.

Shallot: a variety of the common onion which forms a spreading clump of new bulbs from a single planted bulb. The bulbs are smaller than those of the onion but are used in the same way. They are milder in flavour and are widely cultivated in many parts of the world, including the tropics, but particularly in Europe and the USA.

Sharks are a group of large sea fishes with a cartilaginous skeleton, which is hardened by deposits of calcium. The upper jaw is not fused with the skull and the teeth, arranged in rows, are not fused with the jaw bones; as each outer row of teeth wears out it is replaced by a new inner row. They lack the air float, or swim bladder, of bony fish. The intestine has a spiral arrangement to increase its absorptive surface. The body of all sharks is tube-shaped like a torpedo, and the upper lobe of the tail fin is longer than the lower. Their skin is covered with tooth-like dermal denticles, which give it a rough texture. Male sharks have pelvic claspers which act as guides for spermatozoa when these are inserted into the female cloaca to fertilize the eggs. Most species give birth to fully formed young, though some such as the *rays lay their eggs in horny capsules. Sharks are predatory, mostly eating other fishes, but a few species (whale shark and basking shark) feed on plankton. They are marine creatures, although a few species enter rivers and lakes in tropical areas, occurring both inshore and in the deep sea; but they are rare in polar regions. About 300 species are known, mostly moderate to large in size, with the whale shark, the largest known fish, growing to 18 m. (60 ft.) in length.

Sharpbill: a bird found in Central America and much of South America, the only species in its family. The size of a large sparrow, it is olive-green above with a small brownish-black crest, which usually conceals some orange-red feathers in the centre of the crown. It has a straight, slightly broad, beak with which it eats fruit. The sharpbill makes a nest of twigs high in forest trees; however, its breeding habits are mostly unknown.

Shearwaters are long-winged *petrels of several genera which have a swooping, wave-skimming flight. They feed on surface-feeding fishes and marine organisms in large mid-ocean or offshore flocks. Some species dabble at the surface while others dive and swim powerfully under water, using their webbed feet as paddles and wings part-extended as hydrofoils. They nest colonially on cliffs and

The young or non-breeding Manx **shearwaters**, *Puffinus puffinus*, spend much time sitting on rocky cliffs. Adults are rarely seen on land. Out at sea during the day, at night they fly directly to their burrows.

screes or in tussock-screened burrows, feeding by day and returning to the nests after dark to avoid aerial predators. The fledglings of some southern species are much in demand for their oil and rich meat (*mutton birds).

Sheathbills are two species of small white pigeon-like birds of sub-Antarctic and southern South American coasts, and are a family on their own. They scavenge on shores and in the breeding colonies of penguins and other seabirds. The yellow-billed sheathbill, *Chionis alba*, inhabits Atlantic coasts, while the black-billed sheathbill, *C. minor*, is found on the most southerly islands of the Indian Ocean. Combative though gregarious, they feed in small groups, eating seaweeds, flotsam, and abandoned eggs and chicks of other birds. A single chick is usually reared.

Sheath-tailed bats are distinguished by a tail which pierces the tail membrane and emerges free on the upper surface. The tail slides easily through the basal sheath so that the shape of the membrane can be altered in flight by extending the hind legs. This allows the bat to be very manoeuvrable. There are fifty species which include the tomb bats, *Taphozous* species, and the sac-winged bats, *Saccopteryx bilineata*

Sheep are species of goat-antelope closely related to goats. They have transversely ribbed horns which tend to curl in spirals, with the points turning outwards as they rise. Wild sheep, of which there are six species in the genus *Ovis*, do not have the thick, woolly coats of domesticated breeds. They include the *bighorn, *mouflon, and the *argali. The domestication of sheep for production of milk, meat, leather, wool, and lanolin goes far back into antiquity. New Zealand has the highest density of sheep in the world. There are more than forty different breeds which have been developed and improved for particular purposes and terrains. The hill breeds, the most numerous, are small and hardy; they develop into good meat and their fleeces can be used for carpets, tweeds, the finest flannel, and knitwear. The lowland breeds are larger and have heavy coats. The ewes usually produce one or two lambs, although five have been recorded; they are born woolly and open-eyed. The gestation period is about twenty-one weeks.

Sheep-keds   *Louse-flies.

Sheep-ticks, or castor bean ticks, are probably the commonest *ticks in temperate areas, occurring mainly on sheep and cattle, but occasionally they bite man and smaller mammals. The eggs hatch in autumn, but the six-legged larva waits until spring to find a host. It then feeds, falls off, and moults into a larger eight-legged larva which survives for a year without feeding. A year later feeding again occurs, leading to maturity. Bean-like adults are found in the third year, mating on sheep and cattle. The males then die, and the females continue the cycle by laying batches of eggs in crevices in the ground. Redwater fever in cattle and louping in sheep may be transmitted by this tick.

Shelducks form a genus of six species of Old World dabbling *ducks. They have a goose-like character, being slower on the wing than typical ducks, and rather heavily built. The plumage of the sexes is nearly the same. The common shelduck, *Tadorna tadorna*, is a large black and white bird banded with chestnut, with a red beak and pink legs. It inhabits estuaries and coasts, where it breeds in burrows in sand dunes. Of the other five species the chestnut-orange ruddy shelduck, *T. ferruginea*, of the southern Palearctic region, is the most widespread.

Shellfish is a confusing term covering all aquatic invertebrates with shells, especially those which we eat. It has no real zoological meaning, since it includes both crustaceans with complete external jointed skeletons, and molluscs, with incomplete but thicker calcified shells from which the soft body protrudes. In the first group are crabs, shrimps, and prawns; in the latter cockles, winkles, mussels, and their relatives. In some countries other groups, like sea-urchins and marine worms, may also be valued as edible shellfish.

Shells are found in many animal groups, with a great range of shapes and structures, some very different from the familiar sea shells formed by molluscs. Some protozoa (foraminifera, radiolarians) construct elaborate shells, and polyps may also secrete external shells to become corals. Sea-urchins also have shells, just beneath their skins. But it is molluscs and arthropods that are particularly renowned for their hard outer coverings. Molluscs have an incomplete shell, into which the soft body can be withdrawn, while the arthropods always have a complete, jointed, armour-plating which covers their bodies.

Shells have many uses. They give protection not only against predators and mechanical damage, but also against extreme temperatures or desiccation. They also provide rigid surfaces against which muscles may pull. Shells are almost essential for invertebrates living on land, where very few truly soft-bodied animals are successful. They have to be strong but not heavy, and resistant to cracking or buckling. Most animal shells are made either of mineral crystals (like calcium carbonate or silica), or of tough fibres (like chitin in cuticle or collagen), with these components embedded in soft cushioning protein.

Even when well adapted to a life-style, shells can be a hindrance. The complete arthropod shell or *exoskeleton makes it necessary for the animal to moult periodically to allow growth, and even the incomplete molluscan shell, which permits growth at the edges, may cause problems by precluding fast predatory movements, so that the *cephalopod molluscs have evolved by reducing or completely losing their shells.

Shieldbugs are shield-shaped *bugs, often brightly coloured or conspicuously marked, some of which produce a pungent odour and are also known as stink bugs. The majority belong to the family Pentatomidae with over 2,500 species, and can be distinguished by the large triangular scutellum (an extension of the thorax, between or covering the leathery wing-bases). Some feed on other insects, but the majority are plant-feeders. The harlequin bug, *Tectocoris diophthalmus*, which is a pest of cabbage and related crops in North America, belongs to another family, the Scutelleridae, which also has shieldbug-type adults. Some species of shieldbug of tropical regions can reach lengths of up to 3 cm. (1¼ in.).

Shield-tail snakes are a family of thirty-five small, burrowing species of snakes occurring in southern India and Sri Lanka. They have small, pointed heads, which are well adapted for being pushed through crevices in the

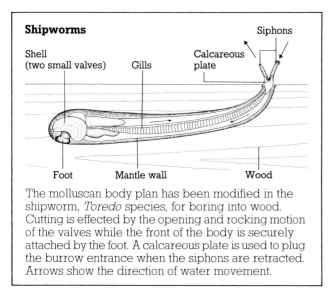

Shipworms

Shell (two small valves)   Gills   Calcareous plate   Siphons

Foot   Mantle wall   Wood

The molluscan body plan has been modified in the shipworm, *Toredo* species, for boring into wood. Cutting is effected by the opening and rocking motion of the valves while the front of the body is securely attached by the foot. A calcareous plate is used to plug the burrow entrance when the siphons are retracted. Arrows show the direction of water movement.

soil. The short tail-tip is covered by one or more spine-bearing scales to which mud adheres. The soil sticking to the tail probably affords the snake some protection from a predator that might follow it down its tunnel. Earthworms are the main food of these snakes.

Shingles is an eruption of small blisters in a localized area of the body, often a band running round one-half of the abdomen. Pain usually precedes, and often outlasts, the eruption. The cause is reactivation of the chicken pox virus, which was dormant in the sensory nerves supplying the skin of the affected area. Early treatment can ease the condition.

Shipworms are not worms at all, but *bivalves. The typical two-piece molluscan shell is much reduced in shipworms, forming the scraping drills whereby the animal burrows into wood. The unprotected wormlike body and siphons follow behind the shell, secreting a chalky lining to the burrow. Shipworms filter food from water via the siphons, but also digest wood while burrowing and therefore pose real problems to wooden boats and pilings.

Shock is, in strictly medical terms, a condition in which blood-flow from the heart is greatly reduced, so that the vital functions of the body are threatened. It occurs as a result of excessive blood loss, a heart attack, or infection in the blood. Urgent treatment is required to save the patient's life. The term is also used in the popular sense for states of extreme anxiety which people may enter after 'shocking' events like robbery or assault. However ill they look, their lives are rarely in danger.

Shoebill *Whale-headed stork.

Shooting star (plant) *Primrose.

Shore birds *Wading birds.

Shore plants are those which, despite being above the high-tide line, are subject to salt spray and a high salt content in the soil. They include a number of cliff plants that are also resistant to high levels of nutrients from the droppings of seabirds. Many are *succulents and often have deep roots. Some vegetables have been derived from shore plants, notably cabbage, spinach beet, sugarbeet, beetroot, and asparagus. So-called scurvy grass (actually a member of the mustard family) is high in vitamin C and was once used by sailors to control scurvy. Shore plants are often found in saltmarshes or in other habitats with a high salt content in the soil.

Short-eared owls: a species of medium-sized, dull-brown *owl with scarcely visible eartufts and rounded wings which span a metre (over 3 ft.). Cosmopolitan in distribution, they prefer rolling country, particularly marshy or wet heath. In such habitats they may be resident throughout the year, but some short-eared owls migrate annually from tundra regions to northern temperate areas. Their food is mostly small mammals and birds, often hunted by day. They nest on the ground, and usually lay between two and six eggs.

Shorthorn grasshoppers have antennae not much longer than the length of their head. They are medium to large insects of the superfamily Acridoidea of *grasshoppers. There are almost 10,000 species world-wide including the *locusts. In most species the males stridulate by rubbing a row of pegs on the hind legs against a thickened vein on the front wings. The sound is a series of chirps, not so long or loud as that of longhorn grasshoppers. The females have short ovipositors and lay pods of eggs by inserting the abdomen deep in the soil. They feed mostly on grasses, although a few eat broad-leaved plants.

Short-tailed bats, of which there are only two species, comprise the family Mystacinidae. They are native to New Zealand, and have many peculiar features, including sharp claws and a leather membrane, within which the wings are folded when they are not flying. This allows the bat to use its arms as limbs and run quickly along the ground. Its tail is similar to that of a *sheath-tailed bat.

Shovellers are dabbling *ducks easily distinguished by their outsize, spoon-shaped beaks. The drake of the com-

The common **shoveller** runs its bill along the surface of the water, using it in a side-to-side movement to filter out small animals or plant particles. Occasionally it may also snap at insects in the air.

mon shoveller, *Anas clypeata*, has a green head, white and chestnut underparts, and a large blue wing-patch. It is chiefly found in shallow freshwater habitats and is widely distributed in the Northern Hemisphere.

Shrews are insectivores belonging to the family Soricidae, although the name is used loosely to describe other insectivores such as *elephant shrews. Most shrews are small, and one, the pygmy white-toothed shrew, *Suncus etruscus*, has the distinction of being the smallest living mammal, weighing only 2 g. (less than $\frac{1}{10}$ oz). Shrews are characterized by a pointed snout, dense fur, and often a strong, musky odour emanating from glands on the flanks. They are usually solitary, burrowing, terrestrial animals, but a few are adapted for an aquatic life. Water shrews, such as the American *Sorex palustris*, have fringes of hair on the feet and tail which help them to swim. Air trapped in the fur gives them buoyancy. These and some other shrews produce venom in the saliva. All shrews are insectivorous or carnivorous and some will take carrion and vegetable matter in their diet. Shrews are noted for their continuous and intense activity, which necessitates almost continuous feeding.

Shrikes belong to a family of about seventy species of birds. The great majority occur in Africa, with some others in the Old World. Two species occur in North America, one of these reaching Central America, but they are absent from South America and Australasia. They vary in size from that of a small thrush to that of a small crow and are often strikingly coloured in blacks, greys, and whites. Some of the tropical species are brilliantly coloured with red, green, and yellow. Their beaks are powerful and slightly hooked. They perch on bushes and pounce on passing insects, small birds, and other small animals. They often impale surplus food on thorns, retiring to their larder later; hence an alternative name of butcher birds. The *pepper shrikes form a family of only two species.

Shrike thrushes/tits   *Whistlers.

Shrimp fishes, of which there are only four species, are found in the tropical Indo-Pacific ocean. They are small, rarely exceeding 15 cm. (6 in.) in length, and are slender and compressed, with a long snout and the second dorsal and tail fins displaced onto the underside of the body. They swim vertically in the water in a head-down posture. One species has been reported to live in underwater caves.

Shrimps are small, laterally compressed *crustaceans with elongate abdomens suited to swimming. They are related to crabs and lobsters (decapods), and the term shrimp is generally used for all free-swimming decapods, especially those trawled for food. But, confusingly, shrimps can also mean *all* elongate swimming crustaceans, so including various *amphipods, *lamp shells, the *euphausid krill shrimps, and *mantis shrimps.

True decapod (ten-legged) shrimps are usually transparent, though deeper-water species may be reddish or luminescent. They swim using special swimming legs and the flicking abdomen, although some prefer life in sandy or coralline burrows. A few shrimps have special behavioural traits, like the snapping-shrimps, which click their pincers together noisily as a threat display, or the cleaner-shrimps, which browse on the surface of fishes to their mutual benefit (symbiosis).

Shrubs are woody plants without the obvious trunk of trees but with a number of basal shoots, giving the plant a bushy appearance. There is no clear disinction between shrubs and trees; species, such as hawthorn, hazel, or birch, may grow from a shrub into a tree. Low-growing shrubs, such as heathers, are difficult to distinguish from herbaceous plants with rhizomes, particularly if these are produced at ground level.

Shrubby vegetation is often found as a belt between grassland and forest, and often precedes forest in its

Sight

Man

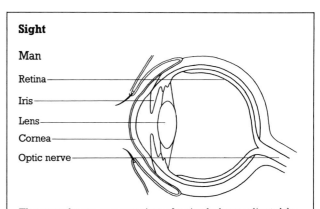

The vertebrate eye consists of a single lens adjustable by muscle action to focus an image on the retina, from which optic nerves carry sight messages to the brain. Iris size can be adjusted to let in more or less light.

Octopus

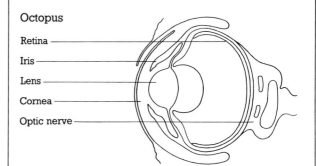

Cephalopod molluscs have independently evolved eyes similar to those of the vertebrates—convergent evolution.

Insect

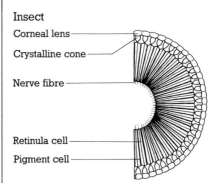

The compound eyes of insects are made up of numerous units, each having a corneal lens, and a crystalline cone which also acts as a lens. The image is passed from the retinal cells along a nerve fibre, as coded impulses, to the brain. Some insects can protect the sensory cells from intense light by moving the pigment cells between the units nearer to the surface.

colonization of abandoned pastures. Such vegetation is commonly referred to as scrub and is regarded as a stage in *succession. Shrubs adapted to grow as scrub are often armed with thorns or spines as a defence against grazing animals. Familiar temperate examples are hawthorn, blackthorn, and gorse.

Shrubs include a number of crops such as tea, coffee, and blackcurrants, and have great importance in horticulture, particularly as hedging plants.

Siamese fighting fish *Fighting fish.

Side-necked turtles are distinguished by their method of retracting their heads into their shells. In contrast to the 'normal' vertical bending of most turtles, side-neck turtles fold the neck laterally; the head thus faces to one side when retracted. There are about fifty species and they occur almost entirely in the Southern Hemisphere. Most of them are entirely aquatic, living in freshwater habitats.

Sight is, at its simplest, the ability of an organism to detect light. In its most elaborate form it permits discrimination of pattern, shape, colour, movement, and depth.

Although most plants respond physically to light they are not generally considered to possess sight. Animals are said to have sight and usually have specialized nerve receptors gathered into *eyes. All such eyes work through the absorption of light by a visual pigment, and this sets up a tiny electrical impulse in *nerves. Snails and worms use their simple eyes to detect light and dark. Arthropods with simple eyes see light and dark and in some instances movement. Those with compound eyes, such as dragonflies, flies, and bees, are able to see small movements, general shape, and colour. Many insects can also see ultraviolet light, which is invisible to man. Octopuses and squids have acute vision, as their eye structure is similar to that of vertebrates, but they lack stereoscopic sight. The well-developed vertebrate eye enables animals to see with high resolution over a range of conditions. Nocturnal animals have good vision at low levels but are often blinded by the light intensities of daytime. Most species of reptiles, amphibians, birds, fishes, and mammals can detect colours to a lesser or greater degree. Sight is the most important sense in primates, including man.

Silk moths are *emperor moths of a number of genera and also include the silkworm moth, *Bombyx mori*, and allied species. The name is used mainly, but not exclusively, for those whose cocoons yield commercially useful silk, and for American counterparts of the *atlas moths. They occur in North, Central, and South America, and in India eastwards to Japan and Australia. Historically, commercial silk production was confined to China, India, and Japan, but has been tried in Europe and North America, using an introduced species which has become established. Silk produced by species other than silkworms is often called shantung or tussore silk.

Silkworms are the caterpillars of the silk moths which extrude silk from mouth glands and use it to weave cocoons within which they pupate. The common silk moth, *Bombyx mori*, a native of China, has been domesticated for centuries and no longer occurs in the wild. The caterpillars are fed on mulberry leaves; adults neither feed nor fly. Before the silk is wound off, the pupae are killed by immersion in hot water. Each cocoon yields up to 900 m. (over half a mile) of thread. Caterpillars of many emperor moths (Saturnidae) also produce silk, but none is as commercially important as that of the common silk moth.

Silky flycatchers belong to a subfamily of some four species of bird which are closely related to waxwings. They are found in the southwestern USA and Central America and have a sparrow-sized body, with a crested head and a longish tail 18–25 cm. (7–10 in.). Their plumage is blackish, grey, or brown with yellow markings, and white in the tail and wings. They live in bushy country and feed on berries and insects.

Silverfish, or bristle-tails, are elongate, tapering, wingless insects, up to 1·2 cm. ($\frac{1}{2}$ in.) long, and covered with silvery scales. They comprise the order Thysanura, with over 550 species distributed world-wide. The antennae are long and there are three long appendages at the tail-end. Silverfish are harmless commensals of man, living in damp places, such as cracks around sinks, and feeding on scraps of food and paper. The eggs are laid in the cracks and the young resemble the adults, except that the scales are not developed. The firebrat, *Thermobia domestica*, is a cosmopolitan example.

Silversides are mostly small, coastal and inshore marine fishes, abundant in tropical and warm temperate seas. A substantial number of the 180 or so species, however, live in fresh water especially in Central America, New Guinea, and Australia. They all belong to the family Atherinidae,

The dense cocoon of the common **silkworm** measures up to 3·8 cm. (1 ft. 6 in.) in length. Here it has been cut open to reveal the caterpillar, which, having already reduced in size somewhat, will soon pupate.

and are recognized by two widely separated dorsal fins, large scales, moderately big eyes, and a mouth opening at the top of the head. All species have a prominent silvery *lateral line. They spawn in shallow water, the eggs of many species having short adhesive threads on their surface. The Californian grunion, *Leuresthes tenuis*, spawns on the shore at high tide and the eggs hatch at subsequent high spring tides.

Silver-Y moth: a medium-sized brown *owlet moth with a silvery Y-shaped mark on the fore-wings. This fast-flying moth is active by night and day, and is a regular migrant throughout the Palearctic region. It can produce several generations in a year and extends its distribution northwards in spring. The adult moths cannot survive the winter of northern regions and often migrate southwards in autumn.

Simpson, Sir James Young (1811–70), Scottish surgeon, introduced the use of chloroform as an anaesthetic. His innovation met with fierce opposition at first, but he won the confidence of Queen Victoria, who was given chloroform during childbirth. He was one of the founders of gynaecology, working out new procedures for diagnosis and treatments.

Sipunculids are a phylum of some 330 species of marine animals which are sometimes called 'peanut worms', and smaller species do resemble unshelled peanuts if they have their proboscis withdrawn. But many are larger than this, up to 75 cm. (30 in.) long, and all have a proboscis which bears tentacles.

Sipunculids mostly live in burrows on muddy shores, gathering food from the mud with their tentacles, though a few bore into corals or live in mollusc shells. These worms have a *coelom, like the *annelids to which they are probably related, but lack segmentation and are slow burrowers.

Sirens are a unique group of *salamanders, though some biologists consider them to belong to a separate order of *amphibians. There are only three species, all found in the southern USA: two in the genus *Siren* and one in the genus *Pseudobranchus*. They are eel-like and permanently aquatic, retaining many larval characters, such as the lack of a pelvic girdle and hind-limbs, and retention of a pair of feathery gills. The fore-limbs are minute and the jaw margins are horny. In dry conditions, when the shallow ditches, swamps, and ponds which they inhabit dry up, they are able to survive by living in mud burrows, protected by a mucous cocoon.

Sisal, or sisal hemps, are useful fibre plants, native to Central America. They are species of *Agave*, members of the monocotyledonous family Agavaceae, and produce the strong, durable fibres named after the plants. Most important are *A. sisalana* and *A. fourcroydes*. Sisal is grown in many tropical countries, including Brazil, Haiti, and several parts of Africa, including the largest producer, Tanzania.

The plant itself is a short-lived perennial, consisting of a rosette of swollen, stout, sword-like leaves, each with a brown, sharply spined tip. The oldest leaves are cut for fibre two to three years after planting and then annually until the plant dies. The leaves are mechanically crushed and scraped to remove the flesh and the washed bundles of fibre are dried in the sun. The uses of sisal fibre include weaving into matting, fishing nets, cordage, and other such products.

Siskin is the common name for sixteen species of small birds of the genus *Carduelis*, members of the finch family. They are widespread in the Northern Hemisphere and also occur in Central and South America. They are mostly small birds, greenish-yellow with bright yellow markings in the wings and tail, resembling their close relatives in the genus *Carduelis*, the greenfinch. The males tend to be brighter than the females and often have black caps or other head markings. They live in forests and eat the seeds of trees. Other species of *Carduelis* include the *goldfinches, which are also seed-eaters.

Sitatunga, or marsh buck: an antelope species related to the kudu, living in the swamps and water courses of central and East Africa. It swims, dives, and even travels considerable distances under water. Its feet are adapted for life near water, for the halves of the slender hoofs spread widely when walking or running, leaving a V-shaped spoor. The male may reach 1·2 m. (4 ft.) at the shoulders, has spiral horns 90 cm. (3 ft.) long, and brown hair with stripes. The female is hornless, and she and the young are bright chestnut in colour with white stripes and spots.

Skate: a species of large *cartilaginous fish growing to a width of 2 m. (6 ft. 6 in.) and a weight in excess of 100 kg. (220 lb.). Living off the coasts of Europe, it is most abundant in depths of 100–200 m. (325–650 ft.), mainly on rocky sea-beds. It feeds on fishes and crustaceans, many of the former being captured in mid-water. The eggs are laid in rectangular capsules about 10 × 15 cm. (4 × 6 in.) with a spike at each corner. A valuable food-fish, and also caught by anglers, the skate has now become rare through over-fishing, for even newly hatched fishes are caught in trawls.

Skeleton   *Bones, *Human skeleton.

Skimmers are three species of large, tern-like birds 40–45 cm. (16–18 in.) long, with a greatly enlarged, flexible lower mandible. They feed by skimming over smooth water with their beak cutting the surface, catching small fish, insects, and crustaceans. Large flocks feed and roost together. Skimmers live in Africa, India, and the Americas.

Skin is the surface covering of all vertebrates and is formed of two layers: the outer surface layer, the epidermis, and the dermis underneath. The epidermis is formed of a single, or multiple, layer of cells which can be replaced if it is damaged or worn away. The epidermis of terrestrial vertebrates consists of many cell layers; those on the surface progressively die and turn into dry scales. These are constantly shed as the result of wear and tear, being replaced from the basal living cells by division. The underlying dermis is formed of a meshwork of tough fibrous tissue interlaced with elastic fibres. It is very strong and resistant to injury. Within the dermis there are numerous blood vessels and nerve endings which when stimulated give rise to the sensations of touch, pressure, warmth, cold, and pain.

The skin covering certain body areas is modified. The epidermis may be greatly thickened to form hoofs, or hu-

Skin

A generalized section through mammalian skin, highly magnified.

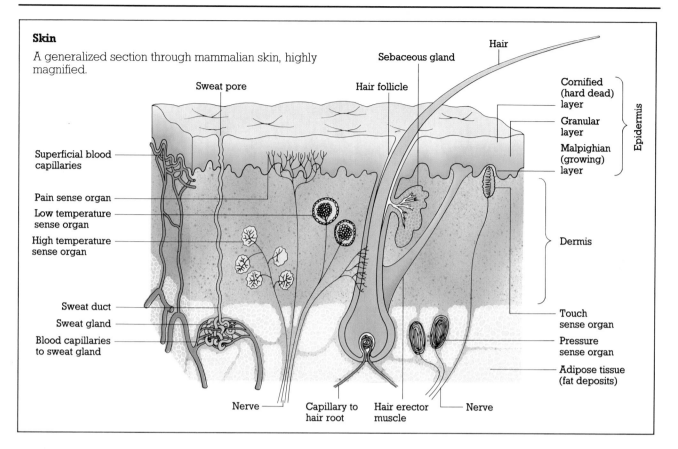

Sweat pore — Hair follicle — Sebaceous gland — Hair

Superficial blood capillaries

Pain sense organ

Low temperature sense organ

High temperature sense organ

Sweat duct

Sweat gland

Blood capillaries to sweat gland

Cornified (hard dead) layer

Granular layer

Malpighian (growing) layer

Epidermis

Dermis

Touch sense organ

Pressure sense organ

Adipose tissue (fat deposits)

Nerve — Capillary to hair root — Hair erector muscle — Nerve

man sole skin. In fishes, horny scales are derived from the epidermis and are very tough. In amphibians, the skin acts as an air-breathing surface. Within the skin of some mammals are sweat glands, which help to cool the body through evaporation. However, other mammals, such as dogs and cats, lose heat via the mouth (panting). Among the *hair follicles in skin lie sebaceous glands, producing an oil which protects the fur and maintains its waterproof qualities.

Skinks are a large family of lizards comprising over 600 species which are very widely distributed but most numerous in the tropical areas. They are mainly ground-dwelling and burrowing lizards. The smaller species usually eat insects, but the larger ones often subsist, at least partly, on plant material. They generally have rather cylindrical bodies, mostly with smooth scales that are reinforced by internal bony plates (osteoderms). Skinks with subterranean habits tend to have reduced, or vestigial limbs, accompanied by tail elongation and thickening which gives them an almost snake-like resemblance.

Skipjacks *Click beetles.

Skipper butterflies are mostly small, moth-like butterflies which beat their wings fast in rapid, darting flight. They make up a superfamily distinct from the rest of the butterflies, with over 3,500 species. Their wings are small in relation to the robust, hairy body; the antennae are widely separated, and often have a narrow hooked tip. Some species rest with the wings flat over the body, but most hold the hind-wings flat and tilt the fore-wings at an angle. The caterpillars, many of which feed on grasses, have a narrow section, like a neck, behind the head. They spin shelters among the leaves of the food plant, within

which they later pupate. They are cosmopolitan in distribution and the butterflies of some tropical species are colourful with long 'tails' to the hind-wings.

Skuas are five species of brown or mottled, gull-like birds which make up their own family. They are found in both hemispheres at high latitudes. Predatory on smaller birds and mammals, they also catch fish and krill at sea, migrating long distances between breeding seasons.

As a defence, the blue-tongued **skink**, *Tiliqua scincoides*, exposes its brightly coloured gape and tongue. It is found over a large part of Australia, and attains a length of some 60 cm. (2 ft.).

Skull: the hollow skeleton of the head in vertebrates, comprising a series of flat, jointed bones, usually immovable except for that of the lower jaw. The bones of the skull can be roughly divided into the cranium, which houses the brain, and the face. It is supported on the top of the spine and moved by the muscles of the neck.

The cranium in humans consists of eight bones; between them they form a 'box' with the mechanical rigidity to resist crushing. Holes in the base and sides allow entry of the spinal cord and blood vessels. The facial bones, of which there are fourteen in humans, have sockets and cavities for the nose, eyes, and ears. In human babies there are membraneous spaces between the bones of the skull vault which allow the head to mould during birth and the brain to reach full maturity before the bones join. Between the main joints, or sutures, are small additional (sutural) bones which suggest that the brain could grow yet larger in the course of evolution.

Skunks are small mammals belonging to the weasel family and well known for their foetid secretions, which can be ejected forcibly from the anal glands towards an attacker. In addition to its strong smell, the secretion has

The two-toed **sloth**, *Choloepus didactylus*, is found in Central America and northern South America. When awake, it lives upside down, but two-thirds of every day is spent asleep, propped on top of a branch.

irritant properties and is an effective deterrent to would-be predators. There are thirteen species of skunk, all of which have bold black and white patterns in the fur, which are clearly warning signals. The commonest species is the striped skunk, *Mephitis mephitis*, whose range extends from Canada to Mexico. The closely related hooded skunk, *M. macroura*, is limited to the southern USA and Central America. Other species are the four species of spotted skunks (*Spilogale*) of North and Central America and the seven species of hog-nosed skunks (*Conepatus*), whose range extends from the southern USA into South America. All skunks feed on insects and small vertebrates, as well as some plant material, and most lie up in burrows during the winter.

Skylarks are two species of bird in the *lark family. The common skylark, *Alauda arvensis*, is widely distributed in Europe and Asia, where it lives in open, grassy country. About the size of a sparrow, it is a streaky brown bird with white outer tail feathers and a very slight crest. Although undistinguished to look at, the skylark is famous for its beautiful song, given when the bird is high in the sky, almost hovering. The second species, restricted to Asia, is the oriental skylark, *A. gulgula*, which is similar to the common skylark in appearance and habits.

Slaters   *Woodlice.

Sleep is a natural state of torpor and a lowering of consciousness associated with changes in the patterns of electrical waves recorded from the brain (electro-encephalograms). The body's relaxation is induced by graded switching off of the nerves which activate postural and other muscles. Sleep consists of alternating periods of 'deep' sleep and a type characterized by rapid eye flickering. It is during the latter phase that dreams occur. The biological need for sleep is not understood, but all species of mammal undergo regular periods of it, though these periods vary with age, species, and, under natural conditions, with the seasonal changes of day length.

Sleeping sickness   *Trypanosomiasis.

Slime moulds are a group of single- to multi-celled organisms, sometimes regarded as fungi but having characteristics of both plants and animals. They reproduce by spores, yet their cells can move like an amoeba and feed by ingesting particles of food.

Slit-faced bats   *Hispid bats.

Sloane, Sir Hans (1660–1753), London physician and naturalist, bequeathed to his country a library and collections which formed the nucleus of the British Museum. Physician to the governor of Jamaica, he returned to Britain with several hundred plants. He succeeded Sir Isaac Newton as President of the Royal Society in 1727, when he also became physician to George II.

Sloths are mammals of the Edentate order, like armadillos. They are fully adapted to tree life. The limbs are long, especially the fore-limbs, and the digits carry hooked claws which allow the animal to hang upside-down from the branch of a tree. As the name suggests, the sloths are very lethargic. They differ from all other mammals in that their body temperature is low and variable between 24 and 37 °C (75–98 °F). Sloths feed upon foliage and their teeth have a grinding surface; the stomach is divided into several compartments like that of the ruminants.

There are two families of sloth. The three-toed sloths, of which there are three species, have three curved claws on their fore-limbs, while the two species of two-toed sloths have only two. Both have three claws on their hind-limbs, are about 60 cm. (2 ft.) long, and have thick coarse hair of a drab brown or grey colour. The presence of algae, growing in their fur, gives them a greenish tinge which helps them blend into their surroundings. A single young is born in the summer months and it clings to the mother's belly.

Slow-worm: a species of limbless *lizard, readily distinguishable from snakes by having movable eyelids. They occur over most of Europe (including Britain but not Ireland) to southwest Asia and parts of North Africa. They are found mainly in thickly vegetated localities, resting by day under stones, logs, and other sun-warmed objects. Their diet comprises mainly small slugs, earthworms, and insect larvae. Largest individuals may attain 50 cm. (1 ft. 7 in.) in total length but most are smaller. Captive specimens have lived over 54 years.

Slugs, the scourge of all gardeners, belong to the same group of *molluscs as snails and periwinkles, but they have mostly lost the typical coiled shell, so they are active only during damp periods when they will not dry out. On dry days their streamlined shape lets them shelter in crevices, protected by foul-tasting slime to deter hungry birds. In hot spells they retreat underground, and can survive the loss of 80 per cent of their body fluid. They feed on vegetable matter, using a ribbon of very fine teeth which rasps plant leaves and stems. Some species also eat earthworms and other species of slugs.

Smallpox is a serious virus infection, once the cause of epidemics throughout the world. It has now officially been eradicated, by a remarkable international effort. Success was possible because the condition is easily recognized, and vaccination gives rapid and almost complete protection. Moreover, it is not 'carried' by humans, and there are no animal reservoirs of the virus. This combination of circumstances may be unique, and it is doubtful that the success can be repeated with any of the other major infective scourges of mankind. Symptoms include a high fever and a rash, with blisters that leave scars on the skin, after bursting and healing.

Smell is an important sense and is used by some animals to help locate food, find a mate, and define territories. Smell is usually considered to involve airborne substances, while taste implies food, but both senses rely on the reception of chemical substances in the surrounding air or food. Molecules arriving at the olfactory organs (often in the *nose) trigger nerves to send electrical messages to the smell centres in the brain.

Some fish, like the carp, have smell receptors on the body surface. In insects, the receptors for smell and taste are often concentrated in the antennae but may occur in almost any other body surface. The elaborate branched antennae of some moths can detect a single molecule of the female's scent or *pheromone. Amphibians and reptiles have smell detectors in the roof of the mouth, where incoming air carries the scents. Snakes extend their forked tongue to pick up airborne molecules, then retract it and insert it into the cavities, where the olfactory organs lie.

Man does not have a well-developed sense of smell but is still able to discriminate several thousand different odours. Other mammals, such as the cats and dogs, make greater use of smells. They mark out their territory with urine and the female signals her sexual receptivity by odour. Some animals, such as salmon, use water-borne scent to locate their stream of origin, to which they return and spawn.

Smellie, William (1697–1763), Scotsman, both practised and taught midwifery, using an artificial model of a woman in labour. He published several books on the subject of childbirth, described more exactly than any previous writer the mechanism of birth, and insisted on careful measurement of the pelvis of expectant mothers.

Smews are the smallest of the *mergansers. The drake is white marked with black, with a white crest; the female chestnut, white, and grey. They breed in the northern Palearctic region, nesting in holes in trees, and spend winter on lakes, ponds, and rivers throughout Europe and central Asia.

Smith, Sir James Edward (1759–1828), English botanist, founded the Linnean Society of London and was author (with James Sowerby as illustrator) of *English Botany* (36 vols., 1790–1814). He did much to popularize botany by writing and lecturing. His best work was his last: *The English Flora* (4 vols., 1824–8).

Smoky bats, or thumbless bats, are small, Central and South American bats called 'smoky' for no good reason apart from their slate-grey colour. They comprise a family of just two species. Their main characteristic is the greatly reduced, non-functioning thumb, which is enclosed within the wing membrane.

Smooth snake: a fairly small, non-venomous species of *colubrid snake from Europe and Asia Minor. Its length only exceptionally exceeds 80 cm. (2 ft. 7 in.). It occurs mainly in dry, sunny localities, where other reptiles, such as lizards, form its main food. In England the smooth snake is rare, being restricted to a few heathland localities in the south, and is a protected species. It gives birth to up to fifteen live young.

Smut fungi are parasites which grow inside plants, often without visible symptoms, and produce a mass of dark, smut-like spores. In 'loose smut' of wheat these spores are produced in place of grain as the ear ripens. Like *rusts, there are numerous species, each with specific hosts.

Snail is the name used for terrestrial gastropod *molluscs, most familiar as garden animals, but the name is also applied to vast numbers of *sea snails and pond snails. All are characterized by the coiled shell which encloses and protects their body. They constitute the largest group of the molluscs; together with a few uncoiled forms like abalones and limpets, and the shell-less slugs, there are 35,000 species of gastropod molluscs.

Most snails creep about on a broad flat foot, leaving trails of mucus, and feed by scraping up algae with a ribbon of renewable tiny teeth. The highly successful land snails breathe with a special lung, and can close the shell with a lid (operculum) to conserve water. Some species

Snails

Snails are hermaphrodite and when two snails come together, in this case Roman snails, *Helix pomatia*, there is a mutual exchange of sperm. Copulation is stimulated by the discharge of a calcareous dart. One mating provides enough sperm to fertilize several batches of round, whitish eggs.

have shells designed to reflect heat and combat desiccation so well that they even survive in deserts.

Snake birds *Darters, *Wrynecks.

Snake-flies, characterized by the elongation of the front of the thorax into a 'neck', form the insect order Megaloptera, along with alder flies. Snake-fly larvae are terrestrial and are found in wooded regions. Females use a long, thin ovipositor to lay eggs in crevices in bark. Adults and larvae are predatory.

Snakehead fish are freshwater fish of the family Channidae, which have a cylindrical body, with a broad head, a wide mouth, and heavy scales, giving the impression of a snake's head. There are about ten species, all native to tropical Africa and Southeast Asia. They have an accessory breathing organ in the upper part of the gill chamber by means of which they can breathe air and they frequently live in swamps. Some species grow to 1 m. (3 ft. 3 in.) long.

Snake-necked turtles are a family of *side-necked turtles, with some thirty species that are found in South America, Australia, and New Guinea. The common name is based on some species, such as the Australian snake-necked turtle, *Chelodina longicollis*, with a neck almost as long as the shell. Several species in the family have shorter necks. A long neck is an advantage for reaching prey and also enables the turtle to breathe air from above the surface of the water while remaining on the bottom of a pool.

Snakes comprise roughly 2,500 species of elongate *reptiles that generally lack all trace of legs, although claw-like hind-limb vestiges occur in some primitive groups. Most snakes have enlarged, wide scales on the underside of the body, but these are reduced or absent in many totally aquatic groups, such as sea snakes, wart snakes, and some burrowing species. Snakes lack movable eyelids, the eyes of most species being covered by a transparent spectacle (brille), and this feature separates them from lizards, which have, in addition, reduced limbs as as adaptation to a subterranean life. In some burrowing species of snake the eye is small and covered by the head-scales. Most snakes have remarkably mobile jaw-bones which, together with stretchable throat skin, enable them to swallow prey that is much wider than their own head.

Three categories (or infra-orders) of snakes may be recognized: Scolecophidia, which includes blind snakes and thread snakes; Henophidia, a group with some primitive characteristics and including boas, pythons, cylinder snakes, sunbeam snake, and shield-tail snakes, and Caenophidia, or higher snakes, including colubrids, elapids (such as cobras, mambas, coral snakes, kraits, and sea snakes), and vipers.

Snake's head (plant) *Fritillaries.

Snapdragon *Antirrhinum.

Snappers are predatory, marine fishes which belong to the family Lutjanidae, containing some 230 species. They occur in all the tropical oceans except the east Pacific and occasionally swim up estuaries. They are moderately deep-bodied fishes, with a rather long, concave profile, and a spiny dorsal fin continuous with a second dorsal fin. They have strong sharp-pointed teeth in the front of the jaws and eat a wide range of fishes, crustaceans, and octopuses. They are important food-fish for man, although in the Caribbean their flesh may be poisonous, due to the build-up of toxins ingested from algal-eating fishes which the snappers in turn have eaten.

Snapping turtles are freshwater, New World species of *turtle with large heads, strong jaws, and rough shells. The alligator snapping turtle, *Macrochelys temmincki*, is exceptionally large, and can weigh more than 90 kg. (200 lb.). The common snapping turtle, *Chelydra serpentina*, is smaller, not often exceeding 25 kg. (55 lb.), but is particularly aggressive, especially when encountered on land. The strong jaws of these turtles, which are endowed with sharp, horny ridges (turtles lack teeth), can sever a man's finger at one go.

Snipe are a group of medium-sized *wading-birds in the family Scolopacidae, and are notable for long, straight beaks with which they locate and swallow underground prey. The position of their eyes gives them an extra-wide field of vision, useful in avoiding predators. They inhabit marshland and, when disturbed, tend to crouch until almost trodden upon before rising. In the breeding season some species, including the widely represented common snipe, *Gallinago gallinago* (which is known as Wilson's snipe in North America), perform spectacular switchback display flights in which bleating or drumming noises are made by outer tail feathers during dives. There are about twenty species of snipe.

Snipe-eels are a family of oceanic eels which have exceptionally long slender bodies and a long, thin beak-like snout. Their eyes are very large, and they are dark brown above and black ventrally. They live in mid-water and feed on small crustaceans. The snipe-eel, *Nemichthys scolapaceus*, is worldwide in distribution and grows to a length of 1 m. (3 ft. 3 in.).

Snipefishes belong to a family of about ten species of marine fishes, which usually live in tropical and sub-tropical seas. They are related to the pipefishes and have a similar long, tubular snout but are deeper-bodied and have a long spine in the first dorsal fin. Most species are small, few growing as long as 30 cm. (12 in.), and generally they are rose-pink in colour.

Snoek: a valuable food-fish which is found off the coasts of South Africa and southern Australia (where it is known as barracouta). It also occurs in the open sea all round the Southern Hemisphere. The snoek, *Thyristes atun*, is a migratory fish travelling in large schools, and feeding on smaller fishes and euphausid crustaceans. Although superficially resembling the mackerels, it belongs to a separate family, which includes mostly open-ocean fishes. Its body is slender, the spiny dorsal fin is long-based, and it has finlets behind both the dorsal and anal fins; its teeth are large, numerous, and sharp. It grows to 1·5 m. (5 ft.) in length.

Snowberries are the eleven or so species of *Symphoricarpos*, native to North America and western China, but the name refers especially to the North American *S. rivularis*. It is a deciduous shrub with simple leaves, small pinkish flowers, and spherical white mushy fruits about 1 cm. (½ in.) in diameter. These shrubs belong to the same family as elders and honeysuckles.

Thick plumage provides the **snowy owl**, *Nyctaea scandiaca*, with excellent insulation. Even the bird's legs and toes are feathered, often almost hiding its claws.

Snowdrops and snowflakes are perennial plants of the daffodil family (Amaryllidaceae), which grow from bulbs. The four species of snowdrop, *Galanthus* species, flower in early spring and are native to woods and scrubland in Europe. Their small pendulous white flowers, held about 18 cm. (7 in.) high, consist of six petals, the three inner ones each having a green patch. The common snowdrop, *G. nivalis*, has been selected by breeders to produce double-flowered forms, which are popular as garden plants. The eight species of snowflake, *Leucojum* species, are native to Europe and are almost twice the height of snowdrops. Their flowers consist of six equal sized petals, and are produced in spring, summer, or autumn, depending upon species.

Snow fleas   *Scorpion-flies.

Snow geese are large, pure white geese with black wing-tips or, in their dark form, grey with a white head and neck. The bill and legs are pink. They breed in the Arctic tundra and move southwards to North America in winter, sometimes visiting Europe accidentally.

Snow leopard: one of the most beautiful species of big *cat, which lives in the Himalayas, above the timber-line, at altitudes of 1,800–5,500 m. (6,000–18,000 ft.). To combat the cold it has a superb, leopard-like coat of deep, soft fur, pale grey or creamy buff in colour, ornamented with large black rosettes or broken rings. It is 60 cm. (2 ft.) tall at the shoulder and 1·9 m. (6 ft. 3 in.) long. It is a shy, nocturnal animal; little is known about it except that wild sheep, goats, marmots, and domestic stock fall prey to it.

Snowy owls are a species of large, white or grey-spotted *owl of the northern tundra, with a wing-span of 1·5 m. (5 ft.). Nesting in the far north, they migrate to sub-arctic and temperate latitudes in winter. Females are darker and

slightly larger than males. They nest on the ground, laying four or five eggs, and their success at raising chicks depends on the availability of rodents. Large numbers of foxes and predatory skuas and gulls also affect breeding success.

Soapwort is a perennial plant related to the carnation. The leaves, when bruised, release a sap which may be used as a substitute for soap.

Social animals are those which, with others of their own species, organize themselves co-operatively. Examples are found among insects and vertebrates, particularly mammals. Insect social behaviour is stereotyped, impersonal, and based upon chemical (*pheromone) communication using largely inherited behaviour patterns (instinct). Ants, honey-bees, and termites are all socially organized. In each case reproduction is controlled by one female, the queen (which will lay thousands of eggs in its lifetime), and the construction and maintenance of the colony is through the activities of the workers. Other members of the colony may have a distinct physical appearance related to their specific tasks, such as the defence role of the termite soldiers.

Vertebrate social animal behaviour may be instinctive or learned. In mammals and birds, it embodies a number

Solenodons

Formerly known only in Haiti, the Hispaniola solenodon was thought to be extinct at the turn of the century, but in 1907 breeding colonies were found in the interior of the neighbouring Dominican Republic. Though by no means safe, this large insectivore now seems in no immediate danger, but clearing and cultivation of its limited, stony forest habitat and further spread of alien predators to these islands would have serious consequences.

of rituals which ensure recognition of some individuals by the rest of the group and establish territory and social dominance. These actions reduce potentially damaging aggression. Birds such as the gannet, which nest in colonies, have elaborate visual signalling systems. Primates such as baboons and man have a large number of learned social behaviours, including complex communication patterns. Their societies are based on personal recognition of other members of the group.

Social animals are thought to gain protection from predators by their co-operation. They may also have increased breeding opportunities, and greater efficiency through division of labour. Many actions of social animals can be regarded as altruistic.

Soft corals are strange, fleshy or leathery structures found in most seas, forming colonies up to 1 m. (3 ft.) across in Indo-Pacific reefs. They differ from other *corals in lacking massive calcium carbonate skeletons, gaining support from their own soft rubbery bodies instead (though there may be some chalky spicules). They are usually irregular in shape, with lobes and projections, and one common type is aptly named dead men's fingers, *Sarcophyton*. They belong to the octocorals, so each *polyp within the coral has eight tiny tentacles which can be withdrawn if the animal is disturbed.

Soft-shelled turtles occur in North America, Africa, Asia, and western Indonesia. Their shells are covered with a layer of leathery skin, unlike the horny, shell covering of most other turtles. They are mainly aquatic in their habits and have a snout which is extended into a snorkel-like proboscis, used for breathing air from the water surface whilst the rest of their body remains submerged. Their lips, which conceal the jaw surfaces, are fleshy and soft. Their main food is crayfish and aquatic insects.

Solan goose is an old name for the northern *gannet.

Solenodons are an insectivore family of only two species from Cuba and Haiti. They resemble large shrews, growing up to 32 cm. (1 ft.) long, excluding the tail, and have long claws, particularly on the front feet, and an elongated, flexible snout that is used to root out small vertebrates and insects from the ground. They are easy prey for introduced dogs and mongooses and, consequently, are becoming rare in their native habits. At least one of the two, the Hispaniola solenodon, *Solenodon paradoxurus*, has a toxin in its saliva which helps to kill its prey.

Soles are a family of *flat-fishes of world-wide distribution in tropical and temperate seas, almost always found in shallow water. They are slender-bodied with the dorsal fin beginning right forward on the snout and continuing almost to the tail fin; the anal fin is almost as long, but the other fins are small. Soles have both eyes on the left side of the head (they undergo a metamorphosis like all flat-fishes) and the mouth is often twisted. On the blind side of the head there are elaborate sensory organs.

The sole, *Solea solea*, also called the Dover sole, is a European fish, living on the sea-bed (although often swimming near the surface at night while migrating) at depths of 10–100 m. (32–325 ft.). It is most common on sandy or muddy bottoms, and usually feeds at night on small crustaceans and worms. It is an important food-fish and grows to a length of 60 cm. (24 in.). Several members of the sole family are found on the Atlantic coast of North America, where the term sole has a much wider application.

Solitaire birds are three species of extinct bird that formerly occupied Rodrigues, an island in the Madagascar group. They were related to the *dodo and like it were completely flightless; in size and general form they resembled a large turkey. By the end of the eighteenth century all three species had become extinct.

Solomon's seal is the common name for some perennial plants with a stout, rhizomatous rootstock, producing annual stems 1 m. (3 ft. 3 in.) in height with clusters of pendent white flowers in the axils of the paired leaves. All belong to the genus *Polygonatum*, itself part of the lily family. Virtually all are native to Europe and Asia.

Song thrush: a species of bird in the thrush family which is widely distributed over Europe and western Asia. Closely related to the redwing, the song thrush is a duller brown, lacking the redwing's conspicuous eyestripe and having pale orange under the wing instead of orange-red. It has a musical song, containing notes or phrases copied from other birds. Each phrase is usually repeated two or three times.

Sorghum is one of the most valuable cereals in the semi-arid areas of the tropics, being more resistant to drought than maize. Of African origin, it has been the staple cereal in parts of Africa and Asia for thousands of years. It is a variable, annual grass up to 4 m. (13 ft.) tall, bearing a large terminal flower head. The round grains vary in colour from white or yellow to red, brown, or even black. The lighter-coloured types are preferred for eating, the more bitter dark-coloured types being used for beer. Sorghum flour is usually made into a porridge. In the southern Great Plains of the USA, a dwarf sorghum with compact heads is widely grown for livestock food.

Sorrel is the common name for several low-growing perennial plants in the widespread genus *Rumex*. They are closely related to *docks and are members of the rhubarb family, or Polygonaceae. The common European sorrel, *R. acetosa*, has arrow-shaped leaves which can be used as a green salad vegetable. A related species, *Oxyria digyna*, known as the mountain sorrel, belongs to the same family.

Soursop is a member of a family of small, tropical American trees that produce compound fruits with a custard-like flavour. Along with the sweetsop, or custard-apple, they belong to the genus *Annona*. Pollination in all of them is carried out naturally by beetles, and ultimately large green, spiny fruits are produced. The flesh of the fruit ripens as a pulp and is eaten also by bats, squirrels, and monkeys. Soursop is cultivated for this fruit, but being very sensitive to cold, it succeeds only in the warmest climates.

Southernwood *Artemisias.

Sow bugs *Woodlice.

Soya beans are an annual *legume native to southwest Asia, and the most important food legume in China, Korea, Japan, and Malaysia. It was first grown commercially in the USA in 1924, and North America now produces 40 per cent or more of the world's output. Because of its long history of cultivation, thousands of varieties are now known, each adapted to a different use. It is one of many beans used in Chinese cookery, either fresh, as whole beans, or fermented to give soy sauce. The dried seeds can be ground into protein-rich flour, useful as an ingredient of ice cream or mixed with wheat flour for baking. The seeds contain up to 50 per cent protein and 25 per cent oil. This oil is used for margarine and cooking oils. The protein-rich residue, left after oil extraction, is an invaluable source of vegetable protein, used mainly as a livestock feed, but increasingly incorporated into human diets as a replacement for animal protein.

Spadefoot toads possess an enlarged, shovel-shaped projection, or 'spade', on the inside of each hind-foot. They usually live in sandy areas and use the spade for digging deep burrows, which are used as retreats during the day or in periods of drought. The name is usually applied specifically to the family Pelobatidae, which includes the spade-footed toads of North America, Europe, and Morocco, but has also been used for the Australian genus, *Neobatrachus*. Spadefoot toads are usually 8–10 cm. (3–4 in.) long and have moist skin, akin to that of a frog.

Spanish flies are blue and green *blister beetles whose adults feed on privet and ash in Mediterranean countries. The dried, bright green wing-cases of one species, *Lytta vesicatoria*, were once used in medicine for raising blisters. If taken internally, they can be dangerously poisonous.

Spanish mackerel: a species of fish of the tuna family which inhabits tropical Atlantic waters. It has a rather slender body with a long-based, spiny, first dorsal fin, and is deep blue on the back, with silvery sides and belly adorned with yellow to orange spots. It grows to 1·2 m. (4 ft.) in length and lives near the surface, often in shallow coastal water, and enters estuaries. It is regarded as a fine game-fish by anglers. The name Spanish mackerel is also used in Britain for a smaller mackerel-like fish, *Scomber japonicus*, which is known as chub mackerel in North America.

Narrow-barred Spanish mackerel, *Scomberomorus commerson*, sometimes called the seer, is an elongate, torpedo-shaped fish with a low, spiny dorsal fin, and numerous wavy, dark bands on the sides. This species is an Indo-Pacific fish, growing to 2·3 m. (7 ft. 6 in.), and forms large schools just outside the coastal reefs. It is an important food- and game-fish.

Spanworms *Loopers.

Sparrow is a name applied to many small birds, especially those which are streaky brown and have short, stout beaks for eating seeds. One main group of 'sparrows' contains about twenty species of the genus *Passer*, belonging to the weaver family. These are Old World in origin, but some, especially the house or English sparrow, *P. domesticus*, have spread widely in man-made habitats around the world. They are streaky brown birds although the males may be quite attractive with patches of rich red-brown and black. They nest in holes or in thick vegetation, building untidy, domed nests of grass and other vegetation. The other main group contains about fifty species of several genera in the bunting family, all occurring in the New World. Most of these are smaller than the house sparrow, but the fox sparrow, *Passerella iliaca*, is almost the size of a thrush. Most are brownish or greyish birds, many heavily streaked below; others have chestnut or black head markings. All are primarily seed-eating, though they may feed insects to their young.

Sparrow-hawks are broad-winged *hawks of the genus *Accipiter*, and the name is used especially for the European sparrow-hawk, *A. nisus*. This species is found in woodland and forests throughout much of Eurasia, and is a stealthy hunter of birds, insects, and other small prey. It uses the cover of woodland edges to swoop on sparrows, pigeons, and other birds feeding on open ground. Twig nests are built in trees and between four and seven eggs are laid, the males feeding the females throughout incubation. Many sparrow-hawks die of insecticide poisoning. The same name is used in America to describe *Falco sparverius*.

Spawn is the name given to the eggs of aquatic animals when they are expelled in a mass. The term is used for certain molluscs and crustaceans, but is best known in fishes and some amphibians, such as frogs and toads. The egg is composed of stored food material in the form of yolk and fat (mostly seen as oil droplets) surrounded by albumin and covered with a thin shell. In fishes the spawn is frequently shed at random in the water, although conditions of temperature and salinity must be suitable. The cod is an example of a fish whose eggs float near the sea's surface. Some species of fishes, such as herring, spawn on the sea-bed, their spawn being laid in a mass sometimes several eggs deep, on stones, shells, and even algae. Many

members of the carp family lay their eggs over plants; their eggs have a sticky outer surface. In other species of fish, spawn is laid in long lacy threads, either in the sea (angler-fish) or wound around vegetation in fresh water (perch). Spawn may also be shed in gravel in a hollow excavated by the parent (trout), or even on the sea-shore at high spring tides (as in the Californian grunion). A few species place their spawn in nests made by the parent (stickleback) or attach their eggs to floating algae in the sea (flying fish).

Spearfishes are three species of the genus *Tetrapturus*, which are relatives of the marlin. Spearfish have a long-based, quite high dorsal fin, a broad tail, and a long snout similar to that of the marlin, and typical of the family. They are found in all tropical seas, living near the surface in the open sea and feeding on fishes and squids.

Species, in *nomenclature in zoology and botany, rank below *genera and form the basic unit of the system. A species is usually considered to include a group of individuals capable of interbreeding with each other to produce fertile offspring, and kept apart from other species by genetic, geographical, or other barriers. A species may include many distinct subspecies, varieties, or races. Cultivars, or named varieties of plants, developed by selection or hybridization, are also included, provided they produce viable offspring. Latin names of species are usually in two parts, as in man's specific name of *Homo sapiens*; a third word may be added to indicate a subspecies or another division below the level of the species.

Speech is the human faculty for the verbal exchange of ideas, using language. Sound is generated by forcing air out of the lungs against the resistance provided by the vocal cords (*larynx). This sets up vibrations in the air-flow that produce sound, which is modified by changing the

Breeding from the end of April, the European **sparrow-hawk** builds a platform nest. The short wings, and longish tail and legs help to give this hawk good manoeuvrability.

position of the lips and tongue to produce the elementary speech sounds. Different sequences of the basic sounds produce words. Learning to do this accurately in childhood requires normal hearing and normal control of the muscles in the throat, mouth, and tongue by the brain.

Speedwell is a name applied to various members of the plant genus *Veronica*, which belongs to the foxglove family. They are annual or perennial plants, often with prostrate or creeping stems, opposite leaves, and spikes of blue or occasionally white flowers. The species are found in a variety of habitats—stream-sides, woodlands, meadows, as weeds in lawns, and at fairly high altitudes on mountains. They are chiefly from the temperate regions of the Northern Hemisphere. The various kinds are identified by suitable prefixes to the name speedwell, for example, fingered, ivy, germander, wall, and thyme-leaved. This genus formerly included a large number of evergreen shrubs, most of them native to New Zealand; these have now been classified as the genus *Hebe*.

Sperm are the male equivalent of the *ovum, and are produced in vast numbers by the *testes. A sperm cell has a head containing the *nucleus with half the normal number of *chromosomes. Attached to the head is a tail which propels the sperm towards the ovum. Only one sperm head penetrates the ovum wall and *fertilizes the ovum, after which development commences.

Sperm whales are the largest of the *toothed whales. The three species range through all the seas, migrating north and south with the seasons. The large, almost square, head is a striking feature of these mammals. Much of the head is taken up by the 'case', a transformed and expanded region of the right nostril, filled with spermaceti. This is liquid at body temperature but forms a white wax when cooled. Spermaceti may be important in controlling buoyancy, particularly during prolonged and deep dives. The sperm whale, *Physeter macrocephalus*, can reach depths of 1,200 m. (4,000 ft.) and has been found entangled in submarine cables at these depths. It can remain below the surface for as long as fifty minutes, after which it spends ten minutes at the surface. It feeds largely on squids and cuttlefish; a large male eats a total of some 99,800 kg. (110 US tons) of these cephalopods in a year. The male can reach a length of 19·8 m. (65 ft.) and is larger than the female. A single calf of up to 4·8 m. (16 ft.) in length is born after a gestation period of about fourteen and a half months; it may be suckled for as long as two years. The two other species, the pygmy sperm whale, *Kogia breviceps*, and dwarf sperm whale, *K. sinus*, are much smaller, the largest reaching 3·4 m. (11 ft.) in length. They occur in warmer seas than the sperm whale.

Sphagnum is a large plant genus of mosses which form large areas of peat bog in wet, upland regions, especially in acid and nutrient-poor conditions in which few other plants can thrive. They have a uniquely spongy structure so that the hummocks, or carpets, of *Sphagnum* plants retain water. The damp, nutrient-poor conditions in which these mosses grow rule out rapid decomposition of dead plant material. Instead, layers of peat develop, gradually building upwards with the *Sphagnum* maintaining wet, boggy conditions. The peat is a useful fuel.

Sphinx moths   *Hawk moths, *Tobacco hornworm.

Spices come from the leaves, flowers, seeds, bark, or roots of many plants. Although there is no clear distinction between them, those of tropical origin tend to be known as spices, whereas 'herbs' such as rosemary, parsley, and thyme are typical of temperate areas. Of importance since the beginning of history, certain spices were highly prized in ancient China, Egypt, Greece, and Rome. Spices are used in food preparation, adding flavour and sometimes colouring. Their addition effectively removes the monotony of many staple diets, and masks the unpleasant odour and flavour of meat that is either dried or not fresh.

Much of the European exploration of the world in the fourteenth and fifteenth centuries was stimulated by the quest for various spices. Such was the esteem of the true pepper in medieval England that rents and taxes could be collected in peppercorns. Columbus's voyage to America arose from Portugal's desire to find a shorter route to the Indies in the search for pepper.

Spider crabs do not resemble the usual flattened ovoid shape of crabs, but have a triangular or pear-shaped body, projecting between the eyes, and very long legs. The superfamily includes the decorator crabs, which camouflage their own shells under a forest of other sessile organisms. The giant of the group is the Japanese spider crab with a leg-span up to 4 m. (13 ft.). Most spider crabs are entirely marine, quite common in rock-pools but always difficult to spot.

Spider monkeys are four species of New World monkey, found in the tropical forests of Central and South America from southern Mexico to Bolivia. They live in the upper branches of trees and can travel at about 6·4 km./hour (4 m.p.h.). They are remarkably acrobatic monkeys, and can take a flying leap of 9 m. (30 ft.). Spider monkeys eat fruit, flowers, insects, and birds. They are easily recognizable by their light, slender body, long spidery limbs, and tail of extraordinary length, 60 cm. (2 ft.), a little longer than the head and body together. The tip of the tail is naked and can grasp firmly anything which it touches, acting almost like a hand. The coarse, woolly fur of the long-haired spider monkey, *Ateles belzebuth*, and the black spider monkey, *A. paniscus*, is entirely black in colour, while in the brown-headed spider monkey *A. fusciceps*, and black-handed spider monkey, *A. geoffroyi*, it can be reddish-brown. There is no fixed breeding season. After birth, the baby is carried and cared for by its mother for some ten months. When being carried, the baby will curl its tail around the mother's tail.

Spiders are the commonest *arachnids, and are found in huge numbers everywhere except in the Antarctic, with up to 2 million in any acre of meadow. There are 35,000 known species, many only 1 mm. ($\frac{1}{25}$ in.) long, but a few tarantulas have bodies 10 cm. (4 in.) in length. Most spiders are carnivores, and are the only invertebrates that trap flying insects. Those which use silk to catch prey employ it as trip-wires, webs, or even lassos! All spiders rely mainly on touch and vibration to sense their prey, as their eyes are simple, and capable only of discriminating between night and day. They all use venom to paralyse prey, though only a few can pierce human skin. Ground-hunting species of jumping spiders have eight prominent eyes and are very agile.

Silk is spun from six glands, of varying thickness and stickiness. It is the strongest material made by any in-

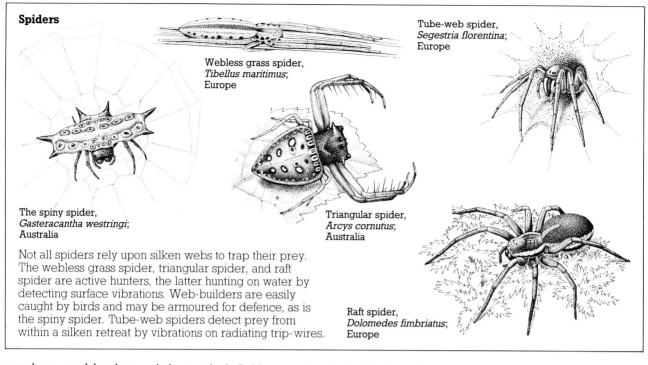

Spiders

Webless grass spider, *Tibellus maritimus*; Europe

Tube-web spider, *Segestria florentina*; Europe

The spiny spider, *Gasteracantha westringi*; Australia

Triangular spider, *Arcys cornutus*; Australia

Raft spider, *Dolomedes fimbriatus*; Europe

Not all spiders rely upon silken webs to trap their prey. The webless grass spider, triangular spider, and raft spider are active hunters, the latter hunting on water by detecting surface vibrations. Web-builders are easily caught by birds and may be armoured for defence, as is the spiny spider. Tube-web spiders detect prey from within a silken retreat by vibrations on radiating trip-wires.

vertebrate, and hardens as it is stretched. Spiders use it not only for trapping but also for insulating their homes, wrapping their eggs, and dispersal by parachuting into the air. A male may use it to wrap gifts for an aggressive spouse, and may even pluck soothing tunes on her web as he approaches, bearing sperm in his *pedipalps. Even so, he may end up as her next meal.

Spiderworts are perennial herbaceous plants, with grass-like leaves and clusters of bright blue flowers composed of three petals. They belong to the monocotyledonous family Commelinaceae and are members of the genus *Tradescantia*. Some species are popular flower-border or greenhouse plants.

Spina bifida is an abnormality in man caused by defective closure of the vertebral column, which normally encases the spinal cord. This is present in the newborn baby. It varies in severity from an imperceptible bony gap, detected only by X-rays, to complete exposure of the nervous tissues. Thus some children have no handicap, while others are paralysed from the spinal defect downwards, sometimes with failure of bladder and bowel control. Initial treatment requires surgery, to close the skin defect and hence prevent infection, and to relieve increased spinal fluid pressure in the brain (hydrocephalus). Prenatal diagnosis of spina bifida is increasingly available, enabling parents to opt for termination of pregnancy.

Spinach is widely cultivated in temperate regions for its leaves, which are are rich in protein. It belongs to the same family as sugar-beat, beetroot, and mangels, and usually refers to the species *Spinacia oleracea*.

Spinach beet is a variety of *Beta vulgaris*, which also includes *beetroot and *sugar-beet. It is a vegetable grown for its abundance of leaves throughout most of the year. Spinach-like, but milder in flavour and also known as perpetual spinach and leaf beet, it has succulent leaves which can be eaten whole, including the long, green stalk.

Spindle trees belong to the cosmopolitan genus *Euonymus*, which comprises some 175 species of deciduous trees and shrubs. The common spindle tree, *Euonymus europaeus*, is typical of the genus in having scarlet fruits like cardinals' hats, which split open to reveal orange seeds. Some other species have winged or spiny fruits. The wood has been used to make spindles, and burnt to produce high-quality charcoal. They give their common name to the spindle-tree family, Celastraceae.

Spine (anatomy), or backbone, is a series of articulating *vertebrae which gives its name to the animal subkingdom called vertebrates. The spinal cord, one of the major parts of the central *nervous system, runs within the spine, inside a cavity of each vertebra. The spine serves to protect the spinal cord and permit the animal's body to twist and turn, and varies from species to species.

In man the spine is a pillar composed of seven cervical, twelve thoracic, and five lumbar vertebrae separated by intervertebral *discs. The vertebrae articulate with one another and are so arranged that spine movements are limited in degree and direction. Cervical vertebrae can move backwards and forwards as well as twist upon one another. Thoracic vertebrae can move forwards, and rotate upon one another to a very limited extent but cannot move backwards. Lumbar vertebrae can only flex and extend upon one another to a limited extent. The human spine is so constructed that it has three natural curvatures. The cervical vertebrae form a forward convex curve, the thoracic a backward curve, and the lumbar a forward convex curve.

Spiny anteater *Echidna.

Spiny dormouse: a single species which, with the Chinese dormouse, *Typhlomys cinereus*, comprises the subfamily of oriental dormice. The spiny dormouse, *Platacanthomys lasiurus*, from southern India, is similar to a dormouse in appearance, with its long bushy tail, and it is about the same size, but it has sharp spines protruding from the fur on its back. The animal lives in trees and feeds on fruits and seeds.

Spiny rats are members of the cavy-like *rodent family Echimyidae; most, but not all, of the fifty-six species have spiny or bristly hairs in the fur. Some have pointed snouts, large ears, and long tails that give them a close similarity to true rats. All are Central or South American. Some are arboreal but others live in burrows. They are herbivorous in diet.

Spiny rock lobsters　*Crawfish.

Spiraea is a genus of some 100 species of deciduous shrubs native to the north temperate zone and extending as far south as Mexico. They have heads of small, white, pink, or reddish flowers, and many species are cultivated as garden plants. As members of the rose family they are closely related to very similar species in the genera *Astilbe*, *Aruncus*, and *Filipendula*. Some of these are commonly referred to as 'spiraea' because of their similar flowers.

Spirogyra is a genus of filamentous green *algae commonly found in fresh water as skeins of fine green threads. The threads consist of individual cells attached end to end: inside each are microscopic granules (chloroplasts) that contain the photosynthetic pigment, chlorophyll. The threads are twisted in the form of a spiral ribbon, hence the

Sexual reproduction seen taking place in **spirogyra**. Dome-shaped protuberances, formed between opposite cells of the two threads in close proximity in the centre, have joined to make a tube. The contents of the cells in the lower thread have moved through it to fuse with those of the upper one. New filaments (zygotes; towards the *top right*) are produced from the resultant structures, after being released from the cells.

name *Spirogyra*. They grow by simple lengthwise division of cells, a process known as asexual reproduction, though occasionally *Spirogyra* undergoes a cycle of sexual reproduction when two threads come to lie alongside each other. One of the strands passes across the contents of each cell through a temporary 'tunnel' into the corresponding cells of the other strand. The donor thread, now an empty shell, breaks away and dies, while the recipient thread undergoes a series of changes before resistant 'spores' are released, eventually to produce new threads.

Spittle-bug　*Frog-hoppers.

Spleen is the name of an abdominal organ concerned with maintaining the proper condition of the blood in most vertebrates. It is a fibrous capsule which encloses developing white blood cells called lymphocytes. It also contains numerous phagocytic cells (capable of engulfing foreign matter or bacteria) and has an extensive, slowly moving blood supply. In mammals, it lies in the upper left part of the abdomen. It acts as a blood purifier and reconditioner, enabling dead or disintegrating red cells to be removed and digested. It also adds lymphocyte defence cells to the circulation. In mammals other than man, the spleen has a muscular capsule which contracts during exercise to augment the blood supply to the muscles, thereby providing them with the extra oxygen they need.

Sponges are unique animals which comprise the phylum Porifera, with 10,000 species. They are so simply organized that they are often regarded as colonial *protozoans, with scarcely any co-operation between their constituent cells. Sponges have no mouth, digestive cavity, nerves, muscles, locomotion, or true behaviour. They do, however, occur as multicellular organisms, often with quite specific shapes and structures, and they are undoubtedly a successful group. With the exception of 150 species in fresh water,

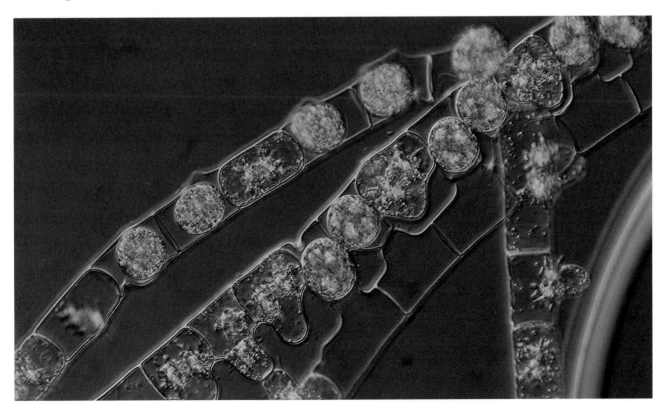

the vast majority of sponges are marine. Most form small and irregular encrusting masses on rocks, although some may reach 1 m. (3 ft.) across. They occur in a wide range of colours.

Sponges live very differently from other animals, surviving on the food particles brought by water currents which enter the colony through pores all over their surface. Within the internal channels, tiny flagellated cells (very like some protozoans) filter out this food and waft the current onwards to an exhalant opening. The whole structure is supported by skeletal spicules of chalk or silica, or by a network of fibres. This last system is apparent in the natural bath sponge when the living cell contents have been dissolved away. Some deep-water sponges have very elaborate and beautiful skeletons, and one form is known as Venus' flower basket.

Spoonbills are large, chiefly freshwater marsh birds akin to ibises, with unusual flattened spoon-tipped beaks. Mainly tropical or subtropical, they breed in colonies, often building stick nests in trees. The five Old World species are white, but the single American species, the roseate spoonbill, *Ajaia ajaia*, is rosy pink.

Spores are produced by *cryptogam plants and microorganisms, including bacteria, fungi, ferns, clubmosses, horsetails, mosses, liverworts, and algae. They are microscopic, or only just visible to the naked eye, but can be seen as clumps on the underside of fern leaves or under the cap of fungi. Unlike seeds, they contain only a few cells and no embryo, but still grow into new individuals.

Most are liberated into the air and travel long distances in air currents, although most algae and some fungi have spores that can swim in water using beating hairs, or flagellae. Other kinds of spores lie dormant where they are produced; dormant spores of bacteria are the longest-surviving living material, resistant to extremes of temperature and drought. Those of the bacterium *Anthrax* can survive up to 50 years.

Sprat: a small species of fish in the herring family which is found in coastal waters around Europe. Its maximum length is 16 cm. (6 in.). It is an abundant, schooling fish, usually found near the surface and is often common in estuaries and bays. Sprats spawn in spring and summer; the eggs and larvae are planktonic and the young fishes drift into coastal waters as they develop. The young are fished for as *whitebait and the adults are eaten either fresh or smoked, or they may be canned as brisling. They superficially resemble small herring, as their back is green, shading to silver on the sides. Sprats have sharply pointed scales along the belly.

Springbok, or springbuck: a species of *gazelle about 76 cm. (30 in.) tall at the shoulders. On its back is a fold of skin which is turned inside out to display an array of white hairs when the animal is startled. This serves as a signal to others in the herd. Huge herds once roamed the plains of South Africa, but persistent hunting has reduced their numbers and range. The common name of this animal comes from its peculiar habit of leaping unexpectedly into the air.

Springhaas, or springhare: an African *rodent which is the sole representative of the family Pedetidae. In appearance it looks like a small bushy-tailed kangaroo and

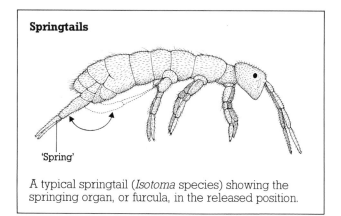

Springtails

A typical springtail (*Isotoma* species) showing the springing organ, or furcula, in the released position.

'Spring'

hops like one. About 4 kg. (9 lb.) in weight, it is prized by the African Bushmen for its flesh. Its distribution is disjunct, with one population in southern Africa and another in East Africa. It lives in dry, sandy country, digging extensive burrow systems, and is nocturnal and herbivorous.

Springtails are minute, wingless insects whose abdomen contains only six segments. They form an order (Collembola) of over 2,000 species. The fourth abdominal segment bears a double appendage which is held at rest under a catch on the third segment. When this catch is released, the springtail jumps high into the air. The first segment of the abdomen bears a long tube which is used either to absorb water, or as an adhesive organ to help springtails climb steep surfaces. They feed mostly on fungi and plant detritus and are the most abundant animals in the upper layers of the soil. A few, including the lucerne flea, *Sminthurus viridis*, live on plant leaves, while others can live on the surface film of fresh or salt water.

Spruces are conifers of the genus *Picea*, which belongs to the pine family. They are evergreen trees with hard, sharp leaves, or needles. Their cones are soft, leathery cylinders which hang from branches and are shed as entire cones. The Norway spruce, *Picea abies*, found from the Pyrenees to the Arctic Circle, provides valuable timber, resin, and turpentine. It is perhaps most familiar as the Christmas tree. There are fifty other species of spruce, of which many are grown for their timber and ornament. The red spruce, *P. rubens*, black spruce, *P. mariana*, and the white spruce, *P. glauca*, of North America, are used for pulp, especially newsprint. A species planted widely in Europe for timber and pulp is the Sitka spruce, *P. sitchensis*.

Spurges comprise the plant genus *Euphorbia*, which has over 1,000 species, widely distributed throughout the temperate and tropical regions of the world and showing great variation in size and form. The succulent kinds (often devoid of true leaves) from southern and southwest Africa include *E. grandidens*, a spiny cactus-like tree up to 12 m. (39 ft.) in height, and *E. globosa* from the Cape Province, a dwarf cylindrical-stemmed succulent only a few centimetres (about an inch) high. The leafy herbaceous species from Europe and the Mediterranean region are often used as garden border plants, while *E. pulcherrima* from Mexico is the popular Christmas pot plant called poinsettia. All have a milky latex which is very poisonous. The family to which this genus belongs, the Euphorbiaceae, contains over 5,000 species of mostly tropical plants.

Squash is the name used, particularly in North America, for the edible fruits of several species and varieties of *Cucurbita*, belonging to the pumpkin family. Summer squash describes those varieties whose young fruits of various shapes and colours are used as vegetables. Winter squash describes the mature fruits of several species which keep well in frost-free conditions and form invaluable food for livestock as well as man. The summer squashes include the vegetables that Europeans refer to as *marrows. They are all varieties of *C. pepo*. The winter squashes are derived from *C. maxima* or *C. moschata* and include a variety which Europeans call a pumpkin.

Squids, the commonest *cephalopods, are sinuous, torpedo-shaped animals usually up to 50 cm. (20 in.) long. They have eight arms and two larger tentacles at the front, all bearing suckers, and a rear fin giving stability as they squirt water out of their siphon for jet propulsion. The shell is reduced to an internal horny 'pen', sometimes found on beaches as 'cuttlebone'. Large schools of squids can eat considerable numbers of shrimp and fishes and may even threaten man's offshore catches.

The rare giant squids, *Architeuthis* species, can be 18 m. (60 ft.) long, and are the largest known invertebrates. There are also 'flying squids', which can glide briefly above the waves.

Squirrel-fishes, of which there are about fifty species, belong to the family Holocentridae, and are spiny marine fishes living in all tropical oceans. They are distinguished by the massive spines in the dorsal, anal, and pelvic fins, rough-edged scales and, in many species, a long spine on the lower edge of the cheek; they have large eyes. Many are red in colour, and most are nocturnal, hiding in crevices and caves in reefs during the day.

Squirrels form a large family (*Sciuridae*) of diurnal *rodents which, in addition to the familiar tree squirrels, includes burrowing and flying forms among the 267 spe-

The European red **squirrel**, *Sciurus vulgaris*, holds its food in the manner of all tree squirrels. Like most of them, its diet is almost entirely vegetarian, although it will occasionally eat insects or small reptiles.

cies. Squirrels are widely distributed throughout much of the world. The typical tree-dwelling squirrels build nests in holes. They have long bushy tails which are characteristically held over the back while at rest. Tree squirrels are very agile, leaping from branch to branch, so that it is not surprising that some have evolved into gliding forms, of which there are many species. The largest, of the genus *Petaurista*, are reputed to glide for up to 450 m. (1,500 ft.) and to be able to change direction in flight. The largest tree-living species is the Indian giant squirrel, *Ratufa indica*, which weighs up to 3 kg. (7 lb.), and the smallest, only 7 cm. (3 in.) long, is the African pygmy squirrel, *Myosciurus pumilio*.

Some squirrels have taken to feeding on the ground and resting in underground burrows. Such *ground squirrels have shorter, less bushy tails, a trend which is carried further in the *marmots and *prairie dogs, whose stumpy tails impair their squirrel-like appearance. These squirrels are unusual in being highly social and living in large underground colonies.

Stable flies are true *flies which bite animals and suck blood. They are related to the *housefly, and are sometimes called the biting housefly. They frequent farm buildings and feed on a variety of warm-blooded animals, using the modified proboscis, which is long, rigid, and non-retractile. The larvae develop in horse dung and stable litter, and pupate in the ground.

Stag beetles are large, heavily built, dark coloured *beetles of the family Lucanidae, whose males have their jaws enlarged and branched to resemble antlers. Their jaw muscles are not enlarged and the jaws are useless for feeding or offence, and may be presumed to be of value in competing with other males for females. Their fat, white larvae develop in rotten wood. The males of the biggest British species, *Lucanus cervus*, which have deep red wing-covers and jaws, reach 5 cm. (2 in.) in length and are the heaviest British insects. Most of the 750 species of stag beetles are tropical insects.

A 7·5 cm. (3 in.) specimen of a male **stag beetle**, *Cyclommatus tarandus*, from Mount Kinabalu, Borneo. This species is known to feed on the sap issuing from the bark of tropical trees, such as *Shorea* species.

Stag's horn ferns are some seventeen species of fern in the genus *Platycerium*, widely distributed in the tropics and subtropics of the Old World. Although they are all *epiphytes, some species also grow on steep cliffs. They are characterized by two types of fronds: an upright, undivided type, adapted to clasp the host; and a spore-bearing divided type (the stag's horn), which droops downwards.

Stamens are the structures in a flower which produce *pollen grains. They usually consist of a stalk (filament) and a terminal head (anther) which contains the pollen-producing cells. They are arranged in one or more whorls, rarely spirals, around the ovary in bisexual flowers.

Stanley cranes, or blue *cranes, are a common species of bird in the open plains of southern Africa. They are pearly grey and have a dark fan of elongated, inner secondary feathers drooping over the tail. Demoiselle cranes are their close relatives.

Starch is the main food storage compound of plants. It is not a pure substance but a mixture of two different compounds both belonging to the polysaccharide group of *carbohydrates. The first substance, called amylose, contains 250–300 glucose subunits linked to form a long right-handed spiral shape, or helix. The second substance, called amylopectin, is very similar except that it has a branching structure, with new branches occurring at intervals of about every twenty-five units. Both substances can be rapidly built up or degraded by enzyme action. Following *photosynthesis, starch grains, consisting of a mixture of amylose and amylopectin, are deposited in the cytoplasm of many plant cells, particularly in seeds, and in storage organs such as potato tubers. The starch is converted to sucrose for transport in plants and, after being further broken down into *glucose, provides a large proportion of the daily energy requirement in animals. Starch is particularly important as a food in third-world countries.

Starfish are the most familiar of the *echinoderms, with over 1,600 species found in marine habits worldwide. As with many of their relatives, they show a five-rayed symmetry. The five arms each bear hundreds of tiny co-ordinated suction feet. Together these may exert enough force to pull open a bivalve. The starfish then everts its own stomach into the prey, releasing digestive enzymes, before sucking up the resulting soup through its ventral mouth.

Even in temperate areas starfish can be brightly coloured, often bearing elaborate spines and miniature claws. In the tropics, where they may reach 1 m. (3 ft.) across, their colours may be quite startling. In some species, such as the cushion stars, the arms are so reduced that the body forms a five-sided pad.

Starlings are members of the family Sturnidae, an Old World family containing about 110 species. Mostly thrush-sized, they tend to have an iridescent sheen on black, purple, or green plumage. This family includes the mynahs, which are well known as mimics and are kept as cage-birds. The common starling, *Sturnus vulgaris*, is a bird which originated in the Old World but which has been introduced to many other parts of the world, including North America, where it is now abundant. About 20 cm.

(8 in.) long, it is blackish with pale speckles and a purple or green sheen. In the breeding season the normally black beak turns yellow. It has been very successful in exploiting man-made habitats and can be a serious pest in orchards, particularly of cherries.

Star of Bethlehem is the name used for several species of the genus *Ornithogalum*. They are all perennials of the lily family, native to Europe and northern Africa. The clusters of green and white flowers are produced in spring and early summer and open only when the sun shines. The bulbs were formerly much valued as a source of food, eaten both raw and cooked.

Stems are the part of a plant which bears the leaves and, unlike leaves, continue to grow throughout the life of the plant. They are usually aerial but may be subterranean, as in rhizomes, bulbs, and corms. Many 'stems' of herbaceous perennials are perhaps best viewed as elaborate flower-heads since they grow only to a certain size and then bear the flowers. Stems are green when in their early stages of growth, and photosynthesize. When older, they are often covered in protective bark, though in leafless succulents they may continue to photosynthesize. With age, they acquire more strengthening tissue and, in seed plants, may accumulate great deposits of wood, forming woody stems and branches, or a thick trunk in trees.

Steppe was originally a term for the great Russian plains, which experience cold winters and slight summer rainfall, but its use has been extended to cover any grassland under such conditions. In some steppes the high evaporation brings salts to the surface, and this land can be colonized only by salt-tolerant plants, such as oraches, otherwise seen as shore plants.

Sternum, or chest bone, is the term applied to the bony anterior part of the chest wall. It articulates with the front ends of the collar-bones, and the rib cartilages. The

Vivid **starfish**, or sea stars, abound in the Red Sea. Movement of the tiny, paired feet on the underside of each arm is effected by an intake of sea water.

The Australian **stick insect**, *Didymura violescens*, crawls over a branch of eucalyptus, showing the effective camouflage of its twig-like legs and leaf-like body, tilted at the same angle as the leaves.

sternum is a flat bone and red, blood-forming, marrow is contained in its spongy bone interior.

Steroids are a group of organic compounds which share the same basic structure, comprising four interlinked rings of carbon atoms. The group includes: cholesterol, a major component of cell membranes; bile acids; *vitamin D (cholecalciferol); and many hormones, including those released from the adrenal cortex and the sex hormones. Some synthetic steroids are used as contraceptives, and others are used in the treatment of asthma and to prevent rejection of a new organ following transplant surgery.

Stick insects are large to very large, elongate, green or brown insects, which with the *leaf insects, make up the order Phasmida. In most species the female is wingless and the males, if they occur, are winged. They live in warm climates and are protectively coloured and camouflaged to look like twigs. Most feed on the leaves of trees and drop their eggs to the ground singly. Those that feed on or near the ground are among the bulkiest insects known, growing up to 30 cm. (12 in.) in length. Many species are parthenogenic, producing fertile eggs without mating and rarely producing males.

Sticklebacks are a family of small fishes confined to the Northern Hemisphere, with species living in fresh, brackish, and sea water. Eight species are known, and most are moderately elongate and scaleless with a series of separate spines (from three to sixteen depending on species) along the back, and a single spine forming each pelvic fin. Best known is the three-spined stickleback, *Gasterosteus aculeatus*, found in fresh water in both North America and Europe, and in the sea in the northern part of their range. It grows to 5 cm. (2 in.) in length and feeds mainly on small crustaceans, breeding in early summer in a nest made by the male; an elaborate courtship precedes egg-laying.

Stifftails are a bizarre and distinctive group of mainly freshwater *bay ducks, possessing stiff tails with which they steer under water. On the surface they often swim with tails cocked. Typical is the podgy little ruddy duck, *Oxyura jamaicensis*, an American bird introduced into Britain: the drake is chestnut, black, and white with a blue beak. Related species inhabit every continent. The black-headed duck, *Heteronetta atricapilla*, of South America lays its eggs in other species' nests, cuckoo-fashion. The largest stifftail is the Australian musk duck, *Biziura lobata*, whose male has a big fleshy lobe beneath its beak, and performs complicated communal courtship displays.

Stilts are nine species of wading birds in the same family as *avocets but with straighter beaks and longer legs. The most common stilt, the black-winged stilt, *Himantopus himantopus*, has an almost cosmopolitan distribution. It is black and white with exaggeratedly gangly, pink legs which trail behind the tail in flight. Australia has the only species of stilt in the genus *Cladorhynchus*, the banded stilt, *C. leucocephalus*.

Sting-rays are a small group of *cartilaginous fishes, most abundant in tropical and warm temperate seas. They have a long, serrated-edged spine at the base of the tail which is used as a defensive weapon. The spine has venom tissue (*poison fishes) in grooves on its surface, which makes wounds very painful; large sting-rays have killed humans who have stepped on them. In Britain the Atlantic sting-ray, *Dasyatis sabina*, is mainly a summer migrant, but it does occur all round the coast, and grows to a length of 1 m. (3 ft. 3 in.). Preferring sandy coasts, it feeds mainly on crustaceans and molluscs; it gives birth to live young. Similar species occur on both the Atlantic and Pacific coasts of North America.

Stink bugs *Shieldbugs.

Stinkhorns are fungi with fruiting bodies which exude a distasteful aroma in some species. The powerful and pervasive smell attracts flies, which carry away the slimy greenish mass of spores. In the common stinkhorn, *Phallus impudicus*, these are borne on a stalk about 10–20 cm. (4–8 in.) high, which develops at remarkable speed from an egg-like structure at ground-level. The dog stinkhorn, *Mutinus caninus*, has a pale finger-like fruiting body.

Stoat, or ermine: a species of carnivore related to weasels. Its range extends throughout Eurasia and North America with little variation in appearance. The stoat is a small carnivore with the male at 230 g. (8 oz.) being very much heavier than the 120 g. (4 oz.) female, but it is capable of overpowering prey, such as rabbits, much bigger than itself. It also takes birds and small mammals. In turn, it is frequently killed by hawks and owls. Stoats are generally solitary, but the sexes keep together while rearing young and the male brings food to the nest. Northern populations of stoats turn white in winter, when the pelt is known as ermine.

Stomach: the first part of the intestines to receive food and start digestion in most animals. The stomach is essentially a holding place for pieces of food where they can be mixed with hydrochloric acid and the protein-digesting enzyme, pepsin. That of fishes, amphibians, and reptiles is tube-like. In birds the stomach's function of food storage may be taken away by the expansion of the lower part of the oesophagus into a crop.

The stomachs of other vertebrates fall into one of two

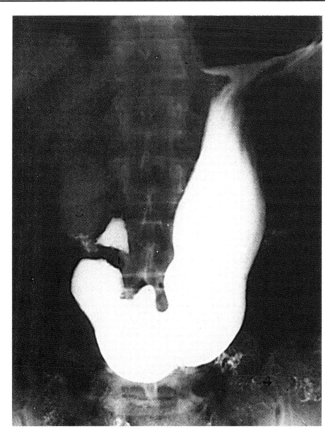

To see if a human **stomach** is functioning correctly, a barium meal may be administered; the processes of digestion may then be monitored by radiography.

general types, those of carnivorous, or omnivorous groups, and those of herbivores, including *ruminants. The ruminant stomach, as typified by the cow, consists of four main chambers. The first, and largest, is the rumen, holding a large quantity of bacteria and protozoa which help to break down the *cellulose in plants. These release some amino acids, proteins, and other useful food substances from the plant material. This mashed-up food is then regurgitated and chewed in the cud, leading to further physical breakdown of the food before it passes back into the other chambers of the stomach. The remaining chambers, the omasum, abomasum, and pylorus, act in the usual manner, secreting enzymes and crushing food.

The human stomach is of the general non-ruminant type, and roughly J-shaped. The external parts are named as the fundus, the body, greater omentum, antrum, and pylorus. A muscular ring, the pyloric sphincter, forces the partly digested food into the first part of the small intestine.

Stonechats are members of the thrush family which occur mainly in Asia. All are small, dark brown or black birds, 12·5–17·5 cm. (5–7 in.) long, with white or chestnut marks and pale or white underparts. They live in open bushy country and sit conspicuously on perches from which they pounce on their insect prey. The best-known species is the common stonechat, *Saxicola torquata*, which is widely distributed across Europe, Asia, and Africa. Its call sounds like two stones being knocked together. Many other species in the thrush family are called chats and include rockchats, cliffchats, and bushchats.

Stone crabs are slightly lopsided *crabs, related to the highly asymmetric hermit crabs but with fully hardened abdomens. They occur mostly in cold oceans and have heavy, sculptured carapaces. The claws are also stout, enabling them to feed on echinoderms and bivalves.

Stonecrops are part of the large family of *succulent plants, Crassulaceae. They all live in extreme environments, some species of *Sedum* and *Sempervivum* (houseleek) are frost-resistant alpines; some species of *Kalanchoe* and *Crassula* live in desert regions. The name stonecrop is usually reserved for species of *Sedum*, some of which are popular as rock-garden plants.

Stone-curlew, also known as thick-knees or dikkops, is the name for nine species of birds. Related to the other *waders, these birds are found in many parts of the Old and New Worlds. They inhabit open, bushy, stony, or sandy country, often away from coasts. They are long-legged, well camouflaged in mottled brown and grey, usually with conspicuous white marks in the wings which are only visible in flight. Most species of stone-curlew are active at dusk or at night, when they feed on a wide range of animal foods.

Stonefish are relatives of the scorpion fishes but belong to the family Synanceiidae. There are several species, confined to the shallow inshore waters of the tropical Indian and west Pacific oceans, which live hidden in coral or among rubble on the bottom. Two species of the genus *Synanceia* are well known for their highly poisonous spines in the dorsal fin. If disturbed the fish erects the spines, which, if trodden upon, automatically inject venom into the wound, which then becomes agonizingly painful. Although deaths are rare from stonefish stings, a wound can lead to amputation of the affected limb.

Stoneflies are small to medium-sized insects, with long antennae and usually a pair of tail filaments. They make up the order Plecoptera, with over 1,600 species found particularly in cooler regions of the Northern and Southern Hemispheres. They rest with the wings flat or wrapped around the body. Their aquatic nymphs, which also have two tail filaments, are found beneath stones; most are herbivorous, some carnivorous.

Stones (medical) may form in many of the ducts in the body, but most commonly in the gall-bladder as *gall-stones, and in the urinary tract or kidneys. There are several types of these but the commonest is due to the precipitation of calcium oxalate. This substance is normally held in supersaturated solution, by means not fully understood, and may precipitate out to form hard, spherical stones under certain conditions. A stone may be symptomless but often causes severe pain especially when naturally ejected from the gall-bladder or kidneys. Small stones may be passed spontaneously, but larger ones need to be surgically removed.

Storks are a family of some seventeen species of birds, closely related to herons. They are widespread in both the Old and New Worlds, though only one, the wood stork, *Mycteria americana*, occurs in North America. Standing up to 1·5 m. (5 ft.) high, they are long-legged, long-necked birds, primarily black and white, often with red legs and beaks. They mostly live in damp, marshy ground and eat

fish and amphibians. Some, such as the European white stork, *Ciconia ciconia*, are migrants, arriving in their northern breeding grounds in spring. This species has for long been associated with the bringing of good luck (or babies!) and for that reason is encouraged to nest on roof tops.

Storm petrels are small, fluttering, brown and white *petrels, 12–25 cm. (5–10 in.) long, which feed in flocks over rough seas, picking small fishes and crustaceans from the surface. Seven species breed in the Southern Hemisphere, and over a dozen in the north, usually in burrows on islands. All wander widely over the oceans of the world.

Strawberries are perennial, trailing plants belonging to the rose family. The wild European wood or alpine strawberry, *Fragaria vesca*, was once cultivated for its richly flavoured, small, dark red fruits. All modern cultivated varieties have been developed from two American species, *F. chiloensis* from the Pacific seaboard, and *F. virginiana* from eastern America, introduced to Europe in the eighteenth century.

The strawberry plant has a crown of leaves, and produces runners which take root, eventually to produce new plants. The 'fruits' consist of swollen red fruit-stalks; the true fruits are the tiny pips that are embedded on their surface.

Strawberry trees are evergreen trees of the genus *Arbutus*, native to Europe, western Asia, and North America. They belong to the heather family and are closely related to cranberry and blueberry. Of the twelve species, the common strawberry tree, *A. unedo*, is perhaps most familiar as a species native to the Mediterranean. It is also found in parts of western Ireland. It has a bitter strawberry-like fruit which is edible but rarely eaten, although used in liqueurs, notably in Portugal. The madrona, *A. menziesii*, of North America produces a useful wood.

Strelitzia *Bird of Paradise flowers.

Strepsiptera are small, strange insects, perhaps related to beetles but placed in the order Strepsiptera with about 370 species. Their young and most females are sac-like internal parasites, usually in solitary bees. The males enjoy a brief free life of active flight, fertilizing the females within their hosts. Females of a few species are briefly free-living, but wingless.

Stress is the experience of adversity or of events largely beyond one's control. It can cause depression or anxiety in vulnerable people. It is also widely believed to cause physical illnesses like high blood pressure, heart attacks, and strokes.

Stroke is a sudden motor or sensory disorder, resulting from loss of blood flow to a region of the brain and usually occurring when a small clot blocks off a blood vessel or the vessel itself bursts. If this happens on the right side of the brain the left side of the body is affected. Speech is also affected if it occurs on the left side of the brain. The age of the patient and the severity of the stroke are among the factors determining the degree of recovery.

Stromatolites are the fossilized remains of algal colonies, or mats. Characteristically, they have a domed or column-like shape and the sediment in which they lie is marked with fine concentric bands. These are very like the structures being produced today by blue-green algae and bacteria in hot, saline environments such as Shark Bay, Australia, where browsing animals cannot survive with the result that algal communities can flourish. The oldest known stromatolites occur in rocks of the Warrawoona Group of Western Australia and are 3,400–3,500 million years old. They reached the peak of their distribution about 2,200 million years ago when blue-green algae and bacteria were the most advanced forms of life.

Sturgeons are an order of fishes which have many primitive features, living in fresh and salt water of the Northern Hemisphere. They are distinguished by the heterocercal tail (in which the vertebrae continue up the upper lobe), a mainly cartilaginous skeleton, and the presence of a spiral valve in the gut (a primitive feature also seen in sharks). Most are large fishes with rows of large bony plates along the back and sides. The beluga, *Huso huso*, of the Black Sea grows to about 5 m. (16 ft.). The European sturgeon, *Acipenser sturio*, reaches a maximum 3·5 m. (11 ft. 6 in.) and migrates up rivers to spawn. It is now a rare fish in western Europe. Several North American species of sturgeon are also threatened with extinction.

Succession is the series of ecological *communities that appear or become extinct as an area changes from bare ground to a mature *ecosystem. Primary succession develops at sites not previously colonized, such as new islands formed by marine volcanoes. Secondary succession takes place where clearance has occurred, often on land cleared by man or fires. The first species to colonize a site are termed pioneers and include annual species of plants and mobile animals. When established, these progressively modify the nutrient status and condition of the soil, enabling other species to colonize; competition usually then excludes the pioneers. Further modification of the site admits more species, and eventually a dynamically stable climax is established with one or two species dominant. The climax vegetation in many parts of the world is forest, but where water is in short supply or grazing animals are abundant this may be grassland. The whole succession (alternatively termed a sere) is directional and predictable.

The first animals and plants to colonize an area typi-

An Atlantic **sturgeon** shows the typical elongated snout with sensory barbels used to detect bottom-living prey. Because of widespread pollution, the only rivers in western Europe to hold a breeding population are the Guadalquivir in Spain and the Gironde in France.

cally produce a lot of seed or offspring, grow quickly, and are small in size. Climax species of plants and animals are usually long-lived, large organisms, which produce few fruits or young, which they may endow with large food reserves to maximize survival.

Succulent plants, including the cacti, are a large group that are adapted to grow successfully in arid or desert regions. They have evolved tissues modified for water storage, which enables them to survive long periods of drought. The modified tissues include stems and leaves. Typical stem-storage species include the cactus-like

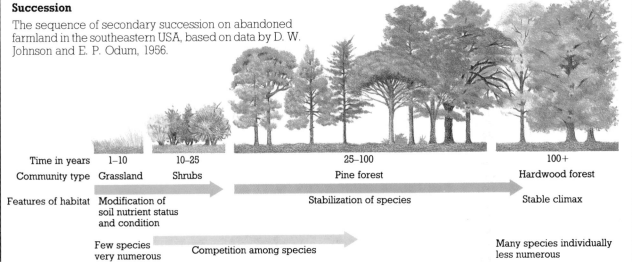

Succession

The sequence of secondary succession on abandoned farmland in the southeastern USA, based on data by D. W. Johnson and E. P. Odum, 1956.

Time in years	1–10	10–25	25–100	100+
Community type	Grassland	Shrubs	Pine forest	Hardwood forest
Features of habitat	Modification of soil nutrient status and condition		Stabilization of species	Stable climax
	Few species very numerous	Competition among species		Many species individually less numerous

Euphorbia canariensis, with swollen spiny stems up to 7 m. (23 ft.) in height. Others include *Stapelia*, the *baobab tree of tropical Africa, and the *Nolina* of Mexico, with a swollen stem base resembling an elephant's foot. Leaf succulents show great variety of form, including the diminutive rounded-leaved sedums, the large fleshy leaves of the African aloes, mesembryanthemums of many shapes, *Cotyledons*, sempervivums, and the sword-like leaves of the agaves of Central and South America.

Sucker-footed bats are similar to *disc-winged bats except that the adhesive discs are not on stalks. In both groups the suckers allow the bat to cling to smooth surfaces. There is only one species, the golden bat, *Myzopoda aurita*, which is confined to Madagascar. It is a rare, little-known animal.

Sucking lice are small insects of the order Siphunculata, which are rarely longer than 5 mm. ($\frac{1}{4}$ in.), but are the largest *lice of mammals. They pierce the skin and feed on blood or cell contents. They are flattened and usually live at the base of hairs, cementing their eggs, or nits, to them. The 300 or so species occur on most mammals, including seals and their relatives, but are absent from mammals such as marsupials, bats, cats, and whales. They are transferred to new hosts mostly by contact between hosts when young, but also by sexual intercourse. They are highly successful parasites. Although a few such as the human louse, *Pediculus humanus*, can carry diseases most are harmless, but intensely irritating. The body louse and head louse are subspecies of *P. humanus*.

Sugar-beet is a variety of *Beta vulgaris*, the species which also gives rise to beetroot. The large, whitish, conical, swollen roots can contain up to 20 per cent of the sugar, sucrose, and are an alternative source to sugar-cane in temperate regions. This crop was developed in about 1800 and large quantities are grown in central and eastern Europe, Britain, and the USA.

Sugarbirds are two South African species thought to be related to the sunbirds and the Australian honeyeaters. Although the body is only the size of a small thrush, the males in particular have long tails, which are up to 45 cm. (18 in.) long. They are greyish-brown birds with flashes of bright yellow underneath the bases of the tail, and long, curved beaks. They specialize in feeding on protea flowers, taking both nectar and insects on the blooms.

Sugar-cane is a perennial grass of the genus *Saccharum*, from Southeast Asia, where it has been cultivated for thousands of years. It was introduced into the West Indies by Spanish explorers of the fifteenth and sixteenth centuries. Its solid stems, up to 6 m. (19 ft. 6 in.) high, are rich in extractable sugar (sucrose). More than half of the world's sugar supply comes from this species. The main areas of production are Brazil, Cuba, Mauritius, the Caribbean, Hawaii, and Australia. Commercial production is concentrated in large plantations, as expensive machinery is needed to culture, harvest, and extract the sugar. Crops are propagated from stem cuttings and are usually harvested after burning to remove unwanted leaves. The raw sugar is light or dark brown in colour, depending upon the amount of molasses, also present in the sap, that is extracted with it. A purple-coloured variety in Asia is more commonly used to provide a sweet drink, rather than to produce raw sugar. Rum is a by-product obtained from sugar-cane.

Sugars are water-soluble *carbohydrates which often have a sweet taste. They include simple sugars such as *glucose, fructose, and ribose; slightly more complex sugars such as maltose, lactose, and sucrose (table sugar); and trisaccharide sugars such as melizitose. Many sugars are basic building units for many natural organic compounds.

Suicide is death deliberately brought about by one's own actions. In Western countries it is most frequently committed by middle-aged, socially isolated people, with depression or alcoholism contributing to their predicament.

Attempted suicide, often by an overdose of drugs, is much more common and occurs in a younger age group. The action is often impulsive, lacking real suicidal intent, and is often called parasuicide; but, of course, it carries the risk of unintended death.

Sulphurs (butterflies)   *Yellows.

Sumacs, or sumachs, are shrubs and trees of the genus *Rhus*, of which there are some 250 species in the subtropics and warm, temperate regions, including the *poison ivy. The leaves of the common sumac, *R. coriaria*, of the Mediterranean, are ground up for use in dyeing and tanning. Familiar in gardens is the fast-growing stag's-horn sumac, *R. typhina*, from North America, a species whose leaves are tinted a deep red in the leaves in the autumn.

Sunbeam snake: a species of blackish to brownish snake which gets its name from the iridescence of its shiny scales. It is found in Southeast Asia, growing to a length of 1·2 m. (4 ft.), and is a harmless semi-burrowing species. Its affinities are obscure but it is thought to be most closely related to snakes of the boa group.

Sunbirds are a family of 105 species of bright, metallic-coloured birds found mainly in Africa, but also in the warmer parts of Asia, with one species in Australia. They are mostly small birds, 9-12·5 cm. (4-5 in.) long, though a few are larger and some have very long tails. They are mostly coloured brilliant metallic blues, greens, or reds, though a few are dull green. Their beak is long and curved for reaching into flowers to obtain nectar and some species also take small insects and spiders.

Sun bittern: a bird that inhabits the forested edges of marshes and streams in tropical South and Central America. It is the sole member of its family and is related to the rails, being about chicken-sized with thinnish legs and a longish beak. The wing, when spread, shows a big circular, sun-like patch of black, chestnut, and white; this is shown in display and is the origin of the bird's name. It feeds on insects and other small animals.

Sundews are a group of *carnivorous plants which, along with the Venus's fly trap (*Dionaea muscipula*), belong to the family Droseraceae. The name is commonly applied to members of the genus *Drosera*. They have rounded or elongated leaves provided with long, stalked glands which secrete a sticky mucilage. Insects are attracted to these leaves and ensnared. Sundews may be plants of bogs or sandy habitats where nitrogen and other nutrients are in

The round-leaved **sundew**, *Drosera rotundifolia*, grows throughout the north temperate region. Seen here after blooming, it has small white flowers in June to September.

short supply, so that the insects which they catch are a valuable supplement to their nutrition. About seventy species of sundew are found throughout the world.

Sunfishes belong to two quite unrelated groups of fishes. The marine sunfishes are members of the family Molidae, giant inhabitants of the open sea and found in all the tropical and temperate oceans. Although rather uncommon, a few of the five species can weigh over 1,000 kg. (2,200 lb.) each. The ocean sunfish, *Mola mola*, is one of the most abnormally shaped fishes. Its dorsal and anal fins are set well back on the body on either side of a much reduced caudal fin. The body is disc-shaped and appears to be all head. The ocean sunfish eats marine invertebrates such as jellyfish.

Freshwater sunfishes belong to the family Centrarchidae, of which there are thirty-two species, confined to the rivers and lakes of North America (although introduced to other parts of the world). Most are deep-bodied, often brightly coloured, with a continuous dorsal fin, the first section of which is spiny. Few of them grow larger than 20 cm. (8 in.) in length, but all build nests in river- or lake-beds, in which the eggs, when laid, are defended by the male. Several species, like the green sunfish, *Lepomis cyanellus*, range from Canada to Mexico, but others are more restricted in range.

Sunflowers are tall annuals reaching 2·5 m. (8 ft.) in height and give their name to the sunflower family, the

Compositae. This includes daisies, asters, ragworts, and many other familiar plants, and is one of the largest families of flowering plants, with over 25,000 species. Sunflowers have large, yellow flower-heads, which produce oil-rich seeds, and are grown in warm temperate areas and at medium altitudes in the tropics. The main producer is the USSR, followed by southern and eastern Europe, and Argentina. Harvested by hand or machine, the seeds, containing up to 40 per cent oil, are threshed out and crushed to yield a high-quality table or cooking oil, used also in the manufacture of margarine. Inferior grades of oil are used for soap, paints, and varnish.

Sun spiders, or wind scorpions, are a unique order of *arachnids of hot dry lands, named for their habit of being active in full sunlight; they are also called wind scorpions because they run extremely fast. They resemble large spiders, but have enormous jaws which can project upwards, and they use only their last three pairs of legs for running, the front pair being sensory. They feed on small animals, captured by the *pedipalps and torn apart by the jaws. Mating necessitates the male's subduing the female by stroking, until he can deposit his sperm, using a delicate action of his jaws.

Sunstar   *Starfish.

Sunstroke, better called heatstroke, is a dangerous rise in body temperature, due to inadequate heat loss by sweating in a hot environment. Excessive exertion, before acclimatization, and dehydration are factors often involved. In extreme cases it may be fatal unless alleviated by giving the patient drinks, or cooling his body.

Surf perch   *Sea perch.

Surgeonfishes   *Unicorn fishes.

Suricates   *Meerkats.

Sunflowers, *Helianthus annuus*, stimulated by sunlight, always face the sun, following it across the sky. Known as heliotropism, this movement also occurs in other plants.

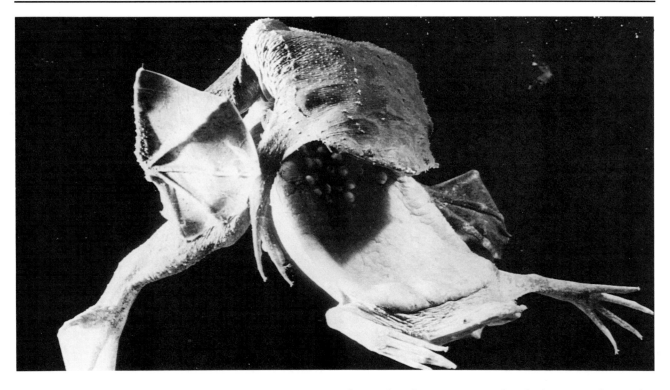

A pair of mating **Surinam toads**, showing the eggs deposited in the spongy tissue of the female's back. Note also the broadly webbed hind-feet adapted for swimming.

Surinam toad: a species of completely aquatic and rather bizarre-looking toad. The body is flattened, the head triangular, the toes of the fore-feet have star-shaped tips, and the hind feet are extensively webbed between the toes. The mating behaviour of the Surinam toad is remarkable and involves the fertilized eggs being pressed into the spongy skin of the female's back by the male. These eggs are then engulfed by the spongy skin until a 'cap' has grown over each. The tadpoles develop inside these pockets on the female's back to emerge as tiny toads 3 to 4 months later.

Swallowing is a sequence of movements of the tongue and throat by means of which food or other objects are pushed from the mouth down the oesophagus. The tongue initiates the process, directing the material to the back of the mouth, where it stimulates sensory nerves. The stimuli reaching the brain trigger a series of reflex actions. These include closure of the passages to the nose and trachea for a brief time and nudging of the food particles down the oesophagus by successive waves of muscular contractions.

Swallows, or *martins, are birds with a worldwide distribution except for cold areas at high latitudes. They are mostly small, less than 15 cm. (6 in.) long; a few species are larger or have long tail feathers. Many are glossy blue-black or greenish above and pale below, and a number have bright reddish-chestnut patches on the breast or head. They are almost exclusively insectivorous, catching their prey on the wing. As a result they cannot survive in most temperate areas in winter because there are no insects available, and so they migrate. Some, such as the common or barn swallow, *Hirundo rustica*, nest in close association with man. Some nest in dense colonies, and in North America complex nesting-boxes with many dif-

ferent chambers are put up for the large purple martin, *Progne subis*.

Swallowtail butterflies include some of the largest and most beautiful butterflies in the world. Many have 'tails' on the hind wings. They are mainly tropical with between 621 and 641 species in the family Papilionidae. They are mainly found in forest, but the abundant citrus swallowtail, *Papilio demodocus*, of Africa and the related chequered swallowtail, *P. demoleus*, of southern Asia and Australia, both intricately patterned in black and yellow, frequent gardens and plantations, where their caterpillars eat citrus leaves, and may be pests. The caterpillars of swallowtails usually have an eversible forked process, often red, behind the head, which releases a pungent aroma. The family Papilionidae also includes birdwings, apollo butterflies of the Palearctic, and the long-tailed dragontail butterflies (*Lamproptera* species) of Southeast Asia.

Swamp eels   *Cuchia.

Swan mussels are freshwater *bivalves, common in rivers and streams. Adults can be 15 cm. (6 in.) long, and they filter-feed, processing 3 litres ($6\frac{1}{2}$ US pints) of water per hour. Young develop within the parental shell, and are eventually released to live for a while as parasites on fish gills, before becoming independent sedentary adults. Their name refers to their being pulled out of the water and eaten by birds such as swans.

Swans are waterfowl in the Anatidae family which also includes geese and ducks. They are generally bigger than geese, the largest being over 1·5 m. (5 ft.) long, and have longer necks. These assist them to feed on underwater vegetation, although they also graze on land. Their ponderous but graceful flight, with slow but regular wing-beats, is characteristic of all swans. The six species of swans have pure white adult plumage, with the exception of the black swan, *Cygnus atratus*, in Australia and the black-

The black **swan** may produce up to four broods a year, each of four to seven downy cygnets. Parents keep in constant communication by powerful trumpeting calls.

necked swan, *C. melanocoryphus*, in South America. The sexes are alike.

Sweat glands are tubular glands of the *skin that supply a watery solution (sweat) whose evaporation serves to cool the body. The rate of sweating is controlled by the brain and related to body temperature. Sweat glands are well developed in primates, horses, and camels. Two million or so of these glands are distributed in human skin.

Swedes, or rutabagas, are one of the two *brassicas commonly grown for their swollen roots (the other is turnip). They are cultivated for both human and livestock food. They may be distinguished from the less frost-resistant turnip by the swollen, ridged neck with its leaf scars. Purple-, white-, or yellow-skinned, the flesh is normally orange-yellow but may be whitish. Swedes are best suited to cool, moist, temperate climates such as parts of northern and western Europe.

Sweet corn is a particular form of *maize, shorter in stature than most other strains, but differing principally in that its grains consist mainly of sugar rather than starch. As the cobs are harvested well before the grains are fully ripe, it can be grown, at least in sheltered areas, in cool, temperate regions such as Britain.

Sweet peas are annual climbing plants of the pea family with large, sweetly scented flowers on long stalks. They have been developed as garden and florists' flowers, by selective breeding from the Italian species *Lathyrus odoratus*. This species was first introduced to England from southern Italy in 1699. Variously coloured varieties were in existence by 1900, when frilled and waved-petal forms appeared and became instantly popular. The genus *Lathyrus* contains many species with 'pea' as part of their common name. They are generally distributed over temperate regions.

Sweet peppers are a variety of *Capsicum annuum*, a species which also gives rise to some paprikas and pimentos. Native to tropical America, it is now widespread in the tropics generally and, like its relative the tomato, is often grown under glass in temperate areas. When ripe the fruits are usually red, up to 25 cm. (10 in.) long, and

vary in their shape from more or less spherical to long and narrow. Varieties differ in their pungency but all are mild compared to the chilli.

Sweet potato: a root crop related to bindweeds and morning glory. It is native to South America and was taken, via sixteenth-century Spain, to Asia where it is now an important crop. Its elongated, swollen tubers have a red or purplish skin and white or yellow flesh rich in starch. The vine-like, climbing sweet potato, *Ipomoea batatas*, seldom flowers, but is propagated from cuttings. In the USA, it is often wrongly referred to as a *yam.

Swift moths differ from most moths in the way the fore- and hind-wings are held together, and in having similar venation in both wings. They comprise the family Hepialidae, with 300 or so species, best represented in Australia, but widely distributed throughout the world. Their caterpillars tunnel into wood or feed on roots of a wide variety of plants, and the pupae have a flexible, spined abdomen with which they work their way to the surface when the adults are ready to emerge. The largest European species is the ghost moth, *Hepialus humili*, which gets its name from the brilliant white upper wings of the male. These dance up and down in the air above vegetation and seem to appear and disappear alternately as their upper-wing surfaces reflect the moonlight. There are many species in Australia; the bent-wing, *Zelotypia stacyi*, has a wing-span of 19 cm. ($7\frac{1}{2}$ in.) and its caterpillars tunnel destructively in eucalyptus trees.

Swifts are a family of sixty-six species of birds with a worldwide distribution except for colder areas at high latitudes. Like the swallows, they are insectivorous, catching all their prey on the wing, so they have to be summer visitors to many temperate areas. They are slightly larger than swallows, and are mostly black or black and white. They often nest in colonies and some, such as the edible-nest swiftlet, *Collocalia fuciphaga*, (the bird whose nest is used for the Chinese bird's nest soup), live in enormous numbers in caves.

Swim bladder: an air-filled sac which lies above the intestine of *bony fishes. Its role is to maintain the buoyancy of the fish by effectively reducing the fish's weight at different depths of water. The swim bladder is either filled with air through a connection with the mouth, or filled with gases passed from the blood system into, or out of, the swim bladder. In lungfishes the swim bladder functions as a lung.

Swine *Pigs.

Swordfish: a species of large *bony fish of worldwide distribution in temperate and warm temperate seas, ranging northwards, or southwards, with the seasons. It occurs off the British coast (rarely) and Nova Scotia (more commonly) in the late summer and early autumn. It is a heavily built fish with a high, but short, first dorsal fin and a broad tail, but its most distinctive feature is the long sword-like snout. This is flattened like a sword, not rounded and rapier-like as in the marlins. Active, strong swimmers, they eat a wide range of smaller fishes and squids, which are captured in the mouth, not impaled on the sword. Living in the upper 1,000 m. (3,250 ft.) of the sea, they are often sighted near the surface. A fine game-fish, the swordfish is also captured for food, but has become scarce in recent years.

Swordtail: a popular tropical aquarium fish, *Xiphophorus helleri*, which originated in Mexico and Guatemala, where it is native to mountain streams as well as still waters on the coastal plains. It is a member of the family Poeciliidae, hence related to guppies, platys, and mollies, and gives birth to well-developed young. The male has a structure called a gonopodium, formed from some of the rays of the anal fin, with which sperm is transferred to the female. Only males have the characteristic, coloured, sword-like, long rays of the tail fin. Females occasionally change sex after producing a few broods, and grow the distinctive sword tail.

Sycamore is a name used for a number of different trees. In North America, it may be used to refer to *planes. In the Bible (the original use) it was a fig, *Ficus sycomorus*. In Britain, it is used for one of the *maples, *Acer pseudoplatanus*, native to southeast Europe but widely introduced elsewhere.

Sydenham, Thomas (1624–89), English physician, was renowned for his systematic study of epidemics, based on seasons, years, and ages of victims. Known as 'the English Hippocrates', he introduced a successful cooling treatment for smallpox; gave clear descriptions of malaria, plague, hysteria, and gout; and was the first to identify scarlatina. He distinguished between acute diseases, thought to be the body's attempts to resist outside influences, and chronic diseases, considered to be caused by unsuitable living and eating habits.

Symbiosis, or mutualism, describes any interspecific association between organisms where both participants, host and symbiont, benefit from the relationship. The degree of co-operation varies. A lichen is the result of an association between a unicellular alga, and a fungus. The alga gains nutrients, protection, and anchorage, and the fungus obtains carbohydrates from its photosynthesizing partner. Another example of symbiosis involves herbivorous animals, such as *ruminants, and the cellulose-digesting micro-organisms which gain protection and food in their alimentary canals. The relationship between leguminous plants (*legumes) and nitrogen-fixing bacteria is a good example of plant symbiosis. Symbiosis is an association related to *commensalism and *parasitism.

Sympathetic nervous system   *Autonomic nervous system.

Syphilis is a contagious, sexually transmitted disease due to the bacterium *Treponema pallidum*. Its importance lies in complications that may affect the nervous system and blood vessels. These complications have become rare because the infection can now be treated effectively with penicillin. The incubation period is two to three weeks. The first sign is a small, painless 'ulcer' on the penis, vulva, or labia. Treatment includes tracing sexual contacts in an attempt to prevent further spread of the disease.

Syringa   *Lilacs.

T

Tachinid flies   *Bristle flies.

Tadpoles are the larval stage of frogs and toads; the name is no longer normally applied to the larvae of other amphibians. Tadpoles have a combined head plus body, usually round or oval in shape, and a well-developed tail; they do not resemble the adult. In other amphibians the larva *is* like the adult except that it has feathery gills.

Not all frogs and toads have a free-living tadpole stage; some undergo direct development, like the *Surinam toad and the West African viviparous toad, both of which produce fully formed young which are miniatures of their parents. In some species the tadpole stage is passed inside the egg.

Taiga: a term originally applied to regions of eastern Siberia. Nowadays it is used to describe scattered patches of coniferous trees, such as larch, which grow at the transition between northern temperate pine-forests and *tundra. Trees of the taiga suffer long, harsh winters, and very short summers.

Tailed frogs are unique among frogs and toads in that the male has a tail-like extension of the cloaca (a common reproductive-excretory tract), which acts as a copulatory organ, preventing sperm loss during mating in the fast-flowing streams of North America where it lives. Tadpole development is slow in the cold water of these streams and between one and three years may pass before the tadpoles are transformed into juvenile frogs. The tailed frog, *Ascaphus truei*, belongs to a family of primitive frogs with normal body-shape. The remaining three species of this family live in New Zealand.

Taipan: a relatively slender-bodied, rather *mamba-like, snake up to 3·4 m. (11 ft.) long. It occurs in coastal areas of northern and northeastern Australia and southern parts of New Guinea. One of the world's most highly venomous snakes, when threatened, a taipan is capable of striking at an intruder with considerable speed and ferocity. It eats small mammals.

Takahe   *Notornis.

Takin: an unusual-looking species of antelope which combines features of the ox, goat, and antelope. It is a close relative of the musk ox, and characteristically has eyes high in its head, and curving horns at the top of the forehead. It lives in the exceedingly rough mountainous country of the Himalayas, western China, and Korea at elevations of 2,500–4,300 m. (8,000–14,000 ft.). A heavy-looking animal, some 90 cm. (3 ft.) high at the shoulder, it has a shaggy coat of yellowish hair. The day is spent in thickets, near the timber-line, from which it emerges at dusk and dawn to feed on the grassy slopes. It is a rare animal about which little is known.

Talipot palm is the common name for *Corypha umbraculifera*, a *palm from India, which can grow to 24 m. (75 ft.) tall. It then produces a terminal inflorescence several metres (yards) in height, at the top of the palm, after which it dies. The large leaves have been used as umbrellas and for thatching, and also as a writing material.

Tamanduas *Ant-eaters.

Tamarind: a tree species of the pea family, *Tamarindus indica*, or its dried fruit. Although it grows in seasonally dry areas throughout the tropics, its origin is unknown. The tart fruit, consisting of brown pods, is used as a flavouring and a food preservative, and is also used in medicine; the timber is of some value.

Tamarisks are shrubs and trees of the genus *Tamarix* with some ninety species throughout Eurasia, many of them growing in soils with high levels of salt. They give their name to the tamarisk family, which has some 120 species. Most are rather wispy shrubs with fine, hair-like leaves and spikes of tiny flowers. *T. mannifera*, when punctured by scale insects, exudes the manna used by the bedouin.

Tanagers are a large family of about 235 species of birds, all of which are native to the New World, mainly to Central and South America. A few are summer migrants to North America. Most are sparrow- to thrush-sized, though one or two species are slightly larger. They are like sparrows in shape, with stoutish beaks, and many are brilliantly coloured red, yellow, or blue. They live in a

The seven-coloured **tanager**, *Tangara fastuosa*, endemic to the forests of northeast Brazil, is now an endangered species. This vivid bird is typical of the perching birds, with three toes pointing forwards as well as one well-developed hind-toe to aid balance.

wide variety of habitats from rain forest to open scrub and eat a wide range of fruits and insects.

Tangerine *Mandarin.

Tansy *Chrysanthemum.

Tapaculos are a family of twenty-nine species of small birds restricted to Central and South America. They are mostly brown, grey, or black in colour and most have short tails. They live in a wide variety of habitats from dense grassland to forest scrub, although most are confined to mountain forests.

Tapeworms are highly specialized creatures, which live inside the guts of vertebrates, and reach 12 m. (40 ft.) in length in some species. They form one of the three major classes (Cestoda) of the phylum Platyhelminthes. The head of a tapeworm bears suckers and hooks with which to hold on, while the rest of the body consists of flat, square sections linked like a segmented tape. Each of these sections is packed with reproductive organs. They have no gut, as they can absorb the host's food directly across their specialized body wall. Mature segments rupture and release eggs via the host's faeces. These eggs are eaten in contaminated food by other animals, often a specific secondary host. The larvae grow within the secondary hosts, which can be vertebrates or arthropods, and return to their main host when the secondary host is eaten by the larger main host. Hence, eating badly cooked meats, which harbour tapeworm larvae, can cause debilitating tapeworm infestation in man. Many species have complex life-cycles with unique adaptations to ensure that some larvae eventually reinfect their main host.

Tapioca *Cassava.

Tapirs

Young Brazilian tapirs have a reddish-brown coat broken up by prominent white spots and stripes, which serves as an effective camouflage amongst the dappled shade of the forest undergrowth. The young of all species show this cryptic coloration, which is lost at around six to eight months.

Tapirs are found in northern South America and Malaya. They are large, brown, or black and white ungulate (hoofed) mammals which comprise a family of four species. The snout and upper lip are elongated to form a short, flexible trunk, with terminal nostrils. The trunk is used to draw twigs, leaves, and branches into the mouth. These animals are quite unique and almost living fossils, as they have apparently changed little throughout their evolution.

Three species live in the New World. The South American or Brazilian tapir, *Tapirus terrestris*, is the smallest with a body length of 1·8 m. (6 ft.). It has a short, stiff mane and inhabits swampy regions of hardwood forests at altitudes of up to 1,200 m. (4,000 ft.). The mountain tapir, *T. pinchaque*, is found at altitudes of 3,000 m. (10,000 ft.) or higher in forests. Its population is almost certainly small. The Central American or Baird's tapir, *T. bairdi*, is found from sea-level up to 1,800 m. (6,000 ft.). The head and body may reach a length of 1·95 m. (6 ft. 5 in.) and it has a tail 7·5 cm. (3 in.) long. The Malay tapir, *T. indicus*, the only species found in the Old World, is the most distinctive, with white back and sides contrasting with the black of the rest of the animal. It stands 1·06 m. (3 ft. 6 in.) at the shoulders and has no mane. It lives in lowlands, especially swampy forests. A shy, inconspicuous, solitary animal, it wanders at night along well-used trails to water, feeding on lush vegetation and fruit. After a gestation period of thirteen months a single young is born, with distinctive white spots and streaks running along the sides of the body.

Tarantulas, or bird-eating spiders, are notorious for their size and diet. The body is fist-sized, with long, hairy legs, and they can prey on small vertebrates, but normally subsist upon invertebrates. North American species are not venomous, unlike species in the Southern Hemisphere, but most have irritant hairs. They live in burrows and do not spin webs, but hunt nocturnally, relying mainly upon vibration to locate their host. Pet tarantulas can live for twenty-five years.

Tardigrades   *Water bears.

Taro is a valuable crop of the wet, humid, tropical areas of Asia and Polynesia, where it is grown for its short, swollen, underground stems (corms) that yield a highly digestible starchy food. It belongs to the genus *Colocasia*,

part of the monocotyledonous family Araceae. Also known as dasheen, true taro, *C. esculenta*, is a native of Asia, and is related to several similar species within the same genus. It has large leaves similar in size and shape to those of its relative the arum lily. The corms are peeled before cooking, when they lose their rather bitter flavour. The young leaves may be used as a vegetable, and the fermented corms are sometimes canned and marketed as 'poi' in Hawaii.

Tarpon: a huge oceanic fish living in the tropical Atlantic and occasionally moving northwards in warm seasons. It grows to at least 2·4 m. (7 ft. 10 in.) in length. A similar species in the Pacific Ocean, *Megalops cyprinoides*, grows to only 1·5 m. (5 ft.). Both are members of the family Megalopidae. They are large silvery fishes with big scales and a single dorsal fin, the last ray of which forms a streamer. They live and breed in the upper waters of the open sea, and are incredibly prolific: a 1·8 m. (6 ft.) female contains 12 million eggs. The young, however, are found close inshore in lagoons, estuaries, and often in the oxygen-poor waters of mangrove swamps.

Tarragon is a member of genus *Artemesia* and belongs to the sunflower family, the Compositae, which also includes

The Mexican red-kneed spider, *Brachypelma smithii*, is a **tarantula** which grows up to 7·6 cm. (3 in.) long. The spider is shown moulting, a process that involves the carapace of the head and thorax lifting off like a lid.

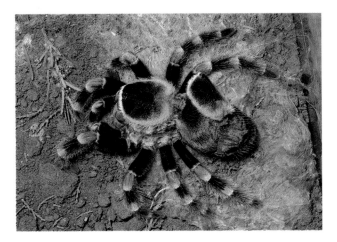

some common garden plants such as the daisy. This southern European plant, which spreads by underground runners, has a distinctive flavour quite unlike that of any other culinary herb.

Tarsiers are three species of nocturnal prosimian *primates that live in the East Indies region and which are placed in a separate family. The adults are 15 cm. (6 in.) long and have tails 25 cm. (10 in.) long, which are hairless except for a tuft at the tip. The head is round, with a short face and comparatively large, rounded ears. The close-set, forward-facing eyes are enormous, as in most nocturnal creatures. The tarsiers hunt insects among the trees where they live. On finding an insect, the tarsier stares at it for some time, then slowly bends forward and finally makes a sudden rush to seize it with both hands and take it to its mouth. Adhesive pads on the fingers and toes allow these animals to cling firmly to smooth surfaces as they leap vertically from branch to branch. When on the ground, they travel with the same type of movements. The female bears a single young.

Tasmanian devil: a species of carnivorous *marsupial that once roamed over much of Australia but now survives only in Tasmania. Despite its name, it is no more fearsome than any other carnivorous mammal. Rather bear-like in appearance, it is about 90 cm. (3 ft.) long and has a black coat with irregular blotches of white hair. A land animal, it haunts the edges of rivers and beaches in search of food. It will take crabs, frogs, and small mammals, but can attack and kill animals larger than itself. At night it calls with a low, yelling growl followed by a snarling cough. Like all marsupials, the female has a pouch in which the young are carried until they are too large; the mother then builds a nest for them.

Tasmanian wolf, or thylacine: a species of *marsupial that has been eliminated in most of Australia but may still

Tasmanian wolf

The Tasmanian wolf, *Thylacinus cynocephalus*, was persecuted as vermine by sheep farmers, encouraged by a substantial Government bounty from 1888 to 1909. After 1909 there was an additional sudden decline in numbers, probably due to distemper or a similar disease. The species is now probably extinct, though a few individuals may still be found in isolated, rugged sectors of western Tasmania.

survive in Tasmania, although it was last seen there in 1936; man is its enemy. It is the largest of the flesh-eating marsupials, being some 165 cm. (65 in.) long, including the tail of 50 cm. (20 in.). Brown in colour, it has sixteen to eighteen chocolate-brown stripes across its back. The head is rather dog-like with a long, pointed muzzle, and broad, rounded ears. During the day it remains in its lair in the rocks, waiting until dusk before venturing out in search of wallabies or other marsupials to prey upon. The female has a pouch large enough for four babies.

Tasting is the sensation experienced when specialized nerve endings in the tongue and pharynx, called taste buds, are activated by substances dissolved in the saliva. The taste buds react selectively to various substances and some animals have taste buds that respond to pure water. Humans possess four basic tastes, sweet, sour, bitter, and salt. All acids taste sour and, in addition to common salt, many inorganic substances taste salty. A number of chemical compounds, as well as sugar, taste sweet. Saccharin, a protein, is 100,000 times sweeter than sucrose (table sugar).

Tawny owls: a species of *owl common in woodlands throughout much of Eurasia. They have also taken readily to parks, nesting as successfully on buildings as in hollow trees and dense undergrowth. They are small, heavily barred, brown owls with rounded wings and tail. They emerge at dusk and fly until dawn, seldom appearing during the day. Their main prey consists of rodents and other small mammals up to the size of baby rabbits, though birds are taken when abundant. Clutches of four or more spherical white eggs are laid; males do most of the hunting throughout the incubation and brooding periods.

Tayberry: a soft fruit which is a hybrid obtained by crossing the American blackberry, variety Aurora, with a raspberry. The new fruit was introduced to cultivation in about 1980. The deep purple fruits, up to 4 cm. ($1\frac{1}{2}$ in.) long, have a rich flavour. The quick growing, prickly plants are cultivated like blackberries or loganberries.

Tea is drunk regularly by half of the world's population, particularly in India, Southeast Asia, and the Far East, where it has been cultivated for thousands of years. Although tea was not introduced to Europe until after coffee drinking had become established, it found particular favour in Britain, which now imports nearly one third of the total production. India is the main producer, followed by Sri Lanka and China.

The tea plant is an evergreen shrub of the genus *Camellia*, related to the ornamental camellias and suited to hill slopes in the tropics with good rainfall and acid soils. Two forms are grown, the hardier narrow-leaved China type, and the larger-leaved Assam type. The two terminal leaves of the young shoots yield the highest-quality tea, though up to four such leaves may be harvested per shoot. Picking is done usually by hand every seven to fourteen days, but increasingly by machine. The leaves are steamed, rolled, and dried rapidly to produce the green tea popular in the Far East, or withered, fermented to release the drug caffeine, and carefully but rapidly dried to develop the flavour and colour of the more familiar black tea most popular in Britain and the USA. Most commercial teas are blends of leaves of different qualities from various sites.

Teak: a tall tree of seasonally dry forest in Burma, Thailand, Malaysia, Indonesia, and India. It has been planted commercially elsewhere, particularly in Africa and the West Indies. The timber is hard and durable, rather oily, and is used in ship-building and for furniture. The term teak is strictly applied only to the timber from the tree *Tectona grandis*, but is also used for a number of timbers with similar qualities but derived from quite unrelated kinds of trees. *Tectona* belongs to the family Verbenaceae, a group of mostly small plants and shrubs, some of which, such as species of *Lantana*, have showy flowers.

Teal is the name of various species of *duck, mostly small dabblers. Typical is the green-winged teal, *Anas crecca*, a species noted for its small size and rapid flight. Tight flocks of this species are formed outside the breeding season, and manoeuvre in the air with the agility of waders.

Tears form from the clear, salty fluid produced by glands in the eye-socket of terrestrial vertebrates. The fluid lubricates the eye-ball and washes away dust, bacteria, and other foreign particles onto the skin of the cheeks, or through a drain, the tear duct, into the nose. Tears contain an enzyme capable of destroying some forms of bacteria. Their flow can be increased by chemical or mechanical irritants, pain, and, in man, strong emotions.

Teasels are biennial or perennial plants with tall spiny stems and leaves. In the common teasel, *Dipsacus fullonum*, the leaf bases form a cup in which water collects, preventing climbing insects such as ants from pillaging the nectar and pollen of the flowers. The fuller's teasel, *D. sativus*, is a cultivated kind with hooked spines on the flower-heads. These are used in the textile industry to tease or raise the nap of cloth. They belong to the same family as scabious.

Teeth are situated in the jaws. In the course of evolution the teeth have become highly specialized. In some fishes they are merely modified and replaceable skin scales called dermal denticles used to hold prey before it is swallowed alive. Fossilized teeth can inform palaeontologists whether the animal (for example a dinosaur) to which they once belonged was herbivorous or carnivorous. Fossilized teeth have also been of great value in the determination of the place of apes and early man on the evolutionary tree.

In man there are only two sets of teeth and the permanent teeth do not grow after they have reached maturity. In rodents teeth grow throughout life and are worn down during mastication. The teeth of mammals are covered by a tough enamel layer overlying a bone-like material called dentine. The tooth root has a hollow canal along which nerves and blood vessels pass. In a living animal the tooth is attached by a spongy bone-like substance referred to as cement, and only a small part of the tooth protrudes above the gums.

Temperatures (body) of individual bodies vary among animal groups. Most have body temperatures which change according to that of their surroundings. These are called cold-blooded animals and include all invertebrates, amphibians and reptiles, and fishes. Certain animals, the homeotherms, or warm-blooded animals, regulate their body temperatures to an accurate setpoint by controlling their heat production and loss. Mammals and birds have setpoints of 37–38 °C and 40 °C respectively. These tem-

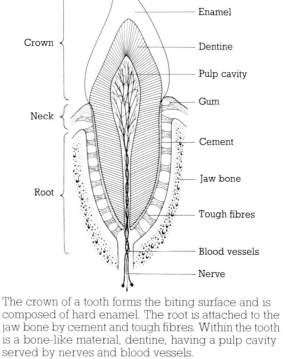

Teeth: structure of a mammalian tooth

Crown — Enamel
— Dentine
— Pulp cavity
— Gum
Neck
— Cement
— Jaw bone
Root
— Tough fibres
— Blood vessels
— Nerve

The crown of a tooth forms the biting surface and is composed of hard enamel. The root is attached to the jaw bone by cement and tough fibres. Within the tooth is a bone-like material, dentine, having a pulp cavity served by nerves and blood vessels.

peratures are those at which the chemical reactions in living tissue work most efficiently. Both warm and cold-blooded animals have temperature sensors in their skin and bodies. These tell the animal whether it needs to cool down or warm up, a process called thermoregulation. Warm-blooded animals produce heat as a by-product of their *metabolism, and cool down by sweating, panting, or behavioural means. Cold-blooded animals either are adapted to work efficiently at low temperatures, or bask in sunshine as a way of heating up. The latter type of animal includes reptiles and many insects, each working optimally at a body temperature much higher than that of

Body **temperature** fluctuates naturally according to the time of day, but in any case it is not uniform all over the body. In these thermograms of a man, a woman, and a child, the white patches represent the warmest areas and dark blue-black the coolest.

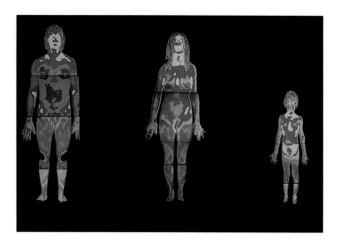

their surroundings. Basking animals do maintain a certain body temperature, like their warm-blooded counterparts, usually by moving out of sunshine when too warm, and vice versa when cold. Many large vertebrates maintain a higher core (deep body) temperature compared to that of the brain, which is cooled by a specific heat-exchange mechanism.

Tench: a member of the *carp family, widely distributed in Europe and central Asia. It is a deep greeny-brown in colour with bronze sides, and has small scales and rounded fins. A bottom-living fish, it prefers muddy lakes and rivers with dense vegetation, and can survive with little oxygen.

Tenrecs belong to a family of insectivores confined to Madagascar and the nearby Comoro Archipelago. In the absence of other insectivores, the tenrecs have evolved into a range of some thirty-four species that resemble insectivores elsewhere. This tendency is most marked in the 'hedgehog' tenrecs, such as the greater hedgehog tenrec, *Setifer setosus*, but there are also shrew-like tenrecs, others like water shrews, and three species of rice tenrecs, *Oryzarictes*, which have the habits and something of the appearance of a mole. The common tenrec, *Tenrec ecaudatus*, resembles the New World opossums.

Tent caterpillars, such as those of lackey moths, *Malocosoma neustria*, live gregariously in or on silken webs,

North American eastern **tent caterpillars** on their web. The communal silken tent protects them against predators and extremes of temperature. The caterpillars progressively denude their host tree.

which are spun over the bushes or trees which serve as their food plant. Two species are abundant and regarded as pests in North America. The eastern tent moth, *M. americana*, builds conspicuous tents and feeds particularly on orchard trees; the forest tent moth, *M. disstria*, constructs webs on forest trees, such as oak, which are used as platforms for resting.

Terman, Lewis Madison (1877-1956), American psychologist, during the First World War created psychological and intelligence tests for soldiers, which he later developed for application to schoolchildren. He had a lifelong interest in gifted children, and he revised the *Binet–Simon intelligence tests.

Termites, or white ants, are social insects totally unrelated to true ants, and forming the order Isoptera with 1,900 species. They are found in all warm countries, where no wooden structure is safe from their feeding activities. Foraging workers move in soil-covered runways, as protection from desiccation. Most species eat wood or deadwood and harbour *symbiotic, flagellate protozoa in their guts, which digest cellulose. Other species eat vegetable matter or debris; a few attack live plants. Many species eat soil, digesting its organic component and building above their nests the excreted clay into rock-hard mounds, which may be several metres (yards) tall and form a familiar part of African and South American grasslands.

Colonies consist usually of a king and queen, sterile, wingless workers and soldiers, and developing nymphs. All castes include males and females. Periodically, winged reproductives are produced and leave the nest in swarms, to mate and found new colonies. Colony composition is regulated chemically by the exchange of regurgitated food and saliva; the queen's faeces also pass among the colony and inhibit the production of reproductives. Some queens live for ten years, grow enormously fat, and lay an egg every two seconds.

Workers of fungus-garden termites, such as *Odontotermes transvalensis*, grow fungus combs on masses of their own faeces, composed largely of wood fragments; they eat the swellings that develop, and pass them on, partially digested, to the rest of the colony.

Terns are slender gull-like birds with narrow wings and forked tails. They hover over water, picking up small fish with forceps bills. Their subfamily includes thirty species of black-capped true terns, and four species of *noddies.

Terrapin is a name of North American origin, usually applied to edible, emydid *turtles, especially the diamond back terrapin, *Malaclemys terrapin*. This species occurs along the east coast of North America, from Massachusetts to Mexico, in salt or brackish water.

In Britain the term 'terrapin' is applied to virtually any small, mainly freshwater, turtle, especially those imported from America as pets, such as the red-eared, or pond, terrapin, *Pseudemys scripta*. Its common name derives from the bright red stripes on the sides of the head.

Territory (in ecology) is the space defended by an organism or a group, using *display or *aggression, against others of the same species. It may be either temporary or permanent. Many birds establish territories for nesting, as do some fishes, such as the stickleback. Territory may be large or small. A pride of lions may occupy territories

This South American **termite** nest has been broken open by an intruder, leaving the gravid queen termite exposed to the air. Workers and soldiers urgently scurry to move her under cover and protect the nest.

several square kilometres in extent. The gannet in large breeding colonies maintains a territory of only about 1·5 m. square (5 ft. sq.), though in this case food is taken from outside the territory. When the young of a species become mature they are usually driven off to establish their own territories.

Testes are the male organs that produce *sperm, and are present in most vertebrates as a pair. In adult male mammals they lie outside the abdomen in the scrotum. They develop inside the abdominal cavity, where they always remain in mammals such as the elephant, but in other mammals they descend through the abdominal wall into the scrotum at around birth. Within the scrotum in the adult the testicles lie in small isolated cavities. It has been shown that testicles in the scrotum are at a lower temperature than that of the body core. This is important, for if the testicles fail to extend into the scrotum, spermatogenesis fails to occur normally.

Tetanus, or lockjaw, is an infectious disease due to the effect on the spinal cord of a toxin produced by the bacterium *Clostridium tetani*. This is most commonly contracted through the contamination of wounds by spores of the bacterium, naturally present in soil. The resistant spores may remain latent for long periods in soil. The condition, which is characterized by painful cramps, is always serious and often fatal. Inoculation with modified toxin gives good protection.

Tetra is a name used mainly by aquarists for a group of fishes belonging to the *characins, so popular as aquarium fishes. The word derives from the now obsolete genus name *Tetragonopterus*. Tetras mostly live in tropical South America (although other characins also occur in Africa) and are small, active, and beautifully coloured. They all have an adipose fin on the back just in front of the tail. Possibly the best-known species is the neon tetra, *Paracheirodon innesi*, from the upper Amazon. Most tetras belong to the genus *Hyphessobrycon*.

Thalassaemia is an important and widespread group of hereditary disorders, in which production of *haemoglobin is impaired. A single thalassaemia *gene causes no disability and may give some protection against malaria by reducing the food value of haemoglobin to the malarial parasite. A double dose produces a severe anaemia, often causing death in childhood. Transfusion will keep the patient alive, but the consequent accumulation of iron in the body is itself damaging. This hereditary disease is widespread in parts of Africa, Asia, and the Mediterranean region.

The Congo **tetra**, *Micralestes interruptus*, from the Congo Basin, swims in shoals, feeding on small fishes and invertebrates. At about 8 cm. (3 in.), the male, here, is longer and more colourful than the female.

Theophrastus of Eresos (*c.*372–287 BC), Greek philosopher, wrote a *History of Plants* and *Causes of Plants*, which influenced botanical science into the Middle Ages. He bequeathed his garden for the benefit of his students.

Thickheads are a family of birds containing about forty-six species, related to the Old World flycatchers and fantails. All are found in wooded areas of India, Southeast Asia, and Australasia. Most are small, less than 20 cm. (8 in.), though some are up to 25 cm. (10 in.) long. The majority are dull grey and brown, or may, like the golden whistler, *Pachycephala pectoralis*, have bright yellow underparts. They are mainly insectivorous. Over half of the group are known as *whistlers, while others include the crested *bell-bird, pitohuis, and New Zealand thrush.

Thick-knees (birds)   *Stone-curlews.

Thirst is a powerful sensation aroused by the need for water, which is essential for the survival of many terrestrial vertebrates. Its function is to stimulate an urgent search for water to replace the deficit in the body fluids. Thirst cannot be referred to a particular sense organ or part of the body, although dryness of the mouth, tongue, and throat usually accompanies it. Thirst is extreme in certain medical conditions, such as haemorrhage, diabetes, or cholera, when the body's fluid volume is severely reduced.

Thistles are spiny-leaved, annual or perennial plants of the sunflower family. Genera within this family include *Cirsium* and *Cnicus*, and to the former belong the very troublesome weeds spear and creeping thistle. Others, such as cotton, milk, and globe thistles, are used as garden plants. Others are used as animal fodder in the young and non-spiny stage.

Thorn apple: one of ten species of poisonous plants, shrubs, or small trees, belonging to the plant genus *Datura*. They are related to the potato and occur widely in the tropics and warm temperate regions, especially the New World. Several American species are naturalized throughout the world and include the thorn apple, *Datura stramonium*. This species has long, pinkish, trumpet-shaped flowers and spiny fruit capsules. Some tropical *Datura* have large pendent flower trumpets and are called angel's trumpets or moon flowers (on account of their strong scent at night).

Thornbills   *Warblers.

Thorndike, Edward Lee (1874–1949), American pioneer in animal psychology and in the psychology of learning, formulated the 'trial and error' concept of learning. He devised methods for measuring *intelligence and introduced the use of identical twins in intelligence and learning experiments.

Thrashers belong to the *mockingbird family. All live in the New World. Most species resemble rather short-winged, long-tailed, thrushes. Their beaks are often curved, and are used to collect insects and seeds. They inhabit woodland and are essentially ground-feeders.

Thread-fins are mainly marine and brackish-water fishes found in all tropical and subtropical seas; a few of the thirty-five species are found in rivers. They are

Called Jimson weed in the USA, the **thorn apple** is an annual plant up to 1 m. (3 ft. 3 in.) tall. Its flowers are about 8 cm. (3 in.) long, and the erect fruit capsules—both ripe and unripe ones can be seen here—are about 5 cm. (2 in.) in diameter.

distinguished by their two-part pectoral fins, each of which has long anterior rays separate from the main fin. Bottom-living, they feed on crustaceans and fishes; most are relatively small fishes but the Asiatic thread-fin, *Eleutheronema tetradactylum*, grows to 1·8 m. (6 ft.).

Thread snakes, of which there are approximately fifty species, occur in Africa, parts of Asia, and the New World. They are similar in external appearance and burrowing habits to *blind snakes, but differ in having teeth only on the lower jaws. They are short (the family contains some of the shortest known snakes), and very slender, and feed mainly on termites.

Thrips, or thunder flies, are very small insects of the order Thysanoptera, most of the 3,000 species having narrow wings fringed with long hairs. Generally they are plant-feeders with sucking mouthparts. The females either have a delicate ovipositor, or a long cylindrical terminal segment to the abdomen, and no ovipositor. Those that are wingless live in the surface of soils, while the winged ones are abundant in flowers. Many species cause considerable damage and are important vectors of virus diseases.

Thrombosis is the formation of a blood clot in an abnormal site. Transformation of liquid blood to a solid form normally helps stop bleeding, but in an intact blood vessel it can block the circulation with serious consequences, for example heart attack or stroke. Thrombosis in a vein can lead to clot fragments reaching the lungs via the circulation, known as pulmonary embolism. Cure of established thrombosis can be difficult, so prevention is extremely important. Arteriosclerosis or hardening of the arteries can lead to thrombosis and is made worse by smoking. Venous thrombosis is common after prolonged confinement to bed.

Thrush (disease), or candidiasis, is an infection by yeast-like fungus of the genus *Candida*. It affects the surface membranes of the mouth and throat, lungs, intestine, vagina, skin, or nails. It produces typical white spots on the palate and inside of the cheeks, and causes some soreness. It is usually a trivial condition, but in some patients the infection may spread and become serious.

Thrushes belong to the family Turdidae, which contains about 304 species of birds with a worldwide distribution. They vary in size from species about the size of the European robin, 14 cm. (5 in.), to those 32 cm. (13 in.) in length. They tend to be active, conspicuous, and strikingly coloured, many of the smaller species having brown, grey, and white plumage, although some have red, blue, or yellow markings. They include the bluebird, nightingale, redstart, stonechat, and wheatear. They live in a wide variety of habitats from open desert to thick forest. Many are primarily insectivorous, though they also take berries and fruits.

The true thrushes belong to the subfamily Turdinae. These are mostly among the larger species of the family, such as the American robin, *Turdus migratorius*, and the European blackbird, *T. merula*, and song thrush, *T. philomelos*. Many tend to be brownish or greyish above and heavily streaked below, though the large Himalayan whistling thrush, *Myiophoneus caeruleus*, is blackish with a bright blue gloss. They are mostly birds of wooded country or rocky alpine areas, though some are found in flocks in more open country outside the breeding season. They eat a wide range of invertebrates and fruits.

Thunder flies *Thrips.

Thyme is the common name of plants with distinctive flavours and odours in the Old World genus *Thymus*. A few are grown as culinary herbs and several more as ornamental plants. Garden thyme, *T. vulgare*, also known as common or French thyme, is a low, evergreen, bushy species from the mountain slopes of the western Mediterranean region, with strongly aromatic leaves. Lemon thyme, *T. citriodorus*, has a distinct but milder lemon scent. Variegated thymes with grey or golden leaves are grown for their decorative qualities, while creeping thymes, with red, pink, or white flowers, are cultivated as garden plants.

Thymus gland is an organ found at the base of the neck of vertebrates. Consisting mainly of lymphoid tissue similar to the spleen, tonsils, and *lymph glands, it enlarges from birth to puberty, shrinking gradually during adult life. It plays an important but obscure part in the development of an individual's *immunity to foreign substances and tolerance of its own tissue components.

Thyroid glands are *endocrine glands in the neck of vertebrates, containing two kinds of secreting tissue. Thyroxine, a hormone containing iodine, is secreted at a rate controlled by a *pituitary gland hormone called thyrotrophin. It is required for the development and growth of the nervous system, and also increases heat production in most metabolizing tissues. Deficiencies in hormone output cause *goitre, while excesses cause overheating and extreme restlessness. The second hormone, calcitonin, is secreted in young animals when blood calcium levels are higher than normal.

Ticks are among the largest representatives of the *mite family, which has a total of 25,000 species. The ticks are entirely parasitic on vertebrates, though they attach to the host only when feeding. They have specialized, hooked mouthparts, and suck out the host's blood, adding an anticoagulant to stop it clotting. Their own body expands enormously as they feed. Their cuticle is quite elastic, and the body becomes a bloated oval bag, filled with blood. After feeding, ticks usually drop off the host, rest for a while, and moult. As much as a year may pass before they need another meal, when they climb up vegetation to find another passing host.

Ticks mate while on the host, and females deposit waxy egg-masses after they leave their host once more. The newly-hatched, six-legged larvae must feed twice, often on different hosts, before they achieve maturity as eight-legged adult ticks. In some areas ticks are important carriers of disease in livestock and sometimes even in man; the common *sheep-tick can cause serious disorders of sheep and cattle. Heavy tick infestations can have a direct debilitating effect on the host, known as tick paralysis.

Tiger beetles are medium-sized, colourful, *ground beetles, with strong jaws. The wing-covers are often metallic blues, reds, or greens. Their larvae live in vertical burrows in the ground and have specialized mouthparts which resemble gin-traps. Both larvae and adults are ferocious predators and usually occur in dry habitats. There are 2,000 or so species throughout the warmer regions of the world.

Tigerfishes are large African *characins. They are rather slender-bodied with a high, first dorsal fin and a separate adipose fin, and have large silvery scales but with a more or less distinct dusky line along each row. Their teeth are large and fang-like, and form a single row on each side of the mouth; they are so large as to be visible when the jaws are closed. Tiger-fishes are avid predators on smaller fishes eat large numbers of practically all the active mid-water fishes in their habitats. Two species are widespread in Africa: the giant tigerfish, *Hydrocynus goliath*, which grows to 1·8 m. (6 ft.) and may attain 70 kg. (154 lb.) in weight, and *H. vittatus* which grows to 1 m. (3 ft. 3 in.) and is found in the Nile, Niger, Congo, and Zambezi rivers.

Tiger moths are stout-bodied, brightly coloured, and conspicuously patterned moths, many of which are combinations of yellow, red, orange, black, and white. They are unpalatable or poisonous to predators, deriving some toxins from the food plants of the caterpillar and manufacturing others. Together with footmen moths, they belong to the family Arctiidae, which is world-wide, with about 3,500 species. The garden tiger moth, *Arctia caja*, of Europe, Asia, and North America, has black and white fore-wings and blue-spotted, red hind-wings, which it displays if attacked. Its hairy caterpillars, often called woolly bears, feed on such plants as nettle, dandelion, and dock.

Tigers resemble lions more than any other member of the *cat family. The single species is similar in size, but the tiger lacks the mane of the lion, and has a striped coat.

This coat provides camouflage among the shadows and sunbeams of the forests and dense underbush where they live. Tigers can swim and take readily to water, and can leap 4 m. (15 ft.) in one bound. Under cover of darkness, they hunt prey, which include deer, antelope, wild pigs, and livestock such as sheep, goats, and cattle. Occasionally a tiger kills a man and may become a persistent menace.

There is no fixed breeding season, and two to four offspring are born about 100 days after mating. The young are blind at birth and remain so for two weeks. The mother takes great care of her kittens and they are suckled until, at about six weeks, they begin to travel with the mother while they learn the elements of hunting. At six months they are of sufficient size and strength to hunt small game. There are eight subspecies of tigers, including the Indian or Bengal, the Siberian, and the Caucasian tigers.

Tiger snake: a species of swamp-loving snake, up to about 2 m. (6 ft. 6 in.) in length, from southern parts of Australia. Coloration is variable but cream or yellow

After a kill the **tiger**, *Panthera tigris*, remains near by and feeds at its leisure. A week may pass before it needs to kill again. It is not a particularly effective hunter, however; on average, only one in every two of its ambushes is successful.

cross-bands are commonly present. It is highly venomous, and if threatened it flattens its neck and body and may strike aggressively.

Tilapia is the name used for many of the *cichlid fishes native to Africa. Scientifically, it is applied to those cichlids in which the eggs are not brooded in the mouth of the male, but it is also used as a general name for all edible species of the African genus *Tilapia*. They have been widely introduced to other parts of the tropics, where they are intensively farmed. All are freshwater fishes growing to a length of 25–30 cm. (10–12 in.).

Timothy grass is a strong-growing European grass species known botanically as *Phleum pratense*. It is a valuable perennial meadow plant for grazing and hay production, and, on this account, it has been widely distributed as a cultivated plant.

Tinamous are a family of forty-six species of birds which are found in South and Central America. Varying from quail- to chicken-size, they look rather like gamebirds, but are not closely related to them. They are well camouflaged in greys and browns, usually barred or spotted, with powerful legs and very short tails. They live on the ground in forests or bushy grassland and eat mainly vegetable matter, though they take some insects.

Tissues (biology) are groups of *cells and associated nerves and blood vessels which perform a particular function. Together these make up a living creature. For example, in vertebrates, skin, muscles, and bones are all tissues making up a limb. The extent to which any tissue will grow is determined both genetically and as the result of environmental influences. Exercise can encourage the development of larger muscles than would form if little were taken.

Tits, or chickadees, include three separate families of birds. The first is the Aegithalidae (the long-tailed tits and bushtits), a family of seven species of very small, longish-tailed birds which occur in Europe, Asia, and North and Central America. The largest are only about 14 cm. (5½ in.) long of which more than half is tail; they weigh about 6–8 g. (⅕–¼ oz.). Most are black, grey, brown, and white, although the common long-tailed tit, *Aegithalos caudatus*, has pink in its plumage. They are primarily insectivorous and build complex, domed nests of feathers, lichens, and mosses held together with spiders' web.

The second family is the Remizidae (the penduline tits), a primarily Old World family of nine species, four of which are found only in Africa and one, the verdin, *Auriparus flaviceps*, in North and Central America. They are only slightly larger than the long-tailed tits, mostly dull greenish or greyish in colour. They also build domed nests; the South African species close the entrance to the nest after they have left to make it more difficult for predators to find a way in.

The third family is the Paridae, the true tits or chickadees, a group of about forty-six species with a worldwide distribution except for Australia and South America. Most are small, less than 14 cm. (5½ in.) long, though the glossy blue, black, and yellow sultan tit, *Melanochlora sultanea*, from Southeast Asia is 20 cm. (8 in.) long. Most are brown, grey, and white (some have blue and yellow markings), often with a white-cheeked pattern on the head; a few have crests. Agile, active birds, some are common in gardens at bird-feeders. They readily use nesting-boxes as they naturally nest in holes. They eat insects in the summer, but will eat seeds in winter.

Toadfishes belong to the family Batrachoididae, which contains about fifty species of mainly marine fishes occurring in shallow tropical seas. They are blunt-headed fishes with a large mouth, and two dorsal fins, the first of which has two or three sharp spines (in some species these spines are hollow and have venom glands). Most toadfishes are dull coloured but a few are brightly marked. Some species, such as the American midshipmen, *Porichthys* species, have luminous organs on their sides; others make loud noises underwater.

Toads are *amphibians which in popular usage have stout bodies, short limbs, and a dry, warty skin. This holds true only if the word is used in its narrowest sense, for members of the genus *Bufo*, the true toads. The distinction between frogs and toads becomes blurred if generalizations are made about most families of tail-less amphibians. Members of the genus *Bufo* are found in all continents of the world with the exception of Antarctica. Among the better-known species are the common toad, *B. bufo*, the natterjack toad, *B. calamita*, and the giant or marine toad, *B. marinus*, introduced into Australia to control the cane beetle, an agricultural pest which attacks sugar-cane.

In more general usage, the name is also applied to other species like the *midwife toad, the *spadefoot toads, the *Surinam toad, and the *Mexican burrowing toad.

Toadstools are the spore-bearing structures, or fruiting bodies, of certain *fungi belonging to the order called Agaricales. A toadstool, like a mushroom, has a stalk and a cap, on the underside of which are the spore-producing gills or pores. There is no biological difference between mushrooms and toadstools, but inedible or poisonous fungi are usually called toadstools.

Tobacco is an annual or short-lived perennial plant, *Nicotiana tabacum*, producing a spiral of large leaves on a stout stem about 2 m. (6 ft. 6 in.) tall. It belongs to the same family as the potato, tomato, and sweet pepper. Native to tropical America, it is now cultivated throughout the world, with the USA being the largest producer. Seedlings raised from the extremely tiny seeds are transplanted and the terminal bud pinched out when about twenty leaves have been produced. This prevents flowering and promotes further development of the existing leaves. These are harvested and dried before being air-cured, flue-cured (with artificial heat), or fire-cured (over the smoke of slow fires). Selected leaves are aged in a long fermentation process, during which the aroma develops and levels of the drug, nicotine, decrease. One other species of *Nicotiana* is cultivated for its high nicotine content (up to 9 per cent of the leaf), which is extracted for use as an insecticide.

Tobacco smoke, particularly from cigarettes, is a direct cause of lung and other cancers, and smoking is the single most important avoidable cause of cancer in man.

Tobacco hornworm: a North American *hawk moth, or sphinx moth, whose large caterpillars feed on tobacco and other plants of the same family, and so are regarded as pests. The green caterpillar has a red horn, and seven oblique white bands on each side. The moth, with a wing-span up to 12 cm. (5 in.), is grey with banded hind-wings and six orange-yellow spots along each side of the abdomen. The tomato hornworm, *Manduca quinquemaculata*, with the same feeding habits, has eight bands and a black horn, and the moth has five spots on the abdomen. Both species are kept under control as pests by trapping the adults.

Tobacco mosaic virus is typical of the many viruses that attack plants in that it contains only *RNA as its chromosomal material. As its name suggests, it attacks tobacco and other plants of the family Solanaceae and causes dead patches on their leaves. The viral particles of tobacco mosaic virus are rod-like and can be transmitted by insects, which feed on infected plants.

Tobacco plant   *Nicotiana.

Todies are a family of five species of bird of the genus *Todus*, found only on Caribbean islands. Small birds, about 10 cm. (4 in.) long, they are brilliant green above, and pale greyish or white below with a red chin patch. They live in wooded country, making sallies from perches to catch flying insects. They nest in holes in banks and lay two to five white eggs.

Tomatoes are annual plants with weak, trailing stems. They arose from the species *Lycopersicon esculentum* in tropical Central and South America. They belong to the family Solanaceae, along with potato, tobacco, sweet pepper, and aubergine. Grown for the round or egg-shaped, red or yellow, smooth-skinned fruits, a wide range of types has been produced and cultivated. In the USA a variety suitable for mechanical harvesting has recently been developed. Tomatoes are grown throughout the warmer areas of the world, or in heated glasshouses in cool temperate regions.

Tongue: a complex mass of muscle arising from the floor of the mouth in all vertebrates, except bony fishes. It is very flexible and has the capacity to change shape for exploration within and without the mouth. It is covered with mucous membrane whose upper surface is roughened by projections, the filiform papillae. In mammals, other papillae are found around the sides and back of the tongue to support taste-buds. In man these determine the recognition of salt, sweet, bitter, and sour. The tongue is not only used for the mastication of food but, for instance in the case of chameleons and frogs, can be projected out of the mouth to catch prey. It is also used to modify sound produced by the voice box, so helping to develop a complex system of communication.

Tongue-worms are unusual parasites of vertebrate respiratory tracts, particularly of tropical lizards. They comprise a small phylum (Pentastomida) of some seventy species. Some species can grow to 15 cm. (6 in.) long, with a five-clawed 'head' and thin annulated body covered with an arthropod-like cuticle. Each female produces thousands of eggs, which hatch as mite-like larvae, which may be transmitted by fish or rodents.

Tonsils are two masses of lymphoid (*lymph gland) tissue situated in the pharynx at the back of the mouth of all mammals. They are part of the body's defence system against infection. In human children, they are comparatively large but unless diseased do not merit removal.

Toothcarps *Killifish.

Toothed whales are smaller and more varied in form than the whalebone whales. The adults have as many as 180 triangular, rather shark-like teeth, a dorsal fin, and two longitudinal grooves on the throat, and the *beaked whales have a distinct beak. Best known of these whales is the killer whale, *Orcinus orca*, distinguished by its black and white markings. It travels in schools of up to forty animals which will attack and kill other, smaller, whales, seals, and dolphins. Other toothed whales feed on fish, squids, and octopuses. The largest of the sixty-six species is the *sperm whale. Others include the beluga, narwhal, beaked whale, dolphin, and porpoise.

Toothwort is the common name for several perennial plant species in the genus *Lathraea*, without normal leaves or chlorophyll; they are parasitic on the roots of shrubs or trees. The common toothwort, *L. squamaria*, parasitizes the roots of hazel, elm, and beech. A plant of woodlands and hedgerows throughout Europe and Asia, it appears above ground in spring as short spikes of dull white flowers. It belongs to the broomrape family which has some 180 species, found mainly in temperate Eurasia, of which all are parasitic on the roots of other plants.

Top-minnows *Killifish.

Torch thistle: one of the largest species of cacti which sometimes grows to 21 m. (69 ft.) high. Indigenous to Mexico and Arizona, it has branching stems which are occasionally used for torches.

Tormentil is the common name for some species within the plant genus *Potentilla*, itself part of the rose family.

Most *Potentilla* are perennial plants native to Europe and Asia. Common tormentil, *P. erecta*, is a plant of open situations on heaths, fens, and mountains, with a thick rootstock, leaves composed of four to five leaflets, and slender flower stems with yellow flowers. Common tormentil has long been valued for its medicinal properties for stomach disorders; it was also formerly used for tanning or staining leather. In the United States, this name is occasionally used for the spotted cranesbill, *Geranium maculatum*.

Torpedo fish: the largest of the electric *rays, growing to a length of 1·8 m. (6 ft.). It occurs on both sides of the North Atlantic and off the coasts of South Africa, mainly on sand and mud bottoms. It has powerful electric organs, which generate a current of 8 amps at 220 volts, but the power of the shocks lessens if repeated quickly. It feeds on fishes, often quite active ones, which are stunned by the electric current and then eaten. It gives birth to young fishes about 25 cm. (10 in.) in length, mainly in summer. The back is smooth and usually dark brown in colour, while below it is white. It has two small dorsal fins and a broad tail fin.

Torrey, John (1796–1873), American professor of chemistry, classified plant specimens and became recognized as a botanical taxonomist. With his pupil Asa *Gray, he wrote *A Flora of North America* (2 vols., 1838–43); and he was instrumental in setting up the New York botanic garden.

Tortoises are largely herbivorous *turtles that occur, usually in dry habitats, in most warm regions of the world. The greatest range of species occurs in Africa and none at

Giant **tortoises**, although of the same genus (*Testudo*) as European species, are much larger and have a rather different pattern of 'plates' in the carapace. Some, like this *T. elephantopus* from the Galapagos Islands, may eventually have a carapace 1·5 m. (5 ft.) long.

all in Australia. Although many true tortoises are known as fossils in North America, only the four species of burrowing gopher tortoises, *Gopherus* species, occur in the region now.

Tortoises have rather short, broad feet and elephantine hind legs that are cylindrical in shape. Their shells have hard, horny coverings. As additional protection, the front legs have thick scales over the part that remains exposed when the tortoise retracts into its shell; in the hind region, only the tail and the soles of the feet are incompletely retracted. Some of the largest living tortoises are found on oceanic islands, especially the Galapagos Islands in the eastern Pacific, and Aldabra in the western Indian Ocean. Each of the bigger islands in the Galapagos had its own distinctive population of tortoises at one time. On islands with sparse food rather small forms evolved, with long legs and a characteristic saddle-backed shell (elevated above the neck) which enabled them to reach and browse on taller vegetation.

Tortoiseshell butterflies　*Brush-footed butterflies.

Tortrix moths, or bell moths, are small moths, with broad, square-ended fore-wings. These are held flat over the rounded hind-wings at rest, giving an outline like a bell. The caterpillars of many species feed within rolled-up leaves, securing them with silk. Several species, such as the codlin moth, are pests on fruit, and others eat seeds. The family Tortricidae is found worldwide and contains about 4,000 species. The green oak tortrix, *Tortrix viridana*, found throughout Europe, northern Africa, and western Asia, is a leaf-roller, and is sometimes so abundant in oak woodland as to defoliate the trees.

Toucans belong to a family of about thirty-eight species of birds, all of which live in forests in Central and South America. Their body-size is similar to that of a crow, and they are usually black or green, with white, red, or yellow patches. They are famous for their enormous and brightly coloured beaks, which they use for reaching fruit at the extremities of branches. They also eat large insects, lizards, and other small animals. They nest in holes in trees.

Touch supplies detailed information about the surfaces in contact with all or parts of an animal's body and the nature of objects examined by hands, lips, or snout. To blind people, nocturnal animals, and dwellers in burrows it is a particularly valuable sense. Sense organs activated by minute deformations are called mechanoreceptors and several specialized forms of these are present in animals. The most sensitive consist of single hairs with a nerve fibre attachment. A combination of sensory information from different types contribute to the discrimination of the sensations of touch, pressure, and vibrations.

Touracos, turacos, or plantain-eaters, are a family of about twenty-two species of birds related to the cuckoos. All occur in Africa south of the Sahara, and most live in forest, though a few are found in open savannah. Their body-size is similar to that of a crow, but they have long tails. Most species are glossy blue-black or green, with red and yellow markings, but some are greyish-brown and white. They eat mainly fruits, but also take some insects.

Toxicology is the study of adverse effects of drugs used in therapy, and of chemicals used in the home, in industry, or found in the environment. Adverse drug reactions are very common and understandably cause widespread interest. Overdosage toxicity results when drugs are given in excess of the recommended dose, whereas drug side-effects are unwanted but predictable effects occurring with normal doses. Unpredictable and potentially serious toxic effects include allergy, bleeding, anaemia, liver damage, and kidney failure.

Prevention of toxicity depends on careful trials of all new drugs before release for public use, along with close monitoring of adverse reactions by an impartial central body, high standards of occupational health care, and surveillance of industrial disease.

Trace elements are chemical elements needed in the body of an organism in very small amounts, usually 100 parts per million or less. They are essential components of certain *enzymes and other proteins, and include copper, iodine, iron, cobalt, chromium, silicon, manganese, and zinc. Copper is needed for the enzyme cytochrome oxidase, which forms part of the electron transport chain. Iron is needed during chlorophyll manufacture in plants, for cytochrome proteins, and as part of haemoglobin. Many trace elements are toxic in larger quantities and are usually supplied by a well-balanced diet.

Trace fossils are the signs of animal activity that have been preserved in hardened sedimentary rock. They include trails, burrows, and tracks such as footprints. Because similar traces can be made by quite different animals, they are classified according to their type rather than the specific animals which may have caused them. From trace fossils geologists can calculate such things as the abundance of life in a particular sediment and rates of sedimentation. The study of trace fossils is called ichnology.

Trachea, or wind-pipe: the tube which starts below the *larynx in mammals and ends in the thorax as it divides into two bronchi. It is lined with hair-like cilia and moistened by mucus. The cilia beat so that foreign particles are wafted outwards, towards the pharynx. The tube is kept open, and resists crushing, by means of rings of *cartilage in its walls.

Translocation in botany is the transport of organic molecules, mainly sugars and amino acids, about the plant. Food is manufactured by photosynthesis in leaves, but is frequently needed for growth elsewhere, such as in roots or buds. The process of translocation requires energy and occurs in specialized tissue called *phloem, which forms part of the plant's vascular system. The compounds moved must be soluble; the most frequently involved compound is a sugar called sucrose.

Transpiration is the evaporation of water from the aerial parts of plants, mostly from the leaves, but also from stems, flowers, and fruits. The impermeable *cuticle covering these structures is designed to prevent desiccation, but must be perforated by pores (stomata, or lenticels) to allow gases through for *photosynthesis and *respiration. Evaporation also occurs through these pores, generating a continuous upward flow from the roots to replace water lost. Transpiration does, however, have advantages; it helps to cool the plant in hot weather, and minerals are carried up from the roots in the transpiration stream. In conditions of drought, stomata close, reducing transpiration. Warmth and wind speed up the process.

Trapdoor spiders construct silk-lined burrows, covered by a snugly fitting, sandy or mossy lid. Here the spider may live for many years, always lying in wait just below ground. Often silk drag-lines radiate from the burrow, so that passing prey can be detected as they trip over them. The trapdoor spider jumps out and pierces the animal with its jaws, injecting poison, then drags the paralysed victim back to the burrow to feed on at leisure. Trapdoor spiders belong to the same primitive group of spiders as the tarantulas.

Trauma is a Greek word meaning wound or injury. It is used in this sense by doctors describing physical damage to the body. Its use has been extended in psychoanalysis to mean a disturbing event or experience that may have affected a person's emotional development and happiness.

Tree boas are, as their name suggests, snakes of the *boa group that spend much of their life in trees. Like many other tree-living animals, they have prehensile (grasping) tails. They also have heat-sensitive pits (rather like those of pythons) on their lips, which help them to detect their prey, which often consists of roosting birds or bats. They occur in South America and Madagascar.

Treecreeper is a name given to five species of birds of the family Certhiidae which live in forests in the cooler parts of the Northern Hemisphere, and to six species of the Australian family Climacteridae. All are small, streaky brown birds 12·5–17 cm. (5–7 in.) long, with pale underparts. They have slender, longish, curved beaks. They creep up trees (or rocks) supporting themselves on their legs and the stiffened tail while they probe into cracks for

The grizzled **tree kangaroo**, *Dendrolagus inustus*, grasps branches with its shorter fore-paws and steps along a branch or trunk with alternate hind-feet when climbing. Ground-based kangaroos use their hind-feet in unison.

their insect food. The Australian treecreepers have a longer tail than their namesakes, and are considered to be related to nuthatches.

Tree ferns are true *ferns which look like trees because the crown of fronds is borne on top of a trunk-like stem which may be up to 25 m. (82 ft.) high and 30 cm. (12 in.) in diameter. The stem is covered in leaf scars or dead leaf bases where old fronds have been shed, giving it a hairy appearance; it is also thickened by roots which emerge from it high above ground and grow down through the tangle of leaf bases. Tree ferns are found almost exclusively in the tropics and warm temperate regions.

Tree frogs are species of tail-less *amphibians, which include species which would popularly be called toads. They live not only in trees but also amongst reeds and sedges. Possibly the best-known species belong to the family Hylidae, and include the European green tree frog, the common tree frog of North America, and the Australian tree frogs, which belong to a separate genus (*Litoria*). They are well adapted to a climbing existence, having long limbs with adhesive discs on the fingers and toes. Some species, the so-called flying frogs, have developed the ability to glide between trees.

Tree hoppers resemble *frog-hoppers, but the top of the first segment of the thorax is enlarged backwards, often into bizarre shapes. They comprise the family of Membracidae, with over 2,500 species throughout the world. They are plant-feeders, most common in warm climates, and many species are closely associated with ants. The females may provide parental care, guarding the young if disturbed.

Tree kangaroos are seven species of forest *kangaroo in the genus *Dendrolagus*, at home in the mountains and

A glade of the soft **tree fern**, *Cyathea smithii*, in the Westland district of the South Island, New Zealand. Mature plants have stems 8 m. (26 ft.) or more in height. The leafy parts of the old fronds of the specimen in the left foreground have withered, leaving behind the harder midribs and stalks, which will also be shed.

high table-land of northern Queensland. They have long hind-limbs but live in the branches of trees. They are about 1·2 m. (4 ft.) long, half of which is the tail. During the day they sleep curled up in the crotch of a tree, several individuals often occupying the same grove. After sunset they descend backwards to the ground to visit a water-hole. They feed on leaves, ferns, creepers, and fruit.

Tree of heaven: one of ten species of *Ailanthus*, a small genus of tall, deciduous trees belonging to the family Simaroubaceae. *A. altissima* is native to northern China but is grown as an ornamental tree in parks in eastern North America and in parts of Europe. It is rapid growing and can reach 30 m. (100 ft.) in height. The long, compound leaves, composed of between fifteen and thirty oval leaflets, are up to 46 cm. (18 in.) in length. The greenish flowers, borne in clusters, have male and female flowers on separate trees, and the reddish-brown fruits are shaped rather like the 'keys' of ash trees.

Tree peonies   *Peonies.

Trees are woody plants with a single readily recognized trunk but in some cases they cannot be distinguished from *shrubs. The trunk may be unbranched, as in most palms and many conifers, or, more often, may bear a crown of branches and twigs on which the leaves are produced. Flowers and fruits may be produced on the twigs or on the major branches or trunk or, in some species, even on the leaves. The tallest trees in both temperate and tropical regions are conifers.

The tree form appears to have evolved independently in a number of unrelated plant groups, including the ferns, club-mosses, and horsetails. All of these tree forms save a few tree ferns are now extinct, though in their prime these forests of spore-bearing plants covered wide areas and their fossils have become the coal of today. They were followed by *gymnosperm forests of trees with seeds, which probably arose independently from a number of fern groups and have given rise to the modern *phanerogams. Modern trees are predominantly conifers and *dicotyledons, the former still being dominant in colder parts of the world. Herbaceous plants seem to have evolved from trees independently in many groups of dicotyledons, but the conifers are remarkable for having no herbaceous representatives.

Because of their size, trees dominate much of the world's land vegetation, forming forests or woodlands, but they can not tolerate the climatic extremes experienced by grasses and *cryptogams in the polar regions. Their principal uses to man have been as timber and pulp for construction and paper respectively, though many fruit-trees, shade-trees, and ornamentals are also grown.

Tree-shrews comprise an order of primitive animals that are intermediate in anatomy between primates and insectivores. On the basis of their relatively large brain, enclosed eye sockets, and a permanent scrotum, they used to be classified with the primates, but are now recognized as a unique order. They are squirrel-like in habits and live in tropical forests of Southeast Asia. Some of the eighteen species are shrew-like in appearance, up to 20 cm. (8 in.) in length with long, furry tails.

Tree sparrow is a name applied to two birds. *Passer montanus* is a member of the weaver family closely related

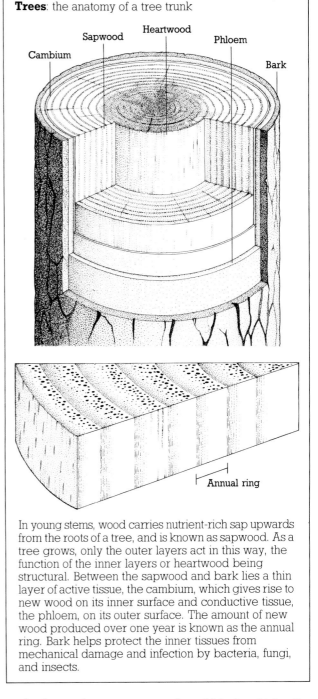

Trees: the anatomy of a tree trunk

Cambium   Sapwood   Heartwood   Phloem   Bark

Annual ring

In young stems, wood carries nutrient-rich sap upwards from the roots of a tree, and is known as sapwood. As a tree grows, only the outer layers act in this way, the function of the inner layers or heartwood being structural. Between the sapwood and bark lies a thin layer of active tissue, the cambium, which gives rise to new wood on its inner surface and conductive tissue, the phloem, on its outer surface. The amount of new wood produced over one year is known as the annual ring. Bark helps protect the inner tissues from mechanical damage and infection by bacteria, fungi, and insects.

to the house *sparrow, a species which is widely distributed and abundant over much of Europe and Asia. It is sparrow-coloured with a russet cap and a white cheek with a black spot on it. It usually nests in holes in trees or buildings. The other bird, *Spizella arborea*, is a member of the bunting family which is widely distributed in North America. It too is sparrow-like above with an orange-brown crown; it is pale below with a black spot on its breast. It lives in open country away from trees and builds a cup-shaped nest of grass on the ground.

Tree tomato: a small tree from Peru, growing up to 6 m. (20 ft.) tall, and belonging to the same family as the tomato. It is cultivated throughout the Andes at moderate altitudes. The fruits are egg-shaped, reddish-yellow or purple, and resemble acid tomatoes in flavour. They are eaten fresh or, more frequently, stewed. A crop of only minor importance, the tree tomato has been introduced to several Pacific islands and is especially popular in New Zealand.

Trepang *Sea cucumbers.

Triceratops, or the horned dinosaur, was one of the bird-hipped *dinosaurs. It was a quadrupedal plant-eater, and its head bore three powerful horns. A frill of bone extended from the skull back over the neck region for protection. It was among the last living dinosaurs, becoming extinct some 65 million years ago.

Trichinella spiralis is a minute, parasitic *roundworm which infests pigs, causing 'measley' pork. If this is consumed without adequate cooking, the larval parasites may become established in the gut. The adult worms can live in the small intestine of man and produce larvae which bore through the intestine wall to cause damage to the body tissue.

Trigger-fishes belong to a family containing perhaps 120 species of mainly tropical marine fishes. They have deep, compressed bodies with heavy, smooth scales, and two dorsal fins, the first of which contains strong spines. The large front spine is blunt and strong and is locked into position by the third spine (which has to be depressed first before the front spine is lowered, thus acting as a trigger). Some species of trigger-fishes wedge themselves into crevices using these strong spines. Most live in shallow water close to reefs, but a few have an oceanic life-style.

Trilobites are an extinct group of arthropods that were common from the Cambrian to Permian eras (570–245 million years ago). They had a single head shield, and some species had insect-like compound eyes and antennae. This was followed by more than twenty short body segments, each bearing a pair of jointed legs on the underside. They lived in marine habitats and were between 2 and 30 cm. (1–12 in.) in length.

Trogons are thirty-five species of birds comprising the family Trogonidae, which inhabit the tropical forests of the Americas, Africa, and Asia. These birds are about the size of a thrush or small crow except for the very long-tailed *quetzal. They are mostly metallic green above with yellow or red underparts. They sit very still beneath the forest canopy and fly out after insects or to pluck fruit. They nest in hollow trees.

Tropic-birds, or boatswain birds, are three species of seabirds of the genus *Phaethon*, widespread over warm tropical and subtropical seas, usually wherever there are flying-fish, their main food. Distinguished by very long central tail feathers, they fly low over the sea singly or in pairs, plunging for fish and squid. They nest on trees or in burrows, usually on islands, rearing single chicks which both parents tend.

Trout are fishes of the salmon family and are native across Europe and in Atlantic coastal waters, but have been introduced to many other parts of the world as a sporting and food-fish. Brown trout, *Salmo trutta*, are dark coloured and heavily spotted, and live in rivers and lakes. Sea trout,

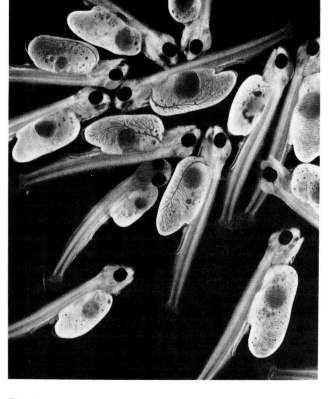

The elongated yolk sac of these brown **trout** alevins (larvae) indicates that they are between two and three days old. At this stage they burrow into the gravel to emerge some three weeks later as fully finned young fry.

which are a form of *S. trutta*, are silvery with few spots, and migrate to the sea to feed. They spawn in winter, the eggs being laid in redds hollowed out in river gravel and hatching in six to eight weeks. Young trout eat insects, insect larvae, and crustaceans; large trout eat fishes too. They require cool water with high oxygen levels. In North America the name 'trout' is used for the rainbow trout, *S. gairdneri*, and for several other species, including members of the genus *Salvelinus*, often known as *charrs.

Trout perch: a freshwater fish, living in the back waters of lowland rivers and in shallow lakes in parts of North America. It grows to about 20 cm. (8 in.) in length and is slender-bodied, with weak spines in the dorsal and anal fins and a small adipose fin. It feeds on aquatic insects, small crustaceans, and molluscs, and spawns over gravel-beds in late spring and early summer.

Truffles are *fungi, belonging to the order Tuberales, and live entirely underground. The fruiting body, often referred to as the truffle, is usually round and pitted, and 1–7 cm. ($\frac{1}{2}$–3 in.) in diameter. Several species are highly esteemed delicacies, particularly the Perigord truffle, *Tuber melanosporum*, and are usually found under oak or beech trees. Many have a strong smell and attract animals such as rabbits and squirrels, which may help disperse them by digging them up. In France, trained dogs or pigs are used to find them by their smell.

Trumpeter swans, named after their bugle-like calls, are the largest of the *swans. They are also the rarest, and have only recently been saved from extinction by conservationists. Except for their black bills they are very like whooper swans and inhabit northwestern America.

Trumpet-fish is the name for three species in the family Aulostomidae, all distantly related to the pipefishes. They occur in the Indo-Pacific, and parts of the Atlantic. All are long-bodied with long tubular snouts and a series of separate spines along the back. They are slow swimmers and live on reefs and in shallow water, catching fishes and shrimps by stealth.

Trunk-fishes are members of the family Ostraciontidae, and are related to the trigger-fishes and puffer-fishes. They are also called box-fishes because of the specialized scales which cover the head and body in a shell-like covering. They are found in all tropical oceans, mainly in shallow water. Slow swimmers, moving mainly by paddle-like motions of the fins, they are protected from predators by their box-like armour and by a toxin which they secrete.

Trypanosomiasis is infection with one of several species of protozoa of the genus *Trypanosoma*. There are two main forms: African sleeping sickness, a fatal disease, which is conveyed to man by the tsetse fly from infected animals, and Chaga's disease, the South American form, which is transmitted by reduviid bugs. These parasites have two stages to their life cycle, one inside an insect, and the other in the blood of a vertebrate. The acute infection is relatively mild, but there can be complications affecting the heart and the gut.

Tsetse flies are dark brown or yellowish blood-sucking flies which make up the family Glossinidae, and the true tsetse flies (*Glossina* species) occur only in tropical Africa. The adults act as vectors for the protozoan parasites that cause sleeping sickness (*trypanosomiasis) in man and nagana in cattle and horses. The word tsetse means 'a fly that kills cattle'. They can be distinguished from houseflies because the wings are folded flat exactly on top of each other, and the horny, piercing proboscis projects in front of the head. Females produce one fully grown larva at a time, dropping it on the ground, where it burrows and immediately pupates.

They are woodland insects, each of the twenty-one species of *Glossina* requiring particular conditions of temperature and humidity to provide suitable conditions for mating, breeding, feeding, and sheltering. The spread of savannah woodland has extended the range of some, including the species that makes it no longer possible to keep horses in Freetown, Sierra Leone. Large wild mammals are their usual source of blood. Slaughter of wild animals and bush clearance are the principal means of control. Ironically, their presence has hitherto saved much of Africa from over-grazing and erosion caused by cattle.

Tuatara: an unusual lizard-like reptile, which lives only on some rocky islands off the coast of New Zealand, where it is protected by law. It is related to, but more primitive than, lizards, snakes, and *amphistaenians. The tuatara, *Sphenodon punctatus*, is the sole survivor of a group of lizards known only as fossils from the late Triassic Period (about 200 million years ago). It represents a major order of reptiles once quite common, but now extinct.

The **tuatara's** staple food is insects but they also eat snails, worms, and lizards. Primitive creatures, they have an archaic bone structure and the vestiges of a third eye (pineal eye) under the skin of the forehead.

It is active at lower body temperatures than other reptiles and has a very slow rate of growth so that it is not sexually mature until it is twenty years old and may live more than a hundred years. The tuatara lives in burrows which it either constructs itself or takes over from excavating birds such as petrels and shearwaters. Their eggs take twelve to fifteen months to hatch.

Tube-mouth fish: a species of fish related to the wrasses, and forming the family Odacidae. The tube-mouth fish resembles a pipefish, being extremely long and slender. It lives among seaweeds on the coasts of southern Australia and grows to a length of 40 cm. (16 in.).

Tuberculosis is an infectious disease caused by the bacterium *Mycobacterium tuberculosis*. Symptoms include fever, weight loss, and blood in spittle. Pulmonary tuberculosis, also known as consumption, affects the lungs, and was once a major cause of death. In developed countries the incidence of tuberculosis has steadily fallen over the last hundred years, and with drug treatment nearly all cases can now be cured. It is still, however, a major cause of death in many parts of the world. Of all infections, it is the one most exacerbated by poor nutrition.

Tubers are swollen plant organs, usually produced underground and derived from roots, as in the dahlia, or from underground stems, as in the potato, so that there is no clear distinction between them and underground stems such as *rhizomes. They act as storage organs and several have been exploited for food.

Tuco-tucos are a family of cavy-like *rodents with a wide distribution in southern South America. They are some thirty-three species found from sea-level to 4,000 m. (13,000 ft.) up the Andes. They are gopher-like in habit, constructing extensive burrows. They weigh up to 700 g. (1½ lb.) and are entirely herbivorous, feeding, mostly at night, on roots as well as leaves and seeds.

Tufted duck: an Old World species of *pochard. *Aythya fuligula* is very common in Europe, where it nests on islands in lakes and ponds, and forms large flocks outside the breeding season. The drake is black and white, and has a thin tuft drooping behind its head, unlike its relative the scaup.

Tulips, of which there are many different kinds, belong to the plant genus *Tulipa*. They are perennial plants, growing from bulbs, and part of the *lily family. With the onset of the growing period the bulbs produce variously shaped leaves or leafy stems which terminate in showy flowers. Over 100 species are known, from Europe, Asia, and North Africa, ranging from small, multiflowered alpines to species with large and colourful solitary flowers. They have long been popular as garden and greenhouse plants ever since the introduction of *T. suaveolens* to Europe in 1572. Dutch plant breeders were mainly responsible for the early varieties, which resulted in the tulip mania of the 1630s, when vast sums were paid for single bulbs of new hybrids. Modern varieties are now divided into several classes according to habit and flowering time.

Tulip-trees are two species of *Liriodendron* which are members of the magnolia family. The tulip-tree, *L. tulipifera*, is native to North America from Nova Scotia to Florida, and the Chinese tulip-tree, *L. chinense*, is native to central China. Both are deciduous trees, which may reach 60 m. (190 ft.) in height, and have tulip-like flowers as their name suggests. The American tulip-tree has fine-grained, light yellow timber which is used in carpentry.

Tumours   *Cancer.

Tunas are a family of medium to large marine fishes which has seventy-five species, including the mackerel. Tunas are oceanic fishes, and most abundant in tropical and warm temperate seas, but they migrate into cooler waters in the warm seasons. They live near the surface and are active, predatory fishes, feeding on a wide range of surface-living fishes and squids. They all have a spindle-

Parts of the **tundra** are inhabited by this dwarf, mat-forming *Arctostaphylos*, a shrub in the heather family, seen here near Churchill on Hudson Bay, Canada. Its evergreen leaves turn a fiery red in the autumn.

shaped body with a pointed head, two dorsal fins, and a broad tail fin. The first dorsal fin slots into a groove on the back, and the second, like the anal one, ends in a series of small finlets. Tunas are immensely powerful swimmers, and have a body-temperature higher than that of the surrounding water.

The most common species in the Atlantic is the blue-fin tunny, or tuna, *Thunnus thynnus*, which grows up to 4 m. (13 ft.) in length; it migrates northwards into Norwegian waters in summer. The albacore, or long-finned tunny, *T. alalunga*, is common worldwide and is an important food-fish.

Tundra is a treeless vegetation type lying between *taiga and the ice-caps, notably in the Arctic lowlands. It is dominated by lichens and mosses which can tolerate the extreme growing conditions. For much of the year the tundra is snow-covered and frozen solid. During the short summer the surface ice and snow melts, but cannot drain away due to the frozen sub-soil, or permafrost. Thus tundra is boggy and swamp-like for its short growing season.

Tunicates   *Sea squirts.

Tunny   *Tunas.

Tupelos are deciduous trees or shrubs of the genus *Nyssa*, native to North America and the Far East. Several species are grown for their durable wood. The trunk of the black gum, or tupelo, *N. sylvatica*, may reach 30 m. (100 ft.) in length. It is native to swampy and other poorly drained ground in eastern North America. The water tupelo or cotton gum, *N. aquatica*, of the southeastern USA, withstands flooding of up to 2 m. (6 ft. 6 in.) or more of water.

Turbot: a species of *flat-fish with a very broad, large-headed body, and large bony tubercles in the skin instead of scales. Widely distributed in European seas, it lives on gravel and coarse sand from the shoreline to depths of about 80 m. (260 ft.). It feeds mostly on fishes.

Turkeys are related to other gamebirds and belong to one of two species of the family Meleagridae. They formerly inhabited woodland in North and Central America but nowadays, because of over-hunting, their natural ranges are much reduced. They are powerfully built birds with dark greenish-grey feathers edged with black. The male of the common turkey, *Meleagris gallopavo*, has a large protuberance on its neck. The bird grows to 1·25 m. (4 ft.), and has been extensively domesticated, with many breeds available. The turkey was almost certainly introduced direct to western Europe by the Spanish, very soon after Columbus discovered America, and did not come by way of Turkey as its name might suggest.

Turkey-vultures are heavily built, vulture-like birds of North and South America, distributed sparsely but widely over mountains and woodlands. They are dark green-brown with a bare red head and neck, and have a wingspan of 2 m. (6 ft. 6 in.). In flight they use upcurrents to soar while they search out carrion, but farmers destroy them for their predation on lambs and sheep. Two chicks are raised in caves or tree nests.

Turk's cap lilies   *Lily.

Turmeric is a plant closely related and similar in growth to the ginger plant. The spice for which it is cultivated is, like ginger, obtained from swollen underground stems. The strong yellow colour of curry is principally provided by turmeric, which also yields a yellow-orange dye used in Asia and Europe to dye natural fibres. An ancient Asian crop, it is grown in India, China, Indonesia, and the West Indies, wherever the climate is hot and moist.

Turnip: a *brassica root crop similar to swedes and equally useful as a vegetable and for livestock feed. Grown mainly for their swollen roots, some varieties (stubble turnips) are particularly fast-growing and can be grazed while the roots are fairly small by sheep or cattle. The leaves are bright green and rough in contrast to the smooth bluish-green leaves of the swede. The round roots are white- or yellow-skinned with either a green- or purple-tinged top. Turnips are an ancient European crop, known since prehistoric times and now widespread throughout the world.

Turtles are members of an order of *reptiles which are distinguished by a relatively short, broad body, enclosed in a box comprised of bony plates. These plates are commonly covered in a horny material or, in some families, such as soft-shelled turtles, a layer of leathery skin. They are known, from fossils, as far back in time as the late Triassic Period (200 million years ago).

All living forms lack teeth; instead they have sharp-

edged, horny beaks. These are modified into broader ridged beaks in species which crush their food. All species are egg-layers, even the most aquatic turtles still needing to return to land for this purpose.

Tusks are formed by the elongation of *teeth, usually of the incisor or canine teeth. Perhaps best known, because of their size, are those of the elephant, where the incisor and the canine teeth have become reduced to a single pair of upward-curving tusks. These may be up to 3·2 m. (10 ft. 6 in.) long in an African elephant. The mammoths had even larger tusks, up to 5 m. (16 ft. 6 in.). The male narwhal has a single tooth which grows continuously, to form a spirally twisted tusk up to 2·7 m. (8 ft. 10 in.) long. This tusk may be used in sparring with other males. Pigs have persistently growing canine teeth which form tusks in the male. The males of some deer also have tusks developed from canine teeth.

Tussock-grass is a name applied to a range of grasses with a tufted, compact habit of growth. It is particularly applied to tufted-hair grass, *Deschampsia flexuosa*, a perennial species with a wide distribution in Europe, Asia

Pacific Ridley **turtles**, *Lepidochelys olivacea*, bury their eggs on a beach in Costa Rica. How so many individuals co-ordinate this behaviour is not fully understood but their numbers may provide some protection.

Minor, Japan, and North America. The species name describes the long, wavy flowering stems, which arise from dense tufts of leaves raised above ground-level. It is a plant of wet meadows, acid heaths, moors, and open woodlands and has little agricultural value.

Tussock moths are moderate-sized moths of the family Lymantridae whose hairy caterpillars bear prominent clumps or tussocks of coloured hairs on their backs. The hairs are irritant, the bright colours advertising this to potential predators, and are incorporated into the cocoon at pupation. Adults never feed, having vestigial mouthparts. The family includes the gold-tail, *Euproctis similis*, and vapourer moths, *Orygia* species. The gold-tail and the closely related brown-tail, *E. chrysorrhoea*, can reach pest proportions and their larval hairs can cause skin irritation to humans. The gypsy moth is another important pest species of trees.

Twite: the name of a species of bird in the *finch family. It occurs in northwest Europe and high mountains of west and central Asia where it is found on open ground or moors. It is a small, streaky brown bird with a pinkish rump. Its beak is yellow in winter, but brown in summer.

Typhoid fever is a specific infection by the bacterium *Salmonella typhi*, characterized by *septicaemia and intestinal ulceration. It generates high fever, a rash of red spots, and in severe cases the intestinal ulcers may perforate. The organism is excreted in the faeces, and transmitted through infected food or water. The illness used to be prolonged, incapacitating, and frequently fatal, but antibacterial treatment has much improved the outlook. Vaccination gives moderate protection against infection.

Typhus, or spotted fever, is an acute infection with micro-organisms called rickettsiae, which, unlike a bacterium, can grow only within cells. The illness is marked by fever, rash, and some loss of consciousness. It is spread from an infected person by parasites such as *lice or *ticks.

Tyrannosaurs are one of the groups of reptile-hipped *dinosaurs. They were the largest of the carnivores, standing 6 m. (20 ft.) high, and were completely bipedal, the front legs being reduced. Their jaws were large and powerful, and armed with dagger-like teeth, which indicate that they were carnivores.

Ulcers are regions where the layer which normally covers either the external or internal surface of the body has been lost and healing has not taken place. One cause is the replacement of normal cells by cancer cells, as, for example, in a malignant ulcer of the rectum. Ulcers commonly occur in the digestive tract. They are caused by the action of acid and *enzymes, such as pepsin, or bile from the liver. These substances act on the lining of the stomach or upper part of the intestine, and the ulcers produced give rise to vomiting, pain, and bleeding. Ulcers in the stomach are often called gastric ulcers, those of the duodenum duodenal ulcers. The term peptic ulcer, which includes gastric, duodenal, and less common types, is used for ulcers that occur in the presence of the enzyme pepsin and acid. Most types are apparently due to local loss of normal resistance to digestive enzymes. They are very common, and mostly heal spontaneously. Healing can be accelerated by reducing the acid output of the stomach.

Umbilical cord: that tissue which connects the embryo to the *placenta in placental mammals. Blood passes from the embryo to the placenta along two arteries, propelled by the embryo's heart. It returns from the placenta via a large umbilical vein to the heart for recirculation. These vessels forming the cord are twisted around one another and resemble a rope. In humans the cord is clamped and cut after birth. Animals bite through it; their saliva contains an enzyme which stops bleeding from the cord after it has been severed.

Umbrella birds are three species of bird which belong to the cotinga family. They occur from Costa Rica southwards down South America, living in tropical forest. They are the largest members of the family. All are black in colour and get their name from the umbrella-like thatch of raised feathers on the crown. All the males have a throat wattle, covered in black feathers in two species; in the male long-wattled umbrella bird, *Cephalopterus penduliger*, the wattle may be up to 35 cm. (14 in.) long. They eat fruits and insects. Like other members of the family, they build a flimsy nest of twigs in the top of a small tree and lay a single egg; however, their breeding behaviour is poorly known.

Underwing moths are *owlet moths which have brightly coloured hind-wings that are exposed in flight but concealed by the camouflaged fore-wings at rest. The name is used for a number of different types of moth, but particularly for the genera *Catocala* and *Noctua* of the family Noctuidae. There are about a hundred species of *Catocala* in North America, but only five in the British Isles, including the red underwing, *Catacola nupta*. There are six British species of yellow underwing (*Noctua*), the most common of which is the large yellow underwing, *Noctua pronuba*, whose caterpillars are *cutworms, which eat roots, stems, and leaves of a variety of low-growing plants.

Unicorn fishes are members of the genus *Naso*, within the family Acanthuridae. All species of *Naso* live in the

Umbrella birds

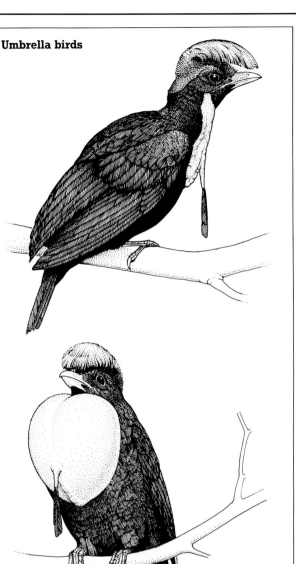

The male bare-necked umbrella bird, *Cephalopterus glabricollis*, has a crest of feathers that spreads like an umbrella over its eyes and beak. It has a throat sac of wrinkled, scarlet skin from which hangs a single, feather-tipped wattle (*top*). The sac inflates (*above*) to produce a loud, booming call which resonates through the forest canopy.

tropical waters of the Indo-Pacific and have a protuberant horn on the front of the head when they are adult. They also have two (occasionally one or three) fixed spines on the tail, with their sharp end pointed forwards. They feed on the leafy, large algae which grow around the bases of coral or on broken coral.

In the same family are the 100 or so species of surgeon-fish whose name is derived from the sharp spines on the tail. Most species reach a length of 30 cm. (12 in.) and feed on plants and animals living on rocks or coral.

Upas trees appeared in travellers' tales as trees which poisoned all the ground around them for up to 25 km. (15 miles) in the Asiatic tropics: contact with them was fatal.

Upas is Malay for poison and, although the upas trees of the tales do not exist, there *is* a tree which contains an acute blood-poison used in arrow poisons in the Malay archipelago. It is *Antiaris toxicaria*, a tree of the same family as figs and mulberries.

Urea is a natural organic compound, molecular formula CH_4ON_2, which is produced in the liver of mammals following the breakdown of amino acids. It circulates in the blood and is excreted by the kidneys, forming 2–4 per cent of urine. Industrially produced urea is used as a fertilizer, and for making plastics.

Urine is the fluid produced by the *kidneys of vertebrates. In aquatic freshwater vertebrates it is produced in large quantities and contains the waste nitrogen-rich gas called ammonia, in addition to salts and other substances. In birds and reptiles, it is a stiff white paste, due to the inclusion of waste nitrogen in the form of insoluble uric acid. Terrestrial vertebrates produce urine rich in urea, and salts. The composition and daily volume of urine are adjusted to keep the salt and water content of the body fluids neutral and constant, varying according to an animal's habitat.

Urticaria, hives, or nettle rash, is an itchy skin eruption of raised flat spots on a reddened background. The usual cause is an allergy, often to an unusual food. When recurrent, and if no provoking cause can be found, it can usually be alleviated by antihistamine drugs.

Uterus, or womb: part of the female reproductive system in mammals. It consists of a thick wall of involuntary muscle surrounding a small cavity which connects with the *ovaries via the *fallopian tubes and with the exterior through a canal in the cervix, a small conical piece of uterus which projects into the vagina. The uterus is basically a tube in which the young develop. In many mammals it has two parts which together enter the birth canal. Embryos develop in both parts, giving multiple pregnancies, such as is seen in pigs. In other mammals, the uterus is a single tube and usually contains only one developing offspring. In most mammals the uterus is inactive until the mother comes into season, when the lining membrane increases in thickness. If fertilization does not occur the lining membrane decreases in thickness, being reabsorbed by the blood supply.

In sexually mature females of some apes, and humans, the lining of the uterus undergoes a series of monthly changes related to ovulation. The membrane thickens and the blood supply increases in readiness to receive a fertilized ovum. In the absence of fertilization the membrane is shed, resulting in *menstruation. If fertilized, the ovum embeds in the uterine wall, which gradually enlarges to accommodate the developing baby.

The number and size of the muscle fibres in the uterine wall increase during pregnancy. At birth the wall contracts and expels the baby together with the afterbirth. The number and size of the uterine wall muscle fibres decrease rapidly after birth and the uterus soon returns to its original size.

Vaccination was originally inoculation with vaccinia or cowpox to protect against smallpox. The term is now used to cover other measures to protect against infections by stimulating immunity. Vaccines may consist of modified bacterial toxins, such as diphtheria and tetanus toxoid, or of killed organisms, as in typhoid vaccine, or of live organisms that have lost their virulence, for example BCG against tuberculosis, and poliomyelitis vaccine. In most countries of the world a scheme of vaccination against major diseases is organized for all children.

Vagina, or birth canal: the lowest part of the female reproductive system in mammals. It opens to the exterior via the vulval lips. In virgin females its opening is partially covered by the hymen, a thin, almost bloodless membrane which is easily ruptured by direct force. The vagina is a muscular tube whose wall is collapsed except when it contains an erect penis or a baby on its way to birth. A thick, round extension of the *uterus, the cervix, projects into the upper part of the vagina. The interior of the vagina is lubricated by secretions of glands which are stimulated to secrete by sexual arousal.

Valerians comprise a family of some 400 species of perennial herbaceous plants with a stoloniferous root-system, divided leaves arranged in pairs, and, typically, heads of pink flowers. The common valerian, *Valeriana officinalis*, is native to Europe and Asia. The root of the plant has long been valued for its medicinal properties, chiefly as a

The common **vampire bat**, *Desmodus rotundus*, prefers to feed upon the blood of comparatively docile, captive animals like this Venezuelan sow. By day it rests and digests its meal in a communal retreat, often a cave.

sedative. It is particularly attractive to cats and other furry animals and is reputed to be the attraction used by the Pied Piper of Hamelin to coax the rats from the town. The name is also used as part of the common name for several other species within the family Valerianaceae, and also in North America for a ground orchid, *Cypripedium pubescens*.

Vampire bats have a fearsome, but not altogether deserved, reputation as blood-suckers. They feed on the blood of vertebrates at night, either by landing on them or crawling towards them after alighting on the ground. The very sharp incisor teeth cut a shallow groove and blood is sucked up through the tongue. The wound is almost painless, but the bat's saliva contains an anticoagulant which causes protracted bleeding. Vampires are dangerous not because of the loss of blood they cause, but because they can transmit serious diseases, such as rabies. There are three species in the family Desmodontidae, all confined to the New World.

Vanilla comes from the dried pods of a climbing *orchid, *Vanilla planifolia*, native to tropical Central America. Vanilla has a gentle, subtle taste and is used particularly in milk puddings, chocolate, and ice cream. The pods need very careful, slow drying, the characteristic odour being absent from the unfermented pod. Most is now produced in Madagascar, but Mexico and some other Central American countries are exporters.

Vapourer moths are common species of moths in the *tussock moth family. They have wingless females, and are found throughout Europe, temperate Asia, and North America. Their colourful caterpillars, clothed in irritant hairs, feed on a variety of trees and shrubs, and are often so abundant in city parks as to cause some alarm.

Varicose veins are dilated, tortuous, superficial veins, most common in the legs. Man's upright posture imposes considerable hydrostatic pressure on his leg veins. Normally, valves in the veins prevent reflux of blood, but

various factors, including pregnancy, prolonged standing, and previous *thrombosis, can cause valve failure. Once developed, varicose veins may lead to eczema and skin ulceration. Elastic support and leg elevation prevent progression, but severe varicosities require surgical treatment.

Varying hare   *Jackrabbits.

Vegetative reproduction   *Reproduction in nature.

Veins are blood vessels carrying deoxygenated blood back to the heart. An exception to this rule is the pulmonary vein, which carries oxygenated blood to the heart from the lungs. The walls of veins are thinner than those of *arteries as they carry blood at a lower pressure. They connect with arteries via fine capillaries.

Velvet ants, or mutillid wasps, are *wasps of warm, arid areas. They are black or reddish in colour, usually with some yellow or silvery velvety markings, and grow up to 3 cm. (1¼ in.) long. The females are always wingless, but the males are usually winged. Their eggs are laid in the nests of solitary or social bees, wasps, or ants, and the young prey on the host larvae, not on the food stores.

Venereal disease   *Sexually transmitted disease.

Venus's fly-trap, or *Dionaea muscipula*: a *carnivorous plant which Darwin described as 'one of the world's most wonderful plants'. It is native to boglands in North and South Carolina. The modified leaves have two lobes, centrally hinged to form a trap, the margins of which are provided with long spines. The trap is sprung when insects, attracted by the colour and nectar-like secretions, touch trigger-hairs on the inner faces of the trap. These cause extremely rapid bursts of growth in the hinge cells which culminate in the closing of the trap. The marginal spines prevent escape even before the lobes are completely closed. Glands secrete enzymes which break down the softer parts of the insect and, by absorbing the 'digested' insect juices, the plant is provided with nutrients such as nitrogen which are difficult to obtain in bogland habitats.

Verdin   *Tits.

Veronica   *Speedwell.

Vertebrae are the individual bones which together form the spine, or backbone, of vertebrates. Each vertebra consists of a circular, thick disc of bone which extends on one side into an arch of bone called the neural arch, along which the spinal cord passes. From this arch extend three bony processes to which muscles are attached. The single dorsal arch and two lateral processes act as sites for muscle attachment. The lateral processes are also interconnected by ligaments and help control movement of the backbone. Differing slightly in shape one to another, the vertebrae are connected via *discs which act to cushion shock waves, and prevent jarring of the complete spine.

Vertebrates are animals with backbones whose *nervous system is differentiated anteriorly into an elaborate brain, housed in a cranium or skull case. These animals can be divided into a series of classes. The class of *jawless fishes (Agnatha) consists of the earliest vertebrates, characterized by the absence of jaws. The earliest group

with jaws was the Placodermi but these are known only from fossils. The rest of the living fishes are divided into two classes: the *cartilaginous fishes and the *bony fishes.

Bridging the gap between aquatic and terrestrial vertebrates are the *amphibians, which are the most primitive land vertebrates, and the first group to emerge from water. They have four limbs although there is reduction in some cases. Most amphibians have to return to water to breed, and the larvae typically have gills for breathing. The *reptiles are virtually all terrestrial, with the exception of crocodiles, alligators, turtles, and iguanas. They are related to the *birds in many respects but are still regarded as a distinct class. The *mammals are the most successful class of the land vertebrates. The brain is much enlarged compared to other classes, there is hair on the body, and the young are suckled after birth using special mammary glands.

Vesalius, Andreas (1514–64), Flemish anatomist, wrote *De humani corporis fabrica* ('On the fabric of the human body', 1543), which was based on actual dissection and examination, and became the foundation of modern anatomy. He was later made physician to the Emperor Charles V, but died at the age of forty-nine as the result of a shipwreck.

Vestigial organs are the atrophied, functionless remains of what were once, in an earlier evolutionary time, ordinary, functional organs. For example, the eyes of many animals that live in caves are vestigial. Different species of cave-dwelling shrimps and fishes show all gradations from almost complete eyes to total loss of them, although even in extreme cases some vestige remains. When an organ becomes useless to its possessor after a change of environment, it will gradually be lost in *evolution. The loss might be for one of two main reasons. Natural selection might actively favour the loss of an organ if a lot of energy is used to produce or maintain it. If an organ is not used, it might be slowly destroyed by the accumulation of *mutations. It is often not known in particular cases which process has operated.

Vetches, or tares, are weakly climbing plants with end-leaflets modified into tendrils. They are *legumes of temperate regions belonging to the genus *Vicia*, best suited to

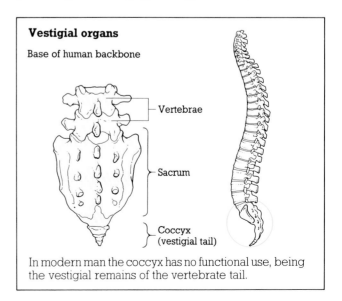

Vestigial organs

Base of human backbone

Vertebrae

Sacrum

Coccyx
(vestigial tail)

In modern man the coccyx has no functional use, being the vestigial remains of the vertebrate tail.

Two decades ago the **vicuña**, *Vicugna vicugna*, was in danger of becoming extinct. Today it is a protected species and there are about 85,000 in total, mostly in Peru. Here, groups congregate safely at a water-hole.

chalky soils. They are sometimes grown, either on their own or in a mixture with cereals, to be cut for silage or hay. A close relative, known as winter vetch in the USA, is useful where the winters are too cold for common vetch. Many vetches occur as native plants throughout temperate regions, with pink, blue, purple, or yellow flowers.

Viburnum   *Guelder rose.

Victoria water-lily   *Royal water-lily.

Vicuña: a species of *llama related to the alpaca, and guanaco. It is less than 90 cm. (3 ft.) high at the shoulders, and is tawny in colour, with a white bib on the lower neck. The coat is not excessively luxuriant, but because of its silkiness the wool is greatly prized and commands a high price. The vicuña lives at very high altitudes on the slopes of the High Andes, travelling in small herds of six to twelve females, with a lone male as leader. He keeps watch and utters a shrill whistle at the first hint of danger, covering the rear of the retreating herd. Although still found in the wild, it is also domesticated.

Vinegar flies   *Fruit flies.

Violets are perennial plants of the genus *Viola*, which has over 400 species widely distributed in temperate regions. Among the numerous species referred to as violets are white-, purple-, and yellow-flowered kinds. Several species are sweetly scented, including *V. odorata*, the sweet violet of Europe. Violets give their name to the family Violaceae, which has some 900 species including the genus *Viola*. This family contains annual and perennial species and many natural or artificially induced hybrids, such as the garden pansy. Plants of the violet family are characterized by heart-shaped or kidney-shaped leaves, and flowers with five unequally shaped petals.

Vipers are a major family of venomous *snakes comprising approximately 170 species. There are three main groups of vipers: Fea's viper (a primitive species from mountain regions of Burma, Tibet, and southern China); pit vipers (including copperhead and rattlesnakes); and viperines (Old World vipers such as the adder, gaboon viper, and puff adder). They are almost worldwide in distribution but are notably absent from the Australian region. They occur in a wide variety of habitats; most are ground-dwelling, many others live in trees, and relatively few are aquatic.

The venom production and injection system is highly developed in these snakes. Long, tubular fangs, through which the venom flows, are stored horizontally (parallel with the roof of the mouth) but can be rotated to a vertical position when the snake opens it jaws prior to striking.

Virchow, Rudolf Carl (1821–1902), German pathologist, discovered leukaemia. His *Cellular Pathologie* (1858) described the body as a state in which every cell is a citizen, and disease as a civil war between cells.

Vireos are birds native to the New World, which make up a subfamily of some thirty-two species in the family Vireonidae, all living in forests. They are small birds, 9–15 cm. (4–6 in.) long; most are greenish, brownish, or greyish above and paler, usually yellow or white, below. Many species have conspicuous pale eye-stripes and white wingbars. They feed primarily on insects, but also take some fruit. Most of the twelve species which spend the summer in North America migrate southwards for the winter.

Virginia creepers are tall deciduous climbers, clinging to their support by discs at the ends of the branches of branched tendrils. They belong to the same family as the grape-vine and include some fifteen species of the genus *Parthenocissus* in Asia and North America. The true virginia creeper, *P. quinquefolia*, has three or usually five leaflets to each leaf and is native to central and eastern North America. It has been largely replaced in gardens by the Japanese creeper or Boston ivy, *P. tricuspidata*, native to China and Japan. This is often called virginia creeper but can be distinguished by lobed leaves which are not divided into leaflets. Several other species are cultivated in gardens.

Viruses are extremely small micro-organisms, visible only under an electron microscope. They form one of the three major groupings of living organisms, and have been described as being between true living cells and organic molecules. They cannot reproduce outside cells of animals, plants, or other micro-organisms such as bacteria. They consist of a short strand of either DNA or RNA, enclosed within a protein shell, or coat. Inside their host cell they take over control of the cell nucleus, directing it to produce viral proteins and DNA/RNA. As a result, severe viral attacks lead to destruction of tissue. Sometimes a piece of viral nucleic acid is incorporated in the chromosomal make-up of the host cell, altering its character. In this way viruses are one of the factors which lead to cancerous transformation of a cell.

Among the diseases in humans caused by viruses are measles, the common cold, mumps, smallpox, and poliomyelitis. The only real defence is through vaccination, which acts by sensitizing the body's natural defence systems to the viral proteins. Many economically important plant diseases, such as tobacco mosaic virus, are due to viral attacks and are often transmitted by plant-feeding insects. Bacteria are attacked by a special type of virus, called a bacteriophage.

Viscachas are large cavy-like *rodents from South America, which belong to the same family as the chinchillas. The plains viscacha, *Lagostomus maximus*, lives in underground colonies on the Argentinian pampas, emerging at night to feed. The mountain viscachas (*Lagidium* species), of which there are three species, are diurnal and live in rocky country. They have long ears and tails and hop like rabbits. Numbers are declining through human persecution and competition from introduced animals.

Vitamin A (retinol) is a fat-soluble vitamin found most plentifully in fish, fish-liver oils, and butter, and in a more complex form as carotenoid pigments. It forms an essential part of rhodopsin, the visual pigment in the retina of the eye. This pigment responds to light and initiates impulses in the optic nerve, thus providing the basis of vision. Deficiency in vitamin A leads to poor night vision and to xerophthalmia, a disease in which the conjunctiva becomes dry and liable to infection. Vitamin A cannot be excreted from the body but is stored in the liver; large amounts are toxic and cause an illness known as hypervitaminosis.

Thiamine, part of the **vitamin B complex**, is commonly known as vitamin B_1 and can be isolated in crystalline form (seen here × 80). Common sources in normal diets are egg yolk, bacon, liver, yeast, and legumes; the daily requirement depends on the total food intake.

Vitamin B complex: a group of water-soluble vitamins obtained from many foods, but notably yeast, liver, and wheat. Most of them function as *coenzymes. The group includes thiamine (vitamin B_1), riboflavin (vitamin B_2), pyridoxine (vitamin B_6), cyanocobalamin (vitamin B_{12}), and nicotinic acid. Thiamine plays an important role in *glycolysis and its absence leads to the disease beri-beri, characterized by nerve and heart disorders. Other B-complex vitamins play similarly important parts in other biochemical reactions throughout the body.

Vitamin C (ascorbic acid) is a water-soluble vitamin found abundantly in citrus fruits and fresh vegetables such as cabbage and potatoes. Its functions have not been fully determined but include some role in the manufacture of collagen proteins, found in skin, blood vessels, and tendons. Deficiency of the vitamin causes scurvy, characterized by swollen and bleeding gums, painful joints, and slow healing of wounds. It is easily destroyed during storage or prolonged cooking.

Vitamin D (cholecalciferol) is a fat-soluble vitamin found mainly in dairy products and liver; it is also produced in the skin following exposure to sunlight. It is essential for normal development of the bones, and acts by stimulating calcium and phosphate absorption from the small intestine. Lack of the vitamin in children causes the disease rickets, characterized by bowed legs and other examples of defective bone growth. Like vitamin A, it cannot be excreted and is toxic in large amounts.

Vitamin E (tocopherol) is a fat-soluble vitamin found principally in wheatgerm and vegetable oils. Its functions in the human body are not fully understood, but lack of it is known to cause infertility and wasting away of muscle tissue in experimental animals, including rabbits, rats, and guinea pigs.

Vitamin K (phylloquinone) is a fat-soluble vitamin present in the leaves of plants and normally produced by some species of bacteria living in the intestines of humans and other mammals. It is needed for making prothrombin, an essential blood-clotting factor, so that deficiency leads to defective blood clotting and possible haemorrhage.

Vitamins are carbon compounds needed in small amounts by all organisms for their normal growth and development. They are a very diverse group of substances, not related chemically or structurally. All are needed as *coenzymes in chemical reactions within organisms. Green plants and most micro-organisms synthesize their own vitamins, but *heterotrophic organisms must obtain most types of vitamin in their diets. Absence or short supply of any of these substances causes a specific vitamin *deficiency disease. Two groups are distinguished: water-soluble vitamins include the vitamin B complex, and vitamin C; and fat-soluble vitamins include vitamins A, D, E, and K.

Voice: consists of a range of complex sounds, varied in pitch and loudness, uttered by animals, including man, for the purposes of communication. Song-birds, dolphins, and porpoises show comparable, though more limited, vocal skills to those of man. The voice is produced in the *larynx, an enlarged cavity at the top of the wind-pipe enclosed by walls of cartilage. Male and female voices in

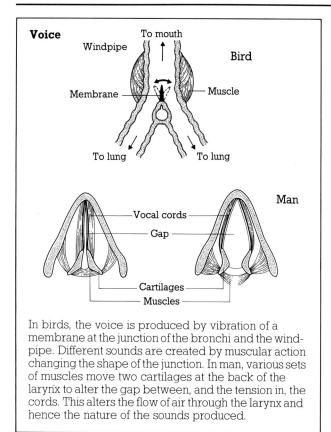

Voice

In birds, the voice is produced by vibration of a membrane at the junction of the bronchi and the windpipe. Different sounds are created by muscular action changing the shape of the junction. In man, various sets of muscles move two cartilages at the back of the larynx to alter the gap between, and the tension in, the cords. This alters the flow of air through the larynx and hence the nature of the sounds produced.

humans differ in pitch because the vocal cords are longer in males with growth stimulated by the male sex-hormones released during adolescence.

Voles are small *rodents related to hamsters, rats, and mice. Along with the lemmings, they comprise a distinct subfamily (Microtinae), which is distributed throughout the cooler regions of the Northern Hemisphere. Their size-range is about the same as that of rats and mice, but they can be distinguished by their shorter tails and rounded noses. The largest of the ninety-nine species of vole is the *muskrat of America, and the largest Old World species is the *water vole. Both show structural adaptations for an aquatic life.

Voles are very numerous over much of their range and, because of their herbivorous habits, frequently become pests of agricultural crops and forest plantations. Many voles, like the lemmings, show periodic fluctuations in numbers, and such changes are of great ecological significance because voles form the staple food of many mammalian and bird predators.

Vultures include fifteen species in Africa, Asia, and Europe and seven species in America. All are heavily built birds, with broad wings and remarkably graceful flight, and a predatory or scavenging habit. Lammergeyers, *Gypaetus barbatus*, and Andean condors, *Vultur gryphus*, are among the largest, spanning almost 3 m. (10 ft.) and weighing 4–5 kg. (10–12 lb.).

Vultures nest in crags or tall trees. Mostly dark-plumaged, they have a generally bare head and neck, a strongly curved beak, a capacious crop and powerful talons. Soaring high on thermals in search of carrion, they follow each other's movements, often landing in flocks to tear a mammal carcase to pieces.

Wading birds, or waders, are in British usage twelve families in the order Charadriiformes, the biggest of which has a worldwide distribution and includes plovers and sandpipers. Typical waders live on the shores of seas, lakes, and rivers, or in marshes, and nest on the ground. They are small or medium in size, some no larger than sparrows, while the biggest curlews are about 60 cm. (2 ft.) long including the beak. They are brown, or black and white in colour. They are powerful fliers, many migrating vast distances and, when not breeding, highly gregarious. Their beaks are variously adapted to different ways of collecting food: many of them are very long, for probing mud and sand; some are straight; others curved up or down. In some groups, the legs too are long, to facilitate wading, the stilts being the extreme example.

In North America these birds are all known as shore birds and the term waders is more usually applied to the Ciconiiformes. These are mostly large marsh- and water-birds with long beaks and legs and include herons, storks, ibises, and flamingos.

Wagtails make up the family Motacillidae, a group of fifty-four species of birds which includes the *pipits. The name is used more specifically for the ten or so species of birds of the genus *Motacilla*, an Old World group of birds. They are mostly small, about 18 cm. (7 in.) long including a long tail, though the Indian large pied wagtail, *M. maderaspatensis*, is larger , being 21 cm. (8½ in.) long. Several species are black with white face markings; others are greenish or greyish above with yellow underparts. The yellow wagtail, *M. flava*, is unusual in that subspecies found in different areas have differently coloured crowns, from greenish through bluish-grey to black. Most species are primarily insectivorous. Several live in wet or damp areas, and many migrate during the northern winter. They wag their tails as they walk—hence the name.

Wainscot moths are *owlet moths typically found in marshy places; some are coastal species, and a few occur in areas of rough grassland. Most have pale yellowish-grey or yellowish-brown fore-wings faintly streaked like an old-fashioned wainscot, and lighter hind-wings. At rest with the wings closed, on reeds and grasses, they are well camouflaged. The caterpillars feed on reeds or grasses, either on the leaves or, in many species, on the pith inside the stems. The straw-coloured common wainscot, *Mythimna pallens*, is widespread in the Palearctic and North America, and its caterpillars feed on grasses. Caterpillars of the larger and browner bulrush wainscot, *Nonagria typhae*, feed and pupate inside stems of reed-mace.

Walking-fern, or walking-leaf: a fern, *Camptosorus rhizophyllus*, native to North America. Its long, thin fronds taper to points which readily root themselves to produce new plants. It is unusual among ferns in being adapted to dry conditions in that the small plant which germinates from a spore (the prothallus) is resistant to drought.

Walking sticks (insects)   *Stick insects.

Walking worms are intriguing animals of the phylum Onychophora, with some sixty-five species which are perhaps the 'missing' link between soft *annelid worms and the stiff-legged, cuticle-covered *arthropods. They have legs and a cuticle, yet the body is soft and flexible and the legs are short, stumpy, and unjointed. These little creatures, of which the best known is *Peripatus*, live only in the tropics, on humid forest floors, breathing through tracheae. Their life on land is aided by sticky defensive secretions, and by giving birth to live young.

Wallabies are a large and diverse assemblage of *marsupials within the *kangaroo family, but are smaller than most kangaroos. They have large hind-feet, strong hind-limbs, and a long tail, and move quickly by jumping. The short-tailed wallaby, or quokka, *Setonix brachyurus*, was once widespread, but is now mostly restricted to two islands off Australia; it is mainly nocturnal, emerging to graze on ground vegetation or browse in the trees after climbing up onto the branches.

The young wallabies are born after a gestation period of a month or so. The single newborn young is very small, about 2·5 cm. (1 in.) long. After climbing into the pouch and attaching itself to a teat, the joey remains there for four to six months. It is not weaned until nine or ten months old. The female can mate again the day after birth. If fertilization occurs the resulting embryo develops for a few days but then remains dormant while the pouch is occupied by a previous young. Once the pouch is empty the embryo resumes development and is born twenty to twenty-five days later. There are some thirty-six species of marsupial called wallaby. They include the tammar, *Macropus eugenii*, spectacled hare, *Lagrostrophus conspicillatus*, and black wallaby, *Wallabia bicolor*.

Wallace, Alfred Russel (1823–1913), was an English naturalist whose theory of natural selection was almost identical to that of *Darwin, but not so thoroughly researched or substantiated. Unlike Darwin, he could not accept that the human mind was the product of evolutionary processes. His main contribution to biology was his study of the distribution of living organisms, now called *biogeography. On a visit to the Malay archipelago (1854–62) he discovered that some islands had a predominantly oriental fauna and other islands had an Australian fauna. He postulated an imaginary faunal boundary between the two, known as Wallace's line, and published his findings in *The Geographical Distribution of Animals* (1876).

Wallflowers are perennial plants related to cabbages, aubrietas, and other species of the mustard family. The familiar garden varieties in white, yellow, orange, and red have been developed by selective breeding from the European species *Cheiranthus cheiri*. Although strictly perennial, for garden use they are treated as biennials, being sown in early summer to flower in the spring of the following year. The Siberian wallflower, with yellow or orange flowers, is a hybrid variety which can be treated as a true perennial.

Walnuts are tall, deciduous trees, native to southeastern Europe, Central Asia, America, and China, and prized for their dark-coloured timber. They belong to the same family as hickories and pecan-nut trees and include fifteen species of the genus *Juglans* from America and Asia. The

irregularly shaped, wrinkled nut of the common, English, or Persian walnut *J. regia* is contained in a hard shell, which in turn is surrounded by a fleshy green layer. High in protein (18 per cent) and fat (60 per cent), these nuts are commercially produced in China, southern Europe, and California, where they are preferred to the stronger flavoured, exceedingly hard-shelled nuts of the native American black walnut, *J. nigra*.

Walrus: a species of seal which is the sole member of the family Odobenidae. They are found in shallow water round the circumpolar Arctic coasts. They are pinnipeds (fin-footed) like seals and sea-lions, and like the latter are able to bring their hind-limbs forward underneath their body, enabling them to move on land. Adult males may reach a length of 3·7 m. (12 ft.); females are a little smaller. The head is somewhat square in profile, the eyes are small, and external ears are absent. Tusks (the upper canine teeth) are present in both sexes, and grow throughout the life of the animal, sometimes reaching a length of 1 m. (3 ft. 3 in.) in the male. They are used to dig for clams and other molluscs in the sea-bed. The female reaches sexual maturity at six years and the male a year or two later. Pregnancy occurs probably in every second or third year.

Walruses commonly crowd together in large herds, as shown here on Round Island, western Alaska. Those just emerged from the sea are characteristically pale.

Gestation lasts about a year and the young is born in April or May. Lactation lasts at least sixteen months and the young remains with the mother for two or three years. Walruses still provide much of the meat eaten by Eskimos, but are the subject of conservation measures.

Walton, Izaak (1593–1683), English writer, is remembered chiefly for his *Compleat Angler* (1653). It is mostly in the form of a dialogue between a hunter and a fisherman (the author), who expounds the art of freshwater angling as they fish together along the banks of the river Lea, near London. Walton himself was a countryman, though he lived in London, where he kept an ironmonger's shop.

Wapiti, or American elk: a species of *deer that originally ranged over most of the USA and southern Canada but today is found only in forests of the Rocky Mountain region and far west. Closely related to the *red deer, it is pale fawn in colour with head and maned neck of dark chestnut-brown, and stands 1·62 m. (5 ft. 4 in.) at the shoulder. The bull makes a clear bugle-call in the rutting season, a challenge to other bulls to battle. Usually one calf is born in May or June; it can stand and usually walk within an hour of birth.

Warble flies   *Bot flies.

Warbler is the name applied to a wide variety of small, insectivorous birds. The Sylviidae or Old World Warblers form a very large family, including about 340 species, a few of which, despite the name, occur in the New World. Most are small, measuring 15 cm. (6 in.) or less, though a

few are a little larger. Most are dull-coloured in browns, greens, and muted yellows, though the so-called scrub warblers of the genus *Sylvia* are quite brightly patterned with black and rich orange-browns. Most are largely insectivorous and many of the species which breed in temperate regions migrate southwards for the winter. The group includes the blackcap, chiffchaff, sedge warbler, white-throat, willow warbler, and wood warbler.

The Acanthizidae, or Australian warblers, include scrub-wrens and thornbills, and comprise a family of about sixty species, the majority from Australia and New Guinea. Most species are small, dull green, grey, or brown in colour. The Parulidae, or New World warblers, is a family of 113 species, restricted to the New World, the members of which are often brightly marked with yellow, white, orange, red, or blue. They live in a wide range of habitats from forest to bushy grassland and are not related to the Old World Warblers. Most of those which breed in North America migrate south to warmer areas for the winter.

Warning coloration is a conspicuous pattern of colour, characteristic of a particular species, showing that it is unpleasant to touch or taste, or is well defended. The display may be continuous, as in many bees and wasps, or suddenly revealed as a form of *flash coloration. Many poisonous animals have warning coloration. The European fire salamander is black with vivid yellow markings. Its skin secretions are poisonous, as are those of many brightly coloured South American frogs. Highly poisonous coral snakes are brightly banded along their length in yellow, orange, and black. In this and other cases, *mimicry of warning coloration is common.

Wart-hog: a species of the *pig family and is found in Africa, inhabiting savannah and open plains. It is grey-brown in colour, its hide sparsely covered with hair, and

The **warning coloration** of the European fire salamander advertises that it is poisonous. The poison is concentrated in the parotid glands behind each eye.

it has warts around the eyes. The long snout of a wart-hog bears two pairs of tusks, the upper of which can reach 63 cm.(25 in.) in large males. The lower tusks, although only 6 cm. (2½in.), are sharp and deadly weapons. The male, or boar, stands 77 cm. (30 in.) at the shoulders; the female is smaller. The birth of the young coincides with the rainy season, usually four in each litter, after a gestation period of 175 days. A vegetarian, it feeds largely on roots and grass rhizomes which it digs up with its tusks.

Warts are benign, localized tumours of the skin, due to infection with a virus. The wart usually grows to a certain size, remains unchanged for perhaps months, and then suddenly disappears, thereby giving anecdotal support to a variety of cures. A wart is sometimes known as a verucca, and, when on the sole of the foot, as a plantar wart.

Wart snakes are aquatic snakes which range from Southeast Asia to northern Australia. They comprise three species, having robust bodies covered with small, wart-like scales. They are non-venomous, mainly nocturnal forms, that eat fish and give birth to live young. The largest species, the Javan file snake, or elephant-trunk snake, grows to about 2·5 m. (8 ft. 2 in.) and has peculiarly baggy skin which is known, in the leather trade, as karung.

Wasps are, in the broadest sense, insects of the order Hymenoptera which are carnivorous and have a thin 'waist', unlike bees or sawflies. In common usage, the name refers specifically to wasps of the family Vespidae, which are stinging insects usually with black and yellow warning colours. These are also known as social wasps, and include *hornets; they overwinter as queens, which often hibernate in houses. They make horizontal combs from chewed-up wood, with a surrounding case of paper and a hole at the bottom. The young are fed on masticated insects and, when the first workers emerge, the queens devote themselves to reproduction while the workers forage and enlarge the nest. New queens and males appear in the autumn and mate before the males die off. The old queen also dies when the cold weather starts, and the colony breaks up.

In New Zealand, introduced European wasps may omit the winter resting stage and produce enormous colonies with many queens. Other solitary wasps of other families may show the beginnings of a social life or, like *mason wasps, may lead a totally solitary life.

Water bears, or tardigrades, are particularly charming tiny freshwater animals of the phylum Tardigrada, with short, plump bodies and four pairs of stumpy legs. Most of the 350 species live in the water-film surrounding mosses and lichens, although a few species live in marine or freshwater habitats. The terrestrial species have a remarkable capacity to survive drought by shrivelling up completely. As soon as it rains again they revive, and resume their feeding on plant cells. They grow by moulting, before reproducing rapidly and often by *parthenogenesis to give large numbers of females.

Water beetles are those beetle species which live, both as larvae and adults, in fresh water. Most are carnivorous; for example the great diving beetle, *Dytiscus marginalis*, eats small fish. To keep afloat it keeps a bubble of air under its wing covers, which is replenished by pushing the tail-end of the body through the water surface. It is about

A common carnivorous **water beetle**, *Colymbetes fuscus*, from a pond in Oxfordshire, England. This species grows to 15 mm. (⅗ in.) long. Note the air supply in the form of an air bubble enclosed by the wing cases.

4 cm. (1½ in.) long. The silver water beetle, *Hydrophilus piceus*, is even larger, 5 cm. (2 in.) long, but is a plant-feeder. It keeps air on its under-side, replenishing it from the head-end. Whirligigs, beetles of the family Gyrinidae, are smaller and spend much time gyrating in groups on the water surface. Water beetles include representatives from several families.

Water boatmen are freshwater *bugs of the families Notonectidae and Corixidae, which swim by using their oarlike hind legs. Greater water boatmen, *Notonecta glauca*, also known as back swimmers, swim upside down, are fierce carnivores, and feed on tadpoles and young fish. All water boatmen fly actively. Lesser water boatmen, *Corixa* species, of the family Corixidae are plant-feeders which swim with their backs up.

Watercress is a hardy, aquatic perennial belonging to the mustard family, native to Britain. It is used as a salad vegetable, its fresh, young shoots having the pungent flavour of mustard and cress, and being rich in vitamins. There are two main types grown, the hardier bronze-green varieties and the frost-susceptible dark green varieties. Watercress needs careful cultivation as it thrives only in clean, running water, as found in chalk streams.

Water fleas   *Daphnia.

Water hyacinths   *Weeds.

Water-lilies are a group of perennial aquatic plants in the family Nymphaeaceae. The largest genus is *Nymphaea*, with fifty species found in temperate or tropical areas. The temperate ones are hardy, and through hybridization a

Because much of Australia is dry, its **water-lilies** are confined to a comparatively narrow coastal strip in the north and east of the continent, which receives sufficient rain to provide suitable habitats. This particular species of *Nymphaea* was photographed in Queensland.

large number of varieties for garden use have been introduced, in colours ranging from white through pink and yellow to red shades. The rootstock is a thick rhizome, from which arises long, stalked leaves that usually lie flat on the water surface. The tropical species, mainly found in Africa, tropical America, and Australia, have a greater colour range and include blue and purple flowers. The genus *Nuphar*, often called yellow water-lilies, has twenty-five species native to north temperate regions.

Water melons are native to the sandy, tropical areas of Africa, such as the Kalahari, but are grown throughout the tropics and subtropics. These sprawling, annual relatives of the cucumber and melon, derived from *Citrullus lanatus*, are very variable in fruit size, flavour, and flesh colour. Green-fleshed types are used only as stock feed, but the large, red-fleshed types with a higher sugar content are prized as attractive, thirst-quenching fruits.

Water plants include most *algae, some liverworts and mosses, some ferns, and a number of *angiosperms. Of marine plants, algae predominate and include some of the largest of all 'plants' in the form of kelps. Angiosperms are restricted to shallow seas, and include the sea-grasses grazed on by the marine mammal, the dugong, and the mangrove trees of the tropics, which grow where high tide inundates their lower parts.

There is a wide range of freshwater plants, including green algae, a few *bryophytes, some ferns, and the *horsetails, which push their spore-producing heads above the water surface. Most of the freshwater angiosperms produce their flowers in the air, though some, like the duckweeds, have flowers that are very reduced in

form and others rarely produce any flowers at all. Most freshwater angiosperms root in the mud, their roots obtaining air through a series of spongy shoots which extend above the mud, and often above the water surface too. Their leaves do not need many strengthening veins, being held up by their natural buoyancy in water. They can grow only if sunlight, of sufficient quality, can reach them. This restricts growth to the first few metres below the water surface.

A number are important in oxygenating lakes, while others are notorious weeds of waterways.

Water rats do not really exist, but the name is applied to a variety of rat-like *rodents that live near water. True rats can swim but they are not in any way modified for aquatic life. The *water vole is the species most often called a water rat but it is not a true rat. The *coypu is another species sometimes given the name. Animals which are called water rats in Spanish-speaking Latin America are the marsh rats, *Holochilus* species, of South America. They have webbed feet and spend much of their time in water, but they are relatives of the voles, and are not rats.

Water scorpions are oval-shaped, flattened water *bugs with a long, thin tube at the end of the abdomen and large grasping fore-legs. The tube is used as a 'snorkel' for respiration and, along with the front legs, gives the animals a superficial resemblance to land scorpions. The 150 or so world species are carnivorous and lie in wait for prey, which they seize with their front pair of legs.

Water shrew is the name of several species of *shrew in several different genera. They include the common European water shrew, *Neomys fodiens*, and the American water shrew, *Sorex palustris*, both of which have hairs on the feet and tail that help them to swim. There are also four species of Asiatic water shrews of the genera *Chimarrogale* and *Nectogale*. All live on the land but take much of their prey of insects and small vertebrates under water.

Water voles are about the size of a rat. There are three species distributed throughout the temperate Old World from Europe to Siberia. They burrow into the banks of streams, so causing damage in some cases, and feed mainly on reeds and grasses, although they will take other vegetation, including root crops. They swim well but are not modified for life in water. They include the European water vole, or water rat, *Arcicola terrestris*.

Wattle *Acacias.

Wattlebirds, or wattled crows, are confined to New Zealand and were once represented by three species of ground-dwelling, colourful birds. The Kokako, *Callaeas cinerea*, and saddleback, *Creadion carunculatus*, have only been seen once or twice in the last twenty-five years. They are about crow-sized. The most famous, the huia, *Heteralocha acutirostris*, has not been seen since early this century and is probably extinct; the male had a medium-length straight beak, and the female a much longer, strongly curved one.

Waxbills are small birds related to the finches. They are found in Africa, Southeast Asia, and Australasia. The large majority of the 130 or so species in the family Estrilidae are less than 12·5 cm. (5 in.) in length and many

Weaverbirds

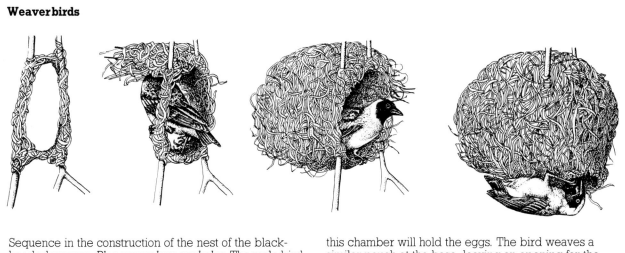

Sequence in the construction of the nest of the black-headed weaver, *Ploceus melanocephalus*. The male bird first weaves a ring of grass fibres, knotting them securely to the two supporting twigs. The bird then builds a cup-like pouch using the original ring as the rim; the floor of this chamber will hold the eggs. The bird weaves a similar pouch at the base, leaving an opening for the entrance hole. In some species the entrance is extended into a long tube to prevent tree snakes from taking eggs and young.

are brightly coloured, often with red or blue. They feed mainly on seeds. They are extremely popular as cage-birds, and one, the Bengalese finch, has been kept in captivity so long that it is no longer like any wild species, though it is probably derived from the white-backed munia, *Lonchura striata*.

Waxwings are three species of birds in the same family as silky flycatchers. They live in temperate areas of the Northern Hemisphere. All are about the size of a small, dumpy thrush with pinkish-brown plumage and a marked crest. The tip of their tail is yellow or red and the inner flight feathers have a curious red waxy tip, hence their name. The European or Bohemian waxwing, *Bombycilla garrulus*, sometimes moves southwards out of its normal range in large numbers when the berries, on which it depends for food, are in short supply. In the past these occasional migrations were thought to be a prelude to some human disaster, such as plague or war.

Weasels are the smallest species of the family Mustelidae, to which they give their name, with a wide distribution in northern temperate regions. The weasels and other species in the genus *Mustela* include the stoat, and the European weasel, *M. nivalis* (which is known as the least weasel in North America). Weasels are wholly carnivorous, feeding on mice, voles, and other small animals. They sometimes kill large prey, such as rabbits, many times their own size.

Weaverbirds, or weavers, belong to a large family of some 140 species of finch-like birds, including some of the sparrows. Most of the species are found in Africa, but some occur in Europe and Asia. The females are generally sparrow-like in plumage while the males are often brightly coloured yellow and black. Many breed in colonies, the males weaving elaborate grass nests before using these as a display ground in an attempt to attract the females. The red-billed quelea, *Quelea quelea*, is a major pest of grain crops in many parts of Africa, where a single colony may contain many millions of birds. Weavers also encompass the buffalo weavers and whydahs.

Web spinners make up an order (Embioptera) of at least 300 species of small, elongate insects of warm climates. Females are always wingless, and males are usually winged. They live gregariously in silken tubes or tunnels, made from silk spun from glands on the front legs. The females guard their eggs, but do not care for the young.

Webworms   *Grass moths.

Weeds are plants growing where man does not want them. They reduce the yield of desirable plants by competing with them for light, water, and nutrients. Some, like *dodders, are direct parasites of crops. Successful weeds are those with features which are very similar to those of the crop or which make their eradication difficult. The first group includes unwanted grass species in cereal crops, of much the same size as the cereals, so that they are gathered with the crop seed. The second group of successful weeds are those which produce large numbers of seeds which are spread widely and germinate over a period of years. Some may remain dormant in the soil for many decades. Other features favouring persistence include fragmenting roots like those of ground elder and bindweed, and the capacity to withstand herbicides, as in certain species of docks.

Aquatic weeds include species which reproduce rapidly through fragmentation, like the water hyacinth in tropical waterways.

Weever-fishes are the five species of the family Trachinidae, and are all marine fishes living in inshore waters of Europe and West Africa. They are rather long-bodied and compressed with the mouth strongly oblique and the eyes on top of the head. They stay buried in sand or mud during most of the day but emerge at night to forage in shallow water for crustaceans and worms. The spiny first dorsal fin and the spines on the gill-cover have venom glands, and wounds from these are very painful but not fatal to man. In Britain, the most abundant species, the lesser weever, *Echiichthys vipera*, grows up to 14 cm. (5 in.); it lives close inshore and is responsible for stings on the feet of bathers.

A **weevil**, *Rhinastus latesternus*, from the rain forest of Peru, shows its large foot-pads used to grip plants and support the relatively bulky body; specimens reach over 3 cm. (1 in.) in length. The adults lay their eggs in bamboo stems after chewing a small slit, using their powerful mandibles at the end of the long snout.

Weevils are members of the beetle family Curculionidae, with some 60,000 species. This makes them the largest family of insects, and possibly the largest family of any animal. The head of a weevil is elongated at the front to form a snout (rostrum), which may be as long as the rest of the animal. The jaws are at its tip and the antennae, often elbowed and with a terminal club, are about halfway along the snout. They are almost all herbivorous, often feeding on seeds, and the long snout may also be used to bore holes in which to insert eggs. The larvae are legless. Many species, such as the granary weevil, *Sitophilus granarius*, or the cotton boll weevil, *Anthonomus grandis*, are destructive to crops.

Weil, Adolf (1848–1916), German physician, in 1886 described a severe form of infectious jaundice among sewer workers, contracted from the urine of rats. The organism (a bacterium) was later isolated and the disease called leptospirosis. It occurs in mammals such as rodents and dogs, and may be transmitted to humans, causing jaundice or meningitis.

Weismann, August (1834–1914), German biologist, expounded a theory of heredity which assumed the continuity of germ plasm, a substance which Weismann postulated was transmitted via the *gametes from one generation to the next, and which gave rise to body cells. The theory rules out transmission of acquired characteristics.

Weka, or weka rail: a species of *rail confined to New Zealand. It is mainly brown in colour and has lost the power of flight. There are four subspecies, of which only one inhabits both the North and South Islands.

Welch, William Henry (1850–1934), Connecticut-born physician, founded the first teaching laboratory of pathology in the USA. Often called 'the dean of American

medicine', he made notable contributions to bacteriology, and immunology, including the discovery of the bacillus which causes gas gangrene. He founded the *Journal of Experimental Medicine* (1896).

Wellingtonia *Redwoods.

Wetlands include bogs, fens, swamps, and marshes where the water-table is at, or near, the surface of the soil for much of the year. They often contain unique *communities and are frequently highly productive, providing food for a large range of organisms. They may also act as breeding sites for mosquitoes and other disease carriers. Depending upon the nutrient status of the soil and water, wetlands can be classified as acidic, neutral, or base-rich. These chemical factors influence the *succession of wetlands and, unless maintained by regular catastrophic clearance, most wetlands will ultimately revert to dry land. As a habitat wetlands have suffered great reduction in total area, either through reclamation for agricultural land, or drainage to eradicate disease organisms. Conservation measures include making some wetlands nature reserves, and in a few instances (notably in Switzerland) transplanting small wetland areas to new sites where the water and nutrient regime can be maintained.

Whalebone whales, or baleen whales, include the largest known animal, the *blue whale, and are usually larger than the toothed whales. The whalebone whales are characterized by a series of plates, baleen or whalebone, which take the place of teeth in the upper jaws. The frayed inner edges of the whalebone allow small planktonic crustacea, such as *krill, to be filtered from the water. Most of these whales have a dorsal fin and a series of parallel grooves, running longitudinally below the throat, except in the right whales. Other whalebone whales include the common *rorqual, lesser rorqual, Rudolph's rorqual, and *humpbacked.

Whale-headed stork, or shoebill: a bird in a family of its own which is native to central Africa. Related to the *stork family, it stands about 1·3 m. (4 ft.) high and is grey with black wings and tail. Its most distinctive feature is its enormous, broad beak with a hook at the end with which it catches fish and amphibia. It makes a nest of reeds in swamps and raises one or two young.

Whales are aquatic, carnivorous mammals. Their specializations for life in water are greater than those of any other mammals, and their entire lives are spent in water. Adaptations to aquatic life are noticeable in the shape of the animal. The most obvious difference from a typical fish shape is the horizontally placed tail flukes, which move up and down to provide the propulsive thrust. The hind-limbs have been completely lost except for small internal traces. The fore-limbs are reduced to small, paddle-like flippers.

Like all mammals, whales breath air, and their respiratory system shows special developments. The nose or blowhole is high on the head and has valves for closing the nostrils during diving. Air is inhaled and exhaled through the blowhole, which connects directly with the lungs. This allows whales to feed while submerged in water without the possibility of water entering the lungs. Their spouts of water-laden air have characteristic forms, so that whales can be recognized at considerable distances.

Sounds are emitted for such purposes as the attraction of females by males, communication between mother and offspring, and for the avoidance of obstacles in the dark. Some whales produce clicks, while others, like the *humpbacked whale, have a song. These animals display elaborate behaviour involving social life, communication, co-operation between individuals, and learning. Many species migrate and some orientate and navigate by following the shape of the sea-bed. Whales include the largest known animals, either fossil or recent. Many species are endangered and conservation of stocks is needed.

Whale shark: the largest living fish, which attains a length of 18 m. (60 ft.). It occurs in all tropical seas and is usually seen close to the surface, where it feeds on planktonic animals and small fishes. It is thought to give birth to live young, each about 50 cm. (20 in.) long. It is frequently found with schools of tuna, probably because they are attracted to plankton-rich areas. The whale shark is now rare over much of its range.

Wheat is the name of a number of cereal species of the genus *Triticum* that have been cultivated for at least 10,000 years. Einkorn and emmer were the earliest domesticated species, the former probably being one of the ancestors of all cultivated wheats. Emmer was once widely grown, and has been found preserved in Egyptian tombs; it was the chief cereal in the Graeco-Roman period. Both types had fairly small grains and were difficult to thresh. They were eventually replaced by bread wheat, *T. aestivum*, a hybrid between emmer and a grass called goat-face grass, *T. tauschii*. Durum wheat, *T. durum*, is closely related to emmer and is grown in the Mediterranean region to provide the flour for macaroni, spaghetti, and other pasta.

Bread wheat is the species most widely cultivated today and is the major cereal of the world in terms of the amount

The **whale-headed stork** of north-central Africa hunts at night; its prey are chiefly lungfishes, which it digs up from the mud with its hooked, scoop-like bill.

Ripe bread **wheat**, *Triticum aestivum*, in Utah, USA. Extensive breeding programmes have been and still are carried out in wheat-growing countries, in a search for varieties with higher yields, better resistance to drought, and greater ability to withstand the constant attacks of plant diseases and insects.

produced. It contains the protein gluten and, depending on the characteristics of this gluten, the grain is used for bread or biscuit manufacture. Bread-making requires a high-grade gluten wheat known as 'hard wheat' since the gluten makes the dough elastic and enables light, airy bread to be produced. It is grown in the USSR, North and South America, Europe, and Australia, and it exists in two forms, depending on the time of sowing. Spring wheats are suitable for areas with very cold winters and so dominate production in the northern wheat belt of the USA and Canada. The slower-growing but higher-yielding winter wheats are characteristic of the southern wheat belt of the USA.

Wheatears are members of the thrush family which belong to the genus *Oenanthe*. There are about nineteen species, most of which are found in Europe, western Asia, and Africa, though one species, the common wheatear, *Oenanthe oenanthe*, spreads all the way across eastern Asia and over the Bering Straits into Alaska. Like many of the other species, including those that breed in Alaska, the common wheatear spends the winter in Africa. Wheatears inhabit open country, including deserts and rocky hillsides, nesting in holes in the ground. They are insectivorous.

Wheel animalcules   *Rotifers.

Whelks are large and predatory *snails of the sea-shore, responsible for the neat, round holes so often drilled through other molluscs' shells. The whelks rasp the hole using their ribbon of stout teeth, aided by a special acidic secretion, and then eat the shell contents. Some whelks also smother their victims with their large flat foot.

Whelk shells are commonly encrusted with other animals—polyps, bryozoans, and barnacles—and when empty they often form the chosen homes of hermit crabs. Masses of papery whelk egg-cases are also a frequent sight along the sea-shore.

Whip-poor-will   *Nightjars.

Whistlers are, broadly speaking, all birds of the sub-family Pachycephalinae, which includes the shrike-tits, the shrike thrushes, and some other species, about fifty in all. They are found in Southeast Asia and Australasia. In a narrower sense, whistlers are birds of the genus *Pachycephala*, mostly smallish, brown birds, though some have white, black, or chestnut patches and a few, such as the golden whistler, *Pachycephala pectoralis*, are bright yellow with black and white markings. They live in forests and are primarily insectivorous.

Whistling swans are the New World cousins of *Bewick's swans, but the yellow patch on the beak is reduced to a small spot. They are numerous and common winter visitors to Atlantic and Pacific coasts of North America, migrating there from the Arctic, where they nest.

White, Gilbert (1720–93), English clergyman and naturalist, wrote *Natural History and Antiquities of Selborne* (1789), which recorded his observations of an area of countryside around his home in Hampshire. He was interested in many aspects of nature, among them the feeding habits of bats, the evening manoeuvres of rooks, and the improvement of horticultural soil by earthworms.

White admiral butterflies are blackish-brown *brush-footed butterflies with a white band across fore- and hind-wings. There are a number of species in woodlands in Europe, Asia, India, and North America. The European white admiral, *Ladoga camilla*, has a powerful, gliding flight; its caterpillars are green with two rows of brown spines, and feed on honeysuckle.

White ants *Termites.

Whitebait are the young of *herrings and *sprats. Both species spawn offshore, but their young drift inshore, entering estuaries in late summer and forming large, mixed schools. At a length of about 5 cm. (2 in.) they are caught for sale as whitebait.

Whitebeam *Rowans.

White-eyes are a family of about eighty species of birds which occur throughout the warmer areas of the Old World, though most species are found in Southeast Asia.

The frontispiece to Gilbert **White's** much-loved book shows a 'North East view of Selborne from the Short Lythe', an engraving by S. H. Grimm.

A pale **white-eye**, *Zosterops pallida*, of South Africa, acts as a pollinating agent of the bird-of-paradise flower by first perching on the male parts of the flower before drinking its nectar.

They are small, 10–13 cm. (4–5 in.) in length, and fairly uniform in appearance, greyish or yellowish-green with pale or yellow underparts. The striking feature of most species is the white ring around the eye, which gives them their name. They spend much of their time in small flocks and feed primarily on fruits, nectar (having a brush-tipped tongue which can be rolled into a tube for this purpose), and insects.

Whiteflies are tiny plant-sucking *bugs of the family Aleyrodoidea and are related to cicadas and aphids. The wings of adults, which rarely exceed 3 mm. ($\frac{1}{10}$ in.) in length, are covered with white, waxy powder. Adults and young pierce individual plant cells and suck out the sap; they excrete unwanted sugars as honeydew. Newly hatched larvae are active, but after one moult they secrete a scale, which covers their body, within which development proceeds, and they become sessile. They are primarily tropical, and are pests of crops throughout the world.

Whites (butterflies), together with yellows, brimstones, orange-tips, and wood-whites, belong to the family Pieridae; a worldwide family, with most species in the tropics. Unlike many butterflies, all Pierids have three pairs of normal legs used for walking. In almost all, the scales contain white (or yellow) pigments derived from a waste product, uric acid. *Cabbage whites, familiar in many countries, are typical of the group, as is the green-veined white, *Pieris napi*, found in meadow and woodland throughout Europe, across Asia to Japan, in North Africa, and in North America. Its caterpillars feed on cruciferous plants, such as hedge mustard, but not on cabbages. There are around 1,100 species of Pierid butterflies.

Whiting are fish species of the cod family and like their relatives have three dorsal and two anal fins, although they lack a chin barbel. The whiting, *Merlangium merlangus*, is conspicuously silvery in colour. It is an abundant fish in inshore waters of the northern Atlantic, and lives mainly in mid-water and just above the sea-bed. Small fishes live close inshore, but at a length of 3–5 cm. (1–2 in.) they often take shelter among the tentacles of jellyfish. They eat small fishes and crustaceans. Whiting is a valuable food-fish, which grows to a length of 40 cm. (16 in.). In North America the name whiting is used for members of the genus *Merluccius*, known as *hake elsewhere.

Whooper swans resemble mute swans, but their bills are yellow and black, and they swim with their necks held erect. In flight they utter loud whooping calls. Breeding in Iceland, northern Europe, and Asia, they migrate south in winter, visiting Britain regularly.

Whooping cough is an infection of the air passages by the bacterium *Haemophilus pertussis*, causing a distressing and sometimes dangerous illness. The organism adheres to the living cells, and prevents the normal clearance of mucus. The first symptoms, which appear seven to ten days after infection, are those common to any other mild respiratory infection. Then a dry, hacking cough appears and becomes progressively worse. Prolonged bouts may be followed by the characteristic whoop, produced by forced intake of breath through a partially closed part of the vocal cords. The cough may persist for weeks after the apparent disappearance of the organism, but can (very rarely) cause brain damage. Vaccination can prevent infection, or severely reduce its effects.

Whydah birds are related to the *weavers. Nine species are known, all of which live in Africa in open, grassy country. They are the size of small sparrows, but the males of some species have extremely long tails, giving them a total length of 60 cm. (2 ft.). The males display in groups to which the females come for mating. The females lay their eggs in the nests of waxbills.

Wigeon is the name of three species of dabbling *duck: the European, *Anas penolope*, American, *A. americana*, and chiloe, *A. sibilatrix*, of South America. The drake of the European wigeon displays a handsomely patterned plumage of chestnut, blue, grey, white, and black, and utters a distinctive whistling call. The European species is mainly vegetarian and notably gregarious.

Wild boar: a species of the *pig family found in the deciduous woodlands of Europe, North Africa, and throughout southern Asia to the Malay peninsula, the Philippines, Japan, and Formosa. It is large, has long tusks and stiff dark grey-black or brown hair. The snout is typically pig-like and has a mobile, cartilaginous disc. The head and body is about 1·8 m. (5 ft. 10 in.) long and the shoulder height about 1 m. (3 ft. 3 in.). The upper canine teeth turn outwards and upwards and wear against the lower ones, causing sharp edges to form. It feeds on acorns, berries, roots, tubers, and green vegetation, and occasionally worms, insects, reptiles, and birds' eggs. The young are born four months after the rutting season; litters vary from five to eight, and the young at birth have almost continuous light stripes on a brown coat. The wild boar is one ancestral species of the domestic pig. Other species of

This litter of **wild boars**, *Sus scrofa*, will be highly vocal, communicating with their mother in grunts and squeaks. After weaning at about three months, they forage for food as a family unit and remain with the mother until she farrows again.

the wild boar's genus, *Sus*, include the pygmy hog, Javan warty hog, bearded pig, and Celebes wild pig.

Wild cats are species of the genus *Felis* apart from the big *cats such as the lion, tiger, cheetah, and leopard. There are twenty-eight species of small, wild cats, found in all types of terrain from thick jungle to desert, and present in the Old and New Worlds. The majority are slightly larger than the domestic cat but all are similar in shape. One of the best known is the European wild cat, *Felis sylvestris*, the head and body of which is about 63 cm. (25 in.) long and the tail 35 cm. (14 in.) long. The coat is yellowish-grey or grey broken up by darker stripes. The tail is heavy and blunt-ended, with black rings. It lives in areas of heavy cover and the lair is a crevice in the rocks. Most of the year is spent alone, except the spring, when male and female meet to mate. About a month after mating two to four kittens are born, fully furred but blind for the first ten days. They learn to hunt at three months, are weaned at four months, and are fully mature at a year. Prey includes mice and voles and even baby deer.

Other species include lynx, *F. lynx*, the ocelot, *F. pardalis*, the bobcat, *R. rufus*, and serval, *F. serval*.

Wildebeest *Gnu.

Wild ox (extinct) *Auroch.

Willow grouse *Red grouse.

Willow-herbs are perennial herbaceous plants which belong to the genus *Epilobium*. They are related to fuchsias, clarkias, and evening primroses. About 160 species are known, widely distributed in the temperate and alpine regions of the world. The European kinds are identified by descriptive prefixes such as great hairy, small-flowered, spear-leaved, broad-leaved, and square-stemmed. The flowers are produced in the axils of the leaves or in terminal spikes and the petals are usually rose-purple or

The rosebay **willow-herb**, or fireweed, has a wide distribution in North America and northern Eurasia. About 1·2 m. (4 ft.) tall, it blooms between June and September.

white. Rosebay willow-herb, *E. angustifolium*, with its handsome purple flower spikes and downy, wind-blown seeds, is one of the first colonizers of waste-ground and woodlands cleared by fire, and for this reason it is known in North America as fireweed.

Willows belong to the genus *Salix*, of which there are about 300 species. They belong to the same family as poplars and vary from large trees like the European white willow, *S. alba*, to tiny creeping shrubs like the dwarf willow, *S. herbacea*. All willows have deciduous, simple leaves and small catkins, often with long hairy bracts, familiar in the pussy willow. All willow catkins are erect, unlike poplars, and they have nectaries which attract insect pollinators.

Willows are found throughout the world, except in Australasia, but are most abundant in the Northern Hemisphere. Dwarf willows are particularly important in mountain vegetation and in taiga and tundra zones. In other regions willows are principally found beside rivers or streams, or on marshy ground. Shrubby species, such

as sallows, *S. caprea* and *S. cinerea*, can tolerate drier soils and are often scrub species in woodland edges.

Osiers are willows with long, narrow leaves, grown in 'osier beds' for their pliable young shoots. These are used for making baskets and hurdles.

Willows are propagated by cuttings and are famous for the rapid production of roots from them. Most familiar in gardens is the weeping willow, perhaps a hybrid between *S. alba* and *S. babylonica*, the latter of which was the original weeping willow introduced from the Near East. Many other varieties and hybrids exist both naturally, or have been induced by horticulturalists.

Willow warbler: a species of bird in the Old World *warbler family. It is closely related and very similar in appearance to the chiffchaff, but with a much more tuneful song. It breeds over a wide area of Europe and the northern parts of Asia, migrating to Africa for the winter. It lives in woodland, nesting on the ground.

Wilson, Alexander (1766–1813), Scottish-born naturalist, emigrated to America and there became a pioneer in ornithology, touring (mostly on foot) the East, the Ohio and Mississippi valley frontier, and much of the southern USA. The result of his experiences was *American Ornithology* (9 vols., 1808–14).

Wind scorpions   *Sun spiders.

Winkles, or periwinkles, are *molluscs found worldwide on sea-shores. They show a phenomenon known as zonation: the occurrence of different species at successive shore heights. Some small species, highly resistant to desiccation, live in cliff crevices rarely even splashed by waves; other forms occur between tide levels amongst algae and in rock-pools, where they are fully aquatic. Winkles thus show a great range of colour and shape, yet all are herbivores and most species are grazers of seaweeds. They are collected as food by birds or by man, though some species are camouflaged by resembling the 'bladders' of common bladderwrack seaweed.

Wintergreen is the name of two distinct genera of plants: *Pyrola* and *Gaultheria*. The first are creeping evergreen plants of the northern temperate zone, with some twenty species which form the major part of the wintergreen family, the Pyrolaceae. The second includes the plant from which oil of wintergreen is derived: this is *G. procumbens* of North America, a member of the heather family and closely related to cranberry. The oil is rich in salicylic acid, a type of antiseptic, and is used as an effective linament for bruises and pulled muscles.

Wireworms   *Click beetles.

Wisteria is a small genus of deciduous climbers native to the eastern United States and northeast Asia. They are members of the pea family and have divided leaves and trusses of white or pale lilac-blue pea flowers. Most familiar is *W. sinensis* from central China, a beautiful climber for house walls, sometimes used in bonsai in Japan.

Witchetty grubs   *Goat moths.

Witch-hazel is the common name for species of *Hamamelis*, a genus of four or five species of small deciduous trees and shrubs with strap-shaped petals in fours. Most widely seen are forms of Chinese witch-hazel, *H. mollis*, from western China, which produce their scented flowers in midwinter. The popular remedy, witch-hazel, is derived from extracts of the leaves and bark of the common witch-hazel, *H. virginiana*, native to the eastern USA and flowering in autumn. The leaves resemble those of hazel trees and the twigs of witch-hazel were used by early settlers in North America for water divining; hence the popular name. Witch, or 'wych', was an old English term used to denote any tree with pliant branches.

Withering, William (1741–99), was the English physician, botanist, and mineralogist after whom *Witheringia* (a genus of potato) and Witherite (barium carbonate) are named. His chief botanical work was *A Botanical Arrangement of all the Vegetables growing in Great Britain* (1776), although he is best remembered for *An Account of the Foxglove and some of its Medical uses* (1785). One of the drugs derived from foxglove was digitalis, which is still employed in the preparation of medicines used for the treatment of cardiac conditions.

Woad is a biennial or short-lived perennial herb, native to central and southern Europe, which belongs to the mustard family. It has a tap root system, bluish-green leaves, and stems of small, yellow flowers. From the leaves is produced a blue dye, formerly much used to colour fabrics.

Wolff, Caspar Friedrich (1733–94), German embryologist, advanced the theory of epigenesis, which postulated (quite correctly) that the development of an embryo involves the gradual and progressive growth of body organs and tissues. The view held prior to the expounding of Wolff's theory was that a miniature of the animal or plant was present inside the sperm, egg, or seed. Wolffian bodies are structures within vertebrate embryos, which become the kidneys in fishes and amphibians and form part of the kidney of mammals.

Wolf spiders are ground-dwelling hunters with highly developed eyes for detecting and tracking their prey, usually small insects, which they kill with quick-acting poisons. A few species even have venom which can cause unpleasant spreading ulceration in man. These spiders are generally quite small and cryptically coloured, often rather hairy. Some females signal receptiveness to mating by releasing a special scent (or pheromone) to attract males. This occurs at night, when males of some species often lull their potential mates by a song produced by scraping two parts of the leg together. Female wolf spiders often carry their eggs around in a coloured silken sac and the young are carried on the mothers' backs after hatching, clinging on to special knobbed hairs.

Wolverine, also known as glutton because of its supposed insatiable appetite, inhabits the tundra and northern forests of North America and Eurasia. Weighing up to 30 kg. (66 lb.), it is by far the largest relative of the weasel and a fearless predator, often driving wolves and bears from their kills. It is capable of killing animals up to the size of deer, but is mainly a scavenger. It also feeds on grubs and birds' eggs, besides taking some fruit in season. Wolverines are solitary, but there are social attachments between a male and several females.

Wolves belong to the dog family and are carnivores. The grey wolf, *Canis lupus*, once ranged over the entire Northern Hemisphere, even into Arctic regions, the exception being extreme desert areas; today it is more restricted. It is 90 cm. (3 ft.) tall at the shoulders, the head is wide and the ears fairly short, the jaws are long, and it has a long bushy tail. Its coat varies from almost black to near white in colour. A pack of wolves usually consists of perhaps a dozen animals including the parents and young together with adult relatives. Their family ties are strong since male and female mate for life. The male will help to excavate an underground den and tunnel leading to it, often near a hilltop and with an unobstructed view of the surrounding country. Four to fourteen pups are born two months after mating, usually in spring. They are weaned in a few weeks and both parents will regurgitate food for them. The pups are taught to hunt and care for themselves, but the family remains together until they are fully grown. Besides the grey wolf there is also the red wolf, *C. rufus*, once found in the southeast USA, but now thought to be extinct in the wild.

Womb *Uterus.

Wombats are three species of *marsupials found in coastal regions and hills of southeastern Australia, Tasmania, and Flinders Island. They are large, burrowing,

The common **wombat**, *Vombatus ursinus*, is distinguished from the other two species by its hairless nose and larger size. It can reach a total length of about 1·2 m. (4 ft.) and weighs approximately 35 kg. (77 lb.). Like many other burrowing animals, it has poor eyesight but its hearing and sense of smell are excellent.

tail-less animals with rodent-like grinding teeth. In some respects they are equivalent to badgers, using their sturdy limbs and claws to dig burrows up to 30 m. (100 ft.) long with a nest chamber at the end. Wombats are nocturnal, remaining in their burrows by day, emerging in darkness to feed upon grasses, roots, and the inner bark of trees. They are solitary except at mating time; a single young is born and this is then carried in the mother's pouch. The wombat has been known to survive for thirty years in captivity.

Woodchuck and groundhog are other names for the *marmot, *Marmota monax*, of North America. It is a species of squirrel but does not look like one, with its thickset body and short tail. Like all marmots, the woodchuck lives in underground colonies on mountain slopes, coming out by day to feed on surface vegetation. It spends the winter in deep hibernation.

Woodcock are six species of *wading birds belonging to the genus *Scolopax*, which are found over much of Eurasia and North America. They live on the forest floor and, like their relatives the snipe, are beautifully concealed by their leaf-like coloration of mottled browns and blacks. They are solidly built, long-beaked birds up to 35 cm. (14 in.) in length, which probe into soft soil for worms and similar animals. The European, *S. rusticola*, and American woodcock, *S. minor*, are valued as sporting birds.

Woodcreepers are confined to forests of Central and South America. Remarkably like *treecreepers, except for their size, they can be as large as a crow—they are speckled or mottled, brown birds with stiffened tails. They climb up tree trunks in their search for food which is primarily insects. Their beak shape varies markedly among the forty-six species, from medium length and straight to long and strongly curved.

Wood hoopoes are a family of some six species of birds, found in central and southern Africa, where they live in wooded country. They have a body about the size of a small thrush, and a long tail. They are glossy blue-black or greenish-black, with white spots at the tips of the tail feathers and white marks in the wing. The beak is usually red, long, and markedly curved in some species. They feed primarily on insects, and nest in holes in trees.

Woodlice are *crustaceans, common in most terrestrial habitats. They have no waterproof covering, and their life revolves around avoiding desiccation. To this end, they congregate in humid spots, only being active at night when humidity is usually higher. Some species, such as pill bugs, resemble tiny armadillos, and can curl into a ball for protection. Most species of woodlice feed on decaying vegetation or rotten wood and protect their young in a special pouch after birth.

Woodmice, often called fieldmice, are small *rodents widely distributed throughout Europe and non-tropical Asia in a variety of habitats from grasslands to woodlands. There are some thirteen species, of which two occur in Britain—the woodmouse, *Apodemus sylvaticus*, and the yellow-necked mouse, *A. flavicollis*. These mice are distinguished by their large, beady eyes and prominent ears, as well as by their long tails. Their food consists of seeds, berries, roots, and insects. They are of economic im-

portance because they kill tree seedlings and are of great ecological significance both as seed-dispersers and as food for avian and mammalian predators. They are nocturnal, spending the day in burrows, in which they also rear their young. They are very prolific, one female being capable of rearing up to six litters in her one year of life. They do not hibernate, even in the coldest winters, but some store food in the summer for winter use. The name fieldmouse is also given to species of the Australian genus *Gyomys* and South African genus *Akodon*.

The equivalent mice of the New World are the deer mice or white-footed mice. They belong to a different subfamily (Hesperomyinae). There are some forty-nine species, belonging to the genus *Peromyscus*, distributed from the Arctic to northern South America.

The Arizona **woodpecker**, *Picoides arizonae*, inhabits oak and pine/oak woodlands in southeast Arizona, southwest New Mexico, and south to central Mexico. About 20 cm. (8 in.) long, this male, distinguished from the female by its red nape, is removing chick droppings from its nest hole.

Woodpeckers are birds of the family Picidae, containing about 200 species. Their distribution is world wide within wooded habitats, though they are absent from Australasia and most oceanic islands. They range in size from 9 to 55 cm. (3½–22 in.) and in colour from green and yellow to spotted black and brown. Most climb vertical trunks using their strong feet and stiffened tail feathers for support. They feed on insects and seeds, extracting many of the former from timber by drilling holes with their powerful beaks and retrieving their prey by means of their very long tongues. Most also use their beaks to excavate nesting cavities in trees, but a few nest in banks or termite mounds. American sapsuckers of the genus *Sphyrapicus* drill rows of small holes in trees, drink the exuding sap and eat the insects attracted by it. The acorn woodpecker, *Melanerpes formicivorus*, drills rows of holes in the bark of a single tree and stores an acorn in each; these enormous stores are guarded by family parties.

The family includes the wrynecks and the flickers (*Colaptes* species) of North America. Many have distinctive, far-carrying calls, the European green woodpecker, *Picus viridis*, being known in some areas as the 'yaffle', a term supposedly mimicking its call.

Woodpigeon, or ring dove: a species of bird in the *pigeon family which occurs over much of Europe and western Asia. It is a large, pale, bluish-grey bird, 40 cm. (16 in.) in length, with a wine-red throat, and a white band across the wings, which is visible only in flight. A white patch on either side of the neck looks as if the bird has a ring around its neck and so gives rise to the name ring dove, which is more properly applied to the Barbary dove. It lives in wooded areas for much of the year, but feeds readily in fields, where it can cause serious damage to crops such as cabbages and sprouts.

Woodruff is the common name for several species of slender, perennial plants with creeping rootstocks, whorls of narrow leaves, and terminal clusters of small white or blue flowers. They are plants of damp soils in woodlands and hedgerows, widely distributed throughout Europe, Italy, and North Africa. The name is applied to plants, related to bedstraws, of the genera *Asperula* and *Galium*. The common woodruff, *A. odorata*, is a fragrant herb which is used as an ingredient of perfumed snuffs and was once in great demand as a medicinal plant.

Wood swallows are a family of ten species of bird native to India, Southeast Asia, Australia, and some Pacific islands. They live in open wooded country, often near water. They are about the size of a stockily built sparrow, primarily greyish or brownish in colour, with a stoutish beak. Most species catch flying insects by fluttering and swooping like true swallows, or by darting out from a perch.

Woodville, William (1752–1805), English physician and botanist, adopted Jenner's discovery of vaccination against smallpox after initially opposing it. He wrote a *History of Inoculation* (1796) and *Observations on the Cowpox* (1800). While physician at St. Pancras smallpox hospital, he turned two acres of hospital land into a botanic garden and published *Medical Botany* (4 vols., 1790–4).

Wood warbler is a name sometimes applied to the family of New World *warblers but more specifically to *Phylloscopus sibilatrix*, a bird closely related to the willow warbler and chiffchaff. It differs from these in being a little larger, 12·5 cm. (5 in.) in length, with a yellow throat and a white belly. It breeds in Europe and spends the winter in Africa, living in woodland which has little ground cover. It has a trilling song which accelerates towards the end.

Woodwasps are species of large *sawflies, the adults of which are orange and black or uniformly blue, and the females have a long ovipositor. They fly with a loud buzz and insert their eggs into the wood of conifers, where the larvae bore damaging tunnels. Often confused with true wasps, they cannot sting.

Wood-white butterflies are one of the few Palearctic representatives of the largely South American subfamily, the Dismorphiinae, with over 100 species. Unlike most members of this subfamily, the wood-whites are white, delicate butterflies. The South American species of *Dismorphia* are often mimics of the Heliconid butterflies, which are toxic and coloured yellow, orange, and black.

Woolly-bear caterpillars   *Tiger moths.

Woolly rhinoceros: an extinct type of *Pleistocene animal which once roamed over much of Eurasia. It stood about 2 m. (6 ft. 6 in.) high, had two horns, and was well protected from the icy climate of that time by a thick, black coat covered with reddish hairs. Early man hunted it and portrayed it extensively in cave paintings.

Worm-lizards   *Amphistaenians.

Worms are invertebrates with long, thin, soft, and generally pinkish brown bodies, which lack hardened outer coverings. The term is used specifically for the phylum Annelida, but is popularly used for many disparate groups. The Annelida include the familiar *earthworms, *bristle worms, and *fan worms.

A worm-shaped body is very suitable for burrowing, sinuous wriggling, or swimming. It is particularly useful to parasites in the guts of other animals or in plants. It is, however, a poor design for terrestrial life, so worms are at best only semi-terrestrial. Only the *walking-worms, *Peripatus* species, possess simple legs, allowing movement over hard terrain without much friction.

Wormwood   *Artemisia.

Wrasses are a large family of *bony fishes, most abundant in tropical seas but extending into temperate waters. About 400 species are known. In some species females change sex to functional males as they age. Like the related *parrot-fishes, they have well-developed teeth. Several wrasses feed on external parasites infesting other fishes, and most sleep at night by burrowing into sand or hiding under cover. Many species are brilliantly coloured.

Wrens belong to a family of about sixty species of birds which occur primarily in South and Central America, though about ten species extend into North America, and one, the common wren, *Troglodytes troglodytes*, is widespread in Europe and Asia. Generally small, they range up to the size of a thrush. Most are greyish or brownish, usually heavily streaked or barred. They live in a wide variety of habitats, from forest to cactus desert, and build domed nests of twigs and leaves.

Wren thrush: the sole member of its family. It is restricted to dense montane forests above about 1,500 m. (5,000 ft.) in Panama and Costa Rica. It is a small bird, 11 cm. (4 in.) in length, and olive-grey above, except for an orange crown, and grey below. It is primarily insectivorous; however, little is known of its habits.

Wrynecks are two species of birds in the same family as the woodpeckers, but they live in open woodland. They are about the size of a sparrow, finely patterned in greys, buffs, and browns with a shortish, sharply pointed beak. The Eurasian wryneck, *Jynx torquilla*, breeds across large areas of Europe and Asia, but spends the winter in Africa, where the other species, the red-breasted wryneck, *J. ruficollis*, occurs. They nest in holes in trees and eat mainly ants. They get their name from the way in which they twist their heads around in all directions.

The globe-shaped nest of the common **wren** has a high entrance. The waiting chicks, spying out, are fed mainly by their mother, who may have two broods annually.

Xenopus, the African clawed toads are sometimes referred to as frogs as they are not at all toad-like. They are totally aquatic and have a smooth, slippery skin. They have a short head with a semi-elliptical outline, long, pointed fingers with sharp, black claws on the three inner toes, and large, fully webbed feet; they are therefore well adapted to an aquatic life. The best-known, most commonly seen species, *Xenopus laevis*, feeds on worms, fishes, or other small animals. They are famous for their use in pregnancy testing; injected samples of urine from pregnant women will induce gravid females to discharge their eggs. This reaction is a response to hormones, which are present only in the urine of pregnant women.

African clawed toads are native to tropical and southern Africa and belong to a family of short-tongued frogs. Unlike most frogs, they spend virtually all of their lives under water.

Y

Yaffle *Woodpeckers.

Yak: a species of large animal unsurpassed among cattle for its capacity to survive under the bleakest mountain conditions at heights of up to 6,000 m. (20,000 ft.) in the Tibetan highlands. It has a rounded forehead with smooth, round horns curving upwards, outwards and forwards, and a hump over the shoulders. It is covered with long, dark brown hair with a fringe of long black hair on the flanks. Large bulls can be nearly 1·82 m. (6 ft.) tall at the shoulder. In the summer, cows and calves collect together in herds of ten to a hundred; the bulls are more or less solitary except in the rutting season.

Yams are species of the plant genus *Dioscorea*, grown for their starch-rich edible underground tubers. They are found throughout the wetter tropics, especially in West Africa. There are around sixty cultivated species, but only eleven are important crop species, differing in their climatic requirements. They form part of the monocotyledonous family Dioscoreaceae, which comprises some 630 species. The Chinese yam, grown in temperate regions, is unusual in this largely tropical family. All are broad-leaved, climbing plants between 3 and 12 m. (9 ft. 9 in. and 39 ft.) tall, depending on the species. The greater yam of Southeast Asia gives the largest yield of tubers, but the white yam of West Africa accounts for most of the world production. Its tubers may reach 3 m. (9 ft. 9 in.) in length and 60 kg. (150 lb.) in weight, but weights of between 5 and 10 kg. (12 and 25 lb.) are more usual. The lesser yam, cultivated mainly in Asia and the Pacific, produces a cluster of five to twenty small, round or ovoid tubers. Less fibrous than some of the other types, they are more palatable but do not store well, for they bruise easily. Yams have been independently cultivated from wild species throughout Southeast Asia, South and Central America, and West Africa. This makes them the most cosmopolitan of food crops used by man.

As the tubers of most cultivated and all wild species are bitter when fresh, they are eaten after peeling and either boiling or roasting. They are more nutritious than cassava, principally because of their higher protein content (up to 8 per cent). Compounds used in the preparation of artificial cortisone and certain oral contraceptives are obtained from the tubers of certain wild species. In the USA the *sweet potato is sometimes wrongly called a yam.

Yeasts are *fungi that grow in the form of microscopic single cells which multiply by budding or dividing. Several unrelated species of fungi have a yeast-like form of growth; for example, the shadow yeasts which live as *saprophytes on leaves and fruit (bloom), others which are parasites on man, causing the disease called thrush, and the yeast used in brewing and baking. Brewers' yeast ferments sugar, producing alcohol and carbon dioxide. It is this gas which makes dough rise when yeast and sugar are added to it.

Yellow fever is a dangerous virus infection, confined to tropical Africa and Central America. It is spread by mosquitoes and typically causes *jaundice due to liver damage. It also causes damage to the kidneys and often proves fatal if untreated. A vaccine gives good protection.

Yellowhammer, or yellow *bunting: a species of bird in the bunting family which is found over wide areas of Europe and western Asia. It lives in open woodland and is common in European hedgerows. The male is streaky brown above with a rufous rump, while the head and much of the underparts are yellow streaked with brown. The female is not as brightly coloured as the male but is otherwise similar. The distinctive song is usually written 'little-bit-of-bread-and-no-cheese' with the last note emphasized and drawn out.

Yellows (butterflies), or sulphurs, together with whites, brimstones, and orange-tips, belong to the family Pieridae. Many, such as the clouded yellow, *Colias croceus*, have black margins to the wings, or at least black wingtips. In brimstones and many others, females are much paler, almost white. Like other Pierids they lay spindle-shaped, ribbed eggs; the caterpillars are usually slender, green, and not very hairy; and the pupae are attached to a silken pad at the tail end and supported by a silken girdle. Caterpillar food plants are varied but many caterpillars eat plants of the pea family, Leguminosae. Some, including the African migrant butterfly, *Catopsilia florella*, are wide-ranging migrants.

Yellow-tail *Tussock moths.

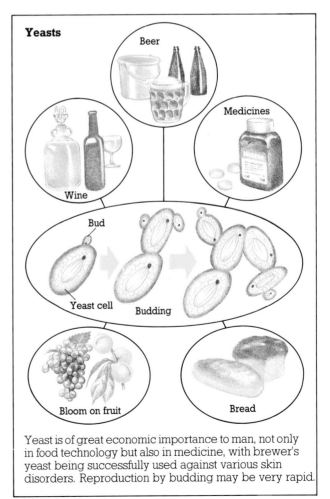

Yeast is of great economic importance to man, not only in food technology but also in medicine, with brewer's yeast being successfully used against various skin disorders. Reproduction by budding may be very rapid.

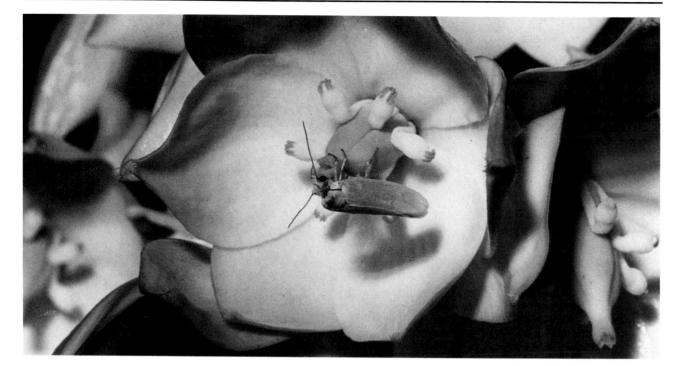

The female **yucca moth**, *Tegeticula maculata*, seen here in a yucca flower, has specially adapted front legs and mouthparts with which it shapes pollen into a ball. When this is inserted into the hollow stigma, it ensures the fertilization of the yucca.

Yellowtails are fishes which belong to the genus *Seriola* of the family Carangidae, and are slender, and streamlined, with deeply forked tails. They occur in coastal waters adjacent to the open ocean in tropical areas and migrate in huge schools, feeding principally on fishes, squids, and crustaceans. They are important food fishes and fine game-fish wherever they occur.

Yews comprise eight or nine species of the genus *Taxus*. They are primitive *gymnosperms allied to the conifers and sometimes included with them. They are evergreen, slow-growing trees casting dense shade and bearing male and female flowers on different trees. The seeds are enclosed in a fleshy type of covering called an aril, which is attractive to birds, but the seed itself, like the rest of the trees, is very poisonous.

The common yew, *T. baccata*, is native to Europe, North Africa, and western Asia and is the longest-lived of all European trees, probably living well over 1,000 years. The tree is widely used for hedging, but is often associated with churchyards and considered a symbol of sadness. Formerly it was the most important wood for the construction of bows in England. In gardens a number of forms are grown for ornament, particularly the Irish Yew, which has a narrow, erect shape, and a number of golden or dwarf forms.

The genus is found in the more humid regions of the northern temperate zone, but reaches as far south as Sulawesi in Indonesia.

Yoga is a set of beliefs and practices developed in India from elements of the Hindu religion. It emphasizes a way of life involving meditation, physical exercises, and self-denial for the religious purpose of union with a supreme spirit. It has become popular in Western countries in a modified form that offers a useful way of relieving anxiety and muscular tension.

Young, Arthur (1741–1820), English writer and exponent of scientific farming, travelled widely in Britain and Ireland studying farming methods, and founded a monthly periodical, *Annals of Agriculture* (1784–1809), much of which he wrote himself. King George III ('Farmer George') once told him: 'Mr Young, I conceive myself more indebted to you than any man in my dominions.'

His *Travels in France* (1792), giving a vivid picture of the *ancien régime* in decline, was notably sympathetic to the French Revolution.

Yucca moths, found in North America and Mexico, have a symbiotic relationship with yucca plants. A female moth collects yucca pollen and uses this to pollinate another flower in the ovary of which she lays eggs. Caterpillars eat some of the developing seeds, but not all; without the moth no seeds are set. These small moths are closely related to the *longhorn moths.

Yuccas belong to a genus of perennial shrubs or medium-sized trees, with some forty species native to Central America and the southern United States. The evergreen, sword-shaped leaves, borne in terminal whorls, are tough, leathery, and often tipped with a sharp spine like that of their other relatives in the family Agavaceae, the sisal hemp and agaves. The white, bell-shaped flowers in dense spikes up to 2 m. (6 ft. 6 in.) in length, have remarkable symbiotic associations with pollinating moths.

Z

Zander *Pike-perch.

Zebras are three species of the horse genus found in eastern and southern Africa. They have a long head with large ears, and an erect mane. The largest and most elegant is Grevy's zebra, *Equus grevyi*, 1·5 m. (5 ft.) tall at the shoulders, the mare and stallion being of similar size. This is an inhabitant of semi-deserts, bush-covered plains, and lowlands. Most numerous is the plains zebra, *E. burchelli*, standing 1·3 m. (4 ft. 2 in.) at the shoulder, found on open plains and hills, and in lightly forested country. The smallest is the mountain zebra, *E. zebra*, 1·2 m (4 ft.) at the shoulders, and confined to the mountains of southern Africa. This zebra has an interesting behavioural adaptation to dry conditions, as it will dig for water in a dried-out river-bed. Zebras move in large herds which are loose gatherings of family units, each unit consisting of a stallion and about six mares with their foals. When alar-

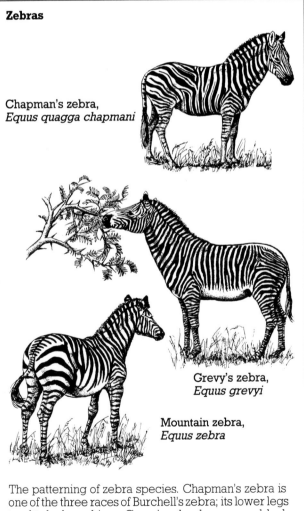

Zebras

Chapman's zebra,
Equus quagga chapmani

Grevy's zebra,
Equus grevyi

Mountain zebra,
Equus zebra

The patterning of zebra species. Chapman's zebra is one of the three races of Burchell's zebra; its lower legs tend to lack markings. Grevy's zebra has narrow black stripes close together and the rare Mountain zebra is distinguished by the grid-iron pattern on its rump.

med, a zebra will emit a yelping bark which sets the whole herd galloping. There is no set breeding season, but the majority of foals are born in the rainy season. The zebra is the favourite prey of the lion and hyena.

Zinnias, of which there are about fifteen species, are plants of the sunflower family, mostly native to Mexico. *Zinnia elegans*, a species with a good deal of natural variation, has been exploited by selective breeding to produce a wide variety of decorative summer-flowering annuals.

Zoogeography *Biogeography.

Zoology is the study of animals and has been a popular science since the early Greek civilization. However, even the first men who hunted animals for food can be considered as zoologists, for they needed to recognize and know something of the habits and occurrence of the animals they hunted. Particular groups of animals are studied in smaller divisions such as entomology (insects), ornithology (birds), and so on. Modern zoology deals with all aspects of the biology of animals, from *biochemistry to *ecology. Subjects peculiar to zoology include animal mechanics, which investigates how animals move; neurology; animal behaviour; and the study of animal social groups.

The concepts and mechanisms derived from zoological research are invaluable to the advancement of modern society. Virtually every technological advance which man 'invents' has some natural parallel. Bats were using 'radar' long before man appeared on earth, and insects had evolved hovering flight virtually before any terrestial animal existed. A sobering lesson to be learned from zoology is that man is part of the animal kingdom.

Zoos, or zoological gardens, began with private collections of animals and were first recorded in Egypt about 1500 BC, when one zoo held monkeys, leopards, and a giraffe. Later in history, an English royal collection was housed in the Tower of London from the thirteenth century until 1834, when the last animals were moved to the new zoo in Regent's Park.

The first public zoo was the Viennese zoo at Schönbrunn, founded as another royal menagerie in 1752 and opened to the public in 1765. Louis XIV's zoo at Versailles began in the seventeenth century; in 1794 it was moved to the Jardin des Plantes in Paris and was supplemented by other collections to become a public amenity. The Zoological Society of London was founded in 1826 and opened its gardens—the first to be called a zoo—in 1828. The oldest North American zoo is the one in Central Park, New York, founded in 1865. There are now about 500 zoos in the world, some with as many as 1,500 species.

From the keeping of animals in cages to amuse and instruct their visitors, the modern zoo has progressed to housing its inhabitants in conditions which resemble their native habitats. 'Open' zoos, set in parkland, supplement more crowded urban ones. The primary purpose of zoological collections is the study of animals and their behaviour. Entertainment value is a by-product, although a useful one in helping to support scientific work and its publication. Many zoos are now active in the conservation of endangered species, and the breeding of rare animals is an important function of zoos.

Zorilla *Polecats.

Acknowledgements

Photographs

Abbreviations: *t* = top; *b* = bottom; *c* = centre; *l* = left; *r* = right.

Bernard D'Abrera, 31*tr*.

Heather Angel, 9*tl*, 13, 66, 71, 78, 83, 85*br*, 99, 143, 160, 210, 276, 293, 331, 346*bl*, 348, 349.

Aquila Photographics / Dennis Green, 259; AP / Wayne Lankinen, 275; AP / Mike Leach, 254; AP / P. J. White, 305; AP / M. C. Wilkes, 51*tr*, 307.

Ardea London, 203; Ardea London / I. R. Beames, 244*br*; Ardea London / R. J. C. Blewitt, 3*br*; Ardea London / Anthony & Elizabeth Bomford, 19, 157; Ardea London / G. K. Brown, 50; Ardea London / John Clegg, 288; Ardea London / Hans D. Dossenbach, 360; Ardea London / Jean-Paul Ferrero, 93, 147*b*, 174, 332; Ardea London / K. W. Fink, 208; Ardea London / Bob Gibbons, 29, 330*obr*; Ardea London / François Gohier, 5, 7; Ardea London / Clem Haagner, 303; Ardea London / Don Hadden, 245; Ardea London / Chuck McDougal, 342; Ardea London / J. L. Mason, 51*bl*, 67, 230; Ardea London / P. Moms, 233*br*; Ardea London / S. Roberts, 159; Ardea London / Peter Steyn, 22; Ardea London / Ron & Valerie Taylor, 302; Ardea London / Valerie Taylor, 150; Ardea London / Richard Vaughan, 295.

C. Ashall, 193.

BBC Hulton Picture Library, 9*br*, 18, 21*br*, 84, 88, 96, 121, 161, 207, 244*tl*.

Biophoto Associates, 52, 132*tl*, 260, 326.

British Library, 107.

Syndics of Cambridge University Library, 138*br*.

Brian Carter, 100.

Bruce Coleman Limited, 14; BCL / Jen & Des Bartlett, 81, 128; BCL / Jane Burton, 117, 124, 272, 339*b*; BCL / B. & C. Calhoun, 165*b*, 350, 370; BCL / R. I. M. Campbell, 44; BCL / Bruce Coleman, 251; BCL / Alain Compost, 31*bl*, 135, 221; BCL / Eric Crichton, 39*bl*, 217; BCL / Gerald Cubitt, 304*tl*; BCL / Peter Davey, 64, 137, 151; BCL / Jack Dermid, 33*tr*, 338; BCL / Inigo Everson, 110; BCL / M. P. L. Fogden, 43, 77*tr*, 91, 179*tr*, 274, 339*tl*; BCL / Jeff Foott, 216, 296, 300*tl*; BCL / Michael Freeman, 239, 255; BCL / Dennis Green, 36; BCL / Anthony Healy, 325; BCL / Pekka Helo, 140; BCL / Udo Hirsch, 241; BCL / David Hughes, 351; BCL / M. P. Kahl, 77*br*, 120; BCL / Jon Kenfield, 24; BCL / Rocco Longo, 102; BCL / L. C. Marigo, 264, 334; BCL / Norman Myers, 1; BCL / Charlie Ott, 368; BCL / Fritz Polking, 179*bl*; BCL / Allan Power, 54; BCL / S. Prato, 309; BCL / Andy Purcell, 314; BCL / Hans Reinhard, 98, 101, 199, 206, 213, 263, 273*br*, 323, 328, 330*tl*, 340, 367; BCL / Leonard Lee Rue III, 26, 94; BCL / Frieder Sauer, 230; BCL / John Shaw, 145; BCL / Kim Taylor, 33*br*, 227, 249, 267*tr*; BCL / Norman Tomalin, 305*t*; BCL / Peter Ward, 73, 194, 273*tl*; BCL / WWF / K. Weber, 69*br*; BCL / Joe van Wormer, 115; BCL / Konrad Wothe, 23, 138*tl*; BCL / G. Ziesler, 70.

E. T. Archive, 366*bl*.

Mary Evans Picture Library, 127, 190, 191*br*.

Gower Scientific Photos, 20.

Robert Harding Picture Library / Peter Arnold Inc, 218; RHPL / Walter Rawlings, 289.

Eric & David Hosking, 108, 177*tl*, 261, 318; E. & D. Hosking / D. P. Wilson, 232*br*.

Jacana / Bertrand Yann Arthus, 191*t*; Jacana / Baranger, 145*t*; Jacana / François Boizot, 234; Jacana / Claude Carré, 321; Jacana / Devez-CNRS, 171*b*; Jacana / Jean-Paul Ferrero, 177*br*, 246*br*; Jacana / François Gohier, 169; Jacana / Nicolas van Ingen, 32; Jacana / Rudolf Konig, 233*t*; Jacana / Jean Michel Labat, 195; Jacana / Jacques Robert, 133, 284; Jacana / Suinot, 247; Jacana / Jean-Philippe Varin, 202, 283; Jacana / Visage Varin, 57; Jacana / Michel Viard, 291; Jacana / Yoff, 175.

Frank Lane Picture Agency / Ray Bird, 40; FLPA / Arthur Christiansen, 365*bl*; FLPA / F. Hartmann, 139; FLPA / Mark Newman, 359; FLPA / L. Robinson, 39*tr*; FLPA / Leonard Lee Rue III, 187; FLPA / H. Schrempp, 304*br*; FLPA / Ronald Thompson, 280*br*; FLPA / Irene Vandermolen, 271; FLPA / Martin B. Withers, 132*b*.

Natural Science Photos / I. Bennett, 189; NSP / P. Bowman, 215; NSP / J. Hobday, 69*tr*, 187*bl*, 346*tr*; NSP / C. Williams, 162; NSP / D. Yendall, 369.

Nature Photographers Ltd / Hugh Clarke, 372; NPL / Michael Gore, 235; NPL / Paul Sterry, 72.

NHPA / Agence Nature, 58; NHPA / Anthony Bannister, 79, 131, 265; NHPA / G. I. Bernard, 361; NHPA / Stephen Dalton, 49, 171; NHPA / Lacz Lemoine, 280; NHPA / Jany Sauvant, 205; NHPA / Philippa Scott, 180; NHPA / James Tallon, 97; NHPA / Bill Wood, 300*b*.

Novosti Press Agency, 201.

OSF / M. J. Adams, 285; OSF / Doug Allan, 188; OSF / G. I. Bernard, 156, 223, 226, 366*tr*; OSF / David Cayless, 173; OSF / Waina Cheng, 248; OSF / J. A. L. Cooke, 163, 164, 269, 286, 296*bl*; OSF / Stephen Dalton, 354; OSF / M. P. L. Fogden, 4, 15, 27, 65*br*, 183, 212, 267*b*, 281, 362, 374; OSF / Laurence Gould, 146; OSF / Rudie H. Kuiter, 299; OSF / G. A. Maclean, 147*tl*; OSF / Godfrey Merlen, 344; OSF / Sean Morris, 211, 243; OSF / Tom Owen Edmunds, 287; OSF / Richard Packwood, 152; OSF / John Paling, 182; OSF / Peter Parks, 257, 298*br*; OSF / Ronald Templeton, 312; OSF / David Thompson, 365*tr*; OSF / Barrie Watts, 17*tl*, 48; OSF / E. J. & J. S. Woolmer, 76.

Partridge Productions / Michael Potts, 268.

Planet Earth Pictures / Charles Bishop, 118; Planet Earth Pictures / Richard Coomber, 279; Planet Earth Pictures / Peter David, 10; Planet Earth Pictures / Walter Deas, 42*b*, 246*tl*, 325*bl*; Planet Earth Pictures / David George, 42*tl*; Planet Earth Pictures / John & Gillian Lythgoe, 129; Planet Earth Pictures / L. P. Madin, 232*tl*; Planet Earth Pictures / Colin Pennycuick, 74; Planet Earth Pictures / Carl Roessler, 85*t*, 114; Planet Earth Pictures / Herwarth Voigtmann, 242; Planet Earth Pictures / Roy Waller, 298*tl*.

Keith Porter, 112, 165.

Ken Preston-Mafham / Premaphotos Wildlife, 12, 17*br*, 46, 47, 87, 125, 128*tl*, 184, 219, 256, 311, 324, 364.

RIDA Photo Library / David Bayliss, 8.

G. R. Roberts, 21*tr*.

Edward S. Ross, 37*tl*, 55, 61, 155, 335.

Science Photo Library, 196; SPL / Dr Jeremy Burgess, John Innes Institute, 262*both*; SPL / R. Clark / M. Goff, 337; SPL / Dr H. A. Davies, 63; SPL / Eric Gravé, 37*br*; SPL / Dr Howells, 126; SPL / David Parker, 357; SPL / K. R. Porter, 34.

Dr D. P. Stribley / Rothamsted Experimental Station, 224.

John Watney, 3*tl*, 65*bl*.

Günter Ziesler, 62, 105, 106, 168, 197, 237, 290, 356.

Picture researcher: Celia Dearing

Artwork

Bruce Coleman Limited / WWF / Helmut Diller, 316, 336.

Andrew Farmer, 95.

Nicholas Hall, 54, 79 (after Tim Halliday), 290, 307.

Richard Lewington, 320.

Anthony Maynard, 308.

Richard Orr, 314.

Oxford Illustrators Limited, 16, 41, 53, 68, 92, 102, 109, 129, 143, 153, 176, 185, 199, 214, 222, 227, 229, 252, 282, 311, 328, 337, 355, 358, 373, end-papers.

John Rignall, 121.

Joyce Tuhill, 161, 363, 375.

Michael Woods, 2, 15, 25, 59, 71, 89, 103, 116, 123, 159, 167, 203, 209, 235, 257, 268, 277, 301, 322, 335, 347, 353.

Publishers' Note

The publishers are grateful to Stanley Remington for his special interest and encouragement.

Classification of life on earth

This chart represents the most familiar groups only

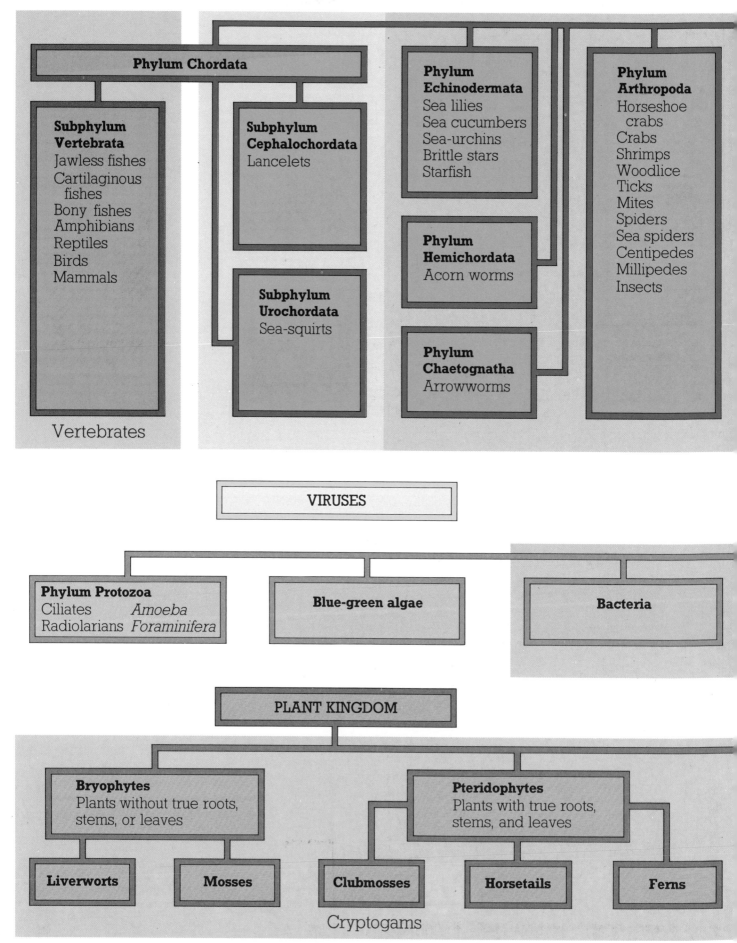

Phylum Chordata

Subphylum Vertebrata
Jawless fishes
Cartilaginous fishes
Bony fishes
Amphibians
Reptiles
Birds
Mammals

Subphylum Cephalochordata
Lancelets

Subphylum Urochordata
Sea-squirts

Vertebrates

Phylum Echinodermata
Sea lilies
Sea cucumbers
Sea-urchins
Brittle stars
Starfish

Phylum Hemichordata
Acorn worms

Phylum Chaetognatha
Arrowworms

Phylum Arthropoda
Horseshoe crabs
Crabs
Shrimps
Woodlice
Ticks
Mites
Spiders
Sea spiders
Centipedes
Millipedes
Insects

VIRUSES

Phylum Protozoa
Ciliates   *Amoeba*
Radiolarians   *Foraminifera*

Blue-green algae

Bacteria

PLANT KINGDOM

Bryophytes
Plants without true roots, stems, or leaves

Pteridophytes
Plants with true roots, stems, and leaves

Liverworts   **Mosses**   **Clubmosses**   **Horsetails**   **Ferns**

Cryptogams